Trends in Mathematics is a series devoted to the publication of volumes arising from conferences and lecture series focusing on a particular topic from any area of mathematics. Its aim is to make current developments available to the community as rapidly as possible without compromise to quality and to archive these for reference.

Proposals for volumes can be sent to the Mathematics Editor at either

Birkhäuser Verlag
P.O. Box 133
CH-4010 Basel
Switzerland

or

Birkhäuser Boston Inc.
675 Massachusetts Avenue
Cambridge, MA 02139
USA

Material submitted for publication must be screened and prepared as follows:

All contributions should undergo a reviewing process similar to that carried out by journals and be checked for correct use of language which, as a rule, is English. Articles without proofs, or which do not contain any significantly new results, should be rejected. High quality survey papers, however, are welcome.

We expect the organizers to deliver manuscripts in a form that is essentially ready for direct reproduction. Any version of $\TeX$ is acceptable, but the entire collection of files must be in one particular dialect of $\TeX$ and unified according to simple instructions available from Birkhäuser.

Furthermore, in order to guarantee the timely appearance of the proceedings it is essential that the final version of the entire material be submitted no later than one year after the conference. The total number of pages should not exceed 350. The first-mentioned author of each article will receive 25 free offprints. To the participants of the congress the book will be offered at a special rate.

Mathematics and Computer Science III

Algorithms, Trees, Combinatorics and Probabilities

Michael Drmota
Philippe Flajolet
Danièle Gardy
Bernhard Gittenberger
Editors

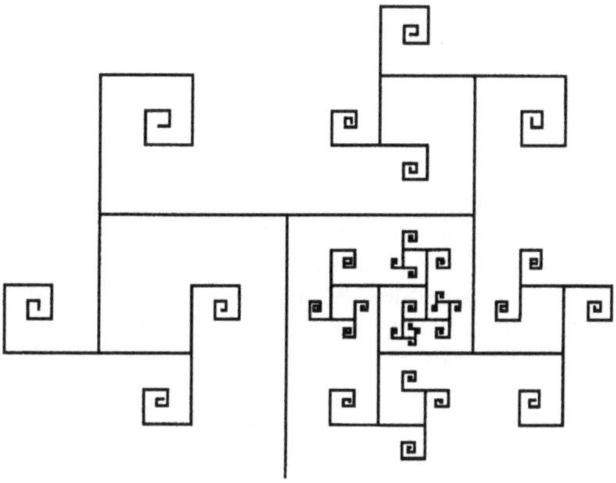

Springer Basel AG

Editors:

Michael Drmota
Vienna University of Technology
Institute of Discrete Mathematics
and Geometry
Wiedner Hauptstrasse 8–10
1040 Wien
Austria
e-mail: michael.drmota@tuwien.ac.at

Danièle Gardy
Université de Versailles-St-Quentin
PRISM
Bâtiment Descartes
45 avenue des Etats-Unis
78035 Versailles Cedex
France
e-mail: gardy@prism.uvsq.fr

Philippe Flajolet
INRIA Rocquencourt
78153 Le Chesnay
France
e-mail: Philippe.Flajolet@inria.fr

Bernhard Gittenberger
Vienna University of Technology
Institute of Discrete Mathematics
and Geometry
Wiedner Hauptstrasse 8–10
1040 Wien
Austria
e-mail: gittenberger@dmg.tuwien.ac.at

2000 Mathematical Subject Classification 05-XX, 60C05, 60Gxx, 68P30, 68Q25, 68Rxx, 68W20, 68W40, 90B15

A CIP catalogue record for this book is available from the Library of Congress, Washington D.C., USA

Bibliographic information published by Die Deutsche Bibliothek
Die Deutsche Bibliothek lists this publication in the Deutsche Nationalbibliografie; detailed bibliographic data is available in the Internet at http://dnb.ddb.de

ISBN 978-3-0348-9620-7 ISBN 978-3-0348-7915-6 (eBook)

DOI 10.1007/978-3-0348-7915-6

The logo on the cover is a binary search tree in which the directions of child nodes alternate between horizontal and vertical, and the edge lengths decrease as 1 over the square root of 2. The tree is a Weyl tree, which means that it is a binary search tree constructed from a Weyl sequence, i.e., a sequence (na) mod 1, $n = 1,2,...$, where a is an irrational real number. The PostScript drawing was generated by Michel Dekking and Peter van der Wal from the Technical University of Delft.

Originally published by Birkhäuser Verlag, Basel – Boston – Berlin in 2004
Softcover reprint of the hardcover 1st edition 2004
Printed on acid-free paper produced from chlorine-free pulp. TCF ∞

ISBN 978-3-0348-9620-7

9 8 7 6 5 4 3 2 1

www.birkhauser-science.com

Foreword

These are the Proceedings of the International Colloquium of Mathematics and Computer Science held at the Vienna University of Technology, September 13–17, 2004. This colloquium is the third one in a now regularly established series following the first two venues in September 2000 and September 2002 in Versailles. The present issue is centered around Combinatorics and Random Structures, Graph Theory, Analysis of Algorithms, Trees, Probability, Combinatorial Stochastic Processes, and Applications. It contains invited papers, contributed papers (lectures) and short communications (posters).

The contributions have been carefully reviewed for their scientific quality and originality by the Scientific Committee chaired by Michael Drmota (Vienna University of Technology, Austria) and composed of Brigitte Chauvin (Université de Versailles, France), Luc Devroye (McGill University, Canada), Danièle Gardy (Université de Versailles, France), Philippe Flajolet (INRIA Rocquencourt, France), Michał Karonski (Adam Mickiewicz University, Poland), Abdelkader Mokkadem (Université de Versailles, France), Helmut Prodinger (University of Witwatersrand, South Africa), J. Michael Steele (University of Pennsylvania, Philadelphia, USA), Brigitte Vallée (Université de Caen, France). We thank them and all anonymous referees for their impressive work.

We also thank the invited speakers: Jean Bertoin (Université Paris VI, France), Mireille Bousquet-Melou (Université Bordeaux 1, France), Hsien-Kuei Hwang (Academia Sinica, Taiwan), Svante Janson (Uppsala University, Sweden), Christian Krattenthaler (Université Lyon, France), Jean-François Marckert (Université de Versailles, France), Boris Pittel (The Ohio State University, USA), Simon Tavaré (University of Southern California, USA), the authors of submitted papers and posters, and the participants for their contribution to the success of the conference.

Finally, we express our acknowledgements to the Institute of Discrete Mathematics and Geometry, the Vienna University of Technology, the Federal Ministry for Education, Science, and Culture, the City of Vienna, the Austrian Research Society (ÖFG), the Austrian Mathematical Society (ÖMG), the Goedel-Society, and the Bank Austria-Creditanstalt for providing generous financial and material support.

The Organizing Committee
Bernhard Gittenberger
Thomas Klausner
Alois Panholzer

Preface

These colloquium proceedings address problems at the interface between Mathematics and Computer Science, with special emphasis on discrete probabilistic models and their relation to algorithms. Combinatorial and probabilistic properties of random graphs random trees, combinatorial stochastic processes (such as random walks) as well as branching processes and related topics in probability are central. Applications are to be found in the analysis of algorithms and data structures, the major application field, but also in statistical theory, information theory, and mathematical logic. This colloquium is the third one in a now regularly established series, following the first two venues in September 2000 and September 2002 in Versailles. The book features a collection of original refereed contributions: contributed papers (lectures) and short communications (posters), supplemented by more detailed articles written by invited speakers (and coauthors): Jean Bertoin (and Christina Goldschmidt), Svante Janson, and Boris Pittel (and Alan Frieze).

During the final preparation of this volume we received the sad news that Rainer Kemp (from Frankfurt, Germany) has passed away. Rainer Kemp was one of the *founding fathers* of the Analysis of Algorithms, a main topic of our conference. His book *Fundamentals of the average case analysis of particular algorithms*, Wiley, 1984, was one of the first books on this subject and had a considerable influence on the development of the field. He was organizer of several meetings on this subject and served the scientific community with many other duties. But first of all we lost a good friend and colleague.

Combinatorics and Random Structures. The starting point of many studies of random discrete models is combinatorics, which often provides us with exact representations in terms of counting generating functions that can also be used for a probabilistic study. Sylvie Corteel, Guy Louchard, and Robin Pemantle work on the common distribution of intervals in pairs of permutations. Next Sylvie Corteel, Jeremy Lovejoy, and Ae Ja Yee provide generating functions for generalized Frobenius partitions. Luca Ferrari, Renzo Pinzani, and Simone Rinaldi present some results on integer partitions. Toufik Mansour considers generating functions for 321-avoiding permutations. Eugenijus Manstavičius proves an iterated logarithm law for the cycle lengths of a random permutation. Martin Rubey provides a sufficient condition for transcendence of generating functions of walks on the slit plane. Michael Schlosser finds some curious q-series expansions, and Klaus Simon presents relations between the numbers of partitions and the divisor functions.

Graph Theory. Graphs are a basic object in discrete mathematics. They are widely used in applications, and algorithms on graphs as well as theoretical questions on graphs have been "modern topics" of research in mathematics and computer science for several decades. Mindaugas Bloznelis uses Hoeffding decomposition to prove asymptotic normality of subgraph count statistics. Robert Cori, Arnaud Dartois, and Dominique Rossin compute so-called "avalanche polynomials" for certain families of graphs. The invited paper by Alan Frieze and Boris Pittel gives a detailed analysis on perfect matchings in random graphs with prescribed minimal degree. Omer Giménez and Marc Noy provide very tight estimates for the growth constant of labelled planar graphs. Finally, Stavros D. Nikolopoulos and Charis Papadopoulos present an algorithm for determining the number of spanning trees in P4-reducible graphs.

Analysis of Algorithms. This field was created by Donald E. Knuth and is concerned with accurate estimates of complexity parameters of algorithms and aims at predicting the behavior of a given algorithm. Javiera Barrera and Christian Paroissin consider specific search cost in random binary search trees. Monia Bellalouna, Salma Souissi, and Bernard Ycart analyze probabilistic bin packing problems. Paweł Hitczenko, Jeremy Johnson, and Hung-Jen Huang consider algorithms for computing the Walsh-Hadamard transform. Tämur Ali Khan and Ralph Neininger analyze the performance of the randomized algorithm to evaluate Boolean decision trees proposed by Smir, in particular they consider the worst case input and provide limit laws and tail estimated. Next, Shuji Kijima and Tomomi Matsui propose a polynomial time perfect sampling algorithm for two-rowed contingency tables. Conrado Martínez and Xavier Molinero combine two generation algorithms to obtain a new efficient algorithm for the generation of unlabelled cycles. Finally, Yuriy A. Reznik and Anatoly V. Anisimov suggest the use of tries for universal data compression.

Trees. Trees are perhaps the most important structure in computer science. They appear as data structures and are used in various algorithms such as data compression. David Auber, Jean-Philippe Domenger, Maylis Delest, Philippe Duchon, and Jean-Marc Fédou present an extension of Strahler numbers to rooted plane trees. Julien Fayolle analyzes mean size and external path length of a suffix tree that is related to the LZ'77 data compression algorithm. Eric Fekete considers two different kinds of external nodes in binary search trees and describes the evolution of this process in terms of martingales. The invited paper by Svante Janson offers an analysis of the number of records in a complete binary tree or equivalently the number of random cutting to eliminate a complete binary tree. Interestingly the distribution is, after normalization, asymptotically a periodic function in $\log n - \log\log n$, where n is the size of the tree. Mehri Javanian and Mohammad Q. Vahidi-Asl consider multidimensional interval trees. Anne Micheli and Dominique Rossin describe a specific distance between unlabelled ordered trees, that is based on deletions and insertions of edges. Katherine Morris determines grand averages on some parameters in monotonically labelled tree structures. Tatiana Mylläri proves local central limit theorems for the number of vertices of a given outdegree in a Galton-Watson forest. And finally, Alois Panholzer gives a precise analysis of the cost distribution for destroying recursive trees in the case of toll functions of polynomial growth.

Probability. Probabilistic methods get more and more important is the analysis of discrete structures: random graphs, random trees, average case analysis of algorithms etc. Margaret Archibald addresses the question of the probability that the maximum in a geometrically distributed sample occurs in the first d positions of a word. The invited paper by Jean Bertoin and Christina Goldschmidt describes the duality between a fragmentation associated to certain Dirichlet distributions and a natural random coagulation. This gives rise to an application to the genealogy of Yule processes. Mykola S. Bratiychuck considers semi-Markov walks in queueing and risk theory. Amke Caliebe characterizes fixed points of linear stochastic fixed point equations as mixtures of infinitely divisible distributions. Peter Jagers and Uwe Rösler describe a systematic approach to find solutions of stochastic fixed points involving the maximum. Arnold Knopfmacher and Helmut Prodinger provide central limit theorems for the number of descents in samples of geometric random variables. Alain Rouault proves a law of large numbers and describes a new large deviation phenomenon for cascades. Christiane Takacs investigates partitioning properties of piecewise constant eigenvectors of matrices describing the mutual positions of points. Vladimir Vatutin and Elena Dyakonova consider branching processes in random environment, find asymptotics of the survival probabilities and prove a Yaglom type limit theorem. Finally, Vladimir Vatutin and Valentin Topchii study the joint distribution of the number of individuals at the origin and outside the origin on a continuous time random walk on the integers.

Combinatorial Stochastic Processes. Random walks are the most prominent representatives of combinatorial stochastic processes. They play a central role in the interplay between combinatorics and probability. Enrica Duchi and Gilles Schaeffer consider a model of particles jumping on a row of cells with general boundary conditions where the stationary distribution is not uniform. Guy Fayolle and Cyril Furtlehner study stochastic deformations of sample paths of random walks. Johannes Fehrenbach and Ludger Rüschendorf show that a Markov chain that is naturally defined on the Eulerian orientation of planar graph converges to uniform distribution. Alexander Gnedin considers regenerative composition structures. Jean Mairesse and Frédéric Mathéus study transient nearest neighbor random walks on groups with a finite set of generators and compute various characteristics such as the drift and the entropy. Finally, Philippe Marchal gives a fractal construction of nested, stable regenerative sets and studies the associated inhomogeneous fragmentation process.

Applications. Random combinatorics interacts with many other areas of science. Eda Cesaratto and Brigitte Vallée consider numeration schemes, defined in terms of dynamical systems and determine the Hausdorff dimension of sets of reals which obey some constraints on their digits. Adriana Climescu - Haulica deals with large deviation analysis of space-time Trellis codes. David Coupier, Agnés Desolneux, and Bernard Ycart provide a zero-one law for first order logic on random images. Nadia Creignou and Hervé Daudé study threshold phenomena for random generalized satisfyability problems. Guy Fayolle, Vadim Malyshev, and Serguei Pirogov introduce new models of energy redistribution in stochastic chemical kinetics with several molecule types and energy parameters. László Györfi discusses Chernoff type large deviations of Hellinger distance on partitions. Nadia Lalam and Christine Jacob address the problem of estimating the offspring mean for a general class of size-dependent branching processes. Małgorzata and Włodzimierz Moczurad deal with the problem of decidability of simple brick codes. And finally, Joel Ratsaby generalizes Sauer's Lemma to finite VC-dimension classes of binary valued functions.

Altogether papers assembled in this volume offer snapshots of current research. At the same time, they illustrate the numerous ramifications of the theory of random discrete structures throughout mathematics and computer science. Many of them, in particular invited lectures, include carefully crafted surveys of their field. We thus hope that the book may serve both as a reference text and as a smooth introduction to many fascinating aspects of this melting pot of continuous and discrete mathematics.

Michael Drmota
Philippe Flajolet
Danièle Gardy
Bernhard Gittenberger

provide central limit theorems for the number of descents in samples of geometric random variables. Alain Rouault proves a law of large numbers and describes a new large deviation phenomenon for cascades. Christiane Takacs investigates [illegible] parameters [illegible] constant [illegible] of matrices describing the [illegible]. Vladimir Vatutin and Elena Dyakonova consider [illegible] three types of branching processes in random environment [illegible] limit theorems [illegible] Vatutin [illegible] reports [illegible] dates as the origin [illegible].

Combinatorial Stochastic Processes. [illegible] walks [illegible] representative [illegible] combinatorial [illegible] in the interplay between combinatorics and probability. [illegible] and Gilles Schaeffer consider a model of particles [illegible] of cells with general boundary conditions when the stationary distribution is not uniform. Guy Fayolle and Cyril Furtlehner study [illegible] walks [illegible] that a Markov chain that is naturally defined on the Eulerian orientations of planar graphs converges to uniform distribution. Alexander Gnedin considers [illegible] composition structures. Jean Mairesse and Frédéric Mathéus study transient nearest neighbor random walks on groups with a finite set of generators and compute various characteristics such as the drift and the entropy. Finally, Philippe Marchal gives a fractal construction of nested stable regenerative sets and studies the associated inhomogeneous fragmentation process.

Applications. Random combinatorics interacts with many other areas of science. Eda Cesaratto and Brigitte Vallée consider numeration schemes, defined in terms of dynamical systems and determine the Hausdorff dimension of sets of reals which obey some constraints on their digits. Adriana Climescu-Haulica deals with large deviation analysis of space-time Trellis codes. David Coupier, Agnès Desolneux, and Bernard Ycart provide a zero-one law for first order logic on random images. Nadia Creignou and Hervé Daudé study threshold phenomena for random generalized satisfiability problems. Guy Fayolle, Vadim Malyshev, and Serguei Pirogov introduce new models of energy redistribution in stochastic chemical kinetics with several molecule types and energy parameters. Laszlo Györfi discusses Chernoff-type large deviations of Hellinger distance on partitions. Nadia Lalam and Christine Jacob address the problem of estimating the offspring mean for a general class of size-dependent branching processes. Malgorzata and Wiesław Moczurad deal with the problem of decidability of simple brick codes. And finally, Joel Ratsaby generalizes Sauer's Lemma to finite VC-dimension classes of binary valued functions.

Altogether papers assembled in this volume offer snapshots of current research. At the same time, they illustrate the numerous ramifications of the theory of random discrete structures throughout mathematics and computer science. Many of them, in particular invited lectures, include carefully crafted surveys of their field. We thus hope that the book may serve both as a reference text and as a smooth introduction to many fascinating aspects of this melting pot of continuous and discrete mathematics.

Michael Drmota
Philippe Flajolet
Danièle Gardy
Bernhard Gittenberger

Contents

PART III. Analysis of Algorithms

PART IV. Trees

Part I

Combinatorics and Random Structures

Trends in Mathematics, © 2004 Birkhäuser Verlag Basel/Switzerland

Common Intervals of Permutations

Sylvie Corteel, Guy Louchard, and Robin Pemantle

ABSTRACT: *An interval of a permutation is a consecutive substring consisting of consecutive symbols. For example, 4536 is an interval in the permutation 71453682. These arise in genetic applications. For the applications, it makes sense to generalise so as to allow gaps of bounded size $\delta - 1$, both in the locations and the symbols. For example, 4527 has gaps bounded by 1 (since 3 and 6 are missing) and is therefore a δ-interval of ****4*5*27**** for $\delta = 2$.*

After analysing the distribution of the number of intervals of a uniform random permutation, we study the number of 2-intervals. This is exponentially large, but tightly clustered around its mean. Perhaps surprisingly, the quenched and annealed means are the same. Our analysis is via a multivariate generating function enumerating pairs of potential 2-intervals by size and intersection size.

1. Introduction

Let $[n]$ denote the set $\{1, 2, \ldots, n\}$. We are interested in counting the common intervals of a pair of permutations. To be precise, if G_A and G_B are two permutations of $[n]$, we are interested in counting the pairs of intervals (I, J) for which $G_A(I) = G_B(J)$. It is equivalent to count intervals I for which $G_B^{-1} G_A(I)$ is also an interval. Accordingly, we define

Definition 1.1. *The interval $I := [i, i+k-1] \subseteq [n]$ is called an* interval of the permutation G *if $G^{-1}(I)$ is an interval, that is, if there is a j such that*

$$G[j, j+k-1] = [i, i+k-1] .$$

The proper intervals are those whose lengths are at least 2 and at most $n-1$.

Here and throughout, we use vector notation for permutations rather than cycle notation, so that $(\sigma_1, \ldots, \sigma_n)$ denotes the permutation $i \mapsto \sigma_i$ rather than the permutation consisting of a single n-cycle.

Example: Let G be the permutation $(3, 1, 2, 4, 5)$. Then the proper intervals of G are $[1, 2], [4, 5], [1, 3]$ and $[1, 4]$.

When G is a random variable, uniformly distributed over all permutations of $[n]$, let X_k denote the number of of intervals of length k of G and let $X = \sum_k X_k$ denote the number of intervals of G. Uno et al [18] compute $\mathbb{E} X_k$; the following easy proposition is proved at the end of this section.

Proposition 1.2. *As $n \to \infty$, the distribution of X converges to a Poisson with mean 2.*

The number of intervals, or runs of a permutation, was studied in the forties by Kaplansky [12] and Wolfowitz [19, 20] from a statistical point of view. See also [14]. Recently several algorithms were designed to efficiently enumerate all common intervals of permutations [10, 18] and their time complexity is $O(n+K)$ where n is the size of the permutation and K the number of intervals. These algorithms were designed because common intervals have several applications. They relate to the consecutive arrangement problem [7]. Genetic algorithms for sequencing problems are based on common intervals [13, 15]. In bioinformatics [4, 5, 9, 10, 11], genomes of prokaryotes can be modelled as a permutation of genes. A common interval is then a set of orthologous genes that appear consecutively, possibly in different orders, in two genomes. Therefore common intervals can be used to detect groups of genes that are functionally associated [10, 11]. As the annotation of genomes is not perfect, the notion of consecutivity in intervals needs to be relaxed. A notion of *gene teams* was defined in [6], where a gene team is a maximal set of orthologous genes, possibly occurring in different orders in the two species, but separated in each case by gaps that do not exceed a fixed threshold, δ. To study these, we consider a generalisation of intervals, namely δ-intervals (the previous case corresponds to $\delta = 1$).

Definition 1.3. *The set $I \subseteq [n]$ is called a δ-*interval *of $[n]$* of length k *if I is a set of integers $\{i_1, \ldots, i_k\}$ with $1 \leq i_{r+1} - i_r \leq \delta$ for each $1 \leq r \leq k-1$. We call I a* δ-interval of length k of G *if both I and $G^{-1}(I)$ are δ-intervals. Proper δ-intervals are again those of cardinality at least 2 and at most $n-1$.*

Example: $G = (3, 1, 2, 4, 5)$ possesses the 2–intervals:

$$\{1,2\}, \{1,3\}, \{1,2,3\}, \{2,3\}, \{1,2,3,4\}, \{1,3,4\}, \{2,3,4\},$$
$$\{2,4,5\}, \{2,4\}, \{2,3,5\}, \{2,3,4,5\}, \{1,3,4,5\}, \{1,2,4,5\}, \{1,2,3,5\}$$

In [6] a polynomial time enumeration algorithm for gene teams is presented. Our notion of δ-intervals removes the maximality constraint, whence the number of these may grow exponentially and it is natural to enumerate asymptotically rather than enumerating exactly.

The main purpose of this note is to investigate the asymptotic properties of $X_k^{(\delta)}$, where this denotes the number of δ–intervals of length k of a uniformly chosen random permutation of $[n]$, and of the total number $X^{(\delta)} := \sum_k X_k^{(\delta)}$ of δ-intervals of a random permutation. We are interested in all $\delta > 1$ but in the present manuscript we examine only the case $\delta = 2$. To reduce the number of superscripts, we let Y and Y_k denote $X^{(2)}$ and $X_k^{(2)}$ respectively.

The number $X^{(\delta)}$ of δ-intervals when $\delta > 1$ behaves very differently from X. Whereas X is $\mathcal{O}(1)$ as $n \to \infty$, with all the contributions coming from short intervals, there will typically be many δ-intervals. In fact a thumbnail computation produces numbers α_k in the unit interval ($\alpha_2 \approx 0.57939$) such that for $k \sim \alpha n$ and $\alpha > \alpha_k$, the random variable X_k will be typically exponentially large: the number of δ-intervals of $[n]$ of size k grows exponentially, the probability of G^{-1} of one of these also being a δ-interval decays exponentially, and the growth overcomes the decay when $\alpha > \alpha_k$.

Seeing that $X^{(\delta)}$ grows exponentially in n, it is natural to look at the rescaled quantity $n^{-1} \log X^{(\delta)}$. In the next section we compute the *annealed* mean, namely $n^{-1} \log \mathbb{E} X^{(\delta)}$. The term "annealed" means that we first take an expectation over the (uniform) measure on permutations. The more interesting quantities are the *quenched* quantities, which refer to the typical, rather than the mean behaviour of $X^{(\delta)}$. Often one has a so-called lottery effect, meaning that the mean of a quantity X comes primarily from an exponentially small number of values that are exponentially larger than the median value, and that consequently, $\mathbb{E} \log X < \log \mathbb{E} X$. For example, when there is a Gaussian limit law, $n^{-1/2}(\log X - n\mu) \to \mathcal{N}(0, \sigma^2)$, then one will typically have a lottery effect. Perhaps surprisingly in light of the discussion in section 4, there is no lottery effect. Our main result, Theorem 4.1 below, is that for $\delta = 2$, we have $\mathbb{E}(X^{(\delta)})^2 = \mathcal{O}(\mathbb{E}X^{(\delta)})^2$. This shows that as $n \to \infty$, the sequence $\overline{X} := X^{(\delta)}/\mathbb{E}X^{(\delta)}$ is tight. The mean of $X^{(\delta)}$ is computed in the next section, with the remaining sections devoted to the computation of the second moment. We start in Section 2 with arguments in the case $\delta = 1$.

2. Intervals

Recall that X_k denotes the number of of intervals of length k of G and that $X = \sum_k X_k$ denotes the number of intervals of G. Uno et al [18] computed

$$\mathbb{E}(X_2) = \frac{2(n-1)}{n} \; ; \; \mathbb{E}(X_3) = \frac{6(n-2)}{n(n-1)} \; ; \; \mathbb{E}(X_k) \leq \frac{24}{n^2} \text{ for } k \geq 4$$

Although this was not explicitly stated in [18], it is not hard to show Proposition 1.2.

Proof of Proposition 1.2: Letting $X' := \sum_{k=3}^{n-1} X_k$, we see that

$$\mathbb{E}X' \leq \frac{6}{n} + (n-4)\frac{24}{n^2} \leq \frac{30}{n}$$

so $X' \to 0$ in probability as $n \to \infty$. Thus it suffices to show that X_2 converges to a Poisson of mean 2. Kaplansky proves this in [12]. A modern approach is to use the Poisson approximation machinery first developed by Chen and Stein, and put in an explicit and usable form in [1]. Given $k \in [n-1]$, let A_k be the event that $G^{-1}\{k, k+1\}$ is an interval. Let $B_k = \{k-1, k, k+1\} \cap [n-1]$. Write p_k for $\mathbb{P}(A_k)$ and $p_{k,l}$ for $\mathbb{P}(A_k \cap A_l)$. Theorem 1 of [2] shows that the total variation distance between X and a Poisson with mean $\sum_k p_k$ is bounded by the sum of three quantities, namely

$$\begin{aligned} b_1 &:= \sum_k \sum_{j \in B_k} p_j p_k \\ b_2 &:= \sum_k \sum_{k \neq j \in B_k} p_{k,l} \\ b_3 &:= \sum_k \mathbb{E} \left| \mathbb{E} \left(\mathbf{1}_{A_k} - p_k | \sigma(A_j : j \notin B_k) \right) \right| . \end{aligned}$$

Bounding b_1 and b_2 by $\mathcal{O}(1/n)$ is straightforward. The same bound on b_3 follows from the identity

$$\begin{aligned} \left| \mathbb{E} \left(\mathbf{1}_{A_k} - p_k | \sigma \right) \right| &= 2 \sup_{H \in \sigma} \left[\mathbb{P}(H \cap A_k) - p_k \mathbb{P}(H) \right] \\ &= 2 p_k || \mu_{A_k} - \mu ||_{TV} , \end{aligned}$$

together with a coupling argument obtaining μ_{A_k} from μ by switching the values of G^{-1} on k and $G(j)$, and on $k+1$ and $G(j+1)$ for a uniformly chosen j. □

More generally, one may consider the distribution of X_k.

$$\mathbb{E}(X_k) = \frac{k!(n-k+1)}{n(n-1)\dots(n-k+2)} = \frac{k(n-k+1)}{\binom{n}{k-1}}. \tag{1}$$

Set

$$X := \sum_{k=2}^{n-1} X_k.$$

We see that, as $n \to \infty$, the dominant terms of $\mathbb{E}(X)$ are given by $k = \mathcal{O}(1)$ and $k = n - \mathcal{O}(1)$. Indeed, by Stirling, and setting $k = \alpha n + 1$, we easily derive

$$\mathbb{E}\left(\sum_{k=\varepsilon n}^{n/2} X_k\right) \sim \int_{\varepsilon}^{1/2} \left[\alpha^{\alpha n}(1-\alpha)^{(1-\alpha)n} \alpha n(1-\alpha)n\sqrt{2\pi\alpha(1-\alpha)n}\right] n d\alpha \quad (2)$$

which goes to zero exponentially as n goes to infinity for fixed ε. We obtain

$$\mathbb{E}(X) \sim 2 + \frac{8}{n} + \frac{36}{n^2} + \frac{228}{n^3} + \ldots$$

As an example of X_k behaviour, let us now turn to X_3 and compute $E_3^2 := \mathbb{E}\left(X_3^2\right)$.

Combinatorial analysis leads to asymptotic series expressions for the moments of X_k and for the probabilities $\mathbb{P}(X_k = j)$. For example,

$$\mathbb{E}X_3 \sim \frac{6}{n} - \frac{6}{n^2} + \cdots \qquad \mathbb{E}X_3^2 \sim \frac{6}{n} + \frac{38}{n^2} + \cdots$$

and

$$\mathbb{P}(X_3 = 0) \sim 1 - \frac{6}{n} + \frac{28}{n^2} + \cdots \qquad \mathbb{P}(X_3 = 1) \sim \frac{6}{n} - \frac{50}{n^2} + \cdots \qquad \mathbb{P}(X_3 = 2) \sim \frac{22}{n^2} + \cdots$$

Details are given in [8]. We turn now to 2-intervals.

3. The mean number of 2-intervals

Let $N(k,n)$ denote the number of 2-intervals that are subsets of $[n]$ and have cardinality k. We will take advantage of the uniformity of G^{-1}. For each of the $N(k,n)$ 2-intervals of cardinality k, its inverse image under G is uniformly distributed on k-subsets of $[n]$. Therefore, the probability is exactly $N(k,n)/\binom{n}{k}$ for any given 2-interval of cardinality k, that its inverse image under G is again a 2-interval. Thus

$$\mathbb{E}Y_k = \frac{N(k,n)^2}{\binom{n}{k}}. \quad (3)$$

To evaluate $N(k,n)$, note that there may be anywhere from 0 to $m_k := \min\{k-1, n-k\}$ "holes" in a 2-interval, where a hole is an element not in the 2-interval but between its endpoints. These may be enumerated by the following procedure. Pick a starting position r with $1 \le r \le n-k-i+1$ and let r be the least element of the 2-interval. Choose any sequence with i occurrences of the word "skip" and $k-1-i$ occurrences of the word "no-skip". If the first word in the sequence is "no-skip" then $r+1$ is the next element of the 2-interval; if the first word is "skip" then $r+2$ is the next element. Continue in this manner until the sequence is used up. This method of enumeration makes it clear that

$$\mathbb{E}Y_k = \frac{N(k,n)^2}{\binom{n}{k}} = \frac{\left(\sum_{i=0}^{m_k}(n+1-k-i)\binom{k-1}{i}\right)^2}{\binom{n}{k}}. \quad (4)$$

The sum may be evaluated exactly for $k/n < 1/2$ and asymptotically for $k/n \geq 1/2$ leading to

$$N(k,n) \sim \begin{cases} \left(n + \frac{3}{2} - \frac{3}{2}k\right) 2^{k-1} & : \text{ if } \frac{k-2n/3}{n^{1/2}} \to -\infty, \\ 2^{k-1} \frac{\sqrt{k}}{2} \Psi_x & : \text{ if } \frac{k-2n/3}{n^{1/2}} \to x \in (-\infty, \infty), \\ \left[\frac{(2k-1-n)^2}{(3k-1-2n)^2}\right] \binom{k-1}{n-k} & : \text{ if } \frac{k-2n/3}{n^{1/2}} \to +\infty, \end{cases} \tag{5}$$

where Ψ_x is the expectation of the positive part of $Z + x$ and Z is a standard normal.

Via equation (4), these asymptotics for $N(k,n)$ lead directly to asymptotics for $\mathbb{E}Y_k$. To obtain asymptotics for $\mathbb{E}Y$, we then sum over k, using a saddle point approximation. The only significant terms are near $k = \alpha_* n$, where α_* will be determined shortly but is evidently greater than 2/3. We therefore use (5) with $n \to \infty$ and $\alpha := k/n$ to obtain

$$\begin{aligned} \frac{N(k,n)^2}{\binom{n}{k}} &\sim \left[\frac{2k-n}{k}\binom{k}{n-k}\left(\frac{2k-n}{3k-2n}\right)^2\right]^2 \binom{n}{k}^{-1} \\ &\sim A(\alpha) n^{-1/2} \exp\left[nF_1(\alpha)\right] \end{aligned} \tag{6}$$

where

$$\begin{aligned} A(\alpha) &:= \frac{(2\alpha-1)^5}{(3\alpha-2)^4}\sqrt{\frac{1}{2\pi\alpha(1-\alpha)}}; \\ F_1(\alpha) &:= 3\alpha\log\alpha - (1-\alpha)\log(1-\alpha) - 2(2\alpha-1)\log(2\alpha-1). \end{aligned}$$

The maximum of F_1 occurs in $[2/3, 1]$ where $F_1'(\alpha)$ vanishes, which occurs when $\alpha = \alpha_* \approx .7840013296$ solves

$$17\alpha^4 - 33\alpha^3 + 24\alpha^2 - 8\alpha + 1 = 0.$$

A saddle point approximation now gives us

$$\begin{aligned} \mathbb{E}Y = \sum_{k=2}^{n-1} \mathbb{E}Y_k &\sim \sum_{k=3n/4}^{n} A(k/n)\exp\left[nF_1(k/n)\right] n^{-1/2} \\ &\sim A(\alpha_*)\exp(nF_1(\alpha_*))\sqrt{\frac{2\pi}{-F_1''(\alpha_*)}} \\ &= \exp(nF_1(\alpha_*))\sqrt{\frac{(2\alpha_*-1)^{11}(3\alpha_*-2)}{(3-2\alpha_*)^8}} \\ &\approx 2.4253\exp(0.40122467n). \end{aligned}$$

4. Counting pairs of 2-intervals

We will now use the second moment method to show that $Y/\mathbb{E}Y$ is tight as $n \to \infty$ by establishing:

Theorem 4.1. $\mathbb{E}Y^2 = \mathcal{O}(\mathbb{E}Y)^2$.

Just as the mean of Y may be obtained from a saddle point analysis of $\mathbb{E}Y_k$ near $k/n = \alpha_*$, we expect the second moment of Y to be dominated by terms $\mathbb{E}Y_k^2$ with k near some α_{**}. Because we have seen from numerical data that the quenched and annealed behaviour are the same, we expect to find, and do find, that $\alpha_{**} = \alpha_*$.

Again we will take advantage of symmetry. This time, if I and I' are 2-intervals, we will need to know the cardinality of their intersection before we can say the probability that $G^{-1}(I)$ and $G^{-1}(I')$ are both 2-intervals. We therefore define $N(k, k', n, \kappa)$ to be the number of pairs of 2-intervals (I, I') of $[n]$ with $|I| = k, |I'| = k'$ and $|I \cap I'| = \kappa$. For the computation of $\mathbb{E}Y_k^2$ we will want to specialise to the case $k = k'$, so we denote $N(k, n, \kappa) := N(k, k, n, \kappa)$. Our computations will now be analogous to the computation F_1 and its argmax, α_*. Specifically, letting

$$\alpha := \frac{k}{n}, \qquad \beta := \frac{k'}{n}, \qquad \rho := \frac{\kappa}{n},$$

we will find a rate function $\mathsf{rate}(\alpha, \beta, \rho)$ such that

$$N(k, k', n, \kappa) \sim A_0(\alpha, \beta, \rho) n^{-3/2} \exp(n \mathsf{rate}(\alpha, \beta, \rho)) \tag{7}$$

for all parameter values in a range containing the dominant contributions to the second moment of Y. To obtain the analogue of (6) for second moments, we need the rate function for total number of pairs of subsets A and B of $[n]$ with $|A| = k, |B| = k'$ and $|A \cap B| = \kappa$. The total number is exactly

$$\binom{n}{\kappa, k-\kappa, k'-\kappa, n-k-k'+\kappa} = \frac{n!}{\kappa!(k-\kappa)!(k'-\kappa)!(n-k-k'+\kappa)!},$$

whence for any pair of sets of respective cardinalities k and k' whose intersection has cardinality $\kappa \le k$, the probability that their union is a specific pair of sets is the reciprocal, P, of this. Let ent denote the exponential rate $\lim n^{-1} \log P$ for this probability:

$$\mathsf{ent} := \rho \log \rho + (\alpha-\rho)\log(\alpha-\rho) + (\beta-\rho)\log(\beta-\rho) + (1-\alpha-\beta+\rho)\log(1-\alpha-\beta+\rho). \tag{8}$$

We may now evaluate the rate function

$$F_2(\alpha, \beta, \rho) := \lim_{n\to\infty, k/n\to\alpha, k'/n\to\beta, \kappa/n\to\rho} N(k, k', n, \kappa)^2 P(k, k', n, \kappa) \tag{9}$$

$$= 2 \cdot \mathsf{rate} + \mathsf{ent}$$

for the expected number of pairs of 2-intervals of respective sizes k and k' with overlap κ. Since $\mathbb{E}Y^2 = \sum_{k,k'} \mathbb{E}Y_k Y_{k'} = \sum_{k,k',\kappa} N(k, k', n, \kappa)^2 P(k, k', n, \kappa)$ is the sum of polynomially many summands, it follows that the exponential order of $\mathbb{E}Y^2$ is the same as the order of the largest summand, namely

$$n^{-1} \log \mathbb{E}Y^2 \to \sup_{\alpha,\beta,\rho} F_2(\alpha, \beta, \rho) := \lambda^{**}. \tag{10}$$

Furthermore, from the inequality $\mathbb{E}Y_k Y_{k'} \le \frac{1}{2}(\mathbb{E}Y_k^2 + \mathbb{E}Y_{k'}^2)$, we see that the maximum of $\mathbb{E}Y_k Y_{k'}$ (for fixed n) can only occur when $k' = k$, and therefore

$$\lambda^{**} = \sup_{\alpha,\rho} F_2(\alpha, \alpha, \rho). \tag{11}$$

Next, consider how $N(k, k', n, \kappa)$ varies with κ for fixed (k, k', n). In other words, enumerate pairs of 2-intervals of fixed sizes k and k' according to the size of their intersection, κ. Observe that $\sum_\kappa N(k,k',n,\kappa)$ counts all pairs of 2-intervals of sizes k and k', so that

$$\sum_\kappa N(k,k',n,\kappa) = N(k,n)N(\kappa,n)\,.$$

Since the number of summands is linear in n, we have at the exponential level that

$$\sup_\rho \mathsf{rate}(\alpha,\beta,\rho) = u(\alpha)u(\beta) \tag{12}$$

where $u(\alpha) = \lim n^{-1}N(k,n)$ as $n \to \infty$ with $k/n \to \alpha$. We will show:

Lemma 4.2. *For α and β in a neighbourhood of α_{**}, this supremum occurs at $\rho = \alpha\cdot\beta$.*

Observe that $-\mathsf{ent}(\alpha,\beta,\cdot)$ has a maximum at $\alpha\cdot\beta$ as well, where it is given by the expression

$$\mathsf{ent}(\alpha,\beta,\alpha\cdot\beta) = h(\alpha) + h(\beta)\,. \tag{13}$$

Here, $h(x) = x\log(x) + (1-x)\log(1-x)$ is the usual entropy function. Both rate and ent are smooth, so we see that the function $F_2 = 2\cdot\mathsf{rate} + \mathsf{ent}$, viewed as a function of the third argument, has a critical point at $\rho = \alpha\cdot\beta$ as well. The next lemma, requiring a four-variable generating function and considerable computation, is:

Lemma 4.3. *The critical point $(\alpha,\beta,\alpha\cdot\beta)$ is in fact a maximum for F_2.*

Corollary 4.4.

$$\lambda_{**} = 2F_1(\alpha_*)\,.$$

Proof. It follows from (12) and (13) that

$$\begin{aligned}\sup_\rho F_2(\alpha,\beta,\rho) &= F_2(\alpha,\beta,\alpha\cdot\beta)\\ &= 2\cdot\mathsf{rate}(\alpha,\beta,\alpha\cdot\beta) + \mathsf{ent}(\alpha,\beta,\alpha\cdot\beta)\\ &= 2u(\alpha) + 2u(\beta) + h(\alpha) + h(\beta)\\ &= F_1(\alpha) + F_1(\beta)\,.\end{aligned}$$

Taking the maximum over α and β then gives the result. □

Theorem 4.1 now follows directly from Corollary 4.4 and:

Lemma 4.5. *1. There is a neighbourhood of (α_*,α_*,ρ) for which*

$$N(k,k',n,\kappa)^2P(k,k',n,\kappa) \sim A(\alpha,\beta,\rho)n^{-3/2}\exp(nF_2(\alpha,\beta,\rho))$$

uniformly as $n\to\infty$ with $(k/n,k'/n,\kappa/n)\to(\alpha,\beta,\rho)$.

2. The Hessian of F_2 here is non-degenerate, whence

$$\sum_{k,k',\kappa} N(k,k',n,\kappa)^2P(k,k',n,\kappa) \sim \frac{A(\alpha_*,\alpha_*,\alpha_*^2)}{\sqrt{(2\pi)^3\mathcal{H}}}\exp(2nF_1(\alpha_*))\,.$$

We will prove Lemmas 4.2, 4.3 and 4.5 in the next section by means of the aforementioned four-variable generating function.

5. The generating function

Recall that $N(k,k',\kappa,n)$ denotes the number of pairs of 2-intervals (I,I') of $[n]$ with $|I|=k, |I'|=k'$ and $|I\cap I'|=\kappa$. Define the generating function

$$F(u_1,u_2,t,z) := \sum_{k,k',\kappa,n} N(k,k',\kappa,n)u_1^k u_2^{k'} t^\kappa z^n$$

so that $N(k,\kappa,n) = [u_1^k u_2^k t^\kappa z^n]F$.

Given a pair of subsets of $[n]$, the positions 1 through n may be divided into three parts:

1. An initial sequence of positions before the first common position;
2. A common position followed by zero or more segments of the form: **a sequence of positions not common to either set, in such a way that no two positions in a row are absent from either set, followed by a common position;**
3. A final sequence of positions after the last common position.

We give generating functions for the three parts separately. To enumerate the second of the three parts, note that each segment between common positions can be one of six possible classes of configuration. We list these here, along with the factor contributed by such a step to the generating function.

- (a) Empty. $f_a = 1$.
- (b) A single position, which can belong to either set or neither, but not both. $f_b = z(1+u_1+u_2)$.
- (c) A positive number of pairs $(j, j+1)$ where $j\in I\setminus I'$ and $j+1\in I'\setminus I$. $f_c = z^2u_1u_2/(1-z^2u_1u_2)$.
- (c') A positive number of pairs $(j,j+1)$ where $j\in I'\setminus I$ and $j+1\in I\setminus I'$. $f_{c'} = f_c$.
- (d) the same as (c) but there is a single position in $I\setminus I'$ at the end. $f_d = zu_1 f_c$.
- (d') the same as (c′) but there is a single position in $I'\setminus I$ at the end. $f_{d'} = zu_2 f_{c'}$.

The generating function for an arbitrary sequence of these is

$$f = \frac{zu_1u_2t}{1-zu_1u_2t(f_a+f_b+(2+z(u_1+u_2)f_c)} . \tag{14}$$

To enumerate the first (or last) of the three parts, we first write down the generating function

$$F_6(u,z) := \frac{uz}{1-(1+z)uz}$$

for that part of a 2-interval between its first and last point. By symmetry, parts 1 and 3 have the same generating function, which is equal to $1/(1-z)$ times the generating function g for the segment to the right of the last common position but to the left of the last position in $I\cup I'$. We may write g as the sum of several cases.

- (e) Empty. $g_e = 1$.
- (f) A position in neither set, followed by a non-empty string of positions, each of which is in neither set or I, with no two in a row not in I. $g_f = zF_6(u_1,z)$.
- (f') A position in neither set, followed by a non-empty string of positions, each of which is in neither set or I', with no two in a row not in I'. $g_{f'} = zF_6(u_2,z)$.
- (g) A string of pairs as in case (c) above, followed by a nonempty sequence as in case (f). $g_g = F_6(u_1,z)/(1-z^2u_1u_2)$.
- (g') A string of pairs as in case (c') above, followed by a nonempty sequence as in case (f'). $g_{g'} = F_6(u_2,z)/(1-z^2u_1u_2)$.
- (h) The same as (g) except with a position in $I'\setminus I$ in the beginning. $g_h = zu_2F_6(u_1,z)/(1-z^2u_1u_2)$.

(h') The same as (g') except with a position in $I \setminus I'$ in the beginning. $g_{h'} = zu_1F_6(u_2,z)/(1-z^2u_1u_2)$.

Summing, we see that the factor from the first and last parts is

$$g = \frac{1}{1-z}\left(1 + zF_6(u_1,z) + zF_6(u_2,z) + \frac{F_6(u_1,z)+F_6(u_1,z)}{1-z^2u_1u_2}\right.$$
$$\left. + \frac{u_2zF_6(u_1,z)+u_1zF_6(u_2,z)}{1-z^2u_1u_2}\right),$$

and finally,

$$F(u_1,u_2,t,z) = fg^2\,. \tag{15}$$

6. Proofs of lemmas

There is only room for sketches here. Details are given in [8].

Let V be the variety where the rational function F blows up, namely the union of the varieties where the denominators of f and g vanish. It simplifies the computation to change variables to $\tau := ztu_1u_2$, $u = zu_1$ and $v = zu_2$. The $[k,k',n,\kappa]$ coefficient of F now becomes the $[k-\kappa, k'-\kappa, n-k-k'+\kappa, \kappa]$ coefficient of the function $\widetilde{F}(x,y,z,\tau)$. Thus

$$\frac{1}{n}\log[n\alpha, n\beta, n, n\rho]F = \frac{1}{n}\log[n\mu, n\nu, n\delta, n\rho]\widetilde{F}$$

when $\alpha = \mu+\rho$, $\beta = \nu+\rho, \rho = \alpha+\beta+\delta-1$ and $\widetilde{F}$ is F under the change of variables. In the new variables,

$$\widetilde{f} = \frac{\tau(1-xy)}{1-xy-\tau(1+x+y+xy+zxy-z)}\,.$$

We let h denote the denominator of this expression. Then V is the union of the variety V_h where h vanishes, together with the varieties where the functions $g_1 := 1-z$, $g_2 := 1-xy$, $g_3 := 1-(1+z)x$ and $g_4 := 1-(1+z)y$ vanish.

Let Dom denote the domain of convergence for the power series for

$$F = \sum_{\mathbf{r}:=(k,k',n,\kappa)} a_{\mathbf{r}}\mathbf{x}^{\mathbf{r}}\,.$$

Let $\log\mathcal{D}$ denote the logarithmic domain, which is always convex:

$$\log\mathcal{D} := \{\log|x_1|,\ldots,\log|x_4|) : \mathbf{x}\in \mathsf{Dom}\}\,.$$

Asymptotics of $a_{\mathbf{r}}$ in the direction $\mathbf{r}/|\mathbf{r}| \to \mathbf{s}$ are "controlled" by corresponding points $\mathbf{x}(\mathbf{s}) \in V$ as described in the following paragraph (see [8] for details).

First, it is shown in [16, Theorem 6.3] that for each vector $\mathbf{s} \in (\mathbb{R}^+)^4$, there is a point $\mathbf{x}(\mathbf{s}) \in V$ with positive real coordinates such that $\mathbf{x} \in \partial\mathsf{Dom}$ and such that the normal to the support hyperplane to $\log\mathcal{D}$ at $\log\mathbf{x}$ is parallel to $\mathbf{s}$. It then follows from [16, Theorem 3.5], provided a certain function has a non-degenerate Hessian, that if $\mathbf{x}(\mathbf{s})$ lies in some smooth component of V for all $\mathbf{s}$ in some neighbourhood $\mathcal{N}$, then there is a function A such that the coefficients of

$$F = \sum_{\mathbf{r}:=(k,k',n,\kappa)} a_{\mathbf{r}}\mathbf{x}^{\mathbf{r}}$$

are uniformly asymptotically approximated on $\mathcal{N}$ by

$$a_{\mathbf{r}} \sim A(\frac{r_1}{r_3}, \frac{r_2}{r_3}, \frac{r_4}{r_3}) n^{-3/2} \exp(-\mathbf{r} \cdot \mathbf{x}(\mathbf{r})) .$$

The key computational step necessary to make use of these results is to verify that the point $\mathbf{x}(\mathbf{r})$ is in the part V_h of the pole set V that corresponds to the vanishing of h. Combinatorially, the factor f in F enumerates pairs of subsets for which the first and last common positions are 1 and n respectively. The other divisors correspond to the cases where one or both intervals begin at $\Theta(n)$ or end at $n - \Theta(n)$. We now give an argument that the pairs of sets beginning at 1 and ending at n count "nearly all" pairs, meaning that the rate $n^{-1} \log N$ is the same.

Let $\widehat{N}(k, k', \kappa, n)$ denote the number of pairs of 2-intervals in which both the subsets contains 1 and n. Suppose, for a given (k, k', κ, n) that

$$\widehat{N}(k, k', \lambda n, \kappa) \geq \widehat{N}(k, k', n, \kappa) \qquad (16)$$

for some fixed $\lambda < 1$ and arbitrarily large n (this can happen, for example when $\mu = \alpha - \rho$ is small). Then

$$\begin{aligned} \widehat{N}(k, k', \lambda n, \kappa)^2 P(k, k', \lambda n, \kappa) &\geq \widehat{N}(k, k', n, \kappa)^2 P(k, k', \lambda n, \kappa) \\ &> \widehat{N}(k, k', n, \kappa)^2 P(k, k', n, \kappa) \end{aligned}$$

by an exponential factor. But $\sup_{\alpha,\beta,\rho} F_2(\alpha, \beta, \rho) > 0$, so if this supremum is achieved at $\alpha = k/n, \beta = k'/n, \rho = \kappa/n$, then the inequality would be reversed. Thus, in a neighbourhood of where the supremum of F_2 is achieved, it is not possible for (16) to hold. Consequently, in this neighbourhood the coefficients of $\tilde{f}$ are, on the exponential scale, as large as those of $\widetilde{F}$, whence the minimal point occurs on V_h.

Proof of Lemma 4.2: Choose a pair of 2-intervals of sizes k and k' uniformly from pairs where both contain 1 and n. Define the random variable $Z_{k,k',n}$ to be n^{-1} times the cardinality of the intersection of these two sets. If $Z_{k,k',n} \to \gamma$ in probability as $n \to \infty$ with $k/n \to \alpha$ and $k'/n \to \beta$, then the supremum of $\mathsf{rate}(\alpha, \beta, \cdot)$ occurs at γ. But we see from the construction of the generating function that the indicator functions of the two sets form two independent Markov chains, both constrained to go from 1 to n in a given number of steps. This implies asymptotically independent statistics, meaning that the limit in probability of $Z_{k,k',n}$ is kk'/n. □

Proof of Lemma 4.3: We argue here only what we need, namely the lemma for $\beta = \alpha$ and α in a neighbourhood of α_*. We have seen that rate and $-\mathsf{ent}$ both have maxima at $(\alpha_*, \alpha_*, \alpha_*^2)$, so this is a matter of checking the second derivative of $2 \cdot \mathsf{rate} + \mathsf{ent}$ for negative definiteness and then checking that this local maximum is in fact a global maximum for F_2. To do this, we evaluate the point $\mathbf{x}$ controlling asymptotics of $F_2(\alpha, \alpha, \rho)$.

Using Maple, we solve for $\mathbf{x} \in V_h$ and

$$(x_1 \partial h/\partial x_1, x_2 \partial h/\partial x_2, x_3 \partial h/\partial x_3, x_4 \partial h/\partial x_4)$$

parallel to $\mathbf{s}$ for which $\alpha = s_1/s_3, \beta = s_2/s_3, \rho = s_4/s_3$. We get

$$u_0 = v_0 = \frac{2\alpha - 1 - R}{2(\alpha + \delta - 1)};$$

$$z_0 = -\frac{(2\alpha - 1 - R)\delta}{4\delta^2 - 6\delta + 2\delta R + 4\delta\alpha - 6\alpha + 3 + 4\alpha^2 - R};$$

$$\tau_0 = -\frac{4\delta^2 - 6\delta + 2\delta R + 4\delta\alpha - 6\alpha + 3 + 4\alpha^2 - R}{2\delta\alpha - \delta - \delta R - 4\alpha + 4\alpha^2 + q + 1 - 2\alpha R};$$

here, $\delta = 1 - 2\alpha + \rho$ and

$$R := \sqrt{8\alpha^2 - 12\alpha + 5 + 8\delta\alpha - 8\delta + 4\delta^2}.$$

We take two derivatives and evaluate at $\rho = \alpha^2$, obtaining a rational function of α. Doubling and adding the second derivative of ent, we may verify negativity of the second derivative in ρ for all $\alpha > 2/3$. This establishes a local maximum at $\rho = \alpha^2$. We have already seen that the global maximum of $F_2(\alpha, \beta, \rho)$ occurs at $\beta = \alpha$. It is straightforward to check numerically that there is no maximum in a set bounded away from the diagonal $\beta = \alpha$ exceeding the value of $F_2(\alpha, \alpha, \alpha^2)$ at $\alpha = \alpha_*$. □

Proof of Lemma 4.5: For this we need to compute the solution $\mathbf{x}(\mathbf{s})$ in all four variables, rather than just when $\beta = \alpha$. The result is rather messy and may be found in [8]. In a neighbourhood of $(\alpha_*, \alpha_*, \alpha_*^2)$ we may use rigorous numerical estimates to verify that the Hessian is non-degenerate. This establishes the second assertion of the lemma. The first follows from [16, Theorem 3.5]. □

References

[1] R. Arratia, L. Goldstein and L. Gordon. Two moments suffice for Poisson approximation: the Chen-Stein method. *Annals of Probability*, 17:9–25, 1989.

[2] R. Arratia, L. Goldstein and L. Gordon. Poisson approximation and the Chen-Stein method. *Statistical Science*, 5:403–424, 1990.

[3] Y. Baryshnikov and R. Pemantle. *Manuscript in preparation.*

[4] A. Bergeron and J. Stoye, On the Similarity of Sets of Permutations and Its Applications to Genome Comparison, COCOON 2003, *Lecture Notes in Computer Science*, 2697: 68-79, (2003).

[5] A. Bergeron, S. Heber and J. Stoye, Common intervals and sorting by reversals: a marriage of necessity. *Proceedings of ECCB 2002*: 54-63, (2002).

[6] A. Bergeron, S. Corteel and M. Raffinot: The Algorithmic of Gene Teams. WABI 2002, *Lecture Notes in Computer Science*, 2452: 464-476, (2002).

[7] K. S. Booth and G. S. Lueker, Testing for the Consecutive Ones Property, Interval Graphs, and Graph Planarity Using PQ-Tree Algorithms. *J. Comput. Syst. Sci.* 13(3): 335-379 (1976)

[8] S. Corteel, G. Louchard and R. Pemantle. Common intervals in permutations (unabridged version). `http://www.math.upenn.edu/~pemantle/louchard/version040122.ps`

[9] G. Didier, Common Intervals of Two Sequences. WABI 2003, *Lecture Notes in Computer Science*, 2812: 17-24 (2003).

[10] S. Heber and J. Stoye, Algorithms for Finding Gene Clusters. WABI 2001, *Lecture Notes in Computer Science*, 2149: 252-263, (2001).

[11] S. Heber and J. Stoye, Finding All Common Intervals of k Permutations. CPM 2001, *Lecture Notes in Computer Science*, 2089: 207-218, (2001).

[12] I. Kaplansky. The asymptotic distributions of runs of consecutive elements. *Annals of Mathematical Statistics*, 16:200–203, 1945.

[13] S. Kobayashi, I. Ono and M. Yamamura, An Efficient Genetic Algorithm for Job Shop Scheduling Problems. *ICGA 1995*: 506-511, (1995).

[14] V.K. Kolchin, A.S. Sevastyanov, and P.C. Chistiakov. *Random Allocations*. Wiley, 1978.

[15] H. Mühlenbein, M. Gorges-Schleuter, and O. Krämer. Evolution algorithms in combinatorial optimization.*Parallel Comput.*, 7:65-85, (1988).

[16] R. Pemantle and M. Wilson Asymptotics of multivariate sequences, part I: smooth points of the singular variety. *J. Comb. Theory, Series A*, 97:129–161, 2001.

[17] R. Pemantle and M. Wilson Asymptotics of multivariate sequences, part II: multiple points of the singular variety. *Combinatorics, Probability and Computing, to appear.*

[18] T. Uno and M. Yagiura, Fast Algorithms to Enumerate All Common Intervals of Two Permutations. *Algorithmica* 26(2): 290-309 (2000).

[19] J. Wolfowitz. Additive partition functions and a class of statistical hypotheses. *Annals of Mathematical Statistics*, 13:247–279, 1942.

[20] J. Wolfowitz. Note on runs of consecutive elements. *Annals of Mathematical Statistics*, 15:97–98, 1944.

Sylvie Corteel
CNRS PRiSM, Université de Versailles Saint-Quentin, 45 Avenue des Etats-Unis, 78035 Versailles France
email: syl@prism.uvsq.fr

Guy Louchard
Université Libre de Bruxelles, Département d'Informatique, CP 212, Boulevard du Triomphe, B-1050 Bruxelles, Belgium
email:louchard@ulb.ac.be

Robin Pemantle
Department of Mathematics, University of Pennsylvania, 209 S. 33rd Street, Philadelphia, PA 19104 USA, *Supported by NSF Grant # DMS-0103635*
email:pemantle@math.upenn.edu

Overpartitions and Generating Functions for Generalized Frobenius Partitions

Sylvie Corteel, Jeremy Lovejoy, and Ae Ja Yee

ABSTRACT: *Generalized Frobenius partitions, or F-partitions, have recently played an important role in several combinatorial investigations of basic hypergeometric series identities. The goal of this paper is to use the framework of these investigations to interpret families of infinite products as generating functions for F-partitions. We employ q-series identities and bijective combinatorics.*

1. Introduction

Let $P_{A,B}(n)$ denote the number of generalized Frobenius partitions of n, i.e., the number of two-rowed arrays,

$$\begin{pmatrix} a_1, a_2, \ldots, a_m \\ b_1, b_2, \ldots, b_m \end{pmatrix}, \tag{1}$$

in which the top (bottom) row is a partition from a set A (B), and such that $\sum(a_i + b_i) + m = n$ [2]. The classical example is the case $P_{D,D}(n)$, where D is the set of partitions into distinct non-negative parts. Frobenius observed that these objects are in one-to-one correspondence with the ordinary partitions of n, giving

$$\sum_{n=0}^{\infty} P_{D,D}(n) q^n = \prod_{n=1}^{\infty} \frac{1}{(1-q^n)}. \tag{2}$$

Andrews [2] later made an extensive study of two infinite families of F-partitions that begin with $P_{D,D}(n)$. He replaced D by D_k or C_k, the set of partitions where parts repeat at most k times and the set of partitions into distinct parts with k colors, respectively. The generating functions are multiple theta series, which in three known cases can be written as an infinite product.

$$\sum_{n=0}^{\infty} P_{D_2,D_2}(n) q^n = \frac{(q^2;q^2)_\infty (-q^3;q^6)_\infty}{(q)_\infty^2 (-q;q^2)_\infty}, \tag{3}$$

$$\sum_{n=0}^{\infty} P_{D_3,D_3}(n) q^n = \frac{(q^6;q^6)_\infty (q^6;q^{12})_\infty^2 (q^2;q^2)_\infty (q;q^2)_\infty^2}{(q)_\infty^3 (q^3;q^6)_\infty^2}, \tag{4}$$

$$\sum_{n=0}^{\infty} P_{C_2,C_2}(n) q^n = \frac{(-q;q^2)_\infty^2}{(q)_\infty (q;q^2)_\infty}. \tag{5}$$

Here we have employed the standard notation

$$(a_1, \ldots, a_j)_\infty := (a_1, \ldots, a_j; q)_\infty := \prod_{k=0}^{\infty} (1 - a_1 q^k) \cdots (1 - a_j q^k). \tag{6}$$

While Andrews' families subsequently received quite a bit of attention [10, 13, 15, 16, 18, 20], other types of Frobenius partitions have recently been turning up as novel interpretations for some infinite products that figure prominently in basic hypergeometric series identities [6, 7, 8, 21]. The combinatorial setting here is that of overpartitions, which are partitions wherein the first occurrence of a part may be overlined. Let O denote the set of overpartitions into non-negative parts. Then it turns out that

$$\sum_{m,n\geq 0} P_{D,O}(m,n)b^m q^n = \frac{(-bq)_\infty}{(q)_\infty} \tag{7}$$

and

$$\sum_{\ell,m,n\geq 0} P_{O,O}(\ell,m,n)a^\ell b^m q^n = \frac{(-aq,-bq)_\infty}{(q,abq)_\infty}. \tag{8}$$

Here $P_{D,O}(m,n)$ denotes the number of F-partitions counted by $P_{D,O}(n)$ that have m non-overlined parts in the bottom row, and $P_{O,O}(\ell,m,n)$ denotes the number of objects counted by $P_{O,O}(n)$ that have ℓ non-overlined parts in the top row and m non-overlined parts in the bottom row.

Since a thorough combinatorial understanding of (7) and (8) has been so useful, we give in this paper a variety of other infinite product generating functions for F-partitions and begin to study them using bijective combinatorics. The first goal is to use restricted overpartitions and a useful property of the ${}_1\psi_1$ summation (see Lemma 2.2) to embed some of (2) – (8) in families of infinite products that generate F-partitions.

Theorem 1.1. *Let O_k be the set of overpartitions where the non-overlined parts occur less than k times. Let $P_{O_k,O_k}(m,n)$ (resp. $P_{O_k,O}(m,n)$) be the number of F-partitions counted by $P_{O_k,O_k}(n)$ (resp. $P_{O_k,O}(n)$) wherein the number of overlined parts on the top minus the number of overlined parts on the bottom is m. Then*

$$\sum_{m,n\geq 0} P_{O_k,O_k}(m,n)b^m q^n = \frac{(-bq)_\infty(-q/b)_\infty(q^k;q^k)_\infty}{(q)^2_\infty(-bq^k,-q^k/b;q^k)_\infty}, \tag{9}$$

$$\sum_{m,n\geq 0} P_{O_k,O}(m,n)b^m q^n = \frac{(-bq)_\infty(-q/b)_\infty(q^k;q^k)_\infty}{(q)^2_\infty(-q^k/b;q^k)_\infty}. \tag{10}$$

Notice that in both instances the case $k\to\infty$ is the case $a=1/b$ of (8), while the case $k=1$ of (9) is Frobenius' example (2), the case $b=1, k=2$ of (9) is Andrews' (5), and the case $k=1$ of (10) is (7).

Our next object is to exhibit more families like those above, but where the base cases are none of (2) – (8). We use the notation AB for the set of vector partitions $(\lambda_A,\lambda_B)\in A\times B$, and D^k for the set of partitions into non-negative parts where each part occurs 0 or k times.

Theorem 1.2.

$$\sum_{n=0}^{\infty} P_{O_k,OD^k}(n)q^n = \frac{(-q)^2_\infty}{(q)^2_\infty}(q^k;q^k)_\infty(q^{2k};q^{4k})_\infty, \tag{11}$$

$$\sum_{n=0}^{\infty} P_{O_k,O_{2k}}(n)q^n = \frac{(-q)^2_\infty(q^k;q^k)_\infty(q^k;q^{2k})^2_\infty}{(q)^2_\infty(q^k,q^{4k};q^{5k})_\infty}. \tag{12}$$

Then, by employing more general q-series identities, we find generating functions with more parameters, like Theorems 1.3 – 1.6 below. The first two contain the $k = 1$ case of (11) and the case $k = 2$ of (10), respectively.

Theorem 1.3. *Let $P_{D,OD}(m,n)$ be the number of F-partitions counted by $P_{D,OD}(n)$ that have m parts in λ_D. Then*

$$\sum_{m,n\geq 0} P_{D,OD}(m,n)y^m q^n = \frac{(-q;q)_\infty(-yq;q^2)_\infty}{(q;q)_\infty}. \tag{13}$$

Theorem 1.4. *Let O^2 denote the number of overpartitions in O where the non-overlined parts repeat an even number of times. Let $P_{O,O^2D}(\ell,m,n)$ denote the number of F-symbols counted by $P_{O,O^2D}(n)$ where ℓ is the number of non-overlined parts in the top row minus the number of parts in λ_D and m is the number of non-overlined parts in the bottom row. Then*

$$\sum_{\ell,m,n\geq 0} P_{O,O^2D}(\ell,m,n)a^\ell b^m q^n = \frac{(-aq)_\infty(-q/a,-ab^2q^2;q^2)_\infty}{(q)_\infty(q,a^2b^2q^2;q^2)_\infty}. \tag{14}$$

The next theorem also contains the case $k = 2$ of (10). We are concerned here with $\bar{D}$, which denotes the set of overpartitions into distinct parts such that parts have to differ by at least two if the bigger is overlined and $\overline{0}$ does not occur. These overpartitions have recently arisen in a number of works [5, 14, 17].

Theorem 1.5. *Let $P_{O\bar{D},O}(\ell,m,n)$ denote the number of overpartitions counted by $P_{O\bar{D},O}(n)$ such that ℓ is the number of non-overlined parts in λ_O plus the number of overlined parts in $\lambda_{\bar{D}}$ and such that m is the number of non-overlined parts on the bottom minus the number of parts in $\lambda_{\bar{D}}$. Then*

$$\sum_{\ell,m,n\geq 0} P_{O\bar{D},O}(\ell,m,n)a^\ell b^m q^n = \frac{(-aq,-bq)_\infty(-q/b;q^2)_\infty}{(q,abq)_\infty(q;q^2)_\infty}. \tag{15}$$

The last example contains (8) and deals with O', the set of overpartitions in O that have no $\overline{0}$.

Theorem 1.6. *Let $P_{OO',O}(k,\ell,m,n)$ denote the number of F-partitions counted by $P_{OO',O}(n)$ where k is the number of non-overlined parts in λ_O plus the number of overlined parts in $\lambda_{O'}$, ℓ is the number of non-overlined parts in the bottom row, and m is the number of parts in $\lambda_{O'}$. Then*

$$\sum_{k,\ell,m,n\geq 0} P_{OO',O}(k,\ell,m,n)a^k b^\ell c^m q^n = \frac{(-aq,-bq,-cq)_\infty}{(q,abq,bcq)_\infty}. \tag{16}$$

Finally, we give bijective proofs for some of the generating functions above. We are able to establish (5), (13), and the case $k = 2$ of (10) in this way.

2. Recollections and Proofs

Given a set A of partitions we denote by $P_A(n,k)$ the number of partitions of n from the set A having k parts. We recall from [2] that

Lemma 2.1. *The generating function for Frobenius partitions is given by*

$$\sum_{n=0}^{\infty} P_{A,B}(n)q^n = [z^0]\sum_{n,k} P_A(n,k)q^n(zq)^k \sum_{n,k} P_B(n,k)q^n z^{-k}, \tag{17}$$

where $[z^k]\sum A_n z^n = A_k$.

We assume enough familiarity with the elementary theory of partitions and overpartitions [1, 8] that we can state generating functions for simple $P_A(n,k)$ without explanation. The following is the key lemma mentioned in the introduction.

Lemma 2.2. *If*

$$[z^0]\frac{(-bzq,-1/bz)_\infty}{(zq,1/z)_\infty}G(z,q) = \frac{(-bq,-q/b)_\infty}{(q)_\infty^2}H(q), \tag{18}$$

then

$$[z^0]\frac{(-bzq,-1/bz)_\infty}{(zq,1/z)_\infty}G(z^k,q^k) = \frac{(-bq,-q/b)_\infty}{(q)_\infty^2}H(q^k). \tag{19}$$

Proof. Let $H(q) = [z^0]F(z,q)G(z,q)$, with $F(z,q) = \sum A_j(q)z^j$ and $G(z,q) = \sum B_j(q)z^j$. If $A_{kj}(q) = A_j(q^k)$, then

$$\begin{aligned}
[z^0]F(z,q)G(z^k,q^k) &= [z^0]\sum A_j(q)z^j \sum B_j(q^k)z^{kj} \\
&= \sum A_{-kj}(q)B_j(q^k) \\
&= \sum A_{-j}(q^k)B_j(q^k) \\
&= H(q^k).
\end{aligned}$$

The proof is finished when we apply the above observation to

$$F(z,q) = \sum \frac{(1+b)q^j}{1+bq^j}z^j. \tag{20}$$

Substituting $a = 1/b$ and $z = bz$ in the ${}_1\psi_1$ summation,

$$\sum_{n=-\infty}^{\infty} \frac{(-1/a)_n(azq)^n}{(-bq)_n} = \frac{(q,abq,-zq,-1/z)_\infty}{(-bq,-aq,azq,b/z)_\infty}, \tag{21}$$

we have

$$F(z,q) = \frac{(q,q,-bzq,-1/bz)_\infty}{(-bq,-q/b,zq,1/z)_\infty}. \tag{22}$$

Then

$$\frac{(-bq,-q/b)_\infty}{(q)_\infty^2}[z^0]F(z,q)G(z^k,q^k) = \frac{(-bq,-q/b)_\infty}{(q)_\infty^2}H(q^k),$$

and the lemma follows. □

Above we have introduced the notation

$$(a)_n := (a;q)_n := \frac{(a;q)_\infty}{(aq^n;q)_\infty}. \tag{23}$$

Proof of Theorem 1.1. For the first part, take $G(z,q) = (zq, 1/z)_\infty$ in Lemma 2.2. By (2), $H(q) = (q)_\infty/(-bq)_\infty(-q/b)_\infty$. Then

$$\begin{aligned}\sum_{n=0}^{\infty} P_{O_k,O_k}(m,n)b^m q^n &= [z^0]\frac{(-bzq,-1/bz)_\infty(z^kq^k,z^{-k};q^k)_\infty}{(zq,1/z)_\infty}\\ &= [z^0]\frac{(-bzq,-1/bz)_\infty}{(zq,1/z)_\infty}G(z^k,q^k)\\ &= \frac{(-bq,-q/b)_\infty}{(q)_\infty^2}H(q^k)\\ &= \frac{(-bq)_\infty(-q/b)_\infty(q^k;q^k)_\infty}{(q)_\infty^2(-bq^k,-q^k/b;q^k)_\infty}.\end{aligned}$$

Similarly, take $G(z,q) = (zq)_\infty$ for the second part. □

In the following we will use the ${}_1\psi_1$ summation (21) or one of its corollaries for the first step of each proof.

Proof of Theorem 1.2. For the first part, take $G(z,q) = (-1/z, zq)_\infty$ in the case $b = 1$ of Lemma 2.2. Then

$$\begin{aligned}H(q) &= \frac{(q)_\infty^2}{(-q,q)_\infty^2}[z^0]\frac{(-zq;q)_\infty(-z^{-1};q)_\infty(-z^{-1};q)_\infty}{(z^{-1};q)_\infty}\\ &= [z^0]\frac{(q;q)_\infty}{(-q;q)_\infty}\sum_{n=-\infty}^{\infty}\frac{z^n q^{\binom{n+1}{2}}}{(-q;q)_n}\sum_{n=0}^{\infty}\frac{z^{-n}q^{\binom{n}{2}}}{(q;q)_n}\\ &= \frac{(-q;q)_\infty}{(q;q)_\infty}\sum_{n=0}^{\infty}\frac{q^{n^2}}{(q^2;q^2)_n}\\ &= \frac{(q;q)_\infty(-q;q^2)_\infty}{(-q;q)_\infty},\end{aligned}$$

by the q-binomial theorem,

$$\sum_{n=0}^{\infty}\frac{(a)_n z^n}{(q)_n} = \frac{(az)_\infty}{(z)_\infty}. \tag{24}$$

For the second part we again apply the case $b = 1$ of Lemma 2.2, this time with $G(z,q) = (z^{-1}, zq)_\infty$. Then

$$\begin{aligned}H(q) &= \frac{(q)_\infty^2}{(-q,q)_\infty^2}[z^0](-zq;q)_\infty(-z^{-1};q)_\infty(-z^{-1};q)_\infty\\ &= \frac{(q)_\infty}{(-q)_\infty^2}[z^0]\sum_{n\in\mathbb{Z}}z^n q^{n(n+1)/2}\sum_{n\geq 0}\frac{z^{-n}q^{n(n-1)/2}}{(q)_n}\\ &= \frac{(q)_\infty}{(-q)_\infty^2}\sum_{n=0}^{\infty}\frac{q^{n^2}}{(q)_n}\\ &= \frac{(q)_\infty}{(-q)_\infty^2(q,q^4;q^5)_\infty},\end{aligned}$$

by the first Rogers-Ramanujan identity,

$$\sum_{n=0}^{\infty} \frac{q^{n^2}}{(q)_n} = \frac{1}{(q,q^4;q^5)_\infty}. \tag{25}$$

□

Proof of Theorem 1.3.

$$\begin{aligned}
\sum_{m,n\geq 0} P_{D,OD}(m,n)y^m q^n &= [z^0]\frac{(-zq;q)_\infty(-z^{-1};q)_\infty(-yz^{-1};q)_\infty}{(z^{-1};q)_\infty} \\
&= [z^0]\frac{(-q;q)_\infty}{(q;q)_\infty}\sum_{n=-\infty}^{\infty}\frac{z^n q^{\binom{n+1}{2}}}{(-q;q)_n}\sum_{n=0}^{\infty}\frac{y^n z^{-n} q^n 2}{(q;q)_n} \\
&= \frac{(-q;q)_\infty}{(q;q)_\infty}\sum_{n=0}^{\infty}\frac{y^n q^{n^2}}{(q^2;q^2)_n} \\
&= \frac{(-q;q)_\infty(-yq;q^2)_\infty}{(q;q)_\infty},
\end{aligned}$$

the final equality being the case $q = q^2, z = -q/a$, and $a \to \infty$ of (24). □

Proof of Theorem 1.4.

$$\sum_{\ell,m,n\geq 0} P_{O,O^2D}(\ell,m,n)a^\ell b^m q^n = [z^0]\frac{(-zq,-1/z,-1/az)_\infty}{(azq)_\infty(b^2/z^2;q^2)_\infty}$$

$$\begin{aligned}
&= [z^0]\frac{(-aq,-bq)_\infty}{(q,abq)_\infty}\sum_{n\in\mathbb{Z}}\frac{(-1/a)_n(azq)^n}{(-bq)_n}\sum_{n=0}^{\infty}\frac{(1/ab)_n(-b/z)^n}{(q)_n} \\
&= \frac{(-aq,-bq)_\infty}{(q,abq)_\infty}\sum_{n=0}^{\infty}\frac{(-1/a,1/ab)_n(-abq)^n}{(-bq,q)_n} \\
&= \frac{(-aq)_\infty(-q/a,-ab^2q^2;q^2)_\infty}{(q)_\infty(q,a^2b^2q^2;q^2)_\infty},
\end{aligned}$$

by the q-Kummer identity,

$$\sum_{n=0}^{\infty}\frac{(a,b)_n(-q/b)^n}{(q,aq/b)_n} = \frac{(aq,aq^2/b^2;q^2)_\infty}{(-q/b,aq/b)_\infty(q;q^2)_\infty}. \tag{26}$$

Proof of Theorem 1.5.

$$\sum_{\ell,m,n\geq 0} P_{O\bar{D},O}(\ell,m,n)a^\ell b^m q^n [z^0]\frac{(-zq,-1/z)_\infty}{(azq,b/z)_\infty}\sum_{n=0}^{\infty}\frac{(-aq)_n q^{n(n+1)/2}(z/b)^n}{(q)_n}$$

$$\begin{aligned}
&= [z^0]\frac{(-aq,-bq)_\infty}{(q,abq)_\infty}\sum_{n\in Z}\frac{(-1/b)_n(b/z)^n}{(-aq)_n}\sum_{n=0}^{\infty}\frac{(-aq)_n q^{n(n+1)/2}(z/b)^n}{(q)_n} \\
&= \frac{(-aq,-bq)_\infty}{(q,abq)_\infty}\sum_{n=0}^{\infty}\frac{(-1/b)_n q^{n(n+1)/2}}{(q)_n} \\
&= \frac{(-aq,-bq)_\infty(-q/b;q^2)_\infty}{(q,abq)_\infty(q;q^2)_\infty},
\end{aligned}$$

the final equality being Lebesgue's identity which is the case $b \to \infty$ and $a = -1/b$ of (26). □

Proof of Theorem 1.6.

$$\begin{aligned}
&\sum_{k,\ell,m,n\geq 0} P_{OO',O}(k,\ell,m,n)a^k b^\ell c^m q^n [z^0]\frac{(-zq,-1/z,-aczq^2)_\infty}{(azq,b/z,czq)_\infty} \\
&= [z^0]\frac{(-aq,-bq)_\infty}{(q,abq)_\infty}\sum_{n\in\mathbb{Z}}\frac{(-1/b)_n(b/z)^n}{(-aq)_n}\sum_{n=0}^{\infty}\frac{(-aq)_n(czq)^n}{(q)_n} \\
&= \frac{(-aq,-bq)_\infty}{(q,abq)_\infty}\sum_{n=0}^{\infty}\frac{(-1/b)_n(bcq)^n}{(q)_n} \\
&= \frac{(-aq,-bq,-cq)_\infty}{(q,abq,bcq)_\infty},
\end{aligned}$$

by the q-binomial theorem (24). □

It should be noted that Theorems 1.4 – 1.6 have k-generalizations like Theorems 1.1 and 1.2 but the combinatorial definition of the F-partitions are less palatable.

3. Bijections

In this section we establish some partitions that explains some of the first cases.

Bijection for (5) We will here give a combinatorial proof of

$$\sum_{n=0}^{\infty} P_{O_2,O_2}(n)q^n = \frac{(-q;q)_\infty(-q;q^2)^2_\infty}{(q;q)_\infty}.$$

We start with a F-partition and add one to each entry of the first row. Let k be the number of overlined parts in the first row minus the number of overlined parts of the second row. Suppose without loss of generality that $k \geq 0$. Then we split the F-partition into two F-partitions : one that contains the overlined parts and one that contains the non-overlined parts. Apply Wright's bijection (see [20]) to both F-partitions and get two ordinary partitions and two triangles $(k, k-1, \ldots, 1)$ and $(k-1, \ldots, 1)$. We keep the first partition, which gives $1/(q;q)_\infty$ and the odd parts of the second partition, which gives $1/(q;q^2)_\infty$. Then we divide the even parts of the second partition by 2. To the left of these parts we put $k, k-1, \ldots, 1$. By applying the reverse of Wright's bijection, we obtain two partitions into distinct parts. We multiply by two and decrease by 1 the parts of the first partition and we multiply by 2 and increase by 1 the parts of the second partition. We get two partitions into distinct odd parts, which is generated by $(-q;q^2)^2$. It is easy to check that the weight is preserved and that every step is reversible.

Bijection for (10). Now let us prove combinatorially that

$$\sum_{n=0}^{\infty} P_{O_2,O}(n)q^n = \frac{(-q;q)_\infty(-q;q^2)_\infty}{(q;q)_\infty(q;q^2)_\infty}$$

This proof is inspired by some ideas of [21]. We start with the Frobenius partition and add one to each entry of the first row. The top row is an overpartition in

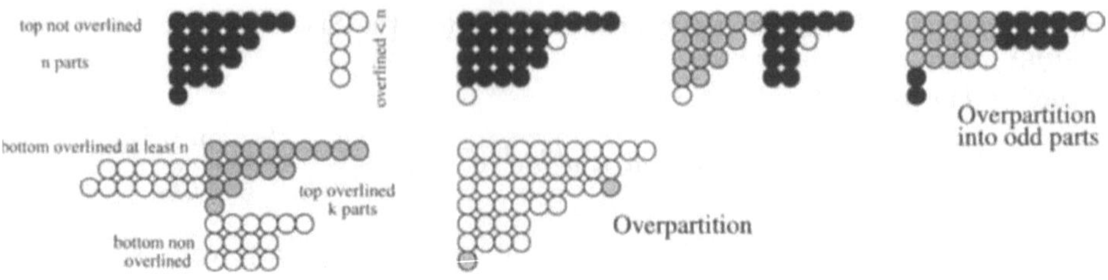

FIGURE 1. Bijection for 10

O_2. Therefore the non-overlined parts form a partition into distinct parts. Let us suppose that this top row has n non-overlined parts and k overlined parts. We separate these into partitions α and β. The bottom row is an overpartition into $n+k$ non-negative parts. Using algorithm Z [4, 22], we can decompose it into a partition δ into parts at most $n+k$ and a partition γ into non-negative distinct parts less than $n+k$.

We take the parts of γ that are less than n, conjugate them, and add them to α to create an overpartition into distinct parts (a part is overlined if the difference with the previous part is at least 2). Then we change α into an overpartition into odd parts generalizing Sylvester's bijection [19]. First we take off $2\lfloor (n-1)/2 \rfloor + 2 - i$ from the ith part and change it to η a $(2\lfloor (n-1)/2 \rfloor + 1) \times \lceil n/2 \rceil$ rectangle. Then we look at the conjugate of what is left. Every odd part is inserted in η and every even part $2i$ is changed to two i parts that are inserted in the conjugate of η. The overlines follow. Note that this part is a bijective proof of Lebesgue's identity.

Now we take the parts of γ that are equal to or greater than n. If the part $n+i-1$ occurs we add it to the ith part of β and take off the overline. After sorting, that creates an overpartition where the non-overlined parts are greater than $n+k$. We add then δ and get an overpartition, which is generated by $(-q;q)_\infty/(q;q)_\infty$.

For example, we start with $\begin{pmatrix} \bar{8}, & 6, \bar{4}, 4, 3, 2, \bar{1}, \bar{0}, 0 \\ \overline{10}, \bar{9}, \bar{5}, 3, 3, \bar{1}, 1, 0, 0 \end{pmatrix}$. We get $\alpha = (7,5,4,3,1)$, $\beta = (9,5,2,1)$, $\gamma = (7,6,4,1)$ and $\delta = (6,4,4)$. Then we apply the mapping and get $\eta = (\overline{11}, 9, \bar{5}, 1, 1)$ and $\mu = (11, \bar{9}, 9, 6, 4, 4, \bar{1})$. See Fig 1.

Bijection for (13) We will here give a combinatorial proof of

$$\sum_{n=0}^{\infty} P_{D,OD}(n) q^n = \frac{(-q;q)_\infty(-q;q^2)_\infty}{(q;q)_\infty}.$$

We start with a Frobenius partition and add one to each entry of the first row. The bottom row is made of an overpartition β and a partition into distinct parts α. Let n be the number of parts of α. We apply a generalization of Wright's bijection to the top row and β. We draw the Ferrers diagram of the top row and we shift the ith part by $i-1$. We draw the Ferrers diagram of the non overlined parts of β and put it at the left of the diagram of the top row. We draw the Ferrers diagram of the overlined parts of β, conjugate it and put it at the bottom right of the diagram of the top row, as shown on the left of Figure 2. Then we break the diagram into two parts: the left and the right of the largest overlined part. On the right we get an ordinary partition, which gives $1/(q;q)_\infty$ and on the left we get a partition λ into distinct parts where all the parts from 1 to n occur.

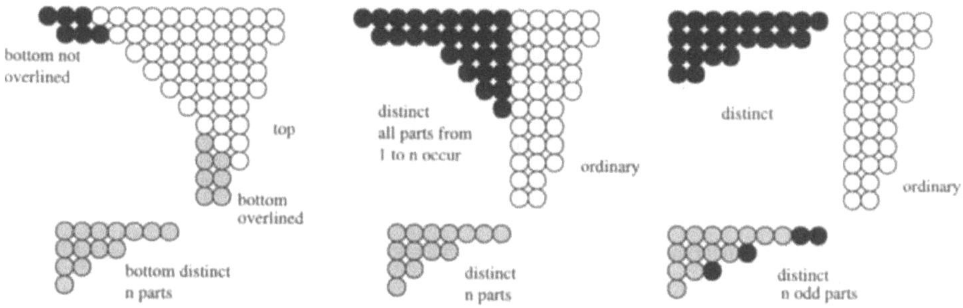

FIGURE 2. Bijection for (13)

Let i be the index of the smallest even part in α. Then we take off i from λ and add it to the conjugate of α. We do that until α has only odd parts. We finally get λ a partition into distinct parts, which is generated by $(-q;q)_\infty$ and α a partition into distinct odd parts, which is generated by $(-q;q)_\infty$. Each step is easily reversible.

For example, starting with $\begin{pmatrix} 11,10,8,7,6,4,3,2,1 \\ (\bar{4},3,3,\bar{2},\bar{0}),(7,4,2,1) \end{pmatrix}$ we get $\beta = (5,5,4,4,4,3,3,3,3,2,2)$, $\lambda = (9,8,4,2)$ and $\alpha = (9,5,3,1)$. See Fig 2.

Bijection for (15). A combinatorial proof of

$$\sum_{n=0}^{\infty} P_{OO_2',O}(n)q^n = \frac{(-q;q)_\infty^2(-q;q^2)_\infty}{(q;q)_\infty^2(q;q^2)_\infty},$$

can be easily done mixing the combinatorial proof of the ${}_1\psi_1$ summation of [21] and a combinatorial proof of the Lebesgue's identity for example the one used in the previous bijection.

Bijection for (16). A combinatorial proof of

$$\sum_{n=0}^{\infty} P_{OO',O}(n)q^n = \frac{(-q;q)_\infty^3}{(q;q)_\infty^3},$$

can be easily done mixing the combinatorial proof of the ${}_1\psi_1$ summation of [21] and a combinatorial proof of the q-binomial identity (see for example [12]).

Acknowledgments

We would like to thank Bruce Berndt for making [15] available to us.

References

[1] G.E. Andrews, The Theory of Partitions, Cambridge University Press, Cambridge, 1998.

[2] G.E. Andrews, Generalized Frobenius partitions, Mem. Amer. Math. Soc. **49** (1984), no. 301.

[3] G.E. Andrews, q-Series: Their Development and Application in Analysis, Number Theory, Combinatorics, Physics, and Computer Algebra, **CBMS 66**, American Mathematical Society, Providence, 1986.

[4] G.E. Andrews and D.M. Bressoud, Identities in combinatorics, III: Further aspects of ordered set sorting, *Discrete Math.* **49** (1984), 223-236.

[5] D. Corson, D. Favero, K. Liesinger, and S. Zubairy, Characters and q-series in $\mathbb{Q}(\sqrt{2})$, preprint.

[6] S. Corteel, Particle seas and basic hypergeometric series, *Adv. Appl. Math.*, **31** (2003), 199-214.

[7] S. Corteel and J. Lovejoy, Frobenius partitions and the combinatorics of Ramanujan's ${}_1\psi_1$ summation, *J. Combin. Theory Ser. A* **97** (2002), 177-183.

[8] S. Corteel and J. Lovejoy, Overpartitions, *Trans. Amer. Math. Soc*, to appear.

[9] N.J. Fine, Basic Hypergeometric Series and Applications, American Mathematical Society, Providence, RI, 1988.

[10] F. Garvan, Partition congruences and generalizations of Dyson's rank, PhD Thesis, Penn State, 1986.

[11] G. Gasper and M. Rahman, Basic Hypergeometric Series, Cambridge University Press, Cambridge, 1990.

[12] J.T. Joichi and D. Stanton, Bijective proofs of basic hypergeometric series identities, *Pacific J. Math.* **127** (1987), 103-120.

[13] L.W. Kolitsch, Some analytic and arithmetic properties of generalized Frobenius partitions, PhD thesis, Penn State, 1985.

[14] J. Lovejoy, Gordon's theorem for overpartitions, *J. Combin. Th. Ser. A* **103** (2003), 393-401.

[15] Padmavathamma, Studies in generalized Frobenius partitions, Ph. D. Thesis, Univ. of Mysore, 1985.

[16] J. Propp, Some variants of Ferrers diagrams. *J. Combin. Theory Ser. A* 52 (1989), no. 1, 98–128.

[17] J.P.O. Santos and D.V. Sills, q-Pell sequences and two identities of V.A. Lebesgue, *Disc. Math.* **257** (2002), 125 - 143.

[18] J. Sellers, New congruences for generalized Frobenius partitions with two or three colors, *Discrete Math.* **131** (1994), 367–373.

[19] J.J. Sylvester, A construtive theory of partitions in three acts, an interact and an exodion, in Collected Math. Papers, vol. 4, pp. 1-83, Cambridge Univ. Press, London and New York, 1912; reprinted by Chelsea, New York, 1974.

[20] A.J. Yee, Combinatorial proofs of generating function identities for F-partitions, *J. Combin. Theory Ser. A* **102** (2003), 217–228.

[21] A.J. Yee, Combinatorial proofs of Ramanujan's ${}_1\psi_1$ summation and the q-Gauss summation, preprint.

[22] D. Zeilberger, A q-Foata proof of the q-Saalschutz identity, *European J. Combin.* **8** (1987), 461-463.

Sylvie Corteel
CNRS, PRiSM, *Email:* syl@prism.uvsq.fr

Jeremy Lovejoy
CNRS, LaBRI, *Email:* lovejoy@labri.fr

Ae Ja Yee
Pennsylvania State University, *Email:* yee@math.psu.edu

Trends in Mathematics, © 2004 Birkhäuser Verlag Basel/Switzerland

Enumerative Results on Integer Partitions Using the ECO Method

Luca Ferrari, Renzo Pinzani, and Simone Rinaldi

ABSTRACT: *In this paper we apply the ECO method to the study of some enumerative properties of integer partitions. In so doing, we both give an original description of some known constructions regarding partitions and propose some results, especially in the context of generalized hook partitions (i.e., partitions whose Ferrers diagrams fit inside of a suitable hook shape).*

1. Introduction

The main goal of this work is to present an alternative approach to the study of enumerative properties of integer partitions. Namely, we are going to apply the ECO[1] method in order to effectively construct integer partitions. After briefly recalling the basics of the ECO method, with particular emphasis on those notions we will need in the sequel, we give a general ECO construction for partitions. We then modify such a construction to deal with restricted classes of partitions. Next we propose a second (less classical) construction of partitions and make use of it to propose a bijective approach to Lecture Hall partitions, leading to a challenging open question. Finally, we state some new results on generalized hook partitions, also proposing an open problem on partitions fitting into the intersection of two hooks having different shapes.

Our notations will be quite classical. In the partition $\lambda = (p_1, \ldots, p_l)$ the *parts* are in decreasing order, that is $p_1 \geq p_2 \geq \cdots \geq p_l$. If λ is a partition of n, then we write $\lambda \vdash n$ or $|\lambda| = n$. The whole class of integer partitions will be denoted $\mathcal{P}$.

2. Background

We call *ECO system* a purely formal system of the kind

$$\Omega : \left\{ \begin{array}{l} (a) \\ (k) \rightsquigarrow (e_1(k))(e_2(k)) \cdots (e_k(k)) \end{array} \right. , \qquad (1)$$

where $a, k, e_1(k), \ldots, e_k(k)$ are positive integers. (a) is called the *axiom* of the ECO system; the second row is just a shortcut in order to express a set of *productions* of the denoted form, for k running over a (finite or infinite) set of positive integers. Thus, for example, the following are typical examples of ECO systems:

$$(*) \quad : \quad \left\{ \begin{array}{l} (2) \\ (k) \rightsquigarrow (2)(3)(4) \cdots (k)(k+1) \end{array} \right. ,$$

$$(**) \quad : \quad \left\{ \begin{array}{l} (2) \\ (k) \rightsquigarrow (k)^{k-1}(k+1) \end{array} \right. , \qquad (***) : \left\{ \begin{array}{l} (2) \\ (1) \rightsquigarrow (2) \\ (2) \rightsquigarrow (1)(2) \end{array} \right. . \qquad (2)$$

[1]We recall that ECO stands for Enumeration of Combinatorial Objects.

Here we tacitly assume that the ECO systems (*) and (**) have, as *labels*, all the positive integers (greater than 2), whereas the ECO system (***) contains the only labels (1) and (2).

Among the many ways of interpreting an ECO system, a well-known and pictorial one is that of drawing its *generating tree*, which is, by definition, the infinite, rooted, labelled tree whose root is labelled (a) (the axiom) and such that every node labelled (k) produces exactly k sons, labelled respectively $(e_1(k)), \ldots, (e_k(k))$. For instance, the generating tree related to the ECO system $(*)$ looks as follows:

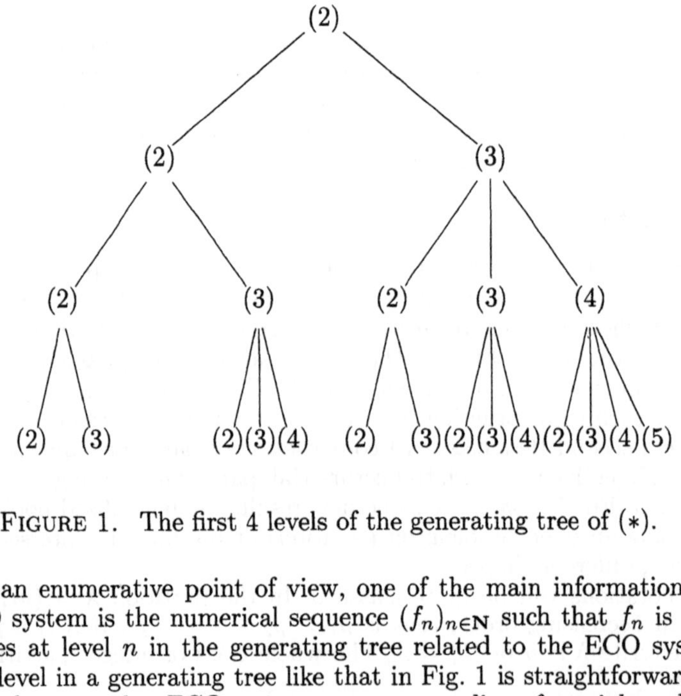

FIGURE 1. The first 4 levels of the generating tree of $(*)$.

From an enumerative point of view, one of the main information encoded by an ECO system is the numerical sequence $(f_n)_{n \in \mathbf{N}}$ such that f_n is the number of nodes at level n in the generating tree related to the ECO system (the concept of level in a generating tree like that in Fig. 1 is straightforward). Actually, it often happens that ECO systems appear as coding of special combinatorial constructions, in which the objects of a given class are recursively constructed by performing a sort of local expansion on them. This kind of combinatorial construction is the core of the ECO method, for which we refer the reader to the paper [BDLPP] introducing it.

Throughout our work we will be mainly concerned with some modifications of the classical concept of ECO system (or succession rule, as it is often called) given above. The first generalization we need is that of *mixed succession rule*.

We start by introducing the concept of *mixed generating tree*: it is a rooted, labelled tree whose edges can have different "lengths" (whereas in the classical case the length of each edge is equal to 1). The *lengthened level* (briefly, level) of a node in a mixed generating tree is defined as the sum of the lengths of the edges connecting the root to that node.

Now consider a *set of production rules* (instead of a single one, as we do in a classical ECO system) and make it act on every label, *each rule producing its sons at a different level.* What we get is precisely a mixed generating tree where a set of h (possibly different) production rules is involved: every node produces sets of sons at h different successive levels, each set according to one of the defining succession rule. A simple example of a mixed succession rule is the following. Consider the two production rules:

$$(k) \rightsquigarrow (2)(3)\cdots(k)(k+1) \qquad , \qquad (k) \rightsquigarrow (1)(2)\cdots(k-1)(k)$$

One can define the following mixed succession rule:

$$\begin{cases} (1) & & \\ (k) & \overset{2}{\rightsquigarrow} & (2)(3)\cdots(k)(k+1) \\ & \overset{3}{\rightsquigarrow} & (1)(2)\cdots(k-1)(k) \end{cases} .$$

The above formal system must be interpreted as follows: in its mixed generating tree a node at level n produces 2 sets of sons, one set at level $n+2$ (according to the first production rule) and the other at level $n+3$ (according to the second production rule). From the point of view of the ECO method, it is not difficult to see that the above mixed succession rule arises from a natural construction of generalized Motzkin paths using an horizontal step of length 3.

The theory of mixed succession rules has not yet been developed in its full generality, but some specific cases have been considered. Here we only mention the paper [FPPR], in which a particular kind of mixed succession rules (called *jumping succession rules*) is systematically studied.

Another generalization of the concept of succession rule we will need in the sequel is the notion of *ECO system with multiple labels.* More precisely, it is an ECO system in which the labels are not simply positive integers, but rather couples (more generally, h-tuples) of positive integers. Also this concept has occasionally surfaced in previous works (see, for instance, [GPP]), but the theory behind it remains largely undeveloped.

3. A general succession rule for integer partitions

In this section we will rediscover a classical construction of integer partitions giving it an ECO interpretation.

For any $k \in \mathbf{N}$, $k \neq 0$, consider the following mixed succession rule:

$$\Omega_k : \begin{cases} (k) & & \\ (h) & \overset{1}{\rightsquigarrow} & (1) \\ & \overset{2}{\rightsquigarrow} & (2) \\ & \cdots & \\ & \overset{h}{\rightsquigarrow} & (h) \end{cases} . \tag{3}$$

If we denote by $F_k(x)$ the generating function of the numerical sequence associated with the rule Ω_k, we easily have:

$$F_1(x) = \frac{1}{1-x}, \qquad F_2(x) = \frac{1}{(1-x)(1-x^2)}.$$

For a generic k, a look at the generating tree of the rule suggests the following equality:

$$F_k(x) = 1 + xF_1(x) + x^2F_2(x) + \ldots + x^kF_k(x),$$

whence, using a simple induction argument:

$$F_k(x) = \prod_{i=1}^{k} \frac{1}{1-x^i}. \tag{4}$$

If we let k tend to infinity, we obtain the following generating function:

$$F(x) = \lim_{k\to\infty} F_k(x) = \prod_{k\geq 1} \frac{1}{1-x^k}. \tag{5}$$

Finding an ECO system Ω associated with the generating function $F(x)$ is possible, provided that we allow the use of the symbol ∞ as a possible label. In fact, we get for Ω the following expression:

$$\Omega : \begin{cases} (\infty) & \\ (h) & \overset{1}{\rightsquigarrow} \quad (1) \\ & \overset{2}{\rightsquigarrow} \quad (2) \\ & \dots \\ & \overset{h}{\rightsquigarrow} \quad (h) \end{cases}, \tag{6}$$

where the unique node labelled (∞) (which is the axiom of the system) produces, by convention, a node labelled (i) at level i, for every $i \geq 1$ (so that it has an infinite number of sons). The first levels of the generating tree of Ω are given in Fig. 2.

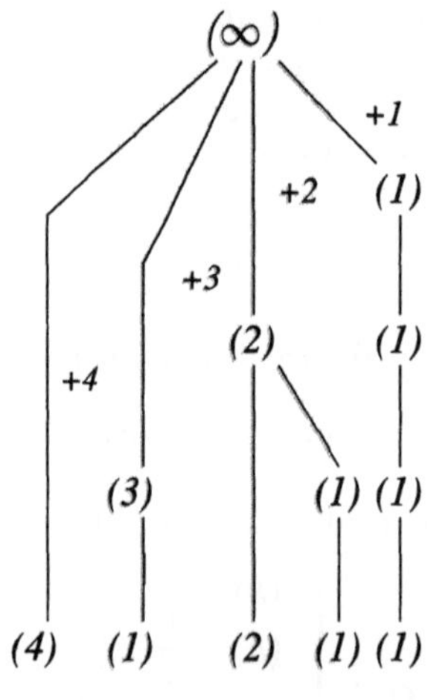

FIGURE 2. The first 4 levels of the generating tree of Ω.

Using a terminology taken from [FPPR], we can informally say that $\Omega = \lim_{k\to\infty} \Omega_k$. The rule Ω enumerates precisely integer partitions, since $F(x)$ is the well-known generating function of them. Every integer partition $\lambda = (p_1, \dots, p_l)$ has a simple graphical representation in terms of *Ferrers diagrams* made of l rows such that the i-th row is made of p_i cells.

There is an ECO construction of Ferrers diagrams of partitions encoded by Ω; it can be described as follows:

i) the empty object (Ferrers diagram without cells, corresponding to the void partition of 0) has label (∞) and produces a single row having h cells at each level h (corresponding to the partition (h) of h);

ii) a Ferrers diagram representing a partition of n, whose bottom row is made of h cells, has label (h) and produces h Ferrers diagrams representing partitions of $n+1, \dots, n+h$, respectively, and having labels $(1), \dots, (h)$, as it is shown in Fig. 3.

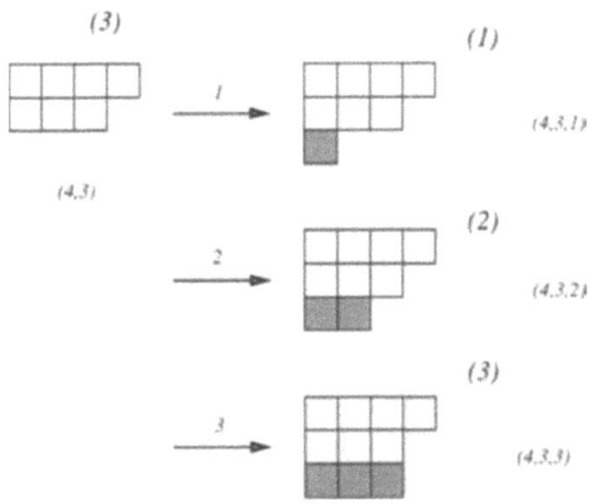

FIGURE 3. The ECO construction for partitions.

An obvious consequence of this construction is that, for any $k \in \mathbf{N}$, the rule Ω_k enumerates partitions $\lambda = (p_1, \dots, p_l)$ where each part p_i is less than or equal to k.

4. A generalization and some examples

The previous rule for integer partitions can be suitably generalized. Suppose to consider only those partitions using a prescribed type of parts. More precisely, a sufficiently general setting consists of taking partitions each of whose parts depends on the previous one. So, given a partition $\lambda = (p_1, \dots p_l) \vdash n$, we require that $p_i \in \Pi(p_{i-1}) \subseteq \mathbf{N}$ (i.e., the set $\Pi(p_{i-1})$ only depends on p_{i-1}). In particular, we define $\Pi(\infty)$ to be the (possibly infinite) set of allowed "maximum parts" for the partitions we are taking into consideration. Therefore, a construction for the above described class of partitions can be obtained by suitably modifying the general ECO construction for partitions given above, so getting to the following succession rule:

$$\begin{cases} (\infty) & & \\ (\infty) & \overset{i_1}{\rightsquigarrow} & (a_1) \\ & \cdots & \\ & \overset{i_n}{\rightsquigarrow} & (a_n) \\ & \cdots & \\ (k) & \overset{j_1}{\rightsquigarrow} & (k_1) \\ & \cdots & \\ & \overset{j_k}{\rightsquigarrow} & (k_k) \end{cases} . \tag{7}$$

In the above ECO system the set $\{(a_n)\}_n$ is the set of labels of the parts belonging to the set $\Pi(\infty) = \{i_n\}_n$, whereas, if (k) is the label of a part p of a partition, then $\{(k_1), \dots (k_k)\}$ is the set of labels of the parts belonging to the set

$\Pi(p) = \{j_1, \ldots, j_k\}$. Some interesting examples fall into this framework and can be obtained by specializing the quite general succession rule in (7).

4.1. Partitions into distinct parts and partitions into odd parts

Let $\mathcal{D}$ be the set of partitions into distinct parts and $\mathcal{O}$ the set of partitions into odd parts. The two classes are easily proved to be enumerated by the same generating function:

$$\sum_{\lambda \in \mathcal{D}} x^{|\lambda|} = \prod_{i \geq 1} (1 + x^i) = \prod_{i \geq 1} \frac{1 - x^{2i}}{1 - x^i} = \prod_{i \geq 0} \frac{1}{1 - x^{2i+1}} = \sum_{\lambda \in \mathcal{O}} x^{|\lambda|}. \tag{8}$$

The ECO construction previously proposed for integer partitions can be slightly modified in order to describe an ECO construction for the classes $\mathcal{O}$ and $\mathcal{D}$.

A construction for $\mathcal{D}$.: Partitions of integers into distinct parts are represented by Ferrers diagrams whose rows all have different lengths. The ECO construction is the following:

i) the empty partition of 0 has label (∞) and produces a single row made of h cells at each level h (corresponding to the partition (h) of h); so $\Pi(\infty) = \mathbf{N}^* = \mathbf{N} \setminus \{0\}$;

ii) a Ferrers diagram representing a partition of n, whose bottom row is made of $h+1$ cells, is labelled (h) and produces h Ferrers diagrams representing partitions of $n+1, \ldots n+h$ and having labels $(0), \ldots, (h-1)$, respectively, as it is shown in Fig. 4; therefore, $\Pi(h+1) = \{1, 2, \ldots, h\}$.

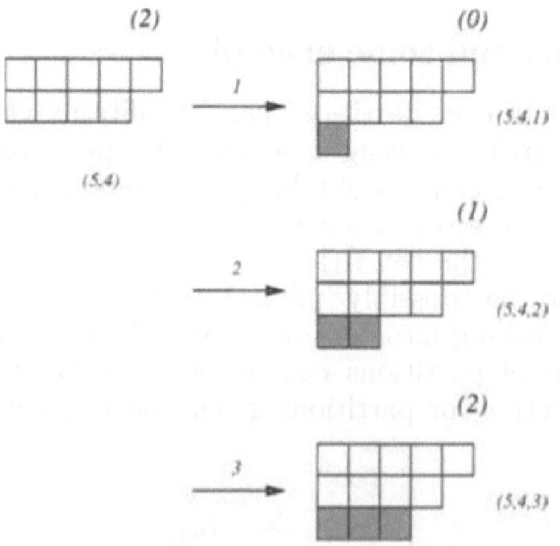

FIGURE 4. The ECO construction for partitions into distinct parts.

The succession rule associated with this construction is the following:

$$\Omega^{\mathcal{D}} : \left\{ \begin{array}{lll} (\infty) & & \\ (h) & \stackrel{1}{\rightsquigarrow} & (0) \\ & \stackrel{2}{\rightsquigarrow} & (1) \\ & \cdots & \\ & \stackrel{h}{\rightsquigarrow} & (h-1) \end{array} \right. \tag{9}$$

In the generating tree of $\Omega^{\mathfrak{D}}$, a node labelled (0) produces no son. The generating function for $\Omega^{\mathfrak{D}}$ can be obtained in an analogous way to that for the generating function of Ω, getting again the series $F_{\mathfrak{D}}(x) = \prod_{i\geq 0}(1+x^i)$ in (8).

A construction for $\mathfrak{O}$.: An ECO construction for the class of partitions into odd parts can be obtained by specializing the above described general setting to the case in which $\Pi(\infty)$ is the set of odd positive integers and, for every odd p, $\Pi(p) = \{q \mid q \text{ odd}, q \leq p\}$ (see Fig. 5).

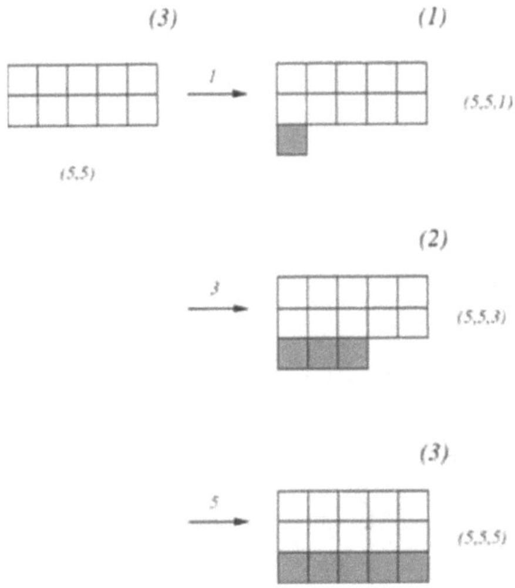

FIGURE 5. The ECO construction for partitions into odd parts.

The associated succession rule is:

$$\Omega^{\mathfrak{O}}: \begin{cases} (\infty) & \\ (h) & \stackrel{1}{\rightsquigarrow} (1) \\ & \stackrel{3}{\rightsquigarrow} (2) \\ & \dots \\ & \stackrel{2h-1}{\rightsquigarrow} (h) \end{cases} .$$

Also in this case it is easy to determine the generating function by an inductive argument, so obtaining $F_{\mathfrak{O}}(x) = \prod_{i\geq 0} \frac{1}{1-x^{2i+1}}$, as it is well-known.

This last example is particularly nice, since $\Pi(\infty)$ is the set of all possible parts appearing in the ECO system (odd positive integers) and $\Pi(p)$ is just the subset of $\Pi(\infty)$ whose elements are less than or equal to p. More generally, if $\Pi(\infty) = \{a_n\}_n$ and $\Pi(a_n) = \{a_k \mid k \leq n\}$, the usual inductive argument leads to the generating function

$$F(x) = \prod_n \frac{1}{1 - x^{a_n}}.$$

This is the classical result concerning the enumeration of partitions whose parts belong to a fixed set.

5. An alternative ECO system for integer partitions

The ECO approach suggests an alternative construction for integer partitions. Starting from a given partition $\lambda = (p_1, \ldots, p_l)$ of n, we define two new partitions of $n+1$ and $n+p_1$, respectively, which are precisely $\lambda_1 = (p_1+1, p_2, \ldots, p_l) \vdash n+1$ and $\lambda_2 = (p_1, p_1, p_2, \ldots, p_l) \vdash n+p_1$. Fig. 5 graphically describes this construction on Ferrers diagrams.

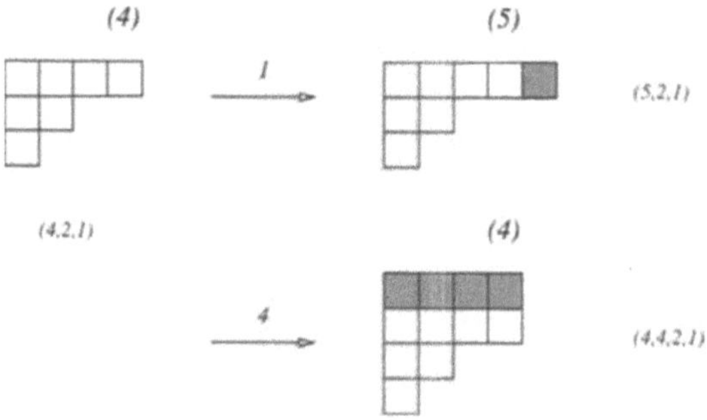

FIGURE 6. An alternative ECO construction for integer partitions.

Such a construction is immediately seen to be associated with the succession rule:

$$\Omega_* : \begin{cases} (1) \\ (h) \stackrel{1}{\rightsquigarrow} (h+1) \\ \stackrel{h}{\rightsquigarrow} (h) \end{cases} . \tag{10}$$

Rule Ω_* in (10) does not satisfy the consistency principle typical of ECO systems, i.e. labels do not denote the number of sons; in particular, each node in the generating tree of Ω_* produces exactly two sons. According to the construction suggested by Fig. (5), each label (h) corresponds to a partition whose maximum part is h.

6. Some applications and open problems

6.1. Lecture Hall partitions

The theory of Lecture Hall partitions has been initiated in [BME1], which is the basic article we refer the reader to concerning this topic.

Only in this section, we change our notation for partitions: if $\lambda = (p_1, \ldots, p_l)$ then we assume that $p_1 \leq p_2 \leq \cdots \leq p_l$. For $k \geq 1$, let $\mathfrak{L}_k$ be the following set of partitions (having possibly some empty parts):

$$\mathfrak{L}_k = \left\{ (p_1, \ldots, p_k) \mid 0 \leq \frac{p_1}{1} \leq \frac{p_2}{2} \leq \cdots \leq \frac{p_k}{k} \right\}.$$

The elements of $\mathfrak{L}_k$ are called *Lecture Hall partitions* of length k. We also denote by $\mathfrak{D}_k$ the set of all partitions of $\mathfrak{L}_k$ with empty parts removed. For example, the partition $(2,3)$ of 5 belongs to $\mathfrak{D}_3$ but not to $\mathfrak{D}_2$. It is clear that, for any given λ, there exists a minimum k such that $\lambda \in \mathfrak{D}_k$: this will be called the *minimum length* of λ as a Lecture Hall partition. The concept of Lecture Hall partition allows to give a finite version of the well-known result (due to Euler) that the number of partitions of n into odd parts equals the number of partitions of n into distinct parts. More precisely, in [BME1] it is shown that the number of partitions of n into odd parts less than or equal to $2k-1$ is equal to the number of Lecture Hall partitions of length k of n.

The first bijective proof of (a refined version of) this result appears in [BME2]; however, the authors themselves admit that such a proof finds its origin in the algebraic context of Coxeter groups. Some bijective proofs have been recently given in [E, Y]. The present approach to integer partitions suggests a possible way to find a new natural bijection proving the Lecture Hall theorem in a purely combinatorial way. The idea is to give two distinct combinatorial interpretations to the generating tree associated with a given ECO construction.

A possible ECO construction for $\mathcal{O}$ can be obtained by suitably modifying the one given in section 5 for unrestricted partitions. It is not difficult to see that the associated succession rule is the following:

$$\Omega_*^{\mathcal{O}} : \begin{cases} (1) & \\ (h) & \stackrel{2}{\rightsquigarrow} \ (h+2) \\ & \stackrel{h}{\rightsquigarrow} \ (h) \end{cases} . \qquad (11)$$

It is clear that the set of labels of this ECO system is the set of odd positive integers. The first levels of its generating tree are the following:

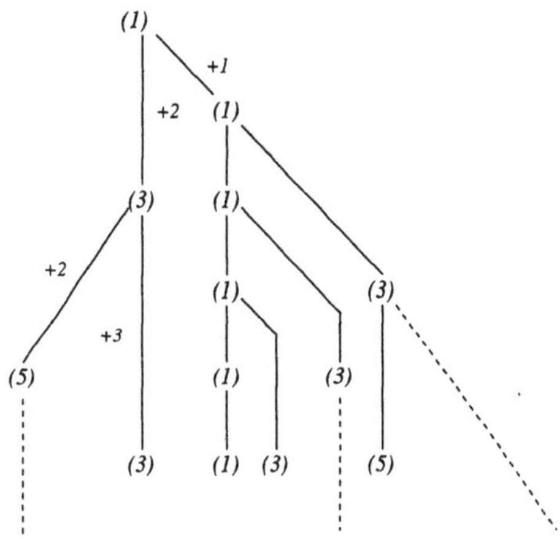

FIGURE 7. The first levels of the generating tree of the rule $\Omega_*^{\mathcal{O}}$.

We conjecture that the ECO system in (11) provides a construction also for partitions into distinct parts. This would lead to a presumably new bijection between $\mathcal{O}$ and $\mathcal{D}$ from which it would be immediate to deduce an explicit bijection proving the Lecture Hall theorem. Indeed, in the conjectured interpretation of the above generating tree, it seems natural to think that the label of a partition is strictly related to its (minimum) length as a Lecture Hall partition: more precisely, a node labelled $2k-1$ represents a Lecture Hall partition of minimum length k.

Concerning this problem, we cite the remarkable paper [P], where the author introduces a truly nice setting to deal with bijective questions on partitions which could be useful in this context.

6.2. Generalized Hook partitions

In the paper [BR], Berele and Regev show how the representation theory of Lie superalgebras heavily relies upon the knowledge of the combinatorics of partitions fitting inside of a hook shaped figure (we will briefly call them *generalized hook partitions*). The above article has been followed by many others, such as [R1, R2], and the study of this kind of partitions is still object of investigations in combinatorics and algebra.

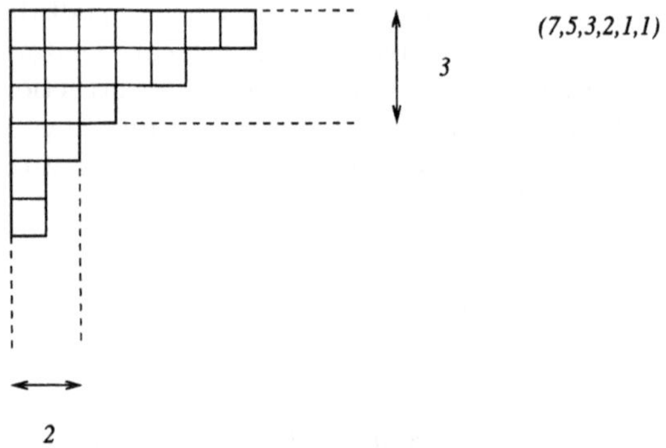

FIGURE 8. A hook partition of shape (2, 3).

Let $\mathcal{H}_{h,k}$ be the set of *generalized hook partitions of shape* (h, k), that is, by definition, the set of all partitions that fit inside a hook shape of k rows and h columns. Our aim is to restrict the general ECO construction for partitions to the set $\mathcal{H}_{h,k}$. This will lead us to the determination of the generating function of $\mathcal{H}_{h,k}$.

We consider two disjoint subsets of $\mathcal{H}_{h,k}$.

i) Partitions having $j < k$ parts; a partition in this set has label (l, j), where l is the number of cells in the last row of its Ferrers diagram. The ECO construction applied to such a partition works exactly like in the general case, leading to the following production:

$$\left\{\begin{array}{lll} (l,j) & \overset{1}{\rightsquigarrow} & (1,j+1) \\ & \cdots & \\ & \overset{l}{\rightsquigarrow} & (l,j+1) \end{array}\right. . \tag{12}$$

Observe that (12) is a production with double labels, since we have to take into account also the number of parts. However, if $j = k-1$, we choose to delete the second label, for reasons that will be explained below.

ii) All the remaining partitions. In this case, because of the hook shape constraint, the ECO construction is made by adding to a Ferrers diagram only rows having at most h cells. Since the last production described in i) produces simple labels, here we can avoid the use of multiple labels. The associated production is then the following:

$$\left\{\begin{array}{lll} (l) & \overset{1}{\rightsquigarrow} & (1) \\ & \cdots & \\ & \overset{m}{\rightsquigarrow} & (m). \end{array}\right. , \tag{13}$$

where $m = \min(l, h)$.

Let $H_{h,k}(x)$ be the generating function for the class $\mathcal{H}_{h,k}$. For any label (l,j), with $j < k$, denoting by $H_{h,k}^{(l,j)}(x)$ the generating function of the ECO system having (l,j) as axiom and the above ones as production rules, we can deduce the following recursion:

$$H_{h,k}^{(l,j)}(x) = 1 + xH_{h,k}^{(1,j+1)}(x) + \ldots + x^l H_{h,k}^{(l,j+1)}(x). \tag{14}$$

Analogously, if $j \geq k$ we have:

$$H_{h,k}^{(l)}(x) = 1 + xH_{h,k}^{(1)}(x) + \ldots + x^m H_{h,k}^{(m)}(x). \tag{15}$$

Starting from this general setting we are able to compute the generating function of generalized hook partitions of any fixed shape.

For instance if we consider hook partitions of shape $(2,3)$, we get to the following succession rule:

$$\left\{\begin{array}{lll} (\infty,0) & & \\ (l,1) & \stackrel{1}{\rightsquigarrow} & (1,2) \\ & \cdots & \\ & \stackrel{l}{\rightsquigarrow} & (l,2) \\ (l,2) & \stackrel{1}{\rightsquigarrow} & (1) \\ & \cdots & \\ & \stackrel{l}{\rightsquigarrow} & (l) \\ (1) & \stackrel{1}{\rightsquigarrow} & (1) \\ (l) & \stackrel{1}{\rightsquigarrow} & (1) \\ & \stackrel{2}{\rightsquigarrow} & (2) \end{array}\right. , \tag{16}$$

leading to the generating function:

$$\begin{aligned} H_{2,3}(x) &= H_{3,2}(x) \\ &= \frac{1}{(1-x)(1-x^2)(1-x^3)}\left(1+\frac{x^4}{1-x}+\frac{x^8}{(1-x)(1-x^2)}\right). \end{aligned}$$

More generally, we have:

$$\begin{aligned} H_{h,k}(x) &= H_{k,h}(x) \\ &= \prod_{i\leq h}\frac{1}{1-x^i}\left(1+\frac{x^{h+1}}{1-x}+\ldots+\frac{x^{k(h+1)}}{(1-x)\ldots(1-x^k)}\right). \end{aligned} \tag{17}$$

Observe that the equality $H_{h,k}(x) = H_{k,h}(x)$, which is immediate by a combinatorial point of view, is by no means obvious from an algebraic one.

We point out here that the problem of determining the generating functions $H_{h,k}(x)$ was previously considered in [OZ], where the authors find an explicit formula for them. However, their expression involves rather complicated quantities, whereas formula (17) is quite easy to read. It would be instead interesting to extend this result to the case of partitions fitting inside of the intersection of two hooks of different shapes. Also this problem arises in the study of representation of Lie superalgebras, in connection with the module decomposition of supersymmetric power of matrices [S], and it is still open.

References

[BDLPP] E. Barcucci, A. Del Lungo, E. Pergola, R. Pinzani, *ECO: A methodology for the enumeration of combinatorial objects*, J. Differ. Equations Appl. 5 (1999) 435-490.

[BR] A. Berele, A. Regev, *Hook Young diagrams with applications to combinatorics and to representations of Lie superalgebras*, Adv. in Math. 64 (1987) 118-175.

[BME1] M. Bousquet-Mélou, K. Eriksson, *Lecture Hall partitions*, Ramanujan J. 1 (1997) 101-111.

[BME2] M. Bousquet-Mélou, K. Eriksson, *A refinement of the Lecture Hall theorem*, J. Combin. Theory Ser. A 86 (1999) 63-84.

[DFR] E. Deutsch, L. Ferrari, S. Rinaldi, *Production matrices*, (submitted).

[E] N. Eriksen, *A simple bijection between Lecture Hall partitions and partitions into odd integers*, proceedings of FPSAC 2002, Melbourne.

[FPPR] L. Ferrari, E. Pergola, R. Pinzani, S. Rinaldi, *Jumping succession rules and their generating functions*, Discrete Math. 271 (2003) 29-50.

[FP] L. Ferrari, R. Pinzani, *A linear operator approach to succession rules*, Linear Algebra Appl. 348 (2002) 231-246.

[GPP] O. Guibert, E. Pergola, R. Pinzani, *Vexillary involutions are enumerated by Motzkin numbers*, Ann. Comb. 5 (2001) 153-174.

[OZ] R. C. Orellana, M. Zabrocki, *Some remarks on the characters of the general Lie superalgebra*, arXiv:math.CO/0008152v1, 2000.

[P] I. Pak, *Partition identities and geometric bijections*, Proc. Amer. Math. Soc., to appear.

[R1] J. B. Remmel, *The combinatorics of (k,l)-hook Schur functions*, Combinatorics and Algebra (Boulder, Colorado, 1983), 253-287, Contemp. Math., 34, Amer. Math. Soc., Providence, RI, 1984.

[R2] J. B. Remmel, *A bijective proof of a factorization theorem for (k,l)-hook Schur functions*, Linear and Multilinear Algebra 28 (1990) 119-154.

[S] T. Seeman private communication, 2003.

[Y] A.J.Yee *On the refined lecture hall theorem*, Discrete Math. 248 (2002) 293-298.

Luca Ferrari
Dipartimento di Scienze Matematiche ed Informatiche, Pian dei Mantellini, 44, 53100, Siena, Italy
ferrari@math.unifi.it

Simone Rinaldi
Dipartimento di Scienze Matematiche ed Informatiche, Pian dei Mantellini, 44, 53100, Siena, Italy
rinaldi@unisi.it

Renzo Pinzani
Dipartimento di Sistemi e Informatica, via Lombroso 6/17, 50135 Firenze, Italy
pinzani@dsi.unifi.it

Trends in Mathematics, © 2004 Birkhäuser Verlag Basel/Switzerland

321-Avoiding Permutations and Chebyshev Polynomials

Toufik Mansour

ABSTRACT: *In* [6] *it was shown that the generating function for the number of permutations on n letters avoiding both* 321 *and* $(d+1)(d+2)\dots k12\dots d$ *is given by* $\frac{2tU_{k-1}(t)}{U_k(t)}$ *for all* $k \geq 2$, $2 \leq d+1 \leq k$, *where* U_m *is the* m*th Chebyshev polynomial of the second kind and* $t = \frac{1}{2\sqrt{x}}$. *In this paper we present three different classes of* 321*-avoiding permutations which are enumerated by this generating function.*

Let $\alpha \in S_n$ and $\tau \in S_k$ be two permutations. Then α *contains* τ if there exists a subsequence $1 \leq i_1 < i_2 < \dots < i_k \leq n$ such that $(\alpha_{i_1}, \dots, \alpha_{i_k})$ is order-isomorphic to τ; in such a context τ is usually called a *pattern*; α *avoids* τ, or is τ*-avoiding*, if α does not contain such a subsequence. The set of all τ-avoiding permutations in S_n is denoted by $S_n(\tau)$. For a collection of patterns T, α *avoids* T if α avoids all $\tau \in T$; the corresponding subset of S_n is denoted by $S_n(T)$.

While the case of permutations avoiding a single pattern has attracted much attention, the case of multiple pattern avoidance remains less investigated. In particular, it is natural to consider permutations avoiding pairs of patterns τ_1, τ_2. This problem was solved completely for $\tau_1, \tau_2 \in S_3$ (see [8]), for $\tau_1 \in S_3$ and $\tau_2 \in S_4$ (see [10]). Several recent papers [1, 2, 4, 5, 6, 7] deal with the case $\tau_1 \in S_3$, $\tau_2 \in S_k$ for various pairs τ_1, τ_2, e.g. in [1] it was found by using transfer matrices that the generating function for the number of permutations in $S_n(321, [k,k])$ is given by

$$R_k(x) = \frac{2tU_{k-1}(t)}{U_k(t)}, \quad t = \frac{1}{2\sqrt{x}}$$

where $U_m(\cos\theta) = \sin(m+1)\theta/\sin\theta$ is the mth Chebyshev polynomial of the second kind and $[d,k] = d(d+1)\dots k12\dots(d-1)$. Later, in [6] Mansour and Vainshtein proved a natural generalization for this theorem.

Theorem 1. For any $k \geq 2$ and $2 \leq d+1 \leq k$, the generating function for the number of permutations in $S_n(321, [d+1,k])$ is given by $R_k(x)$.

Recently, Mansour and Stankova [3] presented an exact enumeration for the case 321-k-gon-avoiding permutations in S_n which generalizes the methods in [9] and [6]. In particular they proved the following result:

Theorem 2. For any $k \geq 4$ and $2 \leq d \leq k-2$, the generating function for the number of permutations in $S_n(321, (d+1)(d+2)\cdots(k-1)1k23\cdots d)$ is given by $R_k(x)$.

Let us define there patterns:

$$\alpha_{d,k} = d(d+2)(d+3)\dots k12\dots(d-1)(d+1),$$
$$\beta_{d,k} = d(d+2)(d+3)\dots(k-1)1k23\dots(d-1)(d+1),$$
$$\gamma_{d,k} = d(d+2)(d+4)\dots k12\dots(d-1)(d+1)(d+3).$$

The main theorem of the paper is formulated as follows.

Theorem 3.

(i) Let $k \geq 4$ and $2 \leq d \leq k-2$. Then the generating function for the number of permutations which avoid both 321 and $\alpha_{d,k}$ is given by $R_k(x)$.

(ii) Let $k \geq 6$ and $3 \leq d \leq k-3$. Then the generating function for the number of permutations which avoid both 321 and $\beta_{d,k}$ is given by $R_k(x)$.

(iii) Let $k \geq 6$ and $2 \leq d \leq k-4$. Then the generating function for the number of permutations which avoid both 321 and $\gamma_{d,k}$ is given by $R_k(x)$.

Our proof of the Theorem 3 is based on finding a recursion for the numbers in question by purely analytical means. In particular, we generalize the methods and extend the results in [1, 3, 6].

In spite of the paradigm formulated in [2], that any enumeration problem leading to Chebyshev polynomials is related to Dyck paths, it would be tempting to find a proof that exploits such a relation.

References

[1] T. Chow and J. West, Forbidden subsequences and Chebyshev polynomials, *Discr. Math.* **204** (1999) 119–128.

[2] C. Krattenthaler, Permutations with restricted patterns and Dyck paths, *Adv. in Applied Math.* **27** (2001) 510–530.

[3] T. Mansour and Z. Stankova, 321-polygon-avoiding permutations and Chebyshev polynomials, *Elect. J. Combin.* **9**:2 (2003) #R5.

[4] T. Mansour and A. Vainshtein, Restricted permutations, continued fractions, and Chebyshev polynomials, *Elect. J. Combin.* **7** (2000) #R17.

[5] T. Mansour and A. Vainshtein, Restricted 132-avoiding permutations, *Adv. Appl. Math.* **126** (2001) 258–269.

[6] T. Mansour and A. Vainshtein, Layered restrictions and Chebyshev polynomials, *Ann. of Combin.* **5** (2001) 451–458.

[7] T. Mansour and A. Vainshtein, Restricted permutations and Chebyshev polynomials, *Sém. Lothar. de Combin.* **47** (2002) Article B47c.

[8] R. Simion, F.W. Schmidt, Restricted Permutations, *Europ. J. Combin.* **6** (1985) 383–406.

[9] Z. Stankova and J. West, Explicit enumeration of 321-hexagon-avoiding permutations, *Disc. Math.*, to appear.

[10] J. West, Generating trees and forbidden subsequences, *Discr. Math.* **157** (1996) 363–372.

Toufik Mansour
Department of Mathematics, University of Haifa, 31905 Haifa, Israel
toufik@math.haifa.ac.il

Trends in Mathematics, © 2004 Birkhäuser Verlag Basel/Switzerland

Iterated Logarithm Laws and the Cycle Lengths of a Random Permutation

Eugenijus Manstavičius

ABSTRACT: *We are concerned with the iterated logarithm laws for mappings defined on the symmetric group. For the sequences of the cycle lengths and the different cycle lengths appearing in the decomposition of a random permutation, such laws provide asymptotical formulas valid uniformly in a wide region for the sequence parameter. The main results are analogues to Feller's and Strassen's theorems proved for partial sums of independent random variables.*

1. Weak convergence of distributions

Let $\sigma \in \mathbf{S}_n$ be an arbitrary permutation and

$$\sigma = \kappa_1 \cdots \kappa_w, \qquad w = w(\sigma), \tag{1}$$

be its unique up to the order expression by the product of the independent cycles κ. Denote by $\nu_n(\dots) = (n!)^{-1}|\{\sigma \in \mathbf{S}_n : \dots\}|$ the uniform probability measure on $\mathbf{S}_n$. In what follows we assume that $n \to \infty$. Despite to the long list of asymptotic results (see [1], [2], [7–10], [12–15], and other publications), we examine the ordered statistics

$$1 \leq J_1(\sigma) \leq \cdots \leq J_w(\sigma) \leq n$$

and

$$1 \leq j_1(\sigma) < \cdots < j_s(\sigma) \leq n, \qquad s = s(\sigma),$$

consisting of the cycle lengths and the different cycle lengths appearing in decomposition (1). In this remark, from these two very close cases we always choose that requiring less technical details.

Let us remind a few of classical results. In 1942 Goncharov [8] (see also [9]) found the limiting distribution, say, $F_k(x)$, of $J_{w-k}(\sigma)/n$ for arbitrary fixed $k \geq 0$. In particular, for the longest cycle, we have

$$\begin{aligned} &\nu_n\left(J_w(\sigma) < xn\right) \to F_0(x) \\ &= 1 + \sum_{1 \leq l < 1/x} \frac{(-1)^l}{l!} \int_1^\infty \cdots \int_1^\infty \mathbf{1}\{u_1 + \cdots + u_l \leq 1/x\} \frac{du_1}{u_1} \cdots \frac{du_l}{u_l} \end{aligned} \tag{2}$$

if $0 < x < 1$ and $F_0(x) = 1$ if $x \geq 1$. Here and in what follows $\mathbf{1}\{\cdot\}$ denotes the indicator function. The moments of $F_k(x)$ were calculated in [13]. As it is stated in [9] (see also [10]),

$$\nu_n\left(\log J_m < x\sqrt{m} + m\right) \to \Phi(x)$$

provided that $m = \alpha \log n + o(\sqrt{\log n})$ and $0 < \alpha < 1$. Here and in what follows $\Phi(x)$ denotes the distribution function of the standard normal law.

In 1972 A.M. Vershik and A.A. Schmidt [14] announced (all details were furnished in [15]) several results on the asymptotic distribution of the random element

$$(J_w(\sigma)/n, J_{w-1}(\sigma)/n, \dots J_1(\sigma)/n, 0, \dots)$$

under the measure ν_n in the simplex

$$\Gamma := \{(x_1, x_2, \dots) : 1 \geq x_1 \geq x_2 \geq \dots \geq 0, \, x_1 + x_2 + \dots = 1\}.$$

Applying continuous mappings of Γ to other spaces, they derived some fairly interesting corollaries. In particular, they proved that

$$\lim_{i\to\infty} \lim_{n\to\infty} \nu_n \left(\log(J_{w-i}(\sigma)/n) < x\sqrt{i} - i \right) \to \Phi(x).$$

Our questions are simple:

What asymptotic results on $\{J_m(\sigma)\}$ and $\{j_m(\sigma)\}$ are valid uniformly in $m \in [a_n, b_n]$, where $[a_n, b_n]$ is a fortiori given subinterval of $[1, n]$? What type of convergence can give such information?

First, we indicate an idea that the functional limit theorems can give some approach to the problem. Let $k_j(\sigma)$, $1 \leq j \leq n$, denote the number of cycles of length j in (1). By the functional limit theorem for additive functions on the symmetric group [4] or [3], the processes

$$W_n(\sigma; t) := \frac{1}{\log n} \left(w(\sigma; n^t) - t \log n \right), \qquad w(\sigma; y) := \sum_{j \leq y} k_j(\sigma), \quad t \in [0, 1],$$

weakly converge to the standard Brownian motion $W(t)$. Taking the maximum functional, we obtain

$$\nu_n \left(\max_{t\in[0,1]} |W_n(\sigma; t)| < x \right) \to P \left(\max_{t\in[0,1]} |W(t)| < x \right)$$
$$= \frac{1}{\sqrt{2\pi}} \sum_{l\in\mathbf{Z}} (-1)^l \int_{-x}^{x} \exp\left\{ -\frac{(u - 2lx)^2}{2} \right\} du =: V(x).$$

The maximum under consideration can be attained at the points $t = \log J_m(\sigma) / \log n$, $1 \leq m \leq w$ or at $t = 1$. Applying this together with $w(\sigma; J_m(\sigma)) = m$, we derive

$$\nu_n \left(\max_{m\leq w} |\log J_m(\sigma) - m| < x\sqrt{\log n}, \, |w(\sigma) - \log n| < x\sqrt{\log n} \right) = V(x) + o(1). \tag{3}$$

Further, observing that by (2)

$$\nu_n \left(|\log J_w(\sigma) - \log n| \geq \varepsilon\sqrt{\log n} \right) \qquad = 1 - F_0(\exp\{\varepsilon\sqrt{\log n}\})$$
$$+ F_0(\exp\{-\varepsilon\sqrt{\log n}\}) + o(1) = o(1)$$

for each $\varepsilon > 0$, we can drop the second event in the frequency of equality (3). So, we obtain the following result.

Theorem 1.1. *We have*

$$\nu_n \left(\max_{m\leq w} |\log J_m(\sigma) - m| < x\sqrt{\log n} \right) = V(x) + o(1). \tag{4}$$

Most probably, the last assertion could also be derived from the aforementioned results proved in [14] and [15].

2. Strong convergence

The next result related to our question involves stronger convergence than the weak convergence of distributions. Actually, we examine an analog of the convergence with probability one. To give a sense for this notion on a sequence of probability spaces, we have to consider convergence in distribution of the "tails" of sequences of random variables. In [12] we proved the following result.

Theorem 2.1. *We have*

$$\lim_{n_1\to\infty}\overline{\lim_{n\to\infty}}\,\nu_n\left(\max_{n_1\le m\le s}\frac{|\log j_m(\sigma)-m|}{(2m\log\log m)^{1/2}}\ge 1+\delta\right)=0 \qquad (5)$$

and

$$\lim_{n_1\to\infty}\underline{\lim}_{n\to\infty}\,\nu_n\left(\min_{n_1\le m\le s}\left|\frac{\log j_m(\sigma)-m}{(2m\log\log m)^{1/2}}-b\right|<\delta\right)=1 \qquad (6)$$

for each $b\in[-1,1]$ *and* $\delta>0$.

In connection to (4), (5), and (6), it is worth to recall that the mean values of $w=w(\sigma)$ and $s=s(\sigma)$ are asymptotically equivalent to $\log n$. The estimate of $|\log j_m(\sigma)-m|$ following from (5) still has the error $\delta(2m\log\log m)^{1/2}$. In this direction we now have the following improvement.

Denote $Lu=\log\max\{u,e\}=L_1u,\ldots,L_ku=L(L_{k-1}u)$ for $u\in\mathbf{R}$. For $\delta>0$ and $k\ge 2$, set

$$\beta_{mk}(1\pm\delta)=\left(2m\left(L_2m+\frac{3}{2}L_3m+L_4m+\cdots+(1\pm\delta)L_km\right)\right)^{1/2}.$$

Theorem 2.2. *For arbitrary* $0<\delta<1$ *and* $k\ge 2$, *we have*

$$\lim_{n_1\to\infty}\overline{\lim_{n\to\infty}}\,\nu_n\left(\max_{n_1\le m\le s}\frac{|\log j_m(\sigma)-m|}{\beta_{km}(1+\delta)}\ge 1\right)=0$$

and

$$\lim_{n_1\to\infty}\underline{\lim}_{n\to\infty}\,\nu_n\left(\max_{n_1\le m\le s}\frac{|\log j_m(\sigma)-m|}{\beta_{km}(1-\delta)}\ge 1\right)=1.$$

Thus, answering our question we may say that "for almost all $\sigma\in\mathbf{S}_n$"

$$|\log j_m(\sigma)-m|\le\beta_{km}(1+\delta)$$

uniformly in m, $n_1\le m\le s(\sigma)$, where $n_1\to\infty$ arbitrarily slowly. This assertion is sharp in the sense that we can not change δ by $-\delta$.

The numbers $j_m(\sigma)$ are just the success epochs of the following partial sum function

$$s(\sigma;y):=\sum_{j\le y}\mathbf{1}\{k_j(\sigma)\ge 1\},\quad 1\le y\le n.$$

By virtue of $s(\sigma;j_m(\sigma))=m$ but using a bit complicated calculations, one can show that Theorem 2.2 is equivalent to the next result. For $\delta>0$ and $k\ge 3$, set

$$\gamma_{km}(1\pm\delta)=\left(2(Lm)\left(L_3m+\frac{3}{2}L_4m+L_5m+\cdots+(1\pm\delta)L_km\right)\right)^{1/2}.$$

Theorem 2.3. *Let $0 < \delta < 1$ and $k \geq 3$ be arbitrary. We have*

$$\lim_{n_1\to\infty} \overline{\lim_{n\to\infty}} \nu^+_{n_1 n} := \lim_{n_1\to\infty} \overline{\lim_{n\to\infty}} \nu_n \left(\max_{n_1 \leq m \leq n} \frac{|s(\sigma;m) - \log m|}{\gamma_{km}(1+\delta)} \geq 1 \right) = 0$$

and

$$\lim_{n_1\to\infty} \underline{\lim}_{n\to\infty} \nu^-_{n_1 n} := \lim_{n_1\to\infty} \underline{\lim}_{n\to\infty} \nu_n \left(\max_{n_1 \leq m \leq n} \frac{|s(\sigma;m) - \log m|}{\gamma_{km}(1-\delta)} \geq 1 \right) = 1.$$

A proof will be given in the last section.

3. Strassen's law of iterated logarithm

Adopting our experience obtained in probabilistic number theory [11], we can propose further generalization of the results of Section 2. We now deal with sequences of functions defined on the symmetric group. Set

$$S_m(\sigma;t) = \frac{s(\sigma;m^t) - t\log m}{\sqrt{2LmL_3m}}, \qquad t \in [0,1],\ 1 \leq m \leq n.$$

The trajectories of these random processes lay in the space $\mathbf{D} = \mathbf{D}[0,1]$ of the right continuous functions on $[0,1]$ having left-hand limits at each point. We assume that $\mathbf{D}$ is endowed with the Skorokhod metrics. We also introduce the linearized version $G_m(\sigma;t)$ of $S_m(\sigma;t)$. For $t_l := (\log l)/\log m$ and $t \in [t_l, t_{l+1})$, we set

$$G_m(\sigma;t) = S_m(\sigma;t_l) + (S_m(\sigma;t_{l+1}) - S_m(\sigma;t_l))\,\frac{t-t_l}{t_{l+1}-t_l},$$

where $1 \leq l \leq m-1$. Our problem now is to describe the set of limit curves of $\{S_m(\sigma;t)\}$ and $\{G_m(\sigma;t)\}$ "for almost all $\sigma \in \mathbf{S}_n$" in the sense of Section 2. Again, to overcome some obstacles appearing in the sequence of probability spaces, we have to introduce more general notions.

Let $(\mathbf{U}, d)$ be a separable metric space and $Y, Y_1, \ldots, Y_n$ be $\mathbf{U}$-valued random elements all defined on the probability space $\{\Omega_n, \mathcal{F}_n, P_n\}$, $\mathcal{P} = \{P_n\}$, where $n = 1, 2, \ldots$. Let, as usual, $d(X,A) = \inf\{d(X,Z) : Z \in A\}$, where $A \subset \mathbf{U}$. We say that Y_m converges to Y $\mathcal{P}$-almost surely ($\mathcal{P} - a.s.$) if for each $\varepsilon > 0$

$$\lim_{n_1\to\infty} \overline{\lim_{n\to\infty}} P_n \left(\max_{n_1 \leq m \leq n} d(Y_m, Y) \geq \varepsilon \right) = 0$$

Thus, a compact set $A \subset \mathbf{U}$ such that, for each $\varepsilon > 0$ and each $X \in A$,

$$\lim_{n_1\to\infty} \overline{\lim_{n\to\infty}} P_n \left(\max_{n_1 \leq m \leq n} d(Y_m, A) \geq \varepsilon \right) = 0 \tag{7}$$

and

$$\lim_{n_1\to\infty} \underline{\lim}_{n\to\infty} P_n \left(\min_{n_1 \leq m \leq n} d(Y_m, X) < \varepsilon \right) = 1 \tag{8}$$

may be called a cluster set of the sequence $\{Y_m\}$ $\mathcal{P} - a.s.$. In what follows we denote the relations (7) and (8) by

$$Y_m \hookrightarrow A \qquad (\mathcal{P} - a.s.). \tag{9}$$

Denoting (8) by $Y_{m'} \to X \quad (\mathcal{P} - \text{a.s.})$ we have in mind that m' can be a random increasing subsequence.

Let $\mathbf{C} = \mathbf{C}[0,1]$ be the space of continuous functions on the interval $[0,1]$ endowed with the supremum distance $\rho(.,.)$. We recall that the Strassen set $\mathcal{K}$ agrees with the set of absolutely continuous functions g such that $g(0) = 0$ and

$$\int_0^1 (g'(t))^2 \, dt \leq 1.$$

These definitions are applicable to $\{G_m(\sigma,\cdot)\}$ and $\{S_m(\sigma,\cdot)\}$, where $1 \leq m \leq n$. For the family of probabilities, we now take $\nu := \{\nu_n\}$.

Theorem 3.1. *We have*

$$G_m(\sigma;\cdot) \hookrightarrow \mathcal{K} \quad (\nu - a.s.) \tag{10}$$

in the space $\mathbf{C}$ *and*

$$S_m(\sigma;\cdot) \hookrightarrow \mathcal{K} \quad (\nu - a.s.) \tag{11}$$

in the space $\mathbf{D}$.

The idea of the proof is similar to that of Theorem 2.3. The details will be exposed in a forthcoming paper. Applying the same argument as in the last section, we can verify that

$$\nu_n \left(\max_{n_1 \leq m \leq n} \sup_{t \in [0,1]} |S_m(\sigma;t) - G_m(\sigma;t)| \geq \varepsilon \right) = o(1)$$

for every $\varepsilon > 0$ as $n \to \infty$ and $n_1 \to \infty$. So, (10) and (11) are equivalent. The following lemma is very useful in various applications of the last result.

Lemma 3.2. *Let* $(\mathbf{U}, d)$ *and* $(\mathbf{U}_1, d_1)$ *be separable metric spaces and let* $f : \mathbf{U} \to \mathbf{U}_1$ *be a continuous map into* $\mathbf{U}_1$. *If* A *is compact subset of* $\mathbf{U}$ *and* $Y_m \hookrightarrow A \quad (\mathcal{P} - a.s.)$ *in* $(\mathbf{U}, d)$, *then* $f(Y_m) \hookrightarrow f(A) \quad (\mathcal{P} - a.s.)$ *in the second space* $(\mathbf{U}_1, d_1)$.

By virtue of this lemma, Theorem 3.1 implies Theorem 2.3 which assertion can now be rewritten as

$$G_m(\sigma;1) \hookrightarrow [-1,1] \quad (\nu - a.s.).$$

Going along this path, we can list more consequences.

Corollary 3.3. *The following relations hold* $(\nu - a.s.)$:

- $(G_m(\sigma;1/2), G_m(\sigma;1)) \hookrightarrow \mathcal{L} := \{(u,v) : u^2 + (v-u)^2 \leq 1/2\}$;
- $G_m(\sigma;1/2) \hookrightarrow [-\sqrt{2}/2, \sqrt{2}/2]$;
- *if* m' *is the subsequence for which* $G_{m'}(\sigma;1/2) \to \sqrt{2}/2$, *then we have* $G_{m'}(\sigma;\cdot) \to g_1$, *where*

$$g_1(t) = \begin{cases} t\sqrt{2}, & \text{if} \quad 0 \leq t \leq 1/2, \\ \sqrt{2}/2, & \text{if} \quad 1/2 \leq t \leq 1; \end{cases}$$

- *if* m' *is the subsequence for which* $G_{m'}(\sigma;1/2) \to 1/2$ *and* $G_{m'}(\sigma;1) \to 0$, *then we have* $G_{m'}(\sigma;\cdot) \to g_2$, *where*

$$g_2(t) = \begin{cases} t, & \text{if} \quad 0 \leq t \leq 1/2, \\ 1-t, & \text{if} \quad 1/2 \leq t \leq 1. \end{cases}$$

These sophisticated examples of continuous functionals f can be found in [6]. Using the fact that the cycle lengths $\jmath_m(\sigma)$ are the counts of the partial sums $s(\sigma;m)$, as in Section 2, we could convert the relations of this Corollary to that for $\jmath_m(\sigma)$. Nevertheless, we now prefer the functional form of doing that.

First, we introduce a new sequence of processes. Set $j(\sigma;u)=1$ if $0\le u<1$ and $j(\sigma;u)=j_{[u]}(\sigma)$ if $1\le u\le s=s(\sigma)$. Define

$$\Psi_m(\sigma;t)=\frac{\log j(\sigma;tm)-tm}{\sqrt{2m\log\log m}},\quad 0\le t\le 1,\ 3\le m\le s=s(\sigma).$$

Theorem 3.4. *We have*

$$\Psi_m(\sigma;\cdot)\hookrightarrow \mathcal{K}\quad (\nu-a.s.)$$

in the space **D**.

To derive this claim from Theorem 3.1, we can use the generalized inverses. Let $\mathbf{D}_0$ denote the subspace of $\mathbf{D}$ consisting of nonnegative nondecreasing functions. For $X\in\mathbf{D}_0$, we define $X^{-1}\in\mathbf{D}_0$ by

$$X^{-1}(t)=\inf\{u\in[0,1]: X(u)>t\}$$

with the agreement $X^{-1}(t)=1$ for $X(1)\le t\le 1$. A useful auxiliary result for the proof has been provided by W. Vervaat [16].

Lemma 3.5. *Let $X_m\in\mathbf{D}_0$ and δ_m be a sequence of positive numbers, $\delta_m\to 0$. If $g=g(t)\in\mathbf{C}$, then as $m\to\infty$ the following relations are equivalent*

$$\sup\left\{\left|\frac{X_m(t)-t}{\delta_m}-g(t)\right| : t\in[0,1]\right\}\to 0$$

and

$$\sup\left\{\left|\frac{X_m^{-1}(t)-t}{\delta_m}+g(t)\right| : t\in[0,1]\right\}\to 0.$$

For the proof of Theorem 3.4, it suffices to apply (11) and Lemma 3.5 with $g\in\mathcal{K}$,

$$X_m=X_m(\sigma,t)=s(\sigma,m^t)/\log m,$$

and $\delta_m=(2L_3m)/\log m)^{1/2}$. In the style of Corollary of Theorem 3.1, we have, for instance,

$$\Psi_m(\sigma;1/2)\hookrightarrow[-\sqrt{2}/2,\sqrt{2}/2],\ \Psi_m(\sigma;1)-\Psi(\sigma;1/2)\hookrightarrow[-\sqrt{2}/2,\sqrt{2}/2]\quad(\nu-a.s).$$

To check the second relation here, we can use the two-dimensional convergence pointed out in the first item of Corollary 3.3. The last two relations show some fascinating symmetry between the behaviour of $\log j(\sigma;tm)$ and $\log(j(\sigma;m)/j(\sigma;(1-t)m))$ at the point $t=1/2$. We have some other observations of a similar symmetry. In what general forms can such phenomenon appear?

4. Proof of Theorem 2.3

The main probabilistic ingredient is found in the classical paper of W. Feller [5]. Let X_n, $n\ge 1$ be independent random variables (r.vs), $\mathbf{E}X_n=0$, $\mathbf{E}X_n^2<\infty$, and $B_n^2:=\mathbf{E}X_1^2+\cdots+\mathbf{E}X_n^2\to\infty$. Here as previously $n\to\infty$. Denote $Y_n=X_1+\cdots+X_n$.

Lemma 4.1. *Let $C>0$ be a constant and λ_n be a positive increasing sequence. In addition, assume that the r.vs X_n, $n\ge 1$, satisfy*

$$|X_n|\le CB_n^2/\lambda_n^3$$

with probability one. If the series

$$\sum_{n\geq 1} \frac{\mathbf{E}X_n^2\lambda_n}{B_n^2} e^{-\lambda_n^2/2} \tag{12}$$

converges, then

$$P\left(Y_n > \lambda_n B_n \quad \textit{infinitely often}\right) = 0.$$

If the series (12) *diverges, then*

$$P\left(Y_n > \lambda_n B_n \quad \textit{infinitely often}\right) = 1.$$

We will need just a corollary for independent Bernoulli r.vs ξ_n, $n \geq 1$, such that $P(\xi_n = 1) = p_n = 1 - P(\xi_n = 0)$. We reformulate it for the sum $\zeta_n := \xi_1 + \cdots + \xi_n$ in a slightly modified form.

Corollary 4.2. *Let the Bernoulli r.vs satisfy the condition $p_n = 1/n + O(1/n^{1+\varepsilon})$, where $\varepsilon > 0$ is arbitrary. Then, for every $0 < \delta < 1$ and $k \geq 3$, we have*

$$\lim_{n_1\to\infty}\lim_{n\to\infty} P_{n_1 n}^{+} := \lim_{n_1\to\infty}\lim_{n\to\infty} P\left(\max_{n_1\leq m\leq n} \frac{|\zeta_m - \log m|}{\gamma_{km}(1+\delta)} \geq 1\right) = 0$$

and

$$\lim_{n_1\to\infty}\lim_{n\to\infty} P_{n_1 n}^{-} := \lim_{n_1\to\infty}\lim_{n\to\infty} P\left(\max_{n_1\leq m\leq n} \frac{|\zeta_m - \log m|}{\gamma_{km}(1-\delta)} \geq 1\right) = 1.$$

Our approach originated in probabilistic number theory (see [11]) has two steps. The first one is based upon the following lemma.

Lemma 4.3 (Fundamental Lemma). *There exist a probability space $\{\Omega, \mathcal{F}, P\}$ and independent Poisson r.vs Z_j, $\mathbf{E}Z_j = 1/j$, $j \geq 1$, such that*

$$\left|\nu_n\left((k_1(\sigma),\ldots,k_r(\sigma)) \in A\right) - P\left((Z_1,\ldots,Z_r) \in A\right)\right| \leq Cr/n$$

with an absolute constant $C > 0$ uniformly in $A \subset \mathbf{Z}^{+r}$ for each $1 \leq r \leq n$.

The assertion of this lemma follows from the Feller coupling. Much more precise estimate of this total variation distance is proved in [2]. It is also shown that the distance does not vanish when the condition $r = o(n)$ is not satisfied. So, as in the probabilistic number theory (see [11]) Fundamental lemma allows to deal with "truncated" up to r additive functions. The remainder appearing in this procedure can be estimated by the following inequality obtained in author's paper [12].

Lemma 4.4. *Let $\{h_j(k)\}$, $k \geq 0$ and $j \geq 1$, be a two-dimensional array of real numbers such that $h_j(0) \equiv 0$. If Z_j, $j \geq 1$, are the Poisson r.vs as in Lemma 4.3, then for arbitrary $x > 0$, $b_m \in \mathbf{R}$, and $1 \leq r \leq m \leq n$, we have*

$$\nu_n\left(\max_{r\leq j\leq n}\left|\sum_{j\leq m} h_j(k_j(\sigma)) - b_m\right| \geq x\right) \leq 32e^2 P\left(\max_{r\leq j\leq n}\left|\sum_{j\leq m} h_j(Z_j) - b_m\right| \geq x/3\right).$$

In particular, for arbitrary $d_r \geq \cdots \geq d_n > 0$ and $a_j \in \mathbf{R}$, $j \geq 1$,

$$\nu_n\left(\max_{r\leq m\leq n} d_m\left|\sum_{j\leq m} a_j k_j(\sigma) - \sum_{j\leq m}\frac{a_j}{j}\right| \geq x\right) \leq \frac{288e^2}{x^2}\left(d_r^2\sum_{j\leq r}\frac{a_j^2}{j} + \sum_{r<j\leq n}\frac{d_j^2 a_j^2}{j}\right).$$

Proof of Theorem 2.3. We start from the observation which actually explains why the results on the sequences $\{w(\sigma;m)\}$ and $\{s(\sigma;m)\}$ are so close. This further leads to the similarity of estimates valid for $\{J_m(\sigma)\}$ and $\{j_m(\sigma)\}$. If $\varepsilon > 0$ and a positive sequence $\psi(n) \nearrow \infty$ are arbitrary, then by Lemma 4.4 we have

$$\begin{aligned} \nu_{n_1 n}(\varepsilon) \quad &:= \nu_n\left(\max_{n_1\le m\le n} \psi(m)^{-1}\left|s(\sigma;m)-w(\sigma;m)\right| \ge \varepsilon\right) \\ &\le 32e^2 P\left(\max_{n_1\le m\le n}\psi(m)^{-1}\left|\sum_{j\le m}(Z_j - \mathbf{1}\{Z_j\ge 1\})\right| \ge \varepsilon/3\right) \\ &\le 32e^2 P\left(\sum_{j\le n}|Z_j - \mathbf{1}\{Z_j \ge 1\}| \ge \varepsilon\psi(n_1)/3\right) = o(1) \end{aligned}$$

as $n \to \infty$ and $n_1 \to \infty$.

Take $r = r(n) = n/Ln \ge n_1$ and some arbitrary $\varepsilon > 0$. By virtue of the last estimate and Lemma 4.4, we obtain

$$\begin{aligned} &\nu_n\left(\max_{r\le m\le n}\gamma_{km}(1+\delta)^{-1}\left|s(\sigma;m)-\log m\right|\ge 1\right) \\ &\le \nu_n\left(\max_{r\le m\le n}\gamma_{km}(1+\delta)^{-1}\left|w(\sigma;m)-\log m\right|\ge 1/2\right) + \nu_{rn}(1/2) \\ &= O\left((LrL_3r)^{-2}\sum_{j\le r}\frac{1}{j} + \sum_{r<j\le n}\frac{1}{jLjL_3j}\right) + o(1) = o(1). \end{aligned}$$

Hence, with such a choice of r we have

$$\nu^+_{n_1 n} \le \nu_n\left(\max_{n_1\le m\le r}\frac{|s(\sigma;m)-\log m|}{\gamma_{km}(1+\delta)}\ge 1\right) + o(1).$$

By the fundamental Lemma this may be rewritten as

$$\nu^+_{n_1 n} \le P\left(\max_{n_1\le m\le r}\gamma_{km}(1+\delta)^{-1}\left|\sum_{j\le m}\mathbf{1}\{Z_j\ge 1\} - \log m\right| \ge 1\right) + o(1)$$

For the Bernoulli r.vs $\hat{\xi}_j := \mathbf{1}\{Z_j \ge 1\}$, $j \ge 1$, we may apply Corollary 4.2. It yields the first assertion of Theorem 2.3.

To prove the second one, it suffices to observe that

$$\nu^-_{n_1 n} \ge \nu_n\left(\max_{n_1\le m\le r}\frac{|s(\sigma;m)-\log m|}{\gamma_{km}(1-\delta)}\ge 1\right) = P^-_{n_1 r} + o(1)$$

and again to use the corollary with $\xi_j = \hat{\xi}_j$. Theorem 2.3 is proved.

References

[1] R. Arratia and S. Tavaré, (1992) Limit theorems for combinatorial structures via discrete process approximations, *Random Structures and Algorithms 3*, 321–345.

[2] R. Arratia, A. Barbour, and S. Tavaré, *Logarithmic Combinatorial Structures: a Probabilistic Approach*, EMS Monographs in Math., The EMS Publishing House, Zürich, 2003.

[3] G.J. Babu and E. Manstavičius, (1999) Brownian motion for random permutations, *Sankhyā A: The Indian J. of Statistics 61*, No. 3, 312–327.

[4] J.M. DeLaurentis and B.G. Pittel, Random permutations and the Brownian motion, *Pacific J. Math. 119* (1985), No. 2, 287–301.

[5] W. Feller, The general form of the so-called law of the iterated logarithm, *Trans. AMS 54* (1943), 373–402.

[6] S. D. Freedman, *Brownian Motion and Diffusion*, Holden–day, San Francisco, 1971.

[7] V.L. Goncharov, On the distribution of cycles in permutations, *Dokl. Acad. Nauk SSSR 35* (1942), 9, 299–301 (Russian).

[8] V.L. Goncharov, Some facts from combinatorics, *Izv. Akad. Nauk SSSR, Ser. Mat. 8* (1944), 3–48 (Russian); On the field of combinatory analysis, *Transl. AMS 19* (1962), 1–46.

[9] V.F. Kolchin, A problem of the allocation of particles in cells and cycles of random permutations, *Teor. Veroyatnost. i Primenen. 16* (1971), No. 1, 67–82 (Russian).

[10] V.F. Kolchin, *Random Mappings*, Optimization Software, Inc. New York, 1986.

[11] E. Manstavičius, Functional approach in the divisor distribution problems, *Acta Math. Hungarica 66* (1995), No. 3, 343–359.

[12] E. Manstavičius, The law of iterated logarithm for random permutations, *Lith. Math. J. 38* (1998), No. 2, 160–171.

[13] L.A. Shepp and S.P. Lloyd, Ordered cycle lengths in a random permutation, *Trans. AMS 121* (1966), 340–357.

[14] A.M. Vershik and A.A. Schmidt, Symmetric groups of high order, *Dokl. Acad. Nauk SSSR 206* (1972), No. 2, 269–272 (Russian).

[15] A.M. Vershik and A.A. Schmidt, Limit measures that arise in the asymptotic theory of symmetric groups I, *Teor. Veroyatnost. i Primenen. 22* (1977), No. 1, 72–88; II, *ibid. 23* (1978), No. 1, 42–54 (Russian).

[16] W. Vervaat, *Success epochs in Bernoulli trials with applications in number theory*, Math. Centrum, Amsterdam, 1972.

Eugenijus Manstavičius
Vilnius University, Department of Mathematics and Informatics
eugenijus.manstavicius@maf.vu.lt

Transcendence of Generating Functions of Walks on the Slit Plane

Martin Rubey

ABSTRACT: *Consider a single walker on the slit plane, that is, the square grid $\mathbb{Z}^2$ without its negative x-axis, who starts at the origin and takes his steps from a given set $\mathfrak{S}$. Mireille Bousquet-Mélou conjectured that – excluding pathological cases – the generating function counting the number of possible walks is algebraic if and only if the walker cannot cross the negative x-axis without touching it. In this paper we prove a special case of her conjecture.*

1. Introduction

Let $\mathfrak{S}$ – the set of *steps* – be a finite subset of $\mathbb{Z}^2$. A *walk on the slit plane* is a sequence $(0,0) = w_0, w_1, \dots, w_n$ of points in $\mathbb{Z}^2$, such that the difference of two consecutive points $w_{i+1} - w_i$ belongs to the set of steps $\mathfrak{S}$ and none of the points but the first lie on the half-line $\{(x,0) : x \le 0\}$. An example for such a walk with set of steps

$$\mathfrak{S} = \{(-1,-2), (-1,1), (-1,2), (1,-2), (1,1), (1,2)\}$$

is shown in Figure 1.

Recall that a generating function $F(t) = \sum_{n\ge 0} f_n t^n$ is *algebraic*, if there is a nontrivial polynomial P in two variables, such that $P(F(t), t) = 0$. Otherwise, it is *transcendental*.

In [2] Mireille Bousquet-Mélou conjectured the following:

Conjecture 1.1. *Consider the generating function for walks in the slit plane with a given set of steps $\mathfrak{S}$, counted according to their length and their end-coordinates:*

$$S(x,y;t) = \sum_{\substack{W \text{ walk on the slit plane} \\ \text{starting at the origin} \\ \text{with steps in } \mathfrak{S}}} t^{\text{length } W} x^{\text{x-final } W} y^{\text{y-final } W}.$$

Suppose that the set of steps is not degenerated and thus all four quadrants of the plane can be reached by some walk, and that the greatest common divisor of the vertical parts of the steps is equal to one.

Then this generating function is algebraic in t, if and only if the height of any step is at most one.

In fact, she proved one part of this conjecture in Section 7 of the above paper, namely, that walks with steps that have height at most one have an algebraic generating function. Furthermore, in Section 8 she proved for one family of step-sets that the corresponding generating functions have to be transcendental. In the present paper, we prove the following:

[1]Research supported by the Austrian Science Foundation FWF, grant S8302-MAT.

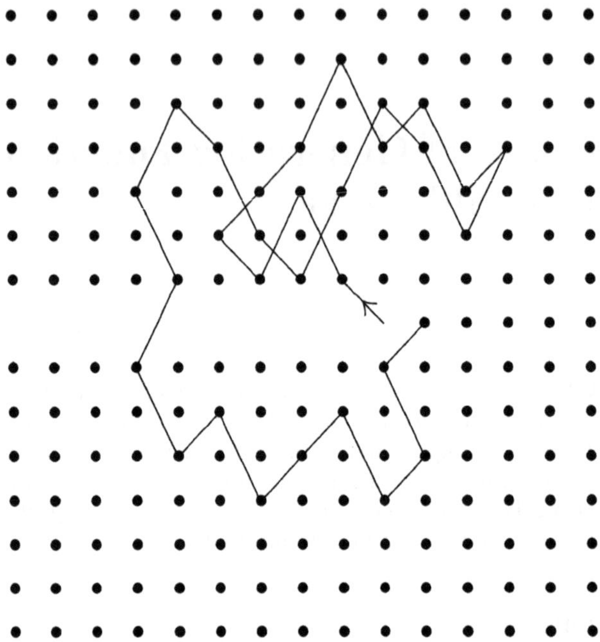

FIGURE 1. A walk on the slit plane

Theorem 1.2. *Let $\mathcal{H}$ and $\mathcal{V}$ be two finite sets of integers, the greatest common divisor of the integers in each set being equal to one. Furthermore, assume that both of the sets $\mathcal{H}$ and $\mathcal{V}$ contain positive and negative numbers, and that $\mathcal{V}$ contains an element with absolute value at least 2. Finally, assume that the minimum of $\mathcal{V}$ is at least -2.*

Let $\mathfrak{S}$ be the Cartesian product of the two sets: $\mathfrak{S} = \mathcal{H} \times \mathcal{V}$, where $\mathcal{H}$ is the horizontal and $\mathcal{V}$ is the vertical part of the steps. Then the following generating functions for walks in the slit plane with set of steps $\mathfrak{S}$ are transcendental in t:

- *the generating function $S_{i,0}(t)$ for walks ending at a prescribed coordinate $(i, 0)$,*
- *the generating function $L(t)$ for loops, i.e., walks that return to the origin,*
- *the generating function $S_0(1;t)$ for walks ending anywhere on the x-axis, and*
- *$S(1,1;t)$, which is the generating function for walks ending anywhere in the slit plane.*

For example, the set of steps of the walk in Figure 1 is the Cartesian product of $\mathcal{H} = \{-1, +1\}$ and $\mathcal{V} = \{-2, +1, +2\}$.

In fact we consider a slightly more general problem: we allow the steps in $\mathcal{H}$ and $\mathcal{V}$ to be weighted with positive real numbers. The weight of a step in the product set $\mathfrak{S} = \mathcal{H} \times \mathcal{V}$ then is the product of the weights of its corresponding vertical and horizontal parts and the weight of a walk is the product of the weights of its individual steps.

As in [2], we will use a special case of the following theorem to determine in which cases the generating function cannot be algebraic:

Theorem 1.3. [4] *Let $F(t)$ be an algebraic function over $\mathbb{Q}$ that is analytic at the origin, then its nth Taylor coefficient f_n has an asymptotic equivalent of the form*

$$f_n = \frac{\beta^n n^s}{\Gamma(s+1)} \sum_{i=0}^{m} C_i \omega_i^n + O(\beta^n n^t),$$

where $s \in \mathbb{Q} \setminus \{-1, -2, -3, \dots\}$, $t < s$; β is a positive algebraic number and the C_i and ω_i are algebraic with $|\omega_i| = 1$.

It follows easily that an algebraic function cannot have an appearance of a negative integer power of n anywhere in the full asymptotic expansion of its Taylor coefficients.

2. An expression for the generating function for walks on the slit plane

The fundamental theorem for walks on the slit plane is the following:

Theorem 2.1. *(Proposition 9 in [2]) Let*

$$B(x;t) = \sum_{\substack{W \text{ walk on } \mathbb{Z}^2 \\ \text{starting at the origin} \\ \text{ending on the } x\text{-axis} \\ \text{with steps in } \mathfrak{S}}} x^{\text{x-final}(W)} t^{\text{length}(W)}$$

be the generating function for bilateral walks, that is, walks that end on the x-axis but are otherwise unconstrained.

For $i \geq 1$, the generating function $S_{i,0}(t)$ for walks on the slit plane ending at $(i,0)$ can be computed by induction on i via the following identity:

$$\sum_{k=1}^{i} \frac{(-1)^{k-1}}{k} \sum_{\substack{i_1+i_2+\dots+i_k=i \\ i_1>0, i_2>0, \dots, i_k>0}} S_{i_1,0}(t) S_{i_2,0}(t) \dots S_{i_k,0}(t) = [x^i] \log B(x;t).$$

Note that it follows that $\log B(x;t)$ has positive Taylor coefficients.

Now we can take advantage of the special structure of the set step $\mathfrak{S}$. Since it decomposes into a horizontal and a vertical part, the generating function for bilateral walks factorises:

$$B(x;t) = \bar{B}(H(x)t)$$

where

$$\bar{B}(t) = \sum_{\substack{W \text{ walk on } \mathbb{Z} \\ \text{from 0 to 0} \\ \text{with steps in } \mathcal{V}}} t^{\text{length}(W)}$$

is the generating function for *bridges* and

$$H(x) = \sum_{h \in \mathcal{H}} x^h$$

is the *step (Laurent-)polynomial* for $\mathcal{H}$.

3. Transcendence of $[x^i]\log B(x;t)$

In this section we show that $[x^i]\log B(x;t)$ cannot be algebraic if the set $\mathcal{V}$ contains an element with absolute value strictly greater than one.

To this end, we consider the asymptotic expansion of $[t^n x^i]\log B(x;t)$. Since $B(x;t)$ factorises, we have

$$\begin{aligned}[t^n x^i]\log B(x;t) &= [t^n x^i]\log \bar{B}\left(H(x)t\right)\\ &= [x^i]\left(H(x)\right)^n [t^n]\log \bar{B}(t).\end{aligned} \tag{1}$$

Therefore, we have divided the problem in two: we will show that the asymptotic expansions of both $[x^i]\left(H(x)\right)^n$ and $[t^n]\log \bar{B}(t)$ contain a term $n^{-k/2}$ for some odd k.

3.1. Asymptotics of the horizontal part

Let min be the minimal integer such that $H(x)x^{min}$ is a polynomial. To determine the asymptotics of $[x^i]\left(H(x)\right)^n = [x^{n\cdot min}]x^{-i}\left(H(x)x^{min}\right)^n$, we can use the following theorem:

Theorem 3.1 ([3, 5]). *Let $g(z)$ be an analytic function of degree d with positive coefficients assumed to be aperiodic and such that $g(0)\neq 0$, and let $a(z)$ be analytic except possibly at zero, where a pole is allowed. Let λ be a positive number of some subinterval $[\lambda_a,\lambda_b]$ of the open interval $]0,d[$. Then, with $N = \lfloor \lambda n\rfloor$, one has uniformly for $\lambda\in[\lambda_a,\lambda_b]$*

$$[z^N]a(z)(g(z))^n = a(\zeta)\frac{(g(\zeta))^n}{\zeta^{N+1}\sqrt{2\pi nR}}(1+o(1)),$$

where ζ is the unique positive root of the equation

$$\zeta\frac{g'(\zeta)}{g(\zeta)} = \lambda$$

and

$$R = \frac{d^2}{d\zeta^2}[\log g(\zeta) - \lambda\log\zeta].$$

Let $p = \gcd\{k + min : k\in\mathcal{H}\}$ and note that in the case $p>1$, every p^{th} coefficient of $\left(H(x)\right)^n$ will be zero. Thus, we take $g(x) = H(x^{1/p})x^{min/p}$, and $a(x) = x^{-i/p}$, and use the theorem to determine $[x^{n\cdot min/p}]a(x)(g(x))^n$. In this situation, $\lambda = min/p$ is constant, therefore ζ and R must be constant, too. Hence, the asymptotic expansion of $[x^i]\left(H(x)\right)^n$ contains an appearance of $n^{-1/2}$.

3.2. Asymptotics of the vertical part

To determine the asymptotic behaviour of $[t^n]\log\bar{B}(t)$ we will use analysis of singularities. In a first step, we have to determine the singularities of the expression. Following the general theory of singularity analysis, all the contributions from these singularities must be added up.

According to the remark after Theorem 2.1 and the factorisation (1), we have that $[t^n]\log\bar{B}(t)$ is always positive. Therefore we can apply Pringsheim's theorem:

Theorem 3.2. [6] *If a function with a finite radius of convergence has Taylor coefficients that are nonnegative, then one of its singularities of smallest modulus – a dominant singularity – is real positive.*

Since the logarithm is singular only at the origin, and $\bar{B}(t)$ is strictly positive for positive real numbers, Pringsheim's theorem implies that one dominant singularity of $\log \bar{B}(t)$ is in fact a singularity – call it ρ – of $\bar{B}(t)$. Of course, there can be other dominant singularities of $\log \bar{B}(t)$ arising from singularities of $\bar{B}(t)$. We will discuss them below.

Furthermore, it might happen – although we believe that it does not – that $\bar{B}(t)$ has zeros on the circle around the origin with radius ρ, thus also making $\log \bar{B}(t)$ singular. However, such singularities can never contribute a summand of order $n^{-k/2}$ for odd k, so we can simply ignore them. Note that $\bar{B}(t)$ cannot vanish for $|t| < \rho$, since ρ is a dominant singularity of $\log \bar{B}(t)$.

Now we want to compute the contribution of the dominant singularities of $\bar{B}(t)$ to the asymptotic expansion of $\log \bar{B}(t)$. To do so, we need a better understanding of $\bar{B}(t)$, which is the generating function for bridges with step set $\mathcal{V}$.

Luckily, this generating function has already been studied. We define the *step (Laurent-)polynomial* for $\mathcal{V}$ as

$$V(y) = \sum_{v \in \mathcal{V}} y^v$$

and the *characteristic curve* determined by $\mathcal{V}$ by the equation

$$1 - tV(y) = 0 \quad \text{or equivalently} \quad y^{min_v} = t(y^{min_v} V(y)) = 0, \tag{2}$$

where $min_v = -\min \mathcal{V}$ is the minimal integer to make the equation polynomial.

We say that the functional equation (2) is *reduced*, if the greatest common divisor of the exponents of the monomials in $V(y)$ is equal to one, which is one of the assumptions in our main Theorem 1.2.

We say that the functional equation (2) has *period* p, if the greatest common divisor of the exponents of the monomials in $y^{min_v} V(y)$ is equal to p.

As is well known, the period is also the number of dominant singularities of $\bar{B}(t)$, which are all conjugate to the real dominant singularity ρ. In our case however, it can be seen ([1, Section 3.3]) that the asymptotic formula for bridges is obtained from the asymptotic expansion derived from the singularity at ρ by multiplying with p. Since we are only interested in the presence or absence of a term $n^{-k/2}$ for some odd k in the asymptotic expansion, we can assume from now on that the functional equation (2) is aperiodic, i.e., has period one.

It can be seen [1, 6] that the solutions of this functional equation organise themselves into "small" and "large" branches. Here, "small" means that the solution $y(t)$ tends to zero as t tends to zero, whereas "large" means that $y(t)$ tends to infinity as t approaches zero.

It is only the set of "small" solutions that is interesting for us, and it can be seen – using a limit case of Pellet's Theorem, see for example [7] – that there are min_v of them. A nice expression for the generating function for bridges is given by the following theorem:

Theorem 3.3. [1, Theorem 1 and proof of Theorem 3] *The generating function for bridges is an algebraic function given by*

$$\bar{B}(t) = t\sum_{j=1}^{min_v} \frac{y_j'(t)}{y_j(t)} = t\frac{d}{dt}\log(y_1(t)y_2(t)\cdots y_{min_v}(t))$$
$$= \frac{1}{2\pi i}\int_{|y|=\tau} \frac{dy}{y(1-tV(y))},$$

where the expressions involve all the small branches $y_1, y_2, \ldots, y_{min_v}$ *of the characteristic curve (2), and* ρ *is the radius of convergence of* $\bar{B}(t)$*. Furthermore, the principal branch* $y_1(t)$*, i.e., the branch with real coefficients, has a square root singularity at* ρ *and the product of all the other small branches is analytic for all* t *with* $|t| \le \rho$*. More precisely,* ρ *is given by* $\rho = 1/V(\tau)$*, where* τ *is the unique positive number with* $V'(\tau) = 0$*.*

Thus, applying the Newton-Puiseux theorem, we can develop $\bar{B}(t)$ around the singularity ρ, setting $\tilde{t} = \sqrt{\rho - t}$:

$$\bar{B}(t) = a_{-1}/\tilde{t} + a_0 + a_1\tilde{t} + a_2\tilde{t}^2 + \cdots$$

Composing this expansion with the Taylor expansion of the logarithm we obtain

$$\log\bar{B}(t) = \log a_{-1}/\tilde{t} + \log(1 + \frac{a_0}{a_{-1}}\tilde{t} + \frac{a_1}{a_{-1}}\tilde{t}^2 + \cdots).$$

Now we want to find a term $n^{-k/2}$ for some odd k in the asymptotic expansion of the coefficients of the above series. Clearly,

$$[t^n]\log a_{-1}/\tilde{t} = [t^n]\log a_{-1}/\sqrt{\rho - t} \sim c_0\rho^{-n}n^{-1},$$

for some constant c_0, is not what we are looking after. However, we have

$$\log(1 + \frac{a_0}{a_{-1}}\tilde{t} + \frac{a_1}{a_{-1}}\tilde{t}^2 + \cdots) = \frac{a_0}{a_{-1}}\tilde{t} + \cdots$$

and

$$[t^n]\frac{a_0}{a_{-1}}\tilde{t} = [t^n]\frac{a_0}{a_{-1}}\sqrt{\rho - t} \sim c_1\rho^{-n}n^{-3/2},$$

for some constant c_1. Provided that a_0 does not vanish, this term will guarantee transcendence of the generating function for walks on the slit plane. Thus we need the constant term in the singular expansion of $\bar{B}(t)$.

Since the product of the non-principal branches $y_2(t)y_3(t)\ldots y_{min_v}(t)$ is analytic and non-zero at ρ, the contribution of $t\frac{d}{dt}\log(y_2(t)\cdots y_{min_v}(t))$ to the constant term in the singular expansion of $\bar{B}(t)$ around ρ is the sum of the residues of $1/y(1-\rho V(y))$ at the zeros of $1-\rho V(y)$ that are strictly smaller than τ in modulus.

To obtain the contribution of $t\frac{d}{dt}\log(y_1(t))$, we proceed as follows:

$$[\tilde{t}^0]t\frac{d}{dt}\log(y_1(t))$$
$$= [\tilde{t}^0](\rho - \tilde{t}^2)(-\frac{1}{2\tilde{t}})\frac{d}{d\tilde{t}}\log(y_1(t))$$
$$= -\frac{\rho}{2}[\tilde{t}]\frac{d}{d\tilde{t}}\log(y_1(t))$$
$$= \rho[\tilde{t}^2]\log(y_1(t)).$$

To obtain the coefficient of $\tilde{t}^2$ in $z = \log(y_1(t))$, we consider the Taylor expansion of $0 \equiv G(t,z) = 1 - tV(e^z)$ around $(\rho, \log\tau)$, where $\tau = y_1(\rho)$. We set $\tilde{z} = z - \log\tau$ and write G short for $G(\rho, \log\tau)$, subscripts denote the partial derivative:

$$\begin{aligned} 0 &= G(t,z) \\ &= G - G_t\tilde{t}^2 + G_z\tilde{z} - G_{t,z}\tilde{t}^2\tilde{z} + \frac{1}{2}G_{z,z}\tilde{z}^2 - \frac{1}{2}G_{t,z,z}\tilde{t}^2\tilde{z}^2 + \frac{1}{6}G_{z,z,z}\tilde{z}^3 + \dots, \end{aligned}$$

since $G_{t,t} \equiv 0$. We have

$$\begin{aligned} G &= 0 \\ G_t &= -V(e^z)|_{(t=\rho, z=\log\tau)} = -\frac{1}{\rho} \\ G_z &= -te^zV'(e^z)|_{(t=\rho,z=\log\tau)} = -\rho\tau V'(\tau) = 0 \\ G_{t,t} &= 0 \\ G_{t,z} &= -e^zV'(e^z)|_{(t=\rho,z=\log\tau)} = 0 \\ G_{z,z} &= -t\left(e^{2z}V''(e^z) + e^zV'(e^z)\right)\big|_{(t=\rho,z=\log\tau)} = -\rho\tau^2V''(\tau) \\ G_{z,z,z} &= -\rho\tau^3V'''(\tau) - \rho\tau^2V''(\tau). \end{aligned}$$

Therefore, substituting into the Taylor expansion $\tilde{z} = \alpha\tilde{t} + \beta\tilde{t}^2 + O(\tilde{t}^3)$ we obtain

$$\begin{aligned} 0 &= -G_t\tilde{t}^2 + \frac{1}{2}G_{z,z}\tilde{z}^2 + \frac{1}{6}G_{z,z,z}\tilde{z}^3 + O(\tilde{t}^4) \\ &= \left(-G_t + \frac{1}{2}G_{z,z}\alpha^2\right)\tilde{t}^2 + \left(G_{z,z}\alpha\beta + \frac{1}{6}G_{z,z,z}\alpha^3\right)\tilde{t}^3 + O(\tilde{t}^4). \end{aligned}$$

Thus

$$\begin{aligned} \alpha &= \sqrt{\frac{2G_t}{G_{z,z}}} = \frac{1}{\rho\tau}\sqrt{\frac{2}{V''(\tau)}} \\ \beta &= -\frac{\alpha^2 G_{z,z,z}}{6G_{z,z}} = -\frac{\tau V'''(\tau) + 3V''(\tau)}{3\left(\rho\tau V''(\tau)\right)^2}. \end{aligned} \tag{3}$$

It is easy to check that β, i.e., the coefficient of $\tilde{t}^0$ in the singular expansion of $t\frac{d}{dt}\log(y_1(t))$ is exactly one half of the residue of $1/y(1-\rho V(y))$ at τ.

In summary, we have shown the following:

Theorem 3.4. *Consider bridges with set of steps $\mathcal{V}$. Let $y_1(t), y_2(t), \dots, y_{min_v}(t)$ be the solutions of the functional equation $1 - t\sum_{v\in\mathcal{V}} y^v = 0$ that tend to zero as t goes to zero, $y_1(t)$ being the branch with real positive Taylor coefficients. Let τ be the unique positive real number with $V'(\tau) = 0$ and let $\rho = \frac{1}{V(\tau)}$. Furthermore, let $\tau_k = y_k(\rho)$ for $k \in \{2, 3, .., min_v\}$.*

Then the constant term in the singular expansion of the generating function for bridges is

$$\frac{1}{2}\operatorname{res}_{y=\tau}\frac{1}{y(1-\rho V(y))} + \sum_{k=2}^{min_v}\operatorname{res}_{y=\tau_k}\frac{1}{y(1-\rho V(y))}. \tag{4}$$

Unfortunately, we were not able to show that this expression does not vanish if the step set contains a step of height strictly greater than one. However, we have the following conjecture, that we are able to prove partially for some special cases:

Conjecture 3.5. *Let $V(y)$ be a Laurent-polynomial with positive coefficients with highest exponent equal to max_v and lowest exponent equal to $-min_v$. Let τ be the unique positive solution of $V'(\tau) = 0$ and*

$$f_<(y) = \prod_{\substack{V(\kappa)=V(\tau)\\ |\kappa|<\tau}} (y-\kappa) = \sum_{k=0}^{min_v-1} a_k z^k$$

and

$$f_>(y) = \prod_{\substack{V(\kappa)=V(\tau)\\ |\kappa|>\tau}} (y-\kappa) = \sum_{k=0}^{max_v-1} b_k z^k.$$

Consider the decomposition

$$\frac{1}{y(V(y)-V(\tau))} = \frac{\alpha+\beta y}{(y-\tau)^2} + \frac{p_<(y)}{f_<(y)} + \frac{p_>(y)}{f_>(y)}, \tag{5}$$

where the degree of $p_<$ is $min_v - 2$ and the degree of $p_>$ is $max_v - 2$.

Then

$$0 < a_0 < a_1 < \cdots < a_{min_v-1} \tag{6}$$
$$0 < b_{max_v-1} < b_{max_v-2} < \cdots < b_0, \tag{7}$$

the leading term of $p_<$ is negative if $min_v > 1$ and the leading term of $p_>$ is positive if $max_v > 1$.

The negativity of the leading term of $p_<$ in the conjecture would already imply that the constant term of the singular expansion of $\bar{B}(t)$ around ρ does not vanish: $-\frac{1}{\rho}(\frac{\beta}{2} + [y^{min_v-2}]p_<(y))$ is exactly the value given by (4). Since replacing y by $1/y$ in the Laurent-polynomial $V(y)$ changes the sign of β in the decomposition (5), we can assume that β is negative or zero. In fact, if $min_v = 1$, it follows from (3) that β is negative. Since $[y^{min_v-2}]p_<(y) < 0$ for $min_v > 1$, the claim follows.

We can prove parts of the conjecture for $\min_v \leq 3$: In general, it is easy to see that the product of $f_<$ and $f_>$ has positive coefficients and that both of their constant terms must be positive.

If $min_v \leq 2$ we can show (6) and (7) by inductive arguments. We were also able to check the case $min_v = 3$ and $max_v \leq 4$.

If $min_v = 2$ we can also show that $[y^{min_v-2}]p_<(y) < 0$: in this case, $p_<(y)$ is constant and equals $1/(\tau_2 V'(\tau_2))$, where τ_2 is the only negative zero of $V(y) = V(\tau)$ which is smaller than τ in modulus. Since $V(y)$ tends to infinity as y approaches $0-$, we have that $V'(\tau_2) > 0$, which implies the claim.

Finally, if $min_v = 2$ and the coefficients of V are either zero or one, we can also conclude that β is negative: in this case the numerator of (3) equals

$$-\tau V'''(\tau) - 3V''(\tau) + 3\tau^{-1}V'(\tau) = -3a_{-1}\tau^{-3} + 3a_1\tau^{-1} - 15a_3\tau + \ldots$$

If $a_1 = 0$ then the above expression is trivially negative. Otherwise we have to show that $3\tau^{-1} < 15\tau$. We show that $\tau > \frac{1}{2}$, which is sufficient: We have

$$\begin{aligned} V'(y) &= -2y^{-3} + \sum_{k \geq -1} k a_k y^{k-1} \\ &\leq -2y^{-3} + \frac{1}{(1-y)^2} \end{aligned}$$

which is negative for $y \leq \frac{1}{2}$.

4. Transcendence

It is now a simple matter to complete the proof of the main Theorem 1.2: Since in the circumstances of the theorem the asymptotic expansion of

$$[t^n x^i] \log B(x;t) = [x^i] (H(x))^n [t^n] \log \bar{B}(t)$$

contains a term n^{-2}, the series $[x^i] \log B(x;t)$ cannot be algebraic. When i is minimal such that there is at least one walk in the slit plane with steps in $\mathfrak{S}$ ending at $(i, 0)$, Theorem 2.1 gives that $[x^i] \log B(x;t)$ is the generating for such walks. To settle the transcendence of $S_{i,0}(t)$ for general i, we only need to note that $[t^n] \log B(x;t) \sim c_0 \rho^{-n} n^{-1}$, where, as we proved in the last section, $c_0 = \sqrt{\frac{2G_t}{G_{z,z}}} = \frac{1}{\rho \tau_k} \sqrt{\frac{2}{V''(\tau_k)}}$, and thus does not vanish. Hence, the leading term of $[t^n x^i] \log B(x;t)$ contains a factor of $1/n^{3/2}$. Thus, in the convolution formula for $S_{i,0}(t)$, the term $1/n^2$ in the asymptotic expansion of $[x^i] \log B(x;t)$ cannot be cancelled by terms of the asymptotic expansion of the product of two or more functions $S_{i_j,0}$.

The proof of the non-D-finiteness of the other functions can be copied verbatim from the proof of Proposition 22, page 282 of [2].

5. Acknowledgements

I would like to thank Michael Drmota, Bernhard Gittenberger, Bernhard Lamel and Bodo Laß for numerous stimulating discussions concerning the nature of the solutions of the functional equation (2). Also I am very grateful to two anonymous referees who pointed out numerous mistakes and a wrong conjecture appearing in the manuscript. And, of course, I would like to thank Mireille Bousquet-Mélou for introducing me to the problem and for a wonderful stay in Bordeaux.

References

[1] Cyril Banderier and Philippe Flajolet, *Basic analytic combinatorics of directed lattice paths*, Theoret. Comput. Sci. **281** (2002), no. 1-2, 37–80, Selected papers in honour of Maurice Nivat. MR **2003g:**05006

[2] Mireille Bousquet-Mélou, *Walks on the slit plane: other approaches*, Advances in Applied Mathematics **27** (2001), no. 2-3, 243–288, Special issue in honor of Dominique Foata's 65th birthday (Philadelphia, PA, 2000). MR **2002j:**60076

[3] Michael Drmota, *A bivariate asymptotic expansion of coefficients of powers of generating functions*, European Journal of Combinatorics **15** (1994), no. 2, 139–152. MR **94k:**05014

[4] Philippe Flajolet, *Analytic models and ambiguity of context-free languages*, Theoretical Computer Science **49** (1987), no. 2-3, 283–309, Twelfth international colloquium on automata, languages and programming (Nafplion, 1985). MR **89e**:68067

[5] Philippe Flajolet and Robert Sedgewick, *The average case analysis of algorithms*, 1994.

[6] Einar Hille, *Analytic function theory. Vol. I, II. 2nd ed. corrected.*, Chelsea Publishing Company, 1973.

[7] Morris Marden, *Geometry of polynomials*, Second edition. Mathematical Surveys, No. 3, American Mathematical Society, Providence, R.I., 1966. MR 37 #1562

Martin Rubey
LaBRI, Université Bordeaux I, Research financed by EC's IHRP Programme, within the Research Training Network "Algebraic Combinatorics in Europe", grant HPRN-CT-2001-00272
martin.rubey@labri.fr

Trends in Mathematics, © 2004 Birkhäuser Verlag Basel/Switzerland

Some Curious Extensions of the Classical Beta Integral Evaluation

Michael Schlosser

ABSTRACT: *We deduce curious q-series identities by applying an inverse relation to a certain identity for basic hypergeometric series. After rewriting some of these identities in terms of q-integrals, we obtain, in the limit $q \to 1$, curious integral identities which generalize the classical beta integral evaluation.*

1. Introduction

Euler's beta integral evaluation (cf. [1, Eq. (1.1.13)])

$$\int_0^1 t^{\alpha-1}(1-t)^{\beta-1}dt = \frac{\Gamma(\alpha)\Gamma(\beta)}{\Gamma(\alpha+\beta)}, \qquad \Re(\alpha), \Re(\beta) > 0, \tag{1}$$

is one of the most important and prominent identities in special functions. In Andrews, Askey and Roy's modern treatise [1], the beta integral (and its various extensions) runs like a thread through their whole exposition.

An unusual extension of (1) was recently found by George Gasper and the present author in [4, Th. 5.1] and reads as follows.

$$\frac{\Gamma(\alpha)\Gamma(\beta)}{\Gamma(\alpha+\beta)} = (c-(a+1)^2)\int_0^1 \frac{(c-a(a+t))^\beta\,(c-(a+1)(a+t))^{\beta-1}}{(c-(a+t)^2)^{2\beta}} \times {}_2F_1\left[\begin{matrix}\alpha-\beta-1,-\beta\\ \alpha\end{matrix}; \frac{(a+t)t}{c-a(a+t)}\right] t^{\alpha-1}\,(1-t)^{\beta-1}\,dt, \tag{2}$$

provided $\Re(\alpha), \Re(\beta) > 0$. It is clear that (2) reduces to (1) when either $c \to \infty$ or $a \to \infty$. Two special cases of (2) where the ${}_2F_1$ in the integrand can be simplified are $\alpha = \beta + 1$ and $\alpha = \beta$. Specifically, we have

$$\frac{\Gamma(\beta)\Gamma(\beta)}{2\,\Gamma(2\beta)} = (c-(a+1)^2)\int_0^1 \frac{(c-a(a+t))^\beta\,(c-(a+1)(a+t))^{\beta-1}}{(c-(a+t)^2)^{2\beta}} \times t^\beta\,(1-t)^{\beta-1}\,dt \tag{3}$$

and

$$\frac{\Gamma(\beta)\Gamma(\beta)}{\Gamma(2\beta)} = (c-(a+1)^2)\int_0^1 \frac{(c-a(a+t))^{\beta-1}\,(c-(a+1)(a+t))^{\beta-1}}{(c-(a+t)^2)^{2\beta}} \times (c-(a-t)(a+t))\,t^{\beta-1}\,(1-t)^{\beta-1}\,dt, \tag{4}$$

where in each case $\Re(\beta) > 0$.

In an early version of [4] we claimed that the integral evaluations (3) and (4), proved by the same procedure as the integral identities in this paper, "seem to be difficult to prove by standard methods". However, after seeing our preprint [4],

Mizan Rahman [7] communicated to us a remarkable proof of (3) which involves a sequence of manipulations of hypergeometric series [2].

Another beta-type integral evaluation which has some similarity to (2), is [4, Th. 5.2]. It reads as follows. Let m be a nonnegative integer. Then

$$\frac{\Gamma(\beta)\Gamma(\beta)}{2\,\Gamma(2\beta)} = (c-(a+1)^2)\int_0^1 \frac{(c-a(a+t))^\beta\,(c-(a+1)(a+t))^{\beta-1}}{(c-(a+t)^2)^{2\beta}}$$
$$\times\,{}_2F_1\left[\begin{matrix}-\beta,-m\\ -2\beta\end{matrix};\frac{c-(a+t)^2}{(c-a(a+t))(1-et)}\right]\left(\frac{1-et}{1-e}\right)^m t^{\beta-m}\,(1-t)^{\beta-1}\,dt, \quad (5)$$

provided $\Re(\beta) > \max(0, m-1)$. Some special cases are considered in [4, Sec. 5].

In this paper, we generalize both identities (2) and (5), see Corollary 5.3 and Theorem 5.1, respectively. While (5) does not extend the classical beta integral evaluation (1), its extension in Theorem 5.1 now does. In order to deduce our results, we apply essentially the same machinery which was utilized in [4] with the difference that our derivation now makes use of a more general basic hypergeometric identity (namely, (6)).

We start with some preliminaries on hypergeometric and basic hypergeometric series, see Section 2. In the same section we also exhibit an explicit matrix inverse which will be crucial in our further analysis. This matrix inverse is applied in Section 3 to derive a new q-series identity which we list together with some corollaries. In Section 4 we rewrite two of the obtained identities in terms of q-integrals. From these we deduce in Section 5, by letting $q \to 1$, new beta-type integral identities by which we generalize the results from [4].

2. Preliminaries

2.1. Hypergeometric and basic hypergeometric series

For a complex number a, define the *shifted factorial*

$$(a)_0 := 1, \qquad (a)_k := a(a+1)\dots(a+k-1),$$

where k is a positive integer. Let r be a positive integer. The *hypergeometric* ${}_rF_{r-1}$ *series* with numerator parameters $a_1,\dots,a_r$, denominator parameters $b_1,\dots,b_{r-1}$, and argument z is defined by

$${}_rF_{r-1}\left[\begin{matrix}a_1,\dots,a_r\\ b_1,\dots,b_{r-1}\end{matrix};z\right] := \sum_{k\geq 0}\frac{(a_1)_k\dots(a_r)_k}{k!\,(b_1)_k\dots(b_{r-1})_k}\,z^k.$$

The ${}_rF_{r-1}$ series terminates if one of the numerator parameters is of the form $-n$ for a nonnegative integer n. If the series does not terminate, it converges when $|z|<1$, and also when $|z|=1$ and $\Re[b_1+b_2+\dots+b_{r-1}-(a_1+a_2+\dots+a_r)]>0$. See [2, 10] for a classic texts on (ordinary) hypergeometric series.

Let q (the "base") be a complex number such that $0<|q|<1$. Define the *q-shifted factorial* by

$$(a;q)_\infty := \prod_{j\geq 0}(1-aq^j) \qquad \text{and} \qquad (a;q)_k := \frac{(a;q)_\infty}{(aq^k;q)_\infty}$$

for integer k. The *basic hypergeometric* ${}_r\phi_{r-1}$ *series* with numerator parameters $a_1,\dots,a_r$, denominator parameters $b_1,\dots,b_{r-1}$, base q, and argument z is defined by

$$ {}_r\phi_{r-1}\left[\begin{matrix} a_1,\ldots,a_r \\ b_1,\ldots,b_{r-1}\end{matrix};q,z\right] := \sum_{k\ge 0} \frac{(a_1;q)_k\ldots(a_r;q)_k}{(q;q)_k(b_1;q)_k\ldots(b_{r-1};q)_k} z^k. $$

The ${}_r\phi_{r-1}$ series terminates if one of the numerator parameters is of the form q^{-n} for a nonnegative integer n. If the series does not terminate, it converges when $|z| < 1$. For a thorough exposition on basic hypergeometric series (or, synonymously, *q-hypergeometric series*), including a list of several selected summation and transformation formulas, we refer the reader to [3].

We list two specific identities which we utilize in this paper.

First, we have the following three-term transformation (cf. [3, Eq. (III.34)]),

$$ {}_3\phi_2\left[\begin{matrix} a,b,c \\ d,e\end{matrix};q,\frac{de}{abc}\right] = \frac{(e/b;q)_\infty(e/c;q)_\infty}{(e;q)_\infty(e/bc;q)_\infty}\,{}_3\phi_2\left[\begin{matrix} d/a,b,c \\ d,bcq/e\end{matrix};q,q\right] $$
$$ + \frac{(d/a;q)_\infty(b;q)_\infty(c;q)_\infty(de/bc;q)_\infty}{(d;q)_\infty(e;q)_\infty(bc/e;q)_\infty(de/abc;q)_\infty}\,{}_3\phi_2\left[\begin{matrix} e/b,e/c,de/abc \\ de/bc,eq/bc\end{matrix};q,q\right], \tag{6} $$

where $|de/abc| < 1$. Further, we need (cf. [3, Eq. (III.9)])

$$ {}_3\phi_2\left[\begin{matrix} a,b,c \\ d,e\end{matrix};q,\frac{de}{abc}\right] = \frac{(e/a;q)_\infty(de/bc;q)_\infty}{(e;q)_\infty(de/abc;q)_\infty}\,{}_3\phi_2\left[\begin{matrix} a,d/b,d/c \\ d,de/bc\end{matrix};q,\frac{e}{a}\right], \tag{7} $$

where $|de/abc|, |e/a| < 1$.

2.2. Inverse relations

Let $\mathbb{Z}$ denote the set of integers and $F = (f_{nk})_{n,k\in\mathbb{Z}}$ be an infinite lower-triangular matrix; i.e. $f_{nk} = 0$ unless $n \ge k$. The matrix $G = (g_{kl})_{k,l\in\mathbb{Z}}$ is said to be the *inverse matrix* of F if and only if

$$ \sum_{l\le k\le n} f_{nk}g_{kl} = \delta_{nl} $$

for all $n, l \in \mathbb{Z}$, where δ_{nl} is the usual Kronecker delta.

The method of applying *inverse relations* [8] is a well-known technique for proving identities, or for producing new ones from given ones.

If $(f_{nk})_{n,k\in\mathbb{Z}}$ and $(g_{kl})_{k,l\in\mathbb{Z}}$ are lower-triangular matrices that are inverses of each other, then

$$ \sum_{n\ge k} f_{nk}a_n = b_k \tag{8a} $$

if and only if

$$ \sum_{k\ge l} g_{kl}b_k = a_l, \tag{8b} $$

subject to suitable convergence conditions. For some applications of (8) see e.g. [6, 8, 9].

Note that in the literature it is actually more common to consider the following inverse relations involving finite sums,

$$ \sum_{k=0}^{n} f_{nk}a_k = b_n \qquad \textit{if and only if} \qquad \sum_{l=0}^{k} g_{kl}b_l = a_k. \tag{9} $$

It is clear that in order to apply (8) (or (9)) effectively, one should have some explicit matrix inversion at hand. The following result, which is a special case of Krattenthaler's matrix inverse [6], will be crucial in our derivation of new identities. It can be regarded as a bridge between q-hypergeometric and certain non-q-hypergeometric identities. (For some other such matrix inverses, see [9].)

Lemma 2.1 (MS [9, Eqs. (7.18)/(7.19)]). *Let*

$$f_{nk} = \frac{(1/b;q)_{n-k}\left(\frac{(a+bq^k)q^k}{c-a(a+bq^k)};q\right)_{n-k}}{(q;q)_{n-k}\left(\frac{(a+bq^k)bq^{k+1}}{c-a(a+bq^k)};q\right)_{n-k}},$$

$$g_{kl} = (-1)^{k-l}\,q^{\binom{k-l}{2}}\,\frac{(c-(a+bq^l)(a+q^l))}{(c-(a+bq^k)(a+q^k))}\,\frac{(q^{l-k+1}/b;q)_{k-l}\left(\frac{(a+bq^k)q^{l+1}}{c-a(a+bq^k)};q\right)_{k-l}}{(q;q)_{k-l}\left(\frac{(a+bq^k)bq^l}{c-a(a+bq^k)};q\right)_{k-l}}.$$

Then the infinite matrices $(f_{nk})_{n,k\in\mathbb{Z}}$ and $(g_{kl})_{k,l\in\mathbb{Z}}$ are inverses of each other.

3. Some curious q-series expansions

Proposition 3.1. *Let a, b, c, d and e be indeterminate. Then*

$$\begin{aligned}
1 = \frac{(bq;q)_\infty}{(b^2q;q)_\infty}\sum_{k=0}^{\infty}&\frac{(c-(a+1)(a+b))}{(c-(a+1)(a+bq^k))}\,\frac{(c-(a+bq^k)^2)}{(c-(a+b)(a+bq^k))}\\
&\times {}_3\phi_2\left[\begin{matrix}e/d,1/b,\frac{(a+bq^k)q^k}{c-a(a+bq^k)}\\ eq^k,1/b^2\end{matrix};q,q\right]\\
&\times\frac{(b;q)_k\,(d;q)_k\left(\frac{(a+bq^k)}{c-a(a+bq^k)};q\right)_k\left(\frac{(a+bq^k)b^2q^{k+1}}{c-a(a+bq^k)};q\right)_\infty}{(q;q)_k\,(e;q)_k\left(\frac{(a+bq^k)bq}{c-a(a+bq^k)};q\right)_\infty}\left(\frac{beq}{d}\right)^k\\
&+\frac{(e/d;q)_\infty\,(1/b;q)_\infty\,(b^2eq;q)_\infty}{(1/b^2q;q)_\infty\,(b^2eq/d;q)_\infty\,(e;q)_\infty}\sum_{k=0}^{\infty}\frac{(c-(a+1)(a+b))}{(c-(a+1)(a+bq^k))}\\
&\times\frac{(c-(a+bq^k)^2)}{(c-(a+b)(a+bq^k))}\,{}_3\phi_2\left[\begin{matrix}bq,b^2eq/d,\frac{(a+bq^k)b^2q^{k+1}}{c-a(a+bq^k)}\\ b^2q^2,b^2eq^{k+1}\end{matrix};q,q\right]\\
&\times\frac{(b;q)_k\,(d;q)_k\left(\frac{(a+bq^k)}{c-a(a+bq^k)};q\right)_\infty}{(q;q)_k\,(b^2eq;q)_k\left(\frac{(a+bq^k)bq}{c-a(a+bq^k)};q\right)_\infty}\left(\frac{beq}{d}\right)^k,
\end{aligned}\tag{11}$$

provided $|beq/d|<1$.

Proof of Proposition 3.1. Let the inverse matrices $(f_{nk})_{n,k\in\mathbb{Z}}$ and $(g_{kl})_{k,l\in\mathbb{Z}}$ be defined as in Corollary 2.1. Then (8a) holds for

$$a_n = \frac{(d;q)_n}{(e;q)_n}\left(\frac{b^2eq}{d}\right)^n$$

and

$$b_k=\frac{(d;q)_k}{(e;q)_k}\left(\frac{b^2eq}{d}\right)^k\frac{(bq;q)_\infty\left(\frac{(a+bq^k)b^2q^{k+1}}{c-a(a+bq^k)};q\right)_\infty}{(b^2q;q)_\infty\left(\frac{(a+bq^k)bq^{k+1}}{c-a(a+bq^k)};q\right)_\infty}\,{}_3\phi_2\left[\begin{matrix}e/d,1/b,\frac{(a+bq^k)q^k}{c-a(a+bq^k)}\\ eq^k,1/b^2\end{matrix};q,q\right]$$
$$+\frac{(e/d;q)_\infty(1/b;q)_\infty(b^2eq^{k+1};q)_\infty\left(\frac{(a+bq^k)q^k}{c-a(a+bq^k)};q\right)_\infty}{(1/b^2q;q)_\infty(b^2eq/d;q)_\infty(eq^k;q)_\infty\left(\frac{(a+bq^k)bq^{k+1}}{c-a(a+bq^k)};q\right)_\infty}$$
$$\times\frac{(d;q)_k}{(e;q)_k}\left(\frac{b^2eq}{d}\right)^k{}_3\phi_2\left[\begin{matrix}bq,b^2eq/d,\frac{(a+bq^k)q^k}{c-a(a+bq^k)}\\ b^2q^2,b^2eq^{k+1}\end{matrix};q,q\right]$$

by (6). This implies the inverse relation (8b), with the above values of a_n and b_k. After performing the shift $k\mapsto k+l$, and the substitutions $a\mapsto aq^l$, $c\mapsto cq^{2l}$, $e\mapsto eq^{-l}$, we get rid of l and eventually obtain (11). □

Corollary 3.2. *Let a, b, c, d and e be indeterminate. Then*

$$1=\sum_{k=0}^{\infty}\frac{(c-(a+1)(a+b))}{(c-(a+1)(a+bq^k))}\frac{(c-(a+bq^k)^2)}{(c-(a+b)(a+bq^k))}\frac{(b;q)_k\,(d;q)_k}{(q;q)_k\,(e;q)_k}$$
$$\times{}_3\phi_2\left[\begin{matrix}1/b,dq^k,\frac{(a+bq^k)q^k}{c-a(a+bq^k)}\\ eq^k,\frac{(a+bq^k)bq^{k+1}}{c-a(a+bq^k)}\end{matrix};q,\frac{b^2eq}{d}\right]\frac{\left(\frac{(a+bq^k)}{c-a(a+bq^k)};q\right)_\infty}{\left(\frac{(a+bq^k)bq}{c-a(a+bq^k)};q\right)_\infty}\left(\frac{beq}{d}\right)^k,\tag{12}$$

provided $|beq/d|<1$ and $|b^2eq/d|<1$.

Proof. Apply (6) to the right-hand side of (11), with respect to the simultaneous substitutions $a\mapsto dq^k$, $b\mapsto 1/b$, $c\mapsto(a+bq^k)q^k/(c-a(a+bq^k))$, $d\mapsto eq^k$, $e\mapsto(a+bq^k)bq^{k+1}/(c-a(a+bq^k))$. □

Corollary 3.3. *Let a, b, c, d and e be indeterminate. Then*

$$\frac{(e;q)_\infty(b^2eq/d;q)_\infty}{(be;q)_\infty(beq/d;q)_\infty}=\sum_{k=0}^{\infty}\frac{(c-(a+1)(a+b))}{(c-(a+1)(a+bq^k))}\frac{(c-(a+bq^k)^2)}{(c-(a+b)(a+bq^k))}$$
$$\times\frac{(b;q)_k\,(d;q)_k}{(q;q)_k\,(be;q)_k}{}_3\phi_2\left[\begin{matrix}1/b,bq,\frac{(a+bq^k)bq}{c-a(a+bq^k)}\\ beq/d,\frac{(a+bq^k)bq^{k+1}}{c-a(a+bq^k)}\end{matrix};q,beq^k\right]\frac{\left(\frac{(a+bq^k)}{c-a(a+bq^k)};q\right)_\infty}{\left(\frac{(a+bq^k)bq}{c-a(a+bq^k)};q\right)_\infty}\left(\frac{beq}{d}\right)^k,\tag{13}$$

provided $|beq/d|<1$ and $|be|<1$.

Proof. Apply (7) to the ${}_3\phi_2$ on the right-hand side of (12), with respect to the simultaneous substitutions $a\mapsto 1/b$, $b\mapsto dq^k$, $c\mapsto(a+bq^k)q^k/(c-a(a+bq^k))$, $d\mapsto(a+bq^k)bq^{k+1}/(c-a(a+bq^k))$, $e\mapsto eq^k$, and divide both sides of the resulting identity by $(be;q)_\infty(beq/d;q)_\infty/(e;q)_\infty(b^2eq/d;q)_\infty$. □

We will make use of Proposition 3.1 and of Corollary 3.3 in our derivation of new beta integral identities.

4. q-Integrals

In the following we restrict ourselves to *real* q with $0 < q < 1$.

Thomae [11] introduced the q-integral defined by

$$\int_0^1 f(t)d_qt = (1-q)\sum_{k=0}^{\infty} f(q^k)q^k. \tag{14}$$

Later Jackson [5] gave a more general q-integral which however we do not need here.

By considering the Riemann sum for a continuous function f over the closed interval $[0,1]$, partitioned by the points q^k, $k \geq 0$, one easily sees that

$$\lim_{q\to 1^-} \int_0^1 f(t)d_qt = \int_0^1 f(t)dt.$$

It is well known that many identities for q-series can be written in terms of q-integrals, which then may be specialized (as $q \to 1$) to ordinary integrals. For instance, the q-binomial theorem (cf. [3, Eq. (II.3)])

$$\sum_{k=0}^{\infty} \frac{(a;q)_k}{(q;q)_k} z^k = \frac{(az;q)_\infty}{(z;q)_\infty}, \qquad |z| < 1, \tag{15}$$

can be written, when $a \mapsto q^\beta$ and $z \mapsto q^\alpha$, as

$$\int_0^1 \frac{(qt;q)_\infty}{(q^\beta t;q)_\infty} t^{\alpha-1} d_qt = \frac{\Gamma_q(\alpha)\Gamma_q(\beta)}{\Gamma_q(\alpha+\beta)}, \tag{16}$$

where

$$\Gamma_q(x) := (1-q)^{1-x} \frac{(q;q)_\infty}{(q^x;q)_\infty} \tag{17}$$

is the q-gamma function, introduced by Thomae [11], see also [1, § 10.3] and [3, § 1.11]. In fact, (16) is a q-extension of the beta integral evaluation (1).

We will rewrite the identities in Proposition 3.1 and in Corollary 3.3 in terms of q-integrals. These will then be utilized in Section 5 to obtain new extensions of the beta integral evaluation.

Starting with (11), if we replace b by q^β, d by $eq^{\beta+1-\alpha}$, and multiply both sides of the identity by

$$\frac{(e;q)_\infty}{(eq^{\beta+1-\alpha};q)_\infty},$$

we obtain the following q-beta-type integral identity:

$$\frac{(e;q)_\infty}{(eq^{\beta+1-\alpha};q)_\infty} = \frac{\Gamma_q(2\beta+1)}{\Gamma_q(\beta+1)\Gamma_q(\beta)} \int_0^1 \frac{(c-(a+1)(a+q^\beta))}{(c-(a+1)(a+q^\beta t))} \tag{18}$$

$$\times \frac{(c-(a+q^\beta t)^2)}{(c-(a+q^\beta)(a+q^\beta t))}\,{}_3\phi_2\left[\begin{matrix} q^{\alpha-\beta-1}, q^{-\beta}, \frac{(a+q^\beta t)t}{c-a(a+q^\beta t)} \\ et, q^{-2\beta}\end{matrix}; q, q\right]$$

$$\times \frac{(qt;q)_\infty\,(et;q)_\infty\left(\frac{(a+q^\beta t)}{c-a(a+q^\beta t)};q\right)_\infty\left(\frac{(a+q^\beta t)q^{2\beta+1}t}{c-a(a+q^\beta t)};q\right)_\infty}{(q^\beta t;q)_\infty\,(eq^{\beta+1-\alpha}t;q)_\infty\left(\frac{(a+q^\beta t)q^{\beta+1}}{c-a(a+q^\beta t)};q\right)_\infty\left(\frac{(a+q^\beta t)t}{c-a(a+q^\beta t)};q\right)_\infty}\,t^{\alpha-1}d_qt$$

$$+\frac{\Gamma_q(-2\beta-1)\,\Gamma_q(\alpha+\beta)}{\Gamma_q(\beta)\,\Gamma_q(\alpha-\beta-1)\,\Gamma_q(-\beta)}\int_0^1\frac{(c-(a+1)(a+q^\beta))}{(c-(a+1)(a+q^\beta t))}$$

$$\times \frac{(c-(a+q^\beta t)^2)}{(c-(a+q^\beta)(a+q^\beta t))}\,{}_3\phi_2\left[\begin{matrix} q^{\beta+1}, q^{\alpha+\beta}, \frac{(a+q^\beta t)q^{2\beta+1}t}{c-a(a+q^\beta t)} \\ q^{2\beta+2}, eq^{2\beta+1}t\end{matrix}; q, q\right]$$

$$\times \frac{(qt;q)_\infty\,(eq^{2\beta+1}t;q)_\infty\left(\frac{(a+q^\beta t)}{c-a(a+q^\beta t)};q\right)_\infty}{(q^\beta t;q)_\infty\,(eq^{\beta+1-\alpha}t;q)_\infty\left(\frac{(a+q^\beta t)q^{\beta+1}}{c-a(a+q^\beta t)};q\right)_\infty}\,t^{\alpha-1}d_qt.$$

Similarly, starting with (13), if we replace b by q^β, d by $eq^{\beta+1-\alpha}$, and multiply both sides of the identity by

$$(1-q)\,\frac{(q;q)_\infty\,(eq^\beta;q)_\infty}{(q^\beta;q)_\infty\,(eq^{\beta+1-\alpha};q)_\infty},$$

we obtain the following q-beta-type integral evaluation:

$$\frac{\Gamma_q(\alpha)\,\Gamma_q(\beta)}{\Gamma_q(\alpha+\beta)}\,\frac{(e;q)_\infty}{(eq^{\beta+1-\alpha};q)_\infty}=\int_0^1\frac{(c-(a+1)(a+q^\beta))}{(c-(a+1)(a+q^\beta t))}$$

$$\times \frac{(c-(a+q^\beta t)^2)}{(c-(a+q^\beta)(a+q^\beta t))}\,{}_3\phi_2\left[\begin{matrix} q^{-\beta}, q^{\beta+1}, \frac{(a+q^\beta t)q^\alpha}{e(c-a(a+q^\beta t))} \\ q^\alpha, \frac{(a+q^\beta t)q^\beta t}{c-a(a+q^\beta t)}\end{matrix}; q, eq^\beta t\right]$$

$$\times \frac{(qt;q)_\infty\,(eq^\beta t;q)_\infty\left(\frac{(a+q^\beta t)}{c-a(a+q^\beta t)};q\right)_\infty\left(\frac{(a+q^\beta t)q^{\beta+1}t}{c-a(a+q^\beta t)};q\right)_\infty}{(q^\beta t;q)_\infty\,(eq^{\beta+1-\alpha}t;q)_\infty\left(\frac{(a+q^\beta t)q^{\beta+1}}{c-a(a+q^\beta t)};q\right)_\infty\left(\frac{(a+q^\beta t)t}{c-a(a+q^\beta t)};q\right)_\infty}\,t^{\alpha-1}d_qt. \qquad (19)$$

5. Curious beta-type integrals

Observe that $\lim_{q\to1^-}\Gamma_q(x)=\Gamma(x)$ (see [3, (1.10.3)]) and

$$\lim_{q\to1^-}\frac{(q^\alpha u;q)_\infty}{(u;q)_\infty}=(1-u)^{-\alpha}$$

for constant u (with $|u|<1$), due to (15) and its $q\to1$ limit, the ordinary binomial theorem.

We thus immediately deduce, as consequences of our q-integral identities from Section 4, new beta integral identities. We implicitly assume that the integrals are well defined, in particular that the parameters are chosen such that no poles occur on the path of integration $t\in[0,1]$ and the integrals converge.

We first consider the beta-type integral identity obtained from multiplying both sides of (18) by

$$\frac{\Gamma(\beta)\,\Gamma(\beta+1)}{\Gamma(2\beta+1)},$$

and letting $q\to1^-$.

Theorem 5.1. *Let $\Re(\alpha),\Re(\beta)>0$. Then*

$$\begin{aligned}
&\frac{\Gamma(\beta)\,\Gamma(\beta)}{2\,\Gamma(2\beta)}\,(1-e)^{\beta+1-\alpha}\\
&\qquad=(c-(a+1)^2)\int_0^1\frac{(c-a(a+t))^\beta\,(c-(a+1)(a+t))^{\beta-1}}{(c-(a+t)^2)^{2\beta}}\\
&\times{}_2F_1\!\left[\begin{matrix}\alpha-\beta-1,-\beta\\-2\beta\end{matrix};\frac{c-(a+t)^2}{(c-a(a+t))(1-et)}\right](1-et)^{\beta+1-\alpha}\,t^{\alpha-1}\,(1-t)^{\beta-1}\,dt\\
&\qquad+\frac{\Gamma(\beta)\,\Gamma(-2\beta-1)\,\Gamma(\alpha+\beta)}{2\,\Gamma(2\beta)\,\Gamma(-\beta)\,\Gamma(\alpha-\beta-1)}\,(c-(a+1)^2)\int_0^1(c-(a+t)^2)\\
&\qquad\times\frac{(c-(a+1)(a+t))^{\beta-1}}{(c-a(a+t))^{\beta+1}}\,{}_2F_1\!\left[\begin{matrix}\beta+1,\alpha+\beta\\2\beta+2\end{matrix};\frac{c-(a+t)^2}{(c-a(a+t))(1-et)}\right]\\
&\qquad\qquad\times(1-et)^{-\alpha-\beta}\,t^{\alpha-1}\,(1-t)^{\beta-1}\,dt. \qquad (20)
\end{aligned}$$

Note that (20) can be further rewritten using Legendre's duplication formula

$$\Gamma(2\beta)=\frac{1}{\sqrt{\pi}}\,2^{2\beta-1}\Gamma(\beta)\Gamma(\beta+\tfrac12),$$

after which the left hand side becomes

$$\frac{\sqrt{\pi}}{4^\beta}\,\frac{\Gamma(\beta)}{\Gamma(\beta+\frac12)}\,(1-e)^{\beta+1-\alpha}.$$

Clearly, (20) reduces to (5) if $\alpha-\beta-1=m$, a nonnegative integer.

Observe that (20) reduces to the classical beta integral evaluation (1) for $e=0$ and $c\to\infty$ due to the *Gauß summation*

$${}_2F_1\!\left[\begin{matrix}A,B\\C\end{matrix};1\right]=\frac{\Gamma(C)\,\Gamma(C-A-B)}{\Gamma(C-A)\,\Gamma(C-B)},$$

where $\Re(C-A-B)>0$, the *reflection formula*

$$\Gamma(z)\,\Gamma(1-z)=\frac{\pi}{\sin\pi z},$$

where z is not an integer, and some elementary identities for trigonometric functions, such as

$$\sin(x+y)+\sin(x-y)=\sin x\,\frac{\sin 2y}{\sin y}.$$

Next, we have the beta-type integral identity obtained from (19) by letting $q\to1^-$.

Theorem 5.2. *Let $\Re(\alpha),\Re(\beta)>0$. Then*

$$\begin{aligned}
&\frac{\Gamma(\alpha)\,\Gamma(\beta)}{\Gamma(\alpha+\beta)}\,(1-e)^{\beta+1-\alpha}=(c-(a+1)^2)\int_0^1\frac{(c-(a+1)(a+t))^{\beta-1}}{(c-(a+t)^2)^\beta}\\
&\times{}_2F_1\!\left[\begin{matrix}-\beta,\beta+1\\\alpha\end{matrix};\frac{(ce-(1+ae)(a+t))t}{c-(a+t)^2}\right](1-et)^{1-\alpha}\,t^{\alpha-1}\,(1-t)^{\beta-1}\,dt. \qquad (21)
\end{aligned}$$

Clearly, (21) reduces to (1) when $e=0$ and $c\to\infty$.

Corollary 5.3. *Let* $\Re(\alpha), \Re(\beta) > 0$. *Then*

$$\frac{\Gamma(\alpha)\,\Gamma(\beta)}{\Gamma(\alpha+\beta)} = (c-(a+1)^2)\int_0^1 \frac{(c-a(a+t))^\beta\,(c-(a+1)(a+t))^{\beta-1}}{(c-(a+t)^2)^{2\beta}} \\ \times {}_2F_1\left[\begin{matrix}-\beta, \alpha-\beta-1\\ \alpha\end{matrix}; \frac{((1+ae)(a+t)-ce)t}{(c-a(a+t))(1-et)}\right] \\ \times \left(\frac{1-et}{1-e}\right)^{\beta+1-\alpha} t^{\alpha-1}\,(1-t)^{\beta-1}\,dt. \tag{22}$$

Proof. We apply the transformation [2, p. 10, Eq. 2.4(1)]

$$(1-z)^{-A}\,{}_2F_1\left[\begin{matrix}A, B\\ C\end{matrix}; \frac{-z}{1-z}\right] = {}_2F_1\left[\begin{matrix}A, C-B\\ C\end{matrix}; z\right], \tag{23}$$

valid for $|z| < 1$ and $\Re(z) < \frac{1}{2}$ (conditions which we implicitly assume), to the ${}_2F_1$ on the right-hand side of (21) and divide both sides by $(1-e)^{\beta+1-\alpha}$. □

Clearly, (22) reduces to (2) for $e = 0$.

As in [4], we observe that by performing various substitutions one may change the form and path of integration of the considered integrals. In particular, using $t \mapsto s/(s+1)$ these integrals then run over the half line $s \in [0, \infty)$.

Acknowledgements

We thank George Gasper for comments.

The author was fully supported by an APART fellowship of the Austrian Academy of Sciences.

References

[1] G. E. Andrews, R. Askey and R. Roy, *Special functions*, Encyclopedia of Mathematics and Its Applications 71, Cambridge University Press, Cambridge, 1999.

[2] W. N. Bailey, *Generalized Hypergeometric Series*, Cambridge University Press, Cambridge, 1935; reprinted by Stechert-Hafner, New York, 1964.

[3] G. Gasper and M. Rahman, *Basic Hypergeometric Series*, Encyclopedia of Mathematics and Its Applications 35, Cambridge University Press, Cambridge, 1990.

[4] G. Gasper and M. Schlosser, *Some curious q-series expansions and beta integral evaluations*, preprint `arXiv:math.CO/0403481`.

[5] F. H. Jackson, "On q-definite integrals," *Quart. J. Pure Appl. Math.* **41** (1910), 193–203.

[6] C. Krattenthaler, "A new matrix inverse," *Proc. Amer. Math. Soc.* **124** (1996), 47–59.

[7] M. Rahman, private communication, April 2004.

[8] J. Riordan, *Combinatorial identities*, J. Wiley, New York, 1968.

[9] M. Schlosser, "Some new applications of matrix inversions in A_r," *Ramanujan J.* **3** (1999), 405–461.

[10] L. J. Slater, *Generalized Hypergeometric Functions*, Cambridge University Press, Cambridge, 1966.

[11] J. Thomae, "Beiträge zur Theorie der durch die Heinesche Reihe ...," *J. reine angewandte Math.* **70** (1869), 258–281.

Michael Schlosser
Institut für Mathematik der Universität Wien, Nordbergstraße 15, A-1090 Wien, Austria
schlosse@ap.univie.ac.at

Trends in Mathematics, © 2004 Birkhäuser Verlag Basel/Switzerland

Divisor Functions and Pentagonal Numbers

Klaus Simon

ABSTRACT: *Let $p(n,m)$ be the number of partitions of n with at most m summands*[1], $\omega(n) = \frac{1}{2}(3n^2 - n)$, $n \in \mathbb{Z}$ *be the pentagonal numbers*[2] *and* $\sigma_j(n) = \sum_{d|n} d^j$, $j \in \mathbb{N}$, *be the divisor functions. Then $\sigma_0(n)$ — the number of the divisors of n — satisfies*

$$\sigma_0(n) = g(n,0) + g(n,1) + \cdots + g(n,n-1) \tag{1}$$

where

$$g(n,m) = p(n,m) - \sum_{i=1}^{\infty}(-1)^{i-1}\Big(p(n-\omega(i),m) + p(n-\omega(-i),m)\Big).$$

Pentagonal numbers are given by a well-known identity due to *Euler*

$$(q)_\infty \stackrel{\text{def}}{=} \prod_{n=1}^{\infty}(1-q^n) = \sum_{n=-\infty}^{\infty}(-1)^n q^{n(3n-1)/2}. \tag{2}$$

They are correlated with the number of partition $p(n)$ of $n \in \mathbb{N}$, generated by

$$\sum_{n=0}^{\infty} p(n)\, q^n = \frac{1}{(q)_\infty}, \tag{3}$$

through the identity

$$1 = \frac{(q)_\infty}{(q)_\infty} = (1 - q^1 - q^2 + q^5 + q^7 - \cdots)\cdot(p(0)\,q^0 + p(1)\,q^1 + \cdots)$$

or equivalently

$$0 = p(n) - \sum_{j=1}^{\infty}(-1)^{j-1}\Big(p(n-\omega(j)) + p(n-\omega(-j))\Big). \tag{4}$$

On the other hand, the pentagonal numbers are connected with the divisor function $\sigma_1(n)$, for instance, by[3]

$$\sigma_1(n) = \sum_{i=1}^{\infty}(-1)^{i-1}\Big(\omega(i)\,p(n-\omega(i)) + \omega(-i)\,p(n-\omega(-i))\Big).$$

---

[1] or equivalent: ... in which no part is greater than m, see G. Andrews, Number Theory, Dover, 1994

[2] $\omega(0), \omega(1), \omega(-1), \omega(2), \omega(-2), \ldots = 0, 1, 2, 5, 7, \ldots$

[3] see G. Walz (ed.), Lexikon der Mathematik, Spektrum Akademischer Verlag, Heidelberg, Berlin, 2002

Our statement (1) is a similar identity for $\sigma_0(n)$. By way of illustration, for $n = 5$ we obtain $-2 = -\sigma_0(5) = g(5,0) + \cdots + g(5,4) = 1+0-1-1-1$. The proof of (1) is based on[1]

$$\lim_{n\to\infty} (n - a_n) = \sum_{i=1}^{\infty} \sigma_0(i)\, q^i \tag{5}$$

where the sequence a_n is defined by $a_0 = 0$ and

$$a_n = 1 + (1 - q^{n-1})\, a_{n-1}. \tag{6}$$

Iterating the recurrence leads to

$$a_n = \prod_{i=1}^{n-1} (1 - q^i) \sum_{j=0}^{n-1} \frac{1}{(1-q)\cdots(1-q^j)}. \tag{7}$$

Now, the product $\big((1-q)\cdots(1-q^m)\big)^{-1} \stackrel{\text{def}}{=} (q)_m^{-1}$ is well-known as generating function of the numbers $p(n,m)$, hence

$$\sum_{n=0}^{\infty} p(n,m)\, q^n = \frac{1}{(1-q)\,(1-q^2)\cdots(1-q^m)}, \quad m \geq 0. \tag{8}$$

With (8) the equation (7) can be written as

$$a_n = \sum_{m=0}^{n-1} \underbrace{(q)_{n-1} \sum_{h=0}^{\infty} p(h,m)\, q^h}_{\stackrel{\text{def}}{=} H_{n,m}(q)}. \tag{9}$$

For $n \to \infty$ we observe

$$\sum_{n=0}^{\infty} g(n,m)\, q^n = \lim_{n\to\infty} H_{n,m}(q) \stackrel{(2)}{=} \sum_{i=-\infty}^{\infty} (-1)^i\, q^{i(3i-1)/2} \sum_{j=0}^{\infty} p(j,m)\, q^j \tag{10}$$

with $g(0,m) = 1$ and

$$g(n,m) = p(n,m) - p(n-1,m) - p(n-2,m) + p(n-5,m) + \cdots, \quad m \geq 1.$$

Note that $p(n,m) = p(n)$, for $m \geq n$. This implies $g(n,m) \stackrel{(4)}{=} 0$, for $m \geq n \geq 1$. Therefore

$$\lim_{n\to\infty} (n - a_n) = \sum_{j=1}^{\infty} \big(g(j,0) + \cdots + g(j,j-1)\big)\, q^j \tag{11}$$

which, together with (5), completes the proof of (1).

Klaus Simon
Swiss Federal Laboratories for Materials Testing and Research (EMPA)
Lerchenfeldstrasse 5, CH-9014 St. Gallen
klaus.simon@empa.ch

[1] see G. Andrews, D. Crippa and K. Simon, q-Series Arising from the Study of Random Graphs, SIAM Journal on Discrete Mathematics, 10:41–56, 1997

Part II

Graph Theory

Trends in Mathematics, © 2004 Birkhäuser Verlag Basel/Switzerland

On Combinatorial Hoeffding Decomposition and Asymptotic Normality of Subgraph Count Statistics

Mindaugas Bloznelis

ABSTRACT: *Given k and n, consider a graph with k vertices and n "blue" edges. We assume that the set of "blue" edges $\{X_1, \ldots, X_n\}$ is uniformly distributed among n-subsets of $N = \binom{k}{2}$ pairs of vertices. Given a graph $\mathcal{G}$, the number $\mathcal{N}_{\mathcal{G}}$ of blue copies of $\mathcal{G}$ is a U-statistic based on random sample $X_1, \ldots, X_n$. We show how the combinatorial Hoeffding decomposition of the random variable $\mathcal{N}_G$ can be applied to establish the asymptotic normality of $\mathcal{N}_G$ as $k, n \to \infty$.*

1. Introduction

Given a complete graph based on k vertices let $\mathcal{X} = \{x_1, \ldots, x_N\}$ denote the set of edges. Let $\{X_1, \ldots, X_n\} \subset \mathcal{X}$ be a random n-subset uniformly distributed over the class of n-subsets of $\mathcal{X}$. Here $n < N$. We paint edges $X_1, \ldots, X_n$ blue. The graph based on k vertices and (random) blue edges is denoted by $G(k, n)$. Given a graph $\mathcal{G}$ let $\mathcal{N}_{\mathcal{G}} = \mathcal{N}_{\mathcal{G}}(k, n)$ denote the number of copies of $\mathcal{G}$ in $G(k, n)$. We are interested when the random variable $(\mathcal{N}_{\mathcal{G}} - \mathbf{E}\,\mathcal{N}_{\mathcal{G}})/\sigma(\mathcal{N}_{\mathcal{G}})$ is asymptotically standard normal as $k, n \to \infty$. Here $\sigma^2(\mathcal{N}_G)$ denotes the variance of $\mathcal{N}_G$.

Another random graph model assumes that edges become blue independently with probability $t \in (0, 1)$. Let $\nu_1, \ldots, \nu_N$ be independent Bernoulli random variables with success probability t, i.e., $\mathbf{P}\{\nu_i = 1\} = 1 - \mathbf{P}\{\nu_i = 0\} = t$ for every i. We paint the edge x_i blue if $\nu_i = 1$. The graph based on k vertices and (random) blue edges is denoted by $G'(k, t)$. It is called Bernoulli random graph. Given a graph $\mathcal{G}$ let $\mathcal{N}'_{\mathcal{G}} = \mathcal{N}'_{\mathcal{G}}(k, t)$ denote the number of copies of $\mathcal{G}$ in $G'(k, t)$. Note that the conditional distribution of $\mathcal{N}'_{\mathcal{G}}$ given the event $\nu_1 + \cdots + \nu_N = n$ coincides with the distribution of $\mathcal{N}_{\mathcal{G}}$. Therefore, the problems of the asymptotic normality of distributions of $\mathcal{N}_{\mathcal{G}}$ and $\mathcal{N}'_{\mathcal{G}}$ are closely related.

The asymptotic normality for $\mathcal{N}'_{\mathcal{G}}$ as $k \to \infty$ was studied by several authors using different methods (method of moments, Stein's method, projection method and martingale limit theorems). An overview of the results and methods is given in the book of Janson, Luczak and Rucinski (2000). Let us mention that the first complete description of the conditions that are necessary and sufficient for the asymptotic normality of $\mathcal{N}'_{\mathcal{G}}$ was given by Ruciński (1988).

The asymptotic normality of $\mathcal{N}_G$ is shown in Janson (1990). He consider the random graph process $\{G(k, t), t \in [0, 1]\}$, where for every t the random graph $G(k, t)$ is defined as above, but with $\nu_i = \nu_i(t) = \mathbf{I}_{\{u_i \le t\}}$. Here $u_1, u_2, \ldots$ denote independent random variables uniformly distributed in $[0, 1]$. For every $k = 1, 2, \ldots,$

the collection of random variables $\{\mathcal{N}'_{\mathcal{G}}(k,t), t \in [0,1]\}$ can be viewed as a random process with sample paths in the Skorokhod space $D[0,1]$. Using a martingale convergence theorem Janson (1990) proved a functional limit theorem for the sequence of random processes $\{\mathcal{N}'_{\mathcal{G}}(k,t), t \in [0,1]\}$. Substitution of the random time $t_{k,n} = \min\{t : \nu_1(t) + \cdots + \nu_N(t) = n\}$ gives the asymptotic normality of $\mathcal{N}'_{\mathcal{G}}(k, t_{k,n})$. Since the distributions of $\mathcal{N}'_{\mathcal{G}}(k, t_{k,n})$ and $\mathcal{N}_{\mathcal{G}}$ coincide, this implies the asymptotic normality of $\mathcal{N}_{\mathcal{G}}$.

The present paper proposes another approach to the asymptotic normality of $\mathcal{N}_G$. We illustrate this approach by showing the asymptotic normality of the simplest subgraph count statistic: the number $\mathcal{N}_{P_2}$ of 2–stars ($\mathcal{G} = P_2$). Denote $n^* = \min\{n, N-n\}$.

Proposition 1.1. *Assume that $\sigma^2(\mathcal{N}_{P_2}) \to \infty$ as $k, n^* \to \infty$. Then the distribution of $(\mathcal{N}_{P_2} - \mathbf{E}\,\mathcal{N}_{P_2})/\sigma(\mathcal{N}_{P_2})$ is asymptotically standard normal.*

The proof combines projection's method and Stein's method. By means of Hoeffding's decomposition, the random variable $\mathcal{N}_{\mathcal{G}}$ is expanded into a sum of mutually uncorrelated U statistics of random variables $X_1, \dots, X_n$. The decomposition enables us to write the characteristic function $f(t)$ of (properly standardized) random variable $\mathcal{N}_{\mathcal{G}}$ in Erdős-Rényi form, see Erdős and Rényi (1959). Finally, we show that

$$f'(t) + tf(t) \to 0 \tag{1}$$

as $k, n \to \infty$. This implies the asymptotic normality, see Stein (1970) and Tikhomirov (1976, 2001).

The remaining part of the paper is organized as follows. In Section 2 we construct Hoeffding's decomposition for three examples of subgraph count statistics: the number of 2–stars ($\mathcal{G} = P_2$), the number of triangles ($\mathcal{G} = K_3$) and the number of 4–cycles ($\mathcal{G} = C_4$). In section 3 we write Erdős-Rényi representation for the characteristic function f (see (11) below) and prove Proposition 1.

2. Decomposition

Let $(X_1, \dots, X_N)$ denote the random permutation of the sequence $x_1, \dots, x_N$. A real function $t(x_{i_1}, \dots, x_{i_n})$ defined on $n-$ subsets $\{x_{i_1}, \dots, x_{i_n}\} \subset \mathcal{X}$ defines the random variable $T = t(X_1, \dots, X_n)$. We assume that the variance $\sigma^2(T) > 0$. Hoeffding's decomposition expands the random variable T in the series of $U-$statistics

$$T = \mathbf{E}\,T + U_1 + \cdots + U_{n^*}, \qquad n^* = \min\{n, N-n\}. \tag{2}$$

Here $U_r = \sum_{1 \le i_1 < \cdots < i_r \le n} g_r(X_{i_1}, \dots, X_{i_r})$. The function g_r is defined on $r-$ subsets of $\mathcal{X}$ and satisfies $\mathbf{E}\,g_r(X_{i_1}, \dots, X_{i_r}) = 0$. Furthermore, for every $s < r$ and every $1 \le i_1 < \cdots < i_r \le N$ and $1 \le j_1 < \cdots < j_s \le N$ we have almost surely

$$\mathbf{E}\left(g_r(X_{i_1}, \dots, X_{i_r}) \middle| X_{j_1}, \dots, X_{j_s}\right) = 0. \tag{3}$$

The kernels $g_r(x_{i_1}, \dots, x_{i_r})$ are linear combinations of conditional expectations $\mathbf{E}\,(T | X_1 = y_1, \dots, X_j = y_j)$ for $1 \le j \le r$ and $\{y_1, \dots, y_j\} \subset \{x_{i_1}, \dots, x_{i_r}\}$. Basic facts about the decomposition and formulas defining g_r can be found in Bloznelis and Götze (2001) and Bloznelis (2003), see also Zhao and Chen (1990). Note that (3) implies that random variables U_r are mutually uncorrelated. This yields the variance decomposition

$$\sigma^2(T) = \sigma^2(U_1) + \cdots + \sigma^2(U_{n^*}).$$

Furthermore using (3) it is easy to show, see Bloznelis and Götze (2001), that

$$\sigma^2(U_r) = \frac{\binom{n}{r}\binom{N-n}{r}}{\binom{N-r}{r}}\sigma_r^2 \sim \frac{N^r}{r!}(pq)^r\sigma_r^2, \tag{4}$$

as $k, n^* \to \infty$. Here

$$p = \frac{n}{N}, \qquad q = \frac{N-n}{N}, \qquad \sigma_r^2 = \mathbf{E}\, g_r^2(X_{i_1}, \dots, X_{i_r}).$$

2.1. Let T denote the number of blue copies of 2–star P_2 in $G(k,n)$. Given two edges $x, y \in \mathfrak{X}$ let L_{xy} be the indicator of the event that x and y are incident. We have $T = \sum_{1\le i<j\le n} L_{X_1X_2}$. Hoeffding's decomposition

$$T = \mathbf{E}\, T + U_1 + U_2, \qquad U_1 \equiv 0, \qquad U_2 = \sum_{1\le i<j\le n} (L_{X_iX_j} - p_L). \tag{5}$$

Here $p_L := \mathbf{E}\, L_{X_iX_j} = 2(k-2)/(N-1)$ and $\sigma_2^2 = p_L(1-p_L)$.

The variance

$$\sigma^2(T) = \sigma^2(U_2) = \frac{\binom{n}{2}\binom{N-n}{2}}{\binom{N-2}{2}}\sigma_2^2 \sim \frac{(pq)^2}{2}k^3 \quad \text{as} \quad k, n^* \to \infty. \tag{6}$$

2.2. Let T denote the number of blue copies of the triangle K_3 in $G(k,n)$. Given three edges $x, y, z \in \mathfrak{X}$, let $a(x,y,z)$ be the indicator of the event that x, y and z make up a triangle. We have $T = \sum_{1\le i_1<i_2<i_3\le n} a(X_{i_1}, X_{i_2}, X_{i_3})$. Hoeffding's decomposition

$$\begin{aligned} T &= \mathbf{E}\, T + U_1 + U_2 + U_3, \\ g_1 &\equiv 0, \qquad g_2(x,y) = \frac{n-2}{N-4}(L_{xy} - p_L), \\ g_3(x,y,z) &= a(x,y,z) - p_a - \frac{1}{N-1}(L_{xy} + L_{xz} + L_{yz} - 3p_L). \end{aligned} \tag{7}$$

Here $p_a := \mathbf{E}\, a(X_1, X_2, X_3) = \frac{\binom{k}{3}}{\binom{N}{3}}$. Using (3), (4) we obtain as $k, n^* \to \infty$

$$\sigma^2(U_2) \sim \frac{p^4q^2}{2}k^3, \quad \sigma^2(U_3) \sim \frac{(pq)^3}{6}k^3, \quad \sigma^2(T) \sim \frac{(pq)^3}{6}(3\frac{p}{q} + 1)k^3. \tag{8}$$

2.3. Let T denote the number of blue copies of the cycle C_4 in $G(k,n)$. Given edges $x, y, z, w \in \mathfrak{X}$, let $b(x,y,z)$ be the indicator of the event that x, y and z make up a path and let $d(x,y,z,w)$ be the indicator of the event that x, y, z and w make up a cycle. We have $T = \sum_{1\le i_1<i_2<i_3<i_4\le n} d(X_{i_1}, X_{i_2}, X_{i_3}, X_{i_4})$. Hoeffding's

decomposition

$$\begin{aligned}
T &= \mathbf{E}T + U_1 + U_2 + U_3 + U_4,\\
g_1 &\equiv 0, \qquad g_2(x,y) = \binom{n-2}{2}\frac{N-2}{N-4}\frac{N-3}{N-5}Q_{\{x,y\}},\\
g_3(x,y,z) &= (n-3)\frac{N-3}{N-6}Q_{\{x,y,z\}},\\
g_4(x,y,z,w) &= d(x,y,z,w) - p_d - \frac{N-3}{N-6}\sum_{A\subset\{x,y,z,w\},|A|=3} Q_A\\
&- \frac{N-2}{N-4}\frac{N-3}{N-5}\sum_{A\subset\{x,y,z,w\},|A|=2} Q_A.
\end{aligned}$$

Here we denote

$$\begin{aligned}
Q_{\{xy\}} &= \frac{k-5}{\binom{N-2}{2}}(L_{xy} - p_L),\\
Q_{\{xyz\}} &= \frac{1}{N-3}\big(b(x,y,z) - p_b\big) - \frac{N-2}{N-4}\sum_{A\subset\{x,y,z\},|A|=2} Q_A.
\end{aligned}$$

Furthermore,

$$p_d := \mathbf{E}\,d(X_1,X_2,X_3,X_4) = 3\frac{\binom{k}{4}}{\binom{N}{4}}, \qquad p_b = \mathbf{E}\,b(X_1,X_2,X_3) = 12\frac{\binom{k}{4}}{\binom{N}{3}}.$$

Using (3), (4) we obtain as $k, n^* \to \infty$

$$\sigma^2(U_2) \sim \frac{p^6q^2}{2}k^5, \qquad \sigma^2(U_3) \sim \frac{p^5q^3}{2}k^4, \qquad \sigma^2(U_4) \sim \frac{p^4q^4}{8}k^4. \tag{9}$$

2.3. Examples suggest that the linear part of the decomposition of $\mathcal{N}_\mathcal{G}$ vanishes, i.e. $U_1 \equiv 0$ almost surely. This can be easily shown for arbitrary $\mathcal{G}$. Furthermore, different $U-$ statistics contributing to the Hoeffding's decomposition (2) can have the same stochastic order even in the case where the parameter $p = p(k)$ is bounded away from 0 and 1, i.e., for some $\varepsilon > 0$,

$$\varepsilon < p(k) < 1 - \varepsilon \qquad \text{as} \qquad k, n^* \to \infty. \tag{10}$$

Thus, we have $\sigma^2(U_j) = \Theta(k^3)$ for $j = 2,3$ in (8) and $\sigma^2(U_j) = \Theta(k^4)$ for $j = 3,4$ in (9). Similarly, Hoeffding's decomposition of the number $\mathcal{N}_{K_4}$ of blue copies of the complete graph K_4, is the sum $\mathbf{E}\,\mathcal{N}_{K_4} + U_2 + \cdots + U_6$, where $\sigma^2(U_i) = \Theta(k^5)$ for $i = 2,3$, and $\sigma^2(U_i) \sim \Theta(k^4)$ for $i = 4,5,6$ provided that (10) holds. We do not present formulas of various parts of the decomposition of $\mathcal{N}_{K_4}$ here because the notation becomes awkward.

One may expect that, for a large class of graphs $\mathcal{G}$, the leading $U-$ statistics of the decomposition ($U-$statistics having largest variances) of $\mathcal{N}_\mathcal{G}$ correspond to the components P_2 and K_3, i.e., they are U_2 and (in the case where $K_3 \subset \mathcal{G}$) U_3.

3. Asymptotic normality

Recall that $\nu_1, \ldots, \nu_N$ denotes a sequence of independent Bernoulli random variables with the same success probability (say) $p = \mathbf{P}\{\nu_i = 1\} = 1 - \mathbf{P}\{\nu_i = 0\}$. We shall assume that this sequence and the random permutation $(X_1, \ldots, X_N)$ are independent.

3.1. Let us write the characteristic function of the distribution of T in the Erdős-Rényi form using the decomposition (2). Denote

$$U_k^\star = \sum_{1 \le i_1 < \cdots < i_k \le N} g_k(X_{i_1}, \ldots, X_{i_k}) \nu_{i_1} \ldots \nu_{i_k}.$$

Replacing the factors ν_i by $w_i = (\nu_i - p)$ in the formula of $U_k^\star$ we obtain the random variable U_k^*. Note that (3) implies $U_k^\star = U_k^*$ almost surely. Denote $S = w_1 + \cdots + w_N$.

The distribution of $T - \mathbf{E}T$ coincides with the conditional distribution of the sum

$$U_1^\star + \cdots + U_{n^*}^\star = U_1^* + \cdots + U_{n^*}^* =: T^*$$

given the event $\{S = 0\}$. Therefore, the characteristic function

$$\mathbf{E} \exp\{it(T - \mathbf{E}T)\} = \frac{1}{2\pi \mathbf{P}\{S = 0\}} \int_{-\pi}^{\pi} \mathbf{E} \exp\{itT^* + isS\} ds. \tag{11}$$

This way of representing the characteristic function of a linear statistic, like U_1, was used by Erdős and Rényi (1959).

3.2. Here we prove Proposition 1. Denote $f(t) = \mathbf{E} \exp\{it(T - \mathbf{E}T)/\sigma(T)\}$. In view of (6) it suffices to show that the relation $p^2q^2k^3 \to \infty$ implies for every $t \in R$ that $f'(t) + tf(t) \to 0$ as $k, n^* \to \infty$.

It follows from (5), (11) that

$$f(t) = \frac{1}{\lambda} \int_{-\pi}^{\pi} \mathbf{E}\, e^{iJ} ds, \qquad J = tH + sS, \qquad \lambda = 2\pi \mathbf{P}\{S = 0\},$$

$$H = \sum_{1 \le i < j \le N} h_{ij}, \qquad h_{ij} = g_{ij} w_i w_j, \quad g_{ij} = \frac{1}{\sigma(T)} (L_{X_i X_j} - p_L).$$

Furthermore, by symmetry, we obtain

$$f'(t) = \frac{i}{\lambda} \int_{-\pi}^{\pi} \mathbf{E}\, H e^{iJ} ds = i \frac{\binom{N}{2}}{\lambda} \int_{-\pi}^{\pi} \mathbf{E}\, h_{12} e^{iJ} ds. \tag{12}$$

Split $S = S_* + S_0$ and $H = h_{12} + H_* + H_0$ where $S_* = w_1 + w_2$ and

$$S_0 = \sum_{3 \le j \le N} w_j, \qquad H_* = \sum_{3 \le j \le N} (h_{1j} + h_{2j}), \qquad H_0 = \sum_{3 \le i < j \le N} h_{ij}.$$

Expanding e^{iJ} in powers of ith_{12}, itH_* and isS_* we show that

$$\frac{\binom{N}{2}}{\lambda} \left| \int_{-\pi}^{\pi} \mathbf{E}\, h_{12} (e^{iJ} - ith_{12} e^{iJ_0}) ds \right| \to 0, \tag{13}$$

where $J_0 = tH_0 + sS_0$. Therefore, we replace $\mathbf{E}\, h_{12} e^{iJ}$ by $itp^2q^2 \mathbf{E}\, g_{12}^2 e^{iJ_0}$ in the right integral (12).

Similarly, expanding e^{iJ} in powers of ith_{12}, itH_* and isS_* we show that

$$\frac{\binom{N}{2}}{\lambda} \left| p^2 q^2 \int_{-\pi}^{\pi} \mathbf{E}\, g_{12}^2 (e^{iJ_0} - e^{iJ}) ds \right| \to 0. \tag{14}$$

Using (13) and (14) we replace $\mathbf{E}\, h_{12}e^{iJ}$ by $itp^2q^2\mathbf{E}\, g_{12}^2e^{iJ}$ in the right integral (12). Furthermore, by symmetry, we can replace $\binom{N}{2}\mathbf{E}\, g_{12}^2e^{iJ}$ by $\mathbf{E}\, a^2e^{iJ} = a^2\mathbf{E}\, e^{iJ}$, since the number $a^2 := \sum_{1\le i<j\le N} g_{ij}^2$ is non-random. Finally, invoking (4) we obtain

$$a^2p^2q^2 = \frac{\mathbf{E}\,(L_{X_1X_2} - p_L)^2}{\sigma^2(T)}\binom{N}{2}p^2q^2 = 1 + O(1/Npq).$$

Since $O(1/Npq\lambda) = O(1/\sqrt{Npq})$ we can replace $p^2q^2a^2\mathbf{E}\, e^{iJ}$ by $\mathbf{E}\, e^{iJ}$. We have shown that

$$i\frac{\binom{N}{2}}{\lambda}\int_{-\pi}^{\pi}\mathbf{E}\, h_{12}e^{iJ}ds = -\frac{t}{\lambda}\int_{-\pi}^{\pi}\mathbf{E}\, e^{iJ}ds + o(1)$$

thus completing the proof.

The proof of (13) and (14) is rather technical and laborious. We do not present it here and refer to an extended version of the paper (Bloznelis 2004). Let us mention that in the proof of (13) and (14) we apply techniques developed in Bloznelis and Gözte (2002) for the analysis of the accuracy of the normal approximation of $U-$ statistics based on samples drawn without replacement, see also Bentkus, Götze and van Zwet (1997), Helmers and van Zwet (1982).

3.3. Note that the orthogonal decomposition (projection method) was used by Janson and Nowicki (1991) to prove limit theorems for subgraph count statistics of Bernoulli random graphs. The present paper can be considered as an attempt to extend these techniques to subgraph count statistics in the random graph model $G(k, n)$. In contrast to Bernoulli graph case the subgraph count statistics studied here have decompositions with vanishing linear part. Therefore, known results on the central limit theorem for asymptotically linear statistics based on samples drawn without replacement (see e.g., Bloznelis and Götze (2002), Zhao and Chen (1990)) are not applicable. We show that Erdős-Rényi (1959) representation (11) combined with Stein method (1) can be used to establish the asymptotic normality.

References

[1] Bentkus, V., Götze, F. and van Zwet, W. R. (1997) An Edgeworth expansion for symmetric statistics, *Ann. Statist.* **25**, 851–896.

[2] Bloznelis, M. (2004) Some results on the orthogonal decomposition and asymptotic normality of subgraph count statistics, *Preprint 2004-10 Faculty of Mathematics and Informatics, Vilnius university.*

[3] Bloznelis, M. (2003) Orthogonal decomposition of symmetric functions defined on random permutations, *Combinatorics, Probability and Computing*, Accepted for publication.

[4] Bloznelis, M. and Götze, F. (2001) Orthogonal decomposition of finite population statistic and its applications to distributional asymptotics, *Ann. Statist.*, **29**, 899–917.

[5] Bloznelis, M. and Götze, F. (2002) An Edgeworth expansion for symmetric finite population statistics, *Ann. Probab.*, **30**, 1238–1265.

[6] Erdős, P. and Rényi, A. (1959) On the central limit theorem for samples from a finite population, *PublMath. Inst. Hungar. Acad. Sci.* **4**, 49–61.

[7] Helmers, R., and van Zwet, W. R. (1982) The Berry–Esseen bound for U-statistics, *Statistical Decision Theory and Related Topics*,III. Vol. 1. (S.S. Gupta and J.O. Berger, eds.), Academic Press, New York, 497–512.

[8] Janson, S. (1990) A functional limit theorem for random graphs with applications to subgraph count statistics *Random. Struct. Algorithms*, **1**, 15–37.

[9] Janson, S., Luczak, T., and Rucinski, A. (2000) *Random graphs*, Wiley-Interscience, New York.

[10] Janson, S., Nowicki, K. (1991) The asymptotic distributions of generalized U-statistics with applications to random graphs, *Probab. Theory Related Fields*, **90**, 341–375.

[11] Ruciński, A. (1988) When are small subgraphs of a random graph normally distributed? *Probab. Theory Related Fields*, **78**, 1–10.

[12] Stein, Ch. (1970) A bound for the error in the normal approximation to the distribution of a sum of dependent random variables *Proc Sixth Berkeley Symp. Math. Stat. Probab.*, **2**, 583–602.

[13] Tikhomirov, A.N. (1976) On the rate of convergence in the central limit theorem for weak dependent variables, *Vestn. Leningr. Univ. (Mat. Mekh. Astron.)*, **7**, 158-159.

[14] Tikhomirov, A.N. (2001) On the central limit theorem, *Vestn. Syktyvkar. Univ., Ser. 1, Mat. Mekh. Inform.*, **4**, 51–76.

[15] Zhao, L. C. and Chen, X. R. (1990) Normal approximation for finite-population U-statistics, *Acta Math. Appl. Sinica*, **6**, 263–272.

Mindaugas Bloznelis
Vilnius University and Institute of Mathematics and Informatics
Mindaugas.Bloznelis@maf.vu.lt

Trends in Mathematics, © 2004 Birkhäuser Verlag Basel/Switzerland

Avalanche Polynomials of Some Families of Graphs

Robert Cori, Arnaud Dartois, and Dominique Rossin

ABSTRACT: *We study the Abelian sandpile model on different families of graphs. We introduced the avalanche polynomial which enumerates the size of the avalanches triggered by the addition of a particle on a recurrent configuration. This polynomial is calculated for several families of graphs. In the case of the complete graph, the result involves some known result on Parking functions* [12, 11].

1. Introduction

Bak, Tang and Wiesenfeld [2] introduced 15 years ago the concept of self organized criticality which allowed to describe a large variety of physical systems like earthquakes [5, 14], forest fires and even some fluctuations in the stock market [1]. One version of this concept is the sandpile cellular automaton model which uses a 2 dimensional lattice; in the sites of this lattice particles are added giving rise to a toppling when their number in a site exceeds a given bound. A toppling on a site may be followed by the toppling on one or more of its neighbors and this sequence of topplings is called an avalanche. Many authors have studied the distribution of the sizes (the number of topplings performed) of the avalanches for this model showing that they obey to power-laws [10, 7, 13].

The sandpile model was also considered by combinatorialists as a game on a graph called the chip firing game [3, 4]. Relationships between the structure of the graph and the recurrent configurations of the physical model were pointed out [6, 9].

Experiments on the distribution of sizes of the avalanches were considered only for the 2 dimensional lattice and for some classes of regular graphs. Very little is known for arbitrary graphs [8]. In this paper, a polynomial, encoding the avalanche sizes obtained by adding a particle to a site in a recurrent configuration, is associated to a graph. We determine this polynomial for various families of graphs.

These families are the trees, the cycles, the complete graphs and the lollipop graphs. For these families of graphs the power law observed for the 2 dimensional grid is no more satisfied. The computation of the avalanche polynomial of the complete graph uses a bijection between recurrent configurations of this graph and the so-called parking functions. This computation allows the determination of the avalanche distributions on the lollipop graphs which shows out the existence of peaks also observed in some other families of regular graphs.

2. Recurrent configurations of the sandpile model

In this section we recall the main results on the sandpile model which are useful in this paper.

In what follows $G = (X, E)$ is a connected multigraph with $n+1$ vertices:

$$X = \{x_1, x_2, \ldots, x_n, x_{n+1}\},$$

vertex x_{n+1} is distinguished and called the *sink*. A *configuration* of the sandpile model in this graph is a sequence of n integers

$$u = (u_1, u_2, \ldots, u_n).$$

For a configuration u, the integer u_i will be considered as a number of *particles* placed on the vertex x_i.

A configuration is *stable* if the integers u_i satisfy $0 \le u_i < d_i$, where d_i is the degree of the vertex x_i. In a configuration which is not stable a vertex x_i with $u_i \ge d_i$ may perform a *toppling* giving a new configuration v such that $v_i = u_i - d_i$ and $v_j = u_j + \varepsilon_{i,j}$, where $\varepsilon_{i,j}$ denotes the number of edges between vertices x_i and x_j in the multigraph G.

We will write the toppling of vertex x_i by:

$$u \quad \rightarrow_i \quad v.$$

An *avalanche* is a sequence of topplings; we will use the notation

$$u \overset{*}{\rightarrow} v.$$

for a configuration v which is reached from u after an avalanche.

The size of the avalanche is the number of topplings performed. In the Figure 1 below is given an avalanche of size 3, the sink is represented by a black vertex.

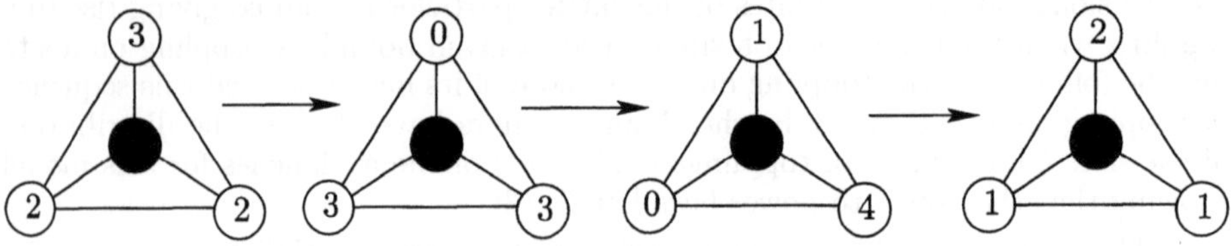

FIGURE 1. An avalanche on K_4

In the sandpile model the main operation consists in:

1. taking a stable configuration
2. adding a particle on one of the vertices
3. performing topplings until a new stable configuration is obtained.

The above example is an illustration of this operation for the graph K_4 and the configuration $(2, 2, 2)$. It is not difficult to prove that the stable configuration obtained after a sequence of topplings from a given unstable configuration does not depend on the order in which these topplings are performed.

The *recurrent* configurations of the sandpile model are the stable configurations which are met infinitely often when performing the above operations 2 and 3. Note that not all the stable configurations are recurrent. The recurrent configurations play a key role in the sandpile model; their number is the tree number of the underlying graph, hence independent of the vertex chosen as the sink.

There are many characterizations of recurrent configurations and many structural results on them; in the sequel we will simply use the following characterization. Let π denotes the configuration such that π_j is the number of edges between vertex x_j and the sink.

Proposition 2.1. *The configuration u is recurrent if and only if it is stable and*

$$\pi + u \overset{*}{\rightarrow} u,$$

where $\pi + u$ denotes the configuration v such that $v_i = u_i + \pi_i$ for $1 \leq i \leq n$.

Moreover in this avalanche every vertex of G topples exactly once.

3. Avalanche polynomials: some simple examples

Let u be a recurrent configuration on the multigraph $G = (X, E)$, we consider the avalanche obtained when adding a particle on site x_i, we will call it a *principal avalanche*, we denote by $ad_G(u, x_i)$ the size of this principal avalanche. Note that a principal avalanche may be of size 0. This is the case when the particle is added on a site i such that $u_i < d_i - 1$.

The three recurrent configurations of the cycle C_3 are given below:

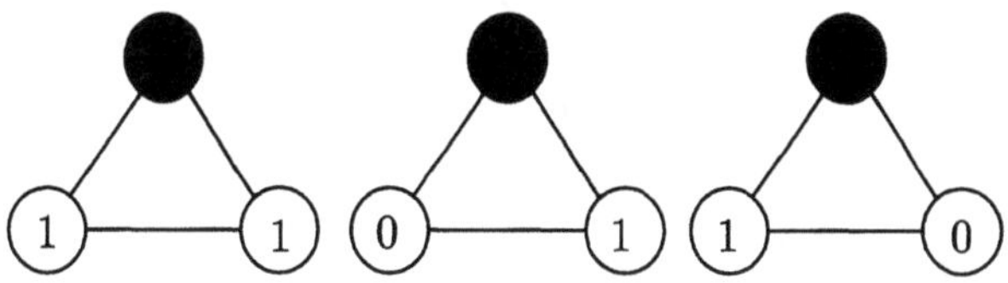

FIGURE 2. Recurrent configurations on C_3.

The sizes of the principal avalanches are 2, 2 for the first recurrent configuration, 1, 0 and 0, 1 for the two other configurations.

To any connected graph $G = (X, E)$ with a sink, we associate a polynomial enumerating the sizes of the principal avalanches. This polynomial is given by:

$$Av_G(x) = \sum \alpha_k x^k,$$

where α_k is the number of principal avalanches of size k.

For instance from the example on Figure 2 we obtain:

$$Av_{C_3}(x) = 2 + 2x + 2x^2.$$

Note that for any graph G, $Av_G(1)$ is equal to n times the tree number of G. Thus it is independent of the sink chosen for G. Note that indeed, the polynomial Av_G is dependent of this sink.

3.1. Avalanche polynomials of trees

If G is a tree T then it has only one recurrent configuration, this configuration u_T is such that $u_{T_i} = d_i - 1$ for each vertex x_i. It is convenient to draw the tree in such a way that the sink is the root of the tree.

Adding a particle on vertex x_i gives a sequence of topplings on all the vertices of the subtree of T rooted at x_i, the avalanche ends there if x_i is a son of the root. If x_i is not a son of the root, then after this first sequence of topplings the father

x_j of x_i in T gets one particle and a new sequence of topplings can be performed. Hence we have:

$$ad_T(u_T, x_i) = ad_T(u_T, x_j) + t_i,$$

where t_i is the number of vertices of the subtree with root x_i.

An example of the computation of avalanche sizes on a tree is given below; on each vertex is indicated the size of the avalanche obtained when adding a particle on it.

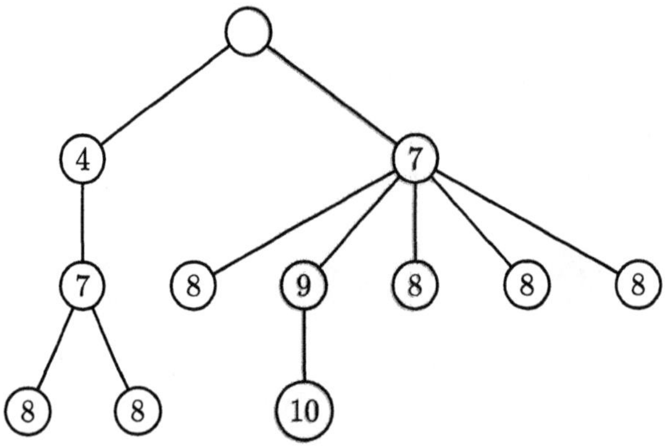

FIGURE 3. A tree T_1 and the sizes of the principal avalanches.

The avalanche polynomial on this tree is:

$$Av_{T_1}(x) = x^4 + 2x^7 + 6x^8 + x^9 + x^{10}.$$

Note that $Av_G(0) = 0$ if and only if G is a tree. Indeed a graph which is not a tree has more than one recurrent configuration and necessarily at least one of them has a vertex x_i with $u_i < d_i - 1$; for this vertex, adding a particle does not produce any toppling.

The avalanche polynomial does not characterize the tree since the avalanche polynomial of the tree T_2 below is also $Av_{T_1}(x)$ [1].

[1]this example is due to Michel Marcus.

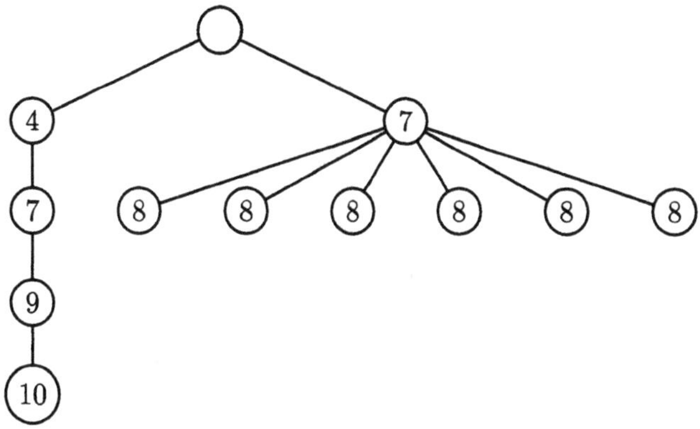

FIGURE 4. A tree T_2 with the same avalanche polynomial as T_1.

When we choose a different root, the avalanche polynomial changes also. From any tree or graph, we can define the set of avalanche polynomials for all different possible roots. This set is neither characteristic of the tree nor the graph. Indeed, the trees T and T' of Figure 6 built from T_1 and T_2 are not isomorphic but admit the same set of avalanche polynomials, when the root spans the set of vertices.

The set $Av^{\mathcal{T}}$ of polynomials P such that there exists a tree T satisfying $Av_T = P$ is the smallest set such that:

$$x \in Av^{\mathcal{T}} \tag{1}$$

$$P, Q \in Av^{\mathcal{T}} \implies P + Q \in Av^{\mathcal{T}} \tag{2}$$

$$P \in Av^{\mathcal{T}} \implies x^a(P+1) \in Av^{\mathcal{T}}, \text{ where } a = P(1) + 1. \tag{3}$$

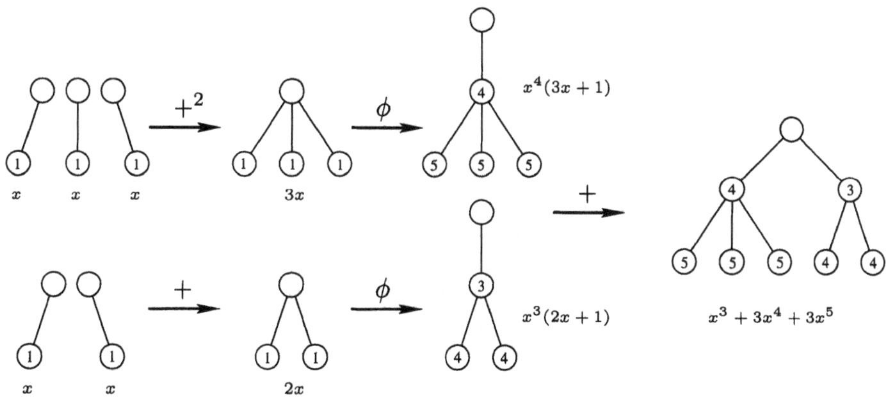

FIGURE 5. Inductive computation of the avalanche polynomial of a tree.

The proof is quite straightforward, since the two operations (2) and (3) correspond to two basic operations on trees (cf Figure 5): sticking two trees by merging their root (+), and adding a new root only connected to the old one (ϕ). The result follows by the fact that any rooted tree can be obtained by these two operations starting from the two-vertices tree (whose avalanche polynomial is x).

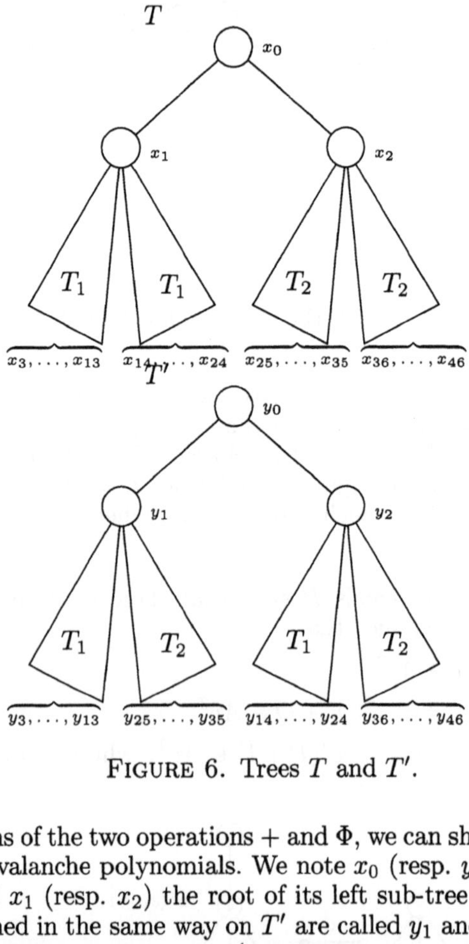

FIGURE 6. Trees T and T'.

By the means of the two operations + and Φ, we can show that T and T' have the same set of avalanche polynomials. We note x_0 (resp. y_0) the root of the tree T (resp. T'), and x_1 (resp. x_2) the root of its left sub-tree (resp. right sub-tree). The vertices defined in the same way on T' are called y_1 and y_2. Among the other vertices of T, we note $x_3, x_4, \ldots, x_{13}$ (resp. $x_{25}, x_{26}, \ldots, x_{35}$) the ones belonging to the first copy of T_1 (resp. T_2) and $x_{14}, x_{15}, \ldots, x_{24}$ (resp. $x_{36}, x_{37}, \ldots, x_{46}$) the ones belonging to the second copy of T_1 (resp. T_2). The corresponding vertices of T' are called y_i, where the index i is defined by the same manner (cf. Figure 6). Then, we can show that the avalanche polynomial of T rooted in x_i equals the avalanche polynomial of T' rooted in y_i for all indices i.

If x_0 is chosen as the root of T, we choose y_0 as root of T'. Then $T = \Phi(T_1+T_1)+\Phi(T_2+T_2)$ and $T' = \Phi(T_1+T_2)+\Phi(T_1+T_2)$. In term of polynomials, we get $Av_T = Av_{T'}$, since $Av_{T_1} = Av_{T_2}$.

By symmetry of (T, T') in regards of T_1 and T_2, it is sufficient to show that $Av_T = Av_{T'}$ when x_i (resp. y_i) is chosen as root of T (resp. T'), only for indices $i = 1$ or $3 \leq i \leq 24$.

If x_1 is chosen as root of T, we choose y_1 as root of T'. This time, $T = T_1+T_1+\Phi^2(T_2+T_2)$ and $T' = T_1+T_2+\Phi^2(T_1+T_2)$. But, in term of polynomials, we still get $Av_T = Av_{T'}$.

Let choose x_i (resp. y_i) as root of T (resp. T'), with $3 \leq i \leq 13$. Then, it exists a sequence s of operations $+$ and Φ that build T_1 rooted in x_i starting from $\{x_1\}$. If we set $Av_A = 0$ for the tree A with one vertex, then, in term of polynomials, the sequence s applied to 0 leads to the avalanche polynomials of T_1 rooted in x_i. Hence we have $T = s(T1 + \Phi^2(T_2 + T_2))$ and also $T' = s(T2 + \Phi^2(T_1 + T_2))$. Consequently, $Av_T = Av_{T'}$. If $14 \leq i \leq 24$, the same argument is possible, and we also get the wanted result.

Hence, T and T' are non-isomorphic but have the same set of avalanche polynomials.

3.2. Avalanche polynomials of cycles

Another simple example on which the avalanche polynomial can be computed easily is that of the cycle C_{n+1}; we have the following result:

Proposition 3.1. *There exists a principal avalanche of size k on C_{n+1} if and only if there exists two integers p, q such that $pq = k$ and $p + q \leq n + 1$. Moreover the number of principal avalanches of size k is equal to twice the number of couples (p, q) such that $pq = k$ and $p + q \leq n$ plus the number of couples such that $pq = k$ and $p + q = n + 1$.*

Proof. The recurrent configurations on the cycle C_{n+1} are the configurations in which all vertices have one particle except possibly one which has no particle; there are $n + 1$ such configurations. The recurrent configuration is determined by the distance from the sink to the vertex with no particle and to obtain the size of the avalanche one has to consider where the particle has been added. Let us denote by 1^n the recurrent configuration consisting of a full sequence of 1's and by 1^p01^q (with $p + q = n - 1$) those containing $n - 1$ 1's and one 0.

It is easy to check that the avalanche starting with $1^{p-1}21^{q-1}$ has size pq and the same is true for the avalanche starting with the configuration $1^{p-1}21^{q-1}01^r$ or with configuration $1^r01^{p-1}21^{q-1}$ giving the result. □

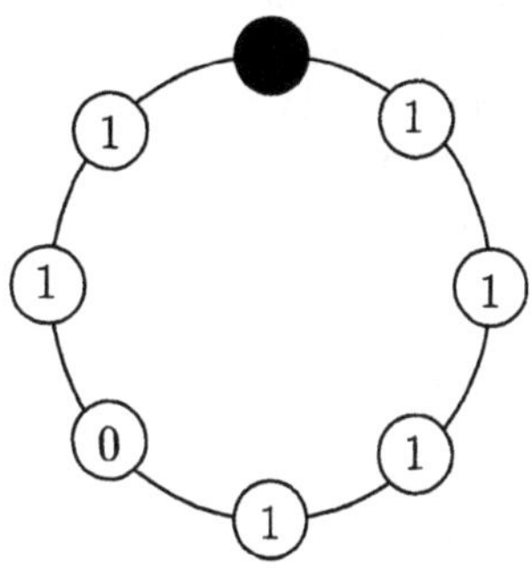

FIGURE 7. A recurrent configuration on C_8.

Note that the number of couples (u, x_i) such that $u_i = 0$ is n, hence the constant term of $Av_{C_{n+1}}$ is n.

The avalanche polynomial of C_8 is given by

$$Av_{C_8} = 7+2x+4x^2+4x^3+6x^4+4x^5+8x^6+2x^7+4x^8+2x^9+4x^{10}+6x^{12}+2x^{15}+x^{16}$$

For instance the number of avalanches of size 6 is 8 since there are 4 couples (p,q) with $pq = 6$ and $p+q \leq 7$ namely: $(1,6),(6,1),(2,3),(3,2)$; the number of avalanches of size 16 is 1 since the couple $(4,4)$ is the only one satisfying $pq = 16$ and $p+q = 8$ and there is no p,q satisfying $pq = 16$ and $p+q \leq 7$.

4. Avalanche polynomials of the complete graph K_{n+1}

In the complete graph K_{n+1} the recurrent configurations are in bijection with parking functions. Recall that a sequence of non negative numbers $(w_1, w_2, \ldots, w_n)$ is an n-parking function if there exists a permutation $a_1, a_2, \ldots, a_n$ of $1, 2, \ldots, n$ such that for all i, $0 \leq w_i < a_i$ (cf [12]). For instance the 16 3-parking functions are the permutations of the following sequences:

$$(0,1,2),(0,0,1),(0,1,1),(0,0,2),(0,0,0).$$

A simple use of Proposition 1 gives

Proposition 4.1. *The configurations $u_1, u_2, \ldots, u_n$ of K_{n+1} is recurrent, if and only if $n-1-u_1, n-1-u_2, \ldots, n-1-u_n$ is a parking function. Then the number of recurrent configurations on K_{n+1} is $(n+1)^{n-1}$.*

Moreover in this correspondence the number of saturated vertices in u is equal to the number of 0 in the associated parking function.

The number of avalanches of size 0 in K_{n+1} is equal to the number of non zero elements in all the n-parking functions. To compute this number we need the following lemma (cf [11]):

Lemma 4.2. *The number of parking functions containing k 0's is :*

$$\binom{n-1}{n-k} n^{n-k} = \frac{k}{n}\binom{n}{k} n^{n-k}.$$

Proof. Consider the set U_n of all sequences u of n integers containing k 0's and $n-k$ numbers $1 \leq u_i \leq n$, clearly there are $\binom{n}{k} n^{n-k}$ such sequences. All parking functions are in U_n but the converse is not true. To determine if $u \in U_n$ is a parking function we apply the following parking algorithm:

Consider a park place with n slots numbered $1, 2, \ldots, n$ lying on a circle. For each i such that $u_i > 0$ put a car on the first free slot starting from position u_i and going around the circle.

The sequence u is a parking function if and only if the slot n is free at the end of the algorithm.

For instance the algorithm applied to the sequence $2,5,0,5,5,0,0$ fills the slots $2,5,6,7$, and the slots $1,3,4$ are free; the sequence is not a 7-parking function.

After the execution of the algorithm there are k free slots, and by symmetry each of the slots $1, 2, \ldots, n$ has an equal probability of being free; hence the probability of the slot n to be free is $\frac{k}{n}$, giving the result. □

From the above Lemma we obtain;

Proposition 4.3. *The number of principal avalanches of size 0 in K_{n+1} is*

$$n(n-1)(n+1)^{n-2}.$$

Proof. It suffices to check that the number of couples (u, x_i) such that u is an n-parking function and $u_i = 0$ is equal to $2n(n+1)^{n-2}$; the result is then obtained by subtracting this value from the total number of couples (u, x_i) which is $n(n+1)^{n-1}$. The number of the couples we are considering is by the Lemma above

$$T_n = \sum_{k=1}^{n-1} k\binom{n-1}{n-k} n^{n-k}.$$

Denote $f_n(x) = x(n+x)^{n-1}$ we have:

$$f_n(x) = \sum_{k=0}^{n-1} x^k n^{n-k}\binom{n-1}{n-k},$$

and

$$T_n = f_n'(1) = (n+1)^{n-1} + (n-1)(n+1)^{n-2},$$

giving the result □

In order to enumerate the number of principal avalanches of positive sizes we have:

Proposition 4.4. *The number of principal avalanches of size $m > 0$ in K_{n+1} is:*

$$\binom{n}{m} m^{m-1}(n-m+1)^{n-m-1} = n\binom{n-1}{m-1} m^{m-2}(n-m+1)^{n-m-1}.$$

Proof. In order to prove this result we associate a subset and two different recurrent configurations to any couple (u, x_i) consisting of a recurrent configuration u and a saturated site x_i which gives a principal avalanche of size $m > 0$ when a particle is added on site x_i:

- a subset J of $m-1$ sites among the $n-1$ sites of K_{n+1} different from the sink and from x_i: these are the sites which perform a toppling during the avalanche triggered by the toppling of x_i,
- a recurrent configuration on K_m: consider the values u_j for $x_j \in J$ and subtract $m-2$ to all these values; it is easy to check using Proposition 1 that it is a recurrent configuration,
- a recurrent configuration on K_{n-m+1}: the values u_k for $x_k \notin J \cup \{x_i\}$ determine a recurrent configuration on this graph.

Conversely it is easy to build a principal avalanche of size $m > 0$ from a subset J of $m-1$ sites, a vertex x_i, a recurrent configuration on K_m and another one on K_{n-m+1}. □

Below is given the sizes of avalanches for the complete graph K_5. The recurrent configurations which differ by a permutation of the number of particles on the sites are considered as equivalent. For K_5, we represent each equivalence class by the configuration satisfying; $u_1 \leq u_2 \leq \ldots \leq u_n$, we also give the class sizes and the number of principal avalanches of each size.

class	config.	0	1	2	3	4
24	3,2,1,0	72	24			
12	3,2,1,1	36	12			
12	3,2,2,0	36	12			
12	3,2,2,1	36	12			
4	3,2,2,2	12	4			
12	3,3,1,0	24		24		
6	3,3,1,1	12		12		
12	3,3,2,1	24				24
6	3,3,2,2	12				12
12	3,3,2,0	24			24	
4	3,3,3,0	24			12	
4	3,3,3,1	4				12
4	3,3,3,2	4				12
1	3,3,3,3					4
		300	64	36	36	64

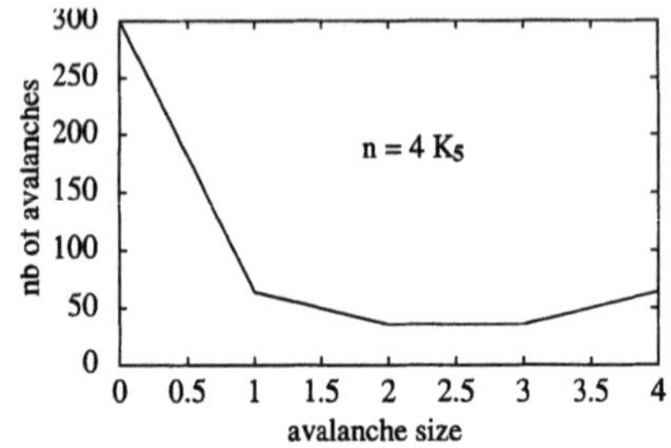

The distribution of avalanches sizes of the graph K_{21} are given in Figure 8.

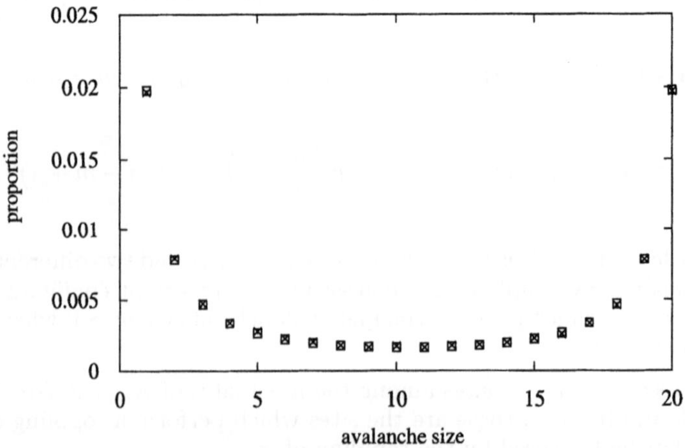

FIGURE 8. Distribution of principal avalanches on K_{21}: theoretical prediction (square) and experiments (cross).

5. Avalanche polynomials of lollipop graphs

In this section we consider avalanches on the lollipop graph with $n+m+1$ vertices. It consists of a path of vertices $x_0, x_1, x_2, \ldots, x_m$ leading to the complete graph K_{n+1} whose vertices are $x_m, x_{m+1}, \ldots, x_{m+n}$. It is more convenient to consider x_0 as the sink. In other words, we apply ϕ^m to K_{n+1}.

A recurrent configuration on this graph consists of a recurrent configuration

$$u_{m+1}, u_{m+2}, \ldots, u_{m+n}$$

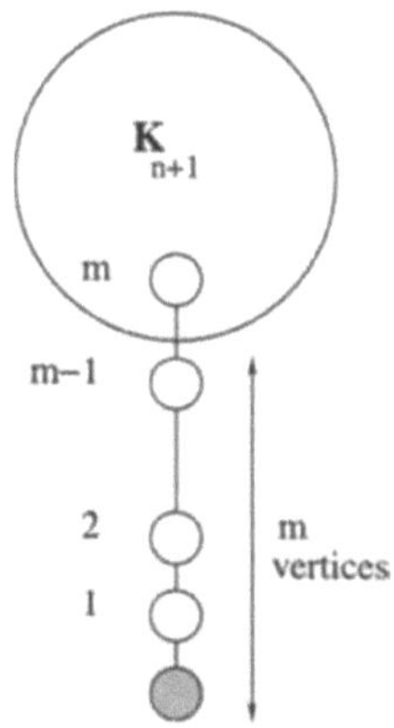

FIGURE 9. Lollipop graph $L_{m,n}$.

on the complete graph, where the other vertices are saturated. This gives:

$$u_m = n, u_{m-1} = 1, \ldots, u_1 = 1.$$

The sizes of the principal avalanches are given by:

Proposition 5.1. *In the lollipop graph the sizes of the principal avalanches of the recurrent configuration u are*

- *0, if the particle is added on a non saturated vertex for u,*
- *$i(m+n) - \frac{i(i-1)}{2}$, if the particle is added to the i-th vertex of the path ($i \leq m$) starting from the sink,*
- *$[\frac{m(2n+m+1)}{2} + 1]ad_{K_{n+1}}(u, x_i)$, if the particle is added in a saturated vertex x_i for u.*

As an example the avalanche polynomial of the lollipop graph $L_{4,3}$ is :

$$24 + 16x^7 + 16x^{13} + 16x^{18} + 16x^{22} + 9x^{23} + 6x^{46} + 9x^{69}.$$

A computation of the sizes of principal avalanches for the lollipop graph $L_{10,20}$ gives the distribution shown in the Figure below:

Let us define the operator $\mathtt{Tr}^k$ for every polynomial $P = \sum_{i=0}^{n}$ as follows:

$$\mathtt{Tr}^k(P)(x) = \sum_{i=0}^{min(k,n)} a_i x^i.$$

Corollary 1. *The avalanche polynomial of the lollipop graph $L_{m,n}$ is:*

$$Av_{L_{m,n}}(x) = Av_{K_{n+1}}(x^{a+1}) + (n+1)^{n-1}\mathtt{Tr}^a(Av_{L_{m+n}})(x),$$

where $a = \frac{m(2n+m+1)}{2}$ and L_k is the line with $k+1$ vertices.

Proof. The first term of the sum is obvious by the precedent proposition. The second term corresponds to principal avalanches for couples (u, x_i), where $i \leq m$, i.e., where x_i is a vertex of the path L_m. The size of this principal avalanche is the same if the vertices x_j for $j > i$ are all in line. Indeed, what matters when a particle is added on x_i is how many vertices are in the subgraph when deleting the path

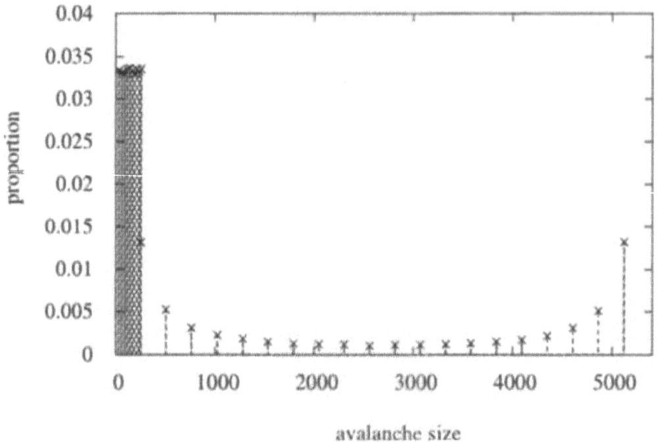

FIGURE 10. Avalanche distribution for $L_{10,20}$: expected result (dashed line), and experiments over 10^6 computations (cross).

$x_0, x_1, \ldots, x_i$. For $i \leq m$, the avalanche size is at most a, thus $\mathrm{Tr}^a(Av_{L_{m+n}})(x)$ gives the good exponents. Since there are $(n+1)^{n-1}$ recurrent configurations, we get the result. □

In fact, this result could be generalized. Let $G = (X, E)$ be a rooted graph. We call *dissipation* of a principal avalanche the number d of particles that the sink (root) receives during the avalanche. We associate a new polynomial enumerating the principal avalanches by size and dissipation to any graph G; this polynomial is given by:

$$\widetilde{Av_G}(x,y) = \sum \alpha_{k,d} x^k y^d,$$

where $\alpha_{k,d}$ is the number of principal avalanches of size k and dissipation d.

If G is a tree, $\widetilde{Av_G}(x,y)$ has a very simple expression: $\widetilde{Av_G}(x,y) = Av_G(x)y$. Indeed, every principal avalanche has dissipation 1.

If G is such that every vertex is connected to the sink, like K_{n+1} for example, $\widetilde{Av_G}(x,y)$ has also a very simple expression: $\widetilde{Av_G}(x,y) = Av_G(xy)$. Every principal avalanche of size m admits m as dissipation, since every toppling gives a particle to the sink.

Then, if we apply ϕ^m to G like we did with K_{n+1} to obtain a lollipop graph, we get a graph G_m, whose avalanche polynomial is:

$$Av_{G_m}(x) = \widetilde{Av_G}(x, x^a) + \frac{Av_G(1)}{n} \mathrm{Tr}^a(Av_{L_{m+n}})(x),$$

where $a = \frac{m(2n+m+1)}{2}$ and $|G| = n+1$. In fact we have:

$$\widetilde{Av_{G_m}}(x,y) = \widetilde{Av_G}(x, x^a y) + \frac{Av_G(1)}{n} \mathrm{Tr}^a(Av_{L_{m+n}})(x)y.$$

For the lollipop graph, it is particularly simple, since $\widetilde{Av_{K_{n+1}}}(x,y) = Av_{K_{n+1}}(xy)$.

6. Concluding remarks

In this paper, we have considered avalanche polynomials which encode the distribution of the sizes of avalanches. We have seen that for the complete graph, we have a huge peak for size 0. Now the tree T_n defined on Figure 11 gives a peak for size n.

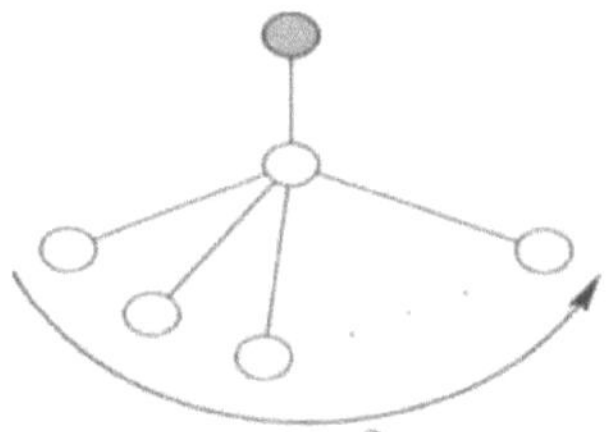

FIGURE 11. Tree T_n.

The construction consisting of merging the sinks of two rooted graphs translates in adding the avalanche polynomials. Applying this construction to the two kind of graphs considered above, we are able to build graphs having any shape of distribution in a certain sense.

This informal remark can be formalized in a more precise result which will be developped in a future work.

Acknowledgement

We thank Michel Marcus for his example.

References

[1] P. Bak. How nature works - the science of soc. *Oxford university press*, 1997.

[2] P. Bak, C. Tang, and K. Wiesenfeld. Self-organized criticality: an explanation of $1/f$ noise. *Phys. Rev. Lett.*, 59, 1987.

[3] N. L. Biggs. Chip-firing on distance-regular graphs. *Tech. Report LSE-CDAM-96-11, CDAM Research Report Series*, June 1996.

[4] N. L. Biggs. Chip-firing and the critical group of a graph. *Journal of Algebraic Combinatorics*, 9(1):25–45, 1999.

[5] K. Chen, P. Bak, and S. P. Obukhov. Self-organized criticality in a crack propagation model of earthquakes. *Physical Review Letters A*, 43:625–30, 1991.

[6] R Cori and D Rossin. On the sandpile group of dual graphs. *European J. Combin.*, 21(4):447–459, 2000.

[7] M. Creutz. Abelian sandpile. *Computers in Physics*, 5:198–203, 1991.

[8] A. Dartois and D. Rossin. Analysis of the distribution of the length of avalanches on the sandpile group of the (n, k)-wheel. *FPSAC*, 2003.

[9] D. Dhar. Self-organized critical state of sandpile automaton models. *Physical Review Letters*, 64:1613–1616, 1990.

[10] D. Dhar, P. Ruelle, S. Sen, and D. Verma. Algebraic aspects of Abelian sandpile models. *Journal of Physics A*, 28:805–831, 1995.

[11] D. Foata and J. Riordan. Mappings of acyclic and parking functions. *Aequationes Math.*, 10:10–22, 1974.
[12] A. G. Konheim and B. Weiss. An occupancy discipline and applications. *SIAM J. Appl. Math.*, 1966.
[13] D. V. Ktitarev, S. Lübeck, P. Grassberger, and V. B. Priezzhev. Scaling of waves in the Bak-Tang-Wiesenfeld sandpile model. *Physical Review E*, 61:81–92, 2000.
[14] A. Sornette and D. Sornette. Self-organized criticality of earthquakes. *Europhys. Lett.*, 9:197–202, 1989.

Robert Cori
LABRI, Université Bordeaux 1
cori@labri.u-bordeaux.fr

Arnaud Dartois
LIX, École Polytechnique
dartois@lix.polytechnique.fr

Dominique Rossin
CNRS and LIAFA, Université Paris 7
rossin@liafa.jussieu.fr

Trends in Mathematics, © 2004 Birkhäuser Verlag Basel/Switzerland

Perfect Matchings in Random Graphs with Prescribed Minimal Degree

Alan Frieze* and Boris Pittel†

ABSTRACT: *We consider the existence of perfect matchings in random graphs with n vertices (or $n+n$ vertices in the bipartite case) and m random edges*, subject to a lower bound on minimum vertex degree. *A random bipartite graph without isolated vertices and $m > n$ edges with high probability* (**whp**) *has a perfect matching iff the average vertex degree is $0.5 \log n + \log \log n + c_n$, $c_n \to \infty$ however slow. A random graph with minimum degree at least two* **whp** *has a matching that matches all the vertices except "odd-man-out" vertices, one per each isolated cycle of odd length, and one for the remaining vertex set if its cardinality is odd. So, for n even,* **whp** *the random graph has a perfect matching iff it does not have isolated odd cycles.*

1. Introduction

To quote from Lovász [17], "the problem of the existence of 1-factors (perfect matchings), the solution of which (the Kőnig-Hall theorem for bipartite graphs and Tutte's theorem for the general case) is an outstanding result making this probably the most developed field of graph theory". Erdős and Rényi ([8], [9]) found a way to use these results for a surprisingly sharp study of existence of perfect matchings in random graphs. For $B_{n,m}$, a random bipartite graph with $n+n$ vertices and $m = n(\ln n + c_n)$ random edges, they proved [8] that

$$\lim_{n\to\infty} \mathbf{Pr}(B_{n,m} \text{ has a perfect matching}) = \lim_{n\to\infty} \mathbf{Pr}(\delta(B_{n,m}) \geq 1) = \begin{cases} 0 & c_n \to -\infty, \\ e^{-2e^{-c}}, & c_n \to c, \\ 1 & c_n \to \infty, \end{cases}$$

where δ denotes minimum degree. Of course, minimum degree at least one is a trivial necessary condition for the existence of a perfect matching. The Hall theorem turned out to be perfectly tailored for use in combination with probabilistic techniques, pioneered in [8] several years earlier. Even though Tutte's theorem for the non-bipartite case is considerably more involved, in [9] Erdős and Rényi managed to extend the analysis to the random graph $G_{n,m}$, a random general graph

*Supported in part by NSF grant ccr-0200945.
†Supported in part by an NSF grant

with n vertices and $m = \frac{n}{2}(\ln n + c_n)$ edges, showing that

$$\lim_{\substack{n\to\infty \\ n \text{ even}}} \mathbf{Pr}(G_{n,m} \text{ has a perfect matching}) = \lim_{n\to\infty} \mathbf{Pr}(\delta(G_{n,m}) \geq 1) = \begin{cases} 0 & c_n \to -\infty, \\ e^{-e^{-c}} & c_n \to c, \\ 1 & c_n \to \infty. \end{cases}$$

In both cases a perfect matching becomes likely as soon as one has sufficiently many random edges for the minimum degree to be at least one with high probability (**whp**). This has led researchers to consider the existence of perfect matchings in models of a random graph in which the minimum degree requirement is always satisfied. Perhaps the first result along these lines is due to Walkup [23]. He considered a κ-out model $B_{\kappa-\text{out}}$ of a random bipartite graph, again with $n+n$ vertices $V_1 + V_2$. Here each vertex $v \in V_i$ "chooses" κ random neighbours in its complementary class V_{3-i}. Walkup showed that

$$\lim_{n\to\infty} \mathbf{Pr}(B_{\kappa-out} \text{ has a perfect matching}) = \begin{cases} 0 & \kappa = 1 \\ 1 & \kappa \geq 2 \end{cases}$$

Frieze [10] proved a non-bipartite version of this result, the argument being based on Tutte's theorem and considerably harder. Very recently Karoński and Pittel [13] have proven **whp** the existence of a perfect matching in what they called the $B_{(1+e^{-1})-out}$ graph, a subgraph of B_{2-out}, obtained from B_{1-out} by letting each of its degree 1 vertices select another random neighbor in the complementary class. Observe that in all of these results [23], [10] and [13] the number of random edges depends linearly on the number of vertices, and the minimum degree has been raised to 2, in a sharp contrast with the case m being of order $n \log n$. Here is why. When there are order $n \ln n$ random edges, there are few vertices of degree 1 and they are far apart. In sparser models, with minimum degree 1, **whp** there will be a linear (in n) number of vertices of degree 1, and some two vertices of degree 1 will have a common neighbor, which rules out a perfect matching. In the case of random regular graphs it turns out that minimum degree 3 is required, Bollobás [2]: Let G_r denote a random r-regular graph on vertex set $[n]$, n even. Then

$$\lim_{n\to\infty} \mathbf{Pr}(G_r \text{ has a perfect matching}) = \begin{cases} 0 & r = 2, \\ 1 & r = 1 \text{ or } r \geq 3. \end{cases}$$

The case $r = 1$ is trivial since then G_r is itself a perfect matching of $[n]$. G_2 is **whp** a collection of $O(\ln n)$ disjoint cycles and they will all have to be even for G_2 to have a perfect matching. The meat of the result is therefore in the case $r \geq 3$ and this follows from r-connectivity and Tutte's theorem.

Another approach was considered by Bollobás and Frieze [5]. Let $\mathcal{G}_{n,m}^{\delta\geq\kappa}$ denote the set of graphs with vertex set $[n]$, m edges and minimum degree at least κ. Let $G_{n,m}^{\delta\geq\kappa}$ be sampled uniformly from $\mathcal{G}_{n,m}^{\delta\geq\kappa}$. By conditioning on minimum degree 1, say, we will need 50% fewer random edges to get a perfect matching **whp**: Let $m = \frac{n}{4}(\ln n + 2\ln\ln n + c_n)$.

$$\lim_{\substack{n\to\infty \\ n \text{ even}}} \mathbf{Pr}(G_{n,m}^{\delta\geq 1} \text{ has a perfect matching}) = \begin{cases} 0 & c_n \to -\infty \text{ sufficiently slowly}, \\ e^{-\frac{1}{8}e^{-c}} & c_n \to c, \\ 1 & c_n \to \infty. \end{cases} \tag{1}$$

The restriction "sufficiently slowly" may seem out of place, but bear in mind that if n is even and $m = n/2$ then the probability of a perfect matching is 1. The precise threshold between $n/2$ and $\frac{1}{4}n \ln n$ for the non-existence of a perfect matching was not determined. Using the approach developed in the present paper for the bipartite case, we have found that "sufficiently slowly" in (1) can be replaced simply by "and $m > n/2$". (For $m = n/2+1$, say, the likely graph, with minimum degree 1 at least, consists of $n/2-3$ isolated edges, and two paths, each consisting of 3 vertices.) The study in [5] was extended in Bollobás, Fenner and Frieze [3] who considered the probability that $G_{n,m}^{\delta\geq\kappa}$ has $\lfloor\kappa/2\rfloor$ disjoint Hamilton cycles plus a further disjoint perfect matching if κ is odd.

In the present paper we continue this line of research. We first consider the bipartite version of (1). Let $\mathcal{B}_{n,m}^{\delta\geq\kappa}$ denote the set of bipartite graphs with vertex set $[n]+[n]$, m edges and minimum degree at least κ. Let $B_{n,m}^{\delta\geq\kappa}$ be sampled uniformly from $\mathcal{B}_{n,m}^{\delta\geq\kappa}$.

Theorem 1. *Let* $m = \frac{n}{2}(\ln n + 2\ln\ln n + c_n)$. *Then*

$$\lim_{n\to\infty} \mathbf{Pr}(B_{n,m}^{\delta\geq 1} \textit{ has a perfect matching}) = \begin{cases} 0 & c_n \to -\infty,\ m > n, \\ e^{-\frac{1}{4}e^{-c}} & c_n \to c, \\ 1 & c_n \to \infty. \end{cases} \tag{2}$$

(As in the case of $\mathcal{G}_{n,m}^{\delta\geq 1}$, we observe that the threshold for m is reduced by the factor of 2, compared to that of the random graph $B_{n,m}$.) The RHS expression in (2) is the limiting probability that no two vertices of degree 1 have a common neighbor. Thus, the probability that a perfect matching exists is (close to) 1 when either $m = n/2$ or c_n is large, and the probability is very small for m everywhere in between, except c_n not far to the left from 0.

The next natural question is: How many random edges are needed if we constrain the minimum degree to be at least 2, so ruling out the possibility of two vertices of degree 1 having a common neighbour?

In this paper we only consider the non-bipartite graphs. To cover both even and odd values of the number of vertices, it is convenient, and natural, to say that a graph $G = (V,E)$ has a perfect matching if $\mu^*(G) = \lfloor |V|/2 \rfloor$, where $\mu^*(G)$ is the maximum matching number of G.

Unlike the bipartite case, with a positive limiting probability the "sparse" graph $G_{n,cn}^{\delta\geq 2}$ may have (short) isolated odd cycles. This observation rules out a "**whp**-type" result for probability of a perfect matching. Let $X(G)$ stand for the total number of odd isolated cycles in G. Clearly

$$\mu^*(G) \leq \nu(G) := \left\lfloor \frac{|V| - X(G)}{2} \right\rfloor.$$

Let $\mu_n^* = \mu^*(G_{n,cn}^{\delta\geq 2})$, $X_n = X(G_{n,cn}^{\delta\geq 2})$ and $\nu_n = \mu(G_{n,cn}^{\delta\geq 2})$.

Theorem 2. *Let* $\liminf c > 1$. *Then*

$$\lim_{n\to\infty} \mathbf{Pr}(\mu_n^* = \nu_n) = 1,$$

and X_n *is, in the limit, Poisson* (λ),

$$\lambda = \lambda_n := \frac{1}{4}\log\frac{1+\sigma}{1-\sigma} - \frac{\sigma}{2}, \quad \sigma := \frac{\rho}{e^\rho - 1},$$

and ρ satisfies

$$\frac{\rho(e^\rho - 1)}{e^\rho - 1 - \rho} = 2c.$$

In particular,

$$\lim_{n\to\infty} \mathbf{Pr}(G_{n,cn}^{\delta\geq 2} \textit{ has a perfect matching}) = \begin{cases} e^{-\lambda}, & \textit{if } n \textit{ even,} \\ e^{-\lambda} + \lambda e^{-\lambda}, & \textit{if } n \textit{ odd.} \end{cases} \quad (3)$$

Thus the subgraph obtained by deletion of isolated odd cycles **whp** has a perfect matching The RHS in (3) is the limiting probability that the total number of isolated odd cycles is 0 (n even), or 1 (n odd). Notice that $c = 1$ corresponds to the random 2-regular (non-bipartite) graph, which typically has $\Theta(\log n)$ isolated cycles, both odd and even. Sure enough, the explicit term in the RHS of (3) approaches zero as $c \downarrow 1$, since $\lambda \to \infty$.

Theorem 2 does leave open the case where the number of edges $m = 2 + o(n)$ and so it is not quite as tight as Theorem 1.

Here is an interesting application of Theorem 2. Consider the Erdős-Rényi random graph $G(n,m)$, $m = cn$, for $\liminf c > 1/2$, i.e. the supercritical phase. By consecutive deletion of the vertices of degree 1 at most, we obtain a 2-core, the largest subgraph of $G(n,m)$ with minimum degree 2 at most. Let ν, μ stand for the number of vertices and the number of edges in the 2-core. Conditioned on ν, μ, and the vertex set, the 2-core of $G(n,m)$ is distributed as $G_{\nu,\mu}^{\delta\geq 2}$. Since **whp** μ, ν are of order n, and μ/ν is bounded away from 1, we see that **whp** the 2-core of the giant component of $G(n,m)$ has a perfect matching.

Among other things, the proof of Theorem 2 is based on an asymptotic analysis [1] of a matching algorithm initially discovered and studied by Karp and Sipser [14]. A related analysis for the bipartite graph $B_{n,m}^{\delta\geq 2}$ is considerably more than a technical extension of that in [1], basically because of some serious complications due to bipartiteness. It is shown in [11] that $B_{n,m}^{\delta\geq 2}$ has a perfect matching **whp**when $m = cn$, $c \geq 2$ constant.

To conclude our discussion, for integer $k \geq 1$, let graph G have property $\mathcal{A}_k$ if G contains $\lfloor k/2 \rfloor$ edge disjoint Hamilton cycles, and, if k is even, a further edge disjoint matching of size $\lfloor n/2 \rfloor$. Bollobás, Cooper, Fenner and Frieze [4] show that for $k \geq 2$, there exists a constant $c_k \leq 2(k+2)^3$ such that if $c \geq c_k$, $G_{n,cn}^{\delta\geq k+1}$ has property $\mathcal{A}_k$. Thus the current paper deals with the property $\mathcal{A}_1$ and proves a sharp result. It is reasonable to conjecture that the true value for c_k is $(k+1)/2$. Note that if $c = (k+1)/2$ and cn is integer then $G_{n,cn}^{\delta\geq k+1}$ is a random $(k+1)$-regular graph and this is known to have the property $\mathcal{A}_k$ **whp**, Robinson and Wormald [22], Kim and Wormald [15].

2. Enumerating some bipartite graphs.

In our probabilistic model, the sample space $\mathcal{B}_{n,m}^{\delta\geq k}$ is the set of all bipartite graphs on the bipartition $[n] + [n]$ with m edges, and the minimum degree at least k. The probability measure is uniform, i.e. each sample graph $B_{n,m}^{\delta\geq k}$ is assigned the same probability, $N_k(n,m)^{-1}$, where $N_k(n,m) = |\mathcal{B}_{n,m}^{\delta\geq k}|$. We will obtain a sharp asymptotic formula for $N_k(n,m)$, as a special case for the number of bipartite graphs meeting more general conditions on vertex degrees.

Let the ν_1-tuple $\mathbf{c} = (c_1, \ldots, c_{\nu_1})$ and the ν_2-tuple $\mathbf{d} = (d_1, \ldots, d_{\nu_2})$ of non-negative integers be given. Introduce $N_{\mathbf{c},\mathbf{d}}(\boldsymbol{\nu}, \mu)$, $\boldsymbol{\nu} = (\nu_1, \nu_2)$, the total number of bipartite graphs with μ edges, such that $a_i \geq c_i$, $(i \in [\nu_1])$, and $b_j \geq d_j$, $(j \in [\nu_2])$. Of course, $N_{\mathbf{c},\mathbf{d}}(\boldsymbol{\nu}, \mu) = 0$ if $\mu < \sum_i c_i$, or $\mu < \sum_j d_j$. So we assume that $\mu \geq \max\{\sum_i c_i, \sum_j d_j\}$

Define

$$G_{\mathbf{c}}(x) = \prod_{i\in[\nu_1]} f_{c_i}(x); \tag{4}$$

$$H_{\mathbf{d}}(y) = \prod_{j\in[\nu_2]} f_{d_j}(y), \tag{5}$$

where

$$f_t(z) = \sum_{\ell\geq t} \frac{z^\ell}{\ell!} = e^z - \sum_{\ell<t} \frac{z^\ell}{\ell!}. \tag{6}$$

The following estimates will be proved in Appendix A, along with some other lengthy computations.

Lemma 2.1. *Suppose that $\nu_1, \nu_2, \mu \to \infty$ are such that that $\nu_1, \nu_2 = O(\mu)$ and $\mu = O(\nu_i \log \nu_i)$, $i = 1, 2$. Let $G_{\mathbf{c}}(x)$ and $H_{\mathbf{d}}(y)$ be defined by (4) and (5).*

(i): *Suppose that $\mu^{-1} \leq_b r_1, r_2 = O(\log \mu)$. Then*

$$N_{\mathbf{c},\mathbf{d}}(\boldsymbol{\nu}, \mu) \sim \mu! \frac{G_{\mathbf{c}}(r_1)H_{\mathbf{d}}(r_2)}{(r_1r_2)^\mu} \left[e^{-\frac{1}{2}\mathbf{E}\lambda(\mathbf{Y})\mathbf{E}\lambda(\mathbf{Z})}\mathbf{Pr}(R=\mu)\mathbf{Pr}(S=\mu) + O(e^{-\Omega(\log^5 \mu)})\right], \tag{7}$$

$$N_{\mathbf{c},\mathbf{d}}(\boldsymbol{\nu}, \mu) \leq_b \mu! \frac{G_{\mathbf{c}}(r_1)H_{\mathbf{d}}(r_2)}{(r_1r_2)^\mu\sqrt{\nu_1\nu_2r_1r_2}} e^{-\frac{1}{2}\mathbf{E}\lambda(\mathbf{Y})\mathbf{E}\lambda(\mathbf{Z})}, \tag{8}$$

the last estimate holding without the condition $\nu_1, \nu_2, \mu \to \infty$, where $Y_i = Po(r_1; \geq c_i)$, $Z_j = Po(r_2; \geq d_j)$ are all independent, and $R = \sum_i Y_i$, $S = \sum_j Z_j$.

(ii): *Suppose also that $\max_i c_i = O(1)$, $\max_j d_j = O(1)$, and $\mu > \max\{\sum_i c_i, \sum_j d_j\}$. Then there exist (unique) positive roots ρ_1, ρ_2 of (100) and (101), and*

$$N_{\mathbf{c},\mathbf{d}}(\boldsymbol{\nu}, \mu) \sim \mu! \frac{G_{\mathbf{c}}(\rho_1)H_{\mathbf{d}}(\rho_2)}{(\rho_1\rho_2)^\mu} e^{-\frac{1}{2}\mathbf{E}\lambda(\mathbf{Y})\mathbf{E}\lambda(\mathbf{Z})} \cdot \mathbf{Pr}(R=\mu)\mathbf{Pr}(S=\mu), \tag{9}$$

where $Y_i = Po(\rho_1; \geq c_i)$, $Z_j = Po(\rho_2; \geq c_j)$.

Furthermore

$$\mathbf{Pr}(R=\mu) \sim \frac{1}{(2\pi\sum_i \mathbf{Var}(Y_i))^{1/2}}, \quad \text{or} \quad \mathbf{Pr}(R=\mu) \sim e^{-\sigma_1}\frac{\sigma_1^{\sigma_1}}{\sigma_1!},$$

dependent upon whether $\sigma_1 := \mu - \sum_i c_i$ approaches infinity or stays bounded, with the analogous formula for $\mathbf{Pr}(S=\mu)$.

Corollary 2.2. *Suppose $n = O(m)$, $kn < m = O(n \log n)$. Then*

$$N_k(n,m) \sim m! \left(\frac{f_k(\rho)^n \mathbf{Pr}\left(\sum_i Y_i = m\right)}{\rho^m}\right)^2 \exp\left(-\frac{n^2}{2m^2}\mathbf{E}^2[(Y)_2]\right), \tag{10}$$

where Y is Poisson $(\rho; \geq k)$ such that $\mathbf{E}Y = r = m/n$. *Note that*

$$\mathbf{E}[(Y)_2] = \begin{cases} r^2, & k = 0, \\ \rho r, & k = 1, . \\ \rho r/(1-e^{-\rho}), & k = 2. \end{cases} \tag{11}$$

Further

$$\mathbf{Pr}\left(\sum_i Y_i = m\right) \sim (2\pi n \mathbf{Var} Y)^{-1/2}, \qquad \text{if } m - kn \to \infty, \tag{12}$$

and

$$\mathbf{Pr}\left(\sum_i Y_i = m\right) \sim e^{-\sigma}\frac{\sigma^\sigma}{\sigma!} \qquad \text{if } m - kn > 0 \text{ is fixed.} \tag{13}$$

As we will see, these results are all we need to evaluate (bound) the probabilities arising in the proofs of Theorem 1 and Theorem 2. We will also need a crude upper bound for the fraction of bipartite graphs in question, with the maximum degree exceeding m^α. This bound is already implicit in the preceding analysis! Indeed, from (90), (92), (93), and the observation that the factor

$$\mathbf{Pr}^2\left(\sum_i Y_i = m\right) \exp\left(-\frac{n^2}{2m^2}\mathbf{E}^2[(Y)_2]\right) \tag{14}$$

in (10) is $\exp(-\Theta(\log^2 n))$, it follows that, for $\alpha' < \alpha < 1/3$, this fraction is $e^{-m^{\alpha'}}$ at most.

One is tempted to call this "overpowering both the conditioning and the fudge factor". Needless to say, this trick would work for the counts (fractions) of other graph classes, as long as the degrees restrictions are so severe that the probability that Y_i, Z_j meet them is negligible compared to the factor in (14).

3. Proof of Theorem 1

We will use Hall's necessary and sufficient condition for the existence of a perfect matching in a bipartite graph to prove (1).

The random graph $B_{n,m}^{\delta\geq 1}$ has no perfect matching iff for some $k \geq 2$ there exists a *k-witness*. A k-witness is a pair of sets $K \subseteq R, L \subseteq C$, or $K \subseteq C, L \subseteq R$, such that $|K| = k, |L| = k-1$ and $N(K) \subseteq L$. Here $N(K)$ denotes the set of neighbours of vertices in K. A k-witness is *minimal* if there does not exist $K' \subset K, L' \subset L$ such that (K', L') is a k'-witness, where $k' < k$. It is straightforward that if (K, L) is a minimal k-witness then every member of L has degree at least two in $B(K \cup L)$, the subgraph of $B_{n,m}^{\delta\geq 1}$ induced by $K \cup L$. Therefore $B(K \cup L)$ has at least $2(k-1)$ edges. We can restrict our attention to $k \leq n/2$ since for $k > n/2$ we can consider $\tilde{C} = C \setminus L, \tilde{R} = R \setminus K$. For $2 \leq k \leq n/2$, let $W_{n,k,\mu}$ denote the random number of minimal k-witnesses, such that $B(K \cup L)$ has μ edges, $\mu \geq 2(k-1)$. Actually, since $k \leq n/2$, we also have $\mu \leq m - n$.

(i) Suppose $m = O(n \log n)$ and $m \geq (1/3 + \epsilon) n \log n$, $\epsilon > 0$. Let us prove that **whp** $B_{n,m}^{\delta\geq 1}$ has no k-witnesses with $k \geq 3$, i.e.

$$\mathbf{Pr}\left(\sum_{k\geq 3, \mu\geq 2(k-1)} W_{n,k,\mu} = 0\right) \to 1, \quad n \to \infty.$$

It suffices to show that

$$\sum_{k\geq 3,\mu\geq 2(k-1)} E_{n,k,\mu}\to 0,\quad E_{n,k,\mu}:=\mathbf{E}W_{n,k,\mu}. \tag{15}$$

Let us bound $E_{n,k,\mu}$. For certainty, suppose that $K\subset R$, $L\subset C$. We can choose a pair (K,L) in $\binom{n}{k}\binom{n}{k-1}$ ways. (K,L) being a witness imposes the above listed conditions on degrees of the subgraph induced by $K\cup L$. If we delete the row set K, we get a remainder graph, which is a bipartite graph with bipartition (R',C), $R'=R\setminus K$; it has $m-\mu$ edges and every vertex in $R'\cup(C\setminus L)$ has degree 1 at least. We bound N_1, the total number of those subgraphs, and N_2 the total number of the remainder graphs using Lemma 2.1 (i), emphasizing the possibility to choose the corresponding parameters r_1,r_2 anyway we want. The product of these two bounds divided by the asymptotic expression for $N_1(n,m)$ in Corollary 2.2 provides an upper bound for the probability that (K,L) is a k-witness with μ edges. Multiplying this bound by $2\binom{n}{k}\binom{n}{k-1}$, we get a bound for $E_{n,k,\mu}$. To implement this program, we consider separately $k\leq m^\beta$ and $k\geq m^\beta$, where $\beta\in(0,1)$ will be specified in the course of the argument.

Let $k\leq m^\beta$. Pick $\alpha'<\alpha=(1-\beta)/2$. From the note following Corollary 2.2, with probability $1-e^{-m^{\alpha'}}$ at least, the maximum vertex degree in the uniformly random bipartite graph is m^α at most. So, backpedaling a bit, we will consider $\mu\leq m^\gamma$, $(\gamma:=(1+\beta)/2)$, only. To bound N_1 we use (85) with $r_1=\mu/k, r_2=\mu/(k-1)$, and to bound N_2 we use (8) with $r_1=r_2=\rho$. Here ρ is the parameter of Y_i in Corollary 2.2, the root of $xf_0(x)/f_1(x)=r$, $r:=m/n$, so that

$$\rho=r(1-e^{-\rho})<r,\quad \rho=r-\Theta(re^{-r}). \tag{16}$$

The r_i for N_1 seem natural, if one interprets them as parameters of Poissons approximating the vertex degrees that should add to μ on either side of the subgraph induced by $K\cup L$. Since k,μ are relatively small, $r_1=r_2=\rho$ should be expected to deliver a good enough bound for N_2. Most importantly, this choice does the job!

After cancellations and trivial tinkering, the resulting bound is

$$\begin{aligned} E_{n,k,\mu} \quad \leq_b \quad & \frac{m\binom{n}{k}\binom{n}{k-1}}{\mu(m-\mu)\binom{m}{\mu}}\cdot\rho^{2\mu}\left(\frac{k}{\mu}\right)^\mu\left(\frac{k-1}{\mu}\right)^\mu \\ \times\quad & \frac{f_1(\mu/k)^k f_1(\mu/(k-1))^{k-1}f_0(\rho)^{k-1}}{f_1(\rho)^{2k-1}} \\ \times\quad & \frac{\exp\left(-\frac{1}{2}\frac{(n-k)\mathbf{E}(Y)_2}{m-\mu}\cdot\frac{(k-1)\rho^2+(n-k+1)\mathbf{E}(Y)_2}{m-\mu}\right)}{\exp\left(-\frac{1}{2}\frac{(n\mathbf{E}(Y)_2)^2}{m^2}\right)}. \end{aligned} \tag{17}$$

Some explanation: $k-1$ vertices from L in the remaining graph have degrees not bounded away from zero, whence the factor $f_0(\rho)^{k-1}=e^{\rho(k-1)}$ in the second line, and $k-1$ usual Poissons (ρ), each with the second factorial moment equal ρ^2, contributing $(k-1)\rho^2$ in the last line. Also, we have used $f_1(\mu/(k-1))^{k-1}$ where we could have used the smaller $f_2(\mu/(k-1))^{k-1}$.

The last line fraction is of order $O(1)$, as $\mathbf{E}(Y)_2 = \Theta(\rho^2)$. Further, since $\log f_1(z) = \log(e^z - 1)$ is concave,

$$k \log f_1\left(\frac{\mu}{k}\right) + (k-1)\log f_1\left(\frac{\mu}{k-1}\right) - (2k-1)\log f_1(\rho) \leq$$
$$(2k-1)\left(\log f_1\left(\frac{2\mu}{2k-1}\right) - \log f_1(\rho)\right) \leq (2k-1)(\log f_1)'(\rho)\left(\frac{2\mu}{2k-1} - \rho\right)$$
$$\leq 2\mu - (2k-1)\rho + 3\mu e^{-\rho}. \quad (18)$$

Using the last observations and $\mu! = \Theta(\mu^{1/2}(\mu/e)^\mu)$, we see that $E_{n,k,\mu}$ is of order $E^*_{n,k,\mu}$ at most, where

$$E^*_{n,k,\mu} = \frac{n^{2k-1}e^{-k\rho}}{k!(k-1)!} \cdot \frac{(m-\mu)!}{m!\mu!} k^{2\mu}\rho^{2\mu} \cdot \exp(3\mu e^{-\rho}). \quad (19)$$

First, since $2(k-1) \leq \mu \leq m^\gamma$,

$$\frac{E^*_{n,k,\mu+1}}{E^*_{n,k,\mu}} = \frac{k^2\rho^2}{(m-\mu)(\mu+1)} = O(m^{-1/3}\log^2 n),$$

so that

$$\sum_{2(k-1)\leq\mu\leq m^\gamma} E^*_{n,k,\mu} \sim E^*_{n,k,2(k-1)}.$$

Second

$$\frac{E^*_{n,k+1,2k}}{E^*_{n,k,2(k-1)}} = \frac{n^2(k+1)^{4k}\rho^4 e^{-\rho}}{k(k+1)(2k-1)2k(m-2k+1)(m-2k+2)k^{4(k-1)}}$$
$$\leq_b \frac{n^2}{m^2}\rho^4 e^{-\rho} = O(\rho^2 e^{-\rho}).$$

Therefore

$$\sum_{3\leq k\leq m^\gamma} E^*_{n,k,2(k-1)} \sim E^*_{n34} \leq_b \frac{n^5}{m^4}\rho^4 e^{-3\rho} \sim ne^{-3m/n} = O(n^{-3\epsilon}),$$

as $m \geq (1/3+\epsilon)n\log n$. In summary,

$$\sum_{\substack{3\leq k\leq m^\gamma \\ 2(k-1)\leq\mu\leq m^\gamma}} E_{n,k,\mu} = O(n^{-3\epsilon}). \quad (20)$$

Consider now $k \geq m^\beta$. This time we use (7) not only for N_1 but for N_2 as well, using for the latter $r_1 = (\rho n - \mu)/(n-k)$ and $r_2 = (\rho n - \mu)/n$. That the latter r_i are positive follows from $\mu \leq m - n/2$ and (16). (For μ, k not being relatively small anymore, the count N_2 of the remaining graphs would hardly be well bounded via the previous choice $r_1 = r_2 = \rho \sim m/n$. What we have chosen turns out to be a working compromise between that old choice and the "naive"

$r_1 = (m-\mu)/(n-k)$, $r_2 = (m-\mu)/n$.) The resulting bound is

$$\begin{aligned} E_{n,k,\mu} &\leq_p \exp\left(\frac{(n\mathbf{E}(Y)_2)^2}{2m^2}\right) E'_{n,k,\mu}, \\ E'_{n,k,\mu} &= \rho^{2m} \frac{\binom{n}{k}^2}{\binom{m}{\mu}} \cdot \left(\frac{k}{\mu}\right)^{2\mu} \left(\frac{n-k}{\rho n-\mu}\right)^{m-\mu} \left(\frac{n}{\rho n-\mu}\right)^{m-\mu} \\ &\times \quad \frac{f_1(\frac{\mu}{k})^k f_1(\frac{\mu}{k-1})^{k-1} f_1(\frac{\rho n-\mu}{n-k})^{n-k} f_1(\frac{\rho n-\mu}{n})^{n-k+1} f_0(\frac{\rho n-\mu}{n})^{k-1}}{f_1(\rho)^{2n}}. \end{aligned} \tag{21}$$

(We use the notation $a_n \leq_p b_n$ to indicate that a_n/b_n is polynomially large, at most.) Using again convexity of $\log f_1(z)$ and denoting $h = (k-1)(\rho n-\mu)/n$, we obtain that the logarithm of the last line fraction is less than

$$2n \log f_1\left(\rho - \frac{h}{2n}\right) - 2n\log f_1(\rho) + h \leq -h((\log f_1)'(\rho) - 1) = -\frac{h}{e^\rho - 1} \leq 0.$$

Thus the fraction is bounded, 1 at most, like its counterpart for $k \leq m^\beta$. (Our search for the proper r_1, r_2 was driven, in fact, by desire to make that fraction bounded again!)

Introduce $x = k/n$, $y = \mu/n$. Using the Stirling formula for factorials, we obtain easily then that

$$E'_{n,k,\mu} \leq_p \exp(nH(x,y)),$$

where

$$H(u,v) = 2r\log\rho + 2H(u) - rH(v/r) + 2v\log u/v + (r-v)\log\frac{1-u}{(\rho-v)^2},$$

($u \in (x_n, 1/2]$, $v \in (0,\rho)$), $x_n := \frac{m^\beta}{n}$, and

$$H(w) = w\log 1/w + (1-w)\log 1/(1-w).$$

It follows that

$$\begin{aligned} H_v(u,v) &= \log\frac{u^2(\rho-v)^2}{v(r-v)(1-u)} - 2\frac{\rho-r}{\rho-v}, \\ H_{vv}(u,v) &= \frac{1}{r-v} - \frac{2}{\rho-v} - \frac{1}{v} - \frac{2(\rho-r)}{(\rho-v)^2} < 0. \end{aligned}$$

So $H_v(u,v)$ decreases with v, and $H_v(u,0+) = \infty$, $H_v(u,\rho-) = -\infty$. Hence, given u, $H(u,v)$ attains its maximum at a unique root $v(u)$ of the equation

$$\frac{(\rho-v)^2u^2}{(r-v)(1-u)v} = \exp\left(\frac{2(\rho-r)}{\rho-v}\right). \tag{22}$$

By (16), $\rho < r$ and $\rho - r = O(re^{-r}) \to 0$; so we should expect $v(u)$ to be close to $v^*(u)$, the root of (22) with ρ replaced by r, i.e.

$$v^*(u) = \frac{u^2}{1-u+u^2}r \leq \frac{r}{3},$$

($u \leq 1/2$). Careful computations reveal that

$$H^*(u, v^*(u)) = r\log(1-u+u^2) + 2H(u),$$

where $H^*(u,v)$ is obtained from $H(u,v)$ by replacing ρ with r. Furthermore, as the RHS of (22) is $1+O(e^{-r})$, it can be shown that

$$v(u) = v^*(u)(1+O(e^{-r})).$$

In this setting, strictly speaking, $v(u)$ is also a function of ρ, and so is $H(u,v(u))$, both explicitly and implicitly, via $v(u)$. Since $H_v(u,v(u))=0$, the derivative of $H(u,v(u))$ with respect to ρ is just the partial derivative, which is

$$\frac{2r}{\rho} - \frac{2(r-v(u))}{\rho - v(u)} = \frac{2v(u)(\rho - r)}{\rho(\rho - v(u))} = O(e^{-r});$$

therefore

$$H(u,v(u)) = H^*(u,v^*(u)) + O(e^{-r}(r-\rho)) = H^*(u,v^*(u)) + O(re^{-2r}).$$

Using

$$\log(1-u+u^2) \le -u/2,\ (u\le 1/2), \quad (1-u)\log(1-u)^{-1} \le u,$$

we see that, for $u\in(m^\beta/n, 1/2]$ (and $m \ge (1/3+\epsilon)n\log n$),

$$\begin{aligned} H^*(u,v^*(u)) &\le -u(r/2-2-2\log(1/u)) \\ &\le -u((1/6+\epsilon/2)\log n - 2(1-\beta)\log n + O(\log\log n)) \\ &\le -cu\log n, \end{aligned}$$

where $c=c(\beta)>0$ if $\beta > 11/12-\epsilon/4$. So, for this choice of β,

$$\begin{aligned} H(u,v(u)) &= H^*(u,v^*(u)) + O(r^{-2r}) \\ &\le -cm^\beta n^{-1}\log n + O(e^{-2m/n}\log n) \\ &\le -m^\beta n^{-1}. \end{aligned}$$

This inequality shows that

$$E'_{n,k,\mu} \le_p \exp(-m^\beta) \Rightarrow E_{n,k,\mu} \le \exp(-0.5m^\beta),$$

as the fudge factor in (21) is only $\exp(O(\log^2 n))$. Consequently

$$\sum_{\substack{m^\beta \le k \le n/2 \\ \mu \le m}} E_{n,k,\mu} \le \exp(-0.4m^\beta). \tag{23}$$

Combining (20) and (23) we obtain

$$\sum_{\substack{3\le k\le m^\beta \\ 2(k-1)\le\mu\le m^\gamma}} E_{n,k,\mu} + \sum_{\substack{m^{1/3}<k\le n/2 \\ \mu\le m}} E_{n,k,\mu} = O(n^{-3\epsilon}),$$

so that

$$\mathbf{Pr}\left(\sum_{k\ge 3}\sum_{\mu\ge 2(k-1)} W_{n,k,\mu} = 0\right) = 1 - O(n^{-3\epsilon}). \tag{24}$$

(ii) Turn now to the 2-witnesses. From (19), it follows that

$$E_{n,2,1} \le_b \frac{n^3e^{-2\rho}\rho^4}{m^2} = O(n\log^2 ne^{-2m/n}) = O(e^{-c_n}), \tag{25}$$

with c_n defined by the notation

$$m = \frac{n}{2}(\log n + 2\log\log n + c_n).$$

Case 1 $c_n \to \infty$.
Assume first that $m \le 2n \log n$. Then (25) shows that, with probability more than $1 - e^{-c_n} \to 1$, there are no 2-witnesses. By (25), with probability $1 - O(n^{-1/2})$ at least, there are no 3-witnesses either. Thus, with probability approaching 1, there exists a perfect matching.

If $m > 2n \log n$ then **whp** $\delta(B_{n,m}) \ge 1$ and $B_{n,m}$ has a perfect matching. The result in this case follows immediately.

Case 2: $c_n \to c \in (-\infty, \infty)$.
We want to prove that $W_{n,2,1}$, the number of 2-witnesses, is, in the limit, Poisson $(e^{-c}/4)$. We do so via the factorial moments method. To evaluate $\mathbf{E}(W_{n,2,1})$ sharply, we notice that in this case $N_1 = 2\binom{n}{2}n$ exactly, and for N_2 we use the part (ii) of Lemma 1 with $r_1 = r_2 = \rho$. So (compare to (17)

$$E_{n,2,1} \sim \frac{2n\binom{n}{2}}{m^2} \cdot \rho^4 e^{-2\rho} \sim \frac{1}{4}e^{-c_n} \to \lambda := \frac{1}{4}e^{-c}. \tag{26}$$

We need to show that, for each $t \ge 2$

$$\lim_{n\to\infty} \mathbf{E}(W_{n,2,1})_t = \lambda^t, \quad \lambda = \frac{1}{4}e^{-c}.$$

To simplify our task, let us consider instead $W^*_{n,2,1}$, the total number of vertex-disjoint 2-witnesses. The difference $W_{n,2,1} - W^*_{n,2,1}$ is **whp** (24) $\mathcal{W}_n$ at most, where $\mathcal{W}_n$ is the total number of subset pairs (K, L), $K \subset R$, $L \subset C$, or $K \subset C$, $L \subset R$, such that $|K| = 3$, $|L| = 1$, and $L = N(K)$. Analogously to (17),

$$\mathbf{E}\mathcal{W}_n \le_b \frac{n\binom{n}{3}\rho^6}{\binom{m}{3}} e^{-3\rho} = O(n^{-1/2}).$$

Therefore, $W_{n,2,1} = W^*_{n,2,1}$ with probability $1 - O(n^{-1/2})$ at least, and it suffices to show that

$$\lim_{n\to\infty} \mathbf{E}(W^*_{n,2,1})_2 = \lambda^t, \quad t \ge 1. \tag{27}$$

This is obviously true for $t = 1$. Let $t \ge 2$. Combinatorially, $(W^*_{n,2,1})_t$ is the total number of ordered t-tuples of (vertex-disjoint) 2-witnesses. Given $r + s = t$, let us compute E_{rs}, the expected number of t-tuples containing r "2 rows, 1 column" (first kind) witnesses, and s "2 columns, 1 row" (second kind) witness. The r vertex-disjoint first kind of witnesses can be chosen in $\binom{n}{2r}\binom{n}{r}(2r-1)!!r!$ ways. (Indeed, once $2r$ rows and r columns are selected, we pair the rows in $(2r-1)!!$ ways and assign the formed r pairs to r columns in $r!$ ways.) Given any such choice, the s 2-nd witnesses, disjoint among themselves *and* from the r first kind witnesses, can be chosen in $\binom{n-r}{2s}\binom{n-2r}{s}(2s-1)!!s!$ ways. There are $t! = (r+s)!$ ways to order all $r + s$ witnesses. Hence $N_1(r, s)$, the total number of the ordered t-tuples of the "alleged" witnesses, is given by

$$\begin{aligned} N_1(r,s) &= \binom{n}{2r}\binom{n}{r}(2r-1)!!r!\binom{n-r}{2s}\binom{n-2r}{s}(2s-1)!!s!(r+s)! \\ &\sim \binom{t}{r}\frac{n^{3t}}{2^t}. \end{aligned}$$

Deleting $2r$ rows and $2s$ columns involved in first kind and second kind witnesses respectively produces a bipartite graph with $m - 2t$ edges that meets the following conditions. (a) Every row (column) vertex not involved in the s 2-nd (in the r

first) kind witnesses has degree at least 1. (b) No edge can be added to one of (just deleted) $r+s$ 2-witnesses to form a pair (K, L), such that $|K|=3, |L|=1$, $K \subset R, L \subset C$, or $K \subset C, L \subset R$, and $N(K)=L$. (This condition is necessary and sufficient for the $(r+s)$ 2-witnesses to be disjoint from all other 2-witnesses.) Denote the total number of such graphs by $N_2(r,s)$. Clearly $N_2(r,s) \leq \mathcal{N}_2(r,s)$, where $\mathcal{N}_2(r,s)$ is the total number of bipartite graphs with the condition (b) dropped. Using (7) with $r_1 = r_2 = \rho$, we have

$$\begin{aligned}\mathcal{N}_2(r,s) &\sim (m-2t)! \cdot \frac{(e^\rho - 1)^{2n-3t} e^{t\rho}}{\rho^{2(m-2t)}} \\ &\times \left(e^{-\frac{\mathbf{E}\lambda(\mathbf{Y})\mathbf{E}\lambda(\mathbf{Z})}{2}} \cdot \mathbf{Pr}(R=m-2t)\mathbf{Pr}(S=m-2t) + O(e^{-\log^5 m})\right).\end{aligned}$$

Here $R = \sum_{i=1}^{n-2r} Y_i$, $S = \sum_{j=1}^{n-2s} Z_j$, $Y_i, Z_j = \text{Po}(\rho; \geq 1)$ for $1 \leq i \leq n-2r-s$, $1 \leq j \leq n-r-2s$, and $Y_i, Z_j = \text{Po}(\rho)$ for $n-2r-s < i \leq n-2r$, $n-r-2s < j \leq n-2s$. Using (A.1) for both local probabilities, we obtain that the second line in the above formula is asymptotic to

$$\exp\left(-\frac{(n\mathbf{E}(Y_1)_2)^2}{2m^2}\right) \cdot \frac{1}{2\pi n \mathbf{Var}(Y_1)}.$$

Thus

$$\frac{\mathcal{N}_2(r,s)}{N_1(n,m)} \sim m^{-2t} \rho^{4t} (e^\rho - 1)^{-3t} e^{t\rho} \sim \left(\frac{e^{-c_n}}{4n^3}\right)^t. \tag{28}$$

Now

$$\mathcal{N}_2(r,s) - N_2(r,s) \leq r(n-2r-s) N_2^{(1)}(r,s) + s(n-r-2s) N_2^{(2)}(r,s);$$

here $N_2^{(1)}(r,s)$ ($N_2^{(2)}(r,s)$ resp.) is the total number of the remaining graphs, such that a particular row (column resp.) vertex is incident to a single column (row resp.) vertex, which happens to be one of the vertices from r first kind (s second kind resp.) witnesses. Consider $N_2^{(1)}(r,s)$. Deleting that row we get a graph with one less number of row vertices and one less number of edges. So, using (7) with $r_1 = r_2 = \rho$ and $e^\rho - 1 \sim e^\rho$, we obtain that

$$\begin{aligned}\frac{N_2^{(1)}(r,s)}{N_1(n,m)} &\leq_b \frac{\mathcal{N}_2(r,s)}{N_1(n,m)} \frac{\rho^2}{me^\rho}, \\ \frac{N_2^{(2)}(r,s)}{N_1(n,m)} &\leq_b \frac{\mathcal{N}_2(r,s)}{N_1(n,m)} \frac{\rho^2}{me^\rho}.\end{aligned}$$

Therefore

$$\frac{\mathcal{N}_2(r,s) - N_2(r,s)}{N_1(n,m)} \leq_b \frac{n\rho^2}{me^\rho} \leq \rho e^{-\rho} \to 0.$$

Collecting the pieces, we obtain that

$$E_{rs} \sim \binom{t}{r} \frac{n^{3t}}{2^t} \cdot \left(\frac{\rho^4}{m^2 e^{2\rho}}\right)^t \to \binom{t}{r} \left(\frac{e^{-c}}{8}\right)^t,$$

i.e.

$$E(W_{n,2,1}^*)_t \to \sum_{r=0}^{t} \binom{t}{r} \left(\frac{e^{-c}}{8}\right)^t = \lambda^t.$$

Thus $W^*_{n,2,1}$ is in the limit Poisson (λ), and then so is $W_{n,2,1}$. Consequently, a perfect matching exists with the limiting probability equal

$$\lim_{n\to\infty} \mathbf{Pr}(W_{n,2,1}=0) = e^{-\lambda} = \exp\left(-\frac{e^{-c}}{4}\right).$$

Case 3: $c_n \to -\infty$, $m > n$.
3a: $m \geq (\frac{1}{3}-\epsilon)n\log n$.
In this case, after with trivial modifications in the above derivation,

$$\mathbf{E}(W^*_{n,2,1})_2 \sim \left(\frac{e^{-c_n}}{4}\right)^2 \to \infty,$$

and, by Chebyshev's inequality,

$$\lim_{n\to\infty} \mathbf{Pr}(W^*_{n,2,1} > 0) = 1.$$

So, **whp**, a perfect matching does not exist.

3b: $m \leq (1/3-\epsilon)n\log n$, $m-n\to\infty$.
Note that $n\rho \to \infty$. Let X_n denote the total number of isolated trees with 2 row vertices and 1 column vertex. ($X_n > 0$ implies that there is no perfect matching.) If the X_n trees are deleted, the remaining graph has $n-2t$ row vertices, $n-t$ column vertices, and $m-2t$ edges, and every vertex has degree 1, at least. Evaluating the number of such graphs by (7), we easily obtain

$$\begin{aligned}\mathbf{E}(X_n)_t &\sim \binom{n}{2t}\binom{n}{t}(2t-1)!!(t!)^2\frac{(m-2t)!}{m!}\frac{\rho^{4t}}{(e^\rho-1)^{3t}}\\ &\sim \left(\frac{m\rho e^{-3\rho}}{2}\right)^t,\end{aligned}$$

using the definition of ρ for the second equality. Also from this definition, $\rho \sim 2(m-n)/n$ if $\rho\to 0$, and $\rho \leq m/n$ always. So, if $\rho \to 0$,

$$m\rho e^{-3\rho} \sim 2m(m-n)/n \geq 2(m-n) \to \infty,$$

and, if $\lim \rho > 0$, then

$$m\rho e^{-3\rho} \geq \rho n e^{-3m/n} \geq \rho n^{3\epsilon} \to \infty.$$

Thus

$$\mathbf{E}(X_n)\to\infty, \quad \mathbf{E}(X_n)_2 \sim \mathbf{E}^2(X_n),$$

so that (Chebyshev's inequality) $\mathbf{Pr}(X_n > 0) \to 1$. That is, **whp** there is no perfect matching.

3c: $\sigma := m-n > 0$ is fixed.
If we form $4n-3m$ isolated edges, the remaining $3(m-n)$ row vertices and $3(m-n)$ column vertices can be partitioned into $2(m-n)$ trees of size 3, half of the trees each containing 2 row vertices and 1 column vertex, and another half - 1 row vertex and 2 column vertices. The total number of such bipartite graphs is

$$\begin{aligned}N^*(n,m) &= \binom{n}{4n-3m}^2(4n-3m)!\cdot\left[\binom{3(m-n)}{2(m-n)}(2(m-n)-1)!!(m-n)!\right]^2\\ &\sim \frac{(n!)^2}{(n-3\sigma)!2^{2\sigma}(\sigma!)^2}.\end{aligned} \tag{29}$$

As for $N_1(n,m)$, the total number of all bipartite graphs, by Corollary 2.2 and (109), it is given by

$$N_1(n,m) \sim m! \left(\frac{f_1(\rho)^n e^{-(m-n)}(m-n)^{m-n}/(m-n)!}{\rho^m} \right)^2 \exp\left(-\frac{(n\mathbf{E}(Y)_2)^2}{2m^2} \right),$$

where, using the definition of ρ,

$$\rho = \frac{2\sigma}{n}\left(1 - \frac{2\sigma}{3n} + O(\sigma^2/n^2)\right).$$

So, after simple computations,

$$N_1(n,m) \sim \frac{m! n^{2\sigma}}{2^{2\sigma}(\sigma!)^2}. \tag{30}$$

Since, for fixed σ,

$$\frac{(n!)^2}{(n-3\sigma)!} \sim m! n^{2\sigma},$$

it follows from (29) and (30) that, with probability approaching 1, the random graph has $2\sigma > 0$ isolated trees of size 3, thus no perfect matching exists.

Theorem 1 is proved completely. □

4. Proof of Theorem 2

We notice upfront that, for $\liminf m/n > 1$, **whp** the random graph $G_{n,m}^{\delta\geq 2}$ has an almost perfect matching, in a sense that

$$\lim_{n\to\infty} \mathbf{Pr}(\mu^*(G_{n,m}^{\delta\geq 2}) < \lfloor n/2 \rfloor - n^{0.2+\beta}) = 0, \tag{31}$$

for every $\beta > 0$. This follows from the analysis of Karp-Sipser matching algorithm (its Phase 2, to be precise) [14], given in [1]. The analysis of [1] shows that at most $n^{0.2+\beta}$ vertices that are left at the end of Phase 1, are not covered by the matching constructed in Phase 2. The random graph at the end of Phase 1 is uniform, subject to the number of vertices ν, and edges μ and $\delta \geq 2$. The analysis is robust with respect to these parameters and implies (31). So, loosely speaking, our task is to get rid of the term $-n^{0.2+\beta}$.

First we prove (Lemma 4.1) that, analogously to the bipartite case, a graph with minimum vertex degree 2 at least, which has no perfect matching, must contain a certain (witness) subgraph. This result is based on the ideas of Edmonds' matching algorithm, [17] (Section 7, Exer. 34). Conditioned by the proof of Theorem 1, one would expect to be able to show that **whp** the random graph in question does not contain such a witness. Indeed, our next Lemma 4.2 rules out (**whp**) all the witnesses of size ϵn at most, $\epsilon > 0$ being sufficiently small. As in the proof of Theorem 2, our argument consists of showing that the expected count of "small" witnesses is exponentially small. However, we have not being able to extend the proof to larger witnesses. Apparently, for sparse graphs in question, the expected count of witnesses can be exponentially large, even though the count itself is zero **whp**.

Not everything is lost however! As the next step we show (Lemma 4.3) that **whp** either the random graph has a perfect matching, or there are an^2 pairs of disjoint vertices such that adding anyone of these pairs to the edge set of the subgraph, obtained by deletion of isolated odd cycles, increases the maximum matching number. This fact and a coupling device, that allows to relate, approximately, the random graphs $G^{\delta\geq 2}_{n,m_1}$ and $G^{\delta\geq 2}_{n,m}$ to each other, enable us to prove certain monotonicity of the distribution of $\mu^*(G^{\delta\geq 2}_{n,m})$ as the function of m, Lemma 4.8. We combine this monotonicity property of the maximum matching number and (31) to complete the proof of Theorem 2.

4.1. Step 1. Profiling and counting the witnesses.

We begin with

Lemma 4.1. *Let $G = (V, E)$ be a graph with $\delta(G) \geq 2$, and with no isolated odd cycles, which does not have a perfect matching, i.e. $\mu^*(G) < \lfloor |V|/2 \rfloor$. For every $x \in V$ which is not covered by at least one maximum matching, there exists a witness $(K, L) = (K(x), L(x))$, $K, L \subset V$, such that*

- **(i):** $|K| = |L| + 1$;
- **(ii):** $N_G(K) = L, \qquad (N_G(K) = \{w \notin K : \exists (v, w) \in E_G, v \in K\})$;
- **(iii):** $|E_G(K \cup L)| \geq |K| + |L| + 1$;
- **(iv):** *each $v \in L$ has at least 2 neighbours in K;*
- **(v):** *for every $y \in K$, there exists a maximum matching that does not cover y;*
- **(vi):** *adding any (x, y), $y \in K(x)$, to E increases the size of a maximum matching.*

Proof. Let $x \in V$ and let M be a maximum matching which does not cover x. Since $\mu^*(G) < \lfloor |V|/2 \rfloor$, there exists $s \neq x$ which is also left uncovered by M. Now let T be a tree of maximal size which is rooted at s and such that for each $v \in T$, the path from s to v in T is alternating with respect to M. Let K, L be the set of vertices at even and odd distance respectively from s in T. For every $y \in K$, we can switch edges on the even path from the root to y to obtain another maximum matching that does not cover y. Furthermore, no leaf of T is in L, since otherwise switching edges along the odd- length path to such a leaf we would have increased the size of the matching. Therefore all the vertices of T, except s, are covered by M. Next, if a neighbor u of a vertex from K is not in $K \cup L$, then u must be covered by M, which contradicts maximality of T. Therefore the pair (K, L) meets all the conditions, except possibly (iii). Using the fact that all the leaves of T are from K and that their neighbors must be in $K \cup L$, *and* $\delta(G) \geq 2$, we can assert that

$$|E_G(K \cup L)| \geq |E(T)| + 1 = |K| + |L|.$$

But if $|E_G(K \cup L)| = |K| + |L|$ then T consists of two even-length path, sharing the root s only, with the leaves forming an edge in G. Thus s has degree 2 in G, and $K \cup L$ induces an odd cycle in G. As there are no isolated odd cycles in G, there must be some edge (v, w), $v \in L$, $w \notin K \cup L$. Since v is covered by M, $(v, w) \notin M$ and, for some $x \notin K \cup L$, we have $(w, x) \in M$. It is easy to see how to alter M solely on $E(K \cup L)$ to obtain a maximum matching M' and the corresponding tree T' rooted at v instead of s. (Draw a picture!) The degree of v in T' is at least 3, so T' has at least 3 leaves and, for the corresponding $K = K(T')$, $L = L(T')$, the condition (iii) is satisfied, too. □

Turn to $G^{\delta\geq 2}_{n,m}$. Given $\epsilon > 0$, let $\mathcal{A}_n(\epsilon)$ denote the event that there exist K, L satisfying (i)–(iv), and such that $|K| \leq \epsilon n$. The following lemma implies that **whp** witnesses must be large.

Lemma 4.2. *Let* $\liminf m/n > 1$. *There exists an* $\epsilon > 0$ *such that* $\mathbf{Pr}(\mathcal{A}_n(\epsilon)) = O(n^{-1})$.

Proof. First of all, using $\liminf m/n > 1$, we have, [21]: $N(n,m)$ the total number of graphs with minimum degree at least 2 is asymptotic to

$$N_0(n,m) = \frac{(2m-1)!!}{\sqrt{2\pi n \mathrm{Var} Z}} \cdot \frac{f_2(\rho)^n}{\rho^{2m}} \cdot \exp(-\hat{\rho}/2 - \hat{\rho}^2/4). \tag{32}$$

Here $\rho, \hat{\rho}$ satisfy

$$\frac{\rho f_1(\rho)}{f_2(\rho)} = \frac{2m}{n}, \quad \hat{\rho} = \frac{\rho f_0(\rho)}{f_1(\rho)}, \tag{33}$$

i.e. ρ is bounded away from 0 and ∞, and Z is Poisson(ρ), conditioned on $Z \geq 2$. In fact, for all $a, b, x > 0$,

$$N(a,b) \leq c^* \frac{(2b-1)!!}{\sqrt{nx}} \cdot \frac{f_2(x)^a}{x^{2b}}, \tag{34}$$

where c^* does not depend on a, b, x. (The attentive reader certainly notices direct analogy between these formulas and their counterparts for the bipartite case in Section 2.) The independent copies $Z_1, \dots, Z_n$ of Z provide an approximation to $\mathbf{deg}(\Gamma)$, the degree sequence of the random graph Γ, in the following sense:

$$\mathbf{Pr}(\mathbf{deg}(\Gamma) \in B) = O\big(n^{1/2}\mathbf{Pr}(\mathbf{Z} \in B)\big), \tag{35}$$

uniformly for all sets B of n-tuples. Consequently, if B is such that $\mathbf{Pr}(\mathbf{Z} \in B)$ is $O(n^{-b})$ for some $b > 1/2$, then $\mathbf{Pr}(\mathbf{deg}(\Gamma) \in B) = O(n^{-(b-1/2)})$, which goes to zero, too! A particular event B, which will come in handy, is defined as follows. Let $d(j) = d(j, \Gamma)$ denotes the j-th largest degree of Γ. Pick $a > e^{5+\rho}(h(\rho)+1)^2$ where $h(\rho) = \frac{2}{f_2(\rho)}$ and define $\ell(n,j) = \lceil \log \frac{e^a n}{j} \rceil$. Let us show that

$$\mathbf{Pr}(\exists j \in [1,n] : d(j) > \ell(n,j)) = O(n^{-1}). \tag{36}$$

To prove this, consider first $Z(j)$, the j-th largest among $Z_1, \dots, Z_n$. Clearly

$$\begin{aligned}
\mathbf{Pr}(Z(j) > \ell(n,j)) &\leq \binom{n}{j} \mathbf{Pr}^j(Z_1 > \ell(n,j)) \\
&\leq \exp\left(j \log \frac{en}{j} + j \log \mathbf{Pr}(Z_1 \geq \ell(n,j))\right),
\end{aligned}$$

and, using the definition of $\ell(n,j)$ and a,

$$\begin{aligned}
\mathbf{Pr}(Z_1 > \ell(n,j)) &\leq h(\rho) \frac{\rho^{\ell(n,j)}}{\ell(n,j)!} \leq \exp\left(\log h(\rho) - \ell(n,j) \log \frac{\ell(n,j)}{e\rho}\right) \\
&\leq \exp\left(-a \log \frac{e^a n}{j}\right).
\end{aligned}$$

Consequently

$$\mathbf{Pr}(Z(j) > \ell(n,j)) \leq \exp\left(-j(a-1)\log \frac{en}{j}\right),$$

so that

$$\mathbf{Pr}(\exists j \in [1,n] : Z(j) > \ell(n,j)) \leq \sum_{j=1}^{n} \mathbf{Pr}(Z(j) > \ell(n,j)) = O(n^{-2}),$$

whence the probability in (36) is $O(n^{-3/2})$. Now, for a given vertex subset S,

$$\sum_{j\in S} d_j \le \sum_{j=1}^{|S|} d(j),$$

and on the event in (36)

$$\sum_{j=1}^{s} d(j) \le \sum_{j=1}^{s} \left\lceil \log \frac{e^a n}{j} \right\rceil \le (2+a)s + s\log\frac{n}{s}.$$

We conclude that

$$\mathbf{Pr}\big(\exists S\subset [n] : \sum_{j\in S} d_j > (2+a)|S| + |S|\log(n/|S|)\big) = O(n^{-1}). \tag{37}$$

This bound will be needed shortly.

Now, given $k\in[2,\epsilon n]$, let T_{μ,ν,ν_1} denote the total number of pairs (K,L) consisting of disjoint subsets $K,L\subset[n]$ such that $|K|=k$, $|L|=k-1$, **(i)–(iv)** hold and μ,ν,ν_1 are given by

$$|E(K)|+|E(L)| = \mu,$$
$$|\{(u,w)\in E(\Gamma) : u\in K, w\in L\}| = \nu,$$
$$|\{(u,w)\in E(\Gamma) : u\in L, w\in (K\cup L)^c\}| = \nu_1.$$

Note that by **(iii)**

$$\nu+\mu\ge 2k. \tag{38}$$

We want to show that

$$\mathbf{Pr}\left(\sum_{2\le k\le \epsilon n}\ \sum_{\mu,\nu,\nu_1} T_{\mu,\nu,\nu_1} > 0\right) = O(n^{-1}),$$

provided that $\epsilon>0$ is sufficiently small. By 37, we may and will confine ourselves to μ,ν,ν_1 such that

$$\mu,\nu,\nu_1 \le A(k+k\log(n/k)), \tag{39}$$

for a large enough constant A. All we need to show is that

$$\sum_{k\le\epsilon n}\ \sum_{\substack{\mu,\nu,\nu_1:\\ (39)\text{ holds}}} E_{\mu,\nu,\nu_1} = O(n^{-1}),\quad E_{\mu,\nu,\nu_1} := \mathbf{E}(T_{\mu,\nu,\nu_1}). \tag{40}$$

By symmetry,

$$E_{\mu,\nu,\nu_1} = \binom{n}{k,k-1,n-2k+1} P_{\mu,\nu,\nu_1}, \tag{41}$$

where P_{μ,ν,ν_1} is the probability that the subsets $K^*=\{1,\dots,k\}$ and $L^*=\{k+1,\dots,2k-1\}$ form such a pair. To bound this probability we need to bound N_{μ,ν,ν_1}, the total number of graphs in question in which the pair $\{K^*,L^*\}$ has the prescribed properties.

Let $(\delta_j)_{j\in K^*}$, $(\delta_j)_{j\in L^*}$ and $(\delta_j)_{j\in(K^*\cup L^*)^c}$ be the degree sequences for subgraphs $G(K^*)$, $G(L^*)$ and $G((K^*\cup L^*)^c)$ respectively. For $j\in K^*$ ($j\in L^*$ resp.) let Δ_j denote the total number of neighbors of j in L^* (in K^* resp.). For $j\in L^*$ ($j\in(K^*\cup L^*)^c$ resp.) let ∂_j denote the total number of neighbors of j in $(K^*\cup L^*)^c$ (in L^* resp.). Then

$$\sum_{j\in K^*\cup L^*}\delta_j = 2\mu, \qquad \sum_{j\in(K^*\cup L^*)^c}\delta_j = 2(m-\mu-\nu-\nu_1),$$
$$\sum_{j\in K^*}\Delta_j = \sum_{j\in L^*}\Delta_j = \nu, \quad \sum_{j\in L^*}\partial_j = \sum_{j\in(K^*\cup L^*)^c}\partial_j = \nu_1. \tag{42}$$

In addition,

$$\delta_j + \Delta_j \ge 2, \quad j\in K^*\cup L^*,$$
$$\delta_j + \partial_j \ge 2, \quad j\in (K^*\cup L^*)^c. \tag{43}$$

It is worth noticing that (43) is a relaxed version of the actual restrictions. Also, lumping together δ_j for $j\in K^*$ and $j\in L^*$, we effectively ignore the fact that the graphs $G(K^*)$ and $G(L^*)$ are disjoint.

Denoting the total number of graphs with the given $\mathbf{D} = (\boldsymbol{\delta}, \boldsymbol{\Delta}, \boldsymbol{\partial})$ by $N(\mathbf{D})$, and using the degree-dependent bounds for the counts of graphs, both general and bipartite, we obtain

$$\begin{aligned}
&N(\mathbf{D}) \le \\
&\left((2\mu-1)!!\prod_{j\in K^*\cup L^*}\frac{1}{\delta_j!}\right)\times\left(\nu!\prod_{j\in K^*\cup L^*}\frac{1}{\Delta_j!}\right)\times\left(\nu_1!\prod_{j\in(K^*\cup L^*)^c}\frac{1}{\partial_j!}\prod_{j\in L^*}\frac{1}{\partial_j!}\right)\\
&\quad\times\left((2(m-\mu-\nu-\nu_1)-1)!!\prod_{j\in(K^*\cup L^*)^c}\frac{1}{\delta_j!}\right)\\
&=(2\mu-1)!!\nu!\nu_1!(2(m-\mu-\nu-\nu_1)-1)!!\cdot\Phi(\mathbf{D});\\
&\Phi(\mathbf{D}) = \prod_{j\in K^*\cup L^*}\frac{1}{\delta_j!\Delta_j!}\cdot\prod_{j\in L^*}\frac{1}{\partial_j!}\cdot\prod_{j\in(K^*\cup L^*)^c}\frac{1}{\partial_j!\delta_j!}.
\end{aligned}\tag{44}$$

Our task now is to evaluate S_{μ,ν,ν_1}, the sum of $\Phi(\mathbf{D})$, for all $\mathbf{D}$ that meet (42) and (43). To do so, let us first determine a *multivariate* generating function of $\Phi(\mathbf{D})$, for $\mathbf{D}$ satisfying (43) only:

$$\begin{aligned}
&\sum_{\mathbf{D}\text{ satisfies (43)}} y_1^{\sum_{j\in K^*\cup L^*}\delta_j}y_2^{\sum_{j\in K^*\cup L^*}\Delta_j}y_3^{\sum_{j\in L^*}\partial_j}y_4^{\sum_{j\in(K^*\cup L^*)^c}\partial_j}\\
&\qquad\qquad\times y_5^{\sum_{j\in(K^*\cup L^*)^c}\delta_j}\cdot\Phi(\mathbf{D})\\
&=\left(\sum_{\delta+\Delta\ge 2}\frac{y_1^\delta y_2^\Delta}{\delta!\Delta!}\right)^{2k-1}\left(\sum_{\partial\ge 0}\frac{y_3^\partial}{\partial!}\right)^{k-1}\left(\sum_{\partial+\delta\ge 2}\frac{y_4^\partial y_5^\delta}{\partial!\delta!}\right)^{n-2k+1}\\
&\qquad= f_2(y_1+y_2)^{2k-1}f_0(y_3)^{k-1}f_2(y_4+y_5)^{n-2k+1};
\end{aligned}\tag{45}$$

So now S_{μ,ν,ν_1} is the coefficient of $y_1^{2\mu}y_2^{2\nu}(y_3y_4)^{\nu_1}y_5^{2(m-\mu-\nu-\nu_1)}$ in the function on the right hand of (45). Using

$$[z_1^{a_1}z_2^{a_2}]F(z_1+z_2) = \binom{a_1+a_2}{a_1}[z^{a_1+a_2}]F(z),$$

and $f_0(y_3)=e^{y_3}$, we obtain then:

$$S_{\mu,\nu,\nu_1} = \binom{2(\mu+\nu)}{2\mu}\binom{2(m-\mu-\nu)-\nu_1}{\nu_1}\frac{(k-1)^{\nu_1}}{\nu_1!} \times [x_1^{2(\mu+\nu)}]f_2(x_1)^{2k-1}\cdot[x_2^{2(m-\mu-\nu)-\nu_1}]f_2(x_2)^{n-2k+1}. \quad (46)$$

Here

$$[x_1^{2(\mu+\nu)}]f_2(x_1)^{2k-1} \le \frac{f_2(x_1)^{2k-1}}{x_1^{2(\mu+\nu)}}, \quad \forall x_1>0,$$

and we will see that a sufficiently small x_1 will do the job. We can write an analogous bound for the last factor in (46), and (in the light of (32) and the relative smallness of our parameters μ,ν,ν_1) $x_2=\rho$ is a natural choice. In fact, we can do a bit better and get an extra factor $n^{-1/2}$, by applying the Cauchy (circular) contour formula, cf. (82), in combination with (83). Using

$$P_{\mu,\nu,\nu_1} = \frac{N_{\mu,\nu,\nu_1}}{N(n,m)},$$
$$N_{\mu,\nu,\nu_1} = \sum_{\mathbf{D}\text{ satisfies (42),(43)}} N(\mathbf{D}),$$

(32),(44), and an inequality

$$\binom{2u-v}{v}(2(u-v)-1)!! \le \frac{(2u-1)!!}{v!},$$

we obtain then

$$P_{\mu,\nu,\nu_1} \le_b (2\mu-1)!!\nu!\binom{2(\mu+\nu)}{2\mu}\frac{(2(m-\mu-\nu)-1)!!}{(2m-1)!!} \times\frac{f_2(x_1)^{2k-1}}{x_1^{2(\mu+\nu)}}\cdot\frac{\rho^{2(\mu+\nu)}}{f_2(\rho)^{2k-1}}\cdot\frac{(\rho(k-1))^{\nu_1}}{\nu_1!}. \quad (47)$$

Then since

$$\sum_{\nu_1\ge 0}\frac{(\rho(k-1))^{\nu_1}}{\nu_1!} = e^{\rho(k-1)} < e^{\rho k},$$

we get the bound (call it $Q_{\mu,\nu}$) for $\sum_{\nu_1} P_{\mu,\nu,\nu_1}$, which is (47) with the last factor replaced by $e^{\rho k}$.

Next, for $\nu\ge\nu_0=\nu_0(\mu):=\max\{2(k-1),2k-\mu\}$, using (39),

$$\begin{aligned}\frac{Q_{\mu,\nu+1}}{Q_{\mu,\nu}} &\leq_b \frac{\rho^2}{x_1^2}\cdot\frac{\mu^2+\nu^2}{\nu m}\\ &\leq_b \frac{\rho^2}{x_1^2}\cdot\left(\frac{k^2+k^2\log^2(n/k)}{km}+\frac{\nu}{m}\right)\\ &\leq_b \frac{\rho^2}{x_1^2}\cdot(\epsilon+\epsilon\log^2(1/\epsilon))\\ &\leq 1/2,\end{aligned}$$

if $k \leq \epsilon n$ and $0 < \epsilon \leq \epsilon_1(x_1)$ is chosen sufficiently small. For this choice of ϵ,

$$\sum_{\nu\geq\nu_0(\mu)} Q_{\mu,\nu} = O(Q_{\mu,\nu_0(\mu)}). \tag{48}$$

Furthermore, if $\mu \geq 2$ then $\nu_0 = \nu_0(\mu) = 2(k-1)$ and we have

$$\begin{aligned}\frac{Q_{\mu+1,\nu_0(\mu+1)}}{Q_{\mu,\nu_0(\mu)}} &\leq_b \frac{\rho^2}{x_1^2}\cdot\frac{\mu^2+k^2}{\mu m}\\ &\leq_b \frac{\rho^2}{x_1^2}\cdot\left(\frac{k}{m}+\frac{k}{m}\log^2(n/k)\right)\frac{k}{\mu}\\ &\leq_b \frac{\rho^2}{x_1^2}\cdot(\epsilon+\epsilon\log^2(1/\epsilon))\cdot\frac{k}{\mu}\\ &\leq 1/2,\end{aligned}$$

if $\mu \geq \epsilon^{1/2}k$, and $0 < \epsilon \leq \epsilon_2(x_1) < \epsilon_1(x_1)$ is chosen sufficiently small. If so,

$$\sum_{\mu\geq 2} Q_{\mu,\nu_0} \leq_b k \max_{\mu}\{Q_{\mu,\nu_0} : 2 \leq \mu \leq k\epsilon^{1/2}\}. \tag{49}$$

To make the last bound explicit, we use $(2a-1)!! = (2a)!/(2^a a!)$ and the Stirling formula for factorials to bound, for $\nu = \nu_0$ and $2 \leq \mu \leq \epsilon^{1/2}k$, the combinatorial factors in (47) as follows:

$$\begin{aligned}(2\mu-1)!! &\leq_b k^2\left(\frac{2\mu}{e}\right)^{\mu-2};\\ \binom{2(\mu+\nu_0)}{2\mu} &\leq_b \exp(O(k\epsilon^{1/2}\log\epsilon^{-1}));\\ \frac{(2(m-\mu-\nu_0)-1)!!}{(2m-1)!!} &\leq_b (2m)^{-(\mu-2)-2k}\exp(O(\epsilon k)).\end{aligned}$$

Using these bounds and $k^2\nu_0! \leq (2k)!$, we obtain: for $2 \leq \mu \leq \epsilon^{1/2}k$ and $x_1 < \lambda$,

$$\begin{aligned}Q_{\mu,\nu_0} \leq_b\, & n^{1/2}(2k)!(2m)^{-2k}\left(\frac{f_2(x_1)}{x_1^2}\right)^{2k}\left(\frac{\rho^2}{f_2(\rho)}\right)^{2k}\\ &\times e^{\rho k}\left(\frac{\rho}{x_1}\right)^{O(\epsilon^{1/2}k)}\exp(O(k\epsilon^{1/2}\log\epsilon^{-1})).\end{aligned} \tag{50}$$

If $\mu = 0$ then $\nu_0 = 2k$, and if $\mu = 1$ then $\nu_0 = 2k-1$. The direct computation shows that the bound (50) holds in these two remaining cases as well. So, collecting the pieces and using

$$\frac{(2k)!}{k!(k-1)!} \le_b k2^{2k},$$

we get

$$\sum_{\mu,\nu,\nu_1} E_{\mu,\nu,\nu_1} \le_b n^{-1/2} q^{2k} \exp(O(k\epsilon^{1/2} \log \epsilon^{-1})), \tag{51}$$

where

$$q = 2\frac{n}{2m} \cdot \frac{f_2(x_1)}{x_1^2} \cdot \frac{\rho^2}{f_2(\rho)} \cdot e^{\rho/2}. \tag{52}$$

Using (33), we transform (52) into

$$q = \frac{\rho}{e^{\rho/2} - e^{-\rho/2}} \cdot \frac{2f_2(x_1)}{x_1^2}.$$

The first fraction is strictly less than, and bounded away from 1. (That's where the condition "$\liminf m/n > 1$" enters!). And the second fraction approaches 1, from above, when $x_1 \downarrow 0$. So we can pick x_1 small enough to make $\rho < 1$. For this choice of x_1, and the corresponding $\epsilon = \epsilon(x_1) < \epsilon_2(x_1)$, we have

$$q^2 \exp(O(\epsilon^{1/2} \log \epsilon^{-1})) \le q_1 := \frac{1+q^2}{2} < 1.$$

Then (51) implies that

$$\sum_{k \le \epsilon n} \sum_{\mu,\nu,\nu_1} E_{\mu\nu\nu_1} \le_b n^{-1/2} \sum_{k \ge 0} q_1^k = O(n^{-1/2}).$$

Thus (40) is completely proved, and so is Lemma 4.2. □

Note. Let X_n denote the total number of isolated odd cycles in $G_{n,m}^{\delta \ge 2}$, and let $\tilde{G}_n$ denote the random subgraph obtained by deletion of all X_n cycles in question. If $\mu(\tilde{G}_n) < \lfloor (n - X_n)/2 \rfloor$ then, by the previous lemma, $\tilde{G}_n$ must contain a witness (K, L). According to the last lemma, **whp** K has to be large, of size ϵn, at least.

4.2. Step 2. Using a witness to gainfully add new edges.

Introduce ∂E, the set of *non-edges* (x, y) of the random (sub)graph $\tilde{G}_n$, such that adding (x, y) to the edge set makes the maximum matching number increase by one.

Lemma 4.3.

$$\mathbf{Pr}(0 < |\partial E| < \epsilon^2 n^2/2) = O(n^{-1}).$$

Proof. Suppose the event $\{|\partial E| > 0\} \cap \mathcal{A}_n(\epsilon)^c$ happens. Then, by Lemma 4.1, for every vertex $x \in \tilde{V} := V(\tilde{G}_n)$, not covered by at least one maximum matching (in $\tilde{G}_n$, needless to say), there exists a vertex set $K \subset \tilde{V}$ of cardinality ϵn or more, such that (1) $x \notin K$, $x \notin N_{\tilde{G}_n}(K)$; (2) adding any (x, y), $y \in K$, to the edge set of $\tilde{G}_n$ increases the maximum matching number; (3) for every vertex $y \in K$ there exists a maximum matching that does not cover y. This implies existence of the vertices $x_1, \dots, x_{\nu_n} \in \tilde{V}$, ($\nu_n := \lfloor \epsilon n \rfloor$), such that for every x_j there is a corresponding vertex subset $Y_j \subset \tilde{V}$ satisfying the conditions: (1) $x_j \notin Y_j$, $x_j \notin N_{\tilde{G}_n}(Y_j)$; (2) for every $y \in Y_j$, adding (x_j, y) to the edge set increases the maximum matching number; (3) $|Y_j| \geq \nu_n$. Consequently the edge set of $\tilde{G}_n$ is missing at least

$$\sum_{j=1}^{\nu_n}(\nu_n - j + 1) = \frac{\nu_n(\nu_n+1)}{2} \geq \frac{\epsilon^2 n^2}{2}$$

pairs (x, y) such that adding any such pair to the edge set would increase the maximum matching. Therefore

$$\{0 < |\partial E| \leq \epsilon^2 n^2/2\} \subseteq \mathcal{A}_n(\epsilon),$$

and the claim follows from Lemma 4.2. □

Note. Paraphrasing Lemma 4.3, with probability $1 - O(n^{-1})$, the subgraph $\tilde{G}_n$ (equal $G_{n,m}^{\delta\geq 2}$ minus all its isolated odd cycles) either has a perfect matching or there are at least an^2 pairs of vertices $(u, v) \in \tilde{V} \times \tilde{V}$, $(u, v) \notin E(\tilde{G}_n)$, such that adding any such (u, v) would increase the matching number.

4.3. Step 3. Counting the isolated odd cycles.

Clearly, we need to determine the limiting distribution of X_n, the total number of isolated odd cycles in $G_{n,m}^{\delta\geq 2}$.

Lemma 4.4. *Let* $\liminf m/n > 1$. *Then* X_n *is, in the limit, Poisson* (λ), *i.e.*

$$\mathbf{Pr}(X_n = j) - e^{-\lambda}\frac{\lambda^j}{j!}, \quad j \geq 0. \tag{53}$$

Here

$$\lambda := \frac{1}{4}\log\frac{1+\sigma}{1-\sigma} - \frac{\sigma}{2}, \quad \sigma = \frac{\rho}{e^\rho - 1}, \tag{54}$$

and ρ *is defined in (33).*

Proof. Let $X_{n,\ell}$ denote the total number of isolated odd cycles of length $\ell \geq 3$. Then, given $L > 3$,

$$\mathbf{E}\left(\sum_{\ell\geq L} X_{n,\ell}\right) = \sum_{\ell\geq L}\binom{n}{\ell}\cdot\frac{(\ell-1)!}{2}\frac{N(n-\ell, m-\ell)}{N(n,m)}. \tag{55}$$

Using (32) and (34) with $x = \rho$, and

$$\prod_{j=0}^{\ell-1}\frac{n-j}{2(m-j)-1} \leq \left(\frac{n}{2m-1}\right)^\ell,$$

we see that the generic term in the sum is of order at most

$$\frac{1}{2\ell}\sqrt{\frac{n}{n-\ell}}\cdot\left(\frac{n}{2m}\cdot\frac{\rho^2}{f_2(\rho)}\right)^\ell = \frac{1}{2\ell}\cdot\sqrt{\frac{n}{n-\ell}}\cdot\sigma^\ell.$$

Since $\sigma < 1$, it easily follows then that $\mathbf{E}\left(\sum_{\ell\geq L} X_{n,\ell}\right) \to 0$ if $L = L(n) \to \infty$ however slowly. Consequently, **whp** there are no isolated odd cycles of length exceeding $L = L(n)$. Introduce

$$X_n^* = \sum_{\substack{\ell\leq L(n)\\ \ell \text{ odd}}} X_{n,\ell}.$$

Then, for every fixed $k \geq 1$,

$$\begin{aligned}
\mathbf{E}[(X_n^*)_k] &= \sum_{\ell \leq kL} R_{n,m}(\ell) \sum_{\substack{\ell_1,\dots,\ell_k \in [3,L] \\ \ell_1+\dots+\ell_k=\ell \\ \ell_1,\dots,\ell_k \text{ odd}}} \prod_{j=1}^{k} \frac{(\ell_j-1)!}{2(\ell_j!)} \\
&= \sum_{\ell \leq kL} R_{n,m}(\ell) \cdot [x^\ell] \left(\sum_{j \in [3,L], j \text{ odd}} \frac{x^j}{2j} \right)^K ; \\
R_{n,m}(\ell) &:= \frac{n! N(n-\ell, m-\ell)}{(n-\ell)! N(n,m)}.
\end{aligned}$$

For $\ell \leq L(n)$ and $L(n) \to \infty$ sufficiently slowly, $N(n-\ell, m-\ell)$ is asymptotic the RHS in (32), with n and m replaced by $n-\ell$ and $m-\ell$. (The point here is that the difference between ρ and $\rho(\ell)$ corresponding to $n-\ell$, $m-\ell$ is of order $O(L/n)$, and this difference leads to an extra factor $\exp(O(L^2/n)) \to 1$, provided that $L = o(n^{1/2})$.) Consequently $R_{n,m}(\ell) \sim \sigma^\ell$, uniformly for $\ell \leq L$. Therefore, using $\sigma < 1$,

$$\begin{aligned}
\mathbf{E}[(X_n^*)_k] \sim \; & \textstyle\sum_{\ell \leq kL} \sigma^\ell \cdot [x^\ell] \left(\displaystyle\sum_{j \in [3,L], j \text{ odd}} \frac{x^j}{2j} \right)^k \\
&= \sum_{\ell \leq kL} [x^\ell] \left(\sum_{j \in [3,L], j \text{ odd}} \frac{(\sigma x)^j}{2j} \right)^k \\
&\sim \sum_{\ell} [x^\ell] \left(\sum_{j \in [3,L], j \text{ odd}} \frac{(\sigma x)^j}{2j} \right)^k \\
&= \left(\sum_{j \in [3,L], j \text{ odd}} \frac{\sigma^j}{2j} \right)^k .
\end{aligned}$$

Thus X_n^* is in the limit Poisson with parameter

$$\sum_{j \in [3,L], j \text{ odd}} \frac{\sigma^j}{2j} = \frac{1}{4} \log \frac{1+\sigma}{1-\sigma} - \frac{\sigma}{2} + o(1) = \lambda + o(1).$$

Since $\mathbf{Pr}(X_n \neq X_n^*) \to 0$, the proof is complete. □

As a brief summary of Steps 1-3, we have established that **whp** $G_{n,m}^{\delta \geq 2}$ contains few isolated odd cycles, and upon deletion of these cycles we end up with a subgraph such that if it has no perfect matching, then there are of order n^2 non-edges, whose individual insertions would increase the maximum matching number.

To capitalize on these results, we need to find a way to compare the maximum matching number for two different values of m. And this is a serious challenge, since —unlike the Erdős-Rényi random graph $G(n,m)$ —we do not know of any construction which would have allowed to consider $G_{n,m_1}^{\delta \geq 2}$ as a subgraph of $G_{n,m_2}^{\delta \geq 2}$, for $m_1 < m_2$. The next step shows how such a coupling can be done asymptotically.

4.4. Step 4. An asymptotic coupling of $G^{\delta\geq 2}_{n,m-\omega}$ and $G^{\delta\geq 2}_{n,m}$.

Let $\omega = \lfloor r\log n\rfloor$ for some constant $r>0$. Consider the bipartite graph Γ with vertex set bipartition $\mathcal{G}^{\delta\geq 2}_{n,m-\omega} + \mathcal{G}^{\delta\geq 2}_{n,m}$ and the edge set $E(\Gamma)$ defined by the condition: for $G\in \mathcal{G}^{\delta\geq 2}_{n,m-\omega}$ and $H\in \mathcal{G}^{\delta\geq 2}_{n,m}$, $(G,H)\in E(\Gamma)$ iff

$$E(G)\subseteq E(H) \text{ and } E(G)\setminus E(H) \text{ is a matching.}$$

(So $(G,H)\in E(\Gamma)$ iff $E(H)$ is obtained by adding to $E(G)$ ω independent edges.)

Consider the following experiment SAMPLE:

- Choose G randomly from $\mathcal{G}^{\delta\geq 2}_{n,m-\omega}$
- Add a random matching M, disjoint from $E(G)$, of size ω to obtain $H\in \mathcal{G}^{\delta\geq 2}_{n,m}$.

This induces a probability measure $\mathbf{Q}$ on $\mathcal{G}^{\delta\geq 2}_{n,m}$. Our task is to show that $\mathbf{Q}$ is *nearly* uniform.

For $v\in \mathcal{G}^{\delta\geq 2}_{n-\omega} + \mathcal{G}^{\delta\geq 2}_{n,m}$, let $d_\Gamma(v)$ denote the degree of v.

Lemma 4.5. *$G\in \mathcal{G}^{\delta\geq 2}_{n,m-\omega}$ implies*

$$\frac{(\binom{n}{2}-m-2\omega n)^\omega}{\omega!} \leq d_\Gamma(G) \leq \binom{\binom{n}{2}}{\omega}.$$

Proof. The RHS is obvious. For the LHS let us bound from below the number of *ordered* sequences $e_1,e_2,\ldots,e_\omega$ of ω edges which are not in $E(G)$, and form a matching. Observe that after choosing $e_1,e_2,\ldots,e_i$ we rule out at most $m-\omega+2in$ choices for e_{i+1}. (The $m-\omega$ edges of G plus the further $\leq 2i(n-2)$ choices of new edges incident with $e_1,e_2,\ldots,e_i$). Thus there are always at least $\binom{n}{2}-m-2\omega n$ choices for e_{i+1}. Dividing by $\omega!$ accounts for removing the ordering. □

Thus for n large and $G,G'\in \mathcal{G}^{\delta\geq 2}_{n,m-\omega}$,

$$\left|\frac{d_\Gamma(G)}{d_\Gamma(G')}-1\right| \leq \frac{4\omega^2}{n}. \tag{56}$$

We need to prove analogous estimates for the degrees $d_\Gamma(H)$, $H\in \mathcal{G}^{\delta\geq 2}_{n,m}$.

To this end, let $\Delta(H)$ denote the maximum vertex degree in H, and let $E_>(H)$ be the edges of H joining vertices of degree at least 3. (Why looking at $E_>(H)$? Well, if (G,H) is an edge in $\mathcal{G}$, and $e\in E(H)\setminus E(G)$, then other edges of H incident to e must already be in $E(G)$. So $E(H)\setminus E(G)\subseteq E_>(H)$.)

Lemma 4.6. *Let*

$$\theta = c^{-1}y^2, \text{ where } \frac{\rho(e^\rho-1)}{e^\rho-1-y}=c,\quad c=\frac{2m}{n}.$$

If H is chosen uniformly at random from $\mathcal{G}^{\delta\geq 2}_{n,m}$ then **qs**[1]

(a)

$$\Delta(H)\leq \log n.$$

(b)

$$|E_>(H)-\theta n| = O(n^{1/2}\log n).$$

[1]A sequence of events $\mathcal{E}_n$ is said to occur *quite surely* if $\mathbf{Pr}(\mathcal{E}_n)=1-O(n^{-K})$ for any $K>0$.

Proof. Let $Z_1, Z_2, \dots, Z_n$ be independent copies of $\mathrm{Po}(\rho; \geq 2)$. Introduce the random set $S_{\mathbf{Z}}$ that contains Z_i (distinguishable) copies of the vertex i, $1 \leq i \leq n$; denote $s_{\mathbf{Z}} = |S_{\mathbf{Z}}|$. Given $S_{\mathbf{Z}}$, we choose uniformly at random "pairing" of all $s_{\mathbf{Z}}$ elements of $S_{\mathbf{Z}}$. A convenient way of generating such a pairing, is to choose a random permutation $\pi = (\pi_1, \dots, \pi_{s_{\mathbf{Z}}})$ of $S_{\mathbf{Z}}$ and to form pairs $(\pi_1, \pi_2), (\pi_3, \pi_4), \dots$. (When $s_{\mathbf{Z}}$ is odd, one vertex in $\sigma_{\mathbf{Z}}$ remains without a partner.) Conditional on $\{s_{\mathbf{Z}} = 2m\} \cap \{\text{pairing is graph- induced}\}$, the uniformly random pairing (permutation π) defines a uniformly random graph $H \in \mathcal{G}_{n,m}^{\delta \geq 2}$. Like the bipartite case in Section 2, we have that

$$\mathbf{Pr}(\mathcal{E}_0) = \mathbf{E}\left(F(\mathbf{Z}) \cdot 1_{\{s_{\mathbf{Z}}=2m\}}\right), \quad s_{\mathbf{Z}} = \sum_{j=1}^{n} Z_j,$$

where

$$F(\mathbf{Z}) = \exp\bigl(-\eta(\mathbf{Z})/2 - \eta^2(\mathbf{Z})/4 + O(\max_j Z_j^4/m)\bigr), \quad \eta(\mathbf{Z}) = \frac{1}{2m}\sum_{j=1}^{n} Z_j(Z_j - 1),$$

cf. [21]. Implicit in (32), (33) is

$$\mathbf{Pr}(\mathcal{E}_0) = (1 + o(1))\frac{\beta}{\sqrt{n}}, \quad \beta = \frac{\exp\left(-\hat{\rho}/2 - \hat{\rho}^2/4\right)}{\sqrt{2\pi \mathrm{Var}(Z)}}, \tag{57}$$

i.e. $\mathbf{Pr}(\mathcal{E}_0)$ is only polynomially small, of order $n^{-1/2}$ exactly. This implies that if $\{\mathbf{Z} \in A\}$, $A \subset \{\{2, 3, \dots\}^n)$, is a **q**sevent, then so is the event $\{\deg(H) \in A\}$, $\deg(H)$ denoting the degree sequence of $H \in \mathcal{G}_{n,m}^{\delta \geq 2}$. The part (a) follows then immediately since, for $L = \log n$,

$$\mathbf{Pr}(\max_j Z_j, \geq L) \leq n\mathbf{Pr}(Z_1 \geq L = \log n) = O(ny^L/L!),$$

which is $O(n^{-K})$ for any $K > 0$.

Turn to (b). Let W be the number of pairs (π_{2i-1}, π_{2i}) in the random permutation π of the multi-set $S_{\mathbf{Z}}$ such that both π_{2i-1} and π_{2i} are copies of the vertices of degree 3 or more. We know that, conditioned on the event $\mathcal{E}_0$, there is $W = E_>(H)$. And it is easy to see that

$$\mathbf{E}(W \mid \mathbf{Z}) = \frac{s_{\mathbf{Z},3}(s_{\mathbf{Z},3} - 1)}{2(s_{\mathbf{Z}} - 1)}, \tag{58}$$

where

$$s_{\mathbf{Z},3} = \sum_i Z_i \mathbf{I}_{\{Z_i > 2\}}.$$

Now

$$\begin{aligned} \mathbf{E}(s_{\mathbf{Z}}) &= n\mathbf{E}Z_1 = n\frac{\rho(e^\rho - 1)}{e^\rho - 1 - \rho} = 2m, & (59)\\ \mathbf{E}(s_{\mathbf{Z},3}) &= n(\mathbf{E}Z_1 - 2\mathbf{Pr}(Z_1 = 2)) = n\left(\frac{\rho(e^\rho - 1)}{e^\rho - 1 - \rho} - \frac{\rho^2}{e^\rho - 1 - \rho}\right) & (60)\\ &= n\rho. & (61) \end{aligned}$$

And $s_{\mathbf{Z}}$, $s_{\mathbf{Z},3}$ are the sums of independent copies of Z and $\tilde{Z} := Z\mathbf{I}_{\{Z>2\}}$, respectively. Using the pgf's

$$\mathbf{E}(x^Z) = \frac{f_2(x\rho)}{f_2(\rho)}, \quad \mathbf{E}(x^{\tilde{Z}}) = \frac{\rho^2/2 + f_3(x\rho)}{f_2(\rho)},$$

(59), (61), in a standard (Chernoff-type) way, we obtain that **qs**

$$|s_{\mathbf{Z}} - 2m| \leq n^{1/2} \log n, \quad |s_{\mathbf{Z},3} - n\rho| \leq n^{1/2} \log n. \tag{62}$$

Denote the event in (62) by $\mathcal{E}_1$. Then $\mathcal{E}_1$ holds **qs**. It follows from (58) that, on the event $\mathcal{E}_1$,

$$\mathbf{E}(W \mid \mathbf{Z}) = \theta n + O(n^{1/2} \log n), \quad \theta := \frac{n}{2m}\rho^2. \tag{63}$$

Next we appeal to the Azuma-Hoeffding inequality to show that, conditional on $\mathbf{Z}$, W is tightly concentrated around $\mathbf{E}(W \mid \mathbf{Z})$. The A-H inequality applies since transposing any two elements of a permutation of $S_{\mathbf{Z}}$ may change W by at most 2, see Appendix B. So, for every $u > 0$,

$$\mathbf{Pr}(|W - \mathbf{E}(W \mid \mathbf{Z})| \geq u \mid \mathbf{Z}) \leq 2e^{-u^2/(8s_{\mathbf{Z}})}.$$

Removing the conditioning on $\mathbf{Z}$, and using the definition of the event $\mathcal{E}_1$, we obtain

$$\mathbf{Pr}(|W - \mathbf{E}(W \mid \mathbf{Z})| \geq u) \leq \mathbf{Pr}(\mathcal{E}_1^c) + 2e^{-u^2/(17m)}.$$

So, substituting $u = n^{1/2} \log n$ and using (63), we see that **qs**

$$|W - \theta n| \leq A n^{1/2} \log n,$$

if a constant A is sufficiently large. Recalling that $W = E_>(H)$ on the event $\mathcal{E}_0$, and that $\mathbf{Pr}(\mathcal{E}_0)$ is of order $n^{-1/2}$, we complete the proof of the part (b). □

Now let $\tilde{\mathcal{G}}$ be the set of $H \in \mathcal{G}_{n,m}^{\delta\geq 2}$ satisfying the conditions of the above lemma i.e.

- The number of edges joining two vertices of degree ≥ 3 is in the range $\theta n \pm A n^{1/2} \log n$
  for some constant $A > 0$.
- The maximum degree $\Delta(H) \leq \log n$.

According to the lemma 4.6,

$$|\mathcal{G}_{n,m}^{\delta\geq 2} \setminus \tilde{\mathcal{G}}| \leq |\tilde{\mathcal{G}}| n^{-K}, \quad \forall K > 0. \tag{64}$$

Note next that

Lemma 4.7. $H \in \tilde{\mathcal{G}}$ *implies*

$$\frac{(\theta n - A n^{1/2} \log n - 2\omega \log n)^{\omega}}{\omega!} \leq d_{\Gamma}(H) \leq \binom{\theta n + A n^{1/2} \log n}{\omega}.$$

Proof. The upper bound follows from the earlier observation, namely that every edge among ω edges added to a graph $G \in \mathcal{G}_{n,m-\omega}^{\delta\geq 2}$ to obtain the graph H must connect two vertices which have degree 3 or more in H, and from the condition $H \in \tilde{\mathcal{G}}$. For the LHS, as in Lemma 4.5, let us bound from below the number of *ordered* sequences $e_1, e_2, \ldots, e_{\omega}$ of ω edges which are contained in $E_>(H)$ and form a matching. Observe that after choosing $e_1, e_2, \ldots, e_i$ we rule out at most $2i\Delta(H)$ choices for e_{i+1}, *since we have restricted ourselves to matchings.* Thus there are always at least $\theta n - A n^{1/2} \log n - 2\omega\Delta$ choices for e_{i+1}. Dividing by $\omega!$ accounts for removing the the ordering. □

So for $H, H' \in \tilde{\mathcal{G}}$,

$$\left|\frac{d_\Gamma(H)}{d_\Gamma(H')} - 1\right| \leq \frac{2A\omega \log n}{\theta n^{1/2}}. \tag{65}$$

Finally, for $H \in \mathcal{G}_{n,m}^{\delta\geq 2} \setminus \tilde{\mathcal{G}}$ and $H' \in \tilde{\mathcal{G}}$,

$$\frac{d_\Gamma(H)}{d_\Gamma(H')} \leq \frac{\binom{m}{\omega}}{\frac{(\theta n - An^{1/2}\log n - 2\omega \log n)^\omega}{\omega!}} \leq \left(\frac{c}{\theta}\right)^\omega, \quad c = \frac{2m}{n}, \tag{66}$$

as the total number of ways to delete a matching of size ω from $H \in \mathcal{G}_{n,m}^{\delta\geq 2}$ is $\binom{m}{\omega}$ at most.

Having proved the bounds (56), (65), (66), we can now show that the distribution $\mathbf{Q}$ on $\mathcal{G}_{n,m}^{\delta\geq 2}$, induced by the uniform distribution on $\mathcal{G}_{n,m-\omega}^{\delta\geq 2}$ *and* the SAMPLE, is nearly uniform itself.

Let $G_0 \in \mathcal{G}_{n,m-\omega}^{\delta\geq 2}$ be fixed. By (56), if $H \in \mathcal{G}_{n,m}^{\delta\geq 2}$ then

$$\begin{aligned}
\mathbf{Q}(H) &= \mathbf{Pr}(\text{SAMPLE chooses } H) \\
&= \frac{1}{|\mathcal{G}_{n,m-\omega}^{\delta\geq 2}|} \times \sum_{G:\,(G,H)\in E(\Gamma)} \frac{1}{d_\Gamma(G)} \\
&= \frac{1 + O(\omega^2/n)}{|\mathcal{B}_{n,m-\omega}^{\delta\geq 2}|} \cdot \frac{d_\Gamma(H)}{d_\Gamma(G_0)}.
\end{aligned} \tag{67}$$

From this relation, (65), and (66), it follows that

$$H, H' \in \tilde{\mathcal{G}} \quad \text{implies} \quad \left|\frac{\mathbf{Q}(H)}{\mathbf{Q}(H')} - 1\right| \leq \frac{3A\omega \log n}{\theta n^{1/2}}, \tag{68}$$

$$H \in \mathcal{G}_{n,m}^{\delta\geq 2} \setminus \tilde{\mathcal{G}},\ H' \in \tilde{\mathcal{G}} \quad \text{implies} \quad \frac{\mathbf{Q}(H)}{\mathbf{Q}(H')} \leq \left(\frac{2c}{\theta}\right)^\omega. \tag{69}$$

Furthermore, invoking also

$$\sum_{G\in\mathcal{G}_{n,m-\omega}^{\delta\geq 2}} d_\Gamma(G) = \sum_{H\in\mathcal{G}_{n,m}^{\delta\geq 2}} d_\Gamma(H),$$

and picking $H' \in \tilde{\mathcal{G}}$, we obtain (see (56), (65)):

$$\frac{d_\Gamma(H')}{d_\Gamma(G_0)} \leq \left(1 + \frac{4A\omega \log n}{\theta n^{1/2}}\right) \frac{|\mathcal{G}_{n,m-\omega}^{\delta\geq 2}|}{|\tilde{\mathcal{G}}|}. \tag{70}$$

Combining (64), (67), (69), and (70), we get: for every $K > 0$,

$$\begin{aligned}
\mathbf{Q}(\mathcal{G}_{n,m}^{\delta\geq 2} \setminus \tilde{\mathcal{G}}) &\leq \mathbf{Q}(H')\left(\frac{2c}{\theta}\right)^\omega \cdot n^{-2K}|\tilde{\mathcal{G}}| \\
&= \frac{1 + O(\omega^2/n)}{|\mathcal{G}_{n,m-\omega}^{\delta\geq 2}|} \cdot \frac{d_\Gamma(H')}{d_\Gamma(G_0)} \left(\frac{2c}{\theta}\right)^\omega n^{-2K}|\tilde{\mathcal{G}}| \\
&= O\big((2c/\theta)^\omega n^{-2K}\big) \\
&\leq n^{-K}.
\end{aligned} \tag{71}$$

(As $\omega = \lfloor r \log n \rfloor$, the last bound holds for $K > K(r)$.) Finally, since $\mathbf{Q}(\mathcal{G}_{n,m}^{\delta\geq 2}) = 1$, from (64), (71) and (68) we deduce that, for $H \in \tilde{\mathcal{G}}$,

$$\left| \mathbf{Q}(H) - \frac{1}{|\mathcal{G}_{n,m}^{\delta\geq 2}|} \right| \leq \frac{1}{|\mathcal{G}_{n,m}^{\delta\geq 2}|} \cdot \frac{5A\omega \log n}{\theta n^{1/2}}. \tag{72}$$

This means that on the graph set $\tilde{\mathcal{G}}$ the probability measure $\mathbf{Q}$ is almost uniform.

4.5. Step 5. Asymptotic monotonicity of the tail distribution of the maximum matching number.

We will use $\tilde{G} = \tilde{G}(G)$ to denote the subgraph of G obtained by deletion of all isolated odd cycles in G, and $X(G)$ to denote the total number of these cycles. Clearly

$$\mu^*(\tilde{G}) \leq \left\lfloor \frac{n - X(G)}{2} \right\rfloor, \quad \text{if } |V(G)| = n.$$

Let M stand for a generic value of the number of edges, and let $\Pr_M$ denote the uniform distribution on $\mathcal{G}_{n,M}^{\delta\geq 2}$. We will use $\mathbf{Q}_M$ to denote the probability distribution induced on $\mathcal{G}_{n,M}^{\delta\geq 2}$ by the uniform distribution $\Pr_{M-\omega}$ on $\mathcal{G}_{n,M-\omega}^{\delta\geq 2}$ via the SAMPLE procedure analyzed in Step 4. To shorten notations and to underscore dependence on M, we will write G_M and $\tilde{G}_M$ instead of $G_{n,M}^{\delta\geq 2}$ and $\tilde{G}(G_{n,M}^{\delta\geq 2})$ respectively, and X_M instead of $X(G_{n,M}^{\delta\geq 2})$. Introduce

$$\epsilon(M,\tau) = \Pr_M \left(\mu^*(G_M) < \left\lfloor \frac{n - X_M}{2} \right\rfloor - \tau \right).$$

Lemma 4.8. *Let $u \in (0,1)$ be so small that $(1-u) \liminf m/n > 1$. Then, for $M \in [(1-u)m, m]$ and $t \leq t_n = n^{1/2} \log^{-3} n$,*

$$\epsilon(M, t-1) \leq \epsilon(M-\omega, t) + \frac{6A\omega \log n}{n^{1/2}}. \tag{73}$$

Proof. Let $H \in \mathcal{G}_{n,M}^{\delta\geq 2}$ be a right-side vertex in the graph Γ on the bipartition $\mathcal{G}_{n,M-\omega}^{\delta\geq 2} + \mathcal{G}_{n,M}^{\delta\geq 2}$. Suppose that

$$\mu^*(H) < \left\lfloor \frac{n - X(H)}{2} \right\rfloor - t + 1.$$

Then, for every $G \in N_\Gamma(H)$, either $\mu^*(G) < \lfloor (n - X(H))/2 \rfloor - t$, or $\mu^*(G) = \lfloor (n - X(H))/2 \rfloor - t$, in which case none of ω non-edges (u,v), added to $E(G)$, is in $\partial E(\tilde{G}(G))$. We know that

$$\Pr_{M-\omega}(1 \leq |\partial E(\tilde{G}_{M-\omega})| \leq an^2) = O(n^{-1}).$$

Besides, conditionally on $G_{M-\omega}$, the probability that none of ω random vertex disjoint non-edges belongs to $\partial E(\tilde{G}_{M-\omega})$, is bounded by

$$\left(1 - \frac{|\partial E(\tilde{G}_{M-\omega})| - \omega}{\binom{n}{2}} \right)^{\omega}.$$

And for $|\partial E(\tilde{G}_{M-\omega})| > an^2$, the last bound is at most $(1-a)^\omega \le n^{-1}$, if we pick r in $\omega = \lfloor r \log n \rfloor$ sufficiently large. Therefore

$$\mathbf{Q}_M\left(\mu^*(H) < \left\lfloor \frac{n - X(H)}{2} \right\rfloor - t + 1\right) \le \Pr_{M-\omega}\left(\mu^*(G_{M-\omega}) < \left\lfloor \frac{n - X_{M-\omega}}{2} \right\rfloor - t\right) + O(n^{-1})$$
$$\le \epsilon(M-\omega, t) + \frac{b}{n},$$

for some absolute constant $b > 0$. So

$$\mathbf{Q}_M\left(\mu^*(H) < \left\lfloor \frac{n - X(H)}{2} \right\rfloor - t + 1 \text{ and } H \in \tilde{\mathcal{G}}\right) \le \epsilon(M-\omega, t) + \frac{b}{n},$$

and then, using (72),

$$\Pr_M\left(\mu^*(H) < \left\lfloor \frac{n - X(H)}{2} \right\rfloor - t + 1 \text{ and } H \in \tilde{\mathcal{G}}\right)$$
$$\le \left(\epsilon(M-\omega, t) + \frac{b}{n}\right)\left(1 + \frac{5A\omega \log n}{\theta n^{1/2}}\right).$$

Therefore, recalling (71),

$$\Pr_M\left(\mu^*(H) < \left\lfloor \frac{n - X(H)}{2} \right\rfloor - t + 1\right) \le \left(\epsilon(M-\omega, t) + \frac{b}{n}\right)\left(1 + \frac{5A\omega \log n}{\theta n^{1/2}}\right) + n^{-K}$$
$$\le \epsilon(M-\omega, t) + \frac{6A\omega \log n}{\theta n^{1/2}}.$$

In other words,

$$\epsilon(M, t-1) \le \epsilon(M-\omega, t) + \frac{6A\omega \log n}{\theta n^{1/2}},$$

for all $M \in [(1-u)m, m]$. □

4.6. Step 6. Completion of the proof of Theorem 2.
Iterating the inequality (73) $t \le t_n$ times from $M = m$ and $t = 1$, we obtain

$$\epsilon(m, 0) \le \epsilon(m - t\omega, t) + t \cdot \frac{7A\omega \log n}{n^{1/2}}. \tag{74}$$

Take $t = \lfloor n^{0.20+\beta} \rfloor$, $\beta > 0$ being small. Then, using the definition of $\epsilon(\cdot,\cdot)$ and (31), we obtain that $\epsilon(m - t\omega, t) \to 0$. And the second term on the RHS of (74) is $O(n^{-0.3+\beta} \log^2 n) = o(1)$, if $\beta < 0.3$.

We conclude that **whp** $\mu^*(G_{n,m})$ equals $\lfloor (n - X_m)/2 \rfloor$, so that $\tilde{G}(G_{n,m}^{\delta \ge 2})$ has a perfect matching. Since X_m is asymptotically Poisson(λ), this proves Theorem 2.

References

[1] J. Aronson, A.M. Frieze and B.G. Pittel, Maximum matchings in sparse random graphs: Karp-Sipser re-visited, *Random Structures and Algorithms* 12 (1998) 111-178.

[2] B. Bollobás, Random graphs, in *Combinatorics* (H.N.V. Temperley Ed.) London Mathematical Society Lecture Note Series 52, Cambridge University Press (1981) 80-102.

[3] B. Bollobás, T. Fenner and A.M. Frieze, Hamilton cycles in random graphs with minimal degree at least k, in *A tribute to Paul Erdős*, edited by A. Baker, B. Bollobás and A. Hajnal, (1988) 59-96.

[4] B. Bollobás, C. Cooper, T. Fenner and A.M. Frieze, On Hamilton cycles in sparse random graphs with minimum degree at least k, *Journal of Graph Theory* 34 (2000) 42-59.

[5] B. Bollobás and A.M. Frieze, On matchings and Hamiltonian cycles in random graphs, *Annals of Discrete Mathematics* 28 (1985) 23-46.

[6] R.Durrett, Probability: Theory and examples, Wadsworth and Brooks/Cole, 1991.

[7] P. Erdős and A. Rényi, On the evolution of random graphs, *Publ. Math. Inst. Hungar. Acad. Sci.* 5 (1960) 17-61.

[8] P. Erdős and A. Rényi, On random matrices, *Publ. Math. Inst. Hungar. Acad. Sci.* 8 (1964) 455-461.

[9] P. Erdős and A. Rényi, On the existence of a factor of degree one of a connected random graph, *Acta. Math. Acad. Sci. Hungar.* 17 (1966) 359-368.

[10] A.M. Frieze, Maximum matchings in a class of random graphs, *Journal of Combinatorial Theory Series B* 40 (1986) 196-212.

[11] A.M. Frieze, Perfect matchings in random bipartite graphs with minimal degree at least 2, to appear.

[12] A.M. Frieze and B. Pittel, Probabilistic analysis of an algorithm in the theory of markets in indivisible goods, *Annals of Applied Probability* 5 (1995) 768-808.

[13] M. Karoński and B. Pittel, Existence of a perfect matching in a random $(1+e^{-1})$-out bipartite graph, *J. Comb. Theory B* 88 (2003) 1-16.

[14] R.M. Karp and M. Sipser, *Maximum matchings in sparse random graphs*, Proceedings of the 22nd Annual IEEE Symposium on Foundations of Computing (1981) 364-375.

[15] J.H. Kim and N.C. Wormald, Random matchings which induce Hamilton cycles, and Hamiltonian decompositions of random regular graphs, *Journal of Combinatorial Theory, Series B* 81 (2001), 20-44.

[16] V.F. Kolchin, Random mappings, *Optimization Software*,New York, 1986.

[17] L. Lovász, Combinatorial problems and exercises, Second edition, North-Holland, 1993.

[18] C.J.H.McDiarmid, *On the method of bounded differences*, Surveys in Combinatorics, 1989, (J.Siemons ed.), Cambridge University Press (1989) 148-188.

[19] B. D. McKay, *Asymptotics for symmetric 0-1 matrices with prescribed row sums*, Ars Combinatoria, 19A (1985) 15-26.

[20] B.G. Pittel, Paths in a random digital tree: limiting distributions, *Advances in Applied. Probability* 18 (1986), 139-155.

[21] B. Pittel and N. Wormald, Counting connected graphs inside out, *J. Comb. Theory B*, to appear.

[22] R.W. Robinson and N.C. Wormald, *Almost all regular graphs are Hamiltonian*, Random Structures and Algorithms 5, (1994) 363-374.

[23] D.W. Walkup, Matchings in random regular bipartite graphs, *Discrete Mathematics* 31 (1980) 59-64.

APPENDIX

A. Enumerating bipartite graphs

Consider the bipartite graphs with vertex bipartition $R \cup C$, (Rows and Columns), $R = [\nu_1]$ and $C = [\nu_2]$. Given μ, the ν_1-tuple $\mathbf{a}$, and the ν_2-tuple $\mathbf{b}$ of nonnegative integers a_i, $i \in [\nu_1]$, b_j, $j \in [\nu_2]$, let $N(\mathbf{a}, \mathbf{b})$ denote the total number of the bipartite graphs with the row degree sequence $\mathbf{a}$ and the column degree sequence $\mathbf{b}$. Using the bipartite version of the pairing model, we see that

$$N(\mathbf{a}, \mathbf{b}) \leq N^*(\mathbf{a}, \mathbf{b}); \quad N^*(\mathbf{a}, \mathbf{b}) := \frac{\mu!}{\prod\limits_{i \in [\nu_1]} a_i! \cdot \prod\limits_{j \in [\nu_2]} b_j!}. \tag{75}$$

The fudge factor, i.e. the ratio

$$F(\mathbf{a}, \mathbf{b}) = \frac{N(\mathbf{a}, \mathbf{b})}{N^*(\mathbf{a}, \mathbf{b})}, \tag{76}$$

is the probability that the uniformly random pairing is graph induced. A sharp asymptotic formula for $F(\mathbf{a}, \mathbf{b})$ has been a subject of many papers. A culmination point is [19] by McKay who proved that if $D^3/\mu \to 0$, D being the maximum degree, then

$$N(\mathbf{a}, \mathbf{b}) = N^*(\mathbf{a}, \mathbf{b}) \exp\left(-\frac{1}{2}\lambda(\mathbf{a})\lambda(\mathbf{b}) + O(D^3/\mu)\right), \tag{77}$$

$$\lambda(\mathbf{a}) := \frac{1}{\mu} \sum_{i \in [\nu_1]} a_i(a_i - 1), \tag{78}$$

$$\lambda(\mathbf{b}) := \frac{1}{\mu} \sum_{j \in [\nu_2]} b_j(b_j - 1), \tag{79}$$

The formulas (75) and (77) are instrumental in asymptotic evaluation (estimation) of the total number of bipartite graphs with a given number of edges and certain restrictions on the degree sequence.

Neglecting for now the fudge factor in (77),

$$N_{\mathbf{c},\mathbf{d}}(\boldsymbol{\nu}, \mu) \leq \sum_{\substack{a_i \geq c_i, b_j \geq d_j \\ \sum_i a_i = \sum_j b_j = \mu}} N^*(\mathbf{a}, \mathbf{b}), \tag{80}$$

In order to rewrite (75) in a more manageable way, we observe that

$$\sum_{\mu_1, \mu_2 \geq 0} x^{\mu_1} y^{\mu_2} \sum_{\substack{a_i \geq c_i, b_j \geq d_j \\ \sum_i a_i = \mu_1, \sum_j b_j = \mu_2}} \frac{1}{\prod\limits_{i \in [\nu_1]} a_i! \cdot \prod\limits_{j \in [\nu_2]} b_j!} = G_{\mathbf{c}}(x) H_{\mathbf{d}}(y);$$

Therefore (75) becomes

$$N_{\mathbf{c},\mathbf{d}}(\boldsymbol{\nu}, \mu) \leq \mu! [x^\mu y^\mu] G_{\mathbf{c}}(x) H_{\mathbf{d}}(y) \tag{81}$$

$$= \mu! (2\pi i)^{-1} \oint\limits_{|x| = r_1} x^{-\mu-1} G_{\mathbf{c}}(x)\, dx \cdot (2\pi i)^{-1} \oint\limits_{|y| = r_2} y^{-\mu-1} H_{\mathbf{d}}(y)\, dy \tag{82}$$

for all $r_1, r_2 > 0$. Using an inequality (Pittel [20])

$$|f_t(z)| \leq f_t(|z|) \exp\left(-\frac{|z| - \text{Re } z}{t+1}\right), \tag{83}$$

(4), (5), and the fact that

$$|z| - \text{Re } z = r(1-\cos\theta) \geq cr\theta^2, \quad \text{when } z = re^{i\theta},\, \theta \in (-\pi, \pi],$$

we see from (82), after a straightforward estimation, that

$$N_{\mathbf{c},\mathbf{d}}(\boldsymbol{\nu}, \mu) \leq_b \mu! \frac{1}{\sqrt{r_1 \sum_i (c_i+1)^{-1}}} \cdot \frac{G_{\mathbf{c}}(r_1)}{r_1^\mu} \times \frac{1}{\sqrt{r_2 \sum_j (d_j+1)^{-1}}} \cdot \frac{H_{\mathbf{d}}(r_2)}{r_2^\mu}. \tag{84}$$

Here and elsewhere $A <_b B$ means that $A = O(B)$, uniformly for all feasible parameters that determine the values of A and B. In the sequel we consider only $\max_i c_i = O(1)$, $\max_j d_j = O(1)$, in which case the bound (84) simplifies to

$$N_{\mathbf{c},\mathbf{d}}(\boldsymbol{\nu}, \mu) \leq_b \mu! \cdot (\nu_1 r_1)^{-1/2} \frac{G_{\mathbf{c}}(r_1)}{r_1^\mu} \cdot (\nu_2 r_2)^{-1/2} \frac{H_{\mathbf{d}}(r_2)}{r_2^\mu}. \tag{85}$$

The task of determining the "best" values of r_1 and r_2 and incorporating the left-out fudge factor will be made easier by looking at the above through probabilistic lenses.

Fix $r_1, r_2 > 0$ and introduce the independent random variables Y_i, Z_j, with the distributions

$$\mathbf{Pr}(Y_i = \ell) = \frac{r_1^\ell/\ell!}{f_{c_i}(r_1)},\ (\ell \geq c_i), \tag{86}$$

$$\mathbf{Pr}(Z_j = \ell) = \frac{r_2^\ell/\ell!}{f_{d_j}(r_2)},\ (\ell \geq d_j), \tag{87}$$

so that, in distribution, Y_i is Poisson (r_1) conditioned on {Poisson$(r_1) \geq c_i$}, and Z_j is Poisson(r_2) conditioned on {Poisson$(r_2) \geq d_j$}. In short, $Y_i = \text{Po}(r_1; \geq c_i)$ and $Z_j = \text{Po}(r_2; \geq d_j)$. Now (81) can be rewritten as

$$\begin{aligned} N_{\mathbf{c},\mathbf{d}}(\boldsymbol{\nu}, \mu) &\leq \mu! \times \frac{G_{\mathbf{c}}(r_1)}{r_1^\mu} [x^\mu] \frac{G_{\mathbf{c}}(xr_1)}{G_{\mathbf{c}}(r_1)} \cdot \frac{H_{\mathbf{d}}(r_1)}{r_1^\mu} [y^\mu] \frac{H_{\mathbf{d}}(yr_1)}{H_{\mathbf{d}}(r_1)} \\ &= \mu! \times \frac{G_{\mathbf{c}}(r_1)}{r_1^\mu} \mathbf{Pr}\left(\sum_i Y_i = \mu\right) \cdot \frac{H_{\mathbf{d}}(r_2)}{r_2^\mu} \mathbf{Pr}\left(\sum_j Z_j = \mu\right). \end{aligned} \tag{88}$$

Now the RHS expressions in (81), (82) and (88) are equal to each other and the RHS of inequality (85) bounds them all. Therefore,

$$\sup_\mu \mathbf{Pr}\left(\sum_i Y_i = \mu\right) \leq_b \frac{1}{\sqrt{\nu_1 r_1}}, \quad \sup_\mu \mathbf{Pr}\left(\sum_j Z_j = \mu\right) \leq_b \frac{1}{\sqrt{\nu_2 r_2}}. \tag{89}$$

Furthermore, (80) becomes equality when $N^*(\mathbf{a},\mathbf{b})$ is replaced by $N(\mathbf{a},\mathbf{b})$. So, analogously to (88),

$$\begin{aligned} N_{\mathbf{c},\mathbf{d}}(\boldsymbol{\nu},\mu) &= \frac{\mu!}{(r_1r_2)^\mu} \sum_{\substack{a_i\ge c_i, b_j\ge d_j\\ \sum_i a_i=\mu_1,\sum_j b_j=\mu_2}} \prod_{i\in[\nu_1]}\frac{r_1^{a_i}}{a_i!}\prod_{j\in[\nu_2]}\frac{r_2^{b_j}}{b_j!} \\ &= \mu!\frac{G_{\mathbf{c}}(r_1)H_{\mathbf{d}}(r_2)}{(r_1r_2)^\mu} \sum_{\substack{a_i\ge c_i, b_j\ge d_j\\ \sum_i a_i=\mu_1,\sum_j b_j=\mu_2}} \mathbf{Pr}(\mathbf{Y}=\mathbf{a},\mathbf{Z}=\mathbf{b})F(\mathbf{a},\mathbf{b}) \\ &= \mu!\frac{G_{\mathbf{c}}(r_1)H_{\mathbf{d}}(r_2)}{(r_1r_2)^\mu}\cdot\mathbf{E}\left(F(\mathbf{Y},\mathbf{Z})\cdot 1_{\{\sum_i Y_i=\mu\}}1_{\{\sum_j Z_j=\mu\}}\right), \end{aligned} \tag{90}$$

where $F(\cdot,\cdot)$ is defined in (76). To make this formula useful, we need to show that, for a proper choice of r_1, r_2, asymptotically we can replace $\lambda(\mathbf{Y})\lambda(\mathbf{Z})$ in the formula (77) by $\mathbf{E}(\lambda(\mathbf{Y}))\mathbf{E}(\lambda(\mathbf{Z})))$.

From now on let us assume that and

$$\mu^{-1}\le_b r_1, r_2 \le_b \log\mu \text{ and that } \nu_1,\nu_2 = O(\mu). \tag{91}$$

Since $\max_i c_i$, $\max_j d_j$ are both $O(1)$, using the definition of Y_i, Z_j and the conditions on ν_1,ν_2 and r_1,r_2, we have: for $0<\alpha'<\alpha$,

$$\begin{aligned} \mathbf{Pr}(\max\{\max_i Y_i, \max_j Z_j\}\ge\mu^\alpha) &\le \sum_i \mathbf{Pr}(Y_i\ge\mu^\alpha)+\sum_j\mathbf{Pr}(Z_j\ge\mu^\alpha) \\ &\le e^{-\mu^{\alpha'}}. \end{aligned}$$

Therefore, for $\alpha<1/3$, $\mathbf{E}_{\boldsymbol{\nu},\mu}$, the expected value in (90), is given by

$$\mathbf{E}_{\boldsymbol{\nu},\mu} = (1+O(\mu^{-1+3\alpha}))\mathbf{E}^*_{\boldsymbol{\nu},\mu}+O(e^{-\mu^{\alpha'}}); \tag{92}$$

$$\mathbf{E}^*_{\boldsymbol{\nu},\mu} = \mathbf{E}\left(F^*(\mathbf{Y},\mathbf{Z})\cdot 1_{\{\sum_i Y_i=\mu\}}1_{\{\sum_j Z_j=\mu\}}\right); \tag{93}$$

$$F^*(\mathbf{a},\mathbf{b}) := \exp\left(-\frac{1}{2}\lambda(\mathbf{a})\lambda(\mathbf{b})\right), \tag{94}$$

see (77). In particular, see (89),

$$\mathbf{E}^*_{\boldsymbol{\nu},\mu}\le_b(\nu_1\nu_2r_1r_2)^{-1/2}.$$

Let us estimate the effect of replacing $\lambda(\mathbf{Y}),\lambda(\mathbf{Z})$ in (94) by their expected values. To this end, let us introduce

$$U_i=(Y_i)_2-\mathbf{E}((Y_i)_2),\quad V_j=(Z_j)_2-\mathbf{E}((Z_j)_2).$$

Simple computation shows that $\mathbf{E}((Y_i)_2)$ is of order $O(1+r_i^2)$, whence of order $O(\log^2\mu)$, and likewise $\mathbf{E}(Z_j(Z_j-1))=O(\log^2\mu)$. From $\lambda(\mathbf{Y})=\mu^{-1}\sum_i(Y_i)_2$, $\lambda(\mathbf{Z})=\mu^{-1}\sum_j(Z_j)_2$, it follows then that $\mathbf{E}(\lambda(\mathbf{Y}))$, $\mathbf{E}(\lambda(\mathbf{Z}))=O(\log^2\mu)$ and that after using the expansion

$$ab-\bar a\bar b=(a-\bar a)(b-\bar b)+\bar a(b-\bar b)+\bar b(a-\bar a)$$

we have

$$\begin{aligned} |\lambda(\mathbf{Y})\lambda(\mathbf{Z})-\mathbf{E}(\lambda(\mathbf{Y}))\mathbf{E}(\lambda(\mathbf{Z}))| &\le_b (\log^2\mu)\Delta(\mathbf{Y},\mathbf{Z})+\Delta^2(\mathbf{Y},\mathbf{Z}), \\ \Delta(\mathbf{Y},\mathbf{Z}) &= |\lambda(\mathbf{Y})-\mathbf{E}(\lambda(\mathbf{Y}))|+|\lambda(\mathbf{Z})-\mathbf{E}(\lambda(\mathbf{Z}))|. \end{aligned} \tag{95}$$

Therefore, if we replace $F^*(\mathbf{Y},\mathbf{Z})$ in (93) by $\exp(-\frac{1}{2}\mathbf{E}(\lambda(\mathbf{Y}))\mathbf{E}(\lambda(\mathbf{Z})))$, then the compensating factor is $\exp(O(\log^2\mu\Delta(\mathbf{Y},\mathbf{Z})+\Delta^2(\mathbf{Y},\mathbf{Z})))$. Furthermore, setting $u=\log^{10}\mu$, we estimate

$$\begin{aligned}\mathbf{Pr}(|U_i|\geq u) &\leq \sum_{\ell^2\geq u}\frac{r_1^\ell/\ell!}{f_{c_i}(r_1)}\\ &\leq_b r_1^{-c_i}\left(\frac{er_1}{\log^5\mu}\right)^{\log^5\mu}\\ &\leq \exp(-\Omega(\log^5\mu)).\end{aligned}$$

Likewise

$$\begin{aligned}\mathbf{E}(U_i;|U_i|\geq u) &= \sum_{\ell:(\ell)_2-\mathbf{E}((Y_i)_2)\geq u}[(\ell)_2-\mathbf{E}((Y_i)_2)]\frac{r_1^\ell/\ell!}{f_{c_i}(r_1)}\\ &\leq_b r_1^{2-c_i}\left(\frac{2er_1}{\log^5\mu}\right)^{\frac{1}{2}\log^5\mu}\\ &\leq \exp(-\Omega(\log^5\mu)).\end{aligned}\tag{96}$$

Let $\mathfrak{U}_i=U_i\mathbf{I}_{\{|U_i|<u\}}$, so that $|\mathfrak{U}_i|\leq u$. Then (Azuma-Hoeffding inequality), for every $t>0$,

$$\mathbf{Pr}\left(\left|\sum_i(\mathfrak{U}_i-\mathbf{E}\mathfrak{U}_i)\right|\geq t\right)\leq 2\exp\left(-\frac{t^2}{2u^2\nu_1}\right).$$

Since $\mathbf{E}U_i=0$, from (96) we have

$$\left|\sum_i\mathbf{E}\mathfrak{U}_i\right|=\left|\sum_i\mathbf{E}(U_i;|U_i|\geq u)\right|\leq\exp(-\Omega(\log^5\mu)).$$

Therefore, for $t\geq 1$,

$$\begin{aligned}\mathbf{Pr}\left(\left|\sum_i U_i\right|\geq t\right) &\leq \sum_i\mathbf{Pr}(|U_i|\geq u)+\mathbf{Pr}\left(\left|\sum_i\mathfrak{U}_i\right|\geq t\right)\\ &\leq \sum_i\mathbf{Pr}(|U_i|\geq u)+\mathbf{Pr}\left(\left|\sum_i(\mathfrak{U}_i-\mathbf{E}\mathfrak{U}_i)\right|\geq t-\left|\sum_i\mathbf{E}\mathfrak{U}_i\right|\right)\\ &\leq_b \exp(-\Omega(\log^5\mu))+\exp\left(-\frac{\left(t-\exp(-\Omega(\log^5\mu))\right)^2}{2u^2\nu_1}\right)\\ &= \exp(-\Omega(\log^5\mu)),\end{aligned}\tag{97}$$

the latter inequality holding if $t=\mu^{1/2}\log^5\mu$. An analogous inequality holds for $\sum_j V_j$. Equivalently

$$\begin{aligned}\mathbf{Pr}\left(|\lambda(\mathbf{Y})-\mathbf{E}(\lambda(\mathbf{Y}))|\geq\mu^{-1/2}\log^{10}\mu\right)&\leq \exp(-\Omega(\log^5\mu)),\\ \mathbf{Pr}\left(|\lambda(\mathbf{Z})-\mathbf{E}(\lambda(\mathbf{Z}))|\geq\mu^{-1/2}\log^{10}\mu\right)&\leq \exp(-\Omega(\log^5\mu)).\end{aligned}$$

Combining these bounds with (95) and (93), and denoting $R = \sum_i Y_i$, $S = \sum_j Z_j$, we get

$$\mathbf{E}^*_{\boldsymbol{\nu},\mu} = (1+O(\mu^{-1/2}\log^{12}\mu))e^{-\frac{1}{2}\mathbf{E}\lambda(\mathbf{Y})\mathbf{E}\lambda(\mathbf{Z})}\mathbf{Pr}\,(R=\mu)\,\mathbf{Pr}\,(S=\mu)+D_{\boldsymbol{\nu},\mu} \quad (98)$$
$$|D_{\boldsymbol{\nu},\mu}| \leq_b \exp(-\Omega(\log^5\mu)). \quad (99)$$

In (98), the exponential factor is $\exp(-O(\log^4\mu))$, and, by (89) and the conditions on μ, ν_i, r_i, the product of the probabilities is of order $(\nu_1 r_1 \nu_2 r_2)^{1/2}$, the latter being $\Omega((\mu\log\mu)^{-1})$. The resulting bound makes the remainder $D_{\boldsymbol{\nu},\mu}$ relatively negligible, so that

$$\mathbf{E}^*_{\boldsymbol{\nu},\mu} \leq_b \frac{e^{-\frac{1}{2}\mathbf{E}\lambda(\mathbf{Y})\mathbf{E}\lambda(\mathbf{Z})}}{(\nu_1 r_1 \nu_2 r_2)^{1/2}}.$$

The power of this bound is due to wide range of the parameters r_i for which it holds. However, we will also need an asymptotic formula for $\mathbf{E}^*_{\boldsymbol{\nu},\mu}$, and this requires asymptotic formulas for the local probabilities, rather than their upper bounds.

Intuitively, we stand a better chance of achieving this goal when the parameters r_1, r_2 are such that the events $\{\sum_i Y_i = \mu\}$ and $\{\sum_j Z_j = \mu\}$ have "sizeable" probabilities. What better candidates than $r_1 = \rho_1$ and $r_2 = \rho_2$ for which

$$\sum_{i=1}^{\nu_1} \mathbf{E}Y_i = \mu, \quad \sum_{j=1}^{\nu_2} \mathbf{E}Z_j = \mu.$$

Explicitly, using (86) and (87), ρ_1 and ρ_2 are the roots of

$$\sum_{i=1}^{\nu_1} \frac{x f_{c_i-1}(x)}{f_{c_i}(x)} = \mu, \quad (100)$$

and

$$\sum_{j=1}^{\nu_2} \frac{x f_{d_j-1}(x)}{f_{d_j}(x)} = \mu, \quad (101)$$

respectively; ($f_t(z) := e^z$, for $t \leq 0$). For $x \to 0$, the LHS of (100) and (101) approach $\sum_i c_i \leq \mu$ and $\sum_j d_j \leq \mu$, respectively. Each LHS is strictly increasing, asymptotic to $\nu_1 x$ and $\nu_2 x$ respectively, as $x \to \infty$. Assuming that $\sum_i c_i < \mu$ and $\sum_j d_j < \mu$, we see that the positive roots ρ_1 and ρ_2 exist uniquely, and that $\rho_i < \mu/\nu_i$. Assuming from now that $\mu = O(\nu_i \log \nu_i)$, $i = 1,2$, we obtain that $\rho_i = O(\log\mu)$ which puts ρ_1 and ρ_2 into the set of feasible (meeting (91)) r_1 and r_2, respectively.

How do the probabilities in (98) behave if $r_1 = \rho_1, r_2 = \rho_2$? For $Y_i = \mathrm{Po}(\rho; \geq 2)$ and $\mu = \nu\mathbf{E}(Y)$, it was proved in [1] that

$$\mathbf{Pr}\left(\sum_{i=1}^{\nu} Y_i = \mu + a\right) = \frac{1+O(a^2(\rho\nu)^{-1})}{(2\pi\nu\mathbf{Var}(Y))^{1/2}},$$

provided that $\rho\nu \to \infty$, and $a^2/(\rho\nu) \to 0$. (The condition $\rho\nu \to \infty$ is equivalent to $\nu\mathbf{Var}(Y) \to \infty$ since $\mathbf{Var}(Y) = \Theta(\rho)$). Suppose that in the present context $\nu_1\rho_1 \to \infty$, i.e. $\sum_i \mathbf{Var}(Y_i) \to \infty$, which is equivalent to $\mu - \sum_i c_i \to \infty$. Only simple modifications of the proof in [1] are needed to prove

Lemma A.1. *If $\sum_i \mathbf{Var}(Y_i) \to \infty$ and the c_i are uniformly bounded, then*

$$\mathbf{Pr}\left(\sum_{i=1}^{\nu_1} Y_i = \mu + a\right) = \frac{1 + O(a^2(\nu_1\rho_1)^{-1})}{(2\pi \sum_i \mathbf{Var}(Y_i))^{1/2}},$$

if $a^2(\nu_1\rho_1)^{-1} \to 0$. An analogous formula holds for $\mathbf{Pr}\left(\sum_j Z_j = \mu + a\right)$.

Proof. Let $W = \sum_\ell Y_\ell$. As usual, we start with the inversion formula

$$\begin{aligned}\mathbf{Pr}(W = \tau) &= \frac{1}{2\pi}\int_{-\pi}^{\pi} e^{-i\tau x}\mathbf{E}\left(e^{ix\sum_\ell Y_\ell}\right)dx \\ &= \frac{1}{2\pi}\int_{-\pi}^{\pi} e^{-i\tau x}\prod_{\ell=1}^{\nu_1}\mathbf{E}(e^{ixY_\ell})^v\,dx, \end{aligned} \tag{102}$$

where $\tau = \mu + a$. Let

$$\Sigma_1 = \sum_\ell \frac{\rho_1}{c_\ell + 1} = \Theta(\nu_1\rho_1)$$

and consider first $|x| \geq \Sigma_1^{-5/12}$. Using inequality (83) we estimate

$$\begin{aligned}&\frac{1}{2\pi}\int_{|x|\geq\Sigma_1^{-5/12}}\left|e^{-i\tau x}\prod_{\ell=1}^{\nu_1}\left(\frac{f_{c_\ell}(e^{ix}\rho_1)}{f_{c_\ell}(\rho_1)}\right)\right|dx \\ &\leq \frac{1}{2\pi}\int_{|x|\geq\Sigma_1^{-5/12}}\prod_{\ell=1}^{\nu_1} e^{\rho_1(\cos x - 1)/(c_\ell+1)}\,dx \leq e^{-\Sigma_1^{1/6}/3}.\end{aligned} \tag{103}$$

For $|x| \leq \Sigma_1^{-5/12}$, putting $\eta = \rho_1 e^{ix}$ and using $\sum_\ell \rho_1 f'_{c_\ell}(\rho_1)/f_{c_\ell}(\rho_1) = \mu$, $d/dx = i\eta d/d\eta$ we expand as a Taylor series around $x = 0$ to obtain

$$\begin{aligned}-i\tau x + \sum_\ell \log\left(\frac{f_{c_\ell}(e^{ix}\rho_1)}{f_{c_\ell}(\rho_1)}\right) &= -iax - \frac{x^2}{2}\sum_\ell \mathcal{D}\left(\frac{\eta f'_{c_\ell}(\eta)}{f_{c_\ell}(\eta)}\right)\Bigg|_{\eta=\rho_1} \\ &\quad - \frac{ix^3}{3!}\sum_\ell \mathcal{D}^2\left(\frac{\eta f'_{c_\ell}(\eta)}{f_{c_\ell}(\eta)}\right)\Bigg|_{\eta=\rho_1} \\ &\quad + O\left[x^4\sum_\ell \mathcal{D}^3\left(\frac{\eta f'_{c_\ell}(\eta)}{f_{c_\ell}(\eta)}\right)\Bigg|_{\eta=\tilde\eta}\right];\end{aligned} \tag{104}$$

here $\tilde\eta = \rho_1 e^{i\tilde x}$, with $\tilde x$ being between 0 and x, and $\mathcal{D} = \eta(d/d\eta)$. Now, the coefficients of $x^2/2$, $x^3/3!$ and x^4 are $\mathbf{Var}(W)$, $O(\mathbf{Var}(W))$, $O(\mathbf{Var}(W))$ respectively, and $\mathbf{Var}(W)$ is of order Σ_1. So the second and the third terms in (104) are $o(1)$ uniformly for $|x| \leq \Sigma_1^{-5/12}$. Therefore

$$\frac{1}{2\pi}\int_{|x|\leq\Sigma_1^{-5/12}} = \int_1 + \int_2 + \int_3, \tag{105}$$

where

$$\int_1 = \frac{1}{2\pi}\int_{|x|\le\Sigma_1^{-5/12}} e^{-iax-\mathbf{Var}(W)x^2/2}\,dx$$

$$= \frac{1}{\sqrt{2\pi\mathbf{Var}(W)}} + O\left(\frac{a^2+1}{\Sigma_1^{3/2}}\right), \tag{106}$$

$$\int_2 = O\left(\sum_\ell \mathcal{D}^2\left(\frac{\rho_1 f'_{c_\ell}(\rho_1)}{f(\rho_1)}\right)\int_{|x|\le\Sigma_1^{-5/12}} x^3 e^{-\mathbf{Var}(W)x^2/2}\,dx\right)$$

$$= O\left(\Sigma_1\int_{|x|\ge\Sigma_1^{-5/12}} |x|^3 e^{-\mathbf{Var}(W)x^2/2}\,dx\right).$$

$$= O(e^{-\alpha\Sigma_1^{1/6}}), \tag{107}$$

($\alpha > 0$ is an absolute constant), and

$$\int_3 = O\left[\Sigma_1\int_{|x|\le\Sigma_1^{-5/12}} x^4 e^{-\mathbf{Var}(W)x^2/2}\,dx\right]$$

$$= O\left(\frac{1}{\Sigma_1^{3/2}}\right). \tag{108}$$

Using (102)-(108), we arrive at

$$\mathbf{Pr}\,(W=\tau) = \frac{1}{\sqrt{2\pi v\mathbf{Var}(W)}}\times\left[1+O\left(\frac{a^2+1}{\Sigma_1}\right)\right].$$

□

Suppose, say, that $\nu_1\rho_1$ is bounded, or equivalently that $\sigma_1 := \mu_1 - \sum_i c_i = O(1)$. Extending an argument in [21] (which covers the case of identically distributed Y_i), we can show that

$$\mathbf{Pr}\,(R=\mu) = (1+O(\nu_1^{-1}))e^{-\sigma_1}\frac{\sigma_1^{\sigma_1}}{\sigma_1!}. \tag{109}$$

An analogous relation holds for $\mathbf{Pr}(S=\mu)$ if $\sigma_2 := \mu - \sum_j d_j = O(1)$. Clearly then, regardless of the behavior of σ_1, σ_2, in (98) the remainder term $D_{\nu,\mu}$ is negligible compared to the explicit term. □

B. Concentration of W.

We need to prove the following result.

Let S be a set with $|S| = N$. Let Ω be the set of $N!$ permutations of S. Let ω be chosen uniformly from Ω.

Let $Z = Z(\omega)$ be such that $|Z(\omega) - Z(\omega')| \le 1$ when ω' is obtained from ω by interchanging two elements of the permutation.

Lemma B.1.

$$\mathbf{Pr}(|Z - \mathbf{E}Z| \ge t) \le 2e^{-2t^2/N}.$$

Proof. For a fixed sequence permutation $(x_1, x_2, \ldots, x_n)$ and $0 \le i \le N$ let

$$Z_i(x_1, x_2, \ldots, x_i) = \mathbf{E}(Z \mid \omega_j = x_j, 1 \le j \le i).$$

Clearly the sequence $Z_0, Z_1, \ldots, Z_N$ is a martingale. To apply the Azuma -Hoeffding inequality, we need to show that

$$|Z_i(x_1, x_2, \ldots, x_i) - Z_i(x_1, x_2, \ldots, x_{i-1}, x_i')| \leq 1 \tag{110}$$

for all i-tuples $(x_1, x_2, \ldots, x_i)$ with distinct components, and $x_i' \neq x_1, \ldots, x_{i-1}$. (Indeed, the inequality (110) readily implies that $|Z_{i+1} - Z_i| \leq 1$.)

Consider

$$\Omega_1 = \{\omega \in \Omega : \omega_j = x_j, 1 \leq j \leq i\}$$

and

$$\Omega_1' = \{\omega \in \Omega : \omega_j = x_j, 1 \leq j \leq i-1, \omega_i = x_i'\}$$

and the map $f : \Omega_1 \to \Omega_1'$ defined as follows. If $\omega = x_1 x_2 \ldots x_{i-1} x_i y_{i+1} \ldots y_N$ and $y_j = x_i'$ then $f(\omega) = x_1 x_2 \ldots x_{i-1} x_i' y_{i+1} \ldots y_{j-1} x_i y_{j+1} \ldots y_N$. Observe that f is a bijection and that

$$|Z_i(x_1, x_2, \ldots, x_i) - Z_i(x_1, x_2, \ldots, x_i')| = \left| \frac{\sum_{y_{i+1}, \ldots, y_N} (Z(\omega) - Z(f(\omega)))}{(N-i)!} \right| \leq 1.$$

□

Alan Frieze
Department of Mathematical Sciences, Carnegie Mellon University, Pittsburgh PA 15213.

Boris Pittel
Department of mathematics, Ohio State University, Columbus OH43210.

Trends in Mathematics, © 2004 Birkhäuser Verlag Basel/Switzerland

Estimating the Growth Constant of Labelled Planar Graphs

Omer Giménez and Marc Noy

ABSTRACT: *Let G_n be the number of labelled planar graphs on n vertices and $\gamma = \lim_{n\to\infty} (G_n/n!)^{1/n}$. It is known that $26.1848 < \gamma < 30.0606$. In this paper we sharpen these bounds to $27.22685 < \gamma < 27.22688$. The proof is based on recent results of Bender, Gao and Wormald [Elec. J. Combinatorics 9 (2002) R43], and on singularity analysis of generating functions.*

1. Introduction

Let G_n be the number of labelled planar graphs on n vertices. We stress the fact that we are counting planar graphs, that is, graphs that can be embedded in the plane, without considering a particular embedding. A subadditivity argument shows that the following limit exists:

$$\gamma = \lim_{n\to\infty} (G_n/n!)^{1/n}.$$

This was first observed by Denise, Vasconcellos and Welsh [5]. Recently, a number of authors have pursued the study of random planar graphs [7, 11, 13]. One of the problems mentioned in these references is to obtain good bounds for the value of the constant γ.

A lower bound results from the work of Bender, Gao and Wormald [1]. They show that, if B_n is the number of 2-connected labelled planar graphs, then $\lim_{n\to\infty} (B_n/n!)^{1/n} \approx 26.1848$; thus γ is at least this number. On the other hand, by means of an encoding scheme for (unlabelled) planar graphs, Bonichon, Gavoille and Hanusse [2] have shown that $\gamma < 2^{5.007} = 32.1556$. Very recently, this upper bound has been reduced by Bonichon at al. [3] to $2^{4.9098} = 30.0606$.

The purpose of this paper is to sharpen these bounds.

Theorem 1.1. *If G_n is the number of labelled planar graphs on n vertices, and $\gamma = \lim_{n\to\infty} (G_n/n!)^{1/n}$, then*

$$27.22685 < \gamma < 27.22688.$$

The proof is based on singularity analysis of generating functions, as described in the forthcoming book by Flajolet and Sedgewick [6]. If C_n is the number of connected labelled planar graphs on n vertices, there are simple equations linking the generating functions $B(x) = \sum B_n x^n/n!$, $C(x) = \sum C_n x^n/n!$ and $G(x) = \sum G_n x^n/n!$. The singularities of $B(x)$ were determined in [1], and from this we are able to compute the dominant singularities of $C(x)$ and $G(x)$, which are both equal to γ^{-1}. Since we do not have explicit analytic expressions for the corresponding singularities, in order to determine them accurately we have to rely on numerical methods.

In Section 2 we review the preliminaries needed for the proof, and we also recall previous work on the enumeration of planar graphs and planar maps. In Section 3 we present the proof of our main result. We conclude with some remarks and open questions.

All graphs in this paper are simple and labelled, so we omit the qualifier from now on. As a rule, variable x in generating function marks vertices, and variable y marks edges.

2. Preliminaries

Let G_n, C_n and B_n denote, respectively, the number of planar graphs, connected planar graphs, and 2-connected planar graphs on n vertices. We introduce the following exponential generating functions:

$$G(x) = \sum_{n\geq 0} G_n \frac{x^n}{n!}, \qquad C(x) = \sum_{n\geq 0} C_n \frac{x^n}{n!}, \qquad B(x) = \sum_{n\geq 0} B_n \frac{x^n}{n!}.$$

These series are related as follows.

Lemma 2.1. *The series $G(x)$, $C(x)$ and $B(x)$ satisfy the following equations:*

$$G(x) = \exp(C(x)), \qquad xC'(x) = x\exp\left(B'(xC'(x))\right),$$

where $C'(x) = \mathrm{d}C(x)/\mathrm{d}x$ and $B'(x) = \mathrm{d}B(x)/\mathrm{d}x$.

Proof. The first equation is standard, given the fact that a planar graph is a set of connected planar graphs.

The second equation follows from a standard argument on the decomposition of a connected graph into 2-connected components; see, for instance, [9, p. 10]. □

Let $B_{n,q}$ be the number of 2-connected planar graphs with n vertices and q edges, and let

$$B(x,y) = \sum B_{n,q} y^q \frac{x^n}{n!}$$

be the corresponding bivariate generating function. Notice that $B(x,1) = B(x)$.

Define the series $M(x,y)$ by means of the expression

$$M(x,y) = x^2y^2\left(\frac{1}{1+xy} + \frac{1}{1+y} - 1 - \frac{(1+U)^2(1+V)^2}{(1+U+V)^3}\right), \tag{1}$$

where $U(x,y)$ and $V(x,y)$ are algebraic functions given by

$$U = xy(1+V)^2, \quad V = y(1+U)^2.$$

The next result from [1] is essential in what follows.

Theorem 2.2. *We have*

$$\frac{\partial B(x,y)}{\partial y} = \frac{x^2}{2}\left(\frac{1+D(x,y)}{1+y} - 1\right), \tag{2}$$

where $D = D(x,y)$ is defined implicitly by $D(x,0) = 0$ and

$$\frac{M(x,D)}{2x^2D} - \log\left(\frac{1+D}{1+y}\right) + \frac{xD^2}{1+xD} = 0. \tag{3}$$

Moreover, the coefficients of $D(x,y)$ are nonnegative.

From the equations above, it is shown in [1] that the radius of convergence of the series $B(x) = B(x, 1)$ is equal to $R \approx 0.0381910976694$. Moreover, R is given by explicit analytic expressions and can be computed to any degree of accuracy.

Let us comment on the previous equations, which are based on the well-developed theory of counting rooted planar maps (see, for instance, [8]). The algebraic generating function M corresponds to 3-connected planar maps, and they were enumerated by Mullin and Schellenberg [12]. The next ingredient is Whitney's theorem: a 3-connected planar graph has a unique embedding in the sphere. Hence, counting 3-connected planar graphs essentially amounts to counting 3-connected planar maps. Finally, the decomposition of a 2-connected graph into 3-connected components (see, for instance, [14]) allows one to connect the generating functions B and M. This decomposition is encoded in equations (2) and (3) and were found by Walsh [15] using the so-called networks.

If we are interested in the bivariate generating functions $C(x, y)$ and $G(x, y)$, where y marks edges, the corresponding equations are

$$G(x, y) = \exp(C(x, y)), \qquad x\frac{\partial}{\partial x}C(x, y) = x\exp\left(\frac{\partial}{\partial x}B(x\frac{\partial}{\partial x}C(x, y), y)\right).$$

The reason is that the parameter "number of edges" is additive under taking connected components and 2-connected components.

Next we turn to some analytical considerations. Let $f(x)$ be a function analytic at 0 with real non-negative Taylor coefficients and such that $f(0) = 0$. Let $\psi(u)$ be its functional inverse, that is, $\psi(f(x)) = x$, and assume that $R > 0$ is the radius of convergence of ψ. In order to determine the radius of convergence ρ of f, there are two cases to consider (see Proposition IV.4 in [6]):

: (i) There exists $\tau \in (0, R)$ (necessarily unique) such that $\psi'(\tau) = 0$. Then $\rho = \psi(\tau)$.

: (ii) We have $\psi'(u) \neq 0$ for all $u \in (0, R)$. Then $\rho = \sup\limits_{0 \leq u < R} \psi(u)$.

The rationale in (i) is that $\psi(u)$ ceases to be invertible at τ since the derivative vanishes; hence the inverse function $f(x)$ ceases to be analytic at $\psi(\tau)$. In case (ii) there is no obstruction to the inversion of $\psi(u)$ and the radius of convergence of $f(x)$ is as stated.

Notice that in both cases we have $\rho = \sup_{0 \leq u < R} \psi(u)$, in the first case because ψ has an absolute maximum at τ; the following easy consequences can be used to bound ρ from above and from below.

Fact 1.: If $h(u)$ is a function such that $\psi(u) \leq h(u)$ for all $u \in (0, R)$, then $\rho \leq \sup\limits_{0 \leq u < R} h(u)$.

Fact 2.: If $\lambda \leq \psi(u)$ for some $\lambda \in \mathbb{R}^+$ and $u \in (0, R)$, then $\lambda \leq \rho$.

3. Proof of Theorem 1.1

From the definition of γ, it follows that γ^{-1} is the radius of convergence of $G(x)$. Since $G(x) = \exp(C(x))$ and $\exp(x)$ is an entire function, $G(x)$ and $C(x)$ have the same radius of convergence; from now on we concentrate on $C(x)$.

In order to simplify the notation, we define a new series $F(x)$ as

$$F(x) = xC'(x).$$

The second equation in Lemma 2.1 now takes the simpler form

$$F(x) = x \exp(B'(F(x))). \tag{4}$$

Notice that the radius of convergence of $B'(x)$ and $F(x)$, are the same, respectively, as those of $B(x)$ and $C(x)$.

We define

$$\psi(u) = ue^{-B'(u)},$$

so that $F(x)$ and $\psi(u)$ are functional inverses. The radius of convergence of ψ is the same as that of $B'(x)$, namely R. Our task is to estimate the radius of convergence $\rho = \gamma^{-1}$ of $F(x)$ from the fact that ψ and F are inverses to each other.

Lower bound. Let $B_{\langle k\rangle}(x) = B_0 + B_1 x + \cdots B_k x^k/k!$ be the truncation of the series $B(x)$ at order k. Since the B_n are non-negative,

$$B'_{\langle k\rangle}(u) \le B'(u) \qquad \text{for all } u \in (0, R).$$

Hence, if we define

$$\psi_k(u) = u \exp(-B'_{\langle k\rangle}(u)),$$

we have

$$\psi(u) \le \psi_k(u) \qquad \text{for all } u \in (0, R).$$

Since $B_{\langle k\rangle}(u)$ is a polynomial, the maximum value of $\psi(u)$ in the interval $[0, R]$ is easily computed: if the unique root τ_k of $\psi'_k(\tau_k) = 0$ is less than R, the maximum is $\psi_k(\tau_k)$; otherwise it is $\psi_k(R)$. For all values of k we have been able to check, it turns out that $\tau_k > R$, hence the second alternative applies.

We have computed the values of B_n up to $n = 25$ (they are also computed in [1]). We have $\psi_{24}(R) \approx 0.03672844872$, and from Fact 1 it follows that $\rho < 0.03672844872$, hence $\gamma = \rho^{-1} > 27.22685$.

Upper bound. Given u in $(0, R)$, we show next how to determine a number β_u such that $\beta_u \ge B'(u)$. If we set $\lambda = u \exp(-\beta_u)$, then

$$\lambda \le u \exp(-B'(u)) = \psi(u)$$

and we can apply Fact 2 to bound ρ from below.

We start by rewriting equation (2) as

$$B(x, y) = \int_0^y P(x, t) dt,$$

where

$$P(x, y) = \frac{x^2}{2}\left(\frac{1 + D(x, y)}{1 + y} - 1\right).$$

Note that, since B is a power series with non-negative coefficients, so its derivatives and the derivatives of $P(x, y) = \partial B(x, y)/\partial y$. Consequently we obtain that

$$\frac{\partial^2 B(x, y)}{\partial x^2} \ge 0, \qquad \frac{\partial^k P(x, y)}{\partial y^k} \ge 0 \quad \text{for all } k,$$

for positive values of x and y.

Next we show how to compute values B_1 and B_2 such that

$$B_1 \le B(u) = \int_0^1 P(u, t) dt,$$

$$B_2 \ge B(u + \epsilon) = \int_0^1 P(u + \epsilon, t) dt,$$

where $\epsilon > 0$ is such that $u + \epsilon < R$.

We compute B_1 and B_2 using numerical integration methods. We apply several Newton-Cotes quadratures as described, for instance, in [10]. These methods relate the error term to the evaluation of a certain derivative of the integrand, and by the positiveness of $\partial^k P/\partial y^k$ for positive values of x and y, we know with certainty the sign of the error term; this is essential for us, as we are looking both for upper and lower bounds for the integrals.

We obtain the value B_1 by applying, for example, the repeated 3-point open rule. The error term $B(u) - B_1$ is the positive number $28h^5 f^{(4)}(\xi)/90$, where $f(y) = P(u, y)$, h is the distance between successive evaluations of the integrand and ξ is in (0,1). To obtain B_2 we apply the repeated 5-point closed rule. The error term $B(u + \epsilon) - B_2$ is now the negative number $-8h^7 f^{(6)}(\xi)/945$, where $f(y) = P(u + \epsilon, y)$.

These methods require the evaluation of $P(u, t)$ and $P(u + \epsilon, t)$ for several values of t. This is not a problem, since D is a function defined implicitly by equations (1) and (3), and we can apply classical methods to obtain evaluations of D up to any precision required.

Finally, we set $\beta_u = (B_2 - B_1)/\epsilon$. This yields an upper bound for $B'(u)$, since the positiveness of $\partial^2 B/\partial x^2$ implies that

$$B'(u) \le \frac{B(u+\epsilon) - B(u)}{\epsilon} \le \frac{B_2 - B_1}{\epsilon} = \beta_u$$

We have performed these computations with the help of Maple using 25 significant digits. For $u = 0.038191096$, $\epsilon = 1.6 \cdot 10^{-9}$ and $h = 1/30000$, for both integration methods. Using this data we have obtained that $\lambda > 0.036728410432$. By applying Fact 2 we obtain $\gamma = \rho^{-1} < 27.22688$.

4. Concluding remarks

We have determined γ with an accuracy of four decimal digits. Next we comment on the problems faced in order to increase the number of correct digits. For the lower bound one needs to compute exact values of B_n for larger values of n, and this is indeed possible. However, we have worked with larger truncations $B_{<k>}(x)$ of the series $B(x)$ and we have observed only a marginal improvement on the lower bound.

To improve the upper bound, we need to estimate $B(u)$ and $B(u + \epsilon)$ for u close enough to R. But the closer u is to R, the smaller ϵ becomes, so we also need to obtain sharper estimations for $B(u)$ and $B(u+\epsilon)$ in order to really improve the upper bound. On the whole this would require to change the integration method we have used, or to evaluate $P(u, y)$ and $P(u+\epsilon, y)$ at even more points; with our present algorithms this seems computationally unfeasible. The main open problem here is whether γ can be determined analytically.

A second, and more important, remark is the following. One of our aims when performing the computations for the upper bound was to obtain $\tau < R$ such that $\phi'(\tau) = 0$, that is, to show that $F(x) = xC'(x)$ corresponds to case (i) of section 2. Then it would follow by singularity analysis that

$$C_n \sim K n^{-5/2} \gamma^n n!,$$

for some constant K. On the other hand, if τ does not exist then we would obtain, from the results in [1] and singularity analysis on $\phi(u)$, that

$$C_n \sim K' n^{-7/2} \gamma^n n!,$$

for some constant K'. Thus we have a dichotomy for the asymptotic behavior of C_n; deciding which is the true situation remains an open question.

With respect to the existence of τ, notice that $\phi'(\tau) = 0$ is equivalent to $\tau B''(\tau) = 1$. If $\tau \in (0, R)$ exists, then $B''(\tau) = \tau^{-1} > R^{-1} = 26.18\ldots$. But our computations for $u \in (0, R - 10^{-9})$ give $B''(u) < 2.09\ldots$ for Hence, if τ exists, then it must be extremely close to the singularity R.

Finally, we remark that, if g_n is the number of unlabelled planar graphs on n vertices, McDiarmid, Steger and Welsh [11] have shown that $\gamma_u = \lim_{n\to\infty} (g_n)^{1/n}$ exists and that $\gamma < \gamma_u$. Thus our result also gives 27.2268 as a new lower bound for γ_u.

5. Acknowledgements

We are grateful to Philippe Flajolet for useful comments on the paper, in particular, for pointing out the dichotomy mentioned in the conclusions. Thanks also to the anonymous referee for his criticisms.

References

[1] E. A. Bender, Z. Gao, N. C. Wormald, The number of 2-connected labelled planar graphs, *Elec. J. Combinatorics* 9 (2002), #43.

[2] N. Bonichon, C. Gavoille, N. Hanusse, An information-theoretic upper bound of planar graphs using triangulations, *Springer Lecture Notes in Computer Science* 2607 (2003), 499–510.

[3] N. Bonichon, C. Gavoille, N. Hanusse, D. Poulalhon, G. Schaeffer, *Planar Graphs, via Well-Orderly Maps and Trees* (preprint).

[4] M. Bodirsky, C. Gröpl and M. Kang, Generating Labeled Planar Graphs Uniformly at Random, in Proc. of ICALP 2003, LNCS 2719, 1095–1107.

[5] A. Denise, M. Vasconcellos, D. J. A. Welsh, The random planar graph, *Congr. Numer.* 113 (1996), 61–79.

[6] P. Flajolet, R. Sedgewick, *Analytic Combinatorics* (book in preparation), preliminary version available at `http://algo.inria.fr/flajolet/`

[7] S. Gerke, C. McDiarmid, On the Number of Edges in Random Planar Graphs, *Comb. Prob. and Computing* 13 (2004), 165–183.

[8] I. P. Goulden, D. M. Jackson, *Combinatorial enumeration*, John Wiley & Sons, Inc., New York (1983).

[9] F. Harary, E. Palmer, *Graphical Enumeration*, Academic Press, New York-London (1973).

[10] F. B. Hildebrand, *Introduction to Numerical Analysis*, Dover, New York (1987).

[11] C. McDiarmid, A. Steger, D. Welsh, *Random Planar Graphs*, preprint (2003).

[12] R. C: Mullin, P. J. Schellenberg, The enumeration of c-nets via quadrangulations, *J. Combin. Theory* 4 (1968), 259–276.

[13] D. Osthus, H. J. Prömel, A. Taraz, On random planar graphs, the number of planar graphs and their triangulations, *J. Combin. Theory Ser. B* 88 (2003), 119–134.

[14] W. T. Tutte, *Graph theory* (reprint of the 1984 original), Cambridge University Press, Cambridge (2001).

[15] T. R. S. Walsh, Counting labelled three-connected and homeomorphically irreducible two-connected graphs, *J. Combin. Theory Ser. B* 32 (1982), 1–11.

Omer Giménez
Universitat Politècnica de Catalunya
omer.gimenez@upc.es

Marc Noy
Universitat Politècnica de Catalunya
marc.noy@upc.es

Trends in Mathematics, © 2004 Birkhäuser Verlag Basel/Switzerland

The Number of Spanning Trees in P4-Reducible Graphs

Stavros D. Nikolopoulos and Charis Papadopoulos

ABSTRACT: *We present an efficient algorithm for determining the number of spanning trees in the class of P_4-reducible graphs, which are perfect graphs and generalize both the well-known class of cographs and the class of quasi-threshold graphs. In particular, for a P_4-reducible graph G on n vertices and m edges, our algorithm computes the number of spanning trees of G in $O(n+m)$ time and space, where the complexity of arithmetic operations is measured under the uniform cost criterion. The algorithm takes advantage of the modular decomposition tree of the input graph which it gradually shrinks in a systematic fashion until it becomes a single vertex while at the same time appropriately updating certain parameters whose product gives the desired number of spanning trees. The correctness of the algorithm is established through the Kirchhoff matrix tree theorem, and is also based on structural and algorithmic properties of the graphs with few P_4s. Our results generalize previous results and extend the family of graphs admitting linear-time algorithms for the number of their spanning trees.*

1. Introduction

The purpose of this paper is to study the problem of finding the number of spanning trees in the class of P_4-*reducible* graphs, which are defined as the graphs for which no vertex belongs to more than one induced P_4 [3]. We take advantage of the modular decomposition tree of the class of P_4-reducible graphs in order to obtain an efficient solution of the above mentioned problem.

The *modular decomposition* of a graph G is a tree denoted as $T(G)$; the leaves of $T(G)$ are the vertices of G and the set of leaves associated with the subtree rooted at an internal node, induce a strong module of G. Thus, the modular decomposition tree $T(G)$ represents all the strong modules of G. An internal node is labeled by either 0 (or p for *parallel* module), 1 (or s for *series* module), or 2 (or n for *neighborhood* module). The module corresponding to a 0-node induces a disconnected subgraph of G, that of a 1-node induces a connected subgraph of G whose complement is disconnected, and that of a 2-node induces a connected subgraph of G whose complement is also connected. A linear-time algorithm for the construction of the tree $T(G)$ can be found in [4].

Let u be an internal node of $T(G)$. We denote by $M(u)$ the module corresponding to u which consists of the set of vertices of G in the subtree of $T(G)$ rooted at node u. Let $\text{ch}(u) = \{u_1, u_2, \ldots, u_p\}$ be the set of the children of u in $T(G)$. Let $G(u)$ be the *representative graph* of the module $M(u)$, which is defined as follows: $V(G(u)) = \{u_1, u_2, \ldots, u_p\}$, and $E(G(u)) = \{(u_i, u_j) \mid (v_i, v_j) \in E(G), v_i \in M(u_i) \text{ and } v_j \in M(u_j)\}$.

Let G be a graph, $T(G)$ its modular decomposition tree and u an internal 2-node of $T(G)$. The graph G is a P_4-reducible graph iff for every u of $T(G)$, $G(u)$ is either a P_4, or a bull graph. Moreover, the vertices of the P_4, or the P_4 of the bull graph are leaf vertices in $T(G)$ (Giakoumakis and Vanherpe [2]).

2. The Algorithm

Let G be a P_4-reducible graph on n vertices and m edges. In order to compute the number of spanning trees of the graph G, we make use of the Kirchhoff matrix tree theorem [1]; that is, we delete an arbitrary vertex $w \in V(G)$ and all the edges incident on vertex w.

Next, we set $s(v) \leftarrow d(v)$ for every vertex v of the graph $G - w$, where $d(v)$ is the degree of the vertex v in G; we call these labels of the vertices their *s-labels*. The algorithm works by contracting in a systematic fashion the contractible nodes of the modular decomposition tree of the graph $G - w$ and by assigning to the leaf that is produced the highest-index vertex of G which is a child of the contracted node. The contractions are done by means of two functions, namely, *Replace_Parallel-Series*() and *Replace_Neighborhood*(), which also update the s-labels of the children/vertices of the contractible node. We note that the s-labels are assumed to be global variables in our algorithm. The algorithm *Spanning_Trees_Number* and the two functions are given in detail below.

Let u be a 0(1)-node such that $\text{ch}(u) = \{v_1, v_2, \ldots, v_p\}$ forms a contracted block of $T(G)$; that is, each $v_i \in \text{ch}(u)$ is a leaf. The function Replace_Parallel-Series()works as follows:

- if u is a 1-node then increase the s-labels $s(v_1), s(v_2), \ldots, s(v_p)$ by 1;
- compute the parameter $e(u) := \sum_{i=1}^{p-1} \frac{1}{s(v_i)}$;
- update the s-labels $s(v_{p-1})$ and $s(v_p)$ as follows: $s(v_{p-1}) := s(v_{p-1}) \cdot (1 + s(v_p) \cdot e(u))$, and $s(v_p) := s(v_p) \ / \ (1 + s(v_p) \cdot e(u))$; if u is a 1-node then decrease $s(v_p)$ by 1;
- make vertex v_p a child of $p(u)$ and delete the vertices $v_1, v_2, \ldots, v_{p-1}$ and node u from $T(G)$.

Let u be a 2-node such that $\text{ch}(u) = \{v_1, v_2, \ldots, v_p\}$ forms a contracted block of $T(G)$. If $G(u)$ is a P_4 then $|\text{ch}(u)| = 4$, or else if $G(u)$ is a bull graph, then $|\text{ch}(u)| = 5$ (see [2]). The function Replace_Neighborhood() works as follows:

- compute the endpoints v_1, v_4 and the midpoints v_2, v_3 of the P_4 $v_1v_2v_3v_4$ of the graph induced by $\text{ch}(u)$, and also the values of d_s and d_k, where $d_s = d(v_1) = d(v_4)$ and $d_k = d(v_2) = d(v_3)$;
- if $|\text{ch}(u)| = 4$, then update the s-labels of the vertices of $\text{ch}(u)$, as follows:

$$s(v_1) = d_s \tag{1}$$

$$s(v_2) = d_k + \frac{d_s - 1}{d_s} \quad \bar{\sigma} \quad \bar{\sigma} \quad \bar{\sigma} \quad \bar{\sigma} \tag{2}$$

$$s(v_3) = \frac{d_s}{\gamma} \cdot (d_k - \frac{d_s + 1}{d_s}) \tag{3}$$

$$s(v_4) = \gamma \tag{4}$$

where

$$\gamma = 2 \cdot (d_s + d_k + 1); \quad \bar{\sigma} \quad \bar{\sigma} \quad \bar{\sigma} \quad \bar{\sigma} \tag{5}$$

- if $|\mathrm{ch}(u)| = 5$, then update the s-labels of the vertices of $\mathrm{ch}(u)$, as follows:

$$s(v_1) = d_s \tag{6}$$

$$s(v_2) = d_k + \frac{d_s - 1}{d_s} \tag{7}$$

$$s(v_3) = \frac{d_s}{\gamma} \cdot \frac{s(v_5) \cdot (d_k - \frac{d_s+1}{d_s}) - 2}{s(v_5) + \frac{d_s}{d_s+1}} \quad \bar{\sigma} \quad \bar{\sigma} \tag{8}$$

$$s(v_4) = \gamma \tag{9}$$

$$s(v_5) = s(v_5) + \frac{d_s}{d_s + 1} \tag{10}$$

where

$$\gamma = 2 \cdot (d_s + d_k + 1) - (d_s - 1) \cdot \frac{2 + \frac{d_s}{d_s+1} \cdot (d_k - \frac{d_s+1}{d_s})}{s(v_5) + \frac{d_s}{d_s+1}}; \tag{11}$$

- make vertex v_3 child of node $p(u)$ and delete the vertices of $\mathrm{ch}(u)$ and node u from the tree $T(G)$.

We next describe an algorithm which computes the number of spanning trees $\tau(G)$ of a P_4-reducible graph G; it works as follows: first it computes the degree $d(v_i)$ of each vertex $v_i \in V(G)$ and assigns $s(v_i) := d(v_i)$, $1 \leq i \leq n$. Then, it computes the graph $G := G - v_n$, where $v_n \in V(G)$, and constructs its modular decomposition tree $T := T(G)$; the resulting graph G has $n-1$ vertices. Recall that, v_i is a leaf of $T(G)$, $1 \leq i \leq n-1$. Next, it repeatedly applies the functions Replace_Parallel-Series(u, T) to each contracted 0(1)-node u, and Replace_Neighborhood(u, T) to each contracted 2-node u, by a bottom-up traversal of T. With this process, it finds the s-labels $s(v_1), s(v_2), \ldots, s(v_{n-1})$ of the vertices of T. Finally, it computes the number of spanning trees $\tau(G) := \prod_{i=1}^{n-1} s(v_i)$.

Theorem 2.1. *The number of spanning trees of a P_4-reducible graph G on n vertices and m edges can be computed in $O(n+m)$ time and space.*

References

[1] N. Biggs, *Algebraic Graph Theory*, Cambridge University Press, London, 1974.

[2] V. Giakoumakis and J-M. Vanherpe, On extended P_4-reducible and P_4-sparse graphs, *Theor. Comput. Sci.* **180** (1997) 269–286.

[3] B. Jamison and S. Olariu, P_4-reducible graphs a class of tree representable graphs, *Studies Appl. Math.* **81** (1989) 79–87.

[4] R.M. McConnell and J. Spinrad, Modular decomposition and transitive orientation, *Discrete Math.* **201** (1999) 189–241.

Stavros D. Nikolopoulos and Charis Papadopoulos
Dept. of Computer Science, University of Ioannina, GR-45110 Ioannina, Greece
stavros@cs.uoi.gr

Part III

Analysis of Algorithms

Trends in Mathematics, © 2004 Birkhäuser Verlag Basel/Switzerland

On the Stationary Search Cost for the Move-to-Root Rule with Random Weights

Javiera Barrera and Christian Paroissin

ABSTRACT: *Consider a binary search tree containing n items updated according to the move-to-root rule as defined first by Allen and Munro [1]. Assume that request probabilities are themselves random. Formula for the expectation is derived from classical result and is compared to the case of a list updated according to the move-to-front rule [2]. The case of Gamma request is then studied.*

1. Introduction and model

Consider a binary search tree containing n items updated as follows: at each unit of discrete time, an item is requested independently of the previous requests and is moved to the root of the binary tree. This way to update a list is called the move-to-root rule and was suggested by Allen and Munro [1]. The aim of this heuristic is to keep a binary search tree in near-optimal form. The associated Markov chain was studied by Dobrow and Fill [3]. One could be interested in the search cost of an item in the binary tree at a given time. The search cost is simply defined as the number of ancestors of the item requested (or equivalently as the depth of the item in the tree minus one). Let $\omega = (w_i)_{i\in\mathbb{N}^*}$ be a sequence of independent strictly positive random variables. For any $i \geqslant 1$, w_i represents the weight of the item i. Let us consider the n first files. One can construct a vector of requested probabilities $\mathbf{p_n} = (p_1, \dots, p_n)$ from weights as follows:

$$\forall i \in \{1,\dots,n\}\,, \quad p_i = \frac{w_i}{W_n} \qquad \text{where} \qquad W_n = \sum_{i=1}^{n} w_i\,.$$

Such construction of random request probabilities was first made by Kingman [4] and is natural to model the unknown in other areas like Bayesian statistics.

2. Expectation of the stationary search cost

Let us denote by $S_n^{\mathfrak{T}}$ the stationary search cost. Using the same technique as [2], one can get the following theorem from classical result:

Theorem 2.1. *For a sequence ω of independent random weights, we have:*

$$\mathbb{E}[S_n^{\mathfrak{T}}] = 2\sum_{\substack{i,j=1\\ i<j}}^{n} \int_0^{\infty}\int_t^{\infty} \phi_i'(u)\phi_j'(u)\prod_{k=1}^{i-1}\phi_k(t)\prod_{k=i+1}^{j-1}\phi_k(u)\prod_{k=j+1}^{n}\phi_k(t)\,dudt\,, \tag{1}$$

where ϕ_i is the Laplace transform of w_i, for all $i \geqslant 1$.

This expected stationary search cost can be compared to the one when the data structure used is a list. Let us denote by $S_n^{\mathcal{L}}$ the stationary search cost of a list updated with the move-to-front rule (see [2]). From equation (1), since ϕ is a decreasing function of the time, one gets (see corollary 1 in [2]):

$$\mathbb{E}[S_n^{\mathcal{T}}] \leqslant 2 \sum_{1\leq i<j\leq n} \int_0^\infty \left(\int_t^\infty \phi_i'(r)\phi_j'(r)\, dr \right) \prod_{\substack{k=1\\ k\neq i,j}}^{n} \phi_k(t)\, dt = \mathbb{E}[S_n^{\mathcal{L}}]\,.$$

This result is not surprising since quickest retrieving data is one of the advantage of binary search tree over list. One could prove the following inequalities:

Proposition 2.2. *For a sequence ω of independent random weights, we have:*

$$\mathbb{E}[S_n] \leqslant H(\mathbf{p_n}) \leqslant 2\log n\,, \tag{2}$$

where $H(\mathbf{p_n})$ is the Shannon entropy.

This bound is not a surprise since $\log n$ is the order of the height of a random binary search tree with n nodes. Moreover the expectation reaches the equality when one consider deterministic and equal weights.

3. Gamma request probabilities

Let us consider a sequence of independent random variables having the Gamma distribution with parameters a_i and 1. Thus one gets:

$$\mathbb{E}[S_n] = 2 \sum_{1\leq i<j\leq n} \frac{a_i a_j}{A_n(a_i + \cdots + a_j + 1)}\,, \tag{3}$$

where $A_n = a_1 + \cdots + a_n$. In the case of iid random variables (all a_i's equals to a), one can compute an asymptotic equivalent: $\mathbb{E}[S_n] \sim 2\log n$. Remark that it does not depend on a. Moreover the asymptotic equivalent equals to the expectation in the case of deterministic and equal weights, which appears as the "worst case".

References

[1] B. Allen and I. Munro. Self-organizing binary search trees. *J. Assoc. Comput. Mach.*, 25:526–535, 1978.

[2] J. Barrera and C. Paroissin. On the distribution of the stationary search cost for the move-to-front with random weights. *J. Appl. Prob.*, 41(1): 250–262, 2004.

[3] R.P. Dobrow and J.A. Fill. On the Markov chain for the move-to-root rule for binary search trees. *Ann. Appl. Probab.*, 5(1):1–19, 1995.

[4] J.F.C. Kingman. Random discrete distributions. *J. R. Statist. Soc.*, B37:1–22, 1975.

Javiera Barrera
CMM, Univ. de Chile, Casilla 170-3, Correo 7, Santiago, Chile
jbarrera@dim.uchile.cl

Christian Paroissin
MODAL'X, Univ. Paris X, 200 av. de la République, 92001 Nanterre Cedex, France
cparoiss@u-paris10.fr

Trends in Mathematics, © 2004 Birkhäuser Verlag Basel/Switzerland

Average-Case Analysis for the Probabilistic Bin Packing Problem

Monia Bellalouna, Salma Souissi, and Bernard Ycart

ABSTRACT: *In the Probabilistic Bin Packing Problem (PBPP), some items are randomly deleted after having been placed into bins. The problem is to rearrange the remaining items, using the a priori solution. The initial arrangement being done with the Next Fit Decreasing heuristic, we consider two procedures. In the first one, the NF algorithm is applied to the new list. In the second one, successive groups of bins are optimally rearranged. In both cases, we prove a law of large numbers and a central limit theorem for the number of occupied bins as the initial number of items tends to infinity.*

1. Introduction

Bin Packing is a classical NP-hard problem of optimization [10]: given items of sizes $(x_1, \ldots, x_n)$, all smaller than 1, one must pack them into bins of size 1, so as to minimize the total number of non-empty bins. Many approximation heuristics have been proposed and studied: see Coffman et al. [5] for a survey. We shall focus here on the Next Fit Decreasing (NFD) heuristic [1]. Firstly the items are ranked in decreasing order (this can be done in $O(n \log(n))$ time). Then they are put into bins according to the the Next Fit algorithm: one bin is open at a time; when a new item has to be placed, either it fits in the open bin or it does not, in which case the current box is closed and a new one is opened (it takes $O(n)$ time to place the items once sorted). In the average case analysis, the item sizes are random variables, and so is the number of non-empty bins. Its distribution has been thoroughly studied, in particular by Csirik et al. [7, 6], Hofri and Kamhi [13] and Rhee [17, 18] (see also section 5.2 of [11] or section 10.3 of [12]).

The idea of so called Probabilistic Combinatorial Optimisation Problems comes from Jaillet [14, 15] who introduced it for the Traveling Salesman Problem (see also [3, 4]). The Probabilistic Bin Packing Problem (PBPP) was first studied in [2]. The idea is the following. Assume that a list of n items has been given, and an *a priori* solution (exact or approximate) has been found for the BPP. Suppose now that some items randomly disappear from the list. Can the knowledge of the a priori solution for the full list be used to construct a solution for the reduced one? Can this be done efficiently without reopening simultaneously too many bins of the a priori solution?

The aim of this paper is to propose an average case analysis of the PBPP, when the a priori solution is obtained through the NFD heuristic. We shall deal with two sources of randomness. Firstly, the initial sizes of the items are independent and identically distributed random variables (i.i.d.r.v.'s) $(X_1, \ldots, X_n)$. Secondly, once they have been sorted in decreasing order and placed into bins by the NF algorithm, a random binary decision is taken: for each $i = 1, \ldots, n$ the item number i remains or disappears. To formalize this, we consider a n-tuple

$(U_1, \ldots, U_n)$ of i.i.d.r.v.'s, uniformly distributed on $[0,1]$. The two random vectors $(X_i)_{i=1,\ldots,n}$ and $(U_i)_{i=1,\ldots,n}$ are independent. The probability for an item to stay in the list may depend on its size: we denote by $\rho(x)$ the probability for an item of size x to stay in the list. It will be convenient to view disappearing items as objects whose size has become null. Thus the new list of sizes is $(Y_1, \ldots, Y_n)$, where for $i = 1, \ldots, n$:

$$Y_i = X_i \, \mathbb{I}_{\{U_i \leq \rho(X_i)\}} ,$$

denoting by $\mathbb{I}_A$ the indicator of an event A.

If the NFD heuristic has been used for the a priori solution, an obvious procedure immediately comes to mind. Since the initial items were ranked in decreasing order, so are the remaining ones, and it is fast and natural to apply again from scratch the NF algorithm to the list of remaining items. The average case analysis of this procedure is proposed in section 2. The total number of bins will be proved to satisfy a law of large numbers and a central limit theorem, and an explicit expression for the asymptotic mean and variance will be given (theorem 2.3).

However, it is not in the spirit of Probabilistic Combinatorial Problems not to use the a priori solution once the items have been randomly deleted. Therefore, we shall study another heuristic. Suppose the a priori NFD solution has been computed, its bins being numbered by order of opening. Once the items have been randomly deleted, vacancy is left in some of the bins. The Group Rearrangement (GR) procedure depends on a fixed integer m which is the number of bins to be opened simultaneously. Here is the algorithm.

1. Open the bins of the a priori solution by groups of m, one group at a time: first bins with numbers 1 to m, then $m+1$ to $2m$, and so on...
2. For each group of m bins, rearrange the remaining items in an optimal way.
3. Eliminate those bins that have been emptied.

The average case analysis of the GR procedure is treated in section 3. Again, a law of large numbers and a central limit theorem for the total number of non-empty bins will be proved (theorem 3.1).

Of course, the GR procedure is neither faster, nor better on average than the NFD heuristic: both run in linear time, and the asymptotic mean number of bins is larger for the former than for the latter. However, numerical evidence shows that the difference is small. We are not able at this point to propose a similar study for the optimal a priori solution. But we consider our NFD results as a reason to believe that local rearrangements inside small sized groups of bins, such as in the GR procedure, may bring a fast and relatively good solution to the PBPP, when starting from an a priori solution, be it optimal or not.

2. PBPP by the NFD heuristic

In this section, we study the asymptotics of the total number of bins filled by the NFD heuristic, for items of random sizes, once some of them have been randomly deleted.

Two independent sequences of i.i.d.r.v.'s are given: $(X_i)_{i\geq 1}$ and $(U_i)_{i\geq 1}$. The X_i's are the sizes of the original items, and the U_i's are the random variables that decide of their deletion. The probability distribution function of the X_i's is denoted by F and the U_i's have uniform distribution on $[0,1]$. A measurable function ρ, from $[0,1]$ into itself is given. If x is an item size, $\rho(x)$ is its probability to remain in the new list. As already pointed out, it is convenient to consider deleted objects as items of size 0. Thus the new list of item sizes after random deletions becomes $(Y_i)_{i\geq 1}$, where for all $i \geq 1$:

$$Y_i = X_i \, \mathbb{I}_{\{U_i \leq \rho(X_i)\}} \, .$$

Notice that the Y_i's are still i.i.d.r.v.'s. Denote by A_n the number of bins used by the NFD algorithm to arrange the n items of sizes $Y_1, \dots, Y_n$. The asymptotic study of A_n requires very little adaptation of the classical proof for uniformly distributed sizes, developped by Csirik et al. [6] (see Hofri [12] section 10.3.1, p.543 ff.). We shall review below the main arguments. We are aware of the more precise approach of Rhee [17], who gives a much better bounding for A_n than that of lemma 2.2. The reason why we chose Csirik et al.'s truncation technique is that it can also be used for the GR procedure, to be treated in section 3.

The first observation is that the number of bins depends more on the *types* of the items than on their actual sizes.

Definition 2.1. *For $k \geq 1$, an item is said to be of type k if its size x is such that*

$$\frac{1}{k+1} < x \leq \frac{1}{k} \, .$$

Thus a bin can accommodate exactly k objects of type k. To account for deletions, we shall agree that an item of size 0 has type 0. With our probabilistic hypotheses, the item types are i.i.d.r.v.'s with values in $\mathbb{N}$. We shall denote by $p = (p_k)_{k\in\mathbb{N}}$ their distribution. For $k \geq 1$, the probability for an item to be of type k is

$$p_k = \int_{\frac{1}{k+1}}^{\frac{1}{k}} \rho(x) \, dF(x) \, ,$$

whereas its probability to be of type 0 (deletion) is

$$p_0 = 1 - \sum_{k=1}^{\infty} p_k = 1 - \int_0^{\infty} \rho(x) \, dF(x) \, .$$

As a particular case, if the original item sizes are uniformly distributed on $[0, 1]$ and the function ρ is constant, one gets $p_0 = 1 - \rho$ and for $k \geq 1$:

$$p_k = \frac{\rho}{k(k+1)} \, .$$

The results that follow only depend on the distribution p.

Since the items are examined in decreasing order of size, all items of type 1 are treated first, and placed alone in as many bins. Then come type 2 items. The first of them possibly fits in the same bin as the last type 1 item, the others are placed 2 by 2 into new bins, and so on... It is intuitively clear that, apart from a few "frontier" bins that may contain items of different types, most bins will host a fixed number of items of the same type. Lemma 2.2 below gives bounds on the number of used bins, in terms of two functions of the item types.

Lemma 2.2. *Let $r \geq 1$ be an integer. Define the two functions ϕ_1 and ϕ_2, from $[0,1]$ into itself by:*

$$\phi_1(x) = \begin{cases} \frac{1}{k} & \text{if } \frac{1}{k+1} < x \leq \frac{1}{k}, \ k = 1, \ldots, r-1, \\ 0 & \text{if } 0 \leq x \leq \frac{1}{r} \end{cases}$$

$$\phi_2(x) = \begin{cases} \frac{1}{k} & \text{if } \frac{1}{k+1} < x \leq \frac{1}{k}, \ k = 1, \ldots, r-1, \\ \frac{1}{r} & \text{if } 0 < x \leq \frac{1}{r}, \\ 0 & \text{if } x = 0. \end{cases}$$

Let $y_1, \ldots, y_n$ be n (possibly null) item sizes. Let a_n be the total number of bins required to arrange those items using the NFD heuristic. Then:

$$\sum_{i=1}^{n} \phi_1(y_i) - (r-1) \leq a_n \leq \sum_{i=1}^{n} \phi_2(y_i) + r.$$

Proof. For $k \geq 1$, let n_k be the number of type k items. The number of bins they will occupy is at least $\lfloor \frac{n_k}{k} \rfloor$, where $\lfloor \cdot \rfloor$ denotes the integer part. Hence the lower bound, neglecting items of size $\leq r$. For the upper bound, all items of type $k \leq r-1$ can be accommodated in at most $\lfloor \frac{n_k}{k} \rfloor + 1$ bins, and all items of type $k \geq r$ in at most $\lfloor \frac{n_k}{r} \rfloor + 1$ bins. □

When the input sizes are random, lemma 2.2 provides bounds on A_n in terms of two sums of random variables:

$$\sum_{i=1}^{n} \phi_1(Y_i) - (r-1) \leq A_n \leq \sum_{i=1}^{n} \phi_2(Y_i) + r. \tag{1}$$

In (1), both $S_{1,n} = \sum \phi_1(Y_i)$ and $S_{2,n} = \sum \phi_2(Y_i)$ are sums of bounded i.i.d.r.v.'s. Their asymptotic behavior (exponential tail inequalities, law of large numbers, central limit theorem) is described by basic results of probability theory (see [8, 9] or [16] as general references). These can be passed to A_n, through a careful choice of the free parameter r.

For $r \geq 1$ denote by q_r the tail probability for the distribution of types:

$$q_r = \sum_{k=r}^{\infty} p_k = 1 - \sum_{k=0}^{r-1} p_k.$$

We will assume that q_r decreases at least as fast as some negative power of r: there exist two positive constants c and α such that for all $r \geq 1$,

$$q_r \leq c r^{-\alpha}. \tag{2}$$

This is actually an assumption on both the behavior of the size distribution function F close to 0, and the function ρ.

Theorem 2.3. *Under the previous hypotheses, denote by μ and σ^2 the following asymptotic mean and variance:*

$$\mu = \sum_{k=1}^{\infty} \frac{p_k}{k}. \tag{3}$$

$$\sigma^2 = \sum_{k=1}^{\infty} \frac{p_k}{k^2} - \mu^2. \tag{4}$$

Then the following results hold for A_n.

1. *Exponential tail inequality: for all* $x > 0$ *and* $n \geq 1$,

$$\mathbb{P}[\,|A_n - n\mu| \geq x\sqrt{n}|\,] \leq \exp\left(-2(x - n^{-1/2} - 2\sqrt{c}n^{-\frac{\alpha}{2(2+\alpha)}})^2\right) . \quad (5)$$

2. *Law of large numbers:*

$$\lim_{n\to\infty} \frac{A_n}{n} = \mu \quad a.s. \quad (6)$$

3. *Central limit theorem:*

$$\lim_{n\to\infty} \mathbb{P}\left[\frac{1}{\sqrt{n\sigma^2}}(A_n - n\mu) \leq x\right] = \Phi(x)\,, \quad (7)$$

where Φ *denotes the standard Gaussian probability distribution function.*

For the particular case of item sizes uniformly distributed on $[0, 1]$ and a constant value of ρ, the asymptotic mean and variance can be expressed in terms of Riemann's Zeta function:

$$\zeta(u) = \sum_{i=1}^{\infty} \frac{1}{n^u} \,.$$

One gets:

$$\mu = \rho\,(\zeta(2) - 1) \simeq 0.645\,\rho\,,$$

and:

$$\sigma^2 = \rho\,(\zeta(3) - \zeta(2) + 1) - \rho^2\,(\zeta(2) - 1)^2 \simeq 0.557\,\rho - 0.416\,\rho^2\,.$$

Proof. For $j = 1, 2$, we shall denote by μ_j the expectation of $\phi_j(Y_i)$, and by σ_j^2 its variance:

$$\mu_1 = \sum_{k=1}^{r-1} \frac{p_k}{k} \quad , \quad \mu_2 = \sum_{k=1}^{r-1} \frac{p_k}{k} + \frac{q_r}{r}\,,$$

$$\sigma_1^2 = \sum_{k=1}^{r-1} \frac{p_k}{k^2} - \mu_1^2 \quad , \quad \sigma_2^2 = \sum_{k=1}^{r-1} \frac{p_k}{k^2} + \frac{q_r}{r^2} - \mu_2^2\,.$$

Obviously as r tends to infinity, μ_1 and μ_2 tend to μ, whereas σ_1^2 and σ_2^2 tend to σ^2. From (1), the following bounds on the PDF of A_n are easily deduced.

$$\begin{aligned}\mathbb{P}[S_{2,n} - n\mu_2 \leq x + r + n\tfrac{q_r}{r}] &\leq \mathbb{P}[A_n - n\mu \leq x] \\ &\leq \mathbb{P}[S_{1,n} - n\mu_1 \leq x - r - n\tfrac{q_r}{r}]\,.\end{aligned} \quad (8)$$

As sums of bounded i.i.d.r.v.'s, both $S_{1,n}$ and $S_{2,n}$ satisfy classical exponential tail inequalities, such as Hoeffding's [16]:

$$\mathbb{P}[|S_{j,n} - n\mu_j| \geq z\sqrt{n}] \leq 2e^{-2z^2}\,.$$

Both also satisfy the central limit theorem:

$$\lim_{n\to\infty} \mathbb{P}\left[\frac{1}{\sqrt{n}}(S_{j,n} - n\mu_j) \leq x\right] = \Phi(\frac{x}{\sigma_j})\,.$$

To derive from (8) the corresponding results for A_n, one needs to let the parameter r tend to infinity, as a function of n. Recall from (2) that $q_r \leq cr^{-\alpha}$. For all $n \geq 1$, we set $r = r(n)$ as follows:

$$r(n) = \lfloor\sqrt{c}n^{\frac{1}{2+\alpha}}\rfloor + 1\,. \quad (9)$$

One readily checks that:

$$n\frac{q_{r(n)}}{r(n)} \leq \sqrt{c}\, n^{\frac{1}{2+\alpha}} .$$

From there, (5) easily follows from Hoeffding's inequality applied to $S_{1,n}$ and $S_{2,n}$. To derive the strong law of large numbers from an exponential inequality such as (5) is an easy application of the Borel-Cantelli lemma. For the central limit theorem (7), one just has to divide variables by $\sqrt{n}$ in (8), and let n tend to infinity: with our choice of $r(n)$, both $\frac{r(n)}{\sqrt{n}}$ and $\frac{\sqrt{n}\,q_{r(n)}}{r(n)}$ tend to 0. □

3. The Group Rearrangement procedure

We now turn to the GR procedure, in which bins are taken by groups of m, each group being optimally rearranged after deletions. As we saw before, in the a priori solution most bins contain items of the same type. Therefore, most groups of bins will be homogeneous in the sense that each bin of the group contains exactly k items of type k; let us call "k-group" such a group of m bins. We need to restrict slightly our assumption on the function ρ: we assume now that it is constant for objects of the same type, and denote by ρ_k the probability for an item of type k to remain in the list.

Consider a k-group. The number of remaining items in all its m bins, has binomial distribution with parameters mk and ρ_k. Let us denote by $\pi_{k,m} = (\pi_{k,m}(i))_{i=0,\dots,m}$ the probability distribution of the number of remaining bins, once the k-group has been rearranged:

$$\pi_{k,m}(i) = \begin{cases} (1-\rho_k)^{mk} & \text{for } i = 0\,, \\ \displaystyle\sum_{l=(i-1)k+1}^{ik} \binom{mk}{l} \rho_k^l (1-\rho_k)^{mk-l} & \text{for } i = 1,\dots,m\,. \end{cases}$$

The expectation and the variance of $\pi_{k,m}$ will be denoted by $e_{k,m}$ and $v_{k,m}$ respectively.

$$e_{k,m} = \sum_{i=0}^{m} i\,\pi_{k,m}(i) \quad \text{and} \quad v_{k,m} = \sum_{i=0}^{m} i^2\,\pi_{k,m}(i) - e_{k,m}^2 \,.$$

Let p_k^* be the *initial* proportion of type k items: for $k \geq 1$,

$$p_k^* = F(\frac{1}{k}) - F(\frac{1}{k+1})\,.$$

We shall make the same assumption on the tail of p^* as we did for p: there exist two positive constants c and α such that for all $r \geq 1$,

$$q_r^* = \sum_{k=r}^{\infty} p_k^* \leq c r^{-\alpha}\,. \tag{10}$$

Let $B_{n,m}$ be the number of remaining bins after the GR procedure. The asymptotics of $B_{n,m}$ is described in the following result.

Theorem 3.1. *Under the previous hypotheses, denote by μ_m and σ_m^2 the following asymptotic mean and variance.*

$$\mu_m = \sum_{k=1}^{\infty} p_k^* \frac{e_{k,m}}{km}, \tag{11}$$

$$\sigma_m^2 = \sum_{k=1}^{\infty} p_k^* \left(\frac{v_{k,m}}{km} + \frac{e_{k,m}^2}{k^2m^2} \right) - \mu_m^2 . \tag{12}$$

Then the following results hold for $B_{n,m}$.

1. *Law of large numbers:*

$$\lim_{n\to\infty} \frac{B_{n,m}}{n} = \mu_m \quad a.s. \tag{13}$$

2. *Central limit theorem:*

$$\lim_{n\to\infty} \mathbb{P}\left[\frac{1}{\sqrt{n\sigma_m^2}}(B_{n,m} - n\mu_m) \le x \right] = \Phi(x) . \tag{14}$$

Essentially, $B_{n,m}$ behaves asymptotically as a sum of n i.i.d.r.v.'s, each having expectation μ_m and variance σ_m^2; it can be viewed as the sum of the individual contributions of the original n items to the final packing. Indeed, for n large, a typical item belongs to a k-group with probability p_k^*. The contribution of km such items (one k-group) to $B_{n,m}$ has expectation $e_{k,m}$ and variance $v_{k,m}$; hence each contribution should have expectation $\frac{e_{k,m}}{km}$, and variance $\frac{v_{k,m}}{km}$. Thus the expected squared contribution of an object of type k should be $\frac{v_{k,m}}{km} + \frac{e_{k,m}^2}{k^2m^2}$. So σ_m^2 can be seen as the variance for the contribution of a typical item to $B_{n,m}$.

Clearly, as m increases, the space wasted by the GR procedure compared to the NFD heuristic diminishes. In particular, the asymptotic expectation μ_m defined by (11) tends to μ. One may wonder what is the difference for small values of m. To get a partial answer, we computed numerically $\mu_m - \mu$, for $m = 2\dots,5$, in the particular case where the item sizes are uniformly distributed on $[0,1]$ and the probability ρ is a constant. Figure 1 shows a plot of $\mu_m - \mu$ as a function of ρ. It turns out that the difference between the global algorithm (NFD) and the local one (GR) is relatively small, even for $m = 2$. In order to understand why, let us fix m and ρ_k, and look at the asymptotic behavior of $e_{k,m}$ as k increases. The law of large numbers implies that $e_{k,m}$ converges to i for all values of ρ_k in the interval $]\frac{i-1}{m}, \frac{i}{m}]$, for i between 1 and m. In other terms, as k increases, $e_{k,m}$ approaches $\lfloor m\rho_k \rfloor + 1$. So μ_m is actually close to the following sum :

$$\mu_m \simeq \sum_{k=1}^{\infty} p_k^* \frac{\lfloor m\rho_k \rfloor + 1}{km},$$

to be compared with

$$\mu = \sum_{k=1}^{\infty} p_k^* \frac{\rho_k}{k} .$$

This also accounts for the modes in $\mu_m - \mu$, plotted as a function of ρ (figure 1).

Proof. Controlling $B_{n,m}$ by sums of independent random variables is not as straightforward as for A_n, though a similar truncation technique will be used. We shall first describe the lower bound, then the upper bound.

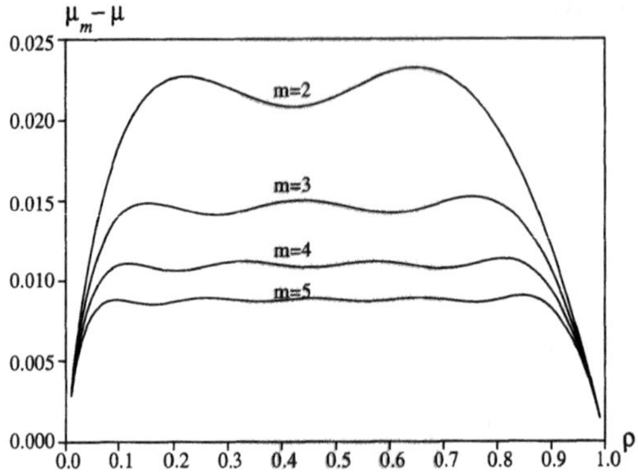

FIGURE 1. Asymptotic difference between the NFD heuristic and the GR procedure, for $m = 2, \ldots, 5$.

To get a lower bound, we restrict ourselves to k-groups, with $k < r$: $B_{n,m}$ is certainly larger than the number of remaining bins after rearrangement of those k-groups, and suppression of all other groups. Denote by N_k the number of items of type k in the original list: the distribution of N_r is binomial with parameters n and p_k^*. If there are N_k items of type k, then the number of k-groups is certainly larger than

$$G_k = \lfloor \frac{1}{m} (\lfloor \frac{N_k}{k} \rfloor - 1) \rfloor - 1 \,. \tag{15}$$

For all $k \geq 1$, consider independent sequences of i.i.d.r.v.'s $(Z_{m,k}^{(l)})_{l \geq 1}$, all independent from the X_i's, where $Z_{m,k}^{(l)}$ has distribution $\pi_{m,k}$. Define $S_{1,n}$ as the sum

$$S_{1,n} = \sum_{k=1}^{r-1} \sum_{l=1}^{G_k} Z_{m,k}^{(l)} \,.$$

The previous reasoning shows that $S_{1,n}$ is smaller than $B_{n,m}$ in the stochastic ordering sense: for all x,

$$\mathbb{P}[S_{1,n} \leq x] \geq \mathbb{P}[B_{n,m} \leq x] \,.$$

As in the proof of theorem 2.3, we need to control the difference between $n\mu_m$ and $\mathbb{E}[S_{1,n}]$. By Wald's theorem, one has

$$\mathbb{E}[S_{1,n}] = \sum_{k=1}^{r-1} \mathbb{E}[G_k] \, e_{k,m} \,.$$

By definition of G_k, one has

$$\mathbb{E}[G_k] \geq \frac{1}{m} \left(\frac{\mathbb{E}[N_k]}{k} - 2 \right) - 2 = \frac{np_k^*}{km} - \frac{2}{m} - 2 \,.$$

Remarking that $e_{k,m} \leq m$ for all k, one gets

$$
\begin{aligned}
n\mu_m - \mathbb{E}[S_{1,n}] &= \sum_{k=1}^{r-1} e_{k,m}\left(\frac{np_k^*}{km} - \mathbb{E}[G_k]\right) + \sum_{k=r}^{\infty} e_{k,m}\frac{np_k^*}{km} \\
&\leq \sum_{k=1}^{r-1} e_{k,m}\left(\frac{2}{m}+2\right) + \sum_{k=r}^{\infty} e_{k,m}\frac{np_k^*}{rm} \\
&\leq (r-1)(2+2m) + \frac{nq_r^*}{r} .
\end{aligned}
$$

Under (10), the same choice of $r(n)$ as in the proof of theorem 2.3 ensures that this difference is small compared to $\sqrt{n}$:

$$
r(n) = \lfloor \sqrt{c}\, n^{\frac{1}{2+\alpha}} \rfloor + 1 . \tag{16}
$$

We also need to check that $Var[S_{1,m}] - n\sigma_m^2 = o(n)$. One has:

$$
Var[S_{1,n}] = \sum_{k=1}^{r-1} \mathbb{E}[G_k]\, v_{k,m} + Var[G_k]\, e_{k,m}^2 + \sum_{k \neq h=1}^{r-1} Cov[G_k, G_h]\, e_{k,m}e_{h,m} .
$$

Using the definition (15) of G_k, one easily gets:

$$
\mathbb{E}[G_k] = \frac{np_k^*}{km} + O(1) \, , \; Var[G_k] = \frac{np_k^*(1-p_k^*)}{k^2m^2} + O(1) \, ,
$$

$$
Cov[G_k, G_h] = -\frac{np_k^*p_h^*}{khm^2} + O(1) .
$$

From this one deduces:

$$
\begin{aligned}
Var[S_{1,n}] &= n\sum_{k=1}^{r-1} p_k^*\left(\frac{v_{k,m}}{km} + \frac{e_{k,m}^2}{k^2m^2}\right) - n\left(\sum_{k=1}^{r-1} p_k^*\,\frac{e_{k,m}}{km}\right)^2 + O(r^2) \\
&= n\sigma_m^2 + o(n) \, ,
\end{aligned}
$$

still using expression (16) for $r(n)$.

Let us now turn to the upper bound. The number of remaining bins $B_{n,m}$ certainly increases if one neglects to rearrange non homogeneous groups. It also increases if all items of size $\leq 1/r$ are replaced by items of size $1/r$ in the original list and none of them disappears. Let M_r denote the number of items of type $\geq r$. The upper bound $S_{2,n}$ is the following:

$$
S_{2,n} = S_{1,n} + (M_r/r + rm) .
$$

One has for all x:

$$
\mathbb{P}[B_{n,m} \leq x] \geq \mathbb{P}[S_{2,n} \leq x] .
$$

As before, one can check that $\mathbb{E}[S_{2,n}] = n\mu_m + o(\sqrt{n})$ and $Var[S_{2,n}] = n\sigma_m^2 + o(n)$.

To finish the proof along the same lines as that of theorem 2.3, we need to check that the law of large numbers and the central limit theorem hold for $S_{1,n}$ and $S_{2,n}$, which are sums of random numbers of r.v.'s. We shall do it for $S_{1,n}$;

similar arguments hold for $S_{2,n}$. The law of large numbers is the easy part. By formula (15), G_k increases a.s. to infinity and

$$\lim_{n\to\infty} \frac{G_k}{n} = \frac{p_k^*}{km} \quad \text{a.s.}$$

Hence:

$$\lim_{n\to\infty} \frac{1}{n}\sum_{l=1}^{G_k} Z_{k,m}^{(l)} = \lim_{n\to\infty} \frac{G_k}{n}\frac{1}{G_k}\sum_{l=1}^{G_k} Z_{k,m}^{(l)} = \frac{p_k^* e_{k,m}}{km} \quad \text{a.s.}$$

Using again the expression (16) for $r(n)$, it follows that

$$\lim_{n\to\infty} \frac{S_{1,n}}{n} = \mu_m \quad \text{a.s.}$$

The central limit theorem is not as straightforward. Here are the main steps.

We first check that the vector $(G_k)_{1\leq k\leq r-1}$ is asymptotically normal. Consider the vector $(N_1, \ldots, N_{r-1}, M_r)$. Its distribution is multinomial, with parameters n and $(p_1^*, \ldots, p_{r-1}^*, q_r^*)$. From there, and formula (15), it follows that the vector $(G_k)_{1\leq k\leq r-1}$ is asymptotically normal (G_k essentially behaves as $N_k/(km)$). More precisely, define for all $k \geq 1$

$$\widetilde{p}_k = \frac{p_k^*}{km} \quad \text{and} \quad H_k = \frac{G_k - n\widetilde{p}_k}{\sqrt{n}} .$$

The distribution of the random vector $(H_k)_{1\leq k\leq r-1}$ converges to the multidimensional Gaussian distribution with null expectation and covariance matrix $C = (c_{k,h})_{1\leq k,h\leq r-1}$, given by:

$$c_{k,k} = \widetilde{p}_k(1-\widetilde{p}_k) \quad \text{and} \quad c_{k,h} = -\widetilde{p}_k\widetilde{p_h}\,, \text{ for } k \neq h .$$

For $k \geq 1$, consider the partial sum

$$\widetilde{S}_{k,m} = \sum_{l=1}^{G_k} Z_{k,m}^{(l)} .$$

Using the classical technique of characteristic functions, one can show that the distribution of the vector

$$\left(\frac{1}{\sqrt{n}}(\widetilde{S}_{k,m} - n\widetilde{p}_k e_{k,m})\right)_{1\leq k\leq r-1}$$

converges to the multidimensional Gaussian distribution with null expectation and covariance matrix

$$D_1 + D_2 C D_2 ,$$

where D_1 and D_2 are the following diagonal matrices.

$$D_1 = Diag((\widetilde{p}_k v_{k,m})_{1\leq k\leq r-1}) \quad \text{and} \quad D_2 = Diag((e_{k,m})_{1\leq k\leq r-1}) .$$

Summing coordinates, it follows that $S_{1,n}$ is asymptotically normal, for any fixed r. There remains to let $r = r(n)$ tend to infinity, using (16). The already given estimates on $\mathbb{E}[S_{1,n}]$ and $Var[S_{1,n}]$, yield that

$$\frac{1}{\sqrt{n\sigma_m^2}}(S_{1,n} - n\mu_m)$$

converges in distribution to the standard Gaussian distribution. □

References

[1] B.S. Baker and E.G. Coffman Jr. A tight asymptotic bound for next fit decreasing bin packing. *SIAM Journal on Algebraic and Discrete Methods*, 2:147–152, 1981.

[2] M. Bellalouna. *Problèmes d'optimisation combinatoire probabilistes.* PhD thesis, Ecole Nationale des Ponts et Chaussées, Paris, 1993.

[3] D. Bertsimas. *Probabilistic Combinatorial Optimisation Problems.* PhD thesis, Massachusetts Institute of Technology, Cambridge, Mass., 1988.

[4] D. Bertsimas, P. Jaillet, and A. Odoni. A priori optimization. *Operations Research*, 38:1019–1033, 1990.

[5] E.G. Coffman, Jr., M.R. Garey, and D.S. Johnson. Approximation algorithms for bin packing - a survey. In D. Hochbaum, editor, *Approximation algorithms for NP-Hard problems*, pages 46–93. PWS Publishing, Boston, 1996.

[6] J. Csirik, J.B.G. Frenk, A.M. Frieze, G. Galambos, and A.H.G. Rinnooy Kan. A probabilistic analysis of the Next-Fit decreasing bin packing heuristic. *Oper. Res. Lett.*, 5(5):233–236, 1986.

[7] J. Csirik and E. Mate. The probabilistic behaviour of the NFD bin packing algorithm. *Acta Cyber.*, 7:241–245, 1986.

[8] W. Feller. *An introduction to probability theory and its applications*, volume I. Wiley, London, 3rd edition, 1968.

[9] W. Feller. *An introduction to probability theory and its applications*, volume II. Wiley, London, 2nd edition, 1971.

[10] M.R. Garey and D.S. Johnson. *Computers and Intractability: A Guide to the Theory of $\mathcal{NP}$-Completeness.* Freeman, San Francisco, 1979.

[11] M. Hofri. *Probabilistic analysis of algorithms.* Springer-Verlag, New York, 1987.

[12] M. Hofri. *Analysis of algorithms – Mathematical methods, computational tools.* Oxford University Press, Oxford, 1995.

[13] M. Hofri and S. Kamhi. A stochastic analysis of the NFD bin packing algorithm. *Journal of Algorithms*, 7:489–509, 1986.

[14] P. Jaillet. A priori solution of a traveling salesman problem in which a random subset of the customers are visited. *Operations Research*, 36:929–936, 1988.

[15] P. Jaillet. Analysis of probabilistic combinatorial optimization problems in Euclidean spaces. *Mathematics of Operation Research*, 18:51–71, 1993.

[16] V.V. Petrov. *Limit theorems of probability theory.* Oxford University Press, Oxford, 1995.

[17] W.T. Rhee. Probabilistic analysis of the next fit decreasing algorithm for bin-packing. *Oper. Res. Lett.*, 6(4):189–191, 1987.

[18] W.T. Rhee. Correction to: Probabilistic analysis of the next fit decreasing algorithm for bin-packing. *Oper. Res. Lett.*, 7:211, 1988.

Monia Bellalouna, Salma Souissi
CRISTAL pôle GRIFT, Ecole Nationale des Sciences de l'Informatique, Tunisia
{monia.bellalouna,salma.souissi}@ensi.rnu.tn

Bernard Ycart
MAP5 CNRS UMR 8145, Université Paris 5, France
ycart@math-info.univ-paris5.fr

Trends in Mathematics, © 2004 Birkhäuser Verlag Basel/Switzerland

Distribution of WHT Recurrences[1]

Paweł Hitczenko, Jeremy R. Johnson, and Hung-Jen Huang

ABSTRACT: *This work explores the performance of a family of algorithms for computing the Walsh-Hadamard transform (WHT), a useful computation in signal and image processing [1, 2, 5]. The algorithms exhibit a wide range of performance and it is non-trivial to determine which algorithm is optimal on a given computer [4]. This paper provides a theoretical basis for the performance analysis. Performance is modelled by a family of recurrence relations that determine the number of instructions required to execute a given algorithm, and are related to standard divide and conquer recurrences [6, 3]. However, since there are a variable number of recursive parts which can grow to infinity as the input size increases, new techniques are required for their analysis. In the full version of this paper, the minimum, maximum, expected values, and variances are calculated and the limiting distribution is obtained.*

The Walsh-Hadamard Transform. The WHT of a signal x, of size $N = 2^n$, is the matrix-vector product $\mathbf{WHT}_N \cdot x$, where

$$\mathbf{WHT}_N = \bigotimes_{i=1}^{n} \mathbf{DFT}_2 = \overbrace{\mathbf{DFT}_2 \otimes \cdots \otimes \mathbf{DFT}_2}^{n}.$$

The matrix

$$\mathbf{DFT}_2 = \begin{bmatrix} 1 & 1 \\ 1 & -1 \end{bmatrix}$$

is the 2-point DFT matrix, and $\otimes$ denotes the tensor or Kronecker product. Let $n = n_1 + \cdots + n_t$ and I_m denote the $m \times m$ identity matrix, then

$$\mathbf{WHT}_{2^n} = \prod_{i=1}^{t} (\mathrm{I}_{2^{n_1+\cdots+n_{i-1}}} \otimes \mathbf{WHT}_{2^{n_i}} \otimes \mathrm{I}_{2^{n_{i+1}+\cdots+n_t}}) \tag{1}$$

This equation provides a family of divide and conquer algorithms for computing the WHT (implemented with a triply nested loop in [4]) and provides a mechanism for exploring different breakdown strategies.

Performance Model for the WHT. Let $\mathcal{W}_{2^n}$ be a WHT algorithm, and let $A(n)$ be the number of times the recursive WHT procedure is called, $A_l(n)$ the number of times the straight-line code for $\mathbf{WHT}_{2^l}$ is called. Finally, let $L_i(n)$, $i = 1, 2, 3$, be the number of times the outermost, middle, and innermost loops are executed.

Then the number of instructions required to execute $\mathcal{W}_{2^n}$ is equal to

$$\alpha A(n) + \sum_l \alpha_l A_l(n) + \sum_{i=1}^{3} \beta_i L_i(n),$$

[1]Supported in part by NSA #MSPF-02G-043 and NSF ITR/NGS #0325687.

where α is the number of instructions for the code in the compiled WHT procedure executed outside the loops, α_i, is the number of instructions in the compiled straight-line code implementations of small WHT's (the largest size considered in practice was 8), and β_i, $i = 1, 2, 3$ is the number of instructions executed in the outer-most, middle, and inner-most loops in the compiled WHT procedure.

The functions $A(n)$, $A_l(n)$, $L_i(n)$ satisfy recurrences that can be derived from (1). Suppose $n = n_1 + \cdots + n_t$ is the composition of n corresponding to the factorization in (1), then the functions $A(n)$, $A_l(n)$, $L_i(n)$, satisfy recurrence relations of the form $F(n) = \sum_{i=0}^{t}\{2^{n-n_i}F(n_i) + f(i)\}$, where $f(i)$ depends on the particular recurrence and is equal to $1/t$, 0, 1, 2^{n-n_i}, $2^{n_1+\cdots+n_{i-1}}$ respectively.

While it is not possible to obtain closed forms for all of these recurrences, it is possible to determine the algorithm with the minimum and maximum number of instructions. It is also possible to determine the expected value and variance and the limiting distribution. In particular, it is possible to show that in the limit the distribution approaches a normal distribution.

Since instruction count by itself does not accurately predict performance on modern processors, where pipeline performance, instruction level parallelism, and the memory hierarchy significantly affect performance, these results can not be used by themselves to model performance and select algorithms. Nonetheless, they give insight into the wide range of performance and the structure of the search space, and the techniques developed to analyze the recurrences provide tools for analyzing divide and conquer recurrences with varying numbers of recursive calls.

References

[1] K.G. Beauchamp. *Applications of Walsh and related functions.* Academic Press, 1984.

[2] D. F. Elliott and K. R. Rao. *Fast Transforms: Algorithms, Analyses, Applications.* Academic Press, 1982.

[3] H-K. Hwang and R. Neininger. Phase change of limit laws in the quicksort recurrence under varying toll functions. *SIAM J. Comput.*, 31:1687–1722, 2002.

[4] J. Johnson and M. Püschel. In Search for the Optimal Walsh-Hadamard Transform. In *Proceedings ICASSP*, volume IV, pages 3347–3350, 2000.

[5] F.J. MacWilliams and N.J. Sloane. *The theory of error-correcting codes.* North-Holland Publ.Comp., 1992.

[6] H. Mahmoud. *Sorting. A Distribution Theory.* Wiley, 2000.

Paweł Hitczenko
Department of Mathematics, Drexel University, Philadelphia, PA19104
phitczenko@cs.drexel.edu

Jeremy R. Johnson
Department of Computer Science, Drexel University, Philadelphia, PA19104
jjohnson@cs.drexel.edu

Hung-Jen Huang
Department of Computer Science, Drexel University, Philadelphia, PA19104
hjhuang@cs.drexel.edu

Trends in Mathematics, © 2004 Birkhäuser Verlag Basel/Switzerland

Probabilistic Analysis for Randomized Game Tree Evaluation

Tämur Ali Khan and Ralph Neininger

ABSTRACT: *We give a probabilistic analysis for the randomized game tree evaluation algorithm of Snir. We first show that there exists an input such that the running time, measured as the number of external nodes read by the algorithm, on that input is maximal in stochastic order among all possible inputs. For this worst case input we identify the exact expectation of the number of external nodes read by the algorithm, give the asymptotic order of the variance including the leading constant, provide a limit law for an appropriate normalization as well as a tail bound estimating large deviations. Our tail bound improves upon the exponent of an earlier bound due to Karp and Zhang, where sub-Gaussian tails were shown based on an approach using multi-type branching processes and Azuma's inequality. Our approach rests on a direct, inductive estimate of the moment generating function.*

1. Introduction

In this note we analyze the performance of the randomized algorithm to evaluate Boolean decision trees proposed by Snir (1985). Given is a complete binary tree of height $2k$, $k \geq 1$, where the root (at depth 0) is labeled $\wedge$ as are all internal nodes with even depth, all internal nodes with odd depth are labeled $\vee$. The $n = 2^{2k}$ external nodes are labeled either 0 or 1 and the objective is to calculate the value of the root. For each node its value is given as the value of the operation labeled at that node applied to the values of its children. The cost for evaluating the Boolean decision tree is measured as the number of external nodes read by the algorithm.

Snir proposed and analyzed the following randomized algorithm to evaluate a Boolean decision tree: At each node one chooses randomly (with probability 1/2) one of its children and calculates its value recursively. If the result allows to identify the value of the node (that is a 0 for a $\wedge$-labeled node and a 1 for a $\vee$-labeled node, respectively) one is done, otherwise also the other child's value has to be calculated recursively in order to obtain the value of the node. Applying this to the root of the tree yields the value of the Boolean decision tree.

The advantage of this algorithm over any deterministic algorithm is that for any input at the external nodes its expected cost is sublinear in n, whereas any deterministic algorithm has linear worst case cost. More precisely, Saks and Wigderson (1986) obtained that the maximum expected cost is of the order $\Theta(n^\alpha)$ with $\alpha = \log_2((1+\sqrt{33})/4) \doteq 0.753$ and showed that this is also a lower bound on the maximum expected cost for any other randomized algorithm to evaluate a Boolean decision tree; see also Motwani and Raghavan (1995, Chapter 2) for an account on this subject. Further analysis was given by Karp and Zhang (1995). For certain regular inputs at the external nodes the cost of the algorithm can be represented via 2-type Galton-Watson processes. Karp and Zhang showed that the normalized cost has sub-Gaussian tails. That argument was based on Azuma's inequality.

We denote the input of 0's and 1's at the external nodes as a vector $v \in \{0,1\}^n$ and the number of external nodes read by the algorithm on input v by $C(v)$. We will see subsequently that for particular $v^\star \in \{0,1\}^n$ not only the expectation of the cost of the algorithm is maximized, i.e., $\mathbb{E}\, C(v^\star) = \max_{v \in \{0,1\}^n} \mathbb{E}\, C(v)$, but also that $C(v^\star)$ is maximal in stochastic order, $C(v) \preceq C(v^\star)$ for all $v \in \{0,1\}^n$. Here, $X \preceq Y$ for random variables X, Y denotes that the corresponding distribution functions F_X, F_Y satisfy $F_X(x) \geq F_Y(x)$ for all $x \in \mathbb{R}$, or, equivalently, that there are realizations X', Y' of the distributions $\mathcal{L}(X), \mathcal{L}(Y)$ of X, Y on a joint probability space such that we pointwise have $X' \leq Y'$.

From this perspective it is reasonable to consider $C(v^\star)$ as the worst case complexity of the randomized algorithm and to analyze its asymptotic probabilistic behavior. Our results for the exact mean of $C(v^\star)$, the asymptotic growth of its variance including the evaluation of the leading constant, a limit law for $C(v^\star)$ after normalization as $k \to \infty$ together with an explicit tail estimate are based on a recursive description of the problem. Since $v^\star$ is a regular input in the sense of Karp and Zhang, also their 2-type Galton-Watson approach applies.

Our main finding is an improvement of the tail bound $\exp(-\text{const}\ t^2)$ for $t > 0$, to $\exp(-\text{const}\ t^\kappa)$, with $1 < \kappa < 1/(1-\alpha) \doteq 4.06$, see Theorem 3.6. This is based on a direct, inductive estimate of the moment generating function. Our approach is also applicable to any regular input as well as to other related problems.

The paper is organized as follows: In section 2 we explain, how a worst case input $v^\star$ is obtained. Section 3 contains the statements of the results. In sections 4 and 5 the 2-type branching process of Karp and Zhang (1995) is recalled and the recursive description of the quantities, that our analysis is based on, is introduced. Section 6 contains the proofs of our results and section 7 has extensions to m-ary Boolean decision trees.

2. Worst case input

In this section we explain how a worst case input $v^\star$ is constructed. We first have a look at the case $k = 1$ and $v \in \{0,1\}^4$ such that the decision tree is evaluated to 1 at the root. Clearly both children of the root have to lead to an evaluation of 1. Now each pair of external nodes attached to the children needs to have at least one value 1. Note that the algorithm reads in both pairs of external nodes until it finds the first one. Hence there will in total be read two 1's no matter how $v \in \{0,1\}^4$ is drawn among the choices that lead to an evaluation of 1 for the decision tree. Clearly, to maximize the number of 0's being read we choose in each pair of external nodes one 0 and one 1. Then both 0's are being read independently with probability 1/2. Hence, $v_1 = (0,1,0,1)$ stochastically maximizes $C(v)$ for all $v \in \{0,1\}^4$ such that the decision tree evaluates 1, see Figure 1.

Analogously look at the case $k = 1$ and $v \in \{0,1\}^4$ such that the decision tree is evaluated to 0. Clearly, one child of the root has to have the value 0, whose external nodes attached need to have both values 0. If we choose also value 0 for the other child of the root, we are lead to $v = (0,0,0,0)$, and the algorithm reads exactly 2 external nodes with values both 0. Therefore, to stochastically maximize $C(v)$ we choose the second child of the root with value 1 and again its external nodes attached with values 0 and 1. Then, $v_0 = (0,0,0,1)$ stochastically maximizes $C(v)$ for all $v \in \{0,1\}^4$ for which the decision tree evaluates to 0, see Figure 1.

Since we have $C(v_0) \preceq C(v_1)$, it follows that $v^\star = (0,1,0,1)$ is a choice with $C(v) \preceq C(v^\star)$ for all $v \in \{0,1\}^4$. For general $k \geq 2$ a corresponding $v^\star = v^\star(k)$ can recursively be constructed from $v^\star(k-1)$ as follows: Each component 0 in $v^\star(k-1)$ is replaced by the block $0,0,0,1$, whereas each 1 is replaced by the block $0,1,0,1$. For example, for $k = 3$, this yields

$$\begin{aligned} v^\star \;=\; & (0,0,0,1,0,0,0,1,0,0,0,1,0,1,0,1,\\ & 0,0,0,1,0,1,0,1,0,0,0,1,0,1,0,1,\\ & 0,0,0,1,0,0,0,1,0,0,0,1,0,1,0,1,\\ & 0,0,0,1,0,1,0,1,0,0,0,1,0,1,0,1). \end{aligned}$$

In Proposition 3.1 we show that this construction yields a $v^\star$ with $C(v) \preceq C(v^\star)$ for all $v \in \{0,1\}^n$ and $k \geq 1$.

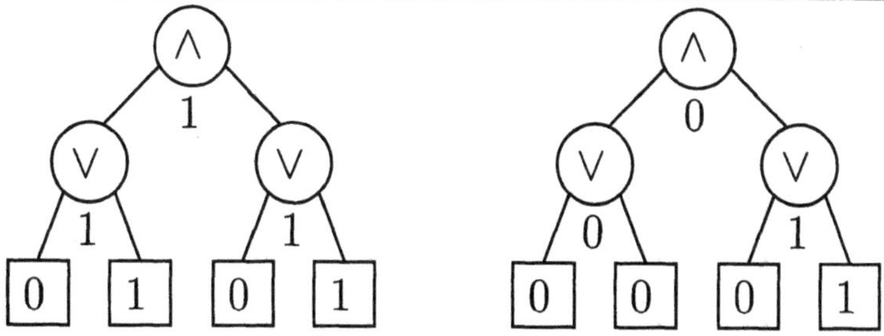

Figure 1: *Shown are decision trees for $k = 1$ evaluating at the root to 1 and 0, respectively, together with a choice for the external nodes that stochastically maximizes the number of external nodes read by the algorithm.*

If we would only want to stochastically maximize the cost over all $v \in R_0(n) \subset \{0,1\}^n$ that evaluate to a 0 at the root, the same recursive construction of replacing digits by corresponding blocks, starting with $v_0 = (0,0,0,1)$, yields a $v_\star \in R_0(n)$ such that $C(v) \preceq C(v_\star)$ for all $v \in R_0(n)$.

3. Results

We assume that we have $n = 2^{2k}$ with $k \geq 1$ and denote by $v^\star \in \{0,1\}^n$ an input as constructed in section 2.

Proposition 3.1. *For $v^\star \in \{0,1\}^n$ as defined in section 2 we have $C(v) \preceq C(v^\star)$ for all $v \in \{0,1\}^n$.*

The stochastic worst case behavior $C(v^\star)$ of the randomized game tree evaluation algorithm has the following asymptotic properties: The subsequent theorems describe the behavior of mean, variance, limit distribution, and large deviations of $C(v^\star)$. For the mean we have:

Theorem 3.2. *The expectation of $C(v^\star)$ is given by $\mathbb{E}\, C(v^\star) = c_1 n^\alpha - c_2 n^\beta$, with*

$$\alpha = \log_2 \frac{1+\sqrt{33}}{4}, \quad \beta = \log_2 \frac{1-\sqrt{33}}{4}, \quad c_1 = \frac{1}{2} + \frac{7}{2\sqrt{33}}, \quad c_2 = c_1 - 1.$$

We denote for sequences $(a_k), (b_k)$ by $a_k \sim b_k$ asymptotic equivalence, i.e., $a_k/b_k \to 1$ as $k \to \infty$. Then we have for the variance of $C(v^\star)$:

Theorem 3.3. *The variance of $C(v^\star)$ satisfies asymptotically* $\mathrm{Var}\, C(v^\star) \sim d\, n^{2\alpha}$ *as $k \to \infty$, where $d \doteq 0.0938$. The constant d can also be given in closed form.*

For random variables X, Y we denote by $X \stackrel{d}{=} Y$ equality in distribution, i.e., $\mathcal{L}(X) = \mathcal{L}(Y)$. Then we have the following limit law for $C(v^\star)$:

Theorem 3.4. *For $C(v^\star)$ we have after normalization convergence in distribution,*

$$\frac{C(v^\star)}{n^\alpha} \longrightarrow C, \quad k \to \infty,$$

where the distribution of C is given as $\mathcal{L}(C) = \mathcal{L}(G_1)$ and $\mathcal{L}(G) = \mathcal{L}(G_0, G_1)$ is characterized by $\mathbb{E}\,\|G\|^2 < \infty$, $\mathbb{E}\, G = (c_0, c_1)$, with $c_0 = 1/2 + 5/(2\sqrt{33})$, and

$$G \stackrel{d}{=} \frac{1}{4^\alpha}\left\{G^{(1)} + G^{(2)} + \begin{bmatrix} B_1 B_2 & 0 \\ 1 - B_2 & 0 \end{bmatrix} G^{(3)} + \begin{bmatrix} 0 & B_1 \\ B_1 & 0 \end{bmatrix} G^{(4)}\right\},$$

with $G^{(1)}, \dots, G^{(4)}$, B_1, B_2 independent with $\mathcal{L}(G^{(r)}) = \mathcal{L}(G)$, for $r = 1, \dots, 4$, and $\mathcal{L}(B_1) = \mathcal{L}(B_2) = B(1/2)$. Here, $B(1/2)$ denotes the Bernoulli$(1/2)$ distribution.

For the estimate of large deviations we rely on Chernoff's bounding technique. We need to follow a bivariate setting for the vector $(C(v^\star), C(v_\star))$ as introduced in section 5. The following bound on the moment generating function is obtained:

Proposition 3.5. *It exists a sequence $(Y_k)_{k\geq 0} = (Y_{k,0}, Y_{k,1})_{k\geq 0}$ of bivariate random variables with marginal distributions $\mathcal{L}((C(v^\star) - \mathbb{E}\, C(v^\star))/n^\alpha)$, $\mathcal{L}((C(v_\star) - \mathbb{E}\, C(v_\star))/n^\alpha)$ such that for all $q > 1/\alpha \doteq 1.33$ there is a $K > 0$ with*

$$\mathbb{E}\, \exp\langle s, Y_k\rangle \leq \exp(K\|s\|^q) \tag{1}$$

for all $s \in \mathbb{R}^2$ and $k \geq 0$. An explicit value for $K = K_q$ is given in (4).

The bound on the moment generating function in the previous proposition implies a large deviation estimate via Chernoff bounds:

Theorem 3.6. *For all $1 < \kappa < 1/(1-\alpha) \doteq 4.06$ there exists an $L > 0$ such that for any $t > 0$ and $n = 2^{2k}$*

$$\mathbb{P}\left(\frac{C(v^\star) - \mathbb{E}\, C(v^\star)}{n^\alpha} > t\right) \leq \exp(-Lt^\kappa). \tag{2}$$

An explicit value for L is given in (5). The same bound applies to the left tail.

The approach of Karp and Zhang (1995) based on Azuma's inequality gives the tail bound $\exp(-L't^2)$ for an explicitly known L'. For $\kappa = 2$ the prefactor $L = L_2$ in Theorem 3.6 can also be evaluated and satisfies $L_2 > 11L'$.

4. Karp and Zhang's 2-type branching process

For the analysis of $C(v^\star)$ note that whenever the algorithm has to evaluate the value of a node at a certain depth that yields a 1, according to the discussion of section 2, the algorithm has to evaluate the values of two nodes of depths two levels below that each yield a 1, and $B_3 + B_4$ nodes of depths two levels below that each yield a 0, cf. Figure 1. Here, B_3, B_4 are independent Bernoulli $B(1/2)$ distributed random variables. Analogously, when the algorithm has to evaluate the value of a node at a certain depth that yields 0, two levels below it has to evaluate B_1 nodes yielding a 1 and $2+B_1B_2$ nodes yielding a 0, where B_1, B_2 are independent $B(1/2)$ distributed random variables. Here, the event $\{B_1 = 1\}$ corresponds to the algorithm first checking the right child of the node to be evaluated and $\{B_2 = 1\}$ to first checking the left child of that child, cf. Figure 1. Since at each node the child being evaluated first is independently drawn from all other choices, this gives rise to the following 2-type Galton-Watson branching process.

We have individuals of type 0 and 1 where the population of the k-th generation corresponds to the number of nodes at depth $2k$ that are read by the algorithm. We consider processes starting either with an individual of type 1 or type 0 and assume that the algorithm is applied to the worst case inputs $v^\star$ and $v_\star$, respectively. Then we have the following offspring distributions: An individual of type 1 has an offspring of 2 individuals of type 1 and $B_3 + B_4$ individuals of type 0. An individual of type 0 has an offspring of B_1 individuals of type 1 and $2 + B_1B_2$ individuals of type 0. We denote the number of individuals of type 0 and 1 in generation k by $(V_n^{(i)}, W_n^{(i)})$, when starting with an individual of type $i = 0, 1$, where $n = 2^{2k}$. Note that for $v^\star, v_\star \in \{0,1\}^n$ we have the representations

$$C(v^\star) \stackrel{d}{=} V_n^{(1)} + W_n^{(1)}, \quad C(v_\star) \stackrel{d}{=} V_n^{(0)} + W_n^{(0)}.$$

This is the approach of Karp and Zhang (1995) for regular inputs like $v^\star, v_\star$. Hence, part of the analysis of $C(v^\star)$ can be reduced to the application of the theory of multi-type branching processes; see for general reference Harris (1963) and Athreya and Ney (1972), and for a survey on the application of branching processes to tree structures and tree algorithms see Devroye (1998).

However, we will also use a recursive description of the problem. This will be given in the next section and enables to use as well results from the probabilistic analysis of recursive algorithms by the contraction method.

5. The recursive point of view

It is convenient to work as well with a recursive description of the distributions $\mathcal{L}(C(v_\star))$ and $\mathcal{L}(C(v^\star))$. For this, we define the distributions of a bivariate random sequence $(Z_n) = (Z_{n,0}, Z_{n,1})$ for all $n = 2^{2k}$, $k \geq 0$ by $Z_1 = (1, 1)$ and, for $k \geq 1$,

$$Z_n \stackrel{d}{=} Z_{n/4}^{(1)} + Z_{n/4}^{(2)} + \begin{bmatrix} B_1B_2 & 0 \\ 1-B_2 & 0 \end{bmatrix} Z_{n/4}^{(3)} + \begin{bmatrix} 0 & B_1 \\ B_1 & 0 \end{bmatrix} Z_{n/4}^{(4)},$$

where $Z_{n/4}^{(1)}, \ldots, Z_{n/4}^{(4)}, B_1, B_2$ are independent, B_1, B_2 are Bernoulli $B(1/2)$ distributed and $\mathcal{L}(Z_{n/4}^{(1)}) = \cdots = \mathcal{L}(Z_{n/4}^{(4)}) = \mathcal{L}(Z_{n/4})$. It can directly be checked by induction that the marginals of Z_n satisfy $\mathcal{L}(Z_{n,0}) = \mathcal{L}(C(v_\star))$ and $\mathcal{L}(Z_{n,1}) = \mathcal{L}(C(v^\star))$. Note that $Z_{n,0}$ and $Z_{n,1}$ become dependent, firstly, since we have coupled the offspring distributions using for the second component of Z_n again B_1 and $1 - B_2$ instead of B_3 and B_4, cf. section 4, and, secondly, since the first component of $Z_{n/4}^{(3)}$ contributes to both components of Z_n. Sequences satisfying recursive equations as (Z_n) are being dealt with in a probabilistic framework, the so called contraction method; see Rösler (1991, 1992), Rachev and Rüschendorf (1995), Rösler and Rüschendorf (2001), and Neininger and Rüschendorf (2004).

6. Proofs

In this section we sketch the proofs of the results stated in section 3.

Proof of Proposition 3.1: *(Sketch)* We denote by $R_0(n), R_1(n) \subset \{0,1\}^n$ the sets of vectors at the external nodes at depth $2k$ that yield an evaluation at the root of the decision tree of value 0 and 1, respectively. From the discussion in section 2 we have

$$C(v) \preceq C(v_\star), \quad v \in R_0(n), \quad \text{and} \quad C(v) \preceq C(v^\star), \quad v \in R_1(n).$$

Hence, it remains to show that $C(v_\star) \preceq C(v^\star)$. This is shown by induction on $k \geq 1$. For $k = 1$ this can directly be checked. For the step $k-1 \to k$ assume that we have $C(v_\star(k-1)) \preceq C(v^\star(k-1))$. It suffices to find realizations of the quantities $(V_n^{(1)}, W_n^{(1)})$ and $(V_n^{(0)}, W_n^{(0)})$ on a joint probability space with $V_n^{(0)} + W_n^{(0)} \leq V_n^{(1)} + W_n^{(1)}$ almost surely, $n = 2^{2k}$.

For this we use B, B', $(V_{n/4}^{(i),j}, W_{n/4}^{(i),j})$ for $i = 1,2$, $j = 1,\dots,4$ being independent for each $i = 0,1$ and with B, B' Bernoulli $B(1/2)$ distributed, $\mathcal{L}(V_{n/4}^{(i),j}) = \mathcal{L}(V_{n/4}^{(i)})$, $\mathcal{L}(W_{n/4}^{(i),j}) = \mathcal{L}(W_{n/4}^{(i)})$ for $i = 1,2$ and $j = 1,\dots,4$. By the induction hypothesis we may assume that we have versions of these random variates with $V_{n/4}^{(0),j} + W_{n/4}^{(0),j} \leq V_{n/4}^{(1),j} + W_{n/4}^{(1),j}$ for $j = 1,\dots,4$. With this coupling we define $(V_n^{(1)}, W_n^{(1)})$ and $(V_n^{(0)}, W_n^{(0)})$ according to the values of B, B': On $\{B = 1, B' = 0\}$ we set

$$\begin{pmatrix} V_n^{(0)} \\ W_n^{(0)} \end{pmatrix} := \begin{pmatrix} V_{n/4}^{(0),2} \\ W_{n/4}^{(0),2} \end{pmatrix} + \begin{pmatrix} V_{n/4}^{(0),3} \\ W_{n/4}^{(0),3} \end{pmatrix} + BB' \begin{pmatrix} V_{n/4}^{(0),4} \\ W_{n/4}^{(0),4} \end{pmatrix} + B \begin{pmatrix} V_{n/4}^{(1),1} \\ W_{n/4}^{(1),1} \end{pmatrix},$$

$$\begin{pmatrix} V_n^{(1)} \\ W_n^{(1)} \end{pmatrix} := B \begin{pmatrix} V_{n/4}^{(0),3} \\ W_{n/4}^{(0),3} \end{pmatrix} + \begin{pmatrix} V_{n/4}^{(1),1} \\ W_{n/4}^{(1),1} \end{pmatrix} + B' \begin{pmatrix} V_{n/4}^{(0),4} \\ W_{n/4}^{(0),4} \end{pmatrix} + \begin{pmatrix} V_{n/4}^{(1),2} \\ W_{n/4}^{(1),2} \end{pmatrix}$$

and obtain $V_n^{(0)} + W_n^{(0)} \leq V_n^{(1)} + W_n^{(1)}$. On the remaining sets $\{B = 0, B' = 0\}$, $\{B = 0, B' = 1\}$, and $\{B = 1, B' = 1\}$ similar couplings of $(V_n^{(0)}, W_n^{(0)})$, $(V_n^{(1)}, W_n^{(1)})$ can be defined with $V_n^{(0)} + W_n^{(0)} \leq V_n^{(1)} + W_n^{(1)}$. ■

Proof of Theorem 3.2: *(Sketch)* Assume that a generation has (w_0, w_1) individuals of type 0 and 1. Then, by the definition on the offspring distribution in section 4, the expected number of individuals in the subsequent generation is given by

$$M \begin{pmatrix} w_0 \\ w_1 \end{pmatrix}, \quad M := \begin{bmatrix} 9/4 & 1 \\ 1/2 & 2 \end{bmatrix}.$$

Since $C(v^\star) = C(v^\star(k))$ is the sum of the individuals at generation k for the process started with an individual of type 1 we obtain

$$\mathbb{E}\, C(v^\star) = (1,1) M^k \begin{pmatrix} 0 \\ 1 \end{pmatrix}.$$

The matrix M has the eigenvalues $\lambda_1 = (17+\sqrt{33})/8$ and $\lambda_2 = (17-\sqrt{33})/8$ and its k-th power can be evaluated to

$$M^k = \frac{1}{2\sqrt{33}} \begin{bmatrix} (\sqrt{33}+1)\lambda_1^k + (\sqrt{33}-1)\lambda_2^k & 8(\lambda_1^k - \lambda_2^k) \\ 4(\lambda_1^k - \lambda_2^k) & (\sqrt{33}-1)\lambda_1^k + (\sqrt{33}+1)\lambda_2^k \end{bmatrix}.$$

From this, $\mathbb{E}\, C(v^\star)$ and various constants needed subsequently can be read off. Note, that $\lambda_1^k = n^\alpha$ with α given in Theorem 3.2 and $n = 2^{2k}$. ■

Before proving Theorem 3.3 it is convenient to first prove Theorem 3.4.

Proof of Theorem 3.4: *(Sketch)* The 2-type branching process defined in section 4 is supercritical, nonsingular, and positive regular. Hence, a theorem of Harris (1963) implies that

$$\frac{1}{n^\alpha}\begin{pmatrix} V_n^{(1)} \\ W_n^{(1)} \end{pmatrix} \longrightarrow Y \begin{pmatrix} \nu_1 \\ \nu_2 \end{pmatrix}$$

almost surely, as $k \to \infty$, where Y is a nonnegative random variable and (ν_1, ν_2) a deterministic vector that could also be further specified. Thus we obtain

$$\frac{C(v^\star)}{n^\alpha} \longrightarrow C$$

in distribution, as $k \to \infty$, with $\mathcal{L}(C) = \mathcal{L}((\nu_1 + \nu_2)Y)$.

On the other hand the recursive formulation of section 5 leads after the normalization $X_n := Z_n/n^\alpha$ to

$$X_n \stackrel{d}{=} \sum_{r=1}^{4} A_r X_{n/4}^{(r)},$$

for $k \geq 1$, where $A_1 = A_2 = (1/4^\alpha)I_2$, with the 2×2 identity matrix I_2, and

$$A_3 = \frac{1}{4^\alpha}\begin{bmatrix} B_1 B_2 & 0 \\ 1 - B_2 & 0 \end{bmatrix}, \quad A_4 = \frac{1}{4^\alpha}\begin{bmatrix} 0 & B_1 \\ B_1 & 0 \end{bmatrix}, \tag{3}$$

where $X_{n/4}^{(1)}, \ldots, X_{n/4}^{(4)}, B_1, B_2$ are independent with $\mathcal{L}(X_{n/4}^{(r)}) = \mathcal{L}(X_{n/4})$ for $r = 1, \ldots, 4$ and $\mathcal{L}(B_1) = \mathcal{L}(B_2) = B(1/2)$. It follows from the contraction method that X_n converges weakly and with all mixed second moments to some G, that can be characterized as in Theorem 3.4. For details, how to apply the contraction method, see Theorem 4.1 in Neininger (2001). Thus, we have $C(v^\star)/n^\alpha \to G_1$ in distribution. ■

Proof of Theorem 3.3: *(Sketch)* As shown in the proof of Theorem 3.4 we have the convergence $X_n = Z_n/n^\alpha \to G$ for all mixed second moments. This, in particular, implies $\operatorname{Var} X_{n,1} \to \operatorname{Var} G_1$. The variances of G_1 can be obtained from the distributional identity for G stated in Theorem 3.4. Then we obtain $\operatorname{Var} C(v^\star) = \operatorname{Var}(n^\alpha X_{n,1}) \sim d n^{2\alpha}$ with $d = \operatorname{Var} G_1$. ■

Proof of Proposition 3.5: For $Y_n = (1/n^\alpha)(Z_n - \mathbb{E}\, Z_n)$ we have marginals $\mathcal{L}(Y_{n,1}) = \mathcal{L}((C(v^\star) - \mathbb{E}\, C(v^\star))/n^\alpha)$ and $\mathcal{L}(Y_{n,0}) = \mathcal{L}((C(v_\star) - \mathbb{E}\, C(v_\star))/n^\alpha)$. The distributional recurrence for Z_n from section 5 implies the relation

$$Y_n \stackrel{d}{=} \sum_{r=1}^{4} A_r Y^{(r)}_{n/4} + b_n, \quad k \geq 1,$$

with $Y^{(1)}_{n/4}, \ldots, Y^{(4)}_{n/4}, B_1, B_2$ independent, $\mathcal{L}(Y^{(r)}_{n/4}) = \mathcal{L}(Y_{n/4})$, for $r = 1, \ldots, 4$, $\mathcal{L}(B_1) = \mathcal{L}(B_2) = B(1/2)$ and $b_n = (1/n^\alpha)(4^\alpha \sum_{r=1}^4 (A_r \mathbb{E}\, Z_{n/4}) - \mathbb{E}\, Z_n)$. The matrices A_r are given in (3).

We prove the assertion by induction on k. For $k = 0$ we have $Y_0 = 0$, thus the assertion is true. Assume the assertion is true for some $n/4 = 2^{2(k-1)}$. Then, conditioning on $(A_1, \ldots, A_4, b_n)$, denoting the distribution of this vector by σ_n, and using the induction hypothesis, we obtain

$$\begin{aligned}
\mathbb{E} \exp\langle s, Y_n\rangle &= \int \exp\langle s, \beta_n\rangle \prod_{r=1}^{4} \mathbb{E} \exp\langle s, a_r Y_{n/4}\rangle d\sigma_n(a_1, \ldots, a_4, \beta_n) \\
&\leq \int \exp\langle s, \beta_n\rangle \prod_{r=1}^{4} \exp(K \|a_r^T s\|^q) d\sigma_n(a_1, \ldots, a_4, \beta_n) \\
&\leq \int \exp\left(\langle s, \beta_n\rangle + K\|s\|^q \sum_{r=1}^{4} \|a_r\|_{\mathrm{op}}^q\right) d\sigma_n(a_1, \ldots, a_4, \beta_n) \\
&= \mathbb{E} \exp(\langle s, b_n\rangle + K\|s\|^q U) \exp(K\|s\|^q),
\end{aligned}$$

with $U := \sum_{r=1}^4 (\|A_r\|_{\mathrm{op}}^q) - 1 = 4^{-\alpha q}(2 + B_1 B_2 + (1 - B_2) + B_1) - 1$ and $\|A\|_{\mathrm{op}} = \sup_{\|x\|=1} \|Ax\|$ for matrices A. Hence, the proof is completed by showing

$$\sup_{k \geq 1} \mathbb{E} \exp(\langle s, b_n\rangle + K\|s\|^q U) \leq 1,$$

for some appropriate $K > 0$. We denote $\xi := -\operatorname{ess\,sup} U = 1 - 4^{1-\alpha q}$, thus $q > 1/\alpha$ implies $\xi > 0$.

Small $\|s\|$: First we consider small $\|s\|$ with $\|s\| \leq c / \sup_{k\geq 1} \|b_n\|_{2,\infty}$ for some $c > 0$, where $\|b_n\|_{2,\infty} := \| \|b_n\| \|_\infty$, the inner norm being the Euclidean norm. Note that throughout we have $n = n(k) = 2^{2k}$. For these small $\|s\|$ we have

$$\mathbb{E} \exp((\langle s, b_n\rangle + K\|s\|^q U) \leq \exp(-K\|s\|^q \xi) \mathbb{E} \exp\langle s, b_n\rangle$$

and, with $\mathbb{E}\,\langle s, b_n\rangle = 0$,

$$\begin{aligned}
\mathbb{E} \exp\langle s, b_n\rangle &= \mathbb{E}\left[1 + \langle s, b_n\rangle + \sum_{k=2}^{\infty} \frac{\langle s, b_n\rangle^k}{k!}\right] \\
&= 1 + \mathbb{E}\,\langle s, b_n\rangle^2 \sum_{k=2}^{\infty} \frac{\langle s, b_n\rangle^{k-2}}{k!} \\
&\leq 1 + \|s\|^2 \mathbb{E}\, \|b_n\|^2 \sum_{k=2}^{\infty} \frac{c^{k-2}}{k!} \\
&= 1 + \|s\|^2 \mathbb{E}\, \|b_n\|^2 \frac{e^c - 1 - c}{c^2}.
\end{aligned}$$

Using $\exp(-K\|s\|^q\xi) \leq 1/(1+K\|s\|^q\xi)$ and with $\Psi(c) = (e^c - 1 - c)/c^2$ we obtain

$$\mathbb{E}\,\exp(\langle s, b_n\rangle + K\|s\|^q U) \leq \frac{1 + \|s\|^2 \mathbb{E}\|b_n\|^2 \Psi(c)}{1 + K\|s\|^q \xi}.$$

Hence, we have to choose K with

$$K \geq \frac{\|s\|^{2-q}\Psi(c)}{\xi} \sup_{k\geq 1} \mathbb{E}\,\|b_n\|^2.$$

With $\|s\| \leq c/\sup_{k\geq 1}\|b_n\|_{2,\infty}$ a possible choice is

$$K = \frac{\sup_{k\geq 1}\mathbb{E}\,\|b_n\|^2}{\sup_{k\geq 1}\|b_n\|_{2,\infty}^{2-q}} \frac{\Psi_q(c)}{\xi},$$

with $\Psi_q(c) = (e^c - 1 - c)/c^q$.

Large $\|s\|$: For general $s \in \mathbb{R}^2$ we have

$$\langle s, b_n\rangle + K\|s\|^q U \leq \|s\|\|b_n\| - \|s\|^q K\xi \leq \|s\|\|b_n\|_{2,\infty} - \|s\|^q K\xi,$$

and this is less than zero if

$$\|s\|^{q-1} \geq \frac{\sup_{k\geq 1}\|b_n\|_{2,\infty}}{K\xi} = \frac{\sup_{k\geq 1}\|b_n\|_{2,\infty}^{3-q}}{\sup_{k\geq 1}\mathbb{E}\,\|b_n\|^2 \Psi_q(c)}.$$

If $\|s\|$ satisfies the latter inequality we call it large. Thus, for large $\|s\|$ we have $\sup_{k\geq 1}\mathbb{E}\,\exp(\langle s, b_n\rangle + K\|s\|^q U) \leq 1$.

In order to overlap the regions for small and large $\|s\|$ we need

$$\Psi_1(c) \geq \frac{\sup_{k\geq 1}\|b_n\|_{2,\infty}^2}{\sup_{k\geq 1}\mathbb{E}\,\|b_n\|^2}.$$

The right hand side of the latter display can be evaluated explicitly for our problem and equals 104/77. Thus, this inequality is true for, e.g., $c = 1.53$. Hence, with the explicit value

$$K := K_q = \frac{\sup_{k\geq 1}\mathbb{E}\,\|b_n\|^2}{\sup_{k\geq 1}\|b_n\|_{2,\infty}^{2-q}} \frac{e^{1.53} - 2.53}{1.53^q(1 - 4^{1-q\alpha})} \tag{4}$$

the proof is completed. ∎

Proof of Theorem 3.6: By Chernoff's bounding technique we have, for $u > 0$ and with Proposition 3.5,

$$\begin{aligned}
\mathbb{P}\left(\frac{C(v^\star) - \mathbb{E}\,C(v^\star)}{n^\alpha} > t\right) &= \mathbb{P}(\exp(uY_{n,1}) > \exp(ut))) \\
&\leq \mathbb{E}\,\exp(uY_{n,1} - ut) \\
&= \mathbb{E}\,\exp(\langle(0,u), Y_n\rangle - ut) \\
&\leq \exp(K_q u^q - ut),
\end{aligned}$$

for all q, K_q as in Proposition 3.5 and (4). Minimizing over $u > 0$ we obtain the bound

$$\mathbb{P}\left(\frac{C(v^\star) - \mathbb{E}\,C(v^\star)}{n^\alpha} > t\right) \leq \exp(-Lt^\kappa),$$

for $1 < \kappa < 1/(1-\alpha)$, with

$$L = L_\kappa = K_{\kappa/(\kappa-1)}^{1-\kappa} \frac{(\kappa-1)^{\kappa-1}}{\kappa^\kappa} \tag{5}$$

and $K_{\kappa/(\kappa-1)}$ given in (4). This completes the tail bound. ■

7. m-ary Boolean decision trees

The analysis can be carried over to the case of m-ary Boolean decision trees. The algorithm visits randomly chosen children and evaluates recursively their values until the value of the root can be identified, the remaining children are discarded afterwards. A worst case input $v^\star \in \{0,1\}^n$ with $n = m^{2k}$ can be constructed similarly. Then we have similar results for $C(v^\star)$:

Theorem 7.1. *For the worst case complexity $C(v^\star)$ of evaluating an m-ary Boolean decision tree we have the following asymptotics:*

$$\begin{aligned}
\mathbb{E}\, C(v^\star) &= c_1^{(m)} n^{\alpha_m} + c_2^{(m)} n^{\beta_m}, \\
\mathrm{Var}\, C(v^\star) &\sim d_m n^{2\alpha_m}, \\
\frac{C(v^\star)}{n^{\alpha_m}} &\to C_m, \\
\mathbb{P}\left(\frac{C(v^\star) - \mathbb{E}\, C(v^\star)}{n^{\alpha_m}} > t \right) &\leq \exp(-L^{(m)} t^\kappa), \quad t > 0,
\end{aligned}$$

with constants $c_1^{(m)}$, α_m, β_m, d_m, $L^{(m)} > 0$, $c_2^{(m)} \in \mathbb{R}$, and $1 < \kappa < \kappa_m = 1/(1-\alpha_m)$.

Numerical values for α_m, d_m and κ_m are listed in Table 1. The distribution of C_m is given as $\mathcal{L}(C_m) = \mathcal{L}(G_1)$ and $\mathcal{L}(G) = \mathcal{L}(G_0, G_1)$ is characterized by $\mathbb{E}\,\|G\|^2 < \infty$, $\mathbb{E}\, G = (c_0^{(m)}, c_1^{(m)})$ and

$$\begin{aligned}
G \stackrel{d}{=} \frac{1}{m^{2\alpha_m}} \Bigg\{ &\sum_{r=1}^{m} G^{(r)} + \sum_{r=1}^{m-1} \begin{bmatrix} 0 & \mathbf{1}_r(U_0) \\ \mathbf{1}_r(U_0) & 0 \end{bmatrix} \bar{G}^{(r)} \\
&+ \sum_{r,\ell=1}^{m-1} \begin{bmatrix} \mathbf{1}_r(U_0)\mathbf{1}_\ell(U_r) & 0 \\ 1 - \mathbf{1}_\ell(U_r) & 0 \end{bmatrix} G^{(r,\ell)} \Bigg\},
\end{aligned}$$

with $\mathcal{L}(G^{(r)}) = \mathcal{L}(\bar{G}^{(r)}) = \mathcal{L}(G^{(r,\ell)}) = \mathcal{L}(G)$ and $G^{(r)}, \bar{G}^{(r)}, G^{(r,\ell)}, U_r$ independent with $\mathcal{L}(U_r) = \mathrm{unif}\{0,\dots,m-1\}$ for all r, ℓ. Here, we denote $\mathbf{1}_i(Y) := \mathbf{1}_{\{i \leq Y\}}$ for integer i and a random variable Y, and we have

$$c_0^{(m)} = \frac{1}{2} + \frac{m+3}{2\sqrt{16m + (m-1)^2}}, \quad c_1^{(m)} = \frac{1}{2} + \frac{3m+1}{2\sqrt{16m + (m-1)^2}}.$$

m	2	3	4	5	6	7	8
α_m	0.754	0.759	0.765	0.769	0.774	0.778	0.781
d_m	0.0938	0.0847	0.0782	0.0731	0.0689	0.0652	0.0619
κ_m	4.060	4.154	4.247	4.336	4.419	4.497	4.571

m	9	10	11	12	13	14	15
α_m	0.785	0.788	0.790	0.793	0.795	0.798	0.800
d_m	0.0590	0.0564	0.0541	0.0519	0.0499	0.0481	0.0464
κ_m	4.641	4.707	4.769	4.829	4.886	4.940	4.993

m	16	17	20	30	40	50	100
α_m	0.802	0.804	0.809	0.821	0.830	0.837	0.856
d_m	0.0448	0.0433	0.0394	0.0304	0.0247	0.0209	0.0117
κ_m	5.043	5.091	5.226	5.596	5.885	6.123	6.928

Table 1: *Numerical values of the quantities α_m, d_m and κ_m appearing in Theorem 7.1 for various values of m.*

References

[1] Athreya, K. B. and Ney, P. (1972) *Branching processes.* Die Grundlehren der mathematischen Wissenschaften, Bd. 196, *Springer-Verlag, New York-Heidelberg.*

[2] Devroye, L. (1998) Branching processes and their applications in the analysis of tree structures and tree algorithms. *Probabilistic methods for algorithmic discrete mathematics*, 249–314, Algorithms Combin., 16, *Springer, Berlin.*

[3] Harris, T. E. (1963) *The theory of branching processes.* Die Grundlehren der Mathematischen Wissenschaften, Bd. 119, *Springer-Verlag, Berlin; Prentice-Hall, Inc., Englewood Cliffs, N.J.*

[4] Karp, R. and Zhang, Y. (1995) Bounded branching process and AND/OR tree evaluation. *Random Structures Algorithms 7*, 97–116.

[5] Motwani, R. and Raghavan, P. (1995) *Randomized algorithms.* Cambridge University Press, Cambridge.

[6] Neininger, R. (2001) On a multivariate contraction method for random recursive structures with applications to Quicksort. *Random Structures Algorithms 19*, 498–524.

[7] Neininger, R. and Rüschendorf, L. (2004) A general limit theorem for recursive algorithms and combinatorial structures. *Ann. Appl. Probab. 14*, 378–418.

[8] Rachev, S. T. and Rüschendorf, L. (1995). Probability metrics and recursive algorithms. *Adv. in Appl. Probab. 27*, 770–799.

[9] Rösler, U. (1991). A limit theorem for "Quicksort". *RAIRO Inform. Théor. Appl. 25*, 85–100.

[10] Rösler, U. (1992). A fixed point theorem for distributions. *Stochastic Process. Appl. 42*, 195–214.

[11] Rösler, U. and Rüschendorf, L. (2001). The contraction method for recursive algorithms. *Algorithmica 29*, 3–33.

[12] Saks, M. and Wigderson, A. (1986) Probabilistic boolean decision trees and the complexity of evaluating game trees. *Proceedings of the 27th Annual IEEE Symposium on Foundations of Computer Science*, 29–38, Toronto, Ontario.

[13] Snir, M. (1985) Lower bounds on probabilistic linear decision trees. *Theoret. Comput. Sci. 38*, 69–82.

Tämur Ali Khan and Ralph Neininger
Department of Mathematics
J.W. Goethe University
Robert-Mayer-Str. 10
60325 Frankfurt a.M.
Germany
{alikhan, neiningr}@ismi.math.uni-frankfurt.de

Trends in Mathematics, © 2004 Birkhäuser Verlag Basel/Switzerland

Polynomial Time Perfect Sampling Algorithm for Two-Rowed Contingency Tables

Shuji Kijima and Tomomi Matsui

ABSTRACT: *This paper proposes a polynomial time perfect (exact) sampling algorithm for $2 \times n$ contingency tables. Our algorithm is a Las Vegas type randomized algorithm and the expected running time is bounded by $O(n^3 \ln N)$ where n is the number of columns and N is the total sum of whole entries in a table. The algorithm is based on monotone coupling from the past (monotone CFTP) algorithm and new Markov chain for sampling two-rowed contingency tables uniformly. We employed the path coupling method and showed the mixing rate of our chain. Our result indicates that uniform generation of two-rowed contingency tables is easier than the corresponding counting problem, since the counting problem is known to be #P–complete.*

1. Introduction

In this paper, we propose a polynomial time perfect (exact) sampling algorithm for $2 \times n$ contingency tables. Our algorithm is a Las Vegas type randomized algorithm and the expected running time is bounded by $O(n^3 \ln N)$ where n is the number of columns and N is the total sum of whole entries in a table. Our result indicates that uniform generation of two-rowed contingency tables is easier than the corresponding counting problem, since the counting problem is known to be #P–complete [11]. The main idea of our algorithm is not the simple rejection sampling but the monotone coupling from the past (monotone CFTP) algorithm proposed by Propp and Wilson [18].

For sampling two-rowed contingency tables, Dyer and Greenhill [10] proposed a fully polynomial time approximately uniform sampler based on the Metropolis-Hastings algorithm with a natural Markov chain. They showed that the mixing time of their chain is bounded by $(1/2)n(n-1)\ln(N\varepsilon^{-1})$ by using the path coupling technique proposed by Bubley and Dyer [5]. We propose a new Markov chain which is obtained by forbidding some moves of Dyer and Greenhill's chain. Although we have shown the bound by using path coupling method also, we need to introduce a preprocedure which is not appeared in Dyer and Greenhill's method. We also introduce a specified partial order on the set of tables and show the monotonicity of our chain. In the paper [18], Propp and Wilson showed that if we have a monotone chain with polynomial time mixing rate, there exists a polynomial time monotone CFTP algorithm. By applying their technique to our rapidly mixing monotone chain we can construct a polynomial time perfect sampling algorithm.

A contingency table is a matrix of nonnegative integers with prescribed positive row and column sums. Contingency tables are used in statistics to store data from sample survey. A test of independence between rows and columns is a statistic interest for contingency tables. Exact test proposed by Fischer [12] is one of the tests for this purpose. Diaconis and Effron also discussed a test for independence in contingency tables [7]. Exact test can be done by systematic enumeration of all tables, but it is hard to enumerate all tables. In practice, Markov chain Monte Carlo (MCMC) method is used for calculation of p value in exact test (see [1, 2] for example). A MCMC method is a Monte Carlo method which uses samples from the stationary distribution of a Markov chain. The problem of using MCMC method, however, is "how many times do we have to simulate the transitions for the purpose of sampling from stationary distribution?" A practical solution for this problem is to use approximate sampler obtained by interrupting transitions in finite time.

There are many works for almost uniform sampling contingency tables using a Markov chain. Diaconis and Saloff-Coste [8] discussed the rate of convergence of a simple Markov chain for 2-dimensional contingency tables. They have shown that the simple chain mixes polynomial time in the table sum when the numbers of rows and columns are fixed. Dyer, Kannan and Mount [11] proposed a different Markov chain for counting the number of 2-dimensional contingency tables. In case of sufficiently large marginal sums, their chain mixes polynomial time in the number of rows and columns. For two-rowed tables, Hernek [14] showed that the mixing time of the simple Markov chain is bounded by a polynomial of table sum and number of columns. Hernek bounded the mixing time of the chain by using coupling theorem shown by Aldous [3]. Dyer and Greenhill [10] proposed a rapidly mixing Markov chain for two-rowed contingency tables. Their chain mixes polynomial time in the number of columns and the logarithm of table sum. They analyzed the mixing rate of their chain by using path coupling technique proposed by Bubley and Dyer [4, 5]. In the paper [17], Matsui, Matsui and Ono extended Dyer and Greenhill's result to $2 \times \cdots \times 2 \times J$ contingency tables. Recently, Cryan, Dyer, Goldberg, Jerrum and Martin [6] showed that 2×2 chain, which is an extension of Dyer and Greenhill's, is rapidly mixing when the number of rows (or columns) is a constant.

Propp and Wilson devised a surprising simple algorithm, called CFTP algorithm (or backward coupling), which produces exact samples from the limit distribution [18, 19]. CFTP algorithm simulate infinite time transitions of a chain in a (probabilistically) finite time, for any finite Markov chain. In CFTP algorithm, however, we need to check the "coalescence condition," by executing the simulations from all the states. Thus CFTP algorithm is not available straightforwardly. A monotone CFTP algorithm is an algorithm for monotone Markov chain, which has a partially ordered state space and a transition rule which preserves the partial order. If the given chain has the monotonicity, it relaxes the difficulty of simulation from all states.

In the next section, we review the (monotone) CFTP algorithm and the theorem proposed by Propp and Wilson in [18]. We propose a new Markov chain for $2 \times n$ contingency tables and a sampling algorithm based on monotone CFTP algorithm in Section 3. In Section 4, we show the monotonicity of our chain. In Section 5, we analyze the expected running time.

2. Review of Coupling From The Past Algorithm

When we simulate an ergodic Markov chain for infinite time, we can gain a sample exactly according to the stationary distribution. Suppose that there exists a chain from infinite past, then a possible state at the present time of the chain for which we can have an evidence of the uniqueness without respect to an initial state of the chain, is a realization of a random sample exactly from the stationary distribution. This is the key idea of CFTP.

Suppose that we have an ergodic Markov chain MC with finite state space Ω and transition matrix P. The transition rule of the Markov chain $X \mapsto X'$ can be described by a deterministic function $\phi : \Omega \times [0,1) \to \Omega$, called *update function*, as follows. Given a random number Λ uniformly distributed over $[0,1)$, update function ϕ satisfies that $\Pr(\phi(x,\Lambda) = y) = P(x,y)$ for any $x, y \in \Omega$. We can realize the Markov chain by setting $X' = \phi(X, \Lambda)$. Clearly, update function corresponding to the given transition matrix P is not unique. The result of transitions of the chain from the time t_1 to t_2 ($t_1 < t_2$) with a sequence of random numbers $\boldsymbol{\lambda} = (\lambda[t_1], \lambda[t_1+1], \ldots, \lambda[t_2-1]) \in [0,1)^{t_2-t_1}$ is denoted by $\Phi_{t_1}^{t_2}(x, \boldsymbol{\lambda}) : \Omega \times [0,1)^{t_2-t_1} \to \Omega$ where $\Phi_{t_1}^{t_2}(x,\boldsymbol{\lambda}) \stackrel{\text{def.}}{=} \phi(\phi(\cdots(\phi(x,\lambda[t_1]),\ldots,\lambda[t_2-2]),\lambda[t_2-1])$. We say that a sequence $\lambda \in [0,1)^{|T|}$ satisfies the *coalescence condition*, when $\exists y \in \Omega$, $\forall x \in \Omega$, $y = \Phi_T^0(x, \boldsymbol{\lambda})$.

With these preparation, standard Coupling From The Past algorithm is expressed as follows.

Algorithm 1. (CFTP Algorithm [18])

Step 1. Set the starting time period $T := -1$ to go back, and set $\boldsymbol{\lambda}$ be the empty sequence.

Step 2. Generate random real numbers $\lambda[T], \lambda[T+1], \ldots, \lambda[\lceil T/2\rceil - 1] \in [0,1)$, and insert them to the head of $\boldsymbol{\lambda}$ in order, i.e., put $\boldsymbol{\lambda} := (\lambda[T], \lambda[T+1], \ldots, \lambda[-1])$.

Step 3. Start a chain from each element $x \in \Omega$ at time period T, and run each chain to time period 0 according to the update function ϕ with the sequence of numbers in $\boldsymbol{\lambda}$. (Here we note that every chain uses the common sequence λ.)

Step 4. [Coalescence check] The state obtained at time period 0 can be denoted by $\Phi_T^0(x, \boldsymbol{\lambda})$.

  (a) If $\exists y \in \Omega$, $\forall x \in \Omega$, $y = \Phi_T^0(x,\boldsymbol{\lambda})$, then return y and stop.

  (b) Else, update the starting time period $T := 2T$, and go to Step 2.

Theorem 2.1. (CFTP Theorem [18]) *Let MC be an ergodic finite Markov chain with state space Ω, defined by an update function $\phi : \Omega \times [0,1) \to \Omega$. If the CFTP algorithm (Algorithm 1) terminates with probability 1, then the obtained value is a realization of a random variable exactly distributed according to the stationary distribution.*

Theorem 2.1 gives a (probabilistically) finite time algorithm for infinite time simulation. However, simulations from all states executed in Step 3 is a hard requirement.

Suppose that there exists a partial order "$\succeq$" on the set of states Ω. A transition rule expressed by a deterministic update function ϕ is called *monotone* (with respect to "$\succeq$") if $\forall \lambda \in [0,1)$, $\forall x, \forall y \in \Omega$, $x \succeq y \Rightarrow \phi(x,\lambda) \succeq \phi(y,\lambda)$. For ease, we also say that a chain is *monotone* if the chain has a *monotone* transition rule.

Theorem 2.2. (monotone CFTP [18, 9]) *Suppose that a Markov chain defined by an update function ϕ is monotone with respect to a partially ordered set of states $(\Omega, \succeq)$, and $\exists x_{\max}, \exists x_{\min} \in \Omega$, $\forall x \in \Omega$, $x_{\max} \succeq x \succeq x_{\min}$. Then the CFTP algorithm (Algorithm 1) terminates with probability 1, and a sequence $\lambda \in [0,1)^{|T|}$ satisfies the coalescence condition, i.e., $\exists y \in \Omega$, $\forall x \in \Omega$, $y = \Phi_T^0(x, \boldsymbol{\lambda})$, if and only if $\Phi_T^0(x_{\max}, \boldsymbol{\lambda}) = \Phi_T^0(x_{\min}, \boldsymbol{\lambda})$.*

When the given Markov chain satisfies the conditions of Theorem 2.2, we can modify Algorithm 1 by substituting Step 4 (a) by

Step 4. (a)′ If $\exists y \in \Omega$, $y = \Phi_T^0(x_{\max}, \boldsymbol{\lambda}) = \Phi_T^0(x_{\min}, \boldsymbol{\lambda})$, then return y.

The algorithm obtained by the above modification is called a monotone CFTP algorithm.

3. Perfect Sampler for $2 \times n$ Contingency Tables

In this section, we introduce our algorithm. We denote the set of real numbers by $\mathbb{R}$ and the set of integers (non-negative, positive integers) by $\mathbb{Z}$ ($\mathbb{Z}_+$, $\mathbb{Z}_{++}$), respectively. Let $\boldsymbol{r} = (r_1, r_2) \in \mathbb{Z}_{++}^2$ and $\boldsymbol{s} = (s_1, \ldots, s_n) \in \mathbb{Z}_{++}^n$ be a pair of vectors satisfying $\sum_{i=1}^2 r_i = \sum_{j=1}^n s_j = N \in \mathbb{Z}_{++}$. The set Ξ of $2 \times n$ contingency tables with row and column sums $(\boldsymbol{r}, \boldsymbol{s})$ is defined by

$$\Xi \stackrel{\text{def.}}{=} \left\{ X \in \mathbb{Z}_+^{2\times n} \;\middle|\; \begin{array}{l} \sum_{j=1}^n X[i,j] = r_i \ (1 \le \forall i \le 2), \\ \sum_{i=1}^2 X[i,j] = s_j \ (1 \le \forall j \le n) \end{array} \right\}$$

where $X[i,j]$ is the value in the cell indexed by ith row and jth column.

We propose a new Markov chain $\mathcal{M}$ with state space Ξ for given $\boldsymbol{r}$ and $\boldsymbol{s}$. For any column index $j \in \{1, \ldots, n-1\}$, we define

$$a_X(j) \stackrel{\text{def.}}{=} X[1,j] + X[1,j+1], \tag{1}$$

$$b_X(j) \stackrel{\text{def.}}{=} X[2,j] + X[2,j+1], \tag{2}$$

$$\theta_X(j) \stackrel{\text{def.}}{=} \min\{a_X(j), b_X(j), s_j, s_{j+1}\} + 1. \tag{3}$$

The transition rule of $\mathcal{M}$ is defined by the following *update function* $\phi : \Xi \times [1,n) \to \Xi$. For a current state $X \in \Xi$, the next state $X' = \phi(X, \lambda) \in \Xi$ with respect to a random number $\lambda \in [1, n)$ is defined by

$$X'[1,j] = \begin{cases} \min\{a_X(j), s_j\} - \lfloor (\lambda - \lfloor \lambda \rfloor)\, \theta_X(j) \rfloor & (j = \lfloor \lambda \rfloor), \\ a_X(j) - \min\{a_X(j), s_j\} + \lfloor (\lambda - \lfloor \lambda \rfloor)\, \theta_X(j) \rfloor & (j = \lfloor \lambda \rfloor + 1), \\ X[1,j] & (\text{otherwise}), \end{cases}$$

$$X'[2,j] = s_j - X'[1,j].$$

Our chain $\mathcal{M}$ is a modification of Dyer and Greenhill's chain ([10]) obtained by restricting to choose only a consecutive pair of columns. Clearly $\mathcal{M}$ is finite, aperiodic and irreducible and so ergodic. The chain has a unique stationary distribution, which is the uniform distribution.

We define two special tables X_{U} and $X_{\mathrm{L}} \in \Xi$ by

$$X_{\mathrm{U}} \stackrel{\text{def.}}{=} \left(X[i,j] \in \mathbb{Z}_+ \;\middle|\; \begin{array}{l} \exists k \in \{1, \ldots, n\},\ r_1 = \sum_{j=1}^k X[1,j] \le \sum_{j=1}^k s_j, \\ X[2,j] = 0 \ (j = 1, \ldots, k-1) \end{array} \right),$$

$$X_{\mathrm{L}} \stackrel{\text{def.}}{=} \left(X[i,j] \in \mathbb{Z}_+ \;\middle|\; \begin{array}{l} \exists l \in \{1, \ldots, n\},\ r_1 = \sum_{j=l}^n X[1,j] \le \sum_{j=l}^n s_j, \\ X[2,j] = 0 \ (j = l+1, \ldots, n) \end{array} \right).$$

Here we note that $X_{\mathrm{U}}, X_{\mathrm{L}}$ are obtained by the North-West corner rule and the North-East corner rule, respectively. Now we describe our sampling algorithm.

Algorithm 2.

Step 1. Set the starting time period $T := -1$ to go back, and set $\boldsymbol{\lambda}$ be the empty sequence.

Step 2. Generate random real numbers $\lambda[T], \lambda[T+1], \ldots, \lambda[\lceil T/2 \rceil - 1] \in [1, n)$, and put $\boldsymbol{\lambda} := (\lambda[T], \lambda[T+1], \ldots, \lambda[-1])$.

Step 3. Start two chains from X_{U} and X_{L}, respectively at time period T, and run them to time period 0 according to the update function ϕ with the sequence of numbers in $\boldsymbol{\lambda}$.

Step 4. [Coalescence check]

(a) If $\exists Y \in \Xi$, $Y = \Phi_T^0(X_{\mathrm{U}}, \boldsymbol{\lambda}) = \Phi_T^0(X_{\mathrm{L}}, \boldsymbol{\lambda})$, then return Y and stop.

(b) Else, update the starting time period $T := 2T$, and go to Step 2.

Theorem 3.1. *With probability 1, Algorithm 2 terminates and returns a table. The table obtained by Algorithm 2 is a realization of a random sample according to the exactly uniform distribution on* Ξ.

Theorem 3.1 guarantees that Algorithm 2 is a perfect sampling algorithm. We prove Theorem 3.1 by showing the monotonicity in the next section.

4. Monotonicity of the Chain

In Section 2, we described two theorems. Thus to prove Theorem 3.1, we only need to show that Algorithm 2 is a monotone CFTP algorithm. For this purpose, in this section, we introduce a partial order on Ξ, and show that X_{U} and X_{L} is a unique pair of maximum and minimum elements of Ξ, and Markov chain $\mathcal{M}$ is monotone.

For any $X \in \Xi$, we define the *cumulative sum vector* $f_X \in \mathbb{Z}_+^{n+1}$ by

$$f_X(i) \stackrel{\text{def.}}{=} \begin{cases} 0 & (i = 0), \\ X[1,1] + \cdots + X[1,i] & (i \in \{1, \ldots, n\}), \end{cases}$$

where $f_X \stackrel{\text{def.}}{=} (f_X(0), f_X(1), \ldots, f_X(n))$. Obviously from the definition, there exists a bijection between Ξ and $\{f_X \mid X \in \Xi\}$. For any pair $X, Y \in \Xi$, we say $X \succeq Y$ if and only if $f_X - f_Y \geq \mathbf{0}$. It is clear that the relation "$\succeq$" is a partial order on Ξ. We can see easily that $X_{\mathrm{U}} \succeq X \succeq X_{\mathrm{L}}$ for any $X \in \Xi$.

We say that a state $X \in \Xi$ *covers* $Y \in \Xi$ (at k), denoted by $X \cdot\!\!\succ Y$ (or $X \cdot\!\!\succ_k Y$), when

$$f_X(i) - f_Y(i) = \begin{cases} 1 & (i = k), \\ 0 & (\text{otherwise}). \end{cases}$$

Note that $X \cdot\!\!\succ_k Y$ if and only if

$$X[1,i] - Y[1,i] = \begin{cases} +1 & (i = k), \\ -1 & (i = k+1), \\ 0 & (\text{otherwise}). \end{cases}$$

Lemma 4.1. *If a pair of distinct states* $X, Y \in \Xi$ *satisfies* $X \succeq Y$, *then* $\exists Z \in \Xi$, $X \cdot\!\!\succ Z \succeq Y$.

Proof is omitted. It is not hard to show the above Lemma.

The following is a key lemma for proving the monotonicity of our chain.

Lemma 4.2. *If a pair of states* $X, Y \in \Xi$ *satisfies* $X \cdot\!\!\succ_k Y$, *then* $\forall \lambda \in [1, n)$, $\phi(X, \lambda) \succeq \phi(Y, \lambda)$.

Proof: We denote $\phi(X,\lambda) = X'$ and $\phi(Y,\lambda) = Y'$ for simplicity. For any index $i \neq \lfloor\lambda\rfloor$, it is clear that $f_{X'}(i) = f_X(i)$ and $f_{Y'}(i) = f_Y(i)$, and so $f_{X'}(i) - f_{Y'}(i) = f_X(i) - f_Y(i) \geq 0$ since $X \succeq Y$. In case that $i = \lfloor\lambda\rfloor$,

$$\begin{aligned} f_{X'}(\lfloor\lambda\rfloor) - f_{Y'}(\lfloor\lambda\rfloor) &= (f_{X'}(\lfloor\lambda\rfloor - 1) + X'[1,\lfloor\lambda\rfloor]) - (f_{Y'}(\lfloor\lambda\rfloor - 1) + Y'[1,\lfloor\lambda\rfloor]) \\ &= \{f_X(\lfloor\lambda\rfloor - 1) - f_Y(\lfloor\lambda\rfloor - 1)\} + (X'[1,\lfloor\lambda\rfloor] - Y'[1,\lfloor\lambda\rfloor]) \\ &= \{f_X(\lfloor\lambda\rfloor - 1) - f_Y(\lfloor\lambda\rfloor - 1)\} \\ &\quad + \min\{a_X, s_{\lfloor\lambda\rfloor}\} - \lfloor(\lambda - \lfloor\lambda\rfloor)\,\theta_X\rfloor - \min\{a_Y, s_{\lfloor\lambda\rfloor}\} + \lfloor(\lambda - \lfloor\lambda\rfloor)\,\theta_Y\rfloor. \\ &= \begin{cases} \Delta\eta + \Delta\theta & (\lfloor\lambda\rfloor \neq k+1), \\ 1 + \Delta\eta + \Delta\theta & (\lfloor\lambda\rfloor = k+1), \end{cases} \end{aligned}$$

where $a_X \stackrel{\text{def.}}{=} a_X(\lfloor\lambda\rfloor)$, $a_Y \stackrel{\text{def.}}{=} a_Y(\lfloor\lambda\rfloor)$, $\theta_X \stackrel{\text{def.}}{=} \theta_X(\lfloor\lambda\rfloor)$, $\theta_Y \stackrel{\text{def.}}{=} \theta_Y(\lfloor\lambda\rfloor)$ (see (1) and (3) for detail), $\Delta\eta \stackrel{\text{def.}}{=} \min\{a_X, s_{\lfloor\lambda\rfloor}\} - \min\{a_Y, s_{\lfloor\lambda\rfloor}\}$ and $\Delta\theta \stackrel{\text{def.}}{=} -\lfloor(\lambda - \lfloor\lambda\rfloor)\,\theta_X\rfloor + \lfloor(\lambda - \lfloor\lambda\rfloor)\,\theta_Y\rfloor$.

1. Consider the case that $\lfloor\lambda\rfloor = k-1$. Then $a_X = a_Y + 1$ and $b_X = b_Y - 1$, where $b_X \stackrel{\text{def.}}{=} b_X(\lfloor\lambda\rfloor)$ and $b_Y \stackrel{\text{def.}}{=} b_Y(\lfloor\lambda\rfloor)$ (see (2) for detail).
   (a) If $a_Y \geq s_{\lfloor\lambda\rfloor}$, then $\Delta\eta = 0$ and $\theta_X \leq \theta_Y$. Thus $\Delta\theta \geq 0$, hence $f_{X'}(\lfloor\lambda\rfloor) - f_{Y'}(\lfloor\lambda\rfloor) \geq 0$.
   (b) If $a_Y < s_{\lfloor\lambda\rfloor}$, then $\Delta\eta = 1$ and $\theta_X \leq \theta_Y + 1$. Thus $\Delta\theta \geq -1$, hence $f_{X'}(\lfloor\lambda\rfloor) - f_{Y'}(\lfloor\lambda\rfloor) \geq 0$.
2. Consider the case that $\lfloor\lambda\rfloor = k+1$. Then $a_X = a_Y - 1$ and $b_X = b_Y + 1$.
   (a) If $a_X \geq s_{\lfloor\lambda\rfloor}$, then $1 + \Delta\eta \geq 1$ and $\theta_X \leq \theta_Y + 1$. Thus $\Delta\theta \geq -1$ and $f_{X'}(\lfloor\lambda\rfloor) - f_{Y'}(\lfloor\lambda\rfloor) \geq 0$.
   (b) If $a_X < s_{\lfloor\lambda\rfloor}$, then $1 + \Delta\eta \geq 0$. Note that $a_X + b_X = a_Y + b_Y = s_{\lfloor\lambda\rfloor} + s_{\lfloor\lambda\rfloor+1}$, $\theta_X \leq \theta_Y$. Thus $\Delta\theta \geq 0$, hence $f_{X'}(\lfloor\lambda\rfloor) - f_{Y'}(\lfloor\lambda\rfloor) \geq 0$.
3. Consider the remained case that $\lfloor\lambda\rfloor \neq k+1$ and $\lfloor\lambda\rfloor \neq k-1$. Then $a_X = a_Y$, $\Delta\eta = 0$, $\Delta\theta = 0$, and $f_{X'}(\lfloor\lambda\rfloor) - f_{Y'}(\lfloor\lambda\rfloor) = 0$.

From the above, we have $f_{X'} \geq f_{Y'}$ and so $\phi(X,\lambda) \succeq \phi(Y,\lambda)$. □

Lemma 4.3. *The Markov chain $\mathcal{M}$ is monotone, i.e.,* $\forall\lambda \in [1,n)$, $\forall X, \forall Y \in \Xi$, $X \succeq Y \Rightarrow \phi(X,\lambda) \succeq \phi(Y,\lambda)$.

Proof: By applying Lemma 4.1 repeatedly, we can show that for any pair of states $X, Y \in \Xi$ satisfying $X \succeq Y$, there exists a sequence $X = Z_0, Z_1, \ldots, Z_R = Y$ with appropriate length such that $Z_i \in \Xi$ $(0 \leq i \leq R)$ and $Z_0 \cdot\!\succ Z_1 \cdot\!\succ \cdots \cdot\!\succ Z_R$. Then Lemma 4.2 implies that $\phi(Z_0,\lambda) \succeq \phi(Z_1,\lambda) \succeq \cdots \succeq \phi(Z_R,\lambda)$ for any $\lambda \in [1,n)$. Thus $\forall\lambda \in [1,n)$, $\phi(X,\lambda) \succeq \phi(Y,\lambda)$. □

Lastly, we show the correctness of our algorithm.

Proof of Theorem 3.1: From Lemma 4.3, the Markov chain $\mathcal{M}$ is monotone, and it is clear that X_{U} and X_{L} is a unique pair of the maximum and minimum elements. Then Algorithm 2 is a monotone CFTP algorithm, and so we can show Theorem 3.1 by using Theorem 2.1 and Theorem 2.2. □

5. Expected Running Time

Here, we discuss the running time of our algorithm. In this section, we assume to introduce a special preprocess and we get the following condition.

Condition 1. *Column sum vector $\mathbf{s}$ satisfies $s_1 \geq s_2 \geq \cdots \geq s_n$.*

We can assume Condition 1 by sorting column sums in $\mathrm{O}(n \ln n)$ time. The following is a main result of this paper.

Theorem 5.1. *Under Condition 1, the expected running time of Algorithm 2 is bounded by $\mathrm{O}(n^3 \ln N)$ where n is the number of columns and N is the total sum of whole entries in a table of Ξ.*

In the rest of this section, we prove Theorem 5.1 by estimating the expectation of *coalescence time* $T_* \in \mathbb{Z}_{++}$ defined by $T_* \stackrel{\text{def.}}{=} \min\{t > 0 \mid \exists y \in \Omega,\ \forall x \in \Omega,\ y = \Phi^0_{-t}(x, \mathbf{\Lambda})\}$. Note that T_* is a random variable.

Given a pair of probabilistic distributions ν_1 and ν_2 on the finite state space Ω, the *total variation distance* between ν_1 and ν_2 is defined by $\mathrm{d_{TV}}(\nu_1, \nu_2) \stackrel{\text{def.}}{=} \frac{1}{2}\sum_{x\in\Omega} |\nu_1(x) - \nu_2(x)|$. The *mixing rate* of an ergodic Markov chain is defined by $\tau \stackrel{\text{def.}}{=} \max_{x\in\Omega}\{\min\{t \mid \forall s \geq t,\ \mathrm{d_{TV}}(\pi, P^s_x) \leq 1/\mathrm{e}\}\}$ where π is the stationary distribution and P^s_x is the probabilistic distribution of the chain at time period $s \geq 0$ with initial state x at time period 0. Path Coupling Theorem is a useful technique for bounding the mixing rate.

Theorem 5.2. (Path Coupling [5]) *Let MC be a finite ergodic Markov chain with state space Ω. Let $G = (\Omega, \mathcal{E})$ be a connected undirected graph with vertex set Ω and edge set $\mathcal{E} \subseteq \binom{\Omega}{2}$. Let $l : \mathcal{E} \to \mathbb{R}$ be a positive length defined on the edge set. For any pair of vertices $\{x, y\}$ of G, the distance between x and y, denoted by $d(x,y)$ and/or $d(y,x)$, is the length of a shortest path between x and y, where the length of a path is the sum of the lengths of edges in the path. Suppose that there exists a joint process $(X, Y) \mapsto (X', Y')$ with respect to MC satisfying that whose marginals are a faithful copy of MC and*

$$0 < \exists\beta < 1,\ \forall\{X, Y\} \in \mathcal{E},\ \mathrm{E}[d(X', Y')] \leq \beta d(X, Y).$$

Then the mixing rate τ of Markov chain MC satisfies $\tau \leq (1-\beta)^{-1}(1+\ln(D/d))$, where $d \stackrel{\text{def.}}{=} \min\{d(x,y) \mid \forall x, \forall y \in \Omega\}$ and $D \stackrel{\text{def.}}{=} \max\{d(x,y) \mid \forall x, \forall y \in \Omega\}$.

The above theorem differs from the original theorem in [5] since the integrality of the edge length is not assumed. We drop the integrality and introduce the minimum distance d. This modification is not essential and we can show Theorem 5.2 similarly.

Now, we show the polynomiality of Algorithm 2. First, we estimate the mixing rate of our chain $\mathcal{M}$ by employing Path Coupling Theorem. In the proof of the following lemma, Condition 1 plays an important role.

Lemma 5.3. *Under Condition 1, the mixing rate τ of our Markov chain $\mathcal{M}$ satisfies $\tau \leq n(n-1)^2(1 + \ln(nN))$.*

Proof: Let $G = (\Xi, \mathcal{E})$ be an undirected simple graph with vertex set Ξ and edge set $\mathcal{E}$ defined as follows. A pair of vertices $\{X, Y\}$ is an edge if and only if $(1/2)\sum_{j=1}^{n} |X[1,j] - Y[1,j]| = 1$. Clearly, the graph G is connected. For each edge $e = \{X, Y\} \in \mathcal{E}$, there exists a unique pair of indices $j_1, j_2 \in \{1, \ldots, n\}$, called the *supporting pair* of e, satisfying

$$|X[1,j] - Y[1,j]| = \begin{cases} 1 & (j = j_1, j_2), \\ 0 & (\text{otherwise}). \end{cases}$$

We define the length $l(e)$ of an edge e by $l(e) \stackrel{\text{def.}}{=} (1/(n-1))\sum_{i=1}^{j^*-1}(n-i)$ where $j^* = \max\{j_1, j_2\} \geq 2$ and $\{j_1, j_2\}$ is the supporting pair of e. Note that $1 \leq \min_{e\in\mathcal{E}} l(e) \leq \max_{e\in\mathcal{E}} l(e) \leq n/2$. For each pair $X, Y \in \Xi$, we define the distance $d(X,Y)$ as the length of the shortest path between X and Y on G. Clearly, the diameter of G, i.e., $\max\{d(X,Y)\}$, is bounded by nN. The definition of edge length implies that for any edge $\{X,Y\} \in \mathcal{E}$, $d(X,Y) = l(\{X,Y\})$.

We define a joint process $(X,Y) \mapsto (X',Y')$ by $(X,Y) \mapsto (\phi(X,\Lambda), \phi(Y,\Lambda))$ with uniform real random number $\Lambda \in [1,n)$, where ϕ is the update function defined in Section 3. Now we show that

$$\mathrm{E}[d(X',Y')] \leq \beta d(X,Y),\ \beta = 1 - 1/(n(n-1)^2), \tag{4}$$

for any pair $\{X,Y\} \in \mathcal{E}$. In the following, we denote the supporting pair of $\{X,Y\}$ by $\{j_1, j_2\}$. Without loss of generality, we can assume that $j_1 < j_2$, and $X[1,j_2] = Y[1,j_2] - 1$. In the following proof, we define $a_X, a_Y, b_X, b_Y, \theta_X, \theta_Y$ in a similar way as in the proof of Lemma 4.2.

1. When $\lfloor\Lambda\rfloor = j_2 - 1$, we show that $\mathrm{E}[d(X',Y') \mid \lfloor\Lambda\rfloor = j_2 - 1] \leq d(X,Y) - (1/2)(n - j_2 + 1)/(n-1)$.
   (a) In case that $j_1 = j_2 - 1$, $X' = Y'$ with conditional probability 1. Hence $d(X',Y') = 0$.
   (b) In case that $j_1 \neq j_2 - 1$ and $\theta_X = \theta_Y$. Condition 1 implies that $s_{j_2-1} \geq s_{j_2}$. Since $\theta_X = \theta_Y$, $a_Y > a_X$ and $b_Y < b_X$, we have $\theta_X = \theta_Y = \min\{s_{j_2-1}, s_{j_2}\} = s_{j_2}$. Thus we have $s_{j_2-1} \geq a_Y > a_X \geq s_{j_2}$, and so $X'[1,j_2-1] = a_X - (\Lambda - \lfloor\Lambda\rfloor)\theta_X$, $Y'[1,j_2-1] = a_Y - (\Lambda - \lfloor\Lambda\rfloor)\theta_Y$ by the definition of the function ϕ. Then $X'[1,j_2-1] = Y'[1,j_2-1] - 1$ since $a_X = a_Y - 1$. Additionally, since $X'[1,j_2] = a_X - X'[1,j_2-1]$ and $Y'[1,j_2] = a_Y - Y'[1,j_2-1]$, we have $X'[1,j_2] = Y'[1,j_2]$. Hence $d(X',Y') = d(X,Y) - (n-j_2+1)/(n-1)$ with conditional probability 1.
   (c) In case that $j_1 \neq j_2 - 1$ and $\theta_X \neq \theta_Y$. Clearly $|\theta_X - \theta_Y| = 1$. First, we discuss the case that $\theta_X = \theta_Y - 1$. We only need to consider two cases, one is the case that $\lfloor(\Lambda - \lfloor\Lambda\rfloor)\theta_X\rfloor = \lfloor(\Lambda - \lfloor\Lambda\rfloor)\theta_Y\rfloor$ and the other is that $\lfloor(\Lambda - \lfloor\Lambda\rfloor)\theta_X\rfloor = \lfloor(\Lambda - \lfloor\Lambda\rfloor)\theta_Y\rfloor - 1$. In the former case, we have $X'[1,j_2-1] = Y'[1,j_2-1]$ and $X'[1,j_2] = Y'[1,j_2] - 1$, and so $d(X',Y') = d(X,Y)$. In the latter case, we have $X'[1,j_2-1] = Y'[1,j_2-1] - 1$ and $X'[1,j_2] = Y'[1,j_2]$, and so $d(X',Y') = d(X,Y) - (n-j_2+1)/(n-1)$. These two cases appear with the same probability 1/2, hence $\mathrm{E}[d(X',Y') \mid \lfloor\Lambda\rfloor = j_2 - 1, j_1 \neq j_2 - 1, \theta_X = \theta_Y - 1] = d(X,Y) - (1/2)(n - j_2 + 1)/(n-1)$. We can show the remained case that $\theta_X = \theta_Y + 1$ in a similar way.
2. When $\lfloor\Lambda\rfloor = j_2$, we show that $\mathrm{E}[d(X',Y') \mid \lfloor\Lambda\rfloor = j_2] \leq d(X,Y) + (1/2)(n-j_2)/(n-1)$.
   (a) In case that $\theta_X = \theta_Y$, we obtain the result that $X'[1,j_2] = Y'[1,j_2] - 1$ and $X'[1,j_2+1] = Y'[1,j_2+1]$ in the same way as Case 1-(b). Hence $d(X',Y') = d(X,Y)$ with conditional probability 1.
   (b) Consider the case that $\theta_X \neq \theta_Y$. In a similar way with Case 1-(c), we can show that $d(X',Y') = d(X,Y)$ with conditional probability 1/2 and $d(X',Y') = d(X,Y) + (n-j_2)/(n-1)$ with conditional probability 1/2. Hence $\mathrm{E}[d(X',Y') \mid \lfloor\Lambda\rfloor = j_2, \theta_X \neq \theta_Y] = d(X,Y) + (1/2)(n - j_2)/(n-1)$.

3. When $\lfloor \Lambda \rfloor \neq j_2 - 1$ and $\lfloor \Lambda \rfloor \neq j_2$, $\{X', Y'\}$ is also an edge of G. It is easy to see that $j_2 = \max\{j'_1, j'_2\}$ where $\{j'_1, j'_2\}$ is the supporting pair of $\{X', Y'\}$. Thus we have $d(X', Y') = d(X, Y)$.

The probability of appearance of Case 1 is equal to $1/(n-1)$, and that of Case 2 is less than or equal to $1/(n-1)$. From the above,

$$\begin{aligned} \mathrm{E}[d(X', Y')] &\le d(X,Y) - \frac{1}{n-1}\frac{1}{2}\frac{n-j_2+1}{n-1} + \frac{1}{n-1}\frac{1}{2}\frac{n-j_2}{n-1} \\ &= d(X,Y) - \frac{1}{2(n-1)^2} \le \left(1 - \frac{1}{2(n-1)^2}\frac{1}{\max_{\{X,Y\}\in\mathcal{E}}\{d(X,Y)\}}\right) d(X,Y) \\ &= \left(1 - \frac{1}{n(n-1)^2}\right) d(X,Y). \end{aligned}$$

Since the diameter of G is bounded by nN, Theorem 5.2 implies that the mixing rate τ satisfies

$$\tau \le n(n-1)^2(1 + \ln(nN)).$$

□

Next, we estimate the coalescence time.

Lemma 5.4. *Under Condition 1, the coalescence time T_* of $\mathcal{M}$ satisfies* $\mathrm{E}[T_*] = \mathrm{O}(n^3 \ln N)$.

Proof: Let $G = (\Xi, \mathcal{E})$ be the undirected graph and $d(X,Y)$, $\forall X, \forall Y \in \Xi$, be the metric on G, both of which are defined in the proof of Lemma 5.3. We define $D_G \stackrel{\text{def.}}{=} d(X_{\mathrm{U}}, X_{\mathrm{L}})$ and $\tau_0 \stackrel{\text{def.}}{=} n(n-1)^2(1 + \ln D_G)$. By using the inequality (4) obtained in the proof of Lemma 5.3, we have

$$\begin{aligned} \Pr(T_* > \tau_0) &= \Pr(\Phi^0_{-\tau_0}(X_{\mathrm{U}}, \mathbf{\Lambda}) \neq \Phi^0_{-\tau_0}(X_{\mathrm{L}}, \mathbf{\Lambda})) \\ &= \Pr(\Phi^{\tau_0}_0(X_{\mathrm{U}}, \mathbf{\Lambda}) \neq \Phi^{\tau_0}_0(X_{\mathrm{L}}, \mathbf{\Lambda})) \\ &\le \sum_{(X,Y)\in\Xi^2} d(X,Y)\Pr(X = \Phi^{\tau_0}_0(X_{\mathrm{U}}, \mathbf{\Lambda}), Y = \Phi^{\tau_0}_0(X_{\mathrm{L}}, \mathbf{\Lambda})) \\ &= \mathrm{E}[d(\Phi^{\tau_0}_0(X_{\mathrm{U}}, \mathbf{\Lambda}), \Phi^{\tau_0}_0(X_{\mathrm{L}}, \mathbf{\Lambda}))] \le \left(1 - \frac{1}{n(n-1)^2}\right)^{\tau_0} d(X_{\mathrm{U}}, X_{\mathrm{L}}) \\ &= \left(1 - \frac{1}{n(n-1)^2}\right)^{n(n-1)^2(1+\ln D_G)} D_G \le \mathrm{e}^{-1}\mathrm{e}^{-\ln D_G} D_G \le \frac{1}{\mathrm{e}}. \end{aligned}$$

By *sub-multiplicativity* of coalescence time ([18]), for any $k \in \mathbb{Z}_+$, $\Pr(T_* > k\tau_0) \le (\Pr(T_* > \tau_0))^k \le (1/\mathrm{e})^k$. Thus

$$\begin{aligned} \mathrm{E}[T_*] &= \sum_{t=0}^{\infty} t\Pr(T_* = t) \le \tau_0 + \tau_0\Pr(T_* > \tau_0) + \tau_0\Pr(T_* > 2\tau_0) + \cdots \\ &\le \tau_0 + \tau_0/\mathrm{e} + \tau_0/\mathrm{e}^2 + \cdots = \frac{\tau_0}{1 - 1/\mathrm{e}} \le 2\tau_0. \end{aligned}$$

Clearly, $D_G \le nN \le N^2$ because $n \le N$. Then we obtain the result that $\mathrm{E}[T_*] = \mathrm{O}(n^3 \ln N)$. □

Lastly, we determine the expected running time of Algorithm 2.

Proof of Theorem 5.1: We denote T_* be the coalescence time of our chain. Note that T_* is a random variable. Put $K = \lceil \log_2 T_* \rceil$. Algorithm 2 terminates when we set the starting time period T to -2^K at $(K+1)$st iteration. Then the total number of generated random numbers in Algorithm 2 is bounded by $2^K \leq 2T_*$ and the total number of simulated transitions is bounded by $2(2^0+2^1+2^2+\cdots+2^K) < 2 \cdot 2 \cdot 2^K \leq 8T_*$. Under the assumption that we can generate a random number in constant time, each transition of a chain is simulated in constant time. Step 4 of Algorithm 2, "Coalescence check," requires $\mathrm{O}(n)$ time. Thus the expectation of total time complexity is bounded by $\mathrm{O}(\mathrm{E}[2T_*]+\mathrm{E}[8T_*]+\mathrm{E}[K+1]n) = \mathrm{O}(\mathrm{E}[T_*]) = \mathrm{O}(n^3 \ln N)$. □

6. Discussions

We proposed a perfect sampling algorithm for $2 \times n$ contingency tables. Our algorithm is based on a new Markov chain which is monotone and rapidly mixing. The rapidity implies the polynomiality of our algorithm. More precisely, Algorithm 2 produces exact samples from the uniform distribution, and the expected running time is bounded by $\mathrm{O}(n^3 \ln N)$ under Condition 1.

Our preliminary computational experience indicates that Condition 1 is an important requirement for rapidity. For example, when we set $\boldsymbol{r} = (500, 500)$, $\boldsymbol{s} = (500, 498, 1, 1)$ and executed monotone CFTP algorithm one thousand times, the average coalescence time was about 20. However, when we substituted $\boldsymbol{s}$ by (500,1,1,498) and executed the algorithm one thousand times, the average coalescence time was about 2 millions.

Even though Algorithm 2 is enough fast but needs to stock the random numbers. We can save the memory storage by using read once algorithm proposed by Wilson in [20]. For detail, please see [20, 13]. Note that the modified algorithm terminates in $\mathrm{O}(n^3 \ln N)$ time and the required memory is bounded by $\mathrm{O}(n \ln N)$.

Our perfect sampling algorithm is applicable to the problem for counting $2 \times n$ contingency tables. In our previous work [16], we modified Dyer and Greenhill's approximate counting scheme [10], and estimated the size of bias of the expectation of approximate solution. When we employ our perfect sampling algorithm in the algorithm proposed in [16], we can show that the total number of required samples halved.

It is easy to extend our monotone CFTP algorithm to $2 \times \cdots \times 2 \times J$ tables discussed in [17]. We can show that the extended algorithm is also a polynomial time algorithm for uniform sampler. In case of conditional multinomial distributions, the existence of polynomial time perfect sampler remains open.

Another remained major open problem is the existence of a monotone Markov chain for $m \times n$ contingency tables.

References

[1] A. Agresti, "A survey of exact inference for contingency tables," *Statistical Science*, 7 (1992), pp. 131–153.

[2] A. Agresti, *Categorical Data Analysis*, John Wiley & Sons, 2002.

[3] D. Aldous, "Random walks on finite groups and rapidly mixing Markov chains," in *Séminarie de Probabilitiés XVII 1981/1982*, vol. 986 of Springer-Verlag Lecture Notes in Mathematics, D. Dold and B. Eckmann, ed., Springer-Verlag, New York, 1983, pp. 243–297.

[4] R. Bubley, *Randomized Algorithms: Approximation, Generation, and Counting*, Springer-Verlag, New York, 2001.

[5] R. Bubley and M. Dyer, "Path coupling : a technique for proving rapid mixing in Markov chains," *Proceedings of the 38th Annual Symposium on Foundations of Computer Science (FOCS 1997)*, pp. 223–231.

[6] M. Cryan, M. Dyer, L. A. Goldberg, M. Jerrum, and R. Martin, "Rapidly mixing Markov chains for sampling contingency tables with constant number of rows," *Proceedings of the 43rd Annual Symposium on Foundations of Computer Science (FOCS 2002)*, pp. 711–720.

[7] P. Diaconis and B. Effron, "Testing for independence in a two-way table: new interpretations of the chi-square statistic (with discussion)," *The Annals of Statistics*, 13 (1985), pp. 845–913.

[8] P. Diaconis and L. Saloff-Coste, "Random walk on contingency tables with fixed row and column sums," *Tech. rep.*, Department of Mathematics, Harvard University, 1995.

[9] X. K. Dimakos, "A guide to exact simulation," *International Statistical Review*, 69 (2001), pp. 27–48.

[10] M. Dyer and C. Greenhill, "Polynomial-time counting and sampling of two-rowed contingency tables," *Theoretical Computer Sciences*, 246 (2000), pp. 265–278.

[11] M. Dyer, R. Kannan, and J. Mount, "Sampling contingency tables," *Random Structures and Algorithms*, 10 (1997), pp. 487–506.

[12] R. A. Fisher, "The logic of inductive inference (with discussion)," *Journal of Royal Statistical Society*, 98 (1935), pp. 39–54.

[13] O. Häggström, *Finite Markov Chains and Algorithmic Application*, London Mathematical Society, Student Texts 52, Cambridge University Press, 2002.

[14] D. Hernek, "Random generation of $2 \times n$ contingency tables," *Random Structures and Algorithms*, 13 (1998), pp. 71–79.

[15] M. Jerrum and A. Sinclair, "The Markov chain Monte Carlo method: an approach to approximate counting and integration," in *Approximation Algorithm for NP-hard Problems*, D. Hochbaum, ed., PWS, 1996, pp. 482–520.

[16] S. Kijima and T. Matsui, "Approximate counting scheme for $m \times n$ contingency tables ," *IEICE Transactions on Information and Systems*, vol. E87-D (2004), pp. 308–314.

[17] T. Matsui, Y. Matsui, and Y. Ono, "Random generate of $2 \times \cdots \times 2 \times J$ contingency tables," *METR* 2003-03, Mathematical Engineering Technical Reports, University of Tokyo, 2003. (available from http://www.keisu.t.u-tokyo.ac.jp/Research/techrep.0 .html)

[18] J. Propp and D. Wilson, "Exact sampling with coupled Markov chains and applications to statistical mechanics," *Random Structures and Algorithms*, 9 (1996), pp. 232–252.

[19] J. Propp and D. Wilson, "How to get a perfectly random sample from a generic Markov chain and generate a random spanning tree of a directed graph," *J. Algorithms*, 27 (1998), pp. 170–217.

[20] D. Wilson, "How to couple from the past using a read-once source of randomness," *Random Structures and Algorithms*, 16 (2000), pp. 85–113.

Shuji Kijima
Department of Mathematical Informatics, Graduate School of Information Science and Technology, University of Tokyo
kijima@misojiro.t.u-tokyo.ac.jp

Tomomi Matsui
Department of Mathematical Informatics, Graduate School of Information Science and Technology, University of Tokyo
http://www.simplex.t.u-tokyo.ac.jp/~tomomi/

Trends in Mathematics, © 2004 Birkhäuser Verlag Basel/Switzerland

An Efficient Generic Algorithm for the Generation of Unlabelled Cycles[1]

Conrado Martínez and Xavier Molinero

ABSTRACT: *In this paper, we combine two recent generation algorithms to obtain a new generic algorithm for the generation of all unlabelled cycles with components from some class $\mathcal{A}$ and total size n. Sawada's algorithm [13] lists all k-ary unlabelled cycles with fixed content, that is, the number of occurrences of each symbol is fixed and given* a priori. *The other algorithm [8], by the authors, generates all multisets of objects with given total size n from any admissible unlabelled class $\mathcal{A}$. By admissible we mean that the class can be specified using the ϵ-class, atomic classes, disjoint unions, products, sequences, (multi)sets, etc. The resulting algorithm, which is the main contribution of this paper, generates all cycles of objects with given total size n from any unlabelled admissible class $\mathcal{A}$. Given the generic nature of the algorithm, it is suitable for inclusion in combinatorial libraries and for rapid prototyping. The new algorithm incurs constant amortized time per generated cycle, the constant only depending in the class $\mathcal{A}$ to which the objects in the cycle belong.*

1. Introduction

The generation of unlabelled cycles (also called *necklaces*) probably poses the most difficult problems if we compare them with the generation of other common combinatorial constructs (see for instance [4, 10, 11]). But in the last few years there has been several notable progress related to the generation of necklaces, as witnessed by the work of Wang and Savage [17], Ruskey and Sawada [12, 14], Cattell *et al.* [1] and Sawada [13].

Recall that a necklace or unlabelled cycle is a sequence of symbols such that it is lexicographically smaller than any of its circular permutations. Thus, abadc is a cycle but adcab is not. Of course, we assume a certain order among the symbols in order to properly define the notion of cycle. When a cycle is aperiodic it is called a *primitive cycle* or *Lyndon* word; thus the cycle abadc is a Lyndon word, but the cycle abab is not.

In our recent works [6, 8, 9] we have shown how to design general algorithms to iterate through all the objects of a given size in labelled and unlabelled *admissible combinatorial* classes, such as those constructed using disjoint unions, products, sets and multisets, sequences, substitutions, etc. We use the adjective *general* above in the sense that the algorithms receive as their input both the size n of the objects and a finite specification of the combinatorial class which the objects belong to. In this paper we present an iteration algorithm for unlabelled

[1]This research was supported by the Future and Emergent Technologies programme of the EU under contract IST-1999-14186 (ALCOM-FT) and the Spanish "Ministerio de Ciencia y Tecnología" programme TIC2002-00190 (AEDRI II).

cycles. Contrary to the other algorithms that we have designed so far, the algorithm for unlabelled cycles is not based upon a suitable recursive decomposition of this combinatorial construction; hence, it cannot be used as the basis for an efficient *unranking* algorithm (that is, given a rank i and a size n, generate the i-th object of size n) for unlabelled cycles, which remains still as an open problem.

The framework presented in [6, 8, 9] follows the approach that was pioneered by Flajolet *et al.* [3] for the random generation of combinatorial objects, and later applied by the authors to the unranking of combinatorial objects [7]. Together with the present paper, these papers show the viability of this elegant approach for an effective and efficient solution of the "big four": counting, random generation, exhaustive/iterative generation and unranking. We have been able to prove that all the proposed iteration algorithms for decomposable objects (sets, sequences, labelled cycles, ...) run in constant amortized time (CAT) per generated object, provided that the class can be finitely specified using ϵ-classes (a single object of size 0), atomic classes (a single object of size 1), disjoint unions, products, sequences, substitutions, sets and multisets. We also provided there a CAT algorithm for labelled cycles, but unlabelled cycles eluded our efforts. The proposed algorithms do not perform better than state-of-the-art algorithms, but do not perform much worse either. And because of their general flavor, they are useful for rapid prototyping and for their inclusion into general combinatorial libraries like `combstruct` for Maple or `MuPAD-combinat` for MuPAD.

This paper is organized as follows. In Section 2 we review a few basic concepts and notation. Sections 3 and 4 present our algorithm, which is based upon our algorithm for unlabelled multisets in [8] and the recent algorithm to generate unlabelled cycles of fixed content of Sawada [13]. We prove that the resulting algorithm has good performance, namely, it is constant amortized time. Finally, in section 5 we discuss our current and future work on this topic and report on our preliminary implementation of the algorithm.

2. Preliminaries

In this paper we consider *unlabelled admissible classes of combinatorial structures* and in particular, unlabelled cycles. Most of the material in this section is standard and can be found elsewhere, see for instance [2, 15, 16]. However, to make the paper more self-contained and to fix notation, we will briefly introduce some basic definitions and concepts. We begin with the formal definition of a combinatorial class.

Definition 2.1. *A* combinatorial class *is a pair* $(\mathcal{A}, |\cdot|)$ *such that* $\mathcal{A}$ *is a finite or infinite denumerable set and* $|\cdot| : \mathcal{A} \to \mathbb{N}$ *is a* size *function such that, for all* $n \geq 0$, $\mathcal{A}_n = \{\alpha \in \mathcal{A} \mid |\alpha| = n\}$ *is finite.*

Shall no confusion arise, we will use the same name for the class and for the set of objects belonging to that class. Also, we use subscripts under a class name to denote the subset of objects of that class with a given size. Typically, complex objects in a given class are composed of smaller units, called *atoms* and generically denoted by Z. Atoms are objects of size 1 and the size of an object is the number of atoms it contains. For instance, a string is composed by the concatenation of symbols, where each of these is an atom, and the size of the string is its length or the number of symbols it is composed of. Similarly, a tree is made up of nodes —its atoms— and the size of the tree is its number of nodes. Objects of size 0 are normally denoted by ϵ.

Two main types of combinatorial classes can be defined depending on whether the atoms that compose a given object can be distinguished or not. In the former case, we say the class is *labelled* whereas in the later we say the class is *unlabelled.*

As it will become apparent, an efficient solution to the problem of counting, namely, given a specification, a class and a size, compute the number of objects of the given size, is fundamental for our solution of the iteration problem. Hence, we turn our attention to *admissible combinatorial classes*, that is, those classes that can be constructed from *admissible constructors.* An admissible constructor is an operation over classes that yields a new class, and such that the number of objects of a given size in the new class can be computed from the number of objects of that size or smaller sizes in the constituent classes.

In order to formalize the notion of admissibility, we need the fundamental notion of *counting generating functions.*

Definition 2.2. *The (counting) generating function of an unlabelled combinatorial class $\mathcal{A}$ is the ordinary generating function for the sequence $\{a_n\}_{n\geq 0}$. That is,*

$$A(z) = \sum_{n\geq 0} a_n z^n = \sum_{\alpha\in\mathcal{A}} z^{|\alpha|}$$

where $a_n = \#\mathcal{A}_n$ is the number of objects in $\mathcal{A}$ of size n. The n-th coefficient of $A(z)$ is denoted $[z^n]A(z)$, i.e., $a_n = [z^n]A(z)$.

Now we can define *admissible operators.*

Definition 2.3. *An operation Ψ over combinatorial classes $\mathcal{A}_1, \mathcal{A}_2, \ldots, \mathcal{A}_k$ is* admissible *if and only if there exists some operator Φ over the corresponding generating functions $A_1(z), \ldots, A_k(z)$ such that*

$$\mathcal{C} = \Psi(\mathcal{A}_1, \ldots, \mathcal{A}_k) \implies C(z) = \Phi(A_1(z), \ldots, A_k(z))$$

where $C(z)$ is the generating function of $\mathcal{C}$.

Finally, we can define admissible specifications.

Definition 2.4. *An* admissible specification *S is a collection of equations of the form*

$$\mathcal{A}_i = \Psi_i(\mathcal{A}^{(i)}_{j_0}, \ldots, \mathcal{A}^{(i)}_{j_i}) \ ,$$

where no two equations have the same left-hand side, each Ψ_i is an admissible operation, and each $\mathcal{A}^{(i)}_{j_r}$ is either an ϵ-class, an atomic class, or there is an equation in the collection with that class as its left-hand side. An ϵ-class is a class that contains a single object of size 0. Each of the classes that appear in the left-hand sides of the equations in S is said to be specified by S.

If a class $\mathcal{A}$ is specified by an admissible specification, then the class itself is called admissible. Admissible classes are also called *decidable* or *well-founded* classes [18]. Table 1 lists some admissible operators and the corresponding ordinary generating functions (OGFs).

Though the collection of operations given above is small, it can be used to describe many important and useful combinatorial classes.

Admissible classes must have a finite number of objects for any size, i.e., $a_n < \infty$ for any $n \in \mathbb{N}$. Hence, some restrictions over the classes are needed. For instance, $\mathsf{Seq}(\mathcal{A})$ and $\mathsf{Set}(\mathcal{A})$ require that $a_0 = 0$; and, $\mathsf{Subst}(\mathcal{A}, \mathcal{B})$ requires that either $b_0 = 0$ or $\mathcal{A}$ is finite.

Class	OGF
$\mathsf{Union}(\mathcal{A},\mathcal{B}) = \mathcal{A}+\mathcal{B}$	$A(z)+B(z)$
$\mathsf{Prod}(\mathcal{A},\mathcal{B}) = \mathcal{A}\times\mathcal{B}$	$A(z)\cdot B(z)$
$\mathsf{Seq}(\mathcal{A})$	$\frac{1}{1-A(z)}$
$\mathsf{PowerSet}(\mathcal{A})$	$\exp\left(\sum_{n>0}(-1)^{n-1}\frac{A(z^n)}{n}\right)$
$\mathsf{Set}(\mathcal{A})$	$\exp\left(\sum_{n>0}\frac{A(z^n)}{n}\right)$
$\mathsf{Cycle}(\mathcal{A})$	$\sum_{n>0}\frac{\phi(n)}{n}\log\left(\frac{1}{1-A(z^n)}\right)$
$\mathsf{Subst}(\mathcal{A},\mathcal{B})$	$A(B(z))$

TABLE 1. Unlabelled admissible combinatorial operators

From now on, by an *admissible class* we mean that the class can be finitely specified using the ϵ class (the class with a single object of size 0), atomic classes (classes that contain a single object of size 1), disjoint unions (Union or '+'), products (Prod or '×'), sequences (Seq), multisets (Set), power sets (PowerSet) and cycles (Cycle) of admissible classes. Furthermore, the techniques and results presented can be easily extended to variants of the cycle operator which are admissible operators, in particular, to cycles with a restricted number of components. As we have already mentioned in the introduction, we have developed in previous works efficient algorithms to cope with all the combinatorial operators mentioned in the list above, except for unlabelled cycles (Cycle). In this paper, we fill that gap.

3. Generating Cycles: The Fundamentals

Let $\mathcal{A}$ be some admissible unlabelled class and assume that we already have an efficient procedure `next` to list all objects of a given size n in $\mathcal{A}$, one at a time. The (very) high-level pseudocode of the algorithm is given in Figure 1. There p points to the root of the tree-like data structure that represents the current combinatorial object; in particular we assume that the current object is an unlabelled cycle of $\mathcal{A}$'s. Each node in the data structure contains a field *size* and a field *class* that contain the size and the class of the object represented by the subtree rooted at the given node, respectively. Similarly, there is a field *count* and a field *rank* on each node that gives the number of objects in the same class with the same size, and the rank of the object within its class. The nodes that represent cycle objects have additional fields, including the fields called *cycle_rank* and *cycle_count*, which give the rank of the cycle within the set of cycles that can be built using the same components as the current cycle and the number of cycles in that set. For instance, the root node of the tree representing the cycle

$$c = \langle b, aa, aaa, ba, aca, aa, b, bb\rangle$$

has *cycle_count*=1260, since that many different cycles can be constructed using the same components as c. The notation $p \to field$ denotes the field named $field$ in the node pointed to by p.

In order to generate all the cycles using the components of the *basis multiset* γ we need to introduce some order among them. If two components a_1 and a_2 have the same size then their order is given by respective ranks in $\mathcal{A}$ (i.e., the order

```
// p points to the root of the tree representing a cycle of A's
function next_cycle(p)
  if p -> rank = p -> count then
    // the object is the last possible cycle; just signal the end
    p -> is_last:= true
    return
  endif
  if p -> cycle_rank = p -> cycle_count then
    // no more cycles can be built with the same
    // (multi)set of components;
    // generate the next multiset
    p:= next_multiset(p)
    // set up for cycle generation
    return init_cycle_from_multiset(p)
  endif
  // Use Sawada's algorithm to generate the next cycle with the same
  // multiset of components
  return next_cycle_from_multiset(p)
endfunction
```

FIGURE 1. The generation of unlabelled cycles of $\mathcal{A}$'s.

in which they are listed by the known `next` procedure). Otherwise, the object of smaller size is considered smaller than the other object. We write $a_1 \preceq a_2$ to denote that a_1 is smaller than or equal to a_2, in the sense above.

On the other hand, the order of multisets could be similarly defined. Let

$$\gamma = \{a_0 \bullet n_0, a_1 \bullet n_0, \ldots, a_{k-1} \bullet n_{k-1}\},$$
$$\gamma' = \{a'_0 \bullet n'_0, a'_1 \bullet n'_1, \ldots, a'_{j-1} \bullet n'_{j-1}\},$$

where the notation $a \bullet n$ indicates that the object a appears $n > 0$ times in the multiset[1]. Assume, furthermore, that each multiset is presented in sorted order, namely, $a_0 \preceq a_1 \preceq \cdots \preceq a_{k-1}$, and similarly γ'. Then the relative order of γ and γ' follows from the obvious lexicographic order; thus $\gamma \prec \gamma'$ if they have a common "prefix" of length $\ell - 1$ and either $a_\ell \prec a'_\ell$ or $a_\ell = a'_\ell$ and $n_\ell > n'_\ell$.

Notice that the procedure `next_cycle` will generate all cycles of size n in $\mathsf{Cycle}(\mathcal{A})$ if we find a way to solve its two main steps; however, the cycles will not be generated in lexicographic order. For instance, take the basis multisets $\gamma = \{a \bullet 3, b, c\}$ and $\gamma' = \{a \bullet 2, b \bullet 3\}$, with $\gamma \preceq \gamma'$. Hence the cycle aacab will be generated before than the cycle aabbb, yet the latter is lexicographically smaller than the former. If we defined the order between multisets in such a way that $\gamma' \preceq \gamma$, we would produce ababb before aaabc, but the second is lexicographically smaller than the first, so the problem is inherent to the structure of our algorithm, not to the way we have defined the order of multisets.

Another key ingredient of the algorithm is a suitable representation of cycles, so that `next_cycle` can be efficiently implemented. The representation of cycles is conditioned by our already existing framework for the generation of other combinatorial objects; some modifications need to be introduced into that framework in order to accommodate the generation of cycles.

[1]We will usually omit the notation $\bullet n$ if $n = 1$, though.

We use binary trees to represent combinatorial objects. We also have a special pointer to mark the last updated node in the object to further speed the generation of the successor of the object. We call the resulting data structure an *iterator*. Leaves in the binary tree contain the atoms and ϵ components of the object; the internal nodes are labelled by operators: '+', '×', etc. Each internal node contains information about the subobject represented by the corresponding subtree. In particular, each node holds the size of the subobject (*size*), the specification of the class that the object belongs to (*class*), the rank of the object within its class (*class*), and the number of objects of that size in the class (*count*). Additional fields are used for some particular nodes, as we describe in next section. Furthermore, each node has pointers to its children and to its parent.

In order to be able to cope with the nested generation of cycles, a 'cycle' node is similar to a 'multiset' node and the object represented by the subtree beneath a 'cycle' is actually the representation for what we have called the basis multiset; the 'cycle' node contains also an auxiliary structure that establishes how the components of the basis multiset are arranged in the current cycle. While generating all the cycles with a common basis multiset, the subtree below the 'cycle' node gets untouched, only the auxiliary structure is modified. When all the cycles with the same basis multiset have been generated, the algorithm to generate the next multiset is applied to the 'cycle' node as if it were a 'multiset' node. Afterwards, we apply `init_cycle_from_multiset` to update the specific auxiliary structure associated to the 'cycle' node.

4. Generating Cycles: The Details

In this section we give a detailed description of the two components of our algorithm. Also, we have a close look at the representation of multiset and cycle objects.

We consider first the representation of multisets and how they are generated. The generation of multisets is based in the following useful recursive decomposition:

$$\mathsf{Set}(\mathcal{A})_n = \bigcup_{n>0} \bigcup_{j=1}^{n} \bigcup_{p=\lfloor n/j \rfloor}^{1} \mathsf{Set}(\mathcal{A}_j, \mathrm{card} = p)_{j\cdot p} \times \mathsf{Set}(\mathcal{A}_{>j})_{n-j\cdot p} \tag{1}$$

where $\mathsf{Set}(\mathcal{A}_j, \mathrm{card} = p)$ is the class of sets with exactly p components which are elements of the class $\mathcal{A}$ of size j, and $\mathsf{Set}(\mathcal{A}_{>j})$ is the class of sets whose components are objects in $\mathcal{A}$ with size strictly larger than j.

Therefore (1) tells us that a multiset consists of a multiset of p components of minimal size, say j, times a multiset of components of larger size. The first part of the multiset is what we call a *multiset block* or *mblock* for short. In our representation, a 'multiset' node has a left son of size ℓ whose root is a 'mblock' node and a right son of size $n - \ell$ whose root is a 'multiset' node. To compute the next multiset, the algorithm `next_multiset` is recursively applied in the right son; but if no more multisets of size $n - \ell$ can be generated, then we apply the `next_mblock` algorithm to the left son and initialize the right son with the smallest multiset of size $n - \ell$ that does not contain components of size smaller or equal to the size of the objects in the 'mblock' node. In order to make this task easier, each 'multiset' node carries an additional field *min_size* which stores the minimal possible size among its components. If there is no 'mblock' object following the one represented by the current left son, then we initialize both the left son and

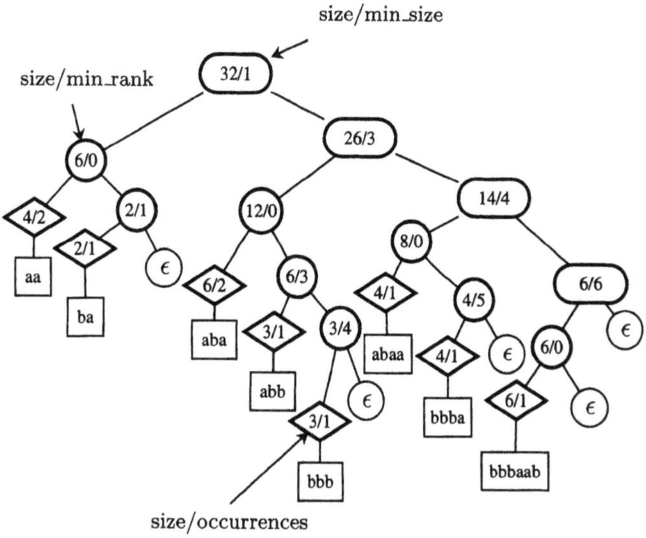

FIGURE 2. Internal representation of a Set-object.

the right son with appropriate 'mblock' and 'multiset' objects of size ℓ' and $n-\ell'$, respectively. We increase the minimal size j for the components of the multiset, and initialize a mblock as large as possible, together with a (eventually empty) multiset whose components are of size strictly larger than the (new) j.

A 'mblock' node represents a set of p components which belong to the same class and have all them the same size j. The mblock recursively decomposes into its component of smallest rank, called its *leader*, occurring a certain number of times, say r, followed by a mblock with $p-r$ components whose ranks are strictly larger than r. Thus a 'mblock' node has a left son which is a 'delta' node representing the smallest component in the mblock and its number of occurrences, and a right son which is a mblock itself. Each mblock node has an additional field *min_rank* which ranges between 0 and $a_j - 1$ (the number of objects in $\mathcal{A}$ of size j) and indicates the minimal possible rank for any component of the mblock.

Figure 2 schematically depicts the internal representation of the multiset

$$\{aa, aa, ba, aba, aba, abb, bbb, abaa, bbba, bbbaab\}$$
$$= \{aa \bullet 2, ba, aba \bullet 2, abb, bbb, abaa, bbba, bbbaab\} \in \mathsf{Set}(\mathsf{Seq}(a+b)).$$

The 'multiset' nodes are represented by oval nodes, together with the values of the *size* and *min_size* attributes. Nodes that represent 'mblocks' are represented by circular nodes with thick border; within each 'mblock' node we write the values of the *size* and *min_rank* attributes. The components of the multiset belong to $\mathsf{Seq}(a+b)$ but we have shown them "collapsed" into single rectangular nodes. Additionally, the diamond-shaped nodes represent 'delta' nodes, and store in the field *occurrences* the number of times that the particular component occurs in the multiset.

Computing the successor of a mblock follows the same pattern as the computation of the successor of a multiset: given a mblock $m = (c, m')$ we recursively try to compute the successor of the mblock m' (represented by the right subtree of m), while maintaining the same leader; if this where not possible, we compute the successor of the leader c and the first mblock with components whose rank is larger that the rank of the new leader replaces m'.

The computation of the next leader (a 'delta' object) is also quite easy. Suppose that the current 'delta' object is $a \bullet k$, of size $\ell = |a| \cdot k$. The 'delta' node has a field *occurrences* to carry the information about the number of repetitions k of the object a, which is represented in the left subtree. The right subtree of the 'delta' node is simply discarded. If a is not the last object in $\mathcal{A}$ of its size and $k > 0$, then the next 'delta' object is $a \bullet k - 1$; but if $k = 0$ then the next object if $a' \bullet n/j$ where a' is the next object of a in $\mathcal{A}$ and n is the size of the mblock.

With this scheme, multisets are not generated in lexicographic order, but as we have discussed, this is not very relevant as the cycles can't be generated in lexicographic order either.

Provided that all objects of a given size in $\mathcal{A}$ can be generated in time proportional to the total number of generated objects, then it can be shown that the algorithm above generates all multisets of size n of $\mathcal{A}$'s in time proportional to the total number of generated multisets [8].

We use Sawada's algorithm [13] to generate all the cycles with a given basis multiset. This recent algorithm is able to produce all possible cycles that contain r_0 occurrences of the symbol 0, r_1 occurrences of symbol 1, etc., i.e., all k-ary cycles of *fixed content.* In our algorithm, the components of the basis multiset are the "symbols" to work with.

To start the generation of cycles, we first initialize the basis multiset $\gamma = \{a_0 \bullet r_0, \dots, a_{k-1} \bullet r_{k-1}\}$ of the 'cycle' object. This is done, except the first time, by applying `next_multiset` algorithm to the root node of the 'cycle' object.

Then the procedure `init_cycle_from_multiset` performs the following tasks. First, it prepares a list L with pointers to the 'delta' objects $a_i \bullet r_i$ and initializes an array R with the number of repetitions r_i of each 'delta' object. Both the list L and the array R are part of the auxiliary structure attached to the 'cycle' node. Because of the requirements of Sawada's algorithm the list L should be filled in reverse order (the first element in the list points to a_{k-1} and so on) and then L should be rearranged to make sure that the first element in L points to the component with the largest value r_i. Of course, the array R should be rearranged accordingly. Sawada's algorithm requires that symbols are indexed to make sure that the largest symbol in lexicographic order has the largest number of occurrences, in order to generate all cycles in (reverse) lexicographic order. Although it would be not too difficult to adapt our algorithm so that the list L and R are arranged as Sawada's algorithm dictates, and undo both the reversing and re-indexing when "printing" the current object, we do not take any further steps on this respect, because we will not be able to generate all cycles of $\mathcal{A}$'s in lexicographic order anyway. So the initialization of L and R is more simple in our setting.

We also have to initialize the rank of the current cycle to 0 (*cycle_rank*) and compute the number of cycles (*cycle_count*) that can be generated with that particular basis multiset [5]:

$$C(N_0, \dots, N_{k-1}) = \frac{1}{N} \sum_{j \mid \gcd(N_0, \dots, N_{k-1})} \phi(j) \frac{(N/j)!}{(N_0/j)! \cdots (N_{k-1}/j)!},$$

where $\phi(x)$ is Euler's totient function (the number of prime divisors of x) and $N = N_0 + N_1 + \cdots + N_{k-1}$ (observe that N is not necessarily the size of the cycle, as the components are not of size 1, in general). Finally, the 'cycle' node contains also an array V of pointers that gives the arrangement of the components within the current cycle, a field *arity* with the number of distinct components that appear in the basis multiset and a field *length* with the number of components of the cycle. All this additional information, in particular L and R, can be easily initialized while initializing the basis multiset, introducing a few minor modifications in the corresponding algorithm. The array V plus the *length* field might be though as the real representation of the cycle, while the other fields are just necessary for the computation of the successor.

With some effort, Sawada's algorithm can be transformed into an iterative version which is more convenient to our purposes, although more involved than the original recursive version. The algorithm works by appending symbols to suitable prefixes; in order to avoid linear searches of the available symbols, the global list L of "available" symbols is used. Also, in the transformation from its original recursive formulation to the iterative version, we have to include additional fields in the 'cycle' node to keep record of the current state of computation. In particular, we need a field *longest_Lyn_pref* with the length of the longest Lyndon prefix of the current cycle.

Recall that when all the cycles with the same basis multiset are exhausted, the `next_multiset` algorithm is applied to obtain a new basis multiset; the list L, the array N and the other fields of the auxiliary structure are updated accordingly. In principle, updating the array R and the list L would need some non-constant amount of work. But when Sawada's algorithm ends, both L and R have recovered their initial contents. Again a minor modification of the `next_cycle` algorithm will allow us to make no more changes to update R and L than to generate the new basis multiset (in the amortized sense). So to speak, most of the times only one or two components of the multiset will be modified and thus only a constant amount of work will be necessary to update R and L.

Sawada's algorithm generates the $C(N_0, \ldots, N_{k-1})$ cycles in constant amortized time per cycle. Denote $C(\gamma)$ the cost of generating all the cycles with basis multiset γ and $\mathfrak{C}(\gamma)$ the set of cycles with basis multiset γ. Thus the cost C_n of our algorithm is given by:

$$\begin{aligned}
C_n &= c \cdot \#\mathsf{Set}(\mathcal{A})_n + \sum_{\gamma \in \mathsf{Set}(\mathcal{A})_n} (C(\gamma) + c') \\
&= (c + c') \cdot \#\mathsf{Set}(\mathcal{A})_n + \sum_{\gamma \in \mathsf{Set}(\mathcal{A})_n} \sum_{v \in \mathfrak{C}(\gamma)} c'' \\
&= c''' \cdot \#\mathsf{Set}(\mathcal{A})_n + c'' \cdot \#\mathsf{Cycle}(\mathcal{A})_n,
\end{aligned}$$

where c, c', c'' and c''' are constants. The constant c is implied by our CAT algorithm for multisets; the constant c'' is implied by Sawada's algorithm for cycles of fixed content, and c' is the constant amount of work that needs to be done to update the auxiliary structure attached to the corresponding 'cycle' node each time that we move from a basis multiset onto the next one. Since $\mathsf{Set}(\mathcal{A})_n \leq \mathsf{Cycle}(\mathcal{A})_n$ if $n > 0$, it follows that C_n is bounded by a constant times the number of generated objects, and thus it is a CAT algorithm too.

5. Final Remarks

We have already conducted a few experiments with a preliminary implementation of the algorithm described in this paper in MAPLE with good results. In particular, we have used the class $\mathcal{N} = \mathsf{Cycle}(\mathsf{Set}(Z, \mathrm{card} \geq 1))$ (cycles of integers) for our experiments. For instance, all cycles of integers of total size 25—there are 1342183 such cycles—are generated in 2269 seconds (0.0016911 seconds/cycle) using a machine equipped with a Pentium processor at 1.7 GHz.

Although the basic ideas behind the algorithms (including the algorithm for multisets and Sawada's algorithm) are rather simple, the implementation details are not. This is because the new algorithm for cycles operates rather differently of the other generation algorithms with which it should be integrated (the algorithm for multisets is prototypical in that respect, with a nice recursive decomposition guiding the algorithm's operation). A suitable recursive decomposition of cycles would allow us to design an iteration algorithm which fits better the framework developed in previous works. Also, such an algorithm would be more amenable to a precise analysis of its performance. Last but not least, if such a recursive decomposition were obtained, an efficient algorithm for the unranking of cycles would suggest itself. Unfortunately, so such decomposition for unlabelled cycles seems to exist.

Other related questions that we are now investigating include variants of the operators (for instance, cycles with restrictions on the number of components) and minor variations of the order in which objects are generated which could improve the overall performance of the process.

References

[1] CATTELL, K., RUSKEY, F., SAWADA, J., AND SERRA, M. Fast algorithms to generate necklaces, unlabeled necklaces, and irreducible polynomials over GF(2). *J. Algorithms 37* (2000), 267–282.

[2] FLAJOLET, P., AND SEDGEWICK, R. The Average Case Analysis of Algorithms: Counting and generating functions. Tech. Rep. 1888, INRIA, April 1993.

[3] FLAJOLET, P., ZIMMERMAN, P., AND CUTSEM, B. V. A calculus for the random generation of combinatorial structures. *Theoret. Comput. Sci. 132*, 1-2 (1994), 1–35.

[4] FREDRICKSEN, H., AND KESSLER, I. An algorithm for generating necklaces of beads in two colors. *Discrete Mathematics 61* (1986), 181–188.

[5] GILBERT, E., AND RIORDAN, J. Symmetry types of periodic sequences. *Illinois J. Mathematics 5* (1961), 657–665.

[6] MARTÍNEZ, C., AND MOLINERO, X. Generic algorithms for the exhaustive generation of labelled objects. In *Proc. of the 4th Workshop on Random Generation of Combinatorial Structures and Bijective Combinatorics (GASCOM'01)* (2001), pp. 53–58.

[7] MARTÍNEZ, C., AND MOLINERO, X. A generic approach for the unranking of labelled combinatorial classes. *Random Structures & Algorithms 19*, 3–4 (2001), 472–497.

[8] MARTÍNEZ, C., AND MOLINERO, X. Generic algorithms for the generation of combinatorial objects. In *Proc. of the 28th Int. Symposium on Mathematical Foundations of Computer Science (MFCS)* (2003), B. Rovan and P. Vojtáš, Eds., vol. 2747 of *Lecture Notes in Computer Science*, Springer-Verlag, pp. 572–581.

[9] MARTÍNEZ, C., AND MOLINERO, X. Efficient iteration in admissible combinatorial classes. *Theoret. Comput. Sci.* (2004). Submitted.

[10] NIJENHUIS, A., AND WILF, H. S. *Combinatorial Algorithms*. Academic Press, 1978.

[11] RUSKEY, F., SAVAGE, C., AND WANG, T. Generating necklaces. *J. Algorithms 13* (1992), 414–430.

[12] RUSKEY, F., AND SAWADA, J. A fast algorithm to generate unlabeled necklaces. In *Proc. of the 11th Annual ACM-SIAM Symposium on Discrete Algorithms (SODA)* (2000), pp. 256–262.

[13] SAWADA, J. A fast algorithm to generate necklaces with fixed content. *Theoret. Comput. Sci. 301*, 1–3 (2003), 477–489.

[14] SAWADA, J., AND RUSKEY, F. An efficient algorithm for generating necklaces with fixed density. In *Proc. of the 10th Annual ACM-SIAM Symposium on Discrete Algorithms (SODA)* (1999), pp. 752–758.

[15] SEDGEWICK, R., AND FLAJOLET, P. *An Introduction to the Analysis of Algorithms*. Addison-Wesley, Reading, MA, 1996.

[16] VITTER, J., AND FLAJOLET, P. Average-case analysis of algorithms and data structures. In *Handbook of Theoretical Computer Science*, J. van Leeuwen, Ed. North-Holland, 1990, ch. 9.

[17] WANG, T., AND SAVAGE, C. A Gray code for necklaces of fixed density. *SIAM J. Discrete Math. 9*, 4 (1996), 654–673.

[18] ZIMMERMANN, P. *Séries génératrices et analyse automatique d'algorithmes*. PhD thesis, École Polytechnique, March 1991.

Conrado Martínez and Xavier Molinero
Departament de Llenguatges i Sistemes Informàtics
Universitat Politècnica de Catalunya
E-08034 Barcelona, Spain.
{conrado,molinero}@lsi.upc.es

Trends in Mathematics, © 2004 Birkhäuser Verlag Basel/Switzerland

Using Tries for Universal Data Compression

Yuriy A. Reznik and Anatoly V. Anisimov

ABSTRACT: *We show how a digital tree (or* trie*) structure can be used for both parsing and encoding (in a Variable-Length to Block (VB) or Variable-Length to Variable-Length (VV) fashion) of sequences of symbols from a stochastic source. As an example, we construct a simple VB code based on a fixed database adaptation model, and derive an asymptotic expression for its* average redundancy rate *for memoryless sources.*

1. Introduction

Given a set of n distinct strings $S = \{s_1, \ldots, s_n\}$ from a binary alphabet $A = \{0,1\}$, a *trie* [1] $T(S)$ can be constructed recursively as follows. If $n = 0$, the trie is *empty*. If $n = 1$, the trie is an *external node* (or *leaf*) containing a pointer to a string in S. If $n \geqslant 2$, the trie is an *internal node*, containing pointers to two *child tries*: $T(S_0)$ and $T(S_1)$, constructed from suffixes of strings in S that begin with symbols 0 and 1 correspondingly: $S_a = \left\{s'_j \,|\, a\, s'_j = s_j,\, a \in A,\, s_j \in S\right\}$.

In a special case, when strings $s_1, \ldots, s_n$ represent successive suffixes of an input sequence $Z = s_1 = z_1 s_2 = \ldots = z_1 \ldots z_{n-1} s_n$, $z_i \in A$, the resulting structure $T(Z)$ is called a *suffix tree* (see Fig. 1.a).

It is well known, that the paths from the root to external nodes in a trie represent *shortest distinguishing prefixes* of strings in S. However, these prefixes may not form a *complete prefix set* (e.g., a trie in Fig. 1.a lacks prefixes 011 and 11). This problem can be trivially solved by adding the missing nodes, or equivalently, by extending the set of strings in a trie to a (smallest possible) complete prefix set. We illustrate this procedure on Fig. 1.b, and call the resulting structure an *extended trie* (or *extended suffix tree*).

In this abstract we propose to use extended tries for construction of non-block (VB or VV structured) source codes. As an example, we develop an algorithm based on a *fixed-database* model of Wyner and Ziv [2].

Both encoder and decoder have access to a *training sequence* (or *database*) Z produced by the source. This sequence is used to construct an extended suffix tree $T(Z)$, which, in turn, is used to define a VB code for this source (see Fig. 1.c). When encoder receives a message, it uses the tree $T(Z)$ to parse the message into words $x_1, \ldots, x_{|T|}$ (corresponding to paths in $T(Z)$), indices of which are then transmitted using $\lceil \log_2 |T| \rceil$ bits, where $|T|$ is the total number of leaves in $T(Z)$.

2. Main Result

Theorem 2.1. *Consider a memoryless source over a binary alphabet $A = \{0,1\}$ with probabilities $p = \Pr\{0\}$, $q = \Pr\{1\}$, $p + q = 1$. Let database Z be a sequence of n symbols produced by this source. Then, the average redundancy rate R_n^{VB} of*

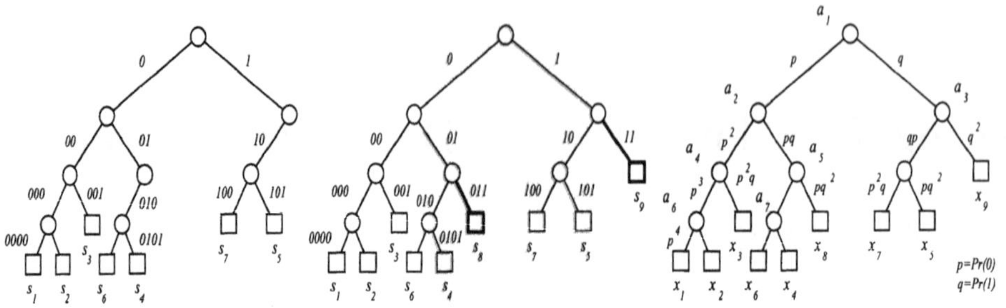

(a) A trie built from 7 suffixes of Z=000010100... (b) An extended version of the same trie. (c) Extended trie used as a VB code for a Bernoulli source.

FIGURE 1. A suffix tree (a), its extended version (b), and the corresponding VB code for a memoryless source (c).

a VB code based on an extended suffix tree $T(Z)$ is (with $n \to \infty$):

$$R_n^{VB} = E\left\{\frac{\lceil \log |T| \rceil}{d(T)}\right\} - h = \frac{C_1 + \delta_1(n,p)}{\log n} + O\left(\frac{1}{\log^2 n}\right), \qquad (1)$$

where: $d(T) = \sum_{i=1}^{|T|} \Pr(x_i)\,|x_i|$ is the average depth of a tree $T(Z)$, $\delta_1(n,p)$ is a fluctuating function of a small amplitude, and C_1 is a constant, which exact formula is

$$C_1 = h\left(\log e\,(1+\gamma) + h - \frac{h_2}{2h} + \log\frac{\log e}{h}\right),$$

where: $h = -p\log p - q\log q$ is the entropy of the source, $h_2 = p\log^2 p + q\log^2 q$, and $\gamma = 0.57721\ldots$ is the Euler's constant. All logarithms have the same base, reflecting the unit of information (e.g. bits or nats) being used.

In other words, we can show (and based on [3], it should be possible to prove it for Markovian sources as well) that the average redundancy of our code is decreasing at a rate of $O(\frac{1}{\log n})$, which is a noticeable improvement over a $O(\frac{\log\log n}{\log n})$ rate of the fixed-database version of a *Lempel-Ziv code* [2]. While a similar improvement has already been reported for a code using two training sequences [4], the present algorithm should be more practical, as it requires only one sequence.

References

[1] D. Knuth, *The Art of Computer Programming. Sorting and Searching. Vol. 3* (Addison-Wesley, Reading MA, 1973).

[2] A.D. Wyner and J. Ziv, Fixed Data Base Version of the Lempel-Ziv Data Compression Algorithm, *IEEE Trans. Inform. Theory*, 37 (1991) 878–880.

[3] P. Jacquet, and W. Szpankowski, Analysis of digital tries with Markovian dependency, *IEEE Trans. Inform. Theory*, 37 (1991) 1470–1475.

[4] A.D. Wyner and A.J. Wyner, Improved Redundancy of a Version of the Lempel-Ziv Algorithm, *IEEE Trans. Inform. Theory*, 41 (1995) 723–732.

Yuriy A. Reznik
RealNetworks, Inc.
Email: yreznik@ieee.org

Anatoly V. Anisimov
Kiev National T. G. Schevchenko University
Email: ava@mi.unicyb.kiev.ua

Part IV

Trees

Trends in Mathematics, © 2004 Birkhäuser Verlag Basel/Switzerland

New Strahler Numbers for Rooted Plane Trees

David Auber, Jean-Philippe Domenger, Maylis Delest,
Philippe Duchon, and Jean-Marc Fédou

ABSTRACT: *In this paper, we present an extension of Strahler numbers to rooted plane trees. Several asymptotic properties are proved; others are conjectured. We also describe several applications of this extension.*

1. Introduction

The Strahler number of binary trees was introduced by the hydrogeologist Horton [11], then refined by Strahler [19] and rediscovered in computer science by Ershov [6] who interpreted it as the minimum number of registers allowing the computation of an arithmetic expression. Numerous results have been obtained from which one can point out explicit expressions for their distribution [8], [12]. In enumerative combinatorics, the research mostly involves bijections mapping Strahler numbers to other known statistics on other types of objects in order to understand their generating series : between binary trees and Dyck paths by Françon [10], between binary trees and forests of plane trees by Zeilberger [25] and more recently between forests of plane trees and Dyck paths by Viennot [23]. The common construction among these results is to redefine the Strahler number as a *pruning number* on rooted plane trees.

A natural question arises, is there a *natural* definition of Strahler numbers for rooted plane trees that extends Ershov's interpretation [6]? This paper proposes such an extension and describes some related properties. This extension is different from the one described and studied on general tries by Nebel [16] or Bourdon *et al.* [5].

Apart from biology and hydrogeology fields, Strahler numbers have been used in computer graphics to give synthetic images of trees and landscapes (see Viennot, Eyrolles, Janey and Arques [22]), in Physics by Vannimenus and Viennot [20]. A survey paper can be found in [21]. Thus, Strahler numbers have been proved efficient in many fields. It is natural to apply them to information visualisation systems that deal with trees and graphs. Here, we present two applications of our definition: navigation [2] and research of "quasi-similar" subtrees [4].

After some definitions, we prove that the computation of these Strahler numbers for all vertices of a tree of size n can be performed in $\Theta(n)$ time and space. Then we prove that the branching ratio r_k (to be defined later) of any simply generated family of trees with finite degree (in the sense of Meir and Moon [14, 13], tends to 4; for binary trees, it was known [18, 15] that $r_k = 4$ for all k. We end with a brief description of applications to information visualization systems.

2. Definitions and notations

A tree is a connected acyclic graph. A rooted tree is defined from a tree by choosing a vertex called the root; edges are oriented such that there exists a path from the root to each vertex of the tree. A vertex with no outgoing edge is called a leaf, other vertices are called internal vertices. A successor of a vertex is called a child of this vertex. The degree (or arity) of a vertex is the number of its children. In this paper, we consider rooted plane trees, i.e. rooted trees where a total order is imposed on the children of each vertex.

Let D be a set of nonnegative integers containing 0. We consider the set of simply generated trees, $\mathcal{T}_D$, in the sense of Meir and Moor [14, 13] : each vertex of a tree has its arity in D. In this setting, d-ary trees correspond to $D = \{0, d\}$; $d = 2$ for binary trees.

The Strahler number σ_b was first introduced on binary trees in some work about the morphological structure of rivers [11, 19]. The recursive definition associates an integer value to each vertex of a binary tree. These values give quantitative information about the complexity of each sub-tree of the original tree. Let s be a vertex of a tree:

- If s is a leaf then $\sigma_b(s) = 1$,
- Else s has two children s_1 and s_2 and

$$\sigma_b(s) = \begin{cases} \max(\sigma_b(s_1), \sigma_b(s_2)) \text{ if } \sigma_b(s_1) \neq \sigma_b(s_2) \\ \sigma_b(s_1) + 1 \text{ otherwise} \end{cases}$$

The Strahler number of a tree is the integer value associated to its root. If T is a tree, $\sigma_b(T)$ will denote the Strahler number of T.

Definition 2.1. *Let T be a tree whose vertices are valued by Strahler numbers. A branch of Strahler order k is a maximal path $(s_0, s_1, \ldots, s_p)$ in T, such that for each $i \in [0..n]$, $\sigma_b(s_i) = k$.*

We denote by $\beta_k(T)$, the total number of branches in T of Strahler order k.

Theorem 2.2. [18, 15] *Let $B_{k,n}$ be the total number of branches of Strahler order k in the set of binary trees with size $2n + 1$, then*

$$r_k = \lim_{n \to +\infty} \left(\frac{B_{k,n}}{B_{k+1}, n} \right) = 4.$$

This ratio is called the branching ratio of binary trees. Horton has shown [11] that for real rivers this ratio is between 3 and 5. A generalization of Strahler numbers to rooted trees is suggested by the nice interpretation by Ershov [6] who proved that the Strahler number of the root of a binary tree is exactly the minimal number of registers needed to compute an arithmetic expression given by the tree (output by the syntactical analysis). Thus, one can define the Strahler number σ on general trees by:

- if s is a leaf then $\sigma(s) = 1$;
- otherwise s has $k + 1$ children s_i such that $\sigma(s_i) \leq \sigma(s_j)$, if $i \leq j$ (i.e. children are reordered for the Strahler computation) and

$$\sigma(s) = \max_{0 \leqslant i \leqslant k} (\sigma(s_i) + i).$$

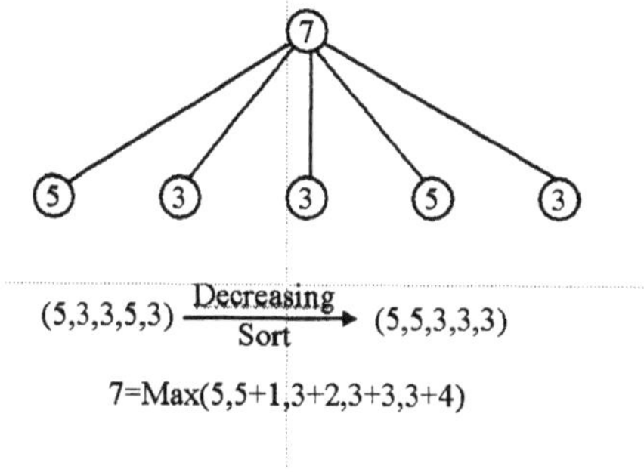

FIGURE 1. Strahler number on one vertex.

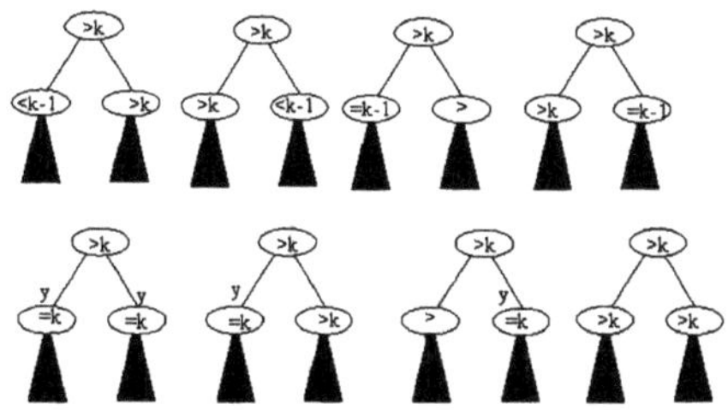

FIGURE 2. Construction of trees of $G_{\{2\}}$.

Equivalently, without requiring the children to be reordered, we have

$$\sigma(s) = \max\{i + j : \#\{\ell : \sigma(s_\ell) > i\} \geq j\}.$$

See an example in Figure 1. As for binary trees, the Strahler number of a tree is the Strahler number of its root. Note that on the set of binary trees, $\mathcal{T}_{\{0,2\}}$, σ and σ_b coincide. The definition of the branching ratio according to the generalization stays unchanged. We conjecture that a weaker (asymptotic) version of Theorem 2.2 holds for any family $\mathcal{T}_D$, and prove it for finite D. Note that if an internal vertex has a Strahler value s then it has at most s children.

3. Complexity of Strahler number computation

Strahler numbers are useful in information visualization as we will show in Section 6. In this field, the effective size of the trees is large : 500000 vertices is average. Thus, the algorithms that compute Strahler numbers must be efficient. In this section we describe a linear time algorithem, and prove that the nodes of any tree can be sorted according to their Strahler numbers in linear time; this is useful for the kind of applications we describe in Section 6.

3.1. Computing the Strahler valuation in linear time

To compute the Strahler number of a node, one has to sort its children in ascending or descending order of their own Strahler numbers. This sorting requirement, together with the *a priori* possibility that some nodes can have a large number of children, make it non obvious that a linear time algorithm exists to compute the Strahler valuation on a tree.

The key to our algorithm is the following observation : if each node of the tree is informed of the Strahler numbers of its children in increasing order, it only needs two counters to compute its own Strahler number; also, if the tree has N nodes, its Strahler number cannot be larger than N.

Thus, we setup an array S of size N; the k-th cell of the array will contain a list of vertices with Strahler number k, before their Strahler number is transmitted to their father. Initially, the first list contains all the leaves of the tree. Then, until all lists are empty, a vertex is extracted from the first nonempty list, and its Strahler number is transmitted to its father; if the father now knows the Strahler numbers of all its children, its Strahler number is then computed, and it is inserted in the appropriate list.

Each node in the tree is examined a bounded number of times overall, so that the total complexity of the algorithm is $\Theta(N)$.

3.2. The set of Strahler values of a tree

In some of the applications described in Section 6, one may have to select a subtree of a given tree, for which the Strahler valuation has already been computed, and sort its nodes according to their Strahler numbers. Thus, it is interesting to know that this sorting of N nodes can be performed in $\Theta(N)$ time. This will be a consequence of the following lemma, which, incidentally, is the best counterpart to the property of Strahler numbers on binary trees stating that a (binary) tree with Strahler number k must have at least $2^k - 1$ vertices.

Lemma 3.1. *Let T be a tree and $E(T)$ denote the set of Strahler values of vertices in T. Then*

$$\sum_{k \in E(T)} k \leqslant |T| \tag{1}$$

Proof: We prove the lemma by induction on the depth of trees. If T has depth 1, then its root has some number k of children, and Strahler number k. Thus equality occurs in (1), unless $k = 1$, in which case the inequality is strict.

Now assume (1) holds for all trees with depth at least h, and consider a tree T with depth $h + 1$. Assume the root of T has p children, which are the roots of subtrees $T_1, \ldots, T_p$ (with decreasing Strahler values). If $\sigma(T) = \sigma(T_1)$, then there is nothing to prove since $E(T) = \cup E(T_i)$. Thus, we assume $\sigma(T) = k > \sigma(T_1)$. By the recursive definition of the Strahler numbers, this ensures there is some integer i, $2 \leq i \leq p$, such that $\sigma(T_1) \geq k - i + 1$. Thus, we have

$$k - 1 \geq \sigma(T_1) \geq \ldots \geq \sigma(T_i) \geq k - i + 1,$$

so that at least two of the subtrees T_j, $j \leq i$, have the same Strahler value $k' \geq k - p + 1$. Also, the value 1 must certainly appear in each of the sets $E(T_i)$, since all leaves have Strahler value 1.

- if $k' > 1$, then we have $\sum_{1 \leqslant i \leqslant p} \sum_{l \in E(T_i)} l \geqslant k' + p - 1 + \sum_{l \in \bigcup_{1 \leqslant i \leqslant p} E(T_i)} l$ which, by induction, proves that inequality in (1) is strict,
- if $k' = 1$, then $k = p$, and we have $\sum_{1 \leqslant i \leqslant p} \sum_{l \in E(T_i)} l \geqslant 1 + \sum_{l \in \cup_i E(T_i)} l$ which also proves (1) by induction.

From this lemma and the fact that any set of k positive integers has a sum at least $\binom{k+1}{2}$, we deduce:

Corollary 3.2. *Let T be a tree then $|E(T)| \leq \sqrt{2|T|}$.*

A bijective proof of this result can be found in [1]. The number of different values of Strahler numbers in a given tree T has a sub-linear upper bound, and these values are in $[1, N]$. Thus the complexity of sorting all nodes becomes linear (say, by using a table similar to the one we described for the Strahler computation, while simultaneously maintaining a balanced tree of the values appearing; each value will have to be inserted only once in this tree, resulting in a total complexity of $O(\sqrt{N} \log N)$ for creating and maintaining it), and we have

Theorem 3.3. *The nodes of any tree of size N, can be sorted according to their Strahler numbers in $\Theta(N)$ time and space.*

4. Branching ratios for binary trees

In this section, we prove a weaker (asymptotic) version of Theorem 2.2 [18, 15] so as to demonstrate the method that we use for more general families. Because D is fixed, we will omit it in all the formulas. We define the following sets :

$$\mathcal{S}_k = \{t \in \mathcal{T}_D : \sigma(t) = k\}, \mathcal{L}_k = \{t \in \mathcal{T}_D : \sigma(t) < k\}, \mathcal{G}_k = \{t \in \mathcal{T}_D : \sigma(t) > k\}.$$

Note that β_k is identically 0 on $\mathcal{L}_k$, and identically 1 on $\mathcal{S}_k$. Next, consider the generating functions

$$F(x) = \sum_{t\in\mathcal{T}_D} x^{|t|}, S_k(x,y) = y\sum_{t\in\mathcal{S}_k} x^{|t|}, L_k(x) = \sum_{t\in\mathcal{L}_k} x^{|t|}, G_k(x,y) = \sum_{t\in\mathcal{G}_k} y^{\beta_k(t)}x^{|t|}.$$

Thus, the variable x always counts size, while y corresponds to the number of segments of Strahler value k. Throughout the rest of this paper, whenever we have a bivariate generating function $F(x,y)$, $F(x)$ denotes the corresponding univariate generating function $F(x,1)$.

The method consists in the following steps.

- Write out an equation for $G_k(x,y)$;
- consider the derivative of $G_k(x,y)$ with respect to y, $\partial G_k/\partial y(x,1) = \sum_{t\in\mathcal{G}_k} \beta_k(t)x^{|t|}$;
- find an asymptotic expression for $\partial G_k \partial y(x,1)$, allowing us to derive asymptotic expressions of the form $B_{k,n} = c_k n^{-1/2}\rho^{-n}(1+o(1))$;
- compute the branching ratio as c_k/c_{k+1}.

In what follows, we will omit the x and y variables in the series as soon as there is no ambiguity. Let us now see what happens on the set of binary trees. In this case, $D = \{0,2\}$. In the first step, write out the equation.

$$G_k = x(2L_{k-1}G_k + 2S_{k-1}G_k + S_k^2 + 2S_kG_k + G_k^2) \tag{2}$$

Each monomial in the right-hand part of (2) corresponds to one of the possible configurations of Strahler values for the two children of a node of Strahler value higher than k (see figure 2). In a second step, we replace $S_k(x)$ by $yS_k(x)$ to account for the number of segments of Strahler value k, and differentiate to get

$$\frac{\partial G_k}{\partial y}(x,1) = \frac{2x\left(G_{k-1}(x,1) - G_k(x,1)\right)^2}{1-2F(x)}. \tag{3}$$

The series $F(x)$ is an algebraic series (this is the well-known Catalan series) with square root type dominant singularities, and each $S_k(x)$ and $L_k(x)$, as a rational series, has only poles as singularities. This implies that the radius of convergence of F is strictly less than that of each L_k or L_k, and, in turn, that $G_k(x) = F(x) - L_{k+1}(x)$ has the same singularity structure as $F(x)$. Furthermore, the denominator in (3) vanishes at the singularities, while the numerator takes a finite, nonzero value.

As a result, the singularities of $\partial G_k/\partial y(x,1)$ are of the inverse square root type, so that the coefficients $B_{k,n}$ have an asymptotic of the form

$$B_{k,n} = c_k n^{-1/2} 4^n (1+o(1)),$$

where c_k is proportional to $(G_{k-1}(1/4,1) - G_k(1/4,1))^2$.

Step 4 consists in proving that $\lim_{k\to\infty} c_{k-1}/c_k = 4$; this is straightforward once one notices that (2) can be rewritten as

$$\frac{G_k(x,1)}{G_{k-1}(x,1)} = 2x\frac{S_k(x,1) - G_k(x,1)}{1-2xF(x)},$$

from where elementary singularity analysis entails that the ratio converges to 1/2 as x goes to $\pm 1/2$.

5. Branching ratio for trees of $\mathcal{T}_D$ with finite D

In this section, we assume that D is any finite set of integers that contains 0 and at least one integer larger than 1. The family of trees we are interested in is the set $\mathcal{T}_D$ of plane rooted trees where the arity of each node lies in D.

Due to space constraints, we only sketch the proofs; the method is very similar to what was used in the previous section.

The generating function for all trees in $\mathcal{T}_D$ thus satisfies the polynomial equation

$$F(x) = x\sum_{d\in D} F(x)^d = x\Phi_D(F(x)). \tag{4}$$

It is a classical result [7, 9] that this series converges as an analytic function inside the complex disc $|x| \le \rho_D$, where $\rho_D = \tau/\Phi_D(\tau)$ and τ is the unique positive real solution to the equation $\Phi_D(\tau) = x\Phi'_D(\tau)$. Furthermore, $F(x)$ has a single[1] dominant singularity of the square root type at ρ_D, and this translates into an asymptotic expression $a_n \sim C.\rho^{-n}n^{-3/2}$ for the number of D-trees of size n.

For each integer $k > 1$, $\mathcal{T}_D$ can be partitioned into 3 sets $\mathcal{S}_k, \mathcal{G}_k, \mathcal{L}_k$ with respective generating functions $S_k(x)$, $G_k(x)$ and $L_k(x)$. Similarly to the binary situation, S_k and L_k are both *rational* power series, so that their poles must all have moduli strictly larger than ρ_D. Thus, *each G_k has a square root singularity at ρ_D*, with the same amplitude as F. This can be interpreted as meaning that, for finite k, *almost all large trees have Strahler number more than k.*

5.1. Generating functions for trees of high Strahler value

Our goal in this paragraph is to write an equation for $G_k(x)$. If a vertex in a tree has d children, then, if we only want to discriminate its Strahler number so as to be able to decide whether it is strictly larger than some value k, we only need to discriminate whether the Strahler value of each child is larger than k, or one of the values between $k-d+2$ and k, or smaller than $k-d+2$; all values lower than $k-d+2$ are equivalent in this regard, because even d children each with Strahler value $k-d+1$ will only result in a Strahler value k for the root node. This means that the generating function for all trees where the root has exactly d children and Strahler value larger than k, can be expressed as x times a *polynomial* $\mathcal{Q}_d$ in the variables $G_k, S_k, \ldots, S_{k-d+2}, L_{k-d+2}$. This polynomial $\mathcal{Q}_d$ can be obtained by expanding (formally) the expression $(G_k + S_k + \cdots + S_{k-d+2} + L_{k-d+2})^d$ into a

[1] In fact, there is a singularity at each complex value of the form $\rho_D.e^{2ik\pi/d'}$, where d' is the greatest common divisor of elements of D; when $d' > 1$, the series $F(x)/x$ is invariant upon the change of variable $x \mapsto xe^{2i\pi/d'}$, and D-trees always have a size of the form $d'k+1$ for some k. The proofs in this paper assume that $d' = 1$, so as to make notations easier; they can be extended to the general case easily.

sum of monomials, and, in the resulting sum, selecting only those monomials that lead to a Strahler number strictly over k.

By performing the global change of variables $S_i = G_{i-1} - G_i$, $L_i = F - G_{i-1}$, we can also express $\mathcal{Q}_d$ as a polynomial $\mathcal{P}_d$ in the variables $G_k, \ldots, G_{k-d+1}, F$; this will later yield somewhat simpler equations. Note that, contrary to $\mathcal{Q}_d$, $\mathcal{P}_d$ has negative coefficients. Once this is done, summing over all possible degrees for the root of a D-tree yields the following equation for the series $G_k(x)$:

$$G_k(x) = x \sum_{d \in D} \mathcal{P}_d(F(x), G_k(x), \ldots, G_{k-d+1}(x)) \tag{5}$$

(with the provision that $G_i(x) = F(x)$ whenever $i \leq 0$)

Our main tool is a recurrence relation between the polynomials $\mathcal{P}_d$ themselves, which enables us to obtain asymptotic results on the branching ratios of any simply generated family of trees with a *finite* set of degrees allowed. We conjecture that similar results should hold for infinite sets of degrees (or at least a wide class of them), but were not able to prove them.

To avoid confusion between the generating functions $F(x)$, $G_k(x)$, $S_k(x)$ and $L_k(x)$ and the variables of polynomials $\mathcal{Q}_d$ and $\mathcal{P}_d$, we will use lowercase letters for the latter, and write

$$\mathcal{Q}_d = \mathcal{Q}_d(g, s_0, \ldots, s_{d-2}, \ell), \mathcal{P}_d = \mathcal{P}_d(f, g_0, \ldots, g_{d-1})$$

We have:

$$\mathcal{Q}_1(g, \ell) = g, \mathcal{Q}_2(g, s_0, \ell) = (g + s_0)^2 + 2g.\ell,$$

or, equivalently,

$$\mathcal{P}_1(f, g_0) = g_0, \mathcal{P}_2(f, g_0, g_1) = g_1^2 - 2g_0g_1 + 2g_0.f,$$

For the first values of d, we obtained $\mathcal{P}_d$ through a computer algebra system. These suggest a recurrence for polynomials:

Lemma 5.1. *For any $d \geq 2$,*

$$\mathcal{P}_d(f, g_0, \ldots, g_{d-1}) = g_{d-1}^d + d \int_{g_{d-1}}^{f} \mathcal{P}_{d-1}(t, g_0, \ldots, g_{d-2})dt \tag{6}$$

and, equivalently,

$$\mathcal{Q}_d(g, s_0, \ldots, s_{d-2}, \ell) = (g + s_0 + \cdots + s_{d-2})^d + d \int_{s_{d-2}}^{s_{d-2}+\ell} \mathcal{Q}_{d-1}(g, s_0, \ldots, s_{d-3}, t)dt \tag{7}$$

Proof: We prove (7); (6) is equivalent under the previously mentioned change of variables. Recall that $\mathcal{Q}_d$ is the enumeration polynomial for the ways a tree with Strahler value higher than k can be constructed with a root and d subtrees whose Strahler values are higher than k (counted by the variable g), or any value between k (counted by s_0) and $k-d+2$ (counted by s_{d-2}), or lower than $k-d+2$ (counted by ℓ).

Now consider the set $\mathcal{L}_d$ of words of length d over the alphabet $\{g, k, k-1, \ldots, k-d+2, \ell\}$ (where g stands for "higher than k", and ℓ stands for "lower than $k-d+2$"), that, if they are interpreted as the sequence of Strahler values for the d children of a node, result in this node having Strahler value higher than k. $\mathcal{Q}_d$ is none other than the commutative image of $\mathcal{L}_d$ when each letter $k-i$ is mapped to the variable s_i. What we need to do is provide a description of $\mathcal{L}_d$ in terms of $\mathcal{L}_{d-1}$ that, under commutation, can be interpreted as (7).

First, note that all words of length d where ℓ does not appear, belong to $\mathcal{L}_d$: having d children, each with Strahler value at least $k-d+2$, is enough for a vertex to have Strahler value at least $k+1$. This justifies the term $(g+s_0+\cdots+s_{d-2})^d$ in (7), and we now turn to words in $\mathcal{L}_d$ where the letter ℓ appears.

Consider a word $w \in \mathcal{L}_d$, with $j+1$ total occurrences of ℓ or $k-d+2$ ($j \geq 0$), and the (multi-)set of $j+1$ words of length $d-1$ obtained by first replacing each $k-d+2$ with an ℓ, then removing one of the ℓ. It is easy to see that each of these words belongs to $\mathcal{L}_{d-1}$. Inversely, for any word $w' \in \mathcal{L}_{d-1}$ with j occurrences of ℓ, the (multi-)set of $d(2^j-1)$ words obtained by first inserting an additional ℓ in any of d positions in w', then replacing each ℓ with either itself or $k-d+2$ – but leaving at least one occurrence of ℓ, since words in $\mathcal{L}_d$ without ℓ are already accounted for – will produce only words of $\mathcal{L}_d$, each word with $j+1$ total occurrences of $k-d+2$ or ℓ being obtained $j+1$ times. Letting all letters commute, and summing over all words of $\mathcal{L}_{d-1}$, we see that each monomial $M\ell^j$ in $\mathcal{Q}_{d-1}$ becomes $\frac{d}{k+1}M((\ell+s_{d-2})^{j+1}-s_{d-2}^{j+1})$ in $\mathcal{Q}_d$, which is exactly symbolic integration with respect to ℓ on the interval $[s_{d-2}, s_{d-2}+\ell]$. Summing all contributions, we get (7).

The following corollary is an easy consequence of the previous lemma and of the expressions for the first polynomials $\mathcal{P}_d$ and $\mathcal{Q}_d$:

Corollary 5.2. *For any $d \geq 1$,*

$$\frac{\partial \mathcal{Q}_d}{\partial g}(g, s_0, \ldots, s_{d-2}, \ell) = d.\,(g+s_0+\cdots+s_{d-2}+\ell)^{d-1} \tag{8}$$

Furthermore, $\mathcal{P}_d$ is homogenous of degree d in its variables, and has $d-1$ in f and degree 1 in g_0. Thus, $\mathcal{P}_d = g_0\mathcal{D}_d + \mathcal{N}_d$, where

$$\mathcal{N}_d(f, g_1, \ldots, g_{d-1}) = g_{d-1}^d + d\int_{g_{d-1}}^{f} \mathcal{N}_{d-1}(t, g_1, \ldots, g_{d-2})dt \tag{9}$$

$$\mathcal{D}_d(f, g_1, \ldots, g_{d-1}) = d\int_{g_{d-1}}^{f} \mathcal{D}_{d-1}(t, g_1, \ldots, g_{d-2})dt \tag{10}$$

$$\mathcal{N}_d = \binom{d}{2} g_1^2 f^{d-2} + O\left(f^{d-3}\right) \tag{11}$$

$$\mathcal{D}_d = df^{d-1} - d(d-1)g_1 f^{d-2} + O\left(f^{d-3}\right) \tag{12}$$

In light of this, (5) solves to

$$G_k(x) = \frac{x\sum_{d\in D}\mathcal{N}_d(F(x), G_{k-1}(x), \ldots, G_{k-d+1}(x))}{1 - x\sum_{d\in D}\mathcal{D}_d(F(x), G_{k-1}(x), \ldots, G_{k-d+1}(x))} \tag{13}$$

Notice that the leading terms (in powers of $F(x)$) in the denominator of (13) exactly cancel the term of 1 when $x = \rho_D$, since

$$x\sum_{d\in D} dF(x)^{d-1} = 1$$

is exactly the equation for the singularities of F. Also, note that (from (11-12) the remaining leading terms (in powers of $F(x)$) in the numerator and denominator are in a ratio that is exactly $G_{k-1}(x)$. Since, for any x, all variables of the polynomials $\mathcal{N}_d$ and $\mathcal{D}_d$ tend to 0 as k tends to $+\infty$, this suggests that the ratio $G_k(\rho_D)/G_{k-1}(\rho_D)$ converges to $1/2$ as k goes to $+\infty$. In fact, the only ingredient missing to complete the proof is to justify using the powers of the various G_{k-i}

variables to select the dominant terms in (13). This is the reason for the following lemma.

Lemma 5.3. *Let D be a finite set of allowed degrees, $0 \in D$, $D \not\subset \{0,1\}$, and let $d' = \max D$. Then, for any $k \geq 1$,*

$$G_{k-1}(\rho_D) \geq G_k(\rho_D) \geq \frac{1}{d'} G_{k-1}(\rho_D). \tag{14}$$

Proof: *(sketch)* The first part is an obvious consequence of the fact that G_{k-1} sums over more trees than G_k; the second part is proved by defining $\mathcal{E}_d = -\mathcal{D}_d + df^{d-1}$, and using (6) to prove, by induction on d, that

$$\mathcal{N}_d(f, g_1, \ldots, g_{d-1}) \geq \frac{1}{d'} g_{d-1} \mathcal{E}_d(f, g_1, \ldots, g_{d-1}) \geq 0 \tag{15}$$

holds as soon as $0 \leq g_1 \leq \cdots \leq g_{d-1} \leq f$. The lemma is then proved by a convex combination of (15), observing that

$$G_k(\rho_D) = \frac{\sum_{d\in D} \mathcal{N}_d(F(\rho_D), G_{k-1}(\rho_D), \ldots, G_{k-d+1}(\rho_D))}{\sum_{d\in D} \mathcal{E}_d(F(\rho_D), G_{k-1}(\rho_D), \ldots, G_{k-d+1}(\rho_D))}.$$

Corollary 5.4. *For any finite D, let γ_k (respectively, σ_k) denote the singular value of G_k: $\gamma_k = G_k(\rho_D)$ (respectively, $\sigma_k = S_k(\rho_D)$). Then,*

$$\lim_{k\to\infty} \frac{\gamma_k}{\gamma_{k-1}} = \frac{1}{2}, \lim_{k\to\infty} \frac{\sigma_k}{\gamma_k} = 1, \lim_{k\to\infty} \frac{\sigma_k}{\sigma_{k-1}} = \frac{1}{2}$$

5.2. The branching ratio

We now turn to the task of evaluating the branching ratio for the Strahler numbers; that is, we must compare the total numbers, in D-trees of size n, of branches of Strahler order k and $k-1$. Let r_k denote the limit, as n tends to infinity, of this ratio $B_{k,n}/B_{k+1,n}$. We will prove that r_k tends to 4 as k tends to infinity – a somewhat weaker result than the previously known fact (for binary trees) that $r_k = 4$. To do this, we consider the bivariate generating function

$$F_k(x,y) = \sum_{T\in\mathcal{T}_D} x^{|T|} y^{\beta_k(T)}.$$

Clearly, $F_k(x,1) = F(x)$. Using standard generating function manipulations and notations, the total number of branches of order k in D-trees of size n is exactly

$$[x^n] \frac{\partial F_k}{\partial y}(x,1).$$

We have $F_k(x,y) = L_k(x) + yS_k(x) + G_k(x,y)$. Using the notations of the previous paragraph,

$$G_k(x,y) = \sum_{d\in D} \mathcal{Q}_d(G_k(x,y), yS_k(x), S_{k-1}(x), \ldots, S_{k-d+2}(x), L_{k-d+2}(x)). \tag{16}$$

Note that (16) proves that $G_k(x,y)$, and by extension $F_k(x,y)$, are *algebraic* series in their two variables, since the various S_i and L_i series are all rational. If we differentiate (16) with respect to y and set $y=1$, we get an equation which we can immediately solve for $\Gamma_k(x) = \partial F_k/\partial y(x,1)$:

$$\Gamma_k(x) = S_k(x) \frac{x\sum_{d\in D} \frac{\partial \mathcal{Q}_d}{\partial s_0}(G_k(x,1), S_k(x), \ldots, S_{k-d+2}(x), L_{k-d+2}(x))}{1 - x\sum_{d\in D} \frac{\partial \mathcal{Q}_d}{\partial g}(G_k(x,1), S_k(x), \ldots, S_{k-d+2}(x), L_{k-d+2}(x))} \tag{17}$$

The denominator in (17) is exactly $1 - x\sum_{d\in D} dF(x)^{d-1} = F(x)/(xF'(x))$. Since this vanishes at the dominant singularity $x = \rho_D$, and the numerator in (17) has a finite, nonzero limit, using a *transfer lemma* [7, 9], we get the following estimates :

- $B_{k,n} \sim b_k \rho_D^{-n} n^{-1/2}$ for some positive constant b_k;
- the branching ratio for branches of Strahler number k is

$$\frac{b_k}{b_{k+1}} = \frac{\sigma_k}{\sigma_{k+1}} \frac{\sum_{d\in D} \frac{\partial\Omega_d}{\partial s_0}(\gamma_k, \sigma_k, \ldots, \sigma_{k-d+2}, L_{k-d+2}(\rho_D))}{\sum_{d\in D} \frac{\partial\Omega_d}{\partial s_0}(\gamma_{k+1}, \sigma_{k+1}, \ldots, \sigma_{k-d+3}, L_{k-d+3}(\rho_D))}$$

The first factor tends to 2 by Corollary 5.4. In the second factor, the variables of the polynomial in the numerator and denominator are in an asymtotic factor of 2 (also by Corollary 5.4), and tend to 0 as k tends to infinity, except the last variable $L_j(\rho_D)$ which tends to $F(\rho_D)$. Since all involved polynomials are homogenous with a nonzero term of degree 1 in the variable ℓ, the whole second factor also tends to 2 as k tends to infinity.

Theorem 5.5. *The branching ratio of trees of any family of finite degree D is asymptotically* 4.

6. Applications to information visualization

Originally, Strahler numbers were designed to highlight river shapes. The extension that we have given is in the same spirit: highlightening information from tree structures. We show here two results. One concerns navigation tools [2] and, the other file system retrieval [4]. When one visualizes large trees on a screen, displaying the elements (vertices and edges) can take more than fifty milliseconds (δ_0). In order to have a high performance visualization system, one needs to accept events coming from the user (mouse movements, keyboard actions) within δ_0. In order to increase performance, the method proposed by Wills [24] consists in predicting the number of elements that can be displayed during δ_0. Thus the system incrementally displays the tree (by parts associated to δ_0) and ensures that the time reaction of the system is less than δ_0 without any request to the clock.

The problem now is to select the *best* elements to display so that the user should not be lost in his data. In our method, we propose to display the elements in decreasing Strahler order of the vertices. The figures show the result for 520000 vertices: Figure 3 shows what happens when *smart* choice is done, and Figure 4 shows the result when applying our method.

An other field of interest is to find *similar* structures in an information system. *Similar* here means not the same but quite the same : the user may choose a "degree of similarity". One real question assigned at the Infovis'03 contest [4] was on file systems (that are trees) : given two views S_1 and S_2 of a same file system taken at two different times, show to the user what parts have changed. Work has already been done based on arity by Zemlyachenko [26] and more recently by Dinitz *et al.* [17]. Their algorithm gives a partition of subtrees into isomorphism equivalence classes. It is proved to be linear. However, it only detects isomorphism and does not provide a measure of similarity for subtrees. In a recent work [4], we used a combination of three parameters (number of vertices, arity, Strahler number) in order to detect *quasi-isomorphic* subtrees. The Figure 5 shows the trees representing the same file system at two different times (S_1 and S_2). Each tree has about 80000 vertices. S_1 and S_2 look similar. Thus roughly, with the "isomorphism

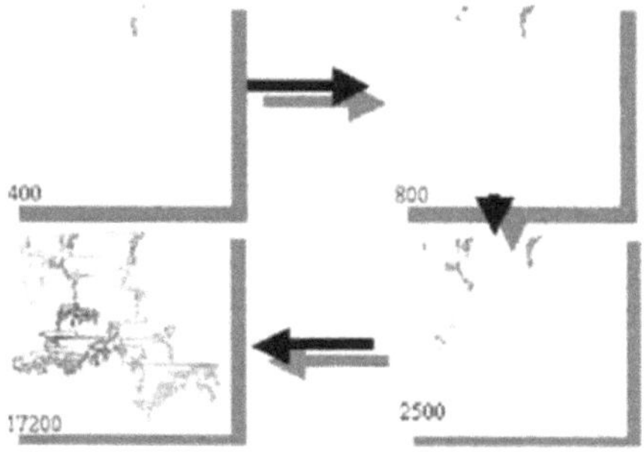

FIGURE 3. Display without Strahler

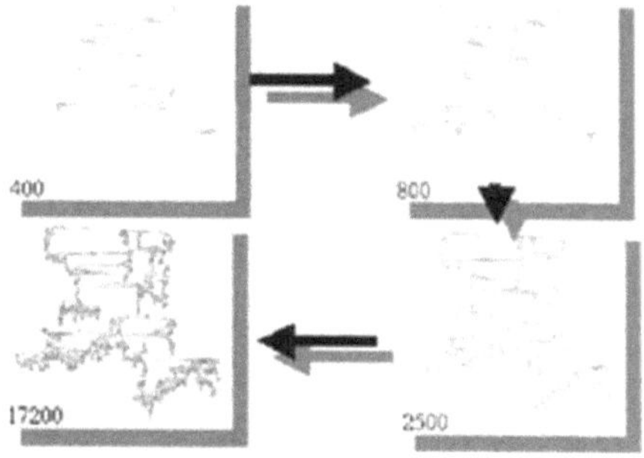

FIGURE 4. Strahler-guided display

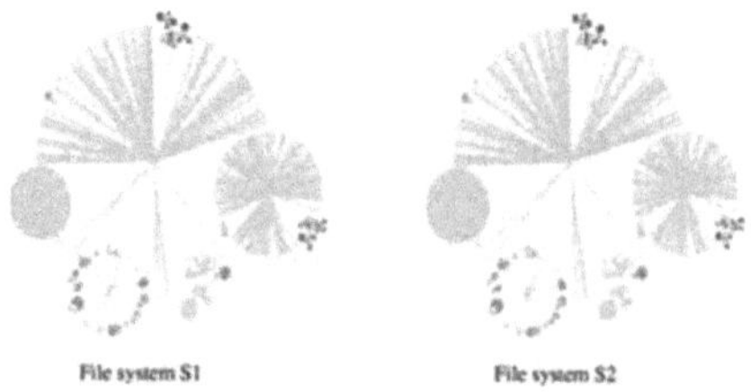

FIGURE 5. Two file systems: overview

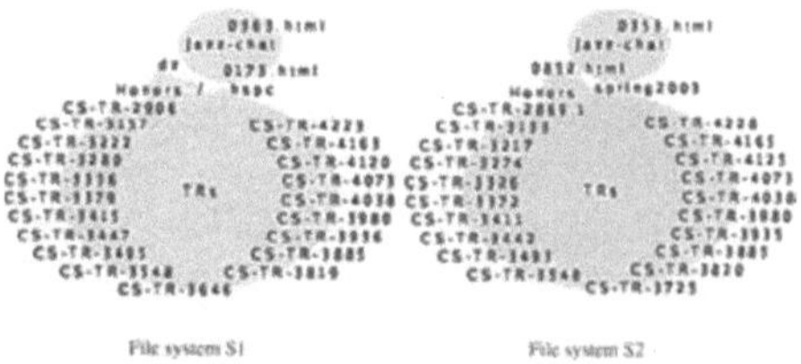

FIGURE 6. Two file systems: first zoom

level" chosen in this view by the user, he can consider that no big changes have occurred during the chosen time. Applying our algorithm recursively, we detect subtrees that are different. One can see in the figure 6 that some smooth changes have occured on the labels even if the drawing of the tree is the same. This is due to the navigation tool that we have described above. In the real version[4], we use only colors because same colors suggest same subtrees. The similar big disks in Figure 6 are different and have 951 and 755 nodes, respectively. Figure 7 was obtained by suppressing the two disks after inspection. We have only displayed a similar part of the two subtrees. All these methods have been implemented in Tulip [3] and work on real examples.

References

[1] D. Auber. Outils de visualisation de larges structures de données. PhD dissertation, Université Bordeaux 1, 2002.

[2] D. Auber. Using strahler numbers for real time visual exploration of huge graphs. In *International Conference on Computer Vision and Graphics*, volume 1-3 of *Journal of WSCG*, pages 56–69, 2002.

[3] D. Auber. Tulip: A huge graph visualisation framework. In P. Mutzel and M. Jünger, editors, *Graph Drawing Softwares*, Mathematics and Visualization, pages 105–126. Springer-Verlag, 2003.

[4] D. Auber, M. Delest, J.P. Domenger, P. Ferraro, and R. Strandh. EVAT : Environment for visualization and analysis of trees. In *IEEE Symposition on Information*

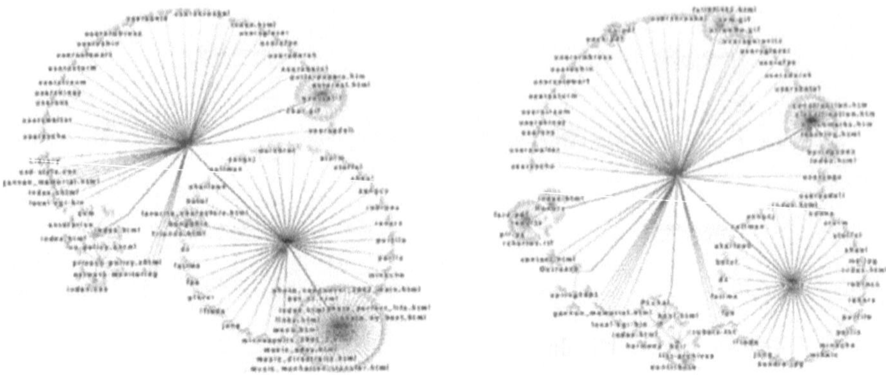

FIGURE 7. A final zoom on both file systems

Visualisation Contest, volume www.cs.umd.edu/hcil/iv03contest/, pages 124–126, 2003.

[5] J. Bourdon, M. Nebel, and B. Vallée. On the stack-size of general tries. *Theoretical Informatics and Applications*, 35(4):163–185, 2000.

[6] A.P. Ershov. On programming of arithmetic operations. *Communication of the A.C.M*, 1(8):3–6, 1958.

[7] P. Flajolet and A. Odlyzko. Singularity analysis of generating functions. *SIAM J. Discrete Math.*, 3:216–240, 1990.

[8] P. Flajolet, J.C. Raoult, and J. Vuillemin. The number of registers required for evaluating arithmetic expressions. *Theoretical Computer Science*, 9:99–125, 1979.

[9] P. Flajolet and R. Sedgewick. The average case analysis of algorithms: Complex asymptotics and generating functions. Rapport de recherche 2026, INRIA, 1993.

[10] J. Françon. Sur le nombre de registres nécessaires à l'évaluation d'une expression arithmétique. *RAIRO Informatique théorique*, 18:355–364, 1984.

[11] R.E. Horton. Erosioned development of systems and their drainage basins, hydrophysical approach to quantitative morphomology. *Bulletin Geological Society of America*, 56:275–370, 1945.

[12] R. Kemp. The average number of registers needed to evaluate a binary tree optimally. *Acta Informatica*, 11:363–372, 1979.

[13] A. Meir and J. W. Moon. Erratum: "On an asymptotic method in enumeration". *J. Combin. Theory Ser. A*, 52(1):163, 1989.

[14] A. Meir and J. W. Moon. On an asymptotic method in enumeration. *J. Combin. Theory Ser. A*, 51(1):77–89, 1989.

[15] W.J. Moon. An extension of Horton's law of stream numbers. Math. Colloq. Univ. cape Town, 1980.

[16] M.E. Nebel. A unified approach to the analysis of Horton-Strahler parameters of binary tree structures. *Random Struct. Algorithms*, 21(3-4):252–277, 2002.

[17] M. Rodeh Y. Dinitz, A. Itai. On an algorithm of Zemlyachenko for subtree isomorphism. *Information Processing Letters*, 703:141–146, 1999.

[18] R. Shreve. Statistical law of stream numbers. *J. Geol.*, 74:178–186, 1966.

[19] A.N. Strahler. Hypsomic analysis of erosional topography. *Bulletin Geological Society of America*, 63:1117–1142, 1952.

[20] J. Vannimenus and X.G. Viennot. Combinatorial analysis of ramified patterns. *J. Stat. Phys.*, 54:1529–1538, 1989.

[21] G. Viennot. Trees everywhere. In A. Arnold, editor, *Colloquium on Trees in Algebra and Programming*, Lecture Notes in Computer Science 431, pages 18–41. Springer-Verlag, 1990.

[22] G. Viennot, G. Eyrolles, N. Janey, and D.Arques. Combinatorial analysis of ramified patterns and computer imagery of trees. In *SIGGRAPH Conference*, volume 23 of *Computer Graphics*, pages 31–40, 1989.

[23] X.G. Viennot. A Strahler bijection between Dyck paths and planar trees. In R. Cori and O. Serra, editors, *11th Formal Power Series and Algebraic Combinatorics*, pages 573–584, 1999.

[24] G.J. Wills. NicheWorks: Interactive visualization of very large graphs. In Giuseppe Di Battista, editor, *5th Symp. Graph Drawing*, volume 1353 of *Lecture Notes in Computer Science*, pages 403–414. Springer-Verlag, 1997.

[25] D. Zeilberger. A bijection from ordered trees to binary trees that sends the pruning order to the Strahler number. *Discrete Math.*, 82:89–92, 1990.

[26] V.N. Zemlyachenko. Determining tree isomorphism. *Seminar on Combinatorial Mathematics*, pages 54–60, 1971.

David Auber
LaBRI, Université Bordeaux 1
auber@labri.fr

Jean-Philippe Domenger
LaBRI, Université Bordeaux 1
domenger@labri.fr

Maylis Delest
LaBRI, Université Bordeaux 1
maylis@labri.fr

Philippe Duchon
LaBRI, Université Bordeaux 1
duchon@labri.fr

Jean-Marc Fédou
I3S, Université de Nice-Sophia Antipolis
fedou@unice.fr

Trends in Mathematics, © 2004 Birkhäuser Verlag Basel/Switzerland

An Average-Case Analysis of Basic Parameters of the Suffix Tree

Julien Fayolle

ABSTRACT: *The* **LZ'77** *algorithm offers one of the best available rates for lossless data compression. It is based on the suffix tree structure. Our aim is to obtain the asymptotics of the mean size and external path length of a suffix tree by comparing them to those of a trie or digital tree. The core problem lies within the set on which we build the suffix tree. This set is correlated, so we cannot use the methods that have proved efficient for the trie. The proof relies on combinatorics, generating functions, and complex analysis.*

1. Introduction

The *trie* or digital tree data structure [7, 8, 9] manages efficiently dictionaries. Queries for an existing word or insertion of a new word in the dictionary can be performed in expected logarithmic time in the number of items stored in the dictionary.

While the **LZ'78** data compression algorithm is based on digital search trees, we will focus on *suffix trees*, a particular kind of trie which lies at the very heart of the popular and efficient **LZ'77** [11] lossless compression algorithm. This algorithm is behind the `gzip` software.

In a groundbreaking article, Jacquet and Szpankowski [6] have developped a sophisticated "string ruler" approach to obtain results on the asymptotics of parameters of suffix trees. This paper uses Jacquet and Szpankowski's lead idea: asymptotically the mean size and external path length for a trie built on n words and for a suffix tree built on n suffixes are very close.

This paper's aim is threefold: first to provide simpler proofs of Jacquet and Szpankowski's results than they do, second to obtain more accurate results, and third to lay the groundwork for a study of suffix trees under a broader model.

1.1. Tries

We first define recursively a trie on a set X of infinite words on the m-ary alphabet $\mathcal{A} = \{a_1, \cdots, a_m\}$ as

$$\text{trie}(X) = \begin{cases} \emptyset & \text{if } |X| = 0, \\ \bullet & \text{if } |X| = 1, \\ \langle \bullet, \text{trie}(X \backslash a_1), ..., \text{trie}(X \backslash a_m) \rangle & \text{else,} \end{cases}$$

where $X \backslash \alpha$ is defined as the set of words starting with the letter α whose first letter is removed.

From now on the alphabet will be binary $\mathcal{A} = \{0, 1\}$, a choice that entails no loss of generality.

1.2. Source model

How do we obtain the infinite words constituting the set X? By a device, called *source*, producing randomly symbols from the binary alphabet regularly in time.

The type of source we are dealing with has two main characteristics: it is *randomised*, symbols are emitted with probabilities; and *memoryless*, the emission of a symbol at a given time is independent of the symbols already emitted. In this work, the probability of occurrence of a symbol is independent of when it is emitted. This specifies the memoryless source model a.k.a. Bernoulli model.

Definition 1.1. *The probability that the source emits a sequence of symbols starting with the pattern w is noted p_w, and called occurrence probability. For a memoryless source, p_w is the product of the probabilities of the letters composing w.*

We note p the probability of emitting the symbol $\mathtt{0}$ and q the probability of emitting $\mathtt{1}$. A source is said to be *symmetric* if $p = q = 1/2$ and *biased* otherwise. For convenience, we adopt the convention that p is the largest of the two probabilities.

1.3. Parameters under the spotlights

For each internal node of a trie, the successive left (encoded by 0) or right (encoded by 1) steps taken to go from the root to the node encode the prefix associated to this node. An internal node exists within the trie (relatively to an infinite complete binary tree) if there are at least two words in X starting by the prefix associated to this node.

For a pattern w, $N_w(X)$ is introduced as the number of words of X starting by w. Sometimes, we note N_w for $N_w(X)$.

Let the parameters S and P denote the size and the external path length. Both can be rewritten in terms of N_w for a trie built on the set X by

$$S(X) := \sum_{w \in \mathcal{A}^*} [\![N_w(X) \geq 2]\!],$$

$$P(X) := \sum_{w \in \mathcal{A}^*} N_w [\![N_w(X) \geq 2]\!],$$

where $[\![P]\!] = 1$ if P is true, 0 else. This is Iverson's bracket notation.

1.4. Suffix trees

Let y be an infinite word on the alphabet $\mathcal{A}$ and Y_n the set of the first n suffixes of y (we consider y to be its own suffix). The suffix tree of index n based on y is nothing but the trie built on Y_n (this operation is valid since Y_n is a set of infinite words over the alphabet $\mathcal{A}$).

Since tries and suffix trees are based on the same recursive decomposition, the expression for the parameters S and P under consideration are identical. For a trie on a set X, N_w means the number of words starting with the pattern w, so for a trie on the set Y_n (suffix tree on y) it coincides with the number of suffixes (amongst the n first of them) of the word y for which w is a prefix, or equivalently, the number of occurrences of the pattern w in the n first positions of the word y. We introduce $\widehat{N}_w(y; n)$ as the number of occurrences of the pattern w in the first n positions of y, and we express the size S and the external path length P of a suffix tree as

$$S(y;n) := \sum_{w\in\mathcal{A}^*} [\![\widehat{N}_w(y;n) \geq 2]\!], \tag{1}$$

$$P(y;n) := \sum_{w\in\mathcal{A}^*} \widehat{N}_w(y;n)[\![\widehat{N}_w(y;n) \geq 2]\!] \tag{2}$$

1.5. Plan

We recall the results obtained by Knuth [7] on the mean size and external path length for a trie built on n strings:

$$\mathbb{E}_n^{\mathrm{t}}(S) = \frac{n}{h}(1+\epsilon'(n)) + O(\log n), \tag{3}$$

$$\mathbb{E}_n^{\mathrm{t}}(P) = \frac{n\log n}{h} + (K+\epsilon(n))n + O(\log n), \tag{4}$$

where ϵ and ϵ' are oscillating function of very small amplitude around 0. These results were proved in a less intricate way in [1].

The purpose of the next section is to obtain, *via* Guibas and Odlyzko's work and complex analysis, an asymptotic expression of the mean size, $\mathbb{E}_n(S)$, and mean external path length, $\mathbb{E}_n(P)$, for suffix trees built on n suffixes. There are two probabilistic models on tries: one can build them on a set of size n (fixed-size model a.k.a. *Bernoulli*) or on a set which size follows a *Poisson* law of parameter z (Poisson model). The difference between mean size for tries under Poisson of parameter n and Bernoulli of parameter n models is fairly small (of order $\log n$), this is also true for mean external path length. Section 3 is dedicated to studying the difference Δ between mean path length for tries under the Poisson model of parameter n, $\mathbb{E}^{\mathrm{t}}_{\mathcal{P}(n)}(P)$, and for a suffix tree built on n suffixes, $\mathbb{E}_n(P)$. Our aim is to show that Δ is small. Section 3 focuses on the external path length but the same techniques can be used for size.

2. Asymptotics

Since we know from the previous part the expression for the size S and the external path length P of a suffix tree, we write down the mean over suffix trees built with n suffixes for both parameters:

$$\begin{aligned}\mathbb{E}_n(S) &= \sum_{w\in\mathcal{M}^*} \mathbb{E}_n([\![\widehat{N}_w \geq 2]\!]) = \sum_{w\in\mathcal{M}^*} \mathbb{P}_n(\widehat{N}_w \geq 2)\\ &= \sum_{w\in\mathcal{M}^*} 1 - \mathbb{P}_n(\widehat{N}_w = 0) - \mathbb{P}_n(\widehat{N}_w = 1),\end{aligned} \tag{5}$$

$$\mathbb{E}_n(P) = \sum_{w\in\mathcal{M}^*} \mathbb{E}_n(\widehat{N}_w[\![\widehat{N}_w \geq 2]\!]) = \sum_{w\in\mathcal{M}^*} \mathbb{E}_n(\widehat{N}_w) - \mathbb{P}_n(\widehat{N}_w = 1). \tag{6}$$

This part is dedicated to finding the asymptotic value for the two probabilities appearing in the formulæ.

2.1. Combinatorics

The expressions for the mean size and external path length of a suffix tree make use of the probabilities $\mathbb{P}_n(\widehat{N}_w = 1)$ and $\mathbb{P}_n(\widehat{N}_w = 0)$, that is the probabilities to obtain one (resp. zero) occurrence of the pattern w in the first $n + |w| - 1$ letters of an infinite word (*i.e.*, an occurrence of w starting at one of the first n letters of the infinite word).

In order to obtain these probabilities, we introduce ordinary generating functions counting the number of texts with a given number of occurrences of the pattern w according to their size: let $O_w(z) = \sum_{n\geq 0} o_n z^n$ be the ordinary generating function (ogf) counting texts with w occurring only once according to their size and $N_w(z)$ the ogf counting texts with no occurrence of the pattern w.

The possible overlap of the pattern w with itself causes problem in the enumeration of occurrences of a pattern in a text. This phenomenon is called *autocorrelation.* The pattern w of size k has an overlap of size j if $1 \leq j \leq k$ and the prefix of size j, P_j, and the suffix of size j, S_j, of w coincide ($P_j = S_j$). Graphically, an overlap of a pattern (white rectangles) looks like this:

,

where the two black boxes are the prefix and suffix matching one above the other. For example w=001001001 has overlaps of size 3 and 6; $|w|$ is always a valid overlap.

The autocorrelation of a pattern is encoded by the autocorrelation polynomials: the combinatorial one is

$$c_w(z) = \sum_{i=0}^{k-1} c_i z^i$$

and, since our symbols are produced by a probabilistic source, the probabilistic version that we need is

$$\widehat{c}(z) = \sum_{i=0}^{k-1} c_i z^i \mathbb{P}(w_{k-i+1} \cdots w_k)$$

where $c_i = [\![S_{k-i} = P_{k-i}]\!]$, meaning there is an overlap of size $k - i$.

For the memoryless sources we are dealing with, the probabilistic autocorrelation polynomial satisfies a neat and useful relation:

Lemma 2.1.

$$\sum_{w\in\mathcal{M}^k} \widehat{c}_w(1) = 2^k + k - 1.$$

Proof: On a binary alphabet there are 2^j patterns of size j for any $j < k$ and from each of these v, one can build a unique w of size k such that $c_{j,w} = 1$ and v is its suffix of size j. So there are *at most* 2^j patterns of size k with a given suffix of size j and satisfying $c_j = 1$. Furthermore, there is no way two different suffixes of size j can create the same word of size k, so there are *exactly* 2^j patterns of size

k satisfying $c_j = 1$ and this for every j between 1 and $k-1$. Hence,

$$\sum_{w\in\mathcal{M}^k} \widehat{c}_w(1) = \sum_{w\in\mathcal{M}^k}\sum_{j=0}^{k-1} c_{j,w}\mathbb{P}(w_{k-i+1}\cdots w_k)$$
$$=\sum_{j=0}^{k-1}\sum_{w\in\mathcal{M}^k:\, c_{j,w}=1} \mathbb{P}(w_{k-i+1}\cdots w_k),$$

but since we just proved that the suffix of size j is enough to obtain a word w of size k with $c_j = 1$ (and $c_{0,w} = 1$ for all $w \in \mathcal{M}^k$),

$$\sum_{w\in\mathcal{M}^k} \widehat{c}_w(1) = 2^k + \sum_{j=1}^{k-1}\sum_{v\in\mathcal{M}^j} \mathbb{P}(v_1\cdots v_j)$$
$$= 2^k + \sum_{j=1}^{k-1} 1 = 2^k + k - 1 \quad \blacksquare$$

In [4, 5], Guibas and Odlyzko devised a combinatorial method based on formal languages in order to obtain the generating functions counting the number of texts with a fixed number of occurrences of a given pattern w according to their size. It leads to the generating function $O_w(z)$ and $N_w(z)$

$$O_w(z) = \frac{z^k}{(c(z)(1-2z)+z^k)^2},$$
$$N_w(z) = \frac{c(z)}{c(z)(1-2z)+z^k},$$

where k is the length of w.

When no ambiguity arises, we abbreviate $O_w(z)$ and $N_w(z)$ by $O(z)$ and $N(z)$. Basing ourself on Guibas and Odlyzko's method, we obtain the probabilistic versions $\mathcal{O}(z)$ and $\mathcal{N}(z)$ of $O(z)$ and $N(z)$, these generating functions count texts with their probability, so

$$\mathcal{O}(z) := \sum_{n\geq 0} \mathbb{P}_n(\widehat{N}_w = 1)z^n = \frac{p_w z^k}{(\widehat{c}(z)(1-z)+p_w z^k)^2}, \tag{7}$$

and

$$\mathcal{N}(z) := \sum_{n\geq 0} \mathbb{P}_n(\widehat{N}_w = 0)z^n = \frac{\widehat{c}(z)}{\widehat{c}(z)(1-z)+p_w z^k}. \tag{8}$$

The next step will be to extract the coefficient of order n in these probabilised generating functions.

2.2. Complex analysis

In order to isolate the dominant pole ρ of the generating functions $\mathcal{O}_w$ and $\mathcal{N}_w$, we use Rouché's theorem. The adequate contour is a circle $\mathcal{C}$ centered at the origin with a radius R depending on the position of the first non-trivial 1 in w's autocorrelation polynomial.

For example, if $c_1 = 1$ we choose $R = 0.5(1+1/p)$, then the disc of radius R contains ρ as unique pole of the generating function. No general formula has yet been devised for the radius, but we are assured of its existence by Pringsheim's theorem.

The contour is also the one we use for an application of Cauchy's theorem. There is only one pole to the generating function $\mathcal{O}(z)$ (or $\mathcal{N}(z)$) inside the disc so that

$$\frac{1}{2i\pi}\int_{\mathcal{C}}\frac{\mathcal{O}(z)}{z^{n+1}}dz = \mathrm{Res}\left(\frac{\mathcal{O}(z)}{z^{n+1}};0\right) + \mathrm{Res}\left(\frac{\mathcal{O}(z)}{z^{n+1}};\rho\right), \tag{9}$$

where $\mathrm{Res}(f,a)$ means the residue of the function f at a.

The modulus of $\mathcal{O}(z)$ can be bounded on the circle $\mathcal{C}$. Furthermore, the residue in 0 of $\mathcal{O}(z)/z^{n+1}$ is the coefficient of order n of $\mathcal{O}(z)$. By developping the generating function in Laurent series near the pole ρ, we obtain the residue at ρ:

$$\mathrm{Res}\left(\frac{\mathcal{O}(z)}{z^{n+1}};\rho\right) = \frac{2^{-k}}{F'(\rho)^3}\rho^{k-n-2}((k-(n+1))F'(\rho) - \rho F''(\rho)),$$

where $F(z) = \widehat{c}(z)(1-z) + p_w z^k$.

2.3. Approximation

This subsection is dedicated to finding a good approximation for the value of the dominant pole ρ. We recall that ρ is defined as the solution of smallest modulus of

$$F(z) = \widehat{c}(z)(1-z) + p_w z^k = 0. \tag{10}$$

Furthermore, due to Pringsheim's theorem [3], we know that ρ is positive real.

Since $F(1) = p_w$ is close to zero on all but a few patterns (the probability decreases exponentially with the size of the word w), and we are dealing with polynomials, ρ is greater than 1 but close enough to it.

Let's introduce α_1 such that $\rho = 1 + \alpha_1$. We know that α_1 is positive, and satisfies

$$\widehat{c}(1+\alpha_1)\alpha_1 + p_w(1+\alpha_1)^k = 0. \tag{11}$$

It is hard to solve this equation to get α_1, so we introduce α such that

$$\widehat{c}(1+\alpha)\alpha + p_w = 0.$$

α is close to α_1 since only small terms have been omitted from (11).

Using Rolle's theorem, we obtain

$$\widehat{c}(1+\alpha) = \widehat{c}(1) + \alpha\widehat{c}'(\beta) = \widehat{c}(1) + p_w\frac{\widehat{c}'(\beta)}{\widehat{c}(1+\alpha)},$$

for $\beta \in]1, 1+\alpha[$. Since the quantity $\widehat{c}'(\beta)/\widehat{c}(1+\alpha)$ is bounded by a constant, and p_w is very small compared to $\widehat{c}(1)$, one has

$$\widehat{c}(1+\alpha) \simeq \widehat{c}(1).$$

Finally, in the residue from the previous section, the ρ^{-n} term becomes

$$\rho^{-n} = (1+\alpha_1)^{-n} \simeq (1+\alpha)^{-n} \simeq \exp(-n\alpha) \simeq \exp\left(-\frac{np_w}{\widehat{c}(1)}\right).$$

3. Splitting in four

From now on, we only deal with the external path length parameter since the methods used for this parameter apply similarly to the size.

Let Δ be the difference between the mean of the external path length P for a trie under Poisson model of parameter n and for a suffix tree on n suffixes:

$$\Delta := \mathbb{E}^{t}_{\mathcal{P}(n)}(P) - \mathbb{E}_n(P).$$

Collecting results from preceding sections, we obtain

$$\Delta = \sum_{w \in \mathcal{M}^*} \mathbb{P}_n(\widehat{N}_w = 1) - \mathbb{P}^{\mathrm{t}}_{\mathcal{P}(n)}(N_w = 1),$$

where $\mathbb{P}^{\mathrm{t}}_{\mathcal{P}(n)}(N_w = 1)$ is the probability of the event $N_w = 1$ taken over all tries built on set of size z, where z follows the Poisson law $\mathcal{P}(n)$ of parameter n. The previous section has provided the asymptotic behavior of $\mathbb{P}_n(\widehat{N}_w = 1)$ and known results from poissonisation [2, 9] allow us to write informally

$$\Delta \simeq \sum_{w \in \mathcal{M}^*} np_w \left(\exp\left(-\frac{np_w}{\widehat{c}(1)} \right) - \exp(-np_w) \right). \tag{12}$$

Remark: Since $\widehat{c}(1) \geq 1$, Δ is a positive quantity, hence asymptotically and on average the path length is longer for a trie than for a suffix tree.

The remainder of this section consists of a delicate subdivision of the set of all patterns in order to control the asymptotic growth that each of these subsets brings to the sum (12). In [6], it was shown that the sum Δ was of order $O(n^{1-\epsilon})$ for some unspecified $\epsilon > 0$, and we want to obtain an explicit bound.

The function $x \to x\exp(-x)$ dominates the behavior of the sum Δ. A perusal of this function's graph induces a three-part splitting of the set of all patterns on whether np_w tends to infinity, to zero or remains "almost constant"; the latter will also be cut into two according to how $\widehat{c}_w(1)$ is close to 1.

3.1. Small sizes

First, we focus on patterns of small sizes, which are in relatively small number. Bounding crudely their contribution to the sum Δ by the product of their number by the worse they grow will be sufficient to prove they do not contribute much in Δ's growth.

We define the small-sized words as those complying with

$$|w| = k \leq \frac{5}{6}\log_{1/q} n = \frac{5}{6} C_q \log n =: k_s(n),$$

for $C_q := (\log 1/q)^{-1}$.

Intuitively, a pattern of small size is one that satisfies $np_w \to \infty$; or in a more quantitative approach

$$np_w \geq n^{1/6}. \tag{13}$$

I call a *slice* of patterns the set of all patterns of a given length. It is more comfortable to handle slices, so the definition for small-sized patterns means those in slices satisfying (13).

We have a binary alphabet, thus the number of patterns of size smaller than $k_s(n)$ is of order $n^{5C_q/6}$ and for any small-sized pattern

$$\begin{aligned} np_w \left(\exp\left(-\frac{np_w}{\widehat{c}(1)} \right) - \exp(-np_w) \right) &\leq n^{1/6} \exp(-n^{1/6}/\widehat{c}(1)) \\ &\leq n^{1/6} \exp(-(1-p)n^{1/6}). \end{aligned}$$

Finally, the patterns of small size contribute to Δ less than

$$n^{5C_q/6} n^{1/6} \exp(-(1-p)n^{1/6}),$$

and due to the dominance of the exponential decrease, to $o(1)$ (this suffices for our goals).

3.2. Large sizes

This part deals with patterns of large size, defined as those whose respective slices satisfies the property $np_w \le \frac{1}{\sqrt{n}}$. The intuition is to catch patterns with $np_w \to 0$, but we refine this condition quantitatively into $np_w \le \frac{1}{\sqrt{n}}$ before resorting to slices.

These patterns indeed are of large sizes: for a symmetric source, for example, $np_w = n2^{-k} \le 1/\sqrt{n}$, implies that k, the length of w, satisfies $k \ge 1.5 \log n$. This definition translates on the length of the patterns into

$$k \ge 1.5 \log_{1/p} n = 1.5 C_p \log n =: k_l(n).$$

With this definition, all large patterns obey $np_w \to 0$; so a Taylor expansion of the function $x \to x\exp(-x)$ near zero yields

$$\sum_{k \ge k_l(n)} \sum_{w \in \mathcal{M}^k} np_w \left(\exp\left(-\frac{np_w}{\widehat{c}(1)} \right) - \exp(-np_w) \right) \simeq \sum_{k \ge k_l(n)} \sum_{w \in \mathcal{M}^k} (np_w)^2 \left(1 - \frac{1}{\widehat{c}(1)} \right).$$

Obviously there are an infinite number of large patterns, which prevents us from using a brute-force majoration like for the small-sized patterns. However, one has

$$\begin{aligned} \sum_{w \in \mathcal{M}^k} p_w^2 &= \sum_{i=0}^{k} \binom{k}{i} (p^i q^{k-i})^2 \\ &= (p^2+q^2)^k =: A_p^k, \end{aligned} \tag{14}$$

for a constant A_p smaller than 1 and depending only on p.

Furthermore we have already seen that $1 \le \widehat{c}(1) \le 1/(1-p)$, hence

$$\sum_{k \ge k_l(n)} \sum_{w \in \mathcal{M}^k} np_w \left(\exp\left(-\frac{np_w}{\widehat{c}(1)} \right) - \exp(-np_w) \right) = O\left(\sum_{k \ge k_l(n)} n^2 A_p^k \right). \tag{15}$$

The largest $\sum_{k \ge k_l(n)} n^2 A_p^k$ can grow is in the case of a symmetric source. This case brings asymptotically an $O(\sqrt{n})$ contribution to the sum Δ; but for other letter occurrence probabilities, we can improve up to a $O(1/n)$ growth.

3.3. Periodic patterns

We introduce $B_k := \{w : |w| = k, \quad \widehat{c}(1) \ge 1 + 2^{-k/2}\}$ as the set of *periodic* patterns of size k. This part aims at patterns of intermediate size (neither small nor large) with the additional constraint they are periodic. We will abusively refer to these patterns as periodic.

A periodic pattern has the first non-trivial 1 in its autocorrelation polynomial for a small index j, and therefore w is formed of repetitions of its suffix of length j. For these patterns, the second term in $\widehat{c}(1)$ is the probability of the suffix of size j of w. But since j is small, the probability is large and $\widehat{c}(1)$ is relatively far from 1.

There are relatively few periodic patterns in B_k:

Lemma 3.1.

$$\#B_k < k2^{k/2}. \tag{16}$$

Proof: We start by partitioning the patterns of size k into two:

$$\sum_{w\in\mathcal{M}^k} \widehat{c}(1) = \sum_{w\in B_k} \widehat{c}(1) + \sum_{w\notin B_k} \widehat{c}(1).$$

For the patterns in B_k, $\widehat{c}(1) \geq 1+2^{-k/2}$ and for the others, $\widehat{c}(1) \geq 1$. Using Lemma 1, we get

$$\sum_{w\in\mathcal{M}^k} \widehat{c}(1) = 2^k + k - 1 \geq \#B_k.(1+2^{-k/2}) + 1.(2^k - B_k) = \#B_k 2^{-k/2} + 2^k \quad \blacksquare$$

From there we bound the contribution of intermediate and periodic patterns

$$\begin{aligned}
\Delta_p := & \sum_{k=k_s(n)}^{k_l(n)} \sum_{w\in B_k} np_w \left(\exp\left(-\frac{np_w}{\widehat{c}(1)}\right) - \exp(-np_w)\right) \\
\leq & \sum_{k=k_s(n)}^{k_l(n)} \#B_k \max_{w\in B_k} \left\{ np_w \exp\left(-\frac{np_w}{\widehat{c}(1)}\right)\right\} \\
< & K \sum_{k=k_s(n)}^{k_l(n)} k2^{k/2},
\end{aligned}$$

where K is any upper bound on the function $x \to x\exp(-x/\widehat{c}(1))$ on positive reals.

We finally obtain a contribution of order $O(n^{0.75C_p}\log n)$ for the periodic patterns. Since we are looking for a sublinear contribution to Δ, this necessitates $0.75C_p < 1$, hence $p < 2^{-0.75} \simeq .5946035575$.

The $2^{-0.75}$ limiting value for p depends on the arbitrary (but smaller than 1) factor defining the bound $k_l(n)$ (here for example this factor is 1.5). We could extend the boundary value of p to $1/\sqrt{2}$ at the expense of a worse error term.

3.4. Aperiodic patterns

The *aperiodic* patterns are those remaining, they are of intermediate sizes (between $k_l(n)$ and $k_s(n)$) and not belonging to the set B_k. For these, $\widehat{c}(1)$ is very close to 1 hence the difference between $np_w\exp(-np_w)$ and $np_w\exp(-np_w/\widehat{c}(1))$ is small.

Since $w \notin B_k$, one has

$$\frac{1}{\widehat{c}(1)} \geq \frac{1}{1+2^{-k/2}} \geq 1 - 2^{-k/2},$$

so that

$$np_w\left(\exp\left(-\frac{np_w}{\widehat{c}(1)}\right) - \exp(-np_w)\right) \leq np_w e^{-np_w}\left(e^{np_w 2^{-k/2}} - 1\right).$$

We are going to use a Taylor expansion of the exponential function near zero, but in order for the expansion to apply we need $np_w 2^{-k/2} \to 0$ for all aperiodic patterns; this leads to the condition $p < p_0 \simeq 0.5469205467$, where p_0 is the unique real solution to

$$\left(\frac{p}{\sqrt{2}}\right)^{5/6} + p - 1 = 0.$$

So, for $p < 0.54$, we can use a Taylor expansion and since $B_k < 2^k$, we derive

$$\begin{aligned}\Delta_a &:= \sum_{k=k_s(n)}^{k_l(n)} \sum_{w \notin B_k} np_w \left(\exp\left(-\frac{np_w}{\widehat{c}(1)}\right) - \exp(-np_w)\right) \\ &\leq \sum_{k=k_s(n)}^{k_l(n)} 2^k (np_w)^2 e^{-np_w} 2^{-k/2} \\ &\leq \sum_{k=k_s(n)}^{k_l(n)} \beta 2^{k/2},\end{aligned}$$

where $\beta = 4e^{-2}$ is the maximum value of $x \to x^2 \exp(-x)$ over the positive reals.

Hence the contribution of the aperiodic patterns to Δ is $O(n^{0.75C_p})$. Similarly to the periodic case, we could increase the upper bound on p up to $2 - \sqrt{2} \simeq 0.5857$ at the expense of a less precise error term.

4. Conclusion

Each subset of patterns contributes less than $O(n^{0.85})$ to the difference Δ. Hence the asymptotic for the external path length (resp. size) of a trie and of a suffix tree only differ by a small quantity. Therefore we have obtained:

Theorem 4.1. *For a suffix tree built on the first n suffixes of a string produced by a memoryless (p,q)-source, and for $p < 0.54$, the mean of the external path length satisfies asymptotically*

$$\frac{n \log n}{h} + (K + \epsilon(n))n + O(n^{0.85}), \tag{17}$$

and the size

$$\frac{n}{h}(1 + \epsilon'(n)) + O(n^{0.85}). \tag{18}$$

where ϵ and ϵ' are oscillating functions of very small modulus centered in 0.

Future research related to this work includes: providing a larger range for the probability p (if possible the whole [0,1] interval); applying this method to other parameters of the suffix tree like the fill-up level, the profile or the height; finally, extending the source model to the powerful dynamical framework introduced by Vallée [10], as it has been done for tries in [2].

References

[1] Clément, J. *Arbres Digitaux et Sources Dynamiques.* Thèse de doctorat, Université de Caen, Sept. 2000.

[2] Clément, J., Flajolet, P., and Vallée, B. Dynamical sources in information theory: A general analysis of trie structures. *Algorithmica 29*, 1/2 (2001), 307–369.

[3] Flajolet, P., and Sedgewick, R. The average case analysis of algorithms: Complex asymptotics and generating functions. Research Report 2026, Institut National de Recherche en Informatique et en Automatique, 1993. 100 pages.

[4] Guibas, L. J., and Odlyzko, A. M. Periods in strings. *Journal of Combinatorial Theory,* Series A *30* (1981), 19–42.

[5] GUIBAS, L. J., AND ODLYZKO, A. M. String overlaps, pattern matching, and non-transitive games. *Journal of Combinatorial Theory. Series A 30*, 2 (1981), 183–208.

[6] JACQUET, P., AND SZPANKOWSKI, W. Autocorrelation on words and its applications: analysis of suffix trees by string-ruler approach. *Journal of Combinatorial Theory. Series A 66*, 2 (1994), 237–269.

[7] KNUTH, D. E. *The Art of Computer Programming*, vol. 3: Sorting and Searching. Addison-Wesley, 1973.

[8] MAHMOUD, H. M. *Evolution of Random Search Trees.* John Wiley, New York, 1992.

[9] SZPANKOWSKI, W. *Average-Case Analysis of Algorithms on Sequences.* John Wiley, New York, 2001.

[10] VALLÉE, B. Dynamical sources in information theory: Fundamental intervals and word prefixes. *Algorithmica 29*, 1/2 (2001), 262–306.

[11] ZIV, J., AND LEMPEL, A. A universal algorithm for sequential data compression. *IEEE Transactions on Information Theory*, IT-23 (1977), 337–343.

Julien Fayolle
INRIA, Projet ALGO
78153 Le Chesnay Cedex
France
julien.fayolle@inria.fr

Trends in Mathematics, © 2004 Birkhäuser Verlag Basel/Switzerland

Arms and Feet Nodes Level Polynomial in Binary Search Trees

Eric Fekete

ABSTRACT: *We define two types of external nodes in binary search trees, the arms and the feet, and we study the vector whose coordinates are the two associated level polynomials. It appears that a so-called growth matrix A describes the evolution of the process. A spectrum analysis of A both provides the average and an almost sure asymptotics of this vector.*

1. Introduction

In the whole paper we are concerned with binary trees whose nodes are labelled by the elements of

$$\mathbb{U} := \maltese \cup \bigcup_{n\geq 1} \{0,1\}^n ,$$

the set of finite words on the alphabet $\{0,1\}$ (with $\maltese$ as empty word). A *complete binary tree* $\mathcal{T}$ is a finite subset of $\mathbb{U}$ such that

$$\begin{cases} \maltese \in \mathcal{T}, \\ \text{if } uv \in \mathcal{T} \text{ then } u \in \mathcal{T}, \\ u1 \in \mathcal{T} \Leftrightarrow u0 \in \mathcal{T}. \end{cases}$$

The elements of $\mathcal{T}$ are called *nodes*, and $\maltese$ *root*; $|u|$, number of letters in u, is the *depth* of u (with $|\maltese| = 0$). We write **BinTree** for the set of complete binary trees.

A tree $\mathcal{T} \in$ **BinTree** can be described by giving the set $\partial\mathcal{T}$ of its *leaves* (also called external nodes), which are the nodes with no descendants. The nodes of $\mathcal{T}\backslash\partial\mathcal{T}$ are called *internal* nodes.

Here, we are concerned with binary search trees. A quick definition is the following: a binary search tree (BST) process is a sequence $(\mathcal{T}_n, n \geq 0)$ of complete binary trees, where $\mathcal{T}_n$ has n internal nodes, which grows by successive insertions of keys, under the so-called random permutation model. Let us describe the dynamics of the sequence of trees. The tree $\mathcal{T}_0$ is reduced to a leaf. The tree $\mathcal{T}_{n+1}$ is obtained from $\mathcal{T}_n$ replacing one of its $n+1$ leaves by an internal node and thus creating two new leaves. The insertion is done uniformly on the leaves, which means with probability $1/(n+1)$ (for a detailed description, see Mahmoud [7]).
The internal nodes contain data and the leaves represent free memory. All the following quantities are random variables on the space **Bintree**, endowed with the probability given by the random permutation model.
To study the shape of these trees, it is usual to define the profile of the tree $\mathcal{T}_n$ by the collection

$$X_{n,k} := \#\{u \in \partial\mathcal{T}_n, |u| = k\} \quad , \quad k \geq 1 ,$$

counting the number of leaves of $\mathcal{T}_n$ at each level. The profile is encoded by the so-called level polynomial defined for $z \in \mathbb{C}$ by

$$W_n(z) := \sum_{k=0}^{n} X_{n,k} z^k .$$

Because of the dynamics of the tree process, this polynomial, renormalized by its expectation, is a $\mathcal{F}_n$-martingale, where $\mathcal{F}_n$ is the σ-field generated by all the events $\{u \in \mathcal{T}_j\}_{j \leq n, u \in \mathbb{U}}$. More precisely, we call BST martingale

$$\mathcal{M}_n(z) = \frac{W_n(z)}{C_n(z)} = \frac{1}{C_n(z)} \sum_k X_{n,k} z^k = \frac{1}{C_n(z)} \sum_{u \in \partial \mathcal{T}_n} z^{|u|}$$

where $C_0(z) = 1$ and

$$C_n(z) := \mathbb{E}(W_n(z)) = \prod_{k=0}^{n-1} \frac{k+2z}{k+1} = (-1)^n \binom{-2z}{n}, \qquad n \geq 1 .$$

It is proved (in Chauvin & al. [1, 2] and Jabbour [6]) that in the supercritical range $z \in (z_c^-, z_c^+)$, this martingale converges in L^1 to a nondegenerate limit $\mathcal{M}_\infty(z)$ and converges a.s. to 0 elsewhere, in particular for the critical values z_c^- and z_c^+. These values are the two positive solutions of equation $2z \log z - 2z + 1 = 0$ ($z_c^- = 0.186...$ and $z_c^+ = 2.155...$). The BST martingale and the level polynomial are the key tools to study the profile of binary search trees.

The motivation of this paper is to refine the study of the profile of binary search trees. Following Dekking & al. [4] we define two kinds of external nodes of a binary search tree as the ones whose brother is an internal node (the *arms*) and the ones whose brother is an external node too (the *feet*). We can already notice that the feet go in pairs and the arms do not. Figure 1 shows a binary search tree whose arms are represented by + and feet by ×. Notice that in computer science a binary search tree is a set of pointers. Each internal node contains 0, 1 or 2 pointers, pointing at each of his descendant only if the corresponding subtree contains at least 2 internal nodes. Thus, every internal node (except the root of the tree) whose descendants are not two feet, receives a pointer from his father. Thus the cost of the tree, which is the total number of pointers, depends on the number of feet.
For all $k \geq 0$ and $n \geq 0$ we first define two random variables $X^A_{n,k}$ and $X^F_{n,k}$ which

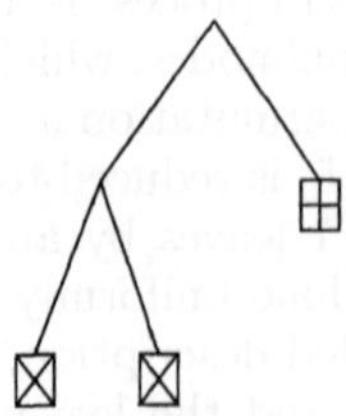

Figure 1.1. *Arms and feet of a binary search tree*

respectively count the number of arms and feet of depth k in a binary search tree of size n. Of course if $k > n$ we have $X^A_{n,k} = X^F_{n,k} = 0$. One can also notice that

$X^A_{n,k} + X^F_{n,k} = X_{n,k}$. We now define two random polynomials, viewed as complex functions, which will help us to study the profile of the binary search tree. We call them respectively arms and feet polynomials:

$$W^A_n(z) = \sum_{k=0}^{n} X^A_{n,k} z^k \quad \text{and} \quad W^F_n(z) = \sum_{k=0}^{n} X^F_{n,k} z^k.$$

Notice that $W^A_n(z) + W^F_n(z) = W_n(z)$.

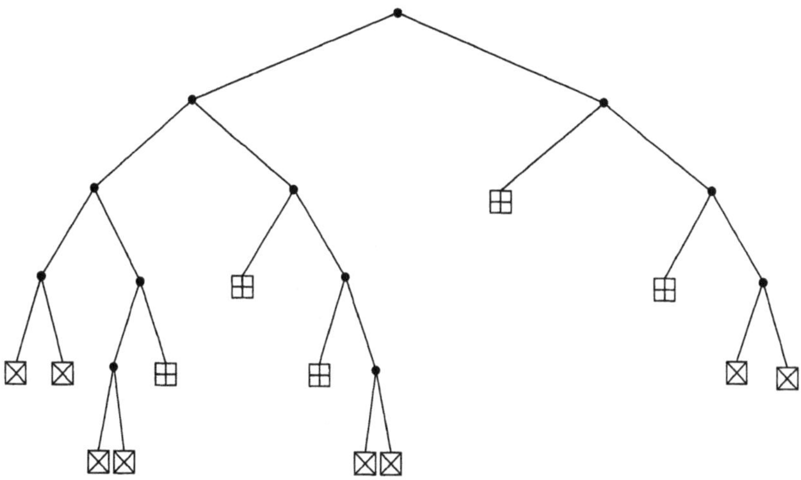

Figure 1.2. *A binary search tree of size 12 with*

$$W^A_n(z) = z^2 + 2z^3 + 2z^4 \text{ and } W^F_n(z) = 4z^4 + 4z^5.$$

We lastly define the vector

$$\mathbf{W}_n(z) = \begin{pmatrix} W^A_n(z) \\ W^F_n(z) \end{pmatrix}$$

whose coordinates are the arms and the feet polynomials defined above and we are interested in the asymptotics of $\mathbf{W}_n(z)$. Remark that $\mathbf{W}_n(z)$ is a powerful object since it permits to count the number of arms and feet nodes for each level. Moreover, taking $z = 1$, the coordinates of $\mathbf{W}_n(z)$ count the total number of arms and feet nodes in the tree.

As in Dekking & al. [4] or Jabbour [6] we use a so-called dynamical or forward method, which means that we take advantage of the evolution of the process between n and $n+1$ (see section 2). This method is complementary to the so-called "divide and conquer" method which takes advantage of similarity and independence between the subtrees of the root. In such a forward method, martingale properties help to find strong limiting results even if limiting distributions are not explicit (it is the case here). For simplicity and for possible generalizations we choose to work in a vectorial frame (in dimension 2 here) instead of working with generating functions as in Dekking & al. [4]. Notice that Dekking & al. in [5] work on more general motives than arms and feet, but using non polynomial vectors and counting the total number of motives instead of counting level by level. It appears in sections 2 and 3 that the spectrum of the matrix driving the evolution of the process plays a crucial role. It is the reason why the study runs along the same lines as in [3] for m-ary search trees. In section 4 we finally prove our main result on $\mathbf{W}_n(z)$:

Theorem 1.1. *For every* $z \in \overline{B}_{\mathbb{C}}(0,1) \cap \{z : \Re(z) > \frac{3}{4}\}$,

$$\frac{\mathbf{W}_n(z)}{C_n(z)} - \mathcal{M}_\infty(z)\mathbf{u}(z) \xrightarrow{n\to\infty} 0 \text{ in } L^2 \text{ and a.s.}$$

where $\mathbf{u}(z) = \frac{1}{2z+1}\begin{pmatrix} 1 \\ 2z \end{pmatrix}$, $\mathcal{M}_\infty(z)$ *is the limit of the BST martingale* $\mathcal{M}_n(z)$, $\Re(z)$ *is the real part of* z *and* $\overline{B}_{\mathbb{C}}(z_0, r)$ *the closed ball with center* z_0 *and radius* r *(*$B_{\mathbb{C}}(z_0,r)$ *is the open corresponding ball).*

Since $X^A_{n,k}$ and $X^F_{n,k}$ are the coefficients of the two polynomials $W^A_n(z)$ and $W^F_n(z)$, these results should help to find the actual profile of arms and feet in binary search trees.

Taking $z = 1$ in the theorem, the coordinates of $\mathbf{W}_n(1)$ respectively count the total number of arms and feet in $\mathcal{T}_n$. We have the following strong law of large numbers:

Corollary 1.2.

$$W^A_n(1) - \frac{1}{3}(n+1) \xrightarrow{n\to\infty} 0 \text{ almost surely}$$

and

$$W^F_n(1) - \frac{2}{3}(n+1) \xrightarrow{n\to\infty} 0 \text{ almost surely.}$$

2. Evolution of the process

To describe the evolution of the process $(\mathcal{T}_n)_{n\in\mathbb{N}}$ we denote by K_n the n^{st} key inserted, D_n the insertion depth of K_n, $D^A_n = D_n.\mathbb{1}_{\{K_n \text{ is inserted in an arm}\}}$ and $D^F_n = D_n.\mathbb{1}_{\{K_n \text{ is inserted in a foot}\}}$.

We recall that in a given binary search tree $\mathcal{T}_n$ with n internal nodes all the leaves have equal probability to receive K_{n+1}. Then

$$\mathbb{P}(D^A_{n+1} = k|\mathcal{T}_n) = \frac{X^A_{n,k}}{n+1} \tag{1}$$

$$\mathbb{P}(D^F_{n+1} = k|\mathcal{T}_n) = \frac{X^F_{n,k}}{n+1}; \tag{2}$$

and we have two cases for the insertion of K_{n+1}:

If K_{n+1} is inserted in an arm node at level k we loose an arm at this level and we win two feet at level $k+1$. If K_{n+1} is inserted in a foot node at level k we loose two feet and we win an arm at this level and we win two feet at level $k+1$. Thus we can describe the evolution of the process as a random walk:

$$\mathbf{W}_{n+1}(z) = \mathbf{W}_n(z) + \mathbf{\Delta}_{n+1}(z) \tag{3}$$

where

$$\mathbf{\Delta}_{n+1}(z) = \begin{cases} \begin{pmatrix} -z^k \\ 2z^{k+1} \end{pmatrix} & if \quad D^A_{n+1} = k \\ \begin{pmatrix} z^k \\ -2z^k + 2z^{k+1} \end{pmatrix} & if \quad D^F_{n+1} = k\,. \end{cases}$$

Calling

$$\delta_1(z) = \begin{pmatrix} -1 \\ 2z \end{pmatrix} \text{ and } \delta_2(z) = \begin{pmatrix} 1 \\ 2z-2 \end{pmatrix}$$

we get

$$\mathbf{\Delta}_{n+1}(z) \quad = \quad \sum_{k=0}^{n} z^k \delta_1(z) \mathbb{1}_{\{D^A_{n+1}=k\}} + z^k \delta_2(z) \mathbb{1}_{\{D^F_{n+1}=k\}}.$$

Taking the conditional expectation in (3) and recalling the evolution rules (1) and (2) one gets

$$\begin{aligned} \mathbb{E}^{\mathcal{F}_n}(\mathbf{W}_{n+1}(z)) &= \mathbf{W}_n(z) + \sum_{k=0}^{n} z^k \delta_1(z) \frac{X^A_{n,k}}{n+1} + z^k \delta_2(z) \frac{X^F_{n,k}}{n+1} \\ &= \mathbf{W}_n(z) + \delta_1(z) \frac{W^A_n(z)}{n+1} + \delta_2(z) \frac{W^F_n(z)}{n+1} \\ &= \mathbf{W}_n(z) + \frac{A(z)}{n+1} \mathbf{W}_n(z), \end{aligned}$$

where $A(z)$ is the endomorphism on $\mathbb{C} \times \mathbb{C}$ defined for any $\mathbf{Y} \in \mathbb{C}^2$ by

$$A(z)\mathbf{Y} = \delta_1(z) y_1 + \delta_2(z) y_2 \,,$$

where (y_1, y_2) are the coordinates of $\mathbf{Y}$ in the canonical basis of $\mathbb{C}^2$. We denote by the same $A(z)$ the endomorphism and its matrix, $A(z) = \begin{pmatrix} -1 & 1 \\ 2z & 2z-2 \end{pmatrix}$.
We finally obtain

$$\mathbb{E}^{\mathcal{F}_n}(\mathbf{W}_{n+1}(z)) = \left(Id + \frac{A(z)}{n+1} \right) \mathbf{W}_n(z) \tag{4}$$

and taking the expectation:

$$\mathbb{E}(\mathbf{W}_n(z)) = \Gamma_n . \mathbf{W}_0(z), \tag{5}$$

where

$$\Gamma_n = \prod_{k=1}^{n} (Id + \frac{A(z)}{k}).$$

Notice that $\mathbf{W}_0(z)$ is nonrandom, one easily gets $\mathbf{W}_0(z) = \begin{pmatrix} 1 \\ 0 \end{pmatrix}$.
At this point it is clear that the properties of the spectrum of $A(z)$ drive the asymptotic behaviour of the expectation $\mathbb{E}(\mathbf{W}_n(z))$ and the expansion of $\mathbf{W}_n(z)$.

3. Average case analysis

The eigenvalues of $A(z)$ are $\lambda_1(z) = 2z - 1$ and $\lambda_2(z) = -2$. We assume from now on $z \neq -\frac{1}{2}$ so that $\lambda_1(z) \neq \lambda_2(z)$ and $A(z)$ is diagonalizable. For each eigenvalue $\lambda(z)$ we denote by $\pi_{\lambda(z)}$ the projection of $\mathbb{C}^2$ on the eigenspace $ker(A(z) - \lambda(z)Id)$ relatively to the decomposition

$$\mathbb{C}^2 = ker(A(z) - \lambda_1(z)Id) \oplus ker(A(z) - \lambda_2(z)Id).$$

The two following relations are consequences of the definition of $\pi_{\lambda_1(z)}$ and $\pi_{\lambda_2(z)}$, for any $X \in \mathbb{C}^2$ we have

$$\begin{aligned} A(z)X &= \lambda_1(z)\pi_{\lambda_1(z)}X + \lambda_2(z)\pi_{\lambda_2(z)}X \qquad (6)\\ X &= \pi_{\lambda_1(z)}X + \pi_{\lambda_2(z)}X . \qquad (7)\end{aligned}$$

Using (6) we have $\Gamma_n = C_1^n(\lambda_1(z)).\pi_{\lambda_1(z)} + C_1^n(\lambda_2(z)).\pi_{\lambda_2(z)}$, where

$$C_{n_0}^n(\lambda) = \prod_{k=n_0}^{n} \left(1 + \frac{\lambda}{k}\right).$$

Remark that $C_1^n(\lambda_1(z)) = C_n(z)$ defined in section 1. We use these two notations as well.
Let us see that the expectation of $\mathbf{W}_n(z)$ only depends on the projection $\pi_{\lambda_1(z)}$: indeed $C_1^n(-2)$ is zero. For the same reason we study $(\mathbf{W}_n(z))_{n\in\mathbb{N}}$ for $z \in \mathbb{C} \setminus \{-\frac{s}{2},\, s \in \mathbb{N}\}$ (nevertheless one can easily get $\mathbf{W}_n(0)$).
From (7) we get the useful relation

$$\mathbf{W}_n(z) = \pi_{\lambda_1(z)}\mathbf{W}_n(z) + \pi_{\lambda_2(z)}\mathbf{W}_n(z). \qquad (8)$$

This relation leads us to study the two projections of $\mathbf{W}_n(z)$. The above remark on $C_1^n(\lambda_2(z))$ indicates this decomposition could coincide with the asymptotic expansion of $\mathbf{W}_n(z)$. The computation of the expectation confirms this impression: using (8), (5), the linearity of the projection on an eigenspace and the fact that for every eigenvalue $\lambda(z)$ of $A(z)$, the projection $\pi_{\lambda(z)}$ commutes with $A(z)$, one gets $\forall n \geq 2$

$$\begin{aligned} \mathbb{E}(\mathbf{W}_n(z)) &= \pi_{\lambda_1(z)}\Gamma_n\mathbf{W}_0(z) + \pi_{\lambda_2(z)}\Gamma_n\mathbf{W}_0(z)\\ &= C_n(z)\pi_{\lambda_1(z)}\mathbf{W}_0(z).\end{aligned}$$

The computation of $\pi_{\lambda_1(z)}\mathbf{W}_0(z)$ is easy, according to (6) and (7) we obtain

$$\mathbf{u}(z) := \pi_{\lambda_1(z)}\mathbf{W}_0(z) = \frac{1}{2z+1}\begin{pmatrix} 1 \\ 2z \end{pmatrix}.$$

Notice that $\mathbf{u}(z)$ is a nonrandom vector.
Moreover, we have (see for example Titchmarsh [9] p56-58)

$$\begin{aligned} C_1^n(\lambda) &= \frac{\Gamma(n+\lambda+1)}{\Gamma(\lambda+1)\Gamma(n+1)}\\ &= \frac{n^\lambda}{\Gamma(\lambda+1)} + \mathcal{O}(n^{\lambda-1}) \qquad (9)\end{aligned}$$

as n tends to infinity.
We have proved the following proposition.

Proposition 3.1. *For $z \in \mathbb{C} \setminus \{-\frac{s}{2},\ s \in \mathbb{N}\}$ we have*

$$\mathbb{E}(\mathbf{W}_n(z)) = C_n(z)\mathbf{u}(z). \tag{10}$$

Asymptotically, when $n \to \infty$

$$\mathbb{E}(\mathbf{W}_n(z)) = n^{2z-1} \times \frac{1}{\Gamma(2z)}\frac{1}{2z+1}\begin{pmatrix} 1 \\ 2z \end{pmatrix}\left(1+\mathcal{O}(\frac{1}{n})\right). \tag{11}$$

In particular, if $z = 1$, then the coordinates of $\mathbf{W}_n(z)$ count respectively the number of arms and feet in $\mathcal{T}_n$. We recover Mahmoud's results [8]:

$$\mathbb{E}(\mathbf{W}_n(1)) = (n+1)\begin{pmatrix} 1/3 \\ 2/3 \end{pmatrix}.$$

In other words, we have on average one arm for two feet.
Notice that the average case is totally described by the first projection of $\mathbf{W}_n(z)$. In order to get the almost sure asymptotics of $\mathbf{W}_n(z)$, let us now see how the projections behave.

4. Almost sure analysis

An elementary computation with $A(z)$ gives the first projection lemma below. It explicitly computes the first projection in terms of the global leaves polynomial $W_n(z)$.

Lemma 4.1. (first projection lemma)
Let $\mathbf{Y}$ be a vector of $\mathbb{C}^2$ and (y_1, y_2) its coordinates in the canonical basis. We have

$$\pi_{\lambda_1(z)}\mathbf{Y} = (y_1 + y_2)\mathbf{u}(z).$$

Using this result on $\mathbf{W}_n(z)$ we obtain

$$\pi_{\lambda_1(z)}\mathbf{W}_n(z) = W_n(z)\mathbf{u}(z),$$

recalling $W_n(z) = \sum_{k=0}^{n}(X^A_{n,k} + X^F_{n,k})z^k$ is the leaves polynomial of the binary search tree. Thus, from (4) and picking up the results of J.Jabbour [6] and Chauvin & al. [2] we get

Proposition 4.2. *Let $z \in \mathbb{C} \setminus \{-\frac{s}{2},\ s \in \mathbb{N}\}$ and $\mathbf{M}_n(z) := \frac{\pi_{\lambda_1(z)}\mathbf{W}_n(z)}{C_n(z)}$. We define*

$$\mathcal{V} := \cup_{1<q<2}\mathcal{V}_q := \cup_{1<q<2}\{z : f(z,q) > 0\}$$

where

$$f(z,q) = 1 + q(2\Re(z) - 1) - 2|z|^q.$$

The process $(\mathbf{M}_n(z), \mathcal{F}_n)_n$ is a martingale and as $n \to \infty$, $\mathbf{M}_n(z)$ converges, a.s. and in L^1, uniformly on every compact subset C of $\mathcal{V}$ to the random variable $\mathcal{M}_\infty(z)\mathbf{u}(z)$.

Figure 4.1 helps to visualize the domain $\mathcal{V}$. For $q = 2$ domain $\mathcal{V}_q$ is the ball $B_{\mathbb{C}}(1, \frac{\sqrt{2}}{2})$. The more q is near 1 the less the domain is thick.

According to this proposition, the almost sure order of magnitude of the first projection of $\mathbf{W}_n(z)$ is clear. We now want to prove that the second projection is negligible comparing to the first one. This leads us to a computation of moments.

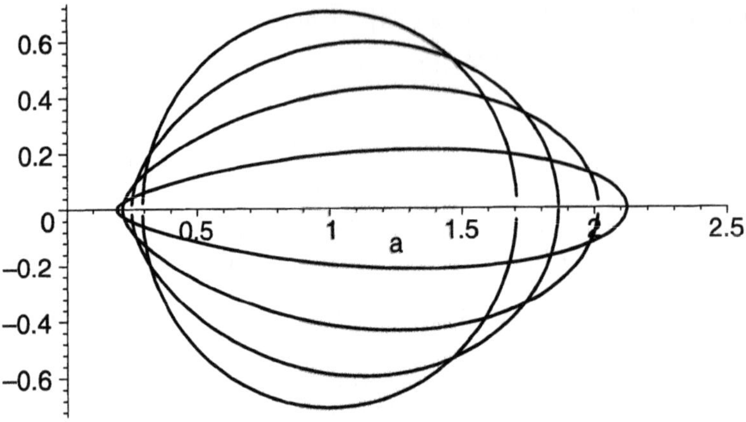

Figure 4.1.

Domains $\mathcal{V}_q$ for $q \in \{1.05, 1.2, 1.5, 2\}$.

Lemma 4.3. (Lemma of moments)
Let $\sigma_i(z)$ denote the real part of the eigenvalue $\lambda_i(z)$ of $A(z)$, $i \in \{1,2\}$ and $< .|. >$ any positive hermitian form on $\mathbb{C}^2$. Let Z be a vector in $\mathbb{C}^2$, $\forall p \in \mathbb{N}$ we have
(i) for every $z \in \overline{B}_{\mathbb{C}}(0,1)$

$$\begin{cases} \mathbb{E}(| < Z|\pi_{\lambda_2(z)}\mathbf{W}_n(z) > |^{2p}) &= \mathcal{O}(n^p) \\ \mathbb{E}(| < Z|\pi_{\lambda_2(z)}\mathbf{W}_n(z) > |^{2p+1}) &= \mathcal{O}(n^{p+1}) \end{cases}$$

(ii) if $z \in \overline{B}_{\mathbb{C}}(0,1)$ and $\mathcal{R}(z) > \frac{3}{4}$ then

$$\begin{cases} \mathbb{E}(| < Z|\pi_{\lambda_1(z)}\mathbf{W}_n(z) > |^{2p}) &= \mathcal{O}(n^{2p\sigma_1(z)}) \\ \mathbb{E}(| < Z|\pi_{\lambda_1(z)}\mathbf{W}_n(z) > |^{2p+1}) &= \mathcal{O}(n^{2p\sigma_1(z)+1}) \end{cases}$$

if $z \in \overline{B}_{\mathbb{C}}(0,1)$ and $\mathcal{R}(z) = \frac{3}{4}$ then

$$\begin{cases} \mathbb{E}(| < Z|\pi_{\lambda_1(z)}\mathbf{W}_n(z) > |^{2p}) &= \mathcal{O}(n^{2p\sigma_1(z)}\log(n)) \\ \mathbb{E}(| < Z|\pi_{\lambda_1(z)}\mathbf{W}_n(z) > |^{2p+1}) &= \mathcal{O}(n^{2p\sigma_1(z)+1}\log(n)) \end{cases}$$

if $z \in \overline{B}_{\mathbb{C}}(0,1)$ and $\mathcal{R}(z) < \frac{3}{4}$ then

$$\begin{cases} \mathbb{E}(| < Z|\pi_{\lambda_1(z)}\mathbf{W}_n(z) > |^{2p}) &= \mathcal{O}(n^p) \\ \mathbb{E}(| < Z|\pi_{\lambda_1(z)}\mathbf{W}_n(z) > |^{2p+1}) &= \mathcal{O}(n^{p+1}) \end{cases}$$

Proof
We prove the results by induction on p.
If $p = 0$ we only have to compute $\mathbb{E}^{\mathcal{F}_n}(| < Z|\mathbf{W}_{n+1}(z) > |)$. Using (3) we almost surely have

$$||\mathbf{W}_{n+1}(z)|| \leq ||\mathbf{W}_n(z)|| + \max_{(i,k)\in\{1,2\}\times\{0..n\}} ||z^k\delta_i(z)||$$

thus assuming $|z| \leq 1$

$$||\mathbf{W}_{n+1}(z)|| \leq ||\mathbf{W}_n(z)|| + \max(||\delta_1(z)||, ||\delta_2(z)||).$$

Therefore, there is some positive constant $c(z)$ such that, almost surely, for every $n \geq 1$,

$$||\mathbf{W}_{n+1}(z)|| \leq c(z)n. \tag{12}$$

The result for $p=0$ follows from this inequality.
Suppose now $p \geq 1$.
Let first write a general result on complex numbers. Let x and y be two complex numbers, then

$$|x+y|^{2p} - |x|^{2p-2}\Re(x\overline{(x+2py)}) = p^2|y|^2|x|^{2p-2} + 2\Re((x^p + px^{p-1}y)\overline{z}) + z\overline{z}$$

where z is a polynomial whose coefficients are nonnegative real numbers and whose degree in x equals $p-2$. Thus bounding above each term of the right part of the equation, we can say that there exists a polynomial P, whose coefficients are nonnegative real numbers and whose degree in the first variable equals $2p-2$, such that

$$|x+y|^{2p} \leq |x|^{2p-2}\Re(x\overline{(x+2py)}) + P(|x|,|y|) \tag{13}$$

Let us start the computation of $\mathbb{E}^{\mathfrak{F}_n}(| < Z|\mathbf{W}_{n+1}(z) > |^{2p})$:

$$\begin{aligned}&\mathbb{E}^{\mathfrak{F}_n}(| < Z|\mathbf{W}_{n+1}(z) > |^{2p})\\ &= \sum_{k=0}^{n} \frac{X_{n,k}^A}{n+1}| < Z|\mathbf{W}_n(z) + z^k\delta_1(z) > |^{2p} + \frac{X_{n,k}^F}{n+1}| < Z|\mathbf{W}_n(z) + z^k\delta_2(z) > |^{2p}.\end{aligned}$$

For every k we apply (13) to $x =< Z|\mathbf{W}_n(z) >$ and $y =< Z|z^k\delta_1(z) >$, afterwards to $x =< Z|\mathbf{W}_n(z) >$ and $y =< Z|z^k\delta_2(z) >$; notice that P is the same one the $2n+2$ times we use (13). We obtain

$$\begin{aligned}\mathbb{E}^{\mathfrak{F}_n}(| < Z|\mathbf{W}_{n+1}(z) > |^{2p}) \leq &\sum_{k=0}^{n} \frac{X_{n,k}^A}{n+1}\Big(| < Z|\mathbf{W}_n(z) > |^{2p-2}\\ &\quad\Re(< Z|\mathbf{W}_n(z) > \overline{(< Z|\mathbf{W}_n(z) > +2p < Z|z^k\delta_1(z) >)})\\ &\quad+P(| < Z|\mathbf{W}_n(z) > |, | < Z|z^k\delta_1(z) > |)\Big)\\ &+\frac{X_{n,k}^F}{n+1}\Big(| < Z|\mathbf{W}_n(z) > |^{2p-2}\\ &\quad\Re(< Z|\mathbf{W}_n(z) > \overline{(< Z|\mathbf{W}_n(z) > +2p < Z|z^k\delta_2(z) >)})\\ &\quad+P(| < Z|\mathbf{W}_n(z) > |, | < Z|z^k\delta_2(z) > |)\Big)\\ \leq\ & | < Z|\mathbf{W}_n(z) > |^{2p-2}\Re(< Z|\mathbf{W}_n(z) > \overline{< Z|(Id + \frac{2p}{n+1}A(z))\mathbf{W}_n(z) >})\\ &+\sum_{k=1}^{n}\Big(\frac{X_{n,k}^A}{n+1}P(| < Z|\mathbf{W}_n(z) > |, | < Z|z^k\delta_1(z) > |)\\ &\quad+\frac{X_{n,k}^F}{n+1}P(| < Z|\mathbf{W}_n(z) > |, | < Z|z^k\delta_2(z) > |)\Big).\end{aligned}$$

Then apply the above inequality to $\pi_{\lambda(z)}^{*}Z$ instead of Z, where f^* denotes the adjoint endomorphism of f relative to $< .|. >$, for any eigenvalue $\lambda(z)$ of $A(z)$ and take the expectation.

$$\begin{aligned}
&\mathbb{E}(|<Z|\pi_{\lambda(z)}\mathbf{W}_{n+1}(z)>|^{2p})\\
\leq\ &(1+\frac{2p\sigma(z)}{n+1})\mathbb{E}(|<Z|\pi_{\lambda(z)}\mathbf{W}_n(z)>|^{2p})\\
&+\sum_{k=1}^{n}\mathbb{E}\Bigg(\frac{X_{n,k}^A}{n+1}P(|<Z|\pi_{\lambda(z)}\mathbf{W}_n(z)>|,|<Z|z^k\pi_{\lambda(z)}\delta_1(z)>|)\\
&\qquad+\frac{X_{n,k}^F}{n+1}P(|<Z|\pi_{\lambda(z)}\mathbf{W}_n(z)>|,|<Z|z^k\pi_{\lambda(z)}\delta_2(z)>|)\Bigg).
\end{aligned}$$

Denoting a_{ij} the coefficients of polynomial P, we have

$$\begin{aligned}
&\sum_{k=1}^{n}\frac{X_{n,k}^A}{n+1}P(|<Z|\pi_{\lambda(z)}\mathbf{W}_n(z)>|,|<Z|z^k\pi_{\lambda(z)}\delta_1(z)>|)\\
=\ &\sum_{k=1}^{n}\frac{X_{n,k}^A}{n+1}\sum_{i=0}^{2p-2}\sum_{j=0}^{2p}a_{i,j}|<Z|\pi_{\lambda(z)}\mathbf{W}_n(z)>|^i|<Z|z^k\pi_{\lambda(z)}\delta_1(z)>|^j\\
=\ &\sum_{i=0}^{2p-2}\sum_{j=0}^{2p}a_{i,j}|<Z|\pi_{\lambda(z)}\mathbf{W}_n(z)>|^i|<Z|\pi_{\lambda(z)}\delta_1(z)>|^j\frac{W_n^A(|z|^j)}{n+1}.
\end{aligned}$$

Assuming $|z|\leq 1$, notice that $\frac{W_n^A(|z|^j)}{n+1}\leq 1$ a.s., and use the non-negativity of the coefficients of polynomial P, so

$$\begin{aligned}
\mathbb{E}(|<Z|\pi_{\lambda(z)}\mathbf{W}_{n+1}(z)>|^{2p}) &\leq (1+\frac{2p\sigma(z)}{n})\mathbb{E}(|<Z|\pi_{\lambda(z)}\mathbf{W}_n(z)>|^{2p})\\
&+\mathbb{E}(P(|<Z|\pi_{\lambda(z)}\mathbf{W}_n(z)>|,|<Z|\pi_{\lambda(z)}\delta_1(z)>|))\\
&+\mathbb{E}(P(|<Z|\pi_{\lambda(z)}\mathbf{W}_n(z)>|,|<Z|\pi_{\lambda(z)}\delta_2(z)>|)),
\end{aligned}$$

and by induction

$$\begin{aligned}
\mathbb{E}(|<Z|\pi_{\lambda(z)}\mathbf{W}_n(z)>|^{2p}) &\leq C_{n_0}^{n}(2p\sigma(z))\Bigg(\mathbb{E}(|<Z|\pi_{\lambda(z)}\mathbf{W}_{n_0}(z)>|^{2p})\\
&+\sum_{k=n_0+1}^{n}\frac{\mathbb{E}(Q(|<Z|\pi_{\lambda(z)}\mathbf{W}_k(z)>|))}{C_{n_0}^{k+1}(2p\sigma(z))}\Bigg), \qquad (14)
\end{aligned}$$

where $n_0=0$ if $\lambda(z)=\lambda_1(z)$ and $n_0=4$ if $\lambda(z)=\lambda_2(z)$ and where Q is a polynomial of degree $2p-2$.

It remains now to be seen whether the series in (14) is convergent or not. Let us detail the end of the computation for $\lambda(z)=\lambda_2(z)$, it is analogous for $\lambda_1(z)$. Assuming the result for all integers $<2p$, the order of magnitude of $\mathbb{E}(Q(|<Z|\pi_{\lambda(z)}\mathbf{W}_k(z)>|))$ is bounded above by the term of highest degree in Q, that is $\mathbb{E}(|<Z|\pi_{\lambda(z)}\mathbf{W}_k(z)>|^{2p-2})$, which is a $\mathcal{O}(k^{p-1})$. Thus according to (9), $C_{n_0}^{k+1}(2p\sigma_2(z))\sim k^{-4p}$ and thus the general term of the series is of size k^{p-1+4p}. The series diverges and this gives the result for the even moments.

For $\lambda_1(z)$ let us just say that the comparison between $\mathcal{R}(z)$ and $\frac{3}{4}$ appears from the comparison between the expectation of $Q(|<Z|\pi_{\lambda_1(z)}\mathbf{W}_k(z)>|^{2p-2})$ and $C_{n_0}^{k+1}(2p\sigma_1(z))$.

Using (12) the result for the odd moments is a consequence of the inequality

$$\mathbb{E}(| < Z|\mathbf{W}_n(z) > |^{2p+1}) \leq \mathbb{E}(| < Z|\mathbf{W}_n(z) > |^{2p}) \times \max_{\Omega} | < Z|\mathbf{W}_n(z) > |,$$

where Ω is the underlying probability space defined as **BinTree** in section 1. □

One can notice that a precise computation can give us the actual size of $\mathbb{E}(| < Z|\pi_{\lambda_1(z)}\mathbf{W}_n(z) > |^2)$ and $\mathbb{E}(| < Z|\pi_{\lambda_2(z)}\mathbf{W}_n(z) > |^2)$. Indeed for $p = 1$ the degree of P in the first variable is zero and we don't have to add the assumption $|z| \leq 1$.

The lemma of moments is used taking (Z_1, Z_2) an orthonormal basis of $\mathbb{C}^2$, so that we get

$$||X||^p \leq | < Z_1|X > |^p + | < Z_2|X > |^p$$

for any integer p, with equality if $p = 2$. We can now prove our main result on $\mathbf{W}_n(z)$.

Proof of Theorem 1

If z belongs to $\overline{B}_{\mathbb{C}}(0,1) \cap \{\mathcal{R}(z) > \frac{3}{4}\}$ then z belongs to $\mathcal{V}$, thus according to proposition 4.2 we have

$$\frac{\pi_{\lambda_1(z)}\mathbf{W}_n(z)}{C_n(z)} = \mathcal{M}_\infty(z)\mathbf{u}(z) + \epsilon_n(z)$$

where $\epsilon_n(z) \xrightarrow{n\to\infty} 0$ almost surely. Furthermore, using lemma 4.3 for $p = 1$, one can prove that $\epsilon_n(z) \xrightarrow{n\to\infty} 0$ in L^2 for all $z \in B_{\mathbb{C}}(1, \frac{\sqrt{2}}{2})$, thus for $z \in \overline{B}_{\mathbb{C}}(0,1) \cap \{\mathcal{R}(z) > \frac{3}{4}\}$. From (8) we get

$$\frac{\mathbf{W}_n(z)}{C_n(z)} - \mathcal{M}_\infty(z)\mathbf{u}(z) = \epsilon_n(z) + \frac{\pi_{\lambda_2(z)}\mathbf{W}_n(z)}{C_n(z)}.$$

We denote $\hat{\epsilon}_n(z) = \frac{\pi_{\lambda_2(z)}\mathbf{W}_n(z)}{C_n(z)}$. We have to prove that $\hat{\epsilon}_n(z) \xrightarrow{n\to\infty} 0$ in L^2 and almost surely.

We obtain the L^2 convergence with lemma 4.3 for $p = 1$ comparing the order of magnitude of $\mathbb{E}(| < Z|\pi_{\lambda_2(z)}\mathbf{W}_n(z) > |^2)$ and $C_n(z)$.

Using Borel-Cantelli lemma one can prove that for any sequence of random variables $(X_n)_{n\in\mathbb{N}}$, if for any $\varepsilon > 0$, $\sum_{n=0}^{\infty} \mathbb{P}(|X_n| > \varepsilon) < \infty$ then $X_n \xrightarrow{n\to\infty} 0$ almost surely. Furthermore the Markov inequality implies that it is sufficient to find an integer $p \geq 1$ such that $\sum_{n=0}^{\infty} |X_n|^{2p}$ is finite.

Using lemma 4.3 we get

$$\mathbb{E}(||\hat{\epsilon}_n(z)||^{2p}) = \mathcal{O}(n^{p(1-2\sigma_1(z))})$$

which is the general term of a convergent series if p is large enough, according to the fact that $1 - 2\sigma_1(z) < 0$ which is equivalent with $\mathcal{R}(z) > \frac{3}{4}$. □

References

[1] B. Chauvin, M. Drmota, and J. Jabbour-Hattab. The profile of binary search trees. *Ann. Appl. Prob.*, 11:1042–1062, 2001.

[2] B. Chauvin, T. Klein, J.F. Marckert, and A. Rouault. Martingales, Embedding and Tilting of Binary Trees. Preprint available at http://fermat.math.uvsq.fr/~ marckert/papers.html, September 2003.

[3] B. Chauvin and N. Pouyanne. m-ary search trees when m > 26: a strong asymptotics for the space requirements, to appear in Random Structures and Algorithm 2004. Preprint available at http://fermat.math.uvsq.fr/~ chauvin.

[4] F. M. Dekking, S. De Graaf, and L. E. Meester. On the node structure of binary search trees. *Trends in Mathematics. Birkhäuser Verlag, Basel*, pages 31–40, 2000.

[5] F. M Dekking and L. E. Meester. An almost sure result for path lengths in binary search trees. *Adv. Appl. Prob.*, 35:363–376, 2003.

[6] J. Jabbour-Hattab. Martingales and large deviations for binary search trees. *Random Structure and Algorithms*, 19:112–127, 2001.

[7] H. Mahmoud. *Evolution of Random Search Trees*. John Wiley, New York, 1992.

[8] H. M. Mahmoud. The expected distribution of degrees in random binary search trees. *Comput. J.*, 29:36–37, 1986.

[9] E. C. Titchmarsh. *Theory of functions.* Oxford university press, 1986 second edition.

Eric Fekete
LAMA Laboratoire de Mathématiques, UMR 8100, Bâtiment Fermat, Université de Versailles Saint Quentin F-78035 Versailles.
fekete@math.uvsq.fr

Trends in Mathematics, © 2004 Birkhäuser Verlag Basel/Switzerland

Random Records and Cuttings in Complete Binary Trees

Svante Janson

ABSTRACT: *We study the number of records in a complete binary tree with randomly labeled vertices or edges. Equivalently, we may study the number of random cuttings required to eliminate a complete binary tree.*

The distribution is, after normalization, asymptotically a periodic function of $\lg n - \lg\lg n$*; thus there is no true asymptotic distribution but a family of limits of different subsequences; these limits are similar to a 1-stable distribution but have some periodic fluctuations.*

1. Introduction

Let each vertex v in a rooted tree T have a random value λ_v attached to it, and assume that these values are i.i.d. with a continuous distribution (so that a.s. there are no ties). Say that a value λ_v is a *record* if it is the smallest value in the path from the root to v. Let $X_v(T)$ denote the (random) number of records. Note that this generalizes the classical record problem (which is the case when T is a path), see for example [9].

Alternatively, we may attach random values to the edges, and let $X_e(T)$ denote the number of edges with record values (along the path from the root).

It is obvious that the choice of common distribution of the labels does not affect the result, and that we as well can count the values that are largest. We can also let the labels be a random permutation of $\{1, \dots, n\}$.

The same random variables appears when we consider random cuttings of the tree T defined as follows, see [6]. Make a random cut by choosing one vertex [edge] at random. Delete this vertex [edge] so that the tree separates into two parts, and keep only the part containing the root. Continue recursively until the root is cut [only the root is left]. Then the (random) number of cuts made is $X_v(T)$ [$X_e(T)$]. (More precisely, these random variables have the same distribution.) This equivalence is shown in [3], where the asymptotic distributions are found for the random trees that can be constructed as conditioned Galton–Watson trees, for example random labelled trees and and random binary trees. See also [6, 1, 7, 8] for earlier results.

We will in this paper study the case of a complete binary tree.

The *complete binary tree* with n vertices has height $m = \lfloor \lg n \rfloor$; it has 2^k vertices of height k, $0 \le k < m$, and $n - 2^m + 1$ vertices of height m, moreover, the vertices of height m have the leftmost positions among the 2^m possible ones, see e.g. [5, page 401]. We denote this rooted tree by T_n, and denote its root by o.

Let $\{x\} := x - \lfloor x \rfloor$ denote the fractional part of a real number x. Further, for a vertex v in a rooted tree, let $h(v)$ be its height (also known as depth), with the root having height 0.

Theorem 1.1. *Suppose that* $n \to \infty$ *such that* $\{\lg n - \lg\lg n\} \to \gamma \in [0,1]$. *Then*

$$\left(X_v(T_n) - \frac{n}{\lg n} - \frac{n \lg\lg n}{\lg^2 n}\right) \Big/ \frac{n}{\lg^2 n} \xrightarrow{\mathrm{d}} -W_\gamma \tag{1}$$

where W_γ *has an infinitely divisible distribution with characteristic function*

$$\mathbb{E}\, e^{\mathrm{i}tW_\gamma} = \exp\Bigl(\mathrm{i}f(\gamma)t + \int_0^\infty \left(e^{\mathrm{i}tx} - 1 - \mathrm{i}tx\mathbf{1}[x<1]\right) d\nu_\gamma(x)\Bigr), \tag{2}$$

where $f(\gamma) := 2^\gamma - 1 - \gamma$ *and the Lévy measure* ν_γ *is supported on* $(0,\infty)$ *and has density*

$$\frac{d\nu_\gamma}{dx} = 2^{\{\lg x + \gamma\}} x^{-2}.$$

The same result holds for $X_e(T_n)$.

We prove Theorem 1.1 in Section 2. The strategy of the proof is to approximate $X_v(T_n)$ and $X_e(T_n)$ by a sum of independent random variables derived from $\{\lambda_v\}$, see Lemma 2.4. We will then apply a classical limit theorem for triangular arrays.

Remark 1.1. *Let* $\widetilde{X}_n$ *denote the left hand side of* (1). *Instead of stating the result for suitable subsequences, we may say that* $\widetilde{X}_n$ *has approximatively the same distribution as* $-W_{\{\lg n - \lg\lg n\}}$ *for large* n; *more precisely, the distance between the two distributions (in for example the Lévy metric) tends to 0 as* $n \to \infty$.

Remark 1.2. *Most records occur at height close to the maximum* $m \approx \lg n$, *simply because almost all vertices are there. On the other hand, it follows from the proof below, see Lemma 2.4 and the proof of Lemma 2.5, that most of the random fluctuations of* $X_v(T_n)$ *or* $X_e(T_n)$ *can be explained by the values at heights close to* $\lg\lg n$. *The explanation is that a few values* λ_v *at these heights will be so small that they significantly reduce the number of records among their descendants. Vertices of smaller height are too few, and there will usually not be any sufficiently small value among them, while vertices of larger height affect only a small proportion of the tree each, and the random effect caused by their values will be wiped out by the law of large numbers.*

Remark 1.3. *It is easy to see* [3] *that* $\mathbb{E}\, X_v(T_n) = \sum_v 1/(h(v)+1)$ *and* $\mathbb{E}\, X_e(T_n) = \sum_{v\neq o} 1/h(v)$; *both sums are easily evaluated as* $n/m + O(n/m^2) = n/\lg n + O(n/\lg^2 n)$. *We see thus from Theorem 1.1 that* $X_v(T_n)$ *and* $X_e(T_n)$ *are concentrated well above their means (at a distance of about* $n \lg\lg n/\lg^2 n$), *so that e.g.* $\mathbb{P}\bigl(X_v(T_n) \le \mathbb{E}\, X_v(T_n)\bigr) \to 0$. *This is connected to the fact that the limit* W_γ *has infinite mean.*

Note also that $\mathbb{E}\, X_e(T_n) - \mathbb{E}\, X_v(T_n) = \sum_{v\neq o}\bigl(h(v)(h(v)+1)\bigr)^{-1} - 1 \sim n/\lg^2 n$, *while there is no similar difference in the limit distribution in Theorem 1.1.*

An explanation of these facts is that the mean is affected by the unlikely event that a vertex close to the root has an extremely small value λ_v, *which would reduce the number of records by a large amount.*

We see that this behaviour makes it impossible to use the method of moments to find the asymptotic distribution in Theorem 1.1, as we did for other trees in [3].

Remark 1.4. *Recall that the Lévy measure* $cx^{-2}\,dx$ *gives a (weakly) 1-stable distribution, see e.g.* [2, XVII.3]; *the measure* ν_γ *is a version of this with periodic fluctuations, so the distribution of* W_γ *is roughly similar to a 1-stable distribution.*

More precisely, we have that if W_γ and W'_γ are independent with the same distribution, then $W_\gamma + W'_\gamma \overset{\mathrm{d}}{=} 2W_\gamma + 2$, as is easily checked from (2), but the corresponding statement for a sum of three copies of W_γ is false.

If we write (2) as $\mathbb{E}\, e^{\mathrm{i}tW_\gamma} = e^{\psi_\gamma(t)}$, it is possible to compute the Fourier coefficients of $\psi_\gamma(t)$ as a function of γ by integrations, using Fubini and some Gamma integrals, and obtain

$$\psi_\gamma(t) = -\frac{\pi}{2}|t| - (\gamma^* - \tfrac{1}{2})\mathrm{i}t \\ - \mathrm{i}t \sum_{n \neq 0} \frac{\Gamma(2\pi \mathrm{i}n/\ln 2 - 1)}{\ln 2 - 2\pi \mathrm{i}n} e^{-\pi^2 n \operatorname{sign} t/\ln 2} |t|^{-2\pi \mathrm{i}n/\ln 2} e^{2\pi n \mathrm{i}\gamma},$$

where γ^* is Euler's constant. We omit the details. This, again, shows the affinity with stable distributions.

The complete binary tree T_n has minimal height among all binary trees with n vertices, but among binary trees with this height, it is maximally unbalanced. The other extreme is the *balanced binary tree* T_n^*, where at each vertex, the two subtrees emanating from it differ in size by at most 1. This tree too has height $m = \lfloor \lg n \rfloor$, and the same number of vertices at each level as T_n. As a companion to Theorem 1.1, we give a similar theorem for T_n^*; note that the results are similar but not identical, which shows that the details of the structure of the tree are important. (If we consider only n of the form $2^k - 1$, $T_n = T$ is a *full binary tree*. Indeed, Theorems 1.1 and 1.2 yield the same result in this case.) In contrast, note that the means of X_v and X_e are the same for T_n and T_n^*, see Remark 1.3.

Theorem 1.2. *Suppose that $n \to \infty$ such that $\{\lg \lg n\} \to \beta \in [0,1]$. Then*

$$\left(X_v(T_n^*) - \frac{n}{\lg n} - \frac{n \lg \lg n}{\lg^2 n}\right) \Big/ \frac{n}{\lg^2 n} \xrightarrow{\mathrm{d}} -W_{1-\beta}$$

where $W_{1-\beta}$ is as in Theorem 1.1. The same result holds for $X_e(T_n^*)$.

The method used below applies also to other binary trees with minimal height, but we leave the details to the reader. Presumably, the method can be used also for a larger class of binary trees, but we have not explored this. In particular, we do not know whether our methods can be used to solve the following problem.

Problem 1.1. *What is the asymptotic distribution of X_v and X_e for a (random) binary search tree?*

2. Proofs

We first treat the case X_v of Theorem 1.1 in detail, and then indicate the small modifications needed for X_e and for T_n^*.

Let $X_n := X_v(T_n)$, and let, for $y > 0$, $X_{n,y}$ be $X_v(T_n) - 1$ conditioned on the root label $\lambda_o = y$, i.e. the number of records in the rest of the tree if we fix the root label (which always is a record).

We will use the notations $m := \lfloor \lg n \rfloor$ (as above) and $l := \lfloor \lg \lg n \rfloor$; we also let $L := \lfloor \frac{3}{2} \lg \lg n \rfloor \approx 3l/2$. We assume that n is so large that $0 < l < L < m$. If a_n are positive numbers and Z_n random variables such that $Z_n/a_n \xrightarrow{\mathrm{p}} 0$ as $n \to \infty$, we write $Z_n = o_p(a_n)$.

In the sequel, we will write T instead of T_n. For a vertex $v \in T$, we let T_v be the subtree of T rooted at v, and let n_v be the number of vertices in T_v.

For later use we note that if we fix $j < m$ and consider the 2^j vertices of height j, labelling them $v_1, \dots, v_{2^j}$ from left to right, then, with $q_j := \lfloor (n-2^m+1)/2^{m-j} \rfloor$,

$$n_{v_i} = \begin{cases} 2^{m+1-j} - 1, & 1 \le i \le q_j, \\ 2^{m-j} - 1 + 2^{m-j}\{(n-2^m+1)/2^{m-j}\}, & i = q_j + 1, \\ 2^{m-j} - 1, & q_j + 1 < i \le 2^j. \end{cases} \tag{3}$$

We will further assume that the labels λ_v have an exponential distribution $\mathrm{Exp}(1)$ with mean 1; as remarked above, this does not affect the distribution of X_n.

Lemma 2.1. *We have*

$$\mathbb{E}\, X_{n,y} = \frac{n-2^m+1}{m}\left(1-e^{-my}\right) + \sum_{k=1}^{m-1} \frac{2^{m-k}}{m-k}\left(1-e^{-(m-k)y}\right) \tag{4}$$

and, uniformly in n and $y > 0$,

$$\operatorname{Var} X_{n,y} = O(m^{-3}n^2).$$

Proof: Fix $y > 0$, and let, for each vertex $v \in T$, I_v be the indicator that λ_v is a minimum, given that $\lambda_o = y$. Thus, $X_{n,y} = \sum_{v \neq o} I_v$. If $h(v) = j$, let $o, v_1, \dots, v_j = v$ be the vertices on the path from the root o to v. Then $I_v = 1$ if and only if $\lambda_{v_j} < y$ and $\lambda_{v_i} > \lambda_{v_j}$ for $i = 1, \dots, j-1$. Hence, since $\lambda_{v_i} \sim \mathrm{Exp}(1)$ are independent,

$$\mathbb{E}\, I_v = \int_0^y \prod_{i=1}^{j-1} \mathbb{P}(\lambda_{v_i} > x) e^{-x}\, dx = \int_0^y e^{-jx}\, dx = \frac{1-e^{-jy}}{j}. \tag{5}$$

Consequently,

$$\mathbb{E}\, X_{n,y} = \sum_{j=1}^{m-1} 2^j \frac{1-e^{-jy}}{j} + (n-2^m+1)\frac{1-e^{-my}}{m},$$

proving (4) by letting $j = m-k$.

To estimate the variance, assume that v and w are two vertices in T of heights $j = h(v)$ and $k = h(w)$, and with their last common ancestor u at height i.

Suppose first $i < j$ and $i < k$. Let $u_0 = o, u_1, \dots, u_i = u$ be the vertices on the path from o to u, and let $Z := \min\{\lambda_{u_s} : 1 \le s \le i\}$. Conditioned on Z, I_v and I_w are independent. Further, since v has height $j-i$ above u, (5) yields

$$\mathbb{E}(I_v \mid Z) = \frac{1-e^{-(j-i)(Z \wedge y)}}{j-i},$$

and similarly for I_w. Consequently, since $Z \sim \mathrm{Exp}(i^{-1})$, being the minimum of i independent Exp(1) variables,

$$\begin{aligned}\mathbb{E}(I_v I_w) &= \mathbb{E}\Big(\frac{1-e^{-(j-i)(Z\wedge y)}}{j-i}\cdot\frac{1-e^{-(k-i)(Z\wedge y)}}{k-i}\Big)\\ &= \frac{1}{j-i}\frac{1}{k-i}\Big(\int_0^y \big(1-e^{-(j-i)z}\big)\big(1-e^{-(k-i)z}\big)ie^{-iz}\,dz\\ &\qquad + e^{-iy}\big(1-e^{-(j-i)y}\big)\big(1-e^{-(k-i)y}\big)\Big) \qquad (6)\\ &= \frac{1}{j-i}\frac{1}{k-i}\Big(1-e^{-iy}-\frac{i}{j}\big(1-e^{-jy}\big)-\frac{i}{k}\big(1-e^{-ky}\big)\\ &\quad+\frac{i}{j+k-i}\big(1-e^{-(j+k-i)y}\big)+e^{-iy}-e^{-jy}-e^{-ky}+e^{-(j+k-i)y}\Big).\end{aligned}$$

Say that the pair (v, w) is *good* if $i \le m/3$ and $j, k \ge 2m/3$, and *bad* otherwise. For a good pair (v, w) we have, by (6) and (5),

$$\begin{aligned}\mathrm{Cov}(I_v, I_w) &= \mathbb{E}\, I_v I_w - \mathbb{E}\, I_v\, \mathbb{E}\, I_w\\ &= \frac{1+O(i/m)}{jk}\Big(1-e^{-jy}-e^{-ky}+e^{-(j+k-i)y}+O(i/m)\Big)\\ &\qquad -\frac{1}{jk}\big(1-e^{-jy}\big)\big(1-e^{-ky}\big) \qquad (7)\\ &= \frac{1}{jk}e^{-(j+k-i)y}\big(1-e^{-iy}\big)+O(i/m^3)\\ &= O(m^{-2}e^{-my}iy)+O(i/m^3)=O(i/m^3).\end{aligned}$$

For given i, j, k, there are at most 2^i choices of u and then at most 2^{j-i} choices of v and 2^{k-i} of w; thus the total number of such pairs is at most 2^{j+k-i}. Hence (7) yields

$$\sum_{\text{good }(v,w)} \mathrm{Cov}(I_v, I_w) = O\Big(\sum_{i=1}^{m}\sum_{j=1}^{m}\sum_{k=1}^{m} 2^{j+k-i} i m^{-3}\Big) = O\big(2^{2m}m^{-3}\big). \qquad (8)$$

The total number of bad pairs is at most

$$\sum_{i>m/3,\ j,k\le m} 2^{j+k-i} + 2\sum_{i\ge 0,\ j<2m/3,\ k\le m} 2^{j+k-i} = O\big(2^{2m-m/3}\big). \qquad (9)$$

For the bad pairs we simply use $\mathrm{Cov}(I_v, I_w) \le \mathbb{E}\, I_v I_w \le 1$, and obtain from (8) and (9)

$$\mathrm{Var}\, X_{n,y} = \sum_{v,w} \mathrm{Cov}(I_v, I_w) = O\big(2^{2m}m^{-3}\big) = O\big(m^{-3}n^2\big).$$

Let $\varphi(n, y) := \mathbb{E}\, X_{n,y}$, given by (4). In the next lemma, we find it useful to be slightly more general than simply requiring $\bar{m} = m$.

Lemma 2.2. *If* $2^{\bar{m}} - 1 \le n \le 2^{\bar{m}+1} - 1$, *then*

$$\varphi(n, y) = \frac{n}{\bar{m}}\big(1-e^{-\bar{m}y}\big) + \frac{2^{\bar{m}+1}}{\bar{m}^2} + O\big(\bar{m}^{-2}e^{-\bar{m}y/4}n + \bar{m}^{-3}n\big).$$

Proof: Let

$$a_k = \frac{2^{m-k}}{m-k}\left(1-e^{-(m-k)y}\right) = \frac{2^{m-k}}{m}\left(1-\frac{k}{m}\right)^{-1}\left(1-e^{-my}-e^{(k-m)y}\left(1-e^{-ky}\right)\right).$$

For $m/2 < k < m$ we use $a_k = O(2^{m/2})$, and for $k \le m/2$

$$a_k = \frac{2^{m-k}}{m}\left(1+\frac{k}{m}+O\Big(\frac{k^2}{m^2}\Big)\right)\left(1-e^{-my}+O\big(kye^{(k-m)y}\big)\right).$$

Summing over k, we see that (4) yields, using $mye^{-my/4} = O(1)$,

$$\begin{aligned}\varphi(n,y) &= \frac{n-2^m}{m}\left(1-e^{-my}\right) + \frac{2^m}{m}\left(1+\frac{2}{m}+O(m^{-2})\right)\left(1-e^{-my}\right)\\ &\qquad + O\Big(m2^{m/2}+\frac{2^m}{m}ye^{-my/2}\Big)\\ &= \frac{n}{m}\left(1-e^{-my}\right)+\frac{2^{m+1}}{m^2}+O(m^{-3}n)+O\big(m^{-2}ne^{-my/4}\big).\end{aligned}$$

This proves the result when $\bar{m} = m$. The only remaining case is $n = 2^{\bar{m}} - 1$ and $m = \bar{m}-1$; the result follows easily in this case too, for example by adding a vertex v at height $\bar{m}$, using the case just considered, and subtracting $\mathbb{E}\, I_v = (1-e^{-\bar{m}y})/\bar{m}$ from (5).

Recall that $L = \lfloor \frac{3}{2}\lg\lg n\rfloor \approx \frac{3}{2}l \approx \frac{3}{2}\lg m$. Let v_i, $1 \le i \le 2^L$, be the 2^L vertices of height L, and let $n_i := n_{v_i}$. Note that $n_i = \Theta(n/2^L)$. Further, let Y_i be the minimum of λ_v along the path $P(v_i) = o\ldots v_i$ from the root to v_i.

Lemma 2.3. *With notations as above,*

$$X_n = \sum_{i=1}^{2^L}\varphi(n_i,Y_i) + o_p(m^{-2}n). \tag{10}$$

Proof: We write the number of records X_n as $V^* + V_1 + \cdots + V_{2^L}$, where V^* is the number of records with height $\le L$ and V_i are the number of records in $T_{v_i} \setminus \{v_i\}$.

If we condition on $\{\lambda_v : h(v) \le L\}$, then V^* and all Y_i become fixed, while V_i, $1 \le i \le 2^L$, become independent random variables with $V_i \overset{\mathrm{d}}{=} X_{n_i,Y_i}$.

Let $\mathcal{F}_L$ be the σ-field generated by $\{\lambda_v : h(v) \le L\}$. Then, by the comments just made, $\mathbb{E}(V_i \mid \mathcal{F}_L) = \mathbb{E}(X_{n_i,Y_i} \mid Y_i) = \varphi(n_i,Y_i)$ and, with $m_i := \lfloor \lg n_i\rfloor = \lg n - L + O(1) \sim m$,

$$\begin{aligned}\mathbb{E}\Big(\Big(X_n - V^* - \sum_{i=1}^{2^L}\varphi(n_i,Y_i)\Big)^2 \,\Big|\, \mathcal{F}_L\Big) &= \mathbb{E}\Big(\Big(\sum_{i=1}^{2^L}(V_i-\varphi(n_i,Y_i))\Big)^2 \,\Big|\, \mathcal{F}_L\Big)\\ &= \sum_{i=1}^{2^L}\mathbb{E}\Big(\big(V_i-\varphi(n_i,Y_i)\big)^2 \mid \mathcal{F}_L\Big) = \sum_{i=1}^{2^L}\operatorname{Var}\big(X_{n_i,Y_i} \mid Y_i\big)\\ &= \sum_{i=1}^{2^L}O(m_i^{-3}n_i^2) = O\big(2^Lm^{-3}2^{-2L}n^2\big) = O\big(m^{-3}2^{-L}n^2\big)\\ &= O\big(m^{-9/2}n^2\big).\end{aligned}$$

Taking the expectation, we find

$$\mathbb{E}\Bigl(X_n - V^* - \sum_{i=1}^{2^L} \varphi(n_i, Y_i)\Bigr)^2 = O\bigl(m^{-9/2} n^2\bigr) = o\bigl(m^{-4} n^2\bigr)$$

and thus

$$X_n - V^* - \sum_{i=1}^{2^L} \varphi(n_i, Y_i) = o_p\bigl(m^{-2} n\bigr).$$

The result follows because also

$$0 \le V^* < 2^{L+1} = O(m^{3/2}) = o\bigl(m^{-2} n\bigr).$$ □

Next, let $\bar{m} := m - L \sim m$. By (3) (with $j = L$), we can apply Lemma 2.2 to each n_i and $\bar{m}$; this yields

$$\varphi(n_i, Y_i) = \frac{n_i}{\bar{m}}\bigl(1 - e^{-\bar{m}Y_i}\bigr) + \frac{2^{\bar{m}+1}}{\bar{m}^2} + O\bigl(m^{-2} e^{-\bar{m}Y_i/4} n_i + m^{-3} n_i\bigr). \tag{11}$$

Since $Y_i \sim \operatorname{Exp}(1/(L+1))$, for every $a > 0$,

$$\mathbb{E}\, e^{-aY_i} = \int_0^\infty (L+1) e^{-ay-(L+1)y}\, dy = \frac{L+1}{L+1+a} = O\Bigl(\frac{L}{a}\Bigr). \tag{12}$$

Hence

$$\begin{aligned}\mathbb{E}\bigl|e^{-\bar{m}Y_i} - e^{-mY_i}\bigr| &= \mathbb{E}\, e^{-\bar{m}Y_i} - \mathbb{E}\, e^{-mY_i} = \frac{L+1}{L+1+\bar{m}} - \frac{L+1}{L+1+m} \\ &= O\Bigl(L\frac{m-\bar{m}}{m^2}\Bigr) = O\Bigl(\frac{L^2}{m^2}\Bigr).\end{aligned} \tag{13}$$

It follows easily from (11), (12) and (13) that

$$\mathbb{E}\Bigl|\varphi(n_i, Y_i) - \frac{n_i}{\bar{m}} + \frac{n_i}{m} e^{-mY_i} - \frac{2^{\bar{m}+1}}{\bar{m}^2}\Bigr| = O\bigl(L^2 m^{-3} n_i\bigr) = o(m^{-2} n_i).$$

Summing over i we find, using Lemma 2.3, since $\sum_i n_i = n - (2^L - 1) = n - O(m^{3/2})$,

$$\begin{aligned} X_n &= \sum_{i=1}^{2^L} \Bigl(\frac{n_i}{\bar{m}} - \frac{n_i}{m} e^{-mY_i} + \frac{2^{\bar{m}+1}}{\bar{m}^2}\Bigr) + o_p(m^{-2} n) \\ &= \frac{n}{m-L} - \frac{1}{m} \sum_{i=1}^{2^L} n_i e^{-mY_i} + 2^L \frac{2^{m-L+1}}{(m-L)^2} + o_p(m^{-2} n) \\ &= \frac{n}{m} + L\frac{n}{m^2} - \frac{1}{m} \sum_{i=1}^{2^L} n_i e^{-mY_i} + \frac{2^{m+1}}{m^2} + o_p(m^{-2} n). \end{aligned} \tag{14}$$

We transform this once more.

Lemma 2.4.

$$X_n = \frac{n}{m} + L\frac{n}{m^2} - \frac{1}{m} \sum_{h(v) \le L} n_v e^{-m\lambda_v} + \frac{2^{m+1}}{m^2} + o_p(m^{-2} n). \tag{15}$$

Proof: We recall that each Y_i is the minimum of the $L+1$ independent variables λ_v, $v \in P(v_i)$; thus e^{-mY_i} is the maximum of the corresponding $e^{-m\lambda_v}$. Let $a = 2\ln m/m$. The probability that at least two λ_v, $v \in P(v_i)$, are less than a is $O(L^2a^2) = O(\ln^4 m/m^2)$; hence the probability that this happens for some i is $O(2^L \ln^4 m/m^2) = o(1)$. With probability tending to 1, there is thus at most one λ_v less than a in each $P(v_i)$, and in this case,

$$0 \leq \sum_{v \in P(v_i)} e^{-m\lambda_v} - e^{-mY_i} \leq Le^{-ma} = L/m^2,$$

and thus,

$$\begin{aligned}\sum_{i=1}^{2^L} n_i e^{-mY_i} &= \sum_{i=1}^{2^L} n_i \sum_{v \in P(v_i)} e^{-m\lambda_v} + O(nL/m^2)\\ &= \sum_{h(v)\leq L} e^{-m\lambda_v} \sum_{i: v\in P(v_i)} n_i + O(nL/m^2)\\ &= \sum_{h(v)\leq L} e^{-m\lambda_v} n_v + O(nL/m^2),\end{aligned}$$

because $n_v - 2^L \leq \sum_{i:v\in P(v_i)} n_i \leq n_v$. Hence,

$$\sum_{i=1}^{2^L} n_i e^{-mY_i} = \sum_{h(v)\leq L} e^{-m\lambda_v} n_v + o_p(n/m),$$

and the result follows from (14).

The sum in (15) is a sum of independent random variables. The proof will be completed by a classical result on convergence of such sums for triangular arrays to infinitely divisible distributions, see e.g. [4, Theorem 15.28].

We write, for convenience, $\xi_v := \frac{mn_v}{n} e^{-m\lambda_v}$. We further write $\alpha_n := \{\lg n\}$ and $\beta_n = \{\lg\lg n\}$; Thus $\lg n = m + \alpha_n$ and $\lg m = \lg\lg n + o(1) = l + \beta + o(1)$. We then have, by Lemma 2.4,

$$\begin{aligned}&\frac{m^2}{n}\Big(X_n - \frac{n}{\lg n} - \frac{n \lg\lg n}{\lg^2 n}\Big)\\ &= m^2\Big(\frac{1}{m} - \frac{1}{\lg n}\Big) + L - m \sum_{h(v)\leq L} \frac{n_v}{n} e^{-m\lambda_v} + \frac{2^{m+1}}{n} - \lg\lg n + o_p(1) \qquad (16)\\ &= \alpha_n + L - l - \beta_n + 2^{1-\alpha_n} - \sum_{h(v)\leq L} \xi_v + o_p(1).\end{aligned}$$

Since $m/\lg n \to 1$, it is thus enough to show that this converges in distribution to $-W_\gamma$ as $n \to \infty$ with $\{\lg n - \lg\lg n\} \to \gamma$.

By considering subsequences, we may assume that the limits $\alpha := \lim \alpha_n$ and $\beta := \lim \beta_n$ exist. Thus $\lg n = m + \alpha + o(1)$ and $\lg m = \lg\lg n + o(1) = l + \beta + o(1)$. Note that $\lg n - \lg\lg n = m - l + \alpha - \beta + o(1)$; thus $\gamma \equiv \alpha - \beta \pmod 1$ and more precisely,

$$\gamma = \begin{cases} \alpha - \beta & \text{if } \alpha > \beta;\\ \alpha - \beta + 1 & \text{if } \alpha < \beta;\\ 0 \text{ or } 1 & \text{if } \alpha = \beta.\end{cases} \qquad (17)$$

Lemma 2.5. *Suppose that $n \to \infty$ such that $\alpha_n \to \alpha$ and $\beta_n \to \beta$ for some α and β in $[0,1]$, and let $h := 2^{\beta-\alpha}$. Then*

(i) $\sup_v \mathbb{P}(\xi_v > x) \to 0$ *for every $x > 0$. (I.e., $\{\xi_v\}$ form a null array.)*
(ii) $\sum_{h(v)\le L} \mathbb{P}(\xi_v > x) \to \nu_\gamma(x,\infty)$ *for every $x > 0$.*
(iii) $\sum_{h(v)\le L} \mathbb{E}(\xi_v \mathbf{1}[\xi_v \le h]) - (L - l + 2^{1-\alpha} + \alpha - \beta) \to \beta - \alpha.$
(iv) $\sum_{h(v)\le L} \mathrm{Var}(\xi_v \mathbf{1}[\xi_v \le h]) \to 3h/2.$

Before proving this lemma, we show how it implies Theorem 1.1. Let $C := L - l + 2^{1-\alpha} + \alpha - \beta$. We apply [4, Theorem 15.28] with $a = 0$ and $b = f(\gamma)$ to $\sum_{h(v)\le L} \xi_v + \sum_{i=1}^{n} \xi_i'$, with $\xi_i' = -C/n$ deterministic. (Note that $C/n \to 0$; thus $\{\xi_v\}\cup\{\xi_i'\}$ is a null array.) We have $d\nu_\gamma/dx = 2^{\{\lg x+\alpha-\beta\}}x^{-2} = 2^{-i+\alpha-\beta}x^{-1}$ when $2^i h < x < 2^{i+1}h$, and thus

$$\int_0^h x^2\, d\nu_\gamma(x) = \sum_{i=-\infty}^{-1} \int_{2^i h}^{2^{i+1}h} 2^{-i+\alpha-\beta} x\, dx = \sum_{i=-\infty}^{-1} \tfrac{3}{2} 2^i h = \frac{3h}{2}.$$

Similarly, if $\beta \le \alpha$ so $1/2 \le h \le 1$, then

$$\int_h^1 x\, d\nu_\gamma(x) = \int_h^1 2^{\alpha-\beta}\, dx = 2^{\alpha-\beta} - 1,$$

while if $\beta \ge \alpha$ so $1 \le h \le 2$, then

$$\int_h^1 x\, d\nu_\gamma(x) = -\int_1^h x\, d\nu_\gamma(x) = 2^{1+\alpha-\beta}(1-h) = 2(2^{\alpha-\beta} - 1).$$

It follows, using (17) and $f(0) = f(1)$, that in both cases

$$f(\gamma) - \int_h^1 x\, d\nu_\gamma(x) = 2^\gamma - 1 - \gamma - \int_h^1 x\, d\nu_\gamma(x) = \beta - \alpha.$$

It is now easy to see from Lemma 2.5 that the conditions of [4, Theorem 15.28] are satisfied, and consequently

$$\sum_{h(v)\le L} \xi_v - (L - l + 2^{1-\alpha} + \alpha - \beta) = \sum_{h(v)\le L} \xi_v + \sum_{i=1}^{n} \xi_i' \xrightarrow{\mathrm{d}} W_\gamma.$$

Theorem 1.1 now follows by (16).

Proof: [Proof of Lemma 2.5] For any $x > 0$,

$$\begin{aligned}\mathbb{P}(\xi_v > x) &= \mathbb{P}\Big(e^{-m\lambda_v} > \frac{nx}{mn_v}\Big) = \mathbb{P}\Big(m\lambda_v < \ln \frac{mn_v}{nx}\Big) \\ &= 1 - \exp\Big(-\frac{1}{m} \ln_+ \frac{mn_v}{nx}\Big).\end{aligned} \tag{18}$$

This shows first that for every $x > 0$,

$$\mathbb{P}(\xi_v > x) < \frac{1}{m} \ln_+ \frac{mn_v}{nx} \le \frac{1}{m} \ln_+ \frac{m}{x} \to 0, \tag{19}$$

which proves (i).

On a given level $j < m$ there are, by (3), $q_j = n2^{j-m} - 2^j + O(1) = (2^{\alpha_n} - 1)2^j + O(1)$ vertices with $n_v = 2^{m+1-j} - 1$, and $2^j - q_j - 1 = (2 - 2^{\alpha_n})2^j + O(1)$ vertices with $n_v = 2^{m-j} - 1$. There is one additional vertex with an intermediate n_v (which could coincide with one of the two main values); for convenience we call such a vertex *bad*. We also call a vertex v with $n_v \geq 2^{m-l/2}$ (which requires $j \leq l/2$) bad. All other vertices v with $h(v) \leq L$ are *good*. The good vertices thus have $n_v = 2^{m-k} - 1$ for some k with $l/2 \leq k \leq L$. For $l/2 \leq k < L$, there are $(2-2^{\alpha_n})2^k + O(1)$ such vertices with $h(v) = k$ and $(2^{\alpha_n} - 1)2^{k+1} + O(1)$ with $h(v) = k+1$; thus together $2^{k+\alpha_n} + O(1)$. For $k = L$, there are only $(2 - 2^{\alpha_n})2^k + O(1)$ such vertices, since we require $h(v) \leq L$. In other words,

$$\#\{v \text{ good} : n_v = 2^{m-k} - 1\} = \begin{cases} 2^{k+\alpha_n} + O(1), & l/2 \leq k < L, \\ (2 - 2^{\alpha_n})2^L + O(1), & k = L. \end{cases} \tag{20}$$

The number of bad vertices is $O\big(L + 2^{l/2}\big) = O\big(m^{1/2}\big)$. By (19), $\mathbb{P}(\xi_v > x) = O(\ln m/m)$ for every fixed $x > 0$. Hence the sum over bad vertices in (ii) is $O(m^{-1/2} \ln m) = o(1)$.

Similarly, using (19) again,

$$\mathbb{E}\big(\xi_v \mathbf{1}[\xi_v \leq h]\big) \leq \frac{1}{m} + h\,\mathbb{P}\Big(\xi_v > \frac{1}{m}\Big) \leq \frac{1}{m} + h\frac{2\ln m}{m} = O\Big(\frac{\ln m}{m}\Big). \tag{21}$$

and

$$\operatorname{Var}\big(\xi_v \mathbf{1}[\xi_v \leq h]\big) \leq \mathbb{E}\big(\xi_v^2 \mathbf{1}[\xi_v \leq h]\big) \leq h\,\mathbb{E}\big(\xi_v \mathbf{1}[\xi_v \leq h]\big) = O\Big(\frac{\ln m}{m}\Big). \tag{22}$$

Consequently, the sum over bad v is $o(1)$ in (ii), (iii) and (iv), so we may in the sequel ignore them and consider only good vertices.

Fix $x > 0$. Then, by (18) and (19),

$$\mathbb{P}(\xi_v > x) = \frac{1}{m} \ln_+\Big(\frac{mn_v}{nx}\Big)\Big(1 + O\Big(\frac{\ln m}{m}\Big)\Big) \tag{23}$$

If $k \geq L$, then $m2^{m-k} \leq 2^{l+1+m-L} < nx$, provided n is large enough. Thus, for large n, by (20) and (23), with all $o(1)$ uniform in k for fixed x,

$$\begin{aligned}
\sum_{v \text{ good}} \mathbb{P}(\xi_v > x) &= (1 + o(1)) \sum_{k=l/2}^{\infty} \big(2^{k+\alpha_n} + O(1)\big) \frac{1}{m} \ln_+\Big(\frac{(2^{m-k} - 1)m}{nx}\Big) \\
&= (1 + o(1)) \sum_{k \geq l/2} 2^{k+\alpha_n - l - \beta_n} \ln_+\big(2^{-k-\alpha_n+l+\beta_n+o(1)} x^{-1}\big) + o(1) \\
&= (1 + o(1)) \sum_{i \leq l/2} 2^{-i+\alpha-\beta} \ln_+\big(2^{i-\alpha+\beta+o(1)} x^{-1}\big) + o(1) \\
&\to F(x) := \sum_{-\infty}^{\infty} 2^{-i+\alpha-\beta} \ln_+\big(2^{i-\alpha+\beta} x^{-1}\big).
\end{aligned}$$

Let $j := \lfloor \lg x + \alpha - \beta \rfloor$; thus $2^{j+\beta-\alpha} \le x < 2^{k+\beta-\alpha+1}$ and

$$\begin{aligned} F(x) &= \sum_{i=j+1}^{\infty} 2^{-i+\alpha-\beta} \ln\bigl(2^{i-\alpha+\beta} x^{-1}\bigr) \\ &= \sum_{k=1}^{\infty} 2^{-k-j+\alpha-\beta}\bigl(k \ln 2 + \ln\bigl(2^{j-\alpha+\beta} x^{-1}\bigr)\bigr) \\ &= 2^{-j+\alpha-\beta}\bigl(2 + \lg\bigl(2^{j-\alpha+\beta} x^{-1}\bigr)\bigr) \ln 2 \\ &= 2^{\alpha-\beta-\lfloor \lg x + \alpha - \beta \rfloor}\bigl(2 - \{\lg x + \alpha - \beta\}\bigr) \ln 2 \\ &= 2^{\gamma - \lfloor \lg x + \gamma \rfloor}\bigl(2 - \{\lg x + \gamma\}\bigr) \ln 2. \end{aligned}$$

Note that $F(x)$ is continuous and decreasing with $F(x) \to 0$ as $x \to \infty$. The derivative is

$$\frac{dF(x)}{dx} = -\frac{1}{x} 2^{\gamma - \lfloor \lg x + \gamma \rfloor} = -x^{-2} 2^{\{\lg x + \gamma\}}.$$

Thus $F(x) = \nu_\gamma(x, \infty)$, which proves (ii).

For (iii) and (iv) we calculate, for $s > 0$,

$$\begin{aligned} \mathbb{E}\bigl(e^{-m\lambda_v} \mathbf{1}[e^{-m\lambda_v} \le s^{-1}]\bigr) &= \int_{m^{-1} \ln_+ s}^{\infty} e^{-mx} e^{-x}\, dx \\ &= \frac{1}{m+1} e^{-(m+1)\frac{1}{m} \ln_+ s} = \frac{1}{m+1} 2^{-(1+1/m) \lg_+ s} \end{aligned} \tag{24}$$

and, similarly,

$$\mathbb{E}\bigl(e^{-2m\lambda_v} \mathbf{1}[e^{-m\lambda_v} \le s^{-1}]\bigr) = \frac{1}{2m+1} 2^{-(2+1/m) \lg_+ s},$$

which gives

$$\operatorname{Var}\bigl(e^{-m\lambda_v} \mathbf{1}[e^{-m\lambda_v} \le s^{-1}]\bigr) = \frac{1}{2m} 2^{-(2+1/m) \lg_+ s}\bigl(1 + O(m^{-1})\bigr). \tag{25}$$

If v is a good vertex with $n_v = 2^{m-k} - 1 = 2^{m-k+o(1)}$, then

$$\frac{mn_v}{nh} = 2^{(l+\beta)+(m-k)-(m+\alpha)-(\beta-\alpha)+o(1)} = 2^{l-k+o(1)},$$

and thus, by (24),

$$\begin{aligned} \mathbb{E}\bigl(\xi_v \mathbf{1}[\xi_v \le h]\bigr) &= \frac{mn_v}{n} \mathbb{E}\Bigl(e^{-m\lambda_v} \mathbf{1}\Bigl[e^{-m\lambda_v} \le \frac{nh}{mn_v}\Bigr]\Bigr) \\ &= \frac{mn_v}{(m+1)n} 2^{-(1+1/m)(l-k)_+ + o(1)} = 2^{-k-\alpha-(l-k)_+ + o(1)}. \end{aligned}$$

Note in particular that if $k \ge l+1$, then $\frac{mn_v}{nh} < 1$ for large n, and thus $\xi_v \le h$ and

$$\mathbb{E}\bigl(\xi_v \mathbf{1}[\xi_v \le h]\bigr) = \mathbb{E}\,\xi_v = \frac{mn_v}{(m+1)n} = 2^{-k-\alpha_n}\Bigl(1 + O\Bigl(\frac{1}{m}\Bigr)\Bigr). \tag{26}$$

It follows from (20), (24) and (26) that, with o and O uniform in k,

$$\begin{aligned}
&\sum_{v \text{ good}} \mathbb{E}(\xi_v \mathbf{1}[\xi_v \le h]) \\
&= \sum_{k=l/2}^{l} 2^{k+\alpha+o(1)} 2^{-k-\alpha-(l-k)_+ + o(1)} + \sum_{k=l+1}^{L-1} (2^{k+\alpha_n} + O(1)) 2^{-k-\alpha_n} (1 + O(m^{-1})) \\
&\quad + \big((2 - 2^{\alpha_n}) 2^L + O(1)\big) 2^{-L-\alpha+o(1)} \\
&= \sum_{k=l/2}^{l} 2^{-(l-k)_+ + o(1)} + \sum_{k=l+1}^{L-1} (1 + O(m^{-1})) + 2^{1-\alpha} - 1 + o(1) \\
&= 2 + L - 1 - l + 2^{1-\alpha} - 1 + o(1) = L - l + 2^{1-\alpha} + o(1).
\end{aligned}$$

Similarly, using (25),

$$\begin{aligned}
\operatorname{Var}(\xi_v \mathbf{1}[\xi_v \le h]) &= \frac{m^2 n_v^2}{n^2} \operatorname{Var}\Big(e^{-m\lambda_v} \mathbf{1}\Big[e^{-m\lambda_v} \le \frac{nh}{mn_v}\Big]\Big) \\
&= \frac{m^2 n_v^2}{2mn^2} 2^{-(2+1/m)(l-k)_+ + o(1)} = 2^{l+\beta-1-2k-2\alpha-2(l-k)_+ + o(1)}
\end{aligned}$$

and

$$\begin{aligned}
\sum_{v \text{ good}} \operatorname{Var}(\xi_v \mathbf{1}[\xi_v \le h]) &= \sum_{k=l/2}^{L} 2^{k+\alpha+o(1)} 2^{l+\beta-1-2k-2\alpha-2(l-k)_+ + o(1)} + o(1) \\
&= \sum_{k=-\infty}^{\infty} 2^{l-k-2(l-k)_+ + \beta-\alpha-1+o(1)} + o(1) \\
&= 3 \cdot 2^{\beta-\alpha-1} + o(1) = 3h/2 + o(1).
\end{aligned}$$

This completes the proof of Lemma 2.5.

We have proved Theorem 1.1 for X_v. For X_e, the only difference is that λ_o is ignored, and thus $Y_i \sim \operatorname{Exp}(1/L)$. The estimates in (12) and (13) remain valid, and thus (15) and (16) still hold, summing over $v \ne o$ only. Since $\xi_0 = me^{-m\lambda_o} \xrightarrow{\text{P}} 0$ by Lemma 2.5(i), this makes no difference for the asymptotics of the distribution. (But note that $\mathbb{E}\,\xi_0 \to 1$, and that the means differ correspondingly, see Remark 1.3.)

For the completely balanced tree (Theorem 1.2), every vertex v with $h(v) = k$ has $2^{-k}n - 2 < n_v \le 2^{-k}n$. We call all vertices with $l/2 \le h(v) \le L$ good, and replace (20) by

$$\#\{v \in T_n^* \text{ good} : n_v = 2^{-k}n + O(1)\} = 2^k, \qquad l \le k \le L. \tag{27}$$

The remaining calculations hold as above, provided we replace α_n and α by 0 and thus γ by $1 - \beta$.

References

[1] P. Chassaing & R. Marchand. In preparation.

[2] W. Feller, *An Introduction to Probability Theory and Its Applications. Vol. II.* Second edition, Wiley, New York 1971.

[3] S. Janson, Random cutting and records in deterministic and random trees. Preprint, 2003. Available from `http://www.math.uu.se/~svante/papers`

[4] O. Kallenberg, *Foundations of Modern Probability.* 2nd ed., Springer-Verlag, New York, 2002.

[5] D.E. Knuth, *The Art of Computer Programming. Vol. 1: Fundamental Algorithms.* 3nd ed., Addison-Wesley, Reading, Mass., 1997.

[6] A. Meir & J.W. Moon, Cutting down random trees. *J. Australian Math. Soc.* **11** (1970), 313–324.

[7] A. Panholzer, Cutting down very simple trees. Preprint, 2003.

[8] A. Panholzer, Non-crossing trees revisited: cutting down and spanning subtrees. *Proceedings, Discrete Random Walks 2003*, Cyril Banderier and Christian Krattenthaler, Eds., *Discr. Math. Theor. Comput. Sci.* **AC** (2003), 265–276.

[9] A. Rényi, (1962). On the extreme elements of observations. *MTA III, Oszt. Közl.* **12** (1962) 105–121. Reprinted in *Collected Works*, Vol III, pp. 50-66, Akadémiai Kiadó, Budapest, 1976.

Svante Janson
Department of Mathematics, Uppsala University, PO Box 480, S-751 06 Uppsala, Sweden
Email: svante.janson@math.uu.se
`http://www.math.uu.se/~svante/`

Trends in Mathematics, © 2004 Birkhäuser Verlag Basel/Switzerland

Multidimensional Interval Trees

Mehri Javanian and Mohammad Q. Vahidi-Asl

ABSTRACT: *The binary interval tree is a random structure that underlies interval division and parking problems. A generalization to trees underlying volume partition is investigated in this paper; the size of the associated tree is studied. In d dimensions (arbitrarily high), the moment generating function of the size of the tree and several pruned variants is shown to satisfy a partial differential equation of order d. So, the situation is considerably more complex than the linear case that arises in one dimension. The paper addresses volume partition by points, and the case can be solved, in contrast with a well-known parking problem, where the partition is done by solid objects; the parking analog remains unsolved, but a variation is shown to be tractable, which may glean research toward an eventual solution to the standard parking problem may. The multidimensional interval tree can be viewed as a continuous analog of the discrete quadtrees.*

1. Introduction

The *interval tree* is a tree associated with repeated division of an interval of volume $x_1 \dots x_d$ until it is partitioned into parts of volume less than 1.

Suppose each x_i is at least 1. The volume $x_1 \dots x_d$ is represented by the root (a distinguished *internal node*) of a 2^d-branching tree. Divide the interval into 2^d random *orthants* by choosing $Q = (U_1, \dots, U_d)$ uniformly at random from the interval. Let us cannonically label the 2^d orthants in any arbitrary way. The ith subtree is then associated with the ith orthant. Each subtree grows recursively on the volume of the interval it represents. The process continues in the 2^d subtrees all the way to the leaves where a leaf stands for an interval with volume less than 1.

The binary interval tree has been introduced in Sibuya and Itoh (1987). Recently, Itoh and Mahmoud (2002) considered incomplete or one-sided variants of the (1-dimensional) interval tree. In d dimensions, the interval tree has 2^d subtrees. We consider here a few such incomplete variants (corner preference, proportionate preference and no preference) to show that the techniques can be extended.

2. Main results

In an interval tree, Suppose at each node, all the 2^d subtrees are pruned, except one specified by a prescribed program. More precisely, let $J_0, J_1, \dots$ be any given *pruning sequence* of independent random variables, each with an arbitrary discrete distribution on the set of integers $\{1, \dots 2^d\}$. At the root node, all the subtrees are pruned, except the J_0th subtree. At the root of the J_0th subtree, all the subtrees are pruned, except the J_1st, and so forth. We call the tree so obtained an *incomplete interval tree grown from the pruning sequence* $J_0, J_1, \dots$.

For example the deterministic sequence $1, 1, 1, \ldots$ corresponds to the consistent choice of the leftmost subtree (corner preference). The pruning process leaves behind the leftmost path (corner preference) in the full tree. All of paths are determined by keeping a random subtree and pruning all the rest. We are interested in the size $S_V(J_0, J_1, \ldots)$ of the incomplete interval tree, grown under the pruning sequence $J_0, J_1, \ldots$. It is sufficient for the purpose to study the length of the path of the corner preference incomplete tree $S_V \equiv S_V(1, 1, \ldots)$, because $S_V(J_0, J_1, \ldots)$ and $S_V(1, 1, \ldots)$ have the same distribution, owing to the symmetry of the 2^d subtrees. We start from a stochastic recurrence for the size of the corner preference incomplete tree S_V.

$$S_V = 1 + S_{V_1}, \qquad \text{for } V \geq 1,$$

where $V_1 = U_1 \ldots U_d$ is the volume of the first orthant. The boundary conditions are $S_1 \equiv 1$, and $S_V \equiv 0$ if $V < 1$. Let $\varphi_V(t)$ be the moment generating function of S_V.

Theorem 2.1. *Let S_V be the size of a random corner preference interval tree grown on the interval of volume V. As $V \to \infty$,*

$$\frac{S_V - \frac{\ln V}{d}}{\sqrt{\ln V}} \xrightarrow{\mathcal{D}} \mathcal{N}(0, \frac{1}{d^2}).$$

Corollary 2.2.

$$\mathbf{E}[S_V] \sim \frac{\ln V}{d}$$

$$\mathbf{Var}[S_V] \sim \frac{\ln V}{d^2}.$$

We have derived the similar results in proportionate and uniform preference incomplete interval trees.

References

[1] Sibuya, M. and Itoh, Y. (1987). Random sequential bisection and its associated binary tree. *Annals of the Institute of Statistical Mathematics*, **39**, 69–84.

[2] Itoh, Y. and Mahmoud, H. (2003+). One-sided variations on interval trees. *Journal of Applied Probability* (accepted).

Mehri Javanian and Mohammad Q. Vahidi-Asl
Department of Statistics, Shahid Beheshti University, Tehran, Iran
javanian_m@yahoo.com

Trends in Mathematics, © 2004 Birkhäuser Verlag Basel/Switzerland

Edit Distance between Unlabelled Ordered Trees

Anne Micheli and Dominique Rossin

1. Definitions

1.1. Sorted permutations

We introduce sorted permutations [1, 2] and show that they are in one-to-one correspondence with ordered trees.

Definition 1.1. *Let $n \in \mathbb{N}$, a sorted permutation of $\{1 \dots n\}$ is a permutation σ such that $\sigma = InJ$ where I and J are sorted permutations on $\{1 \dots p\}$ and $\{p+1 \dots n-1\}$ respectively. Notice that I or J could be empty.*

Theorem 1.2. *Sorted permutations are in one-to-one correspondence with rooted ordered trees. The numbering of the edges corresponds to a postfix Depth First Traversal (DFT). Then the sorted permutation is obtained by a prefix Depth First Traversal of the labeled tree.*

If σ is a sorted permutation , $\mathcal{T}(\sigma)$ denotes the tree associated to σ.

Definition 1.3. *A subsequence of a permutation $\sigma = \sigma_1 \dots \sigma_n$ is a word $\sigma' = \sigma_{i_1} \dots \sigma_{i_k}$ where $i_1, \dots, i_k$ is an increasing sequence of $\{1, \dots, n\}$. Let Φ be the bijective mapping of $\{\sigma_{i_1}, \sigma_{i_2}, \dots, \sigma_{i_k}\}$ on $\{1, \dots, k\}$ preserving the order on σ_{i_l}. The normalized subsequence (pattern) is equal to $\Phi(\sigma')$.*

Remark 1.4. *The sorted permutations are the permutations avoiding the normalized subsequence (pattern)* 231 [3].

1.2. Edit distance

Given two unlabeled trees, the edit distance is the minimal number of operations to transform one into the other. The operations are *Deletion*: this is the contraction of an edge; two vertices are merged. Only one label is kept. *Insertion*: this is the converse operation of deletion.

2. Distance on sorted permutations

A *factor* of a permutation $\sigma = \sigma_1\sigma_2 \dots \sigma_n$ is a *factor* of the word $\sigma_1\sigma_2 \dots \sigma_n$, i.e., a word of the form $\sigma_k\sigma_{k+1} \dots \sigma_{k+l}$. A *compact* factor f is a factor such that its elements are a permutation of an interval of $\mathbb{N}$. A *complete* factor of σ is a compact factor f such that there is no non-empty factor g verifying that fg is compact and the greatest element of fg is equal to the greatest element of f.

Take for example the sorted permutation $\sigma = (1524376)$. The complete factors of σ are $\{1\}$, $\{15243\}$, $\{1524376\}$, $\{5243\}$, $\{524376\}$, $\{2\}$, $\{243\}$, $\{43\}$, $\{3\}$, $\{76\}$, $\{6\}$.

Let $\sigma = \sigma_1 \ldots \sigma_k$ be a word of $\{1 \ldots n\}$ and a be a letter of $\{1 \ldots n\}$. We denote by $[\sigma]_a$ the word $\sigma'_1 \ldots \sigma'_k$ where $\sigma'_i = \begin{cases} \sigma_i \text{ if } \sigma_i < a \\ \sigma_i + 1 \text{ otherwise} \end{cases}$

Definition 2.1. *We define two operations on σ:*

1. *Deletion: Let $1 \leq k \leq n$. The deletion $(k \to \Lambda)$ is the removal of σ_k in σ and the renormalization on S_{n-1} of the result.*
2. *Insertion: $(\Lambda \to \varnothing)$ corresponds to the transformation of $\sigma = \varnothing$ into $\sigma' = (1)$. If $\sigma \neq \varnothing$, let f be a complete factor of σ. Then, $\sigma = ufv$ with u, v factors of σ. The resulting permutation is σ' :*
   (a) *$(\Lambda \to f)$: $\sigma' = [u]_a af[v]_a$, $a = max\{f\} + 1$. This corresponds to the insertion of an inner edge with $\mathcal{T}(f)$ as subtree.*
   (b) *$(\Lambda \xrightarrow{r} f)$: $\sigma' = [u]_a fa[v]_a$, $a = max\{f\} + 1$. This corresponds to the insertion of a free edge as the right brother of $\mathcal{T}(f)$.*
   (c) *$(\Lambda \xrightarrow{l} f)$: $\sigma' = [u]_a a[f]_a[v]_a$, $a = min\{f\}$. This corresponds to the insertion of a free edge as the left brother of $\mathcal{T}(f)$.*

Proposition 2.2. *The Deletion/Insertion algorithm yields a sorted permutation. Moreover insertion and deletion are inverse operations.*

Theorem 2.3. *The edit distance between two sorted permutations σ_1 and σ_2 is the edit distance between the associated ordered trees and is equal to $|\sigma_1| + |\sigma_2| - 2|u|$ where u is a largest normalized subsequence (pattern) of σ_1 and σ_2.*

Corollary 2.4. *Finding the greatest common pattern between two sorted permutations is polynomial.*

In [4], they proved that finding the greatest common pattern between two permutations is NP-complete. We prove here that the problem becomes polynomial when restricting to sorted permutations, ie (132) or (231) avoiding permutations. In fact, the algorithm of Zhang and Shasha [5] on trees solves the problem on sorted permutations because the algorithm outputs not only the distance but also the greatest common subtree.

3. Generating function of the edit distance between sorted permutations and $Id = 1\ 2\ \ldots\ n$

We denote by $S_1(t, q)$ the generating function of sorted permutations where t counts the size of the permutation and q the edit distance between sorted permutations and Id. This is the distance between a tree and the trivial one which is made of n edges and of height 1.

$$S_1(t, q) = \frac{1 + (q^2 - 1)t - \sqrt{(q^2-1)^2t^2 - 2(q^2+1)t + 1}}{2tq^2}$$

Theorem 3.1. *The average edit distance between rooted planar trees with n edges and $n, n-1, \ldots, 2, 1$ is $n - 1$.*

In [6], they determine analytically the average height of a planar tree with n edges which is $\sqrt{\pi n} - \frac{1}{2}$. Thus from [7], we obtain that the the average edit distance is $2(n - \sqrt{\pi n} + \frac{1}{2}) \equiv 2n$.

References

[1] M. Bousquet-Mélou. Sorted and/or sortable permutations. *Disc. Math.*, 225:25–50, 2000.

[2] J. West. *Permutations and restricted subsequences and Stack-sortable permutations.* PhD thesis, M.I.T., 1990.

[3] D.E. Knuth. *The Art of Computer Programming : Fundamental Algorithms*, page 533. Addison-Wesley, 1973.

[4] P. Bose, J.F. Buss, and A. Lubiw. Pattern matching for permutations. *Inf. Proc. Letters*, 65:277–283, 1998.

[5] K. Zhang and D. Shasha. Simple fast algorithms for the editing distance between trees and related problems. *SIAM J. Comput.*, 18(6):1245–1262, Dec. 1989.

[6] N.G. De Bruijn, D.E. Knuth, and S.O. Rice. *Graph theory and Computation*, chapter The average height of planted plane trees. Academic Press, 1972.

[7] E. Roblet and X.G. Viennot. Théorie combinatoire des t-fractions et approximants de Padé en deux points. *Disc. Math.*, 153:271–288, 1996.

Anne Micheli and Dominique Rossin
CNRS, LIAFA, Université Paris 7,
2 Place Jussieu,
75251 PARIS Cedex 05,
FRANCE

On Parameters in Monotonically Labelled Trees

Katherine Morris

ABSTRACT: *Let T be a rooted tree structure with n nodes $t_1, \ldots, t_n$. A function $f : \{t_1, \ldots, t_n\}$ into $\{1 < \cdots < k\}$ is monotone if whenever t_i is a descendant of t_j then $f(t_i) \geq f(t_j)$. Two grand averages, the size of the ancestor tree and the Steiner distance, are determined for some tree structures.*

Binary Trees. We consider binary trees whose nodes are monotonically labelled with $1, 2, \ldots, k$, as described in [4], with generating functions $y_1(z) = 1 + zy_1^2(z), \cdots, y_k(z) = y_{k-1}(z) + zy_k^2(z)$. The asymptotic behaviour of the coefficients of y_k was analyzed in [4]: by determining the singularities q_k of y_k nearest to the origin, given by $q_{l+1} = q_l(1 - q_l)$ with $q_1 = \frac{1}{4}$, we have $y_k(z) = \frac{1}{2q_k} - \frac{c_k}{2q_k}(q_k - z)^{1/2} + \mathcal{O}(q_k - z)$, $c_k = 2(r_2 \ldots r_k)^{-1/4} = 4p_k\sqrt{\pi q_k}$ and $r_k = 1 - 4q_k$. Moreover, from [2], we know that $y_i(q_k) = \frac{q_{k-i}}{q_k}$, $0 \leq i \leq k$.

The first parameter we analyze is the *size of the ancestor tree*. Consider a tree, T, and select p random nodes in it. The ancestor tree is the subtree of T which is spanned by the root and the p chosen nodes.

Theorem 3. *The size of the ancestor tree in binary trees satisfies*
$B(z, u, v) = zv(1 + u)B^2(z, u, v) - zvT^2(z) + T(z)$, *where* $T(z) = \frac{1-\sqrt{1-4z}}{2z}$.

Theorem 4. *The generating function for the size of the ancestor tree in the monotonically labelled binary trees is*
$B_k(z, u, v) = B_{k-1}(z, u, v) + z(1 + u)vB_k^2(z, u, v) + (1 - v)zy_k^2(z)$.

Firstly, the expectation for the size of the ancestor tree is computed. We differentiate $B_k(z, u, v)$ with respect to v, let $v = 1$ and use the substitutions $\frac{\partial B_k(z,u,v)}{\partial v}\big|_{v=1} = \beta_k(z, u)$, $B_k(z, u, 1) = y_k(z(1 + u))$, $y_k(z(1 + u)) = \bar{y}_k(z)$. We obtain $\beta_k(z, u) = \beta_{k-1}(z, u) + \bar{y}_k(z) - \bar{y}_{k-1}(z) + 2z(1 + u)\bar{y}_k(z)\beta_k(z, u) - (y_k(z) - y_{k-1}(z))$, with initial conditions $\bar{y}_0(z) = 0$ and $\beta_0(z, u) = 0$. The solution to this recursion is $\beta_k(z, u) = \sum_{j=1}^{k} \Big[\prod_{i=j}^{k} (1 - 2z(1 + u)\bar{y}_i(z))\Big]^{-1} (\bar{y}_j(z) - \bar{y}_{j-1}(z) - y_j(z) + y_{j-1}(z))$.

Next, we substitute $\bar{y}_k(z)$, $\bar{y}_{k-1}(z)$, $y_k(z)$ and $y_{k-1}(z)$ in the k-th term of β_k with their known expansions for $z \to q_k$. This gives the main term in the recursion $\beta_k \sim \frac{\bar{y}_k - y_k}{1-2z(1+u)\bar{y}_k} \sim \frac{\sqrt{q_k - z} - \sqrt{q_k - z(1+u)}}{2q_k\sqrt{q_k - z(1+u)}}$, and it follows that $[u^p]\beta_k \sim \frac{1}{2^{2p+1}q_k}\binom{2p}{p}\left(1 - \frac{z}{q_k}\right)^{-p}$ and $[z^n u^p]\beta_k \sim \frac{1}{2^{2p+1}q_k^{n+1}}\binom{2p}{p}\frac{n^{p-1}}{\Gamma(p)}$. The latter is normalized with $p_k q_k^{-n} n^{-3/2} \frac{n^p}{\Gamma(p+1)}$, which leads to the expectation for the size of the ancestor tree: $E_{n,p}^{(k)} \sim \frac{p}{2^{2p+1}p_k q_k}\binom{2p}{p}\sqrt{n}$.

An analogous method (differentiate $B_k(z,u,v)$ twice with respect to v, then let $v = 1$) produces the variance for the size of the ancestor tree for monotonically labelled binary trees: $V_{n,p}^{(k)} = \left(\frac{p}{4\pi p_k^2 q_k^2} - \frac{p^2}{2^{4p+2} p_k^2 q_k^2}\binom{2p}{p}^2\right) n + \mathcal{O}(\sqrt{n})$, where $\left(\frac{p}{4\pi p_k^2 q_k^2} - \frac{p^2}{2^{4p+2} p_k^2 q_k^2}\binom{2p}{p}^2\right) \to 0$, as $p \to \infty$.

The second parameter we analyze is the *Steiner distance* which is the size of the subtree spanned by p randomly chosen nodes in a tree.

Theorem 5. *The generating function for the Steiner distance in binary trees is* $S(z,u,v) = \frac{zv(1+u)B^2(z,u,v) - 2zv\,T(z)\,B(z,u,v) + zT^2(z)(v-2) + T(z)}{1 - 2z\,T(z)}$.

Theorem 6. *The Steiner distance in monotonically labelled binary trees has the generating function* $S_k(z,u,v) = \frac{zv(1+u)B_k^2(z,u,v) - 2zvy_k(z)B_k(z,u,v) + zy_k^2(z)(v-2) + y_k(z)}{1 - 2zy_k(z)}$.

The expectation and variance for the Steiner distance in our binary tree model are $E_{n,p}^{(k)} \sim \frac{(p-1)p}{2^{2p}(2p-1)p_k q_k}\binom{2p}{p}\sqrt{n}$ and $V_{n,p}^{(k)} = \left(\frac{(p-1)}{4\pi p_k^2 q_k^2} - \frac{(p-1)^2 p^2}{2^{4p}(2p-1)^2 p_k^2 q_k^2}\binom{2p}{p}^2\right) n + \mathcal{O}(\sqrt{n})$.

***t*-ary Trees.** We consider monotonically labelled t-ary trees with generating functions $y_1(z) = 1 + z\,y_1^t(z), \ldots, y_k(z) = y_{k-1}(z) + z\,y_k^t(z)$.

Theorem 7. *The size of the ancestor tree in the monotonically labelled t-ary trees satisfies* $G_k(z,u,v) = G_{k-1}(z,u,v) + zv(1+u)G_k^t(z,u,v) + (1-v)zy_k^t(z)$. *The size of the ancestor tree has expectation* $E_{n,p}^{(k)} \sim \frac{p}{2^{2p}(t-1)t^{1/(t-1)}p_k q_k^{1/(t-1)}}\binom{2p}{p}\sqrt{n}$.

Ordered Trees. The generating functions for monotonically labelled ordered trees are $y_1(z) = \frac{z}{1-y_1(z)}, \ldots, y_k(z) = y_{k-1}(z) + \frac{z}{1-y_k(z)}$.

Theorem 8. *The size of the ancestor tree in monotonically labelled ordered trees satisfies* $P_k(z,u,v) = P_{k-1}(z,u,v) + \frac{zv(1+u)}{1-P_k(z,u,v)} + \frac{z(1-v)}{1-y_k(z)}$.

The expectation for the size of the ancestor tree is $E_{n,p}^{(k)} \sim \frac{q_k^{1/2} p}{2^{2p+1} p_k}\binom{2p}{p}\sqrt{n}$.

Theorem 9. *The generating function for the Steiner distance in the ordered tree models*

$$S_k(z,u,v) = \left(1 - \frac{z}{(1-y_k(z))^2}\right)^{-1} \times \left(P_k(z,u,v)\left(1 - \frac{zv}{(1-y_k(z))^2}\right) - \frac{z(1-v)y_k(z)}{(1-y_k(z))^2}\right).$$

The formula for the expectation is $E_{n,p}^{(k)} \sim \frac{q_k^{1/2} p(p-1)}{2^{2p}(2p-1)p_k}\binom{2p}{p}\sqrt{n}$.

References

[1] P. Flajolet and A. Odlyzko. Singularity analysis of generating functions. *SIAM Journal of Discrete Mathematics*, 3:216–240, 1990.

[2] P. Kirschenhofer. On the average shape of monotonically labelled tree structures. *Discrete Applied Mathematics*, 7:161–181, 1984.

[3] A. Panholzer and H. Prodinger. Spanning tree size in random binary search trees, *Annals of Applied Probability*, accepted.

[4] H. Prodinger and F. J. Urbanek On monotone functions of tree structures. *Discrete Applied Mathematics*, 5:223-239, 1983.

Katherine Morris
University of the Witwatersrand, Johannesburg, South Africa
kate@maths.wits.ac.za

[illegible]

Trends in Mathematics, © 2004 Birkhäuser Verlag Basel/Switzerland

Number of Vertices of a Given Outdegree in a Galton–Watson Forest

Tatiana Myllári

ABSTRACT: *Galton–Watson forests consisting of N trees and n non-root vertices are considered. The limit distributions of the number of vertices of a given outdegree in such a forest are obtained.*

Let us consider the Galton–Watson forest $\mathfrak{F}_{N,n}$ consisting of N trees and n non-root vertices. This forest can be viewed as a set of all realizations of Galton–Watson process G with N initial particles conditioned to have the total progeny n. Let us assume that the process has the offspring distribution $\mathbf{P}\{\xi = k\} = p_k, \quad k = 1, 2, \dots$, and generating function $F(z) = \sum_{k=0}^{\infty} p_k z^k$. Let us introduce a random variable $\xi^{(r)}$ with the distribution $\mathbf{P}\{\xi^{(r)} = k\} = \mathbf{P}\{\xi = k \mid \xi \neq r\}$, and let $\xi_k,\ k = 1, 2, \dots$, be distributed as ξ, and $\xi_k^{(r)},\ k = 1, 2, \dots$, be distributed as $\xi^{(r)}$. Let us define $S_n = \xi_1 + \dots + \xi_n$ and $S_n^{(r)} = \xi_1^{(r)} + \dots + \xi_n^{(r)}$. Using results of Kolchin [1] we can obtain the following theorem about the distribution of number of vertices with a given outdegree:

Theorem 10. *Let $\Lambda_r(N, n)$ be the number of vertices with outdegree r in the Galton–Watson forest with N trees and n non-root vertices. Then*

$$\mathbf{P}\{\Lambda_r(N,n) = k\} = \binom{N+n}{k} p_r^k \, (1 - p_r)^{N+n-k} \, \frac{\mathbf{P}\{S_{N+n-k}^{(r)} = n - kr\}}{\mathbf{P}\{S_{N+n} = n\}}.$$

This theorem shows that in order to obtain the limit distribution of $\Lambda_r(N, n)$ it suffices to find the limit distribution of the sums of auxiliary independent random variables and to use the normal approximation of the binomial distribution. One can find a proof of Theorem 1 in [2] for $r = 0$ and in [3] for $r > 0$. Using Theorem 1 we prove that the limit distribution of $\Lambda_r(N, n)$ is normal with parameters depending on how N and n approach to infinity. We consider three different cases (see theorems 2-4 below; proofs could be found in [2] and [3]). To formulate these theorems we need some notations and assumptions. Assume that the equation $zF'(z) = F(z)$ has a solution $c > 0$ satisfying $F(c) < \infty$ and $F''(c) < \infty$. Then we can assume, without loss of generality, that $E\xi = 1$. Let the variance of ξ exists and equals B. We assume that there exist at least three nonzero probabilities in the offspring distribution including p_0. Let $j^* := \inf\{k > 0 \ : p_k > 0\}$ and $l^* := \inf\{k > 0 \ : p_{j^*+k} > 0\}$. Denote by d and d_r the span of distributions of ξ and $\xi^{(r)}$ respectively. Let w^* be the smallest nonnegative integer such that $j^* + w^*$ determines the span d of the distribution of ξ, and v_r^* be the smallest nonnegative integer such that $j^* + v_r^*$ determines the span d_r of the distribution of $\xi^{(r)}$.

Without loss of generality we may assume that the offspring distribution is given by $\mathbf{P}\{\zeta = k\} = \lambda^k p_k / F(\lambda)$, where λ is positive number within the circle of convergence of $F(z)$. Let $m = m(\lambda)$ and $\sigma^2 = \sigma^2(\lambda)$ be, respectively, the expectation and the variance of ζ. Introduce

$$\sigma_*^2 = p_r(\lambda)(1 - p_r(\lambda) - \frac{(m-r)^2 p_r(\lambda)}{\sigma^2}). \tag{1}$$

Theorem 11.. *Let $N, n \to \infty$ in such a way that n takes values divisible by d_r, $n/N^2 \to 0$. Let*

$$\lambda^{\max(3l^*,\, v_r^*)} n \to \infty \quad \text{if} \quad r = j^*,$$
$$\lambda^{\max(v_r^*,\, w^*)} n \to \infty \quad \text{if} \quad r = j^* + w^*,\ w^* > 0,$$
$$\lambda^{v_r^*} n \to \infty \quad \text{otherwise},$$

where λ is determined by

$$\frac{\lambda F'(\lambda)}{F(\lambda)} = \frac{n}{N+n}.$$

Then

$$\mathbf{P}(\Lambda_r(N,n) = kd_r/d) = \frac{d_r(1+o(1))}{d\sigma_* \sqrt{2\pi(N+n)}} e^{-u^2/2}$$

uniformly in the integers k such that $u = (kd_r/d - (N+n)p_r(\lambda))/\sigma_* \sqrt{N+n}$ lies in any finite fixed interval.

Theorem 12.. *Let $N, n \to \infty$ in such a way that n takes values divisible by d_r, $n/N^2 \to \gamma$ for some $\gamma > 0$. Then*

$$\mathbf{P}(\Lambda_r(N,n) = kd_r/d) = \frac{d_r(1+o(1))}{d\sigma_* \sqrt{2\pi n}} e^{-u^2/2}$$

uniformly in the integers k such that $u = (kd_r/d - (N+n)p_r + \alpha p_r \sqrt{N+n})/\sigma_* \sqrt{n}$ lies in any finite fixed interval, where $\alpha = (r-1)/B\sqrt{\gamma}$ and σ_* is given as in (1) with $\lambda = 1$.

Theorem 13.. *Let $N, n \to \infty$ in such a way that n takes values divisible by d_r, $n/N^2 \to \infty$. Then*

$$\mathbf{P}(\Lambda_r(N,n) = kd_r/d) = \frac{d_r(1+o(1))}{d\sigma_* \sqrt{2\pi n}} e^{-u^2/2}$$

uniformly in the integers k such that $u = (kd_r/d - (N+n)p_r)/\sigma_* \sqrt{n}$ lies in any finite fixed interval, where σ_* is given as in (1) with $\lambda = 1$.

References

[1] Kolchin, V.F. - *Random mappings*. Springer, New York, 1986.

[2] Mylläri, T. - Limit distributions for the number of leaves in a random forest. *Adv. in Applied Prob.*, vol. 34(4), 2002.

[3] Mylläri, T., Pavlov Yu. - Limit distributions of the number of vertices of a given outdegree in a random forest. To appear in *Journal of Mathematical Sciences*.

Tatiana Mylläri
Åbo Akademi University, Åbo, Finland
tatiana.myllari@abo.fi

Destruction of Recursive Trees

Alois Panholzer

ABSTRACT: *We study, for the family of recursive trees, two procedures that destroy trees by successively removing edges. In both variants, one starts with a tree T of size n and chooses one of the $n-1$ edges at random. Removing this edge costs a toll depending on the size of T, given by the toll function t_n and leads to two subtrees T' and T''. In the one-sided variant, the edge-removal procedure will be iterated with the subtree containing the root, whereas in the two-sided variant it will be iterated with both subtrees. For both variants, we study for toll functions $t_n = n^\alpha$ with $\alpha \geq 0$ the total costs (= sum of the tolls of every step) obtained by completely destroying random recursive trees, where we compute for this quantity the asymptotic behaviour of all moments.*

1. Introduction

In this paper, we are considering two recursive edge-removal procedures **P1** ("one-sided destructions") and **P2** ("two-sided destructions") to destroy (rooted) trees. Both variants start with a tree T of size $|T| = n$, where the size measures as usual the number of nodes of T. If $n = 1$ there are no edges that can be removed and both procedures **P1** and **P2** stop, but we assume that this costs the toll t_1. If $n \geq 2$, then one of the $n-1$ edges in the tree will be chosen and afterwards this edge will be removed from T. We assume now that removing this edge costs a certain toll depending on the size of T and which is given by the toll function t_n. After removing this edge, the original tree T falls into two subtrees T' and T'' with sizes $1 \leq |T'|, |T''| \leq n-1$, where one of them (let us assume T') contains the root of T. In the two-sided variant **P2**, the edge-removal procedure will now be applied recursively to both subtrees T' and T'', whereas in the one-sided variant **P1**, the edge-removal procedure (in [7] called "cutting-down") will only be applied to the subtree T' that contains the root.

Thus the procedure **P2** terminates and T has been destroyed by **P2**, when all $n-1$ edges are removed from T, whereas **P1** terminates and thus T has been destroyed by **P1**, when the root of T has been isolated. We are now interested in the total costs (= sum of the costs of every edge-removal step) $C_1(T)$ resp. $C_2(T)$ that occur when destroying a tree T by **P1** resp. **P2**. Of course, these quantities are for $|T| \geq 2$ given recursively by

$$C_1(T) = C_1(T') + t_{|T|}, \quad \text{resp.} \quad C_2(T) = C_2(T') + C_2(T'') + t_{|T|}, \qquad (1)$$

where T', T'' are the subtrees appearing after the first edge-removal step and T' contains the root of T. If $|T| = 1$ then $C_1(T) = C_2(T) = t_1$. An example for destroying a tree by **P1** resp. **P2** is given in Figure 1.

In this paper, we study for toll functions $t_n = n^\alpha$ with $\alpha \geq 0$ the random variables X_n (resp. Y_n), which measure the total costs that accumulate when destroying a *random* recursive tree of size n by the *random* edge-removal procedures

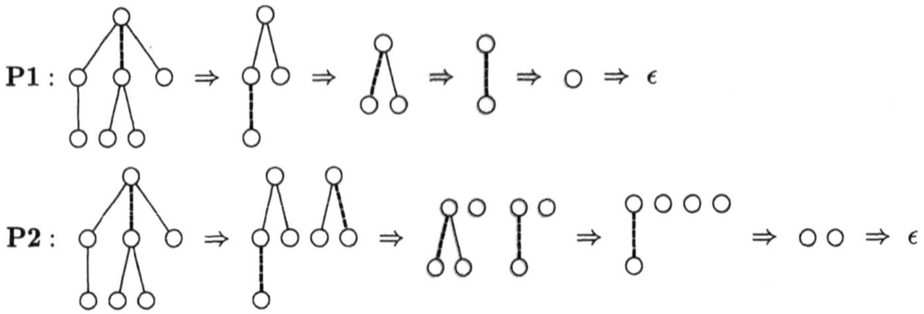

FIGURE 1. Destruction of a tree T of size 7 by the procedures **P1** and **P2**. Using the toll function $t_n = n$ for $n \geq 1$, the one-sided destruction has total costs $C_1(T) = 17$ and the two-sided destruction has total costs $C_2(T) = 28$. Here, ϵ denotes the empty tree.

P1 (resp. **P2**). The tree family considered and the probability model used are described next.

The family of recursive trees can be defined in the following way. A rooted labelled tree T of size n with labels $1, 2, \ldots, n$ is a recursive tree, if the root is labelled with 1, and for each node v holds that the labels of the vertices on the unique path from the root to v form an increasing sequence. It is seen easily that there are $T_n = (n-1)!$ different size-n recursive trees. As the model of randomness we use the random tree model, which means that every recursive tree of size n can be chosen as input for the edge-removal procedures with equal probability $\frac{1}{(n-1)!}$. We speak then about random recursive trees or uniform recursive trees. For a survey of applications and results on random recursive trees see [5].

Further, we will always assume for **P1** resp. **P2** that the removed edges are at each stage chosen at random from the remaining tree. We speak thus about the *random* edge-removal procedures.

If one chooses the toll function $t_n = 1$ for $n \geq 2$ with $t_1 = 0$, then X_n measures exactly the number of edges that are removed by the cutting-down procedure **P1** to destroy a random size-n tree. This quantity was studied by Meir and Moon in [7], where they obtained the following results for the first two moments: $\mathbb{E}(X_n) \sim \frac{n}{\log n}$ and $\mathbb{E}(X_n^2) \sim \frac{n^2}{\log^2 n}$. Choosing the toll function $t_n = 1$, the (corresponding) quantity X_n was also studied for other tree families, see e. g. [6, 8]. For unrooted labelled trees, the (corresponding) quantity Y_n was studied for a few toll functions t_n in [4] and for general toll functions $t_n = n^\alpha$ in [1].

2. Results and mathematical preliminaries

We analyzed the behaviour of the moments $\mathbb{E}(X_n^s)$ resp. $\mathbb{E}(Y_n^s)$ of the total costs when destroying size-n recursive trees with procedures **P1** resp. **P2** for toll functions $t_n = n^\alpha$ with $\alpha \geq 0$ resp. $\alpha > 0$ ($\alpha = 0$ for Y_n is trivial, since then $Y_n = 2n-1$) and have obtained the following results. Here $H_n := \sum_{k=1}^{n} \frac{1}{k}$ denotes as usual the n-th harmonic number and $\Psi(x) := \frac{d}{dx} \log \Gamma(x)$ denotes the Psi-function.

Theorem 2.1. *The s-th moments $\mathbb{E}(X_n^s)$ resp. s-th centered moments $\mathbb{E}([X_n - \mathbb{E}(X_n)]^s)$ of the total costs X_n incurred by one-sided destructions of random recursive trees for toll functions $t_n = n^\alpha$ with $\alpha \geq 0$ are, for s an integer and $n \to \infty$, asymptotically given by*

$$\mathbb{E}(X_n^s) = \frac{1}{(\alpha+1)^s}\frac{n^{s(\alpha+1)}}{\log^s n} + \frac{H_s + s + (\alpha+1)\sum_{l=1}^{s}\Psi\big(l(\alpha+1)\big)}{(\alpha+1)^{s+1}}\frac{n^{s(\alpha+1)}}{\log^{s+1} n} \tag{2}$$
$$+\mathcal{O}\Big(\frac{n^{s(\alpha+1)}}{\log^{s+2} n}\Big),\ s \geq 1,$$

$$\mathbb{E}([X_n - \mathbb{E}(X_n)]^s) = \frac{\frac{(-1)^{s-1}}{(\alpha+1)s} + \sum_{l=0}^{s-1}\binom{s-1}{l}(-1)^{s-1-l}\Psi\big((l+1)(\alpha+1)\big)}{(\alpha+1)^s}\frac{n^{s(\alpha+1)}}{\log^{s+1} n}$$
$$+\mathcal{O}\Big(\frac{n^{s(\alpha+1)}}{\log^{s+2} n}\Big),\ s \geq 2. \tag{3}$$

Theorem 2.2. *The s-th moments $\mathbb{E}(Y_n^s)$ resp. s-th centered moments $\mathbb{E}([Y_n - \mathbb{E}(Y_n)]^s)$ of the total costs Y_n incurred by two-sided destructions of random recursive trees for toll functions $t_n = n^\alpha$ with $\alpha > 0$ are, for s an integer and $n \to \infty$, asymptotically given by*

$$\mathbb{E}(Y_n^s) = \frac{1}{(\alpha+1)^s}\frac{n^{s(\alpha+1)}}{\log^s n} + \gamma_s\frac{n^{s(\alpha+1)}}{\log^{s+1} n} + \mathcal{O}\Big(\frac{n^{s(\alpha+1)}}{\log^{s+2} n}\Big),\ s \geq 1, \tag{4}$$

$$\mathbb{E}([Y_n - \mathbb{E}(Y_n)]^s) = \delta_s\frac{n^{s(\alpha+1)}}{\log^{s+1} n} + \mathcal{O}\Big(\frac{n^{s(\alpha+1)}}{\log^{s+2} n}\Big),\ s \geq 2, \tag{5}$$

where the appearing constants γ_s and δ_s are given by

$$\gamma_s = \frac{\frac{s}{\alpha+1} + \sum_{l=1}^{s}\Psi\big(l(\alpha+1)\big) + \sum_{l=1}^{s}\frac{1}{l(\alpha+1)-1} + \sum_{l=1}^{s}\sum_{j=1}^{l-1}\binom{l}{j}\frac{\Gamma\big(j(\alpha+1)+1\big)\Gamma\big((l-j)(\alpha+1)-1\big)}{\Gamma\big(l(\alpha+1)+1\big)}}{(\alpha+1)^s},$$

$$\delta_s = \frac{1}{(\alpha+1)^s}\Bigg[\sum_{l=0}^{s-1}\binom{s-1}{l}(-1)^{s-1-l}\Psi\big((l+1)(\alpha+1)\big) + \frac{(-1)^{s-1}(s-1)!(\alpha+1)^{s-1}}{\prod_{l=1}^{s}(l(\alpha+1)-1)}$$
$$+\sum_{l=1}^{s}\binom{s-1}{l-1}(-1)^{s-l}\sum_{j=1}^{l-1}\binom{l}{j}\frac{\Gamma\big(j(\alpha+1)+1\big)\Gamma\big((l-j)(\alpha+1)-1\big)}{\Gamma\big(l(\alpha+1)+1\big)}\Bigg].$$

From these theorems follow that for the toll $t_n = n^\alpha$ the r. v. $\frac{(\alpha+1)\log n}{n^{\alpha+1}}X_n$ resp. $\frac{(\alpha+1)\log n}{n^{\alpha+1}}Y_n$ converge in probability to 1 with convergence of all moments. Furthermore, if W_n (resp. $\tilde{W}_n$) denotes a zero-mean and unit-variance normalization of X_n (resp. Y_n), then W_n (resp. $\tilde{W}_n$) has s-th moments growing like $\log^{\frac{s}{2}-1} n$ for $s \geq 2$. This shows that if W_n (resp. $\tilde{W}_n$) has a limiting distribution, this cannot be established by the method of moments. It remains still open whether there exists a limiting distribution for some centered and scaled version of X_n (resp. Y_n). Also it seems surprising, that we do not have a lead-order discrepancy between X_n and Y_n, although it holds of course $Y_n \geq X_n$.

To prove Theorem 2.1 and Theorem 2.2, we use a recursive approach, as it was done for one-sided destructions and the special toll function $t_n = 1$ in [7]. That this recursive approach is indeed permitted is stated in Lemma 3.1. The appearing distribution recurrences (6) lead to recurrences (8) and (12) for the s-th moments. Using generating functions, we obtain differential equations for every s-th moment that can be solved, where the solutions (11) and (15) are composed of the operations differentiation, integration and the Hadamard product of generating functions of the lower moments and the toll function. The Hadamard product $F(z) \odot G(z)$ of generating functions $F(z) = \sum_{n\ge0} f_n z^n$ and $G(z) = \sum_{n\ge0} g_n z^n$ is defined by $F(z) \odot G(z) := \sum_{n\ge0} f_n g_n z^n$. Moreover we use in this paper the abbreviation $F^{\odot s}(z) := \underbrace{F(z) \odot \cdots \odot F(z)}_{s \text{ times}} = \sum_{n\ge0} f_n^s z^n$.

To extract the asymptotic information from the solutions (11) and (15) we cannot use the extension of the "singularity-analysis-toolbox" given in [1], since the theorems shown therein deal only with positive integral powers of logarithmic terms whereas here occur negative powers. Thus we will go back to a quite elementary, but here efficient approach that computes the asymptotic growth directly at the level of the coefficients. This has the advantage that the effect of the operations integration, differentiation and Hadamard product to the growth of the coefficients can be described easily, whereas of course difficulties may arise when computing the growth of the coefficients of the Cauchy product. But for the problem considered here, the two summation formulæ (16) and (17) are sufficient. Using them, the computations for one-sided destructions are done in Section 5 and (only sketched) for two-sided destructions in Section 6.

3. Recurrences for the quantities considered

The basic idea in our approach is to study the distribution recurrences

$$X_n \stackrel{\mathcal{L}}{=} X_{K_n} + t_n,\ n \ge 2,\ X_1 = t_1; \quad Y_n \stackrel{\mathcal{L}}{=} Y_{K_n} + \tilde{Y}_{n-K_n} + t_n,\ n \ge 2,\ Y_1 = t_1, \quad (6)$$

where Y_n and $\tilde{Y}_n$ are identically and independently distributed random variables and K_n is independent of X_n, Y_n and $\tilde{Y}_n$. The random variable K_n will be given by the splitting probabilities $p_{n,k} := \mathbb{P}\{K_n = k\}$, for $1 \le k \le n-1$, where $p_{n,k}$ is the probability that after removing a random edge from a size-n random recursive tree, the subtree containing the root has size k.

To reduce this problem to a study of (6), it is of course necessary that randomness is preserved by cutting-off a random edge. This means that after removing a randomly selected edge from a size-n random recursive tree, the remaining subtrees with sizes k resp. $n-k$ are after natural-order-preserving relabellings of the nodes with labels $\{1,\dots,k\}$ resp. $\{1,\dots,n-k\}$ again *random* recursive trees of sizes k resp. $n-k$. This property of random recursive trees was shown implicitly in [7] when computing the splitting probabilities $p_{n,k}$. Since this is a crucial point in our approach, we will restate their proof. Thus randomness is actually preserved by cutting-off a random edge and the recurrences (6) follow directly from (1).

Lemma 3.1 (Meir and Moon, 1974). *Let us assume that we choose a random recursive tree T of size n and also one of its $n-1$ edges at random, and after removing this edge, the remaining subtrees T' resp. T'' are of sizes k resp. $n-k$, where we further assume that T' contains the root of T. Then it holds that, after an order-preserving relabelling of the nodes, both subtrees are random recursive trees of sizes k resp. $n-k$ and the splitting probabilities $p_{n,k}$ are given by*

$$p_{n,k} = \frac{n}{(n-1)(n-k)(n-k+1)}, \quad \textit{for } 1 \le k \le n-1. \quad (7)$$

Proof. Starting with a size-n recursive tree and removing one of its $n-1$ edges, we obtain a subtree T' of size $1 \leq k \leq n-1$ which contains the root of T and another subtree T'' of size $n-k$. After the order-preserving relabellings, we can consider both subtrees as recursive trees. Now we want to count, how often we can obtain a particular pair (T', T'') of recursive trees with sizes k resp. $n-k$, when removing one edge of recursive trees of size n. It will turn out that this quantity $w(T',T'')$ depends only on the sizes k and n, not on the particular chosen trees T' and T'' and the lemma will be proven. Equivalently we can go the other way around and ask, in how many ways $w(T',T'')$ can we reconstruct size-n recursive trees from the pair (T', T''). Let us assume that the removed edge originally connected the root of T'' with the node with label j in T'. Then all $n-k$ nodes in T'' must have labels larger than j. We have thus $\binom{n-j}{n-k}$ possibilities to select them from $\{j+1,\dots,n\}$ and distribute them order preserving to T'', whereas the remaining $k-j$ labels from $\{j+1,\dots,n\}$ are distributed order preserving to the nodes of T' with labels larger than j. This gives in that case $\binom{n-j}{n-k}$ different size-n recursive trees. By summing up we find that independently from the pair T' and T'', we always have $w(T',T'') = \sum_{j=1}^{k}\binom{n-j}{n-k} = \binom{n}{k-1}$ and therefore randomness is preserved.

Due to the given bijection between pairs of recursive trees with sizes k and $n-k$ and the pairs consisting of a size-n recursive tree and one of its $n-1$ edges, we obtain the equations $\sum_{k=1}^{n-1}\binom{n}{k-1}T_kT_{n-k} = (n-1)T_n$ and $p_{n,k} = \binom{n}{k-1}\frac{T_kT_{n-k}}{(n-1)T_n}$. Thus (7) is also shown.

□

From the distribution recurrence (6), we obtain then the following recurrences for the s-th moments $\mu_n^{[s]} := \mathbb{E}(X_n^s)$ of X_n:

$$\mu_n^{[s]} = \mathbb{E}\big([X_{K_n}+t_n]^s\big) = \sum_{s_1+s_2=s}\binom{s}{s_1}t_n^{s_1}\mathbb{E}\big(X_{K_n}^{s_2}\big) = \sum_{s_1+s_2=s}\binom{s}{s_1}t_n^{s_1}\sum_{k=1}^{n-1}p_{n,k}\mu_k^{[s_2]}.$$

We write them as

$$\mu_n^{[s]} = \sum_{k=1}^{n-1}p_{n,k}\mu_k^{[s]} + r_n^{[s]}, \quad \text{for } n \geq 2, \qquad \mu_1^{[s]} = t_1^s, \tag{8}$$

with $r_n^{[s]} = \sum_{\substack{s_1+s_2=s,\\ s_2<s}}\binom{s}{s_1}t_n^{s_1}\sum_{k=1}^{n-1}p_{n,k}\mu_k^{[s_2]}$.

Introducing the generating functions resp. the common abbreviation

$$\mu^{[s]}(z) := \sum_{n\geq 1}\mu_n^{[s]}\frac{z^n}{n}, \quad t(z) := \sum_{n\geq 1}t_nz^n, \quad r^{[s]}(z) := \sum_{n\geq 2}(n-1)r_n^{[s]}\frac{z^n}{n}, \quad L(z) := \log\frac{1}{1-z},$$

we obtain from (8) by multiplying with $\frac{n-1}{n}z^n$ and summing up:

$$\sum_{n\geq 2}\mu_n^{[s]}\frac{n-1}{n}z^n = z\frac{d}{dz}\mu^{[s]}(z) - \mu^{[s]}(z),$$

$$\sum_{n\geq 2}\sum_{k=1}^{n-1}\frac{1}{(n-k)(n-k+1)}\mu_k^{[s]}z^n = \Big(\frac{d}{dz}\mu^{[s]}(z)\Big)\big(z-(1-z)L(z)\big),$$

$$r^{[s]}(z) = \sum_{\substack{s_1+s_2=s,\\ s_2<s}}\binom{s}{s_1}t^{\odot s_1}(z)\odot\Big[\Big(\frac{d}{dz}\mu^{[s_2]}(z)\Big)\big(z-(1-z)L(z)\big)\Big]. \tag{9}$$

This gives thus the differential equation

$$(1-z)L(z)\frac{d}{dz}\mu^{[s]}(z) - \mu^{[s]}(z) = r^{[s]}(z), \tag{10}$$

where $r^{[s]}(z)$ is given by (9). Solving and adapting to the initial condition $\frac{d}{dz}\mu^{[s]}(z)\big|_{z=0} = \mu_1^{[s]} = t_1^s$, we obtain the solution

$$\mu^{[s]}(z) = L(z)\int_{t=0}^{z} \frac{r^{[s]}(t)}{(1-t)L^2(t)}dt + t_1^s L(z). \tag{11}$$

Analogously we obtain from the distribution recurrence (6) the following recurrences for the s-th moments $\lambda_n^{[s]} := \mathbb{E}(Y_n^s)$ of Y_n:

$$\lambda_n^{[s]} = \mathbb{E}\big([Y_{K_n} + \tilde{Y}_{n-K_n} + t_n]^s\big) = \sum_{s_1+s_2+s_3=s} \binom{s}{s_1, s_2, s_3} t_n^{s_1} \sum_{k=1}^{n-1} p_{n,k}\lambda_k^{[s_2]}\lambda_{n-k}^{[s_3]}.$$

We write them as

$$\lambda_n^{[s]} = \sum_{k=1}^{n-1} p_{n,k}(\lambda_k^{[s]} + \lambda_{n-k}^{[s]}) + r_n^{[s]}, \quad \text{for } n \geq 2, \qquad \lambda_1^{[s]} = t_1^s, \tag{12}$$

with $r_n^{[s]} = \sum_{\substack{s_1+s_2+s_3=s,\\ s_2,s_3<s}} \binom{s}{s_1,s_2,s_3} t_n^{s_1} \sum_{k=1}^{n-1} p_{n,k}\lambda_k^{[s_2]}\lambda_{n-k}^{[s_3]}$.

With the generating function $\lambda^{[s]}(z) := \sum_{n\geq 1} \lambda_n^{[s]}\frac{z^n}{n}$, we obtain from (12) by multiplying with $\frac{n-1}{n}z^n$ and summing up:

$$\sum_{n\geq 2} \lambda_n^{[s]}\frac{n-1}{n}z^n = z\frac{d}{dz}\lambda^{[s]}(z) - \lambda^{[s]}(z),$$

$$\sum_{n\geq 2}\sum_{k=1}^{n-1} \frac{1}{(n-k)(n-k+1)}\lambda_k^{[s]}z^n = \Big(\frac{d}{dz}\lambda^{[s]}(z)\Big)\big(z-(1-z)L(z)\big),$$

$$\sum_{n\geq 2}\sum_{k=1}^{n-1} \frac{1}{(n-k)(n-k+1)}\lambda_{n-k}^{[s]}z^n = \frac{1}{1-z}\int_{t=0}^{z}\lambda^{[s]}(t)dt,$$

$$r^{[s]}(z) = \sum_{\substack{s_1+s_2+s_3=s,\\ s_2,s_3<s}} \binom{s}{s_1,s_2,s_3} t^{\odot s_1}(z) \odot \Big[\Big(\frac{d}{dz}\lambda^{[s_2]}(z)\Big)\int_{t=0}^{z}\lambda^{[s_3]}(t)dt\Big]. \tag{13}$$

This leads to the following equation with $r^{[s]}(z)$ given by (13):

$$(1-z)L(z)\frac{d}{dz}\lambda^{[s]}(z) - \lambda^{[s]}(z) - \frac{1}{1-z}\int_{t=0}^{z}\lambda^{[s]}(t)dt = r^{[s]}(z). \tag{14}$$

By introducing the functions $\tilde{\lambda}^{[s]}(z) := \frac{d}{dz}\lambda^{[s]}(z) = \sum_{n\geq 1}\lambda_n^{[s]}z^{n-1}$, this equation can be transformed to the first order differential equation

$$\frac{d}{dz}\tilde{\lambda}^{[s]}(z) - \frac{2}{1-z}\tilde{\lambda}^{[s]}(z) = \frac{1}{(1-z)^2L(z)}\frac{d}{dz}\big((1-z)r^{[s]}(z)\big),$$

which has by adapting to the initial condition $\tilde{\lambda}^{[s]}(0) = \lambda_1^{[s]} = t_1^s$ the solution

$$\tilde{\lambda}^{[s]}(z) = \frac{1}{(1-z)^2}\int_{t=0}^{z}\frac{1}{L(t)}\frac{d}{dt}\big((1-t)r^{[s]}(t)\big)dt + \frac{t_1^s}{(1-z)^2}. \tag{15}$$

4. Summation lemmata

Essential to our approach are the following auxiliary formulæ to control the asymptotic growth of the appearing convolutions.

Lemma 4.1. *For $\alpha, \beta, p, q \in \mathbb{R}$ with $\alpha, \beta > -1$ and $p, q \geq 0$ the following expansions hold:*

$$\begin{aligned}\sum_{k=2}^{n-2}\frac{k^\beta\,(n-k)^\alpha}{\log^q(k)\,\log^p(n-k)} &= \frac{n^{\alpha+\beta+1}}{\log^{p+q}n}\,\frac{\Gamma(\alpha+1)\Gamma(\beta+1)}{\Gamma(\alpha+\beta+2)}\times\\ &\times\left(1-\frac{q\,\Psi(\beta+1)+p\,\Psi(\alpha+1)-(p+q)\Psi(\alpha+\beta+2)}{\log n}+\mathcal{O}\Big(\frac{1}{\log^2 n}\Big)\right),\end{aligned} \tag{16}$$

$$\sum_{k=2}^{n-2}\frac{(n-k)^\alpha}{k(k-1)\log^p(n-k)} = \frac{n^\alpha}{\log^p n}+\mathcal{O}\Big(\frac{n^{\alpha-1}}{\log^{p-2}n}\Big)+\mathcal{O}\Big(\frac{\log^2 n}{n}\Big). \tag{17}$$

Proof. To obtain (16), we start with the following expansion:

$$\begin{aligned}\sum_{k=2}^{n-2}\frac{k^\beta(n-k)^\alpha}{\log^q(k)\log^p(n-k)} &= \frac{n^{\alpha+\beta}}{\log^{p+q}n}\sum_{k=\lfloor\log n\rfloor}^{n-\lfloor\log n\rfloor}\frac{\left(\frac{k}{n}\right)^\beta\left(1-\frac{k}{n}\right)^\alpha}{\left(1+\frac{\log\frac{k}{n}}{\log n}\right)^q\left(1+\frac{\log(1-\frac{k}{n})}{\log n}\right)^p}\\ &+\sum_{k=2}^{\lfloor\log n\rfloor-1}\frac{k^\beta(n-k)^\alpha}{\log^q(k)\log^p(n-k)}+\sum_{k=n+1-\lfloor\log n\rfloor}^{n-2}\frac{k^\beta(n-k)^\alpha}{\log^q(k)\log^p(n-k)}\\ &= \frac{n^{\alpha+\beta+1}}{\log^{p+q}n}\sum_{k=\lfloor\log n\rfloor}^{n-\lfloor\log n\rfloor}\frac{1}{n}\Big(\frac{k}{n}\Big)^\beta\Big(1-\frac{k}{n}\Big)^\alpha\times\Big[1-\frac{q\log\frac{k}{n}+p\log(1-\frac{k}{n})}{\log n}+\mathcal{O}\Big(\frac{\log^2\frac{k}{n}}{\log^2 n}\Big)\\ &+\mathcal{O}\Big(\frac{\left(\log\frac{k}{n}\right)\left(\log(1-\frac{k}{n})\right)}{\log^2 n}\Big)+\mathcal{O}\Big(\frac{\log^2(1-\frac{k}{n})}{\log^2 n}\Big)\Big]+\mathcal{O}\big(n^\alpha(\log n)^{\beta+1-p}\big)+\mathcal{O}\big(n^\beta(\log n)^{\alpha+1-q}\big).\end{aligned}$$

The appearing sums can be considered as Riemann sums for the Beta integral resp. their derivatives. As examples, we give the following two computations:

$$\begin{aligned}\sum_{k=\lfloor\log n\rfloor}^{n-\lfloor\log n\rfloor}\frac{1}{n}\Big(\frac{k}{n}\Big)^\beta\Big(1-\frac{k}{n}\Big)^\alpha &= \int_{x=0}^{1}x^\beta(1-x)^\alpha dx+\mathcal{O}\Big(\frac{(\log n)^\beta}{n^{\beta+1}}\Big)+\mathcal{O}\Big(\frac{(\log n)^\alpha}{n^{\alpha+1}}\Big)\\ &= \frac{\Gamma(\alpha+1)\Gamma(\beta+1)}{\Gamma(\alpha+\beta+2)}+\mathcal{O}\Big(\frac{(\log n)^\beta}{n^{\beta+1}}\Big)+\mathcal{O}\Big(\frac{(\log n)^\alpha}{n^{\alpha+1}}\Big),\\ \sum_{k=\lfloor\log n\rfloor}^{n-\lfloor\log n\rfloor}\frac{1}{n}\Big(\frac{k}{n}\Big)^\beta\Big(1-\frac{k}{n}\Big)^\alpha\log\frac{k}{n} &= \int_{x=0}^{1}x^\beta(1-x)^\alpha\log x\,dx+\mathcal{O}\Big(\frac{(\log n)^{\beta+1}}{n^{\beta+1}}\Big)\\ +\mathcal{O}\Big(\frac{(\log n)^\alpha}{n^{\alpha+1}}\Big) &= \frac{\partial}{\partial\beta}\Big(\frac{\Gamma(\alpha+1)\Gamma(\beta+1)}{\Gamma(\alpha+\beta+2)}\Big)+\mathcal{O}\Big(\frac{(\log n)^{\beta+1}}{n^{\beta+1}}\Big)+\mathcal{O}\Big(\frac{(\log n)^\alpha}{n^{\alpha+1}}\Big),\end{aligned}$$

where we obtain $\frac{\partial}{\partial\beta}\left(\frac{\Gamma(\alpha+1)\Gamma(\beta+1)}{\Gamma(\alpha+\beta+2)}\right) = \frac{\Gamma(\alpha+1)\Gamma(\beta+1)}{\Gamma(\alpha+\beta+2)}\big(\Psi(\beta+1)-\Psi(\alpha+\beta+2)\big)$. Analogously one can treat the remaining sums which leads eventually to (16).

It remains to prove (17). We start with dissecting the summation interval

$$\sum_{k=2}^{n-2} \frac{(n-k)^\alpha}{k(k-1)\log^p(n-k)} = \sum_{k=2}^{\lfloor \frac{n}{\log n} \rfloor} \frac{(n-k)^\alpha}{k(k-1)\log^p(n-k)} + \sum_{k=\lfloor \frac{n}{\log n} \rfloor+1}^{n-2} \frac{(n-k)^\alpha}{k(k-1)\log^p(n-k)}$$

$$= \frac{n^\alpha}{\log^p n} \sum_{k=2}^{\lfloor \frac{n}{\log n} \rfloor} \frac{(1-\frac{k}{n})^\alpha}{k(k-1)(1+\frac{\log(1-\frac{k}{n})}{\log n})^p} + \mathcal{O}\Big(\frac{n^{\alpha-1}}{\log^{p-2} n}\Big) + \mathcal{O}\Big(\frac{\log^2 n}{n}\Big),$$

where the remainder bounds are coming from the estimate $\frac{(n-k)^\alpha}{\log^p(n-k)} = \mathcal{O}(\frac{n^\alpha}{\log^p n}) + \mathcal{O}(1)$, which combines bounds for $\alpha > 0$ and $\alpha \le 0$ in the considered range $\lfloor \frac{n}{\log n} \rfloor < k \le n-2$. The remaining sum can be evaluated asymptotically which gives the main term in (17):

$$\sum_{k=2}^{\lfloor \frac{n}{\log n} \rfloor} \frac{(1-\frac{k}{n})^\alpha}{k(k-1)(1+\frac{\log(1-\frac{k}{n})}{\log n})^p} = \sum_{k=2}^{\lfloor \frac{n}{\log n} \rfloor} \frac{1}{k(k-1)}\Big(1+\mathcal{O}\big(\frac{k}{n}\big)\Big)$$

$$= \sum_{k=2}^{\lfloor \frac{n}{\log n} \rfloor} \frac{1}{k(k-1)} + \mathcal{O}\Big(\frac{\sum_{k=2}^{\lfloor \frac{n}{\log n} \rfloor} \frac{1}{k-1}}{n}\Big) = 1 + \mathcal{O}\Big(\frac{\log n}{n}\Big).$$

□

5. One-sided destructions

Now we want to prove the results of Theorem 2.1 concerning the asymptotic behaviour of the s-th moments of the total costs X_n when destroying random recursive trees with toll functions $t_n = n^\alpha$, for $n \ge 2$ with $\alpha \ge 0$ and $t_1 = 0$. Choosing $t_1 = 0$ instead of $t_1 = 1$ has of course no influence to the stated asymptotic behaviour ($X_n^* = X_n + 1$, if X_n^* measures the total costs with toll function $t_n = n^\alpha$, for $n \ge 1$), but (11) is then slightly simpler.

In order to reduce extracting coefficients from (11) to an application of formulæ (16) and (17) and to avoid dealing with convolutions of functions growing as n^{-1}, we introduce the generating functions $\tilde{\mu}^{[s]}(z) := \frac{d}{dz}\mu^{[s]}(z) = \sum_{n\ge1} \mu_n^{[s]} z^{n-1}$ and differentiate (11):

$$\tilde{\mu}^{[s]}(z) = \frac{r^{[s]}(z)}{(1-z)L(z)} + \frac{1}{1-z}\int_{t=0}^{z} \frac{r^{[s]}(t)}{(1-t)L^2(t)}dt, \tag{18}$$

where $r^{[s]}(z)$ is given by (9) with there appearing functions $t(z) = \sum_{n\ge2} n^\alpha z^n$.

Now we want to show the expansion (2) by induction, where we additionally use, for $\beta, q > 0$, the asymptotic growth of the coefficients (see [2])

$$[z^n]\frac{1}{(1-z)^\beta (L(z))^q} = \frac{n^{\beta-1}}{\Gamma(\beta)\log^q n}\Big(1 + \frac{q\Psi(\beta)}{\log n} + \mathcal{O}\big(\frac{1}{\log^2 n}\big)\Big). \tag{19}$$

We further use the trivial effect to the growth of the coefficients when differentiating and integrating generating functions $F(z) = \sum_{n\ge2} f_n z^n$ with $f_n = \frac{n^\beta}{\log^q n}$:

$$[z^n]\int_{t=0}^{z} F(t)dt = \frac{n^{\beta-1}}{\log^q n}\Big(1+\mathcal{O}\big(\tfrac{1}{n}\big)\Big), \quad [z^n]\frac{d}{dz}F(z) = \frac{n^{\beta+1}}{\log^q n}\Big(1+\mathcal{O}\big(\tfrac{1}{n}\big)\Big). \tag{20}$$

Since $\mu^{[0]}(z) = \sum_{n\geq 1} \frac{z^n}{n} = L(z)$ and $r^{[1]}(z) = t(z) \odot \left[\frac{1}{1-z} - 1 - L(z)\right]$, we obtain

$$[z^n]t(z) = n^\alpha, \quad [z^n]\Big(\frac{1}{1-z} - 1 - L(z)\Big) = 1 - \frac{1}{n}, \quad [z^n]r^{[1]}(z) = n^\alpha - n^{\alpha-1}, \quad \text{for } n \geq 2.$$

Now we use (16) and (19) to obtain

$$\begin{aligned}
[z^n]\frac{r^{[1]}(z)}{(1-z)L(z)} &= \sum_{k=2}^{n-2} \Big(\tfrac{1}{\log k} + \tfrac{\Psi(1)}{\log^2 k} + \mathcal{O}\big(\tfrac{1}{\log^3 k}\big)\Big)\Big((n-k)^\alpha + \mathcal{O}\big((n-k)^{\alpha-1}\big)\Big) + \mathcal{O}(n^\alpha) \\
&= \sum_{k=2}^{n-2} \frac{(n-k)^\alpha}{\log k} + \Psi(1) \sum_{k=2}^{n-2} \frac{(n-k)^\alpha}{\log^2 k} + \mathcal{O}\Big(\sum_{k=2}^{n-2} \tfrac{(n-k)^\alpha}{\log^3 k}\Big) + \mathcal{O}\Big(\sum_{k=2}^{n-2} \tfrac{(n-k)^{\alpha-1}}{\log k}\Big) + \mathcal{O}(n^\alpha) \\
&= \frac{1}{\alpha+1}\frac{n^{\alpha+1}}{\log n} + \frac{\Psi(\alpha+2)}{\alpha+1}\frac{n^{\alpha+1}}{\log^2 n} + \mathcal{O}\Big(\frac{n^{\alpha+1}}{\log^3 n}\Big)
\end{aligned} \tag{21}$$

and

$$\begin{aligned}
[z^n]\frac{r^{[1]}(z)}{(1-z)L^2(z)} &= \sum_{k=2}^{n-2} \Big(\frac{1}{\log^2 k} + \mathcal{O}\big(\frac{1}{\log^3 k}\big)\Big)\Big((n-k)^\alpha + \mathcal{O}\big((n-k)^{\alpha-1}\big)\Big) + \mathcal{O}(n^\alpha) \\
&= \sum_{k=2}^{n-2} \frac{(n-k)^\alpha}{\log^2 k} + \mathcal{O}\Big(\sum_{k=2}^{n-2} \frac{(n-k)^\alpha}{\log^3 k}\Big) + \mathcal{O}\Big(\sum_{k=2}^{n-2} \frac{(n-k)^{\alpha-1}}{\log^2 k}\Big) + \mathcal{O}(n^\alpha) \\
&= \frac{1}{\alpha+1}\frac{n^{\alpha+1}}{\log^2 n} + \mathcal{O}\Big(\frac{n^{\alpha+1}}{\log^3 n}\Big).
\end{aligned}$$

We proceed with $[z^n]\int_{t=0}^{z} \frac{r^{[1]}(t)}{(1-t)L^2(t)}dt = \frac{1}{\alpha+1}\frac{n^\alpha}{\log^2 n} + \mathcal{O}\Big(\frac{n^\alpha}{\log^3 n}\Big)$ and another application of the convolution formula (16):

$$\begin{aligned}
[z^n]\frac{1}{1-z}\int_{t=0}^{z} \frac{r^{[1]}(t)}{(1-t)L^2(t)}dt &= \frac{1}{\alpha+1}\sum_{k=2}^{n-2} \frac{(n-k)^\alpha}{\log^2(n-k)} + \mathcal{O}\Big(\sum_{k=2}^{n-2} \tfrac{(n-k)^\alpha}{\log^3(n-k)}\Big) \\
&+ \mathcal{O}\Big(\tfrac{n^\alpha}{\log^2 n}\Big) = \frac{1}{(\alpha+1)^2}\frac{n^{\alpha+1}}{\log^2 n} + \mathcal{O}\Big(\frac{n^{\alpha+1}}{\log^3 n}\Big).
\end{aligned} \tag{22}$$

Adding (21) and (22), we obtain finally

$$\mu_{n+1}^{[1]} = [z^n]\tilde{\mu}^{[1]}(z) = \frac{1}{\alpha+1}\frac{n^{\alpha+1}}{\log n} + \frac{2+(\alpha+1)\Psi(\alpha+1)}{(\alpha+1)^2}\frac{n^{\alpha+1}}{\log^2 n} + \mathcal{O}\Big(\frac{n^{\alpha+1}}{\log^3 n}\Big).$$

Substituting $n-1$ for n, it follows that expansion (2) is valid for $s=1$.

Now we assume that for a given $s \geq 2$, all moments $\mathbb{E}(X_n^r) = \mu_n^{[r]}$ with $1 \leq r < s$ have the expansion (2). We want to compute then the asymptotic growth of the coefficients $[z^n]r^{[s]}(z)$. To do this, we use $[z^n]t^{\odot s_1}(z) = n^{s_1\alpha}$ and the expansion given below which follows from (17) and (20):

$$[z^n]\Big(\frac{d}{dz}\mu^{[s_2]}(z)\Big)\big(z - (1-z)L(z)\big) =$$

- $1 + \mathcal{O}\big(\frac{1}{n}\big)$, for $s_2 = 0$,
- $$\sum_{k=2}^{n-2} \frac{1}{k(k-1)}\Big[\frac{1}{(\alpha+1)^{s_2}}\frac{(n+1-k)^{s_2(\alpha+1)}}{\log^{s_2}(n+1-k)} + \frac{H_{s_2} + s_2 + (\alpha+1)\sum_{l=1}^{s_2}\Psi\big(l(\alpha+1)\big)}{(\alpha+1)^{s_2+1}} \times \\ \times \frac{(n+1-k)^{s_2(\alpha+1)}}{\log^{s_2+1}(n+1-k)} + \mathcal{O}\Big(\frac{(n+1-k)^{s_2(\alpha+1)}}{\log^{s_2+2}(n+1-k)}\Big)\Big] + \mathcal{O}\Big(\frac{1}{n^2}\Big)$$

$$= \frac{1}{(\alpha+1)^{s_2}} \frac{n^{s_2(\alpha+1)}}{\log^{s_2} n} + \frac{H_{s_2} + s_2 + (\alpha+1)\sum_{l=1}^{s_2} \Psi\big(l(\alpha+1)\big)}{(\alpha+1)^{s_2+1}} \frac{n^{s_2(\alpha+1)}}{\log^{s_2+1} n}$$
$$+ \mathcal{O}\Big(\frac{n^{s_2(\alpha+1)}}{\log^{s_2+2} n}\Big), \quad \text{for } 1 \le s_2 < s.$$

Thus for $s_1 + s_2 = s$ and the restriction $0 \le s_2 < s$, we can describe the growth of the summands of $[z^n] r^{[s]}(z)$:

$$[z^n] \binom{s}{s_1} t^{\odot s_1} \odot \Big[\big(\frac{d}{dz} \mu^{[s_2]}(z)\big)\big(z - (1-z)L(z)\big) \Big] =$$

- $n^{s\alpha} + \mathcal{O}(n^{s\alpha - 1})$, for $s_2 = 0$,
- $\binom{s}{s_1} \frac{1}{(\alpha+1)^{s_2}} \frac{n^{s\alpha+s_2}}{\log^{s_2} n} + \binom{s}{s_1} \frac{H_{s_2} + s_2 + (\alpha+1)\sum_{l=1}^{s_2} \Psi\big(l(\alpha+1)\big)}{(\alpha+1)^{s_2+1}} \frac{n^{s\alpha+s_2}}{\log^{s_2+1} n}$
  $+ \mathcal{O}\Big(\frac{n^{s\alpha+s_2}}{\log^{s_2+2} n}\Big)$, for $1 \le s_2 < s$.

Therefore the maximum is obtained when choosing $s_2 = s - 1$ and the other contributions are asymptotically negligible. This leads to the required expansion

$$[z^n] r^{[s]}(z) = \frac{s}{(\alpha+1)^{s-1}} \frac{n^{s(\alpha+1)-1}}{\log^{s-1} n}$$
$$+ \frac{s\big[H_{s-1} + s - 1 + (\alpha+1)\sum_{l=1}^{s-1} \Psi\big(l(\alpha+1)\big)\big]}{(\alpha+1)^s} \frac{n^{s(\alpha+1)-1}}{\log^s n} + \mathcal{O}\Big(\frac{n^{s(\alpha+1)-1}}{\log^{s+1} n}\Big).$$

Then by applying (16) and (19), we obtain after some easy manipulations the following expansion:

$$[z^n] \frac{r^{[s]}(z)}{(1-z)L(z)} = \sum_{k=2}^{n-2} \Big(\frac{1}{\log k} + \frac{\Psi(1)}{\log^2 k} + \mathcal{O}\big(\frac{1}{\log^3 k}\big) \Big) \times$$
$$\times \Big[\frac{s}{(\alpha+1)^{s-1}} \frac{(n-k)^{s(\alpha+1)-1}}{\log^{s-1}(n-k)} + \frac{s\big[H_{s-1} + s - 1 + (\alpha+1)\sum_{l=1}^{s-1} \Psi\big(l(\alpha+1)\big)\big]}{(\alpha+1)^s} \times$$
$$\times \frac{(n-k)^{s(\alpha+1)-1}}{\log^s(n-k)} + \mathcal{O}\Big(\frac{(n-k)^{s(\alpha+1)-1}}{\log^{s+1}(n-k)}\Big) \Big] + \mathcal{O}\Big(\frac{n^{s(\alpha+1)-1}}{\log^{s-1} n}\Big) \tag{23}$$
$$= \frac{1}{(\alpha+1)^s} \frac{n^{s(\alpha+1)}}{\log^s n} + \frac{H_s + s - \frac{1}{s} + (\alpha+1)\sum_{l=1}^{s} \Psi\big(l(\alpha+1)\big)}{(\alpha+1)^{s+1}} \frac{n^{s(\alpha+1)}}{\log^{s+1} n} + \mathcal{O}\Big(\frac{n^{s(\alpha+1)}}{\log^{s+2} n}\Big).$$

Analogously we obtain

$$[z^n] \frac{r^{[s]}(z)}{(1-z)L^2(z)} = \frac{1}{(\alpha+1)^s} \frac{n^{s(\alpha+1)}}{\log^{s+1} n} + \mathcal{O}\Big(\frac{n^{s(\alpha+1)}}{\log^{s+2} n}\Big).$$

We proceed with $[z^n] \int_{t=0}^{z} \frac{r^{[s]}(t)}{(1-t)L^2(t)} dt = \frac{1}{(\alpha+1)^s} \frac{n^{s(\alpha+1)-1}}{\log^{s+1} n} + \mathcal{O}\Big(\frac{n^{s(\alpha+1)-1}}{\log^{s+2} n}\Big)$ and obtain again by using the convolution formula the expansion

$$[z^n] \frac{1}{1-z} \int_{t=0}^{z} \frac{r^{[s]}(t)}{(1-t)L^2(t)} dt = \sum_{k=2}^{n-2} \Big[\frac{1}{(\alpha+1)^s} \frac{(n-k)^{s(\alpha+1)-1}}{\log^{s+1}(n-k)} + \mathcal{O}\Big(\frac{(n-k)^{s(\alpha+1)-1}}{\log^{s+2}(n-k)}\Big) \Big]$$
$$+ \mathcal{O}\Big(\frac{n^{s(\alpha+1)-1}}{\log^{s+1} n}\Big) = \frac{1}{s(\alpha+1)^{s+1}} \frac{n^{s(\alpha+1)}}{\log^{s+1} n} + \mathcal{O}\Big(\frac{n^{s(\alpha+1)}}{\log^{s+2} n}\Big). \tag{24}$$

Adding (23) and (24), we obtain

$$\mu_{n+1}^{[s]} = [z^n]\tilde{\mu}^{[s]}(z) = \frac{1}{(\alpha+1)^s}\frac{n^{s(\alpha+1)}}{\log^s n} + \frac{H_s + s + (\alpha+1)\sum_{l=1}^{s}\Psi\big(l(\alpha+1)\big)}{(\alpha+1)^{s+1}}\frac{n^{s(\alpha+1)}}{\log^{s+1} n} + \mathcal{O}\Big(\frac{n^{s(\alpha+1)}}{\log^{s+2} n}\Big).$$

Substituting $n-1$ for n, it follows that expansion (2) is valid also for s.

To show the formula (3) for the centered moments, we plug in the asymptotic expansion (2) for the ordinary moments and get

$$\mathbb{E}([X_n - \mathbb{E}(X_n)]^s) = \sum_{k=0}^{s}\binom{s}{k}(-1)^{s-k}\,\mathbb{E}(X_n^k)\,(\mathbb{E}(X_n))^{s-k} = \left[\sum_{k=0}^{s}\frac{\binom{s}{k}(-1)^{s-k}}{(\alpha+1)^s}\right]\frac{n^{s(\alpha+1)}}{\log^s n}$$
$$+ \left\{\sum_{k=0}^{s}\frac{\binom{s}{k}(-1)^{s-k}}{(\alpha+1)^{s+1}}\Big[H_k + s\big(2+(\alpha+1)\Psi(\alpha+1)\big) - k\big(1+(\alpha+1)\Psi(\alpha+1)\big)\right.$$
$$\left. + (\alpha+1)\sum_{l=1}^{k}\Psi\big(l(\alpha+1)\big)\Big]\right\}\frac{n^{s(\alpha+1)}}{\log^{s+1} n} + \mathcal{O}\Big(\frac{n^{s(\alpha+1)}}{\log^{s+2} n}\Big).$$

Since it holds for $s \geq 1$ (see e. g. [3, p. 187ff]):

$$\sum_{k=0}^{s}\binom{s}{k}(-1)^{s-k}H_k = \frac{(-1)^{s-1}}{s},$$
$$\sum_{k=0}^{s}\binom{s}{k}(-1)^{s-k}\sum_{l=1}^{k}\Psi\big(l(\alpha+1)\big) = \sum_{l=0}^{s-1}\binom{s-1}{l}(-1)^{s-1-l}\Psi\big((l+1)(\alpha+1)\big),$$

and for $s \geq 2$: $\sum_{k=0}^{s}\binom{s}{k}(-1)^{s-k} = \sum_{k=0}^{s}\binom{s}{k}(-1)^{s-k}k = 0$, we obtain (3).

For the special case $\alpha = 0$, where X_n counts the number of removed edges until the root is isolated, we obtain due to further simplifications

$$\mathbb{E}(X_n^s) = \frac{n^s}{\log^s n} + \big((s+1)H_s - \gamma s\big)\frac{n^s}{\log^{s+1} n} + \mathcal{O}\Big(\frac{n^s}{\log^{s+2} n}\Big), \quad \text{for } s \geq 1,$$
$$\mathbb{E}\big([X_n - \mathbb{E}(X_n)]^s\big) = \frac{(-1)^s}{(s-1)s}\frac{n^s}{\log^{s+1} n} + \mathcal{O}\Big(\frac{n^s}{\log^{s+2} n}\Big), \quad \text{for } s \geq 2.$$

Here γ denotes as usual the Euler-constant.

We denote also the case $\alpha = 1$, where removing an edge of a tree T costs exactly the size $|T|$ of T:

$$\mathbb{E}(X_n^s) = \frac{1}{2^s}\frac{n^{2s}}{\log^s n} + \frac{\frac{1}{2}H_s + (2s+1)H_{2s} - (1+2\gamma)s}{2^{s+1}}\frac{n^{2s}}{\log^{s+1} n} + \mathcal{O}\Big(\frac{n^{2s}}{\log^{s+2} n}\Big), \quad \text{for } s \geq 1,$$
$$\mathbb{E}\big([X_n - \mathbb{E}(X_n)]^s\big) = \frac{(-1)^s}{2^{s+1}(s-1)s}\Bigg(1 + \frac{4^{s-1}}{\binom{2s-1}{s}}\Bigg)\frac{n^{2s}}{\log^{s+1} n} + \mathcal{O}\Big(\frac{n^{2s}}{\log^{s+2} n}\Big), \quad \text{for } s \geq 2.$$

We want to remark further that with the approach used here, one can also deal with logarithmic toll functions $t_n = \log^p n$ and compute the asymptotic behaviour of the moments $\mathbb{E}(X_n^s)$, where again one cannot deduce a limiting distribution.

6. Two-sided destructions

Here we sketch the proof of Theorem 2.2 concerning the asymptotic behaviour of the s-th moments of the total costs Y_n when destroying random recursive trees with toll functions $t_n = n^\alpha$, for $n \geq 2$ with $\alpha > 0$ and $t_1 = 0$. Again choosing $t_1 = 0$ instead of $t_1 = 1$ has no influence to the stated asymptotic behaviour ($Y_n^* = Y_n + n$, if Y_n^* measures the total costs with toll function $t_n = n^\alpha$, for $n \geq 1$).

To reduce extracting coefficients from (15) to an application of the convolution formula (16), we write equation (15) as

$$\begin{aligned}\tilde{\lambda}^{[s]}(z) &= \frac{r^{[s]}(z)}{(1-z)L(z)} + \frac{1}{(1-z)^2 L^2(z)} \int_{t=0}^{z} r^{[s]}(t)dt \\ &+ \frac{2}{(1-z)^2} \int_{t=0}^{z} \frac{1}{(1-t)L^3(t)} \int_{u=0}^{t} r^{[s]}(u)du\, dt,\end{aligned} \tag{25}$$

where $r^{[s]}(z)$ is given by (13) with there appearing functions $t(z) = \sum_{n \geq 2} n^\alpha z^n$.

To start the proof of the expansion (4) by induction, we consider $\lambda^{[0]}(z) = \sum_{n \geq 1} \frac{z^n}{n} = L(z)$ and obtain as in Section 5: $r^{[1]}(z) = t(z) \odot \left[\frac{1}{1-z} - 1 - L(z)\right]$ and we can use the expansion for $[z^n]r^{[1]}(z)$ given there. Also (21) is of interest.

Thus $[z^n] \int_{t=0}^{z} r^{[1]}(t)dt = n^{\alpha-1} + \mathcal{O}(n^{\alpha-2})$ and we obtain with (16) and (19)

$$[z^n]\frac{1}{(1-z)^2 L^2(z)} \int_{t=0}^{z} r^{[1]}(t)dt = \frac{n^{\alpha+1}}{\alpha(\alpha+1)\log^2 n} + \mathcal{O}\Big(\frac{n^{\alpha+1}}{\log^3 n}\Big) \tag{26}$$

and by other two applications of the convolution formula also

$$[z^n]\frac{1}{(1-z)^2} \int_{t=0}^{z} \frac{1}{(1-t)L^3(t)} \int_{u=0}^{t} r^{[1]}(u)du\, dt = \mathcal{O}\Big(\frac{n^{\alpha+1}}{\log^3 n}\Big). \tag{27}$$

Adding (21), (26) and (27), we obtain finally

$$\lambda_{n+1}^{[1]} = [z^n]\tilde{\lambda}^{[1]}(z) = \frac{1}{\alpha+1}\frac{n^{\alpha+1}}{\log n} + \Big(\frac{1}{\alpha(\alpha+1)} + \frac{1}{(\alpha+1)^2} + \frac{\Psi(\alpha+1)}{\alpha+1}\Big)\frac{n^{\alpha+1}}{\log^2 n} + \mathcal{O}\Big(\frac{n^{\alpha+1}}{\log^3 n}\Big).$$

Substituting $n-1$ for n, it follows that expansion (4) is valid for $s = 1$.

Now we assume that for a given $s \geq 2$, all moments $\mathbb{E}(Y_n^r) = \lambda_n^{[r]}$ with $1 \leq r < s$ have the expansion (4). To compute the asymptotic growth of the coefficients $[z^n]r^{[s]}(z)$, we use $[z^n]t^{\odot s_1}(z) = n^{s_1 \alpha}$ and the following consequences of (16), (17) and (19), where the γ_r are given in Theorem 2.2:

$$[z^n]\Big(\frac{d}{dz}\lambda^{[s_2]}(z)\Big) \int_{t=0}^{z} \lambda^{[s_3]}(t)dt =$$

- $\frac{\Gamma(s_2(\alpha+1)+1)\Gamma(s_3(\alpha+1)-1)}{(\alpha+1)^{s_2+s_3}\Gamma((s_2+s_3)(\alpha+1))} \frac{n^{(s_2+s_3)(\alpha+1)-1}}{\log^{s_2+s_3} n}\Big(1 + \mathcal{O}\big(\frac{1}{\log n}\big)\Big)$, for $0 < s_2, s_3 < s$,
- $\frac{1}{(\alpha+1)^{s_3}(s_3(\alpha+1)-1)} \frac{n^{s_3(\alpha+1)-1}}{\log^{s_3} n}\Big(1 + \mathcal{O}\big(\frac{1}{\log n}\big)\Big)$, for $s_2 = 0$, $0 < s_3 < s$,
- $\frac{1}{(\alpha+1)^{s_2}} \frac{n^{s_2(\alpha+1)}}{\log^{s_2} n} + \gamma_{s_2} \frac{n^{s_2(\alpha+1)}}{\log^{s_2+1} n}\Big(1 + \mathcal{O}\big(\frac{1}{\log n}\big)\Big)$, for $0 < s_2 < s$, $s_3 = 0$,
- $1 + \mathcal{O}\big(\frac{1}{n}\big)$, for $s_2 = 0$, $s_3 = 0$.

Thus for $s_1+s_2+s_3=s$ and the restrictions $0\le s_2,s_3<s$, we can describe the growth of the summands of $[z^n]r^{[s]}(z)$:

$$[z^n]\binom{s}{s_1,\,s_2,\,s_3}t^{\odot s_1}\odot\Big[\big(\frac{d}{dz}\lambda^{[s_2]}(z)\big)\int_{t=0}^{z}\lambda^{[s_3]}(t)dt\Big]=$$

- $\binom{s}{s_1,\,s_2,\,s_3}\dfrac{\Gamma(s_2(\alpha+1)+1)\Gamma(s_3(\alpha+1)-1)}{(\alpha+1)^{s_2+s_3}\Gamma\big((s_2+s_3)(\alpha+1)\big)}\dfrac{n^{s\alpha+s_2+s_3-1}}{\log^{s_2+s_3}n}\Big(1+\mathcal{O}\big(\frac{1}{\log n}\big)\Big)$, for $0<s_2,s_3<s$,
- $\binom{s}{s_1}\dfrac{1}{(\alpha+1)^{s_3}(s_3(\alpha+1)-1)}\dfrac{n^{s\alpha+s_3-1}}{\log^{s_3}n}\Big(1+\mathcal{O}\big(\frac{1}{\log n}\big)\Big)$, for $s_2=0,\ 0<s_3<s$,
- $\binom{s}{s_1}\dfrac{1}{(\alpha+1)^{s_2}}\dfrac{n^{s\alpha+s_2}}{\log^{s_2}n}+\binom{s}{s_1}\gamma_{s_2}\dfrac{n^{s\alpha+s_2}}{\log^{s_2+1}n}\Big(1+\mathcal{O}\big(\frac{1}{\log n}\big)\Big)$, for $0<s_2<s,\ s_3=0$,
- $n^{s\alpha}\Big(1+\mathcal{O}\big(\frac{1}{n}\big)\Big)$, for $s_2=0,\ s_3=0$.

Therefore the maximum is obtained when choosing $s_3=0$ and $s_2=s-1$, but additionally the instances $s_1=0$ and $0<s_2,s_3<s$ give contributions to the second order term. We get thus the expansion

$$[z^n]r^{[s]}(z)=\frac{s}{(\alpha+1)^{s-1}}\frac{n^{s(\alpha+1)-1}}{\log^{s-1}n}+\tilde{\gamma}_s\frac{n^{s(\alpha+1)-1}}{\log^{s}n}+\mathcal{O}\Big(\frac{n^{s(\alpha+1)-1}}{\log^{s+1}n}\Big),$$

with $\tilde{\gamma}_s:=s\gamma_{s-1}+\sum_{j=1}^{s-1}\binom{s}{j}\frac{\Gamma\big(j(\alpha+1)+1\big)\Gamma\big((s-j)(\alpha+1)-1\big)}{(\alpha+1)^s\Gamma\big(s(\alpha+1)\big)}$. This gives by repeatedly applying (16) and (19) the following expansions:

$$[z^n]\frac{r^{[s]}(z)}{(1-z)L(z)}=\frac{1}{(\alpha+1)^s}\frac{n^{s(\alpha+1)}}{\log^s n}+\Big(\frac{\Psi(s(\alpha+1))}{(\alpha+1)^s}+\frac{1}{(\alpha+1)^{s+1}}+\frac{\tilde{\gamma}_s}{s(\alpha+1)}\Big)\frac{n^{s(\alpha+1)}}{\log^{s+1}n}+\mathcal{O}\Big(\frac{n^{s(\alpha+1)}}{\log^{s+2}n}\Big),$$

$$[z^n]\frac{1}{(1-z)^2L^2(z)}\int_{t=0}^{z}r^{[s]}(t)dt=\frac{1}{(\alpha+1)^s(s(\alpha+1)-1)}\frac{n^{s(\alpha+1)}}{\log^{s+1}n}+\mathcal{O}\Big(\frac{n^{s(\alpha+1)}}{\log^{s+2}n}\Big),$$

$$[z^n]\frac{2}{(1-z)^2}\int_{t=0}^{z}\frac{1}{(1-t)\log^3\frac{1}{1-t}}\int_{u=0}^{t}r^{[s]}(u)\,du\,dt=\mathcal{O}\Big(\frac{n^{s(\alpha+1)}}{\log^{s+2}n}\Big).$$

Combining these results and using (25) we obtain

$$\lambda_n^{[s]}=\frac{1}{(\alpha+1)^s}\frac{n^{s(\alpha+1)}}{\log^s n}+\gamma_s\frac{n^{s(\alpha+1)}}{\log^{s+1}n}+\mathcal{O}\Big(\frac{n^{s(\alpha+1)}}{\log^{s+2}n}\Big),$$

where γ_s satisfies for $s\ge1$ the following recurrence with inital value $\gamma_0:=0$:

$$\gamma_s=\frac{\gamma_{s-1}}{\alpha+1}+\frac{\Psi\big(s(\alpha+1)\big)}{(\alpha+1)^s}+\frac{1}{(\alpha+1)^{s+1}}+\frac{1}{(\alpha+1)^s(s(\alpha+1)-1)}$$
$$+\frac{1}{(\alpha+1)^s}\sum_{j=1}^{s-1}\binom{s}{j}\frac{\Gamma(j(\alpha+1)+1)\Gamma((s-j)(\alpha+1)-1)}{\Gamma\big(s(\alpha+1)+1\big)}.$$

Solving this recurrence leads exactly to the expression for γ_s given in Theorem 2.2. Analogous computations as in Section 5 leads from (4) to the stated result (5) for the centered moments.

Acknowledgment

The author thanks the anonymous referee for many valuable comments.

References

[1] J. A. Fill, P. Flajolet and N. Kapur, Singularity Analysis, Hadamard Products and Tree Recurrences, submitted, 2003. Available at `http://algo.inria.fr/flajolet/Publications/FiFlKa03.pdf`.

[2] P. Flajolet and A. Odlyzko, Singularity Analysis of Generating Functions, *SIAM Journal on Discrete Mathematics* 3, 216–240, 1990.

[3] R. Graham, D. Knuth and O. Patashnik, *Concrete Mathematics (Second Edition)*, Addison Wesley, 1994.

[4] D. Knuth and A. Schönhage, The expected linearity of a simple equivalence algorithm, *Theoretical Computer Science* 6, 281–315, 1978.

[5] H. Mahmoud and R. Smythe, A Survey of Recursive Trees, *Theoretical Probability and Mathematical Statistics* 51, 1–37, 1995.

[6] A. Meir and J. W. Moon, Cutting down random trees, *Journal of the Australian Mathematical Society* 11, 313–324, 1970.

[7] A. Meir and J. W. Moon, Cutting down recursive trees, *Mathematical Biosciences* 21, 173–181, 1974.

[8] A. Panholzer, Non-crossing trees revisited: cutting down and spanning subtrees, *Discrete Mathematics and Theoretical Computer Science*, 2003. `http://dmtcs.loria.fr/proceedings/html/pdfpapers/dmAC0125.pdf`.

Alois Panholzer
Institut für Diskrete Mathematik und Geometrie, Technische Universität Wien, Wiedner Hauptstraße 8–10, A-1040 Wien, Austria.
e-mail: `Alois.Panholzer@tuwien.ac.at`

Part V

Probability

Trends in Mathematics, © 2004 Birkhäuser Verlag Basel/Switzerland

Restrictions on the Position of the Maximum/Minimum in a Geometrically Distributed Sample

Margaret Archibald

ABSTRACT: *We address the question: "What is the probability that the maximum in a geometrically distributed sample occurs in the first d positions of a word?" This can be considered in the strict sense and also when repeats are allowed, both in the first d positions, and in the rest of the word. The same situations are then considered for the minimum. Another generalisation is to consider the smaller probability that the minimum of the first d positions is the maximum of the remaining letters. This can also be interpreted both in the strict and weak sense, as can the last parameter to be discussed, namely: when both the minimum value and the maximum value of all letters in the word are situated in the first d places.*

1. Introduction

Given a word whose letters are natural numbers, we consider these letters to occur independently and with geometric probability. So for $p+q=1$, each letter i appears in the word with probability pq^{i-1}. In this way larger letters occur less frequently than smaller letters. For all of the problems discussed below, we split the word into two parts: the first d letters of the word and the rest. All of these ideas can be interpreted in the strict and weak sense. For the special case $d=1$, a similar concept has been looked at for compositions, see [5]. We make use of generating functions and Rice's method [3, 8, 6, 7, 9, 10, 4, 1].

We first consider the scenario where the largest value in a sequence is situated in the first d letters. (We take words of length n and thus require $d \le n$.) Note that we write $\mathrm{i}=\sqrt{-1}$ to distinguish it from the i we use as an index. We also write $Q=q^{-1}$. By 'strict' we mean that a letter only occurs once and 'weak' means that the letters can occur any number of times (but at least once). We use (s,s) to denote the (strict, strict) cases, (w,s) to denote the (weak, strict) cases etc..

Theorem 1.1. *The probability that the maximum value in a word of length n is in the first d positions is*

$$Max_{(s,s)}(n) \sim Max_{(w,s)}(n) \sim \frac{(1-Q^{-1})d}{Ln}(1+\psi(\log_Q n)) \quad \Big(=\frac{pd}{Ln}(1+\psi(\log_Q n))\Big),$$

where $L=\log Q$ *and* $\psi(x)=\sum_{k\neq 0}\Gamma(1-\chi_k)e^{2k\pi \mathrm{i}x}$. *If we allow this maximum value to repeat in the rest of the word, then the probability is*

$$Max_{(s,w)}(n) \sim Max_{(w,w)}(n) \sim \frac{(Q-1)d}{Ln}(1+\psi(\log_Q n)).$$

We then do the same for minima, only this time there are four different solutions and no fluctuations.

Theorem 1.2. *The probability that the minimum value in a word of length n is in the first d positions is*

$$Min_{(s,s)}(n) = \frac{(Q-1)d}{Q^n - 1}$$

if we allow this minimum to occur only once (strict, strict). If we allow the minimum to recur in the first d positions but not in the rest of the word (weak, strict) we have

$$Min_{(w,s)}(n) = \frac{Q^d - 1}{Q^n - 1} \sim \frac{Q^d - 1}{Q^n}.$$

For the last two we let the min occur any number of times in the rest of the word, but for the first case we allow only one min and for the second, more than one. We get (strict, weak)

$$Min_{(s,w)}(n) = \frac{Q^{n-d}(Q-1)d}{Q^n - 1} \sim \frac{(Q-1)d}{Q^d} \qquad \text{as } n \to \infty$$

or (weak, weak)

$$Min_{(w,w)}(n) = \frac{1 - Q^{-d}}{1 - Q^{-n}} \sim 1 - Q^{-d} \qquad \text{as } n \to \infty.$$

Another generalisation is that the minimum value of the first d letters is greater than (and possibly equal to) all other values in the word.

Theorem 1.3. *If we define $\psi_d(x) := \sum_{k \neq 0} \Gamma(d - \chi_k) e^{2k\pi i x}$, then the probability that the minimum value of the first d is greater than all the remaining letters in the word is*

$$Minmax_{(s,s)}(n) \sim \frac{(Q-1)d!}{Ln^d Q^d} + \frac{(Q-1)d}{Ln^d Q^d} \psi_d(\log_Q n),$$

or

$$Minmax_{(w,s)}(n) \sim \frac{(1 - Q^{-d})(d-1)!}{Ln^d} + \frac{(1 - Q^{-d})}{Ln^d} \psi_d(\log_Q n)$$

depending on whether the minimum is strict or weak (i.e., it occurs once or more than once). The probabilities for the same two cases when we require that the minimum is greater than or equal to all letters in the rest of the word are

$$Minmax_{(s,w)}(n) \sim \frac{(Q-1)d!}{Ln^d} + \frac{(Q-1)d}{Ln^d} \psi_d(\log_Q n)$$

and

$$Minmax_{(w,w)}(n) \sim \frac{(Q^d-1)(d-1)!}{Ln^d} + \frac{(Q^d-1)}{Ln^d} \psi_d(\log_Q n),$$

respectively.

Finally, the probability that a word of length n has both its maximum and minimum in the first d letters is considered. In this case we only list two results, but there are many more cases (see section 5).

Theorem 1.4. *The maximum and minimum values will both occur in the first d letters of a word of length n with probability*

$$Both_{(s)}(n) \sim \frac{(Q-1)^2 d(d-1)}{LQ^{n+1}n}(1+\psi(\log_Q n)),$$

(with $\psi(x)$ as in Theorem 1) if neither is allowed to appear again anywhere in the word. If both are allowed to appear more than once in the word, this probability becomes

$$Both_{(w)}(n) \sim \frac{(Q-1)^2 d(d-1)}{LQn}(1+\psi(\log_Q n)) \quad \text{as } n \to \infty.$$

2. Maximum in the first d positions

This idea can be viewed in four different ways, all of which end up with similar results asymptotically. These asymptotic results are proportional to d as there are d chances that the maximum k will be where we want it to be. We can have (strict, strict) where k can only appear once in the whole word. Next, we can allow k to appear elsewhere in the first d positions, but not in the rest of the word. We will call this the (weak, strict) case. Alternatively, we can allow the letter k to appear any number of times in the rest of the word, but not more than once in the first d places - i.e., the (strict, weak) case. Lastly we can let k be anywhere in the word as long as it occurs at least once in the first d places. We call this case (weak, weak).

For the purposes of keeping the paper a manageable size, we do not include all calculations. Instead only one example of the complete method is given. For all parameters the other cases are similar. For the maximum in the first d positions, the (strict, strict) case is given in detail, and the others are described symbolically.

2.1. Maximum in the first d positions - (strict, strict)

In this case we have strictly only one k in the word, which has to appear somewhere in the first d places. All other letters can only be $1, 2, \dots$ up to $k-1$. Symbolically we have

$$A_k := \bigcup_{i=0}^{d-1} \{1,\dots,k-1\}^i k \{1,\dots,k-1\}^{d-1-i} \qquad \text{in} \qquad \bigcup_{k\geq 1} A_k\{1,\dots,k-1\}^*,$$

and so our generating function becomes

$$\begin{aligned}
F_M^{(s,s)}(z) &:= \sum_{k\geq 1}\sum_{i=0}^{d-1}\Big(\sum_{j=1}^{k-1} zpq^{j-1}\Big)^i zpq^{k-1}\Big(\sum_{j=1}^{k-1} zpq^{j-1}\Big)^{d-1-i} \frac{1}{1-\sum_{j=1}^{k-1} zpq^{j-1}} \\
&= \sum_{k\geq 1}\sum_{i=0}^{d-1} z^d(1-q^{k-1})^i pq^{k-1}(1-q^{k-1})^{d-1-i}\sum_{j\geq 0} z^j(1-q^{k-1})^j.
\end{aligned}$$

Thus

$$\begin{aligned}
[z^n]F_M^{(s,s)}(z) &= \sum_{k\geq 1}\sum_{i=0}^{d-1}(1-q^{k-1})^i pq^{k-1}(1-q^{k-1})^{d-1-i}(1-q^{k-1})^{n-d} \\
&= d\sum_{k\geq 1}\sum_{h=0}^{n-1}\binom{n-1}{h}(-q^{k-1})^h pq^{k-1}
\end{aligned}$$

$$= d\sum_{h=0}^{n-1}\binom{n-1}{h}(-1)^h\frac{Q^{h+1}(1-Q^{-1})}{Q^{h+1}-1}.$$

We can now use complex analysis to evaluate the alternating sum asymptotically. The method is called 'Rice's method' and the lemma we use (which allows us to express a sum such as this as an integral, see [2, 8, 11]) is

Lemma 2.1. *Let $\mathcal{C}$ be a curve surrounding the points $1, 2, \ldots, n$ in the complex plane, and let $f(z)$ be analytic inside $\mathcal{C}$. Then*

$$\sum_{k=1}^{n}\binom{n}{k}(-1)^k f(k) = -\frac{1}{2\pi i}\int_C [n; z] f(z)dz,$$

where

$$[n; z] = \frac{(-1)^{n-1}n!}{z(z-1)\cdots(z-n)} = \frac{\Gamma(n+1)\Gamma(-z)}{\Gamma(n+1-z)}. \qquad \square$$

By extending the contour of integration, it turns out that under suitable growth conditions (see [2]) the asymptotic expansion of our alternating sum is given by

$$\sum \mathrm{Res}([n; z]f(z)) + \text{smaller order terms},$$

where the sum is taken over all poles different from $1, \ldots, n$. Poles that lie more to the left lead to smaller terms in the asymptotic expansion.

In our case, the sum begins at zero and ends at $n-1$, so we start with a contour surrounding points $0, 1, \ldots, n-1$. We use Rice on $f(z) = \frac{Q^{z+1}(1-Q^{-1})}{Q^{z+1}-1}$, and consider poles at $z+1=0$ and $z+1=\chi_k$, where $\chi_k := \frac{2k\pi i}{L}$ for all $k \in \mathbb{Z}\setminus\{0\}$, ($L = \log Q$). To expand $f(z)$ about $z=-1$, we let $\varepsilon = z+1$. Then

$$f(z) = \frac{Q^{\varepsilon}(1-Q^{-1})}{Q^{\varepsilon}-1} = \frac{e^{\varepsilon L}(1-Q^{-1})}{e^{\varepsilon L}-1} \sim \frac{(1-Q^{-1})}{\varepsilon L}$$

and the residue is $\frac{(1-Q^{-1})}{L}$. This is combined with the kernel contribution

$$[n-1; -1] = \frac{(-1)^{n-1-1}(n-1)!}{(-1)(-1-1)\cdots(-1-(n-1))} = \frac{(-1)^{n-2}(n-1)!}{(-1)^n n!} = \frac{1}{n},$$

so that the first pole gives us $\frac{(1-Q^{-1})}{Ln}$. The remainder of the poles use $\varepsilon = z+1-\chi_k$, so (note that $Q^{\chi_k}=1$)

$$f(z) = \frac{Q^{\varepsilon+\chi_k}(1-Q^{-1})}{Q^{\varepsilon+\chi_k}-1} = \frac{Q^{\varepsilon}(1-Q^{-1})}{Q^{\varepsilon}-1} \sim \frac{(1-Q^{-1})}{\varepsilon L},$$

and the residue is again $\frac{(1-Q^{-1})}{L}$. The kernel is

$$[n-1; \chi_k - 1] = \frac{\Gamma(n-1+1)\Gamma(-\chi_k+1)}{\Gamma(n-1+1-\chi_k+1)} = \frac{\Gamma(1-\chi_k)\Gamma(n)}{\Gamma(n-\chi_k+1)}$$

$$\sim \Gamma(1-\chi_k)n^{-(-\chi_k+1)} = \frac{1}{n}\Gamma(1-\chi_k)e^{\chi_k \log n} = \frac{1}{n}\Gamma(1-\chi_k)e^{2k\pi i \log_Q n}$$

which means that the remaining poles give $\frac{1-Q^{-1}}{Ln}\sum_{k\neq 0}\Gamma(1-\chi_k)e^{2k\pi i\log_Q n}$. In total, we have the expected value asymptotic to

$$\frac{(1-Q^{-1})d}{Ln}(1+\psi(\log_Q n)) \qquad \text{where} \qquad \psi(x) := \sum_{k\neq 0}\Gamma(1-\chi_k)e^{2k\pi i x}.$$

2.2. Maximum in the first *d* positions - other cases

For the (weak, strict) scenario we relax the requirements slightly, and allow more than one k in the first d places. This changes the symbolic expression A_k to

$$A_k = \bigcup_{i=0}^{d-1}\{1,\ldots,k-1\}^i k\{1,\ldots,k\}^{d-1-i}.$$

Since we must be sure that k appears at least once, we secure the position of the first k. Again the union is taken over all k - i.e., over all words where k is the maximum, and we include the rest of the word ($\{1,\ldots,k-1\}^*$) as before. Thus the generating function is

$$F_M^{(w,s)}(z) := \sum_{k\geq 1}\sum_{i=0}^{d-1}\Big(\sum_{j=1}^{k-1} zpq^{j-1}\Big)^i zpq^{k-1}\Big(\sum_{j=1}^{k} zpq^{j-1}\Big)^{d-1-i}\frac{1}{1-\sum_{j=1}^{k-1} zpq^{j-1}}.$$

If we swap these requirements round to look at the (strict, weak) case, we assume that k only appears once in the first d positions, but can occur any number of times in the rest of the word. We keep the same A_k as in the (strict, strict) case, but extend the rest of the symbolic equation to include k. The generating function is thus

$$F_M^{(s,w)}(z) := \sum_{k\geq 1}\sum_{i=0}^{d-1}\Big(\sum_{j=1}^{k-1} zpq^{j-1}\Big)^i zpq^{k-1}\Big(\sum_{j=1}^{k-1} zpq^{j-1}\Big)^{d-1-i}\frac{1}{1-\sum_{j=1}^{k} zpq^{j-1}}. \tag{1}$$

For the (weak, weak) case there is at least one k in any position between 1 and d. All other letters in the word can be anything in the alphabet $1,2,\ldots,k$. We express the first d letters symbolically as in the (weak, strict) case, and the rest as in the (strict, weak) case. In terms of generating functions, we thus have the same as in (1) except that the second sum on j ends at k and not $k-1$.

3. Minimum in the first *d* positions

We have four different scenarios here, as we can apply our weak/strict classifications to the first d letters as well as to the rest of the word. We look at the same four cases as highlighted above.

The calculations here are shorter, as the result can be evaluated exactly rather than asymptotically. As a result of this we do not need to use Rice's method and also the solution has no fluctuations. This time only the (weak, strict) case is given in detail, and the others are mentioned more briefly.

3.1. Minimum in the first d positions - (weak, strict)

Suppose we allow the minimum value j of a word of length n to recur (it must occur once) within the first d but not thereafter. Symbolically we express the first d letters of any such word as

$$A_j = \bigcup_{i=0}^{d-1} \{j+1,\dots\}^i j \{j, j+1, \dots\}^{d-1-i},$$

by fixing the first j and allowing further j's to its right. This can be substituted into $\bigcup_{j\geq 1} A_j\{j+1,\dots\}^*$, which symbolises the rest of the word - where no j's are present. We then have

$$F_m^{(w,s)}(z) := \sum_{j\geq 1}\sum_{i=0}^{d-1}\Big(\sum_{k=j+1}^{\infty} zpq^{k-1}\Big)^i zpq^{j-1}\Big(\sum_{k=j}^{\infty} zpq^{k-1}\Big)^{d-1-i}\frac{1}{1-\sum_{k=j+1}^{\infty} zpq^{k-1}}$$

$$= \sum_{j\geq 1}\sum_{i=0}^{d-1}\sum_{k\geq 0} z^{k+d}(q^j)^i pq^{j-1}(q^{j-1})^{d-1-i}q^{jk},$$

and thus

$$[z^n]F_m^{(w,s)}(z) = \sum_{j\geq 1}\sum_{i=0}^{d-1}(q^j)^i pq^{j-1}(q^{j-1})^{d-1-i}q^{j(n-d)} = \sum_{i=0}^{d-1}\frac{p}{q^{d-i}}\frac{q^n}{1-q^n}$$

$$= \frac{Q^d(1-Q^{-1})}{Q^n-1}\sum_{i=0}^{d-1} Q^{-i} = \frac{Q^d-1}{Q^n-1}.$$

3.2. Minimum in the first d positions - other cases

Suppose we consider the probability that the minimum value j appears in the first d positions of the word but never again (i.e., (strict, strict)). Symbolically we express the first d letters of any such word as

$$A_j := \bigcup_{i=0}^{d-1} \{j+1,\dots\}^i j \{j+1, \dots\}^{d-1-i}$$

which differs from the previous A_k only because the last bracket starts from $j+1$. All possible words of this type would be $\bigcup_{j\geq 1} A_j\{j+1,\dots\}^*$. The (strict, weak) case uses the same A_k, but all possible words would be $\bigcup_{j\geq 1} A_j\{j, j+1,\dots\}^*$, so that we have the generating function as

$$\sum_{j\geq 1}\sum_{i=0}^{d-1}\Big(\sum_{k=j+1}^{\infty} zpq^{k-1}\Big)^i zpq^{j-1}\Big(\sum_{k=j+1}^{\infty} zpq^{k-1}\Big)^{d-1-i}\frac{1}{1-\sum_{k=J}^{\infty} zpq^{k-1}}.$$

where J is $j+1$ or j respectively. Similar small changes are needed for the final option - (weak, weak). We let j occur anywhere, as long as it appears at least once in the first d places. Again we have $A_j = \bigcup_{i=0}^{d-1}\{j+1,\dots\}^i j\{j, j+1,\dots\}^{d-1-i}$, so the first j's position is fixed, but others may occur to the right. All such words are represented by $\bigcup_{j\geq 1} A_j\{j, j+1,\dots\}^*$.

Again, for a fixed d and large n, we can see that it is the classification of the 'rest' of the word that takes precedence, i.e., the (strict, strict) and (weak, strict) cases are in a smaller order of magnitude to the (strict, weak) and (weak, weak) cases.

4. The minimum of the first d is the maximum of the rest

We require the minimum value (j) of the first d letters to be either strictly greater than or greater than or equal to the maximum of the rest. Again there are four cases, which are all combinations of the pair 'weak' and 'strict'. The complete working is shown for the (strict, weak) case.

4.1. Min of first d is greater than or equal to the rest - (strict, weak)

If we consider a word in which j is the strict minimum of the first d letters, so that j occurs only once in the first d letters, but that we allow any of $1, 2, \ldots, j$ in the rest of the word then we can represent the first d letters symbolically as

$$A_j = \bigcup_{i=0}^{d-1} \{j+1, \ldots\}^i j \{j+1, \ldots\}^{d-1-i}$$

which is part of the overall symbolic equation

$$\bigcup_{j \geq 1} A_j \{1, \ldots, j\}^*.$$

This translates into the generating function

$$F_d^{(s,w)}(z) := \sum_{j\geq 1} \sum_{i=0}^{d-1} \Big(\sum_{h=j+1}^{\infty} zpq^{h-1} \Big)^i zpq^{j-1} \Big(\sum_{h=j+1}^{\infty} zpq^{h-1} \Big)^{d-1-i} \frac{1}{1 - \sum_{h=1}^{j} zpq^{h-1}}$$

$$= \sum_{j\geq 1} \sum_{i=0}^{d-1} \sum_{h\geq 0} z^{d+h} (q^j)^i pq^{j-1} (q^j)^{d-1-i} (1-q^j)^h,$$

whose coefficients are

$$[z^n] F_d^{(s,w)}(z) = \sum_{j\geq 1} \sum_{i=0}^{d-1} (q^j)^i pq^{j-1} (q^j)^{d-1-i} (1-q^j)^{n-d}$$

$$= \sum_{i=0}^{d-1} \sum_{h=0}^{n-d} \binom{n-d}{h} (-1)^h \frac{p}{q} \sum_{j\geq 1} q^{(d+h)j} = \sum_{i=0}^{d-1} \sum_{h=0}^{n-d} \binom{n-d}{h} (-1)^h \frac{Q-1}{Q^{d+h}-1}.$$

We consider the poles of $f(z) = \frac{Q-1}{Q^{d+z}-1}$ at $z+d=0$ and at $z+d=\chi_k$. Expanding around $\varepsilon := z+d$ gives

$$f(z) = \frac{Q-1}{Q^\varepsilon - 1} = \frac{Q-1}{e^{\varepsilon L}-1} \sim \frac{Q-1}{\varepsilon L}$$

with residue $\frac{Q-1}{L}$, and

$$[n-d; -d] = \frac{(-1)^{n-d-1}(n-d)!}{(-d)(-d-1)\cdots(-d-(n-d))} = \frac{(n-d)!(d-1)!}{n!} = \frac{1}{d\binom{n}{d}}.$$

For $\varepsilon = z + d - \chi_k$,

$$f(z) = \frac{Q-1}{Q^{\varepsilon+\chi_k}-1} = \frac{Q-1}{Q^\varepsilon - 1} \sim \frac{Q-1}{\varepsilon L}$$

so the residue is also $\frac{Q-1}{L}$, and

$$[n-d;\chi_k-d]\sim\Gamma(d-\chi_k)n^{\chi_k-d}=\frac{1}{n^d}\Gamma(d-\chi_k)e^{2k\pi\mathrm{i}\log_Q \mathrm{n}}.$$

Altogether as $n\to\infty$, the expected value tends to

$$\frac{Q-1}{L}\sum_{i=0}^{d-1}\frac{1}{d\binom{n}{d}}+\frac{Q-1}{Ln^d}\sum_{i=0}^{d-1}\sum_{k\neq 0}\Gamma(d-\chi_k)e^{2k\pi\mathrm{i}\log_Q \mathrm{n}}$$
$$\sim\frac{(Q-1)d!}{Ln^d}+\frac{(Q-1)d}{Ln^d}\psi_d(\log_Q n)$$

where $\psi_d(x):=\sum\limits_{k\neq 0}\Gamma(d-\chi_k)e^{2k\pi\mathrm{i}x}$.

4.2. Min of first d is greater than the rest - other cases

For the (strict, strict) case j occurs only once in the first d letters, and is strictly the minimum there. It does not occur in the rest of the word either, and nor does any letter larger than j. We have the same A_j as above, and our total symbolic equation is $\bigcup_{j\geq 1}A_j\{1,\dots,j-1\}^*$. We can translate this into the generating function

$$F_d^{(s,s)}(z):=\sum_{j\geq 1}\sum_{i=0}^{d-1}\Big(\sum_{h=j+1}^{\infty}zpq^{h-1}\Big)^i zpq^{j-1}\Big(\sum_{h=j+1}^{\infty}zpq^{h-1}\Big)^{d-1-i}\frac{1}{1-\sum_{h=1}^{j-1}zpq^{h-1}}.$$

For the (weak, strict) and (weak, weak) cases we allow j any number of times (but at least once) in the first d letters. We have

$$A_j=\bigcup_{i=0}^{d-1}\{j+1,\dots\}^i j\{j,\dots\}^{d-1-i}\quad\text{in all possible words}\quad\bigcup_{j\geq 1}A_j\{1,\dots,J\}^*,$$

where J is $j-1$ or j depending on whether the second restriction is strict or weak. Using the same notation, we translate this into the generating function(s)

$$F_d^{(w,s)}(z):=\sum_{j\geq 1}\sum_{i=0}^{d-1}\Big(\sum_{h=j+1}^{\infty}zpq^{h-1}\Big)^i zpq^{j-1}\Big(\sum_{h=j}^{\infty}zpq^{h-1}\Big)^{d-1-i}\frac{1}{1-\sum_{h=1}^{J}zpq^{h-1}}.$$

5. Min and Max both in first d positions

As in the previous cases, we start by looking at a symbolic equation which in this case represents all words with both the maximum and minimum in the first d places. This is then translated into a generating function whose coefficient of z^n will be the probability that a word of length n has this requirement.

There are sixteen different cases to consider, since we can have strict or weak for the maximum or minimum in the first d places or the rest of the word. The diagram

	First d letters	Remaining letters
Maximum	strict/weak	strict/weak
Minimum	strict/weak	strict/weak

shows why this is so. In this paper we include some details for only the weakest case (see section 5.2 below), where the max and min can appear any number of times, as long as they first occur at least once each in the first d positions. Two other cases are mentioned.

5.1. Max and min occur only once each

We require that $2 \le d \le n$, and that d is fixed. We can express the first d letters of such a word as

$$A_{jk} = \bigcup_{i=0}^{d-2} \bigcup_{h=0}^{d-2-i} \{j+1,\dots,k-1\}^i j\{j+1,\dots,k-1\}^h k\{j+1,\dots,k-1\}^{d-2-i-h}$$

if the minimum occurs first (i.e., to the left of the maximum) and

$$A_{kj} = \bigcup_{i=0}^{d-2} \bigcup_{h=0}^{d-2-i} \{j+1,\dots,k-1\}^i k\{j+1,\dots,k-1\}^h j\{j+1,\dots,k-1\}^{d-2-i-h}$$

if the maximum occurs first. Since we want to include all words (for all values of $j < k$), we must consider

$$\bigcup_{j\ge 1} \bigcup_{k>j} A_B\{j+1,\dots,k-1\}^*,$$

where B could be jk or kj depending on whether the minimum or maximum occurs first. Note that in terms of generating functions, both give the same results due to multiplicative commutativity. Thus translation into generating functions for either case is the same, and so to involve all cases we must include a factor of 2:

$$F^{(s)}(z) := 2\sum_{j\ge 1}\sum_{k>j}\sum_{i=0}^{d-2}\sum_{h=0}^{d-2-i}\Big(\sum_{l=j+1}^{k-1} zpq^{l-1}\Big)^i zpq^{j-1} \cdot \Big(\sum_{l=j+1}^{k-1} zpq^{l-1}\Big)^h zpq^{k-1}\Big(\sum_{l=j+1}^{k-1} zpq^{l-1}\Big)^{d-2-i-h} \frac{1}{1-\sum_{l=j+1}^{k-1} zpq^{l-1}}.$$

Note that the order of the leading term in the first case of Theorem 4 is $\frac{1}{nQ^n}$ - the product of the results in the previous cases, where we had the strict maximum of order $\frac{1}{n}$ and the strict minimum of order Q^n.

5.2. Max and min can occur any number of times

We now look at the other end of the scale in more detail. As long as the max and min occur at least once each in the first d places, they are allowed to appear again any number of times anywhere in the word. Again, we split this into two cases symbolically (according to whether j or k appears first), but one generating function will do for both, if it includes the factor of 2:

$$F^{(w)}(z) := 2\sum_{j\ge 1}\sum_{k>j}\sum_{i=0}^{d-2}\sum_{h=0}^{d-2-i}\Big(\sum_{l=j}^{k} zpq^{l-1}\Big)^i zpq^{j-1} \cdot \Big(\sum_{l=j}^{k} zpq^{l-1}\Big)^h zpq^{k-1}\Big(\sum_{l=j}^{k} zpq^{l-1}\Big)^{d-2-i-h} \frac{1}{1-\sum_{l=j}^{k} zpq^{l-1}}$$

$$= 2\sum_{j\geq 1}\sum_{k>j}\sum_{i=0}^{d-2}\sum_{h=0}^{d-2-i}\sum_{l\geq 0} z^{l+d}(q^{j-1}-q^k)^{d-2+l}p^2q^{j+k-2}$$
$$= d(d-1)\sum_{j\geq 1}\sum_{k>j}\sum_{l\geq 0} z^{l+d}(q^{j-1}-q^k)^{d-2+l}p^2q^{j+k-2}.$$

For the expected value, we are interested in the coefficients:

$$[z^n]F^{(w)}(z) = d(d-1)\sum_{j\geq 1}\sum_{k>j}(q^{j-1}-q^k)^{n-2}p^2q^{j+k-2}$$
$$= d(d-1)\sum_{l=0}^{n-2}\binom{n-2}{l}(-1)^l\frac{p^2}{q^{n-l}}\sum_{j\geq 1}q^{j(n-1-l)}\sum_{k>j}q^{k(l+1)}$$
$$= d(d-1)\sum_{l=0}^{n-2}\binom{n-2}{l}(-1)^l\frac{p^2}{q^{n-l}}\frac{q^{l+1}}{1-q^{l+1}}\frac{q^n}{1-q^n}$$
$$= d(d-1)\frac{Q^n(1-Q^{-1})^2}{Q^n-1}\sum_{l=0}^{n-2}\binom{n-2}{l}(-1)^l\frac{1}{Q^l(Q^{l+1}-1)}.$$

Again, we use Rice's method, with the function $f(z) := \frac{1}{Q^z(Q^{z+1}-1)}$. We consider the simple poles at $z+1=0$ and $z+1=\chi_k$, $k\neq 0$. For the former, let $\varepsilon = z+1$, then

$$f(z) = \frac{1}{Q^{\varepsilon-1}(Q^\varepsilon-1)} = \frac{1}{e^{\varepsilon L}Q^{-1}(e^{\varepsilon L}-1)} \sim \frac{1}{Q^{-1}(\varepsilon L)},$$

so the residue is $\frac{Q}{L}$. The kernel is

$$[n-2;-1] = \frac{(-1)^{(n-2)-1}(n-2)!}{(-1)(-1-1)\cdots(-1-(n-2))} = \frac{(-1)^{n-3}(n-2)!}{(-1)^{n-1}(n-1)!} = \frac{1}{n-1},$$

which means that the whole contribution from the first pole is $\frac{Q}{L(n-1)}$. For the other poles, let $\varepsilon = z+1-\chi_k$,

$$f(z) = \frac{1}{Q^{\varepsilon-1+\chi_k}(Q^{\varepsilon+\chi_k}-1)} = \frac{1}{Q^{\varepsilon-1}(Q^\varepsilon-1)} \sim \frac{1}{Q^{-1}\varepsilon L},$$

thus the residue is again $\frac{Q}{L}$, and the kernel $(n\to\infty)$ is

$$[n-2;\chi_k-1] = \frac{\Gamma(n-2+1)\Gamma(-\chi_k+1)}{\Gamma(n-2+1-\chi_k+1)} = \frac{\Gamma(1-\chi_k)\Gamma(n-1)}{\Gamma(n-\chi_k)}$$
$$\sim \Gamma(1-\chi_k)n^{-1-(-\chi_k)} = \Gamma(1-\chi_k)n^{\chi_k-1}$$
$$= \frac{1}{n}\Gamma(1-\chi_k)e^{\chi_k\log n} = \frac{1}{n}\Gamma(1-\chi_k)e^{2k\pi \mathrm{i}\log_Q n},$$

so altogether, we have

$$\frac{Q}{Ln}\sum_{k\neq 0}\Gamma(1-\chi_k)e^{2k\pi\mathrm{i}\log_Q n}.$$

So the expected value in the 'all weak' case is

$$[z^n]F^{(w)}(z) \sim \frac{(Q-1)^2d(d-1)}{LQn}(1+\psi(\log_Q n)) \quad \text{as } n\to\infty$$

(where $\psi(x) = \sum_{k\neq 0} \Gamma(1-\chi_k)e^{2k\pi ix}$). By comparing the 'all strict' case and the 'all weak' case we can see what a difference it makes to have repeats. The larger case (all weak) is of order $\frac{1}{n}$ whereas the smaller case (all strict) is of order $\frac{1}{nQ^n}$. Since the minimum is likely to occur very often, and the maximum very seldom (due to the geometric probabilities attached to the letters), we would expect the cases with a strict or weak minimum to dominate the various maximums. For interest we therefore look at one more case - where all the maximums are still strong and all the minimums are all weak.

5.3. Max occurs only once and min can recur

The single generating function which expresses this situation is

$$F(z) := 2\sum_{j\geq 1}\sum_{k>j}\sum_{i=0}^{d-2}\sum_{h=0}^{d-2-i}\Big(\sum_{l=j}^{k-1} zpq^{l-1}\Big)^{i} zpq^{j-1}$$
$$\cdot\Big(\sum_{l=j}^{k-1} zpq^{l-1}\Big)^{h} zpq^{k-1}\Big(\sum_{l=j}^{k-1} zpq^{l-1}\Big)^{d-2-i-h}\frac{1}{1-\sum_{l=j}^{k-1} zpq^{l-1}}.$$

The expected value for the 'min weak' and 'max strict' is therefore

$$[z^n]F(z) \sim \frac{(1-Q^{-1})^2 d(d-1)}{Ln}(1+\psi(n)),$$

which is also of order $\frac{1}{n}$ as in the weak case.

Acknowledgements. I would like to acknowledge the continued help, availability and support of my supervisors Prof. A. Knopfmacher and Prof. H. Prodinger.

References

[1] M. Archibald, A. Knopfmacher and H. Prodinger *The number of distinct values in a geometrically distributed sample*, Submitted.

[2] P. Flajolet and R. Sedgewick *Mellin transforms and asymptotics: Finite differences and Rice's integrals*, Theoretical Computer Science, 1995, 144:101-124, Special volume on mathematical analysis of algorithm.

[3] P. Kirschenhofer and H. Prodinger *On the Analysis of Probabilistic Counting*, Lecture Notes in Mathematics, 1990, 1452:117-120.

[4] A. Knopfmacher and H. Prodinger *Combinatorics of geometrically distributed random variables: Value and position of the rth left-to-right maximum*, Discrete Mathematics, 2001, 226:255-267.

[5] A. Knopfmacher and N. Robbins *Compositions with parts constrained by the leading summand.* (To appear in *Ars Combinatoria*).

[6] G. Louchard and H. Prodinger *Ascending runs of sequences of geometrically distributed random variables: A probabilistic analysis*, Theoretical Computer Science, 2003, 304:59-86.

[7] H. Prodinger *Combinatorial problems related to geometrically distributed random variables*, in *Séminaire Lotharingien de Combinatoire (Gerolfingen, 1993)*, Prépubl. Inst. Rech. Math. Av., 1993/34:87-95.

[8] H. Prodinger *Combinatorics of geometrically distributed random variables: Left-to-right maxima*, Discrete Mathematics, 1996, 153:253-270.

[9] H. Prodinger *Combinatorics of geometrically distributed random variables: New q-tangent and q-secant numbers*, International Journal of Mathematics and Mathematical Sciences, 2000, 24:825–838.

[10] H. Prodinger *Combinatorics of geometrically distributed random variables: Inversions and a parameter of Knuth*, Annals of Combinatorics, 2001, 5:241-250.

[11] W. Szpankowski *Average Case Analysis of Algorithms on Sequences*, John Wiley and Sons, New York, 2001.

Margaret Archibald
The John Knopfmacher Centre for Applicable Analysis and Number Theory, School of Mathematics, University of the Witwatersrand, P. O. Wits, 2050 Johannesburg, South Africa.
Email: marchibald@maths.wits.ac.za

Trends in Mathematics, © 2004 Birkhäuser Verlag Basel/Switzerland

Dual Random Fragmentation and Coagulation and an Application to the Genealogy of Yule Processes

Jean Bertoin and Christina Goldschmidt

ABSTRACT: *The purpose of this work is to describe a duality between a fragmentation associated to certain Dirichlet distributions and a natural random coagulation. The dual fragmentation and coalescent chains arising in this setting appear in the description of the genealogy of Yule processes.*

1. Introduction

At a naive level, fragmentation and coagulation are inverse phenomena, in that a simple time-reversal changes one into the other. However, stochastic models for fragmentation and coalescence usually impose strong hypotheses on the dynamics of the processes, such as the branching property for fragmentation (distinct fragments evolve independently as time passes), and these requirements do not tend to be compatible with time-reversal. Thus, in general, the time-reversal of a coalescent process is not a fragmentation process.

Nonetheless, there are a few special cases in which time-reversal does transform a coalescent process into a fragmentation process. Probably the most important example was discovered by Pitman [17]; it is related to the so-called cascades of Ruelle and the Bolthausen-Sznitman coalescent [7], and also has a natural interpretation in terms of the genealogy of a remarkable branching process considered by Neveu, see [4] and [6].

The first purpose of this note is to point out other simple instances of such duality, which rely on certain Dirichlet and Poisson-Dirichlet distributions. Then, in the second part, we shall show that these examples are related to the genealogy of Yule processes.

2. Dual fragmentation and coagulation

2.1. Some notation

For every integer $n \geq 1$, we consider the simplex

$$\Delta_n := \left\{ x = (x_1, \ldots, x_{n+1}) : x_i \geq 0 \text{ for every } i = 1, \ldots, n+1 \text{ and } \sum_{i=1}^{n+1} x_i = 1 \right\} .$$

It will also be convenient to agree that $\Delta_0 := \{1\}$. We shall often refer to the coordinates $x_1, \ldots, x_{n+1}$ of points x in Δ_n as *masses*.

We recall that the n-dimensional Dirichlet distribution with parameter $(\alpha_1, \ldots, \alpha_{n+1})$ is the probability measure on the simplex Δ_n with density

$$\frac{\Gamma(\alpha_1 + \cdots + \alpha_{n+1})}{\Gamma(\alpha_1) \cdots \Gamma(\alpha_{n+1})} x_1^{\alpha_1 - 1} \cdots x_{n+1}^{\alpha_{n+1}-1} .$$

The special case when $\alpha_1 = \ldots = \alpha_{n+1} := \alpha \in]0, \infty[$ will have an important role in this work; it will be convenient to write $\mathrm{Dir}_n(\alpha)$ for this distribution. We recall the following well-known construction: let $\gamma_1, \ldots, \gamma_{n+1}$ be i.i.d. gamma variables with parameters (α, c). Set $\bar{\gamma} = \gamma_1 + \cdots + \gamma_n$, so that $\bar{\gamma}$ has a gamma distribution with parameters $(\alpha(n+1), c)$. Then the $(n+1)$-tuple

$$(\gamma_1/\bar{\gamma}, \ldots, \gamma_{n+1}/\bar{\gamma})$$

has the distribution $\mathrm{Dir}_n(\alpha)$ and is independent of $\bar{\gamma}$.

We also define the (ranked) infinite simplex

$$\Delta_\infty := \left\{ x = (x_1, \ldots) : x_1 \geq x_2 \geq \ldots \geq 0 \text{ and } \sum_{i=1}^{\infty} x_i = 1 \right\}$$

and recall that the Poisson-Dirichlet distribution with parameter $\theta > 0$, which will be denoted by $\mathrm{PD}(\theta)$ in the sequel, is the law of the random sequence

$$\xi := \left(\frac{a_1}{\sum_{i=1}^{\infty} a_i}, \frac{a_2}{\sum_{i=1}^{\infty} a_i}, \ldots \right),$$

where $a_1 \geq a_2 \geq \ldots > 0$ are the atoms of a Poisson random measure on $]0, \infty[$ with intensity $\theta y^{-1} e^{-y} dy$. We also recall that ξ is independent of $\sum_{i=1}^{\infty} a_i$, and that the latter has the gamma distribution with parameters $(\theta, 1)$. By the celebrated Lévy-Itô decomposition of subordinators, we may also rephrase this construction as follows: if $\gamma = (\gamma(t), t \geq 0)$ is a standard gamma process and, for each fixed $\theta > 0$, $\delta_1 \geq \delta_2 \geq \ldots$ denotes the sequence of sizes of the jumps of γ on the time interval $[0, \theta]$, then

$$\left(\frac{\delta_1}{\gamma(\theta)}, \frac{\delta_2}{\gamma(\theta)}, \ldots \right)$$

has the $\mathrm{PD}(\theta)$ distribution and is independent of $\gamma(\theta)$.

2.2. Two dual random transformations

We now define two random transformations:

$$\mathrm{Frag}_k : \Delta_n \to \Delta_{n+k} \quad \text{and} \quad \mathrm{Coag}_k : \Delta_{n+k} \to \Delta_n ,$$

where k, n are integers.

First, we fix $x = (x_1, \ldots, x_{n+1}) \in \Delta_n$ and pick an index $I \in \{1, \ldots, n+1\}$ at random according to the distribution

$$\mathbf{P}(I = i) = x_i , \qquad i = 1, \ldots, n+1 ,$$

so that x_I is a size-biased pick from the sequence x. Let $\eta = (\eta_1, \ldots, \eta_{k+1})$ be a random variable with values in Δ_k which is distributed according to $\mathrm{Dir}_k(1/k)$ and independent of I. Then we split the Ith mass of x according to η and we obtain a random variable in Δ_{n+k}:

$$\mathrm{Frag}_k(x) := (x_1, \ldots, x_{I-1}, x_I \eta_1, \ldots, x_I \eta_{k+1}, x_{I+1}, \ldots, x_{n+1}) .$$

Second, we fix $x = (x_1, \ldots, x_{n+k+1}) \in \Delta_{n+k}$ and pick an index $J \in \{1, \ldots, n+1\}$ uniformly at random. We merge the $k+1$ masses $x_J, x_{J+1} \ldots, x_{J+k}$ to form a single mass $\sum_{i=J}^{J+k} x_i$ and leave the other masses unchanged. We obtain a random variable in Δ_n:

$$\text{Coag}_k(x) = \left(x_1, \dots, x_{J-1}, \sum_{i=J}^{J+k} x_i, x_{J+k+1}, \dots, x_{n+k+1} \right).$$

Remark. Consider the following alternative random coagulation of $x = (x_1, \dots, x_{n+k+1}) \in \Delta_{n+k}$. Pick $k+1$ indices $i_1, \dots, i_{k+1}$ from $\{1, \dots, n+k+1\}$ uniformly at random without replacement, merge the masses $x_{i_1}, \dots, x_{i_{k+1}}$, leave the other masses unchanged and let $\widetilde{\text{Coag}}_k(x)$ be the sequence obtained by ranking the resulting masses in decreasing order. Write also $\text{Coag}_k^{\downarrow}(x)$ for the sequence $\text{Coag}_k(x)$ re-arranged in decreasing order. Then if ξ is exchangeable the pairs $(\xi, \text{Coag}_k^{\downarrow}(\xi))$ and $(\xi, \widetilde{\text{Coag}}_k(\xi))$ have the same distribution. This remark applies in particular to the case when ξ has the law $\text{Dir}_{n+k}(1/k)$, and can thus be combined with forthcoming Proposition 2.1.

The starting point of this work lies in the observation of a simple relation of duality which links these two random transformations via Dirichlet laws.

Proposition 2.1. *Let $k, n \geq 1$ be two integers, and ξ, ξ' two random variables with values in Δ_n and Δ_{n+k}, respectively. The following assertions are then equivalent:*

(i) ξ has the law $\text{Dir}_n(1/k)$ and, conditionally on ξ, ξ' is distributed as $\text{Frag}_k(\xi)$.

(ii) ξ' has the law $\text{Dir}_{n+k}(1/k)$ and, conditionally on ξ', ξ is distributed as $\text{Coag}_k(\xi')$.

It has been observed by Kingman [13] that for $k = 1$, if ξ' is uniformly distributed on the simplex Δ_{n+1} (i.e. has the law $\text{Dir}_{n+1}(1)$), then $\text{Coag}_1(\xi')$ is uniformly distributed on Δ_n. Clearly, this agrees with our statement.

Proof: Let $\gamma_1, \gamma_2, \dots, \gamma_{n+1}$ be independent Gamma$(1/k, 1)$ random variables and set

$$\bar{\gamma} = \sum_{i=1}^{n+1} \gamma_i \quad \text{and} \quad \xi = \left(\frac{\gamma_1}{\bar{\gamma}}, \dots, \frac{\gamma_{n+1}}{\bar{\gamma}} \right),$$

so that ξ has law $\text{Dir}_n(1/k)$ and is independent of $\bar{\gamma}$. Suppose that η is a $\text{Dir}_k(1/k)$ random variable which is independent of the γ_i's, and let $\Phi : \mathbf{R}^{n+k+1} \to \mathbf{R}$ be a bounded measurable function. Let I be an index picked at random from $\{1, \dots, n+1\}$ according to the conditional distribution

$$\mathbf{P}(I = i \,|\, \gamma_1, \dots, \gamma_{n+1}) = \gamma_i / \bar{\gamma}, \qquad i = 1, \dots, n+1,$$

and denote by $\text{Frag}_k(\xi)$ the random sequence obtained from ξ after the fragmentation of its Ith mass according to η. We have

$$\mathbf{E}\left(\Phi(\text{Frag}_k(\xi)), I = i\right) = \mathbf{E}\left[\frac{\gamma_i}{\bar{\gamma}} \Phi\left((\gamma_l/\bar{\gamma})_{l<i}, \gamma_i \eta / \bar{\gamma}, (\gamma_l/\bar{\gamma})_{l>i} \right) \right].$$

Now, using the independence of $\bar{\gamma}$ and ξ and the fact that $\bar{\gamma}$ has the law Gamma$((n+1)/k, 1)$, we see that the last expression is equal to

$$\frac{k}{n+1} \mathbf{E}\left[\gamma_i \Phi\left((\gamma_l/\bar{\gamma})_{l<i}, \gamma_i \eta / \bar{\gamma}, (\gamma_l/\bar{\gamma})_{l>i} \right) \right]$$

$$= \frac{k}{n+1} \mathbf{E} \int_0^\infty x \Phi\left(\frac{(\gamma_l)_{l<i}}{x + \sum_{j \neq i} \gamma_j}, \frac{x\eta}{x + \sum_{j \neq i} \gamma_j}, \frac{(\gamma_l)_{l>i}}{x + \sum_{j \neq i} \gamma_j} \right) \frac{1}{\Gamma(1/k)} x^{1/k-1} \mathrm{e}^{-x} dx$$

$$= \frac{1}{n+1}\mathbf{E}\left[\Phi\left(\frac{(\gamma_l)_{l<i}}{\gamma' + \sum_{j\neq i}\gamma_j}, \frac{\gamma'\eta}{\gamma' + \sum_{j\neq i}\gamma_j}, \frac{(\gamma_l)_{l>i}}{\gamma' + \sum_{j\neq i}\gamma_j}\right)\right]$$

where $\gamma' \sim \text{Gamma}((k+1)/k, 1)$, independently of η and $(\gamma_j)_{j\neq i}$. But then $\gamma'\eta$ is a collection of $k+1$ independent Gamma$(1/k, 1)$ random variables, so $\text{Frag}_k(\xi)$ has the law $\text{Dir}_{n+k}(1/k)$ and is independent of the random index I which is uniformly distributed on $\{1, \ldots, n+1\}$. Since we can recover ξ from $\text{Frag}_k(\xi)$ and I by an obvious coagulation, this completes the proof. □

Next we turn our attention to the infinite ranked simplex and define two random transformations, $\text{Frag}_\infty : \Delta_\infty \to \Delta_\infty$ and $\text{Coag}_a : \Delta_\infty \to \Delta_\infty$, where $a \in [0,1]$ is some parameter. The fragmentation transformation on the infinite simplex simply mimics that on the finite simplex; in this direction, recall that the Poisson-Dirichlet PD(1) arises as the weak limit as $k \to \infty$ of sequence of $\text{Dir}_k(1/k)$ variables after obvious re-ordering. More precisely, given $x = (x_1, \ldots) \in \Delta_\infty$, we pick a mass x_I at random by size-biased sampling and split x_I using an independent variable $\eta = (\eta_1, \ldots)$ with law PD(1). In other words, $\text{Frag}_\infty(x)$ is the ranked sequence with unordered terms $x_1, \ldots, x_{I-1}, x_I\eta_1, x_I\eta_2, \ldots, x_{I+1}, \ldots$.

Next, consider a sequence $U_1, U_2, \ldots$ of i.i.d. uniform random variables and $a \in [0,1]$. Starting again from some fixed $x \in \Delta_\infty$, we merge the masses x_i for which $U_i \leq a$ into a single mass and leave the others unchanged. We denote by $\text{Coag}_a(x)$ the random sequence obtained by putting the resulting masses in decreasing order. We then have the following analogue of Proposition 2.1, which is reminiscent of Corollary 13 of Pitman [17].

Proposition 2.2. *Let ξ, ξ' be two random variables with values in Δ_∞. For every $\theta > 0$, the following assertions are equivalent:*

(i) ξ has the law PD(θ) *and, conditionally on ξ, ξ' is distributed as* $\text{Frag}_\infty(\xi)$.

(ii) ξ' has the law PD$(\theta+1)$ *and, conditionally on ξ', ξ is distributed as* $\text{Coag}_{1/(\theta+1)}(\xi')$.

Proof: Let $\gamma = (\gamma(t), t \geq 0)$ be a standard gamma process and set

$$D_t = \gamma((\theta+1)t)/\gamma(\theta+1),$$

for $0 \leq t \leq 1$, so that $(D_t, 0 \leq t \leq 1)$ is a Dirichlet process of parameter $\theta+1$. (The vector of ordered jumps of this Dirichlet process has the PD$(\theta+1)$ distribution.) Consider the following alternative way of thinking of the random coagulation operator $\text{Coag}_{1/(\theta+1)}$: pick a point V uniformly in $[0,1]$ and define a new process $(D'_t, 0 \leq t \leq 1)$ by

$$D'_t = \begin{cases} D_{\theta t/(\theta+1)} & \text{if } t < V \\ D_{(1+\theta t)/(\theta+1)} & \text{if } t \geq V. \end{cases}$$

As the times of the jumps of D are uniformly distributed on $[0,1]$, this picks a proportion $1/(\theta+1)$ of them and coalesces them into a single jump (say $\beta^* = D_{(1+\theta V)/(\theta+1)} - D_{\theta V/(\theta+1)}$) at V. Let $\beta_1 \geq \beta_2 \geq \ldots > 0$ be the sequence of other jumps of D' and $U_1, U_2, \ldots$ the corresponding jump times. Let $\beta'_1 \geq \beta'_2 \geq \ldots > 0$ be the sequence of jumps of D in the interval $[\theta V/(\theta+1), (1+\theta V)/(\theta+1)]$, so that $\beta^* = \sum_{i=1}^\infty \beta'_i$. We wish to show that D' is a Dirichlet process with parameter θ, so that the vector $(\beta^*, \beta_1, \beta_2, \ldots)$ of its jumps (re-arranged in the decreasing order) has the PD(θ) distribution. We will also show that the mass β^* resulting from the coalescence constitutes a size-biased pick from this vector.

Let

$$\gamma^1(t) = \begin{cases} \gamma(t) & \text{if } t < V\theta \\ \gamma(t+1) - (\gamma(V\theta+1) - \gamma(V\theta)) & \text{if } V\theta \le t \le \theta \end{cases}$$
$$\gamma^2(t) = \gamma(V\theta + t) - \gamma(V\theta) \qquad \text{for } 0 \le t \le 1.$$

Then γ^1 and γ^2 are independent processes with $\gamma^1 \stackrel{d}{=} (\gamma(t), 0 \le t \le \theta)$ and $\gamma^2 \stackrel{d}{=} (\gamma(t), 0 \le t \le 1)$, independently of V. Write $\delta_1 \ge \delta_2 \ge \ldots$ for the ordered sequence of jumps of γ^1 and $T_1, T_2, \ldots$ for the corresponding times of these jumps. Write $\delta'_1 \ge \delta'_2 \ge \ldots$ for the ordered sequence of jumps of γ^2. Then

(i) $U_1 = T_1/\theta, U_2 = T_2/\theta, \ldots$ are i.i.d. U[0, 1],
(ii) $\beta^* = \gamma^2(1)/\gamma(1+\theta)$ and so has a Beta$(\theta, 1)$ distribution,
(iii) $\frac{1}{\beta^*}(\beta'_1, \beta'_2, \ldots) = \frac{1}{\gamma^2(1)}(\delta'_1, \delta'_2, \ldots)$ and so has the PD(1) distribution,
(iv) $\frac{1}{1-\beta^*}(\beta_1, \beta_2, \ldots) = \frac{1}{\gamma^1(\theta)}(\delta_1, \delta_2, \ldots)$ and so has the PD(θ) distribution.

Furthermore, the random variables in (i) to (iv) above are independent. The fact that β^* is a size-biased pick from $(\beta^*, \beta_1, \beta_2, \ldots)$ and the PD(θ) distribution of the latter follow from (i) and (iii) and the stick-breaking scheme (see, for instance, Definition 1 in Pitman and Yor [19]). That D' is a Dirichlet process of parameter θ then follows from (iv) and the independence.

The coagulation operator used here can be re-phrased as follows: starting with $x \in \Delta_\infty$, take a sequence $V, V_1, V_2, \ldots$ of i.i.d. U[0, 1] random variables, merge the masses x_i for which $V_i \in [\theta V/(\theta+1), (1+\theta V)/(\theta+1)]$, leave the other masses unchanged and, finally, rank the resulting sequence in decreasing order. Call this operator $\widetilde{\mathrm{Coag}}_{1/(\theta+1)}$. Then it is clear that whenever ξ' is a random exchangeable sequence in Δ_∞, $(\xi', \mathrm{Coag}_{1/(\theta+1)}(\xi'))$ and $(\xi', \widetilde{\mathrm{Coag}}_{1/(\theta+1)}(\xi'))$ have the same distribution. Our claim follows now readily from these results. □

Remark. It may be interesting to check Proposition 2.2 as follows. Consider Poisson random measure M on $(0, \infty)$ with intensity $\theta x^{-1}\mathrm{e}^{-x}dx$. Let $a_1, a_2, \ldots$ be the atoms of M in decreasing order, so that

$$\left(\frac{a_1}{\sum_{j=1}^{\infty} a_j}, \frac{a_2}{\sum_{j=1}^{\infty} a_j}, \ldots\right)$$

has distribution PD(θ), independently of $\sum_{j=1}^{\infty} a_j$. Let $\eta \sim$ PD(1), independently of M and suppose that $\Phi : \Delta_\infty \to \mathbf{R}$ is any symmetric bounded measurable function. Then if $\xi \sim$ PD(θ), using independence we have that

$$\mathbf{E}\left[\Phi(\mathrm{Frag}_\infty(\xi))\right] = \frac{1}{\mathbf{E}\left[\sum_{j=1}^{\infty} a_j\right]} \mathbf{E}\left[\sum_{i=1}^{\infty} a_i \Phi\left(\frac{a_i \eta}{\sum_{j=1}^{\infty} a_j}, \frac{(a_l)_{l \ne i}}{\sum_{j=1}^{\infty} a_j}\right)\right].$$

By the Palm formula, this is equal to

$$\frac{1}{\theta}\mathbf{E}\int_0^\infty x\Phi\left(\frac{x\eta}{x + \sum_{j=1}^{\infty} a_j}, \frac{(a_l)_{l=1}^{\infty}}{x + \sum_{j=1}^{\infty} a_j}\right)\theta x^{-1}\mathrm{e}^{-x}dx$$
$$= \mathbf{E}\left[\Phi\left(\frac{a'\eta}{a' + \sum_{j=1}^{\infty} a_j}, \frac{(a_l)_{l=1}^{\infty}}{a' + \sum_{j=1}^{\infty} a_j}\right)\right]$$

where $a' \sim \text{Exp}(1)$, independently of M and η. But then $a'\eta$ has the distribution of the atoms of a Poisson random measure with intensity $x^{-1}e^{-x}dx$ arranged in decreasing order and so we see that taking these atoms together with those of M, we obtain a Poisson random measure of intensity $(\theta+1)x^{-1}e^{-1}dx$. Hence, $\text{Frag}_\infty(\xi)$ has the law $\text{PD}(\theta+1)$.

2.3. Dual fragmentation and coagulation chains

The dual fragmentation and coagulation operators that were defined in the preceding section incite us to introduce Markov fragmentation and coagulation chains in duality by time-reversal. Specifically, we consider for each integer $k \geq 1$ a chain

$$X^{(k)}(0), X^{(k)}(1), X^{(k)}(2), \ldots,$$

where $X^{(k)}(n)$ is a random variable with values in Δ_{nk} (in particular $X^{(k)}(0) = 1$), and the conditional distribution of $X^{(k)}(n+1)$ given $X^{(k)}(n) = x$ is the law of $\text{Frag}_k(x)$. We deduce from Proposition 2.1 by induction that for each n, $X^{(k)}(n)$ has the distribution $\text{Dir}_{nk}(1/k)$. The time-reversed coagulation chain

$$\ldots, X^{(k)}(n+1), X^{(k)}(n), \ldots, X^{(k)}(1), X^{(k)}(0)$$

is also Markov; more precisely, this means that the conditional distribution of $X^{(k)}(n)$ given $X^{(k)}(n+1) = x$ is the law of $\text{Coag}_k(x)$. Note that for $k = 1$, this has the distribution of the jump chain of Kingman's coalescent [13].

Analogously, for $k = \infty$, we can define a Markov fragmentation chain on Δ_∞,

$$X^{(\infty)}(0), X^{(\infty)}(1), X^{(\infty)}(2), \ldots,$$

such that the conditional distribution of $X^{(\infty)}(n+1)$ given $X^{(\infty)}(n) = x$ is the law of $\text{Frag}_\infty(x)$. We deduce by induction from Proposition 2.2 that for every $\theta > 0$, if the distribution of the initial state $X^{(\infty)}(0)$ is $\text{PD}(\theta)$ then, for every integer n, $X^{(\infty)}(n)$ has the distribution $\text{PD}(\theta+n)$. Moreover, in this case, the time-reversed coagulation chain

$$\ldots, X^{(\infty)}(n+1), X^{(\infty)}(n), \ldots, X^{(\infty)}(1), X^{(\infty)}(0)$$

is also Markov; more precisely, the conditional distribution of $X^{(\infty)}(n)$ given $X^{(\infty)}(n+1) = x$ is the law of $\text{Coag}_{1/(n+1+\theta)}(x)$.

Remarks. (a) Recall that the parameter θ can be recovered from a sample ξ of a $\text{PD}(\theta)$ random variable as follows:

$$\theta = \lim_{\varepsilon \to 0+} \frac{1}{\log 1/\varepsilon} \max\{n : \xi_n > \varepsilon\}.$$

This shows that the above description for the reversed coagulation chain is indeed Markovian.

(b) There is simple representation for the $k = \infty$ fragmentation chain in terms of compound bridges with exchangeable increments which is inspired by [5]. Let $U_0, U_1, \ldots$ be a sequence of independent uniform variables on $[0,1]$. For each n, we consider the elementary bridge $b_n : [0,1] \to [0,1]$ defined by

$$b_n(t) = \frac{n}{n+1}t + \frac{1}{n+1}\mathbf{1}_{\{t > U_n\}}, \qquad t \in [0,1].$$

Then is is easy to check that for every $n \in \mathbf{N}$, the sequence $b_n \circ b_{n+1} \circ \cdots \circ b_{n+i}$ converges pointwise almost surely as $i \to \infty$ to a bridge with exchangeable increments B_n which has no drift and infinitely many jumps a.s. If we write $\beta_n \in$

Δ_n for the sequence of the sizes of the jumps of B_n ranked the decreasing order, then the chain $(\beta_n, n \in \mathbf{N})$ has the same law as $X^{(\infty)}$. We refer to [5] for the necessary technical background.

3. The genealogy of Yule processes

We shall now show that the dual fragmentation and coagulation chains which we introduced in the preceding section are naturally connected to the genealogy of Yule processes.

3.1. Discrete setting

For every integer $k \geq 1$, we write $Y^{(k)} = \left(Y_t^{(k)}, t \geq 0\right)$ for the Yule process started from $Y_0^{(k)} = 1$: $Y_t^{(k)}$ gives the number of individuals alive a time t in a branching process in which each individual lives for an exponential time of parameter 1 and gives birth at its death to $k+1$ children, which then evolve independently of one another according to the same rules as their parent. We agree to label each child of an individual by an integer in $\{1, \ldots, k+1\}$, which allows us to order individuals at any generation in a consistent way: given two distinct individuals, we may consider their most recent common ancestor. Plainly, two different children of this ancestor are ancestors of exactly one of these two individuals, and the labelling of the children of the most recent common ancestor induces the order of the individuals.

Lemma 3.1. *The process* $\left(\exp(-kt)Y_t^{(k)}, t \geq 0\right)$ *is a uniformly integrable martingale and its limit* $W^{(k)}$ *has the* Gamma$(1/k, 1/k)$ *distribution.*

Proof: A similar limit result is stated in Athreya & Ney [1] on page 130; however, the limiting distribution given there is incorrect and so we shall provide here a detailed proof. The martingale property is classical, so we focus on the distribution of the limit $W^{(k)}$. Define

$$\Phi_t(s) := \mathbf{E}\left(s^{Y_t^{(k)}}\right).$$

The backward equation implies that

$$\frac{\partial}{\partial t}\Phi_t(s) = \Phi_t^{k+1}(s) - \Phi_t(s) \quad , \quad \Phi_0(s) = s.$$

This equation has solution

$$\Phi_t(s) = s\mathrm{e}^{-t}\left(1 - \left(1 - \mathrm{e}^{-kt}\right)s^k\right)^{-1/k}.$$

Hence, for $\theta < 0$,

$$\begin{aligned}\mathbf{E}\left(\exp\left(\theta \mathrm{e}^{-kt}Y_t^{(k)}\right)\right) &= \exp\left(\theta \mathrm{e}^{-kt}\right)\mathrm{e}^{-t}\left(1 - \left(1 - \mathrm{e}^{-kt}\right)\exp\left(\theta k \mathrm{e}^{-kt}\right)\right)^{-1/k} \\ &= \left[\mathrm{e}^{kt}\exp\left(-\theta k\mathrm{e}^{-kt}\right) - \mathrm{e}^{kt} + 1\right]^{-1/k},\end{aligned}$$

and when $t \to \infty$, this quantity converges to

$$(1 - k\theta)^{-1/k} = \left(\frac{1/k}{1/k - \theta}\right)^{1/k},$$

which is the moment generating function of a gamma random variable with parameters $(1/k, 1/k)$. □

We think of $W^{(k)}$ as the size of the terminal population. For every $t \geq 0$, by application of the branching property at time t, we may decompose the terminal population into sub-populations having the same ancestor at time t. Specifically,

$$W^{(k)} = \sum_{i=1}^{Y_t^{(k)}} W_i^{(k)}(t),$$

where $W_i^{(k)}(t)$ is the size of the terminal sub-population descending from the ith individual in the population at time t. Observe that conditionally on $Y_t^{(k)}$, the variables $W_i^{(k)}(t)$ are independent and all have the same law as $\mathrm{e}^{-t}W^{(k)}$.

Finally, we define the genealogical process $G^{(k)} = \left(G^{(k)}(t), t \geq 0\right)$ associated to $Y^{(k)}$ by

$$G^{(k)}(t) = \left(W_1^{(k)}(t), \ldots, W_{Y_t^{(k)}}^{(k)}(t)\right).$$

The genealogical structure of the Yule process can be described in terms of the fragmentation chain $X^{(k)}$ of Section 2.3 as follows.

Theorem 3.1. *Let $N = (N_t, t \geq 0)$ be a standard Poisson process which is independent of the chain $X^{(k)}$. Then for each $w > 0$, the compound chain*

$$\left(wX^{(k)}(N_{wt}), t \geq 0\right)$$

has the same law as the time-changed process

$$\left(G^{(k)}\left(\frac{1}{k}\log(1+kt)\right), t \geq 0\right)$$

conditioned on $W^{(k)} = w$.

Remark. Theorem I of Kendall [12] states that given $W^{(1)}$, $\left(Y^{(1)}_{\log(1+t/W^{(1)})}, t \geq 0\right)$ is a Poisson process with unit parameter. This is clearly an aspect of Theorem 3.1. Moreover, on page 130 of Athreya & Ney [1], it is suggested that no generalization of Kendall's result to a more general continuous-time Markov branching process is known; Theorem 3.1 constitutes a small such generalization.

Proof: Set $\tau(t) := \frac{1}{k}\log(1+kt)$ and let T be the time of the first birth in the Yule process, which is also the time of the first dislocation of $G^{(k)}$. The $k+1$ fragments of $G^{(k)}(T)$ can be written as $\mathrm{e}^{-kT}Z_1, \ldots, \mathrm{e}^{-kT}Z_{k+1}$ where, by the branching property, $Z_1, \ldots, Z_{k+1}$ are i.i.d. Gamma$(1/k, 1/k)$ random variables, independent of T which is Exp(1). Define a change of variables by

$$S = \tau^{-1}(T) = (\mathrm{e}^{kT} - 1)/k$$
$$U_1 = \mathrm{e}^{-kT}Z_1, \quad \ldots, \quad U_k = \mathrm{e}^{-kT}Z_k, \quad W = \mathrm{e}^{-kT}(Z_1 + \cdots + Z_{k+1}).$$

It is straightforward to see that the joint density of $(T, Z_1, \ldots, Z_{k+1})$ is

$$f(t, z_1, \ldots, z_{k+1})$$
$$= \mathrm{e}^{-t}\Gamma(1/k)^{-(k+1)}(1/k)^{(k+1)/k}(z_1z_2\ldots z_{k+1})^{-(k-1)/k}\exp(-(z_1 + \cdots + z_{k+1})/k)$$

and so the joint density of $(S, U_1, \ldots, U_k, W)$ is

$$g(s, u_1, \ldots, u_k, w) = w\mathrm{e}^{-ws} \cdot (1/k)\Gamma(1/k)^{-k}w^{-1/k}(u_1u_2\ldots u_k$$

$$\cdot (w - u_1 - \cdots - u_k))^{1/k-1} \cdot (1/k)^{1/k}\Gamma(1/k)^{-1}w^{-(k-1)/k}\exp(-w/k).$$

Hence, $W \sim \text{Gamma}(1/k, 1/k)$ (as we already knew) and, conditional on $W = w$, we have $S \sim \text{Exp}(w)$ and $(U_1, U_2, \ldots, U_k, W - U_1 - \cdots - U_k) \sim w\text{Dir}_k(1/k)$ independently of S. Thus, the first dislocation has the correct dynamics. But by the branching property, subsequent dislocations are independent for different subpopulations and the total rate of fragmentation is always w. Hence result. □

In the terminology of [2], Theorem 3.1 states that the time-changed genealogical process $G^{(k)} \circ \tau$ is a self-similar fragmentation with index 1, dislocation law $\text{Dir}_k(1/k)$ and erosion coefficient 0. It may be interesting to observe that in the special case $k = 1$, this result can also be derived as follows.

Consider a real Brownian motion B started from 1 and killed when it reaches 0 (at time $T_0 = \inf\{t \geq 0 : B_t = 0\}$). For every $u \in [0, 1[$, let $\widetilde{Y}_u$ denote the number of excursions of B away from 1 which go below level u. Then $(Y^{(1)}_{-\log(1-u)})_{0 \leq u < 1}$ is a version of $(\widetilde{Y}_u)_{0 \leq u < 1}$. To see this, let us consider the evolution of $\widetilde{Y}$. Firstly, $\widetilde{Y}_0 = 1$, corresponding to the single excursion below 1 which reaches 0. Let $D = \sup\{t < T_0 : B_t = 1\}$, the starting time of the final excursion which hits 0, let $U = \inf_{0 \leq t \leq D} B_t$ be the level reached by the deepest excursion below 1 before D and let T_U be the time at which it is reached. Then, by Williams' path decomposition theorem (Theorem VII.4.9 of Revuz and Yor [20]), U is distributed uniformly on $[0, 1[$ and, conditional on U, $(B_t)_{0 \leq t < T_U}$ is a Brownian motion started at 1 and stopped when it first hits level U. By symmetry, $(B_{D-t})_{0 \leq t < D - T_U}$ is another independent Brownian motion started at 1 and stopped when it first hits level U. Thus, $\widetilde{Y}_u$ is equal to 1 on $[0, U[$, $\widetilde{Y}_U = 2$ and $(\widetilde{Y}_{U+v})_{0 \leq v < 1-U}$ evolves as the sum of two independent processes which are the same as $\widetilde{Y}$ except that the times until the first jumps are now uniform on $[0, U[$ rather than on $[0, 1[$. (This is Theorem 8 of Le Gall [16], repeated here for completeness.) Time-changing $Y^{(1)}$ with $u \to -\log(1-u)$ means that its exponential inter-jump times become uniform and so we do, indeed, have $(Y^{(1)}_{-\log(1-u)})_{0 \leq u < 1} \stackrel{d}{=} (\widetilde{Y}_u)_{0 \leq u < 1}$.

A more elegant way of expressing the preceding argument is to say that the Brownian path encodes a continuous-state branching process with quadratic branching mechanism. The local time at level 1, $L^1_{T_0}$, satisfies

$$L^1_{T_0} = \lim_{u \to 1-} 2(1-u)\widetilde{Y}_u.$$

In this context, $\frac{1}{2}L^1_{T_0}$ corresponds to the size of the population at time 1 in the continuous-state branching process generated by a single ancestor conditioned to have descendents up to time 1. The so-called reduced tree associated with the population at time 1 is described up to the deterministic time-change $u \to -\log(1-u)$ by the Yule process $Y^{(1)}$. See, for instance, Section 2.7 in Duquesne and Le Gall [8], and Fleischmann and Siegmund-Schultze [9]. Note that the well-known fact that $\frac{1}{2}L^1_{T_0}$ has an exponential distribution with mean 1 (Proposition VI.4.6 of Revuz and Yor [20]) gives another derivation of the limiting distribution in Lemma 3.1, since

$$W^{(1)} = \lim_{t \to \infty} e^{-t}Y^{(1)}_t = \lim_{u \to 1-} (1-u)Y^{(1)}_{-\log(1-u)} \stackrel{d}{=} \frac{1}{2}L^1_{T_0}.$$

It is known from excursion theory that in the scale of the local time at level 1, the rate of excursions of B away from 1 which reach level $u \in]0, 1[$ but do not exceed $u - du$ is $(1-u)^{-2}du$. Note that the map $s \to 1 - \frac{1}{1+s}$ from $\mathbf{R}_+$ to $[0,1[$ has inverse $u \to \frac{1}{1-u} - 1$ and, thus, transforms Lebesgue measure on $\mathbf{R}_+$ into the measure $(1-u)^{-2}du$ on $[0,1[$. Suppose that we split the local time at level 1 according to the occurrence of excursions exceeding level u, so that we obtain the sequence

$$\widetilde{W}(u) = \left(\widetilde{W}_1(u), \ldots, \widetilde{W}_{\widetilde{Y}_u}(u)\right),$$

where $\widetilde{W}(u)$ is the sequence of the increments of the local time at level 1 on the maximal time intervals such that at the beginning and end of each interval B is at 1 and during the interval remains above level u. Then it follows easily that the time-changed process $\left(\widetilde{W}\left(1 - \frac{1}{1+s}\right), s \geq 0\right)$ is a fragmentation in which each mass, say x, splits at rate x into xU and $x(1-U)$ where U is uniform. In other words, conditionally on $\frac{1}{2}L^1_{T_0} = w$, the process $\left(\widetilde{W}\left(1 - \frac{1}{1+s}\right), s \geq 0\right)$ is distributed as the compound fragmentation chain $\left(wX^{(1)}(N_{ws}), s \geq 0\right)$, where N is an independent standard Poisson process.

Finally, the composition of the two time-changes which appear in this analysis yields

$$s \to -\log\left(1 - \left(1 - \frac{1}{1+s}\right)\right) = \log(1+s)\,, \qquad s \in \mathbf{R}_+\,,$$

and so we recover Theorem 3.1 in the special case $k = 1$. Unfortunately, it does not seem that there are similar interpretations for $k \geq 2$.

Corollary 3.2. *We have that*

$$\left(\frac{1}{W^{(k)}} G^{(k)}\left(\frac{1}{k}\log\left(1 + ke^{-t}/W^{(k)}\right)\right), t \in \mathbf{R}\right)$$

is a time-homogeneous Markov coalescent process which is independent of $W^{(k)}$. For any $n \geq 1$, given that it is in state $x \in \Delta_{nk}$, it waits an exponential time of parameter n and then jumps to a variable distributed as $\mathrm{Coag}_k(x)$, *independently of the exponential time.*

Note that the case $k=1$ of this result gives a variation of Kingman's coalescent. The jump-chains are identical, as we have already noted, but here the rate of coalescence of two blocks depends on the total number of blocks present, whereas in Kingman's coalescent it does not.

Proof: Firstly, we note that by Theorem 3.1,

$$\left(\frac{1}{W^{(k)}} G^{(k)}\left(\frac{1}{k}\log(1 + ke^{-t}/W^{(k)})\right), t \in \mathbf{R}\right)$$

has the same law as

$$\left(X^{(k)}(N_{e^{-t}}), t \in \mathbf{R}\right)$$

and so we will work with the latter process instead. The $k=1$ case is essentially treated in [3] and the proof proceeds in the same way here. The jump chain clearly behaves in the correct manner and so it remains to check that the inter-jump times are as claimed. Let $0 \leq T_1 \leq T_2 \leq \ldots$ be the jump times of $(N_t)_{t \geq 0}$. Then the first instant that $X^{(k)}(N_{e^{-t}})$ has exactly $nk+1$ terms is

$$\inf\{t \in \mathbf{R} : N_{e^{-t}} = n\} = -\log T_{n+1}.$$

The sequence of inter-jump times is

$$\dots, \log T_{n+1} - \log T_n, \log T_n - \log T_{n-1}, \dots, \log T_2 - \log T_1$$

and it is easily shown that this is a sequence of independent exponential random variables with parameters

$$\dots, n, n-1, \dots, 1$$

respectively. □

3.2. Continuous setting

Continuous-state branching processes (or CSBP's) were introduced by Lamperti [14, 15] as limits of rescaled branching processes. Typically, a CSBP is a time-homogeneous Markov process with values in $\mathbf{R}_+$,

$$Z = (Z(t,a),\, t \geq 0 \text{ and } a \geq 0)\,,$$

(where the parameter t refers to time and the parameter a to the starting point i.e. $Z(0,a) = a$ a.s.) which fulfils the branching property: the path-valued process $(Z(\cdot,x), x \geq 0)$ has independent and stationary increments. In particular, if $\widetilde{Z}(\cdot,y)$ is an independent copy of $Z(\cdot,y)$, then $Z(\cdot,x) + \widetilde{Z}(\cdot,y)$ has the law of $Z(\cdot,x+y)$. There is a simple relation connecting CSBP's and Bochner's subordination for subordinators which enables us to define their genealogy; we refer the interested reader to [4] for heuristics, detailed arguments etc.

We call a *continuous state Yule process* a CSBP

$$Y = (Y(t,a),\, t \geq 0 \text{ and } a \geq 0)\,,$$

which evolves as follows: for each $a > 0$, the process $Y(\cdot,a)$ waits an exponential time with parameter a and then jumps to $a+1$. It then evolves independently as if it had been started in state $a+1$. In terms of the genealogy, the sub-population of size 1 which is born at a jump time has a parent which is chosen uniformly at random from the population present before the jump. Note that this genealogy is easy to describe in a consistent manner for different values a of the starting population.

It is immediate that for an integer starting point $a \in \mathbf{N}$, the process $(Y(t,a), t \geq 0)$ is a Yule process $Y^{(1)}$ with 2 offspring, as considered in the preceding section. However, we stress that its genealogy is not the same as that of $Y^{(1)}$, as we are dealing with a continuous population in the first case and a discrete population in the second.

We have the following analogue of Lemma 3.1:

Lemma 3.3. *For every $a \geq 0$, the process $(\mathrm{e}^{-t}Y(t,a), t \geq 0)$ is a uniformly integrable martingale. Its limit, say $\gamma(a)$, viewed as a process in the variable a, has the same finite dimensional laws as a standard gamma process.*

Proof: For $a = 1$, we see from Lemma 3.1 and the identity in distribution $Y(\cdot,1) \stackrel{\mathcal{L}}{=} Y^{(1)}(\cdot)$ that $(\mathrm{e}^{-t}Y(t,1), t \geq 0)$ is a uniformly integrable martingale and that its limit has the standard exponential distribution. The proof is easily completed by an appeal to the branching property. □

Remark. The limiting distribution in Lemma 3.3 is essentially a corollary of Theorem 3 of Grey [10].

Just as in the preceding section, we think of $\gamma(a)$ as the size of the terminal population when the initial population has size a. We can express $\gamma(a)$ as

$$\gamma(a) = \sum_{b \le a} \delta_b ,$$

where $\delta := (\delta_b, b \ge 0)$ is the jump process of γ, which corresponds to decomposing the terminal population into sub-populations having the same ancestor at the initial time. We write $G(0,a)$ for the sequence of the jumps of γ on $[0,a]$, ranked in decreasing order, and we deduce from Lemma 3.3 that conditionally on $\gamma(a) = g$, $G(0,a)/g$ has distribution $\mathrm{PD}(a)$.

More generally, by the branching property, we can decompose the terminal population into sub-populations having the same ancestor at any given time t. This gives

$$\gamma(a) = \sum_{b \le Y(t,a)} \mathrm{e}^{-t} \delta_b^{(t)} ,$$

where $\delta^{(t)} := (\delta_b^{(t)}, b \ge 0)$ is the jump process of a standard gamma process $\gamma^{(t)}$ which is independent of the Yule process up to time t, $(Y(s,c),\ s \in [0,t]$ and $c \ge 0)$. This enables us to define for each $a > 0$ the genealogical process associated to a Yule process $Y(\cdot, a)$,

$$G(\cdot, a) = (G(t,a), t \ge 0) ,$$

where $\mathrm{e}^t G(t,a)$ is the ranked sequence of the sizes of the jumps of the subordinator $\gamma^{(t)}$ on the interval $[0, Y(t,a)]$.

An easy variation of the arguments for the proof of Theorem 3.1 shows that the genealogical structure of the Yule process can be described in terms of the fragmentation chain $X^{(\infty)}$ of Section 2.3 as follows.

Theorem 3.2. *Fix $a, g > 0$ and let the chain $X^{(\infty)}$ have initial distribution* $\mathrm{PD}(a)$. *Introduce a standard Poisson process, $N = (N_t, t \ge 0)$, which is independent of the chain $X^{(\infty)}$. Then the compound chain*

$$\left(gX^{(\infty)}(N_{gt}), t \ge 0\right)$$

has the same law as the time-changed process

$$\left(G\left(\log(1+t), a\right), t \ge 0\right)$$

conditioned on $\gamma(a) = g$.

Likewise, the analogue of Corollary 3.2 is as follows.

Corollary 3.4. *Fix $a > 0$. Then*

$$\left(\frac{1}{\gamma(a)} G\left(\log(1 + \mathrm{e}^{-t}/\gamma(a)), a\right), t \in \mathbf{R}\right)$$

is a time-homogeneous Markov coalescent process which is independent of $\gamma(a)$. Suppose that it is in state $x \in \Delta_\infty$ and recall Remark (a) of Section 2.3. Then if

$$\lim_{\epsilon \to 0+} \frac{1}{\log 1/\epsilon} \max\{i : x_i > \epsilon\} = n + a,$$

the process waits an exponential time of parameter n and then jumps to a variable distributed as $\mathrm{Coag}_{1/(n+a)}(x)$, independently of the exponential time.

References

[1] Athreya, K.B. and Ney, P.E. (1972). *Branching processes.* Springer-Verlag, Berlin-Heidelberg-New York

[2] Bertoin, J. (2002). Self-similar fragmentations. *Ann. Inst. Henri Poincaré* **38**, 319-340

[3] Bertoin, J. (2003). Random covering of an interval and a variation of Kingman's coalescent. To appear in *Random Structures Algorithms.* Also available as Preprint PMA-794 at `http://www.proba.jussieu.fr/mathdoc/preprints/index.html`

[4] Bertoin, J. and Le Gall, J.-F. (2000). The Bolthausen-Sznitman coalescent and the genealogy of continuous-state branching processes. *Probab. Theory Relat. Fields* **117**, 249-266

[5] Bertoin, J. and Le Gall, J.-F. (2003). Stochastic flows associated to coalescent processes. *Probab. Theory Relat. Fields* **126**, 261-288

[6] Bertoin, J. and Pitman, J. (2000). Two coalescents derived from the ranges of stable subordinators. *Elect. J. Probab.* **5**, 1-17. Available via
`http://www.math.u-psud.fr/~ejpecp/ejp5contents.html`

[7] Bolthausen, E. and Sznitman, A.S. (1998). On Ruelle's probability cascades and an abstract cavity method. *Comm. Math. Physics* **197**, 247-276

[8] Duquesne, T. and Le Gall, J.-F. (2002). *Random trees, Lévy processes and spatial branching processes.* Astérisque **281**

[9] Fleischmann, K. and Siegmund-Schultze, R. (1977) The structure of reduced critical Galton-Watson processes. *Math. Nachr.* **79**, 233-241

[10] Grey, D. R. (1974). Asymptotic behaviour of continuous time, continuous state-space branching processes. *J. Appl. Probab.* **11**, 669-677

[11] Kallenberg, O. (1973). Canonical representations and convergence criteria for processes with interchangeable increments. *Z. Wahrsch. verw. Gebiete* **27**, 23-36

[12] Kendall, D. G. (1966). Branching processes since 1873. *J. London Math. Soc.* **41**, 385-406

[13] Kingman, J. F. C. (1982). The coalescent. *Stochastic Process. Appl.* **13**, 235-248

[14] Lamperti, J. (1967). The limit of a sequence of branching processes. *Z. Wahrsch. verw. Gebiete* **7**, 271-288

[15] Lamperti, J. (1967). Continuous-state branching processes. *Bull. Amer. Math. Soc.* **73**, 382-386

[16] Le Gall, J.-F. (1989). Marches aléatoires, mouvement brownien et processus de branchement. *Séminaire de Probabilités XXIII*, Lecture Notes in Math., 1372, Springer, Berlin, 258-274

[17] Pitman, J. (1999). Coalescents with multiple collisions. *Ann. Probab.* **27**, 1870-1902

[18] Pitman, J. (2002). *Combinatorial Stochastic Processes.* Lecture notes for the St Flour summer school. To appear. Available at
`http://stat-www.berkeley.edu/users/pitman/621.ps.Z`

[19] Pitman, J. and Yor, M. (1997). The two-parameter Poisson-Dirichlet distribution derived from a stable subordinator. *Ann. Probab.* **25**, 855-900

[20] Revuz, D. and Yor, M. (1999). *Continuous martingales and Brownian motion*, Third edition. Springer-Verlag, Berlin

Jean Bertoin
Laboratoire de Probabilités et Modèles Aléatoires and Institut universitaire de France,
Université Pierre et Marie Curie,
175, rue du Chevaleret,
F-75013 Paris,
France.

Christina Goldschmidt
Laboratoire de Probabilités et Modèles Aléatoires,
Université Pierre et Marie Curie,
175, rue du Chevaleret,
F-75013 Paris,
France.

Trends in Mathematics, © 2004 Birkhäuser Verlag Basel/Switzerland

Semi-Markov Walks in Queueing and Risk Theory

Mykola S. Bratiychuk

1. Introduction

Let $\{\eta_n(i)\}$, $\{\alpha_n(i)\}$, $i = 1, 2$, $n = 1, 2\dots$ be four sequences of non-negative totally independent random variables having the same distribution within every sequence. Let $\mu_i(t)$, $i = 1, 2$ be the renewal processes generated by $\alpha_n(i)$, $i = 1, 2$ respectively. The process

$$\xi(t) = \sum_{k=1}^{\mu_1(t)} \eta_k(1) - \sum_{k=1}^{\mu_2(t)} \eta_k(2) + \xi(0) \tag{1}$$

is called *the semi-Markov random walk.*

Processes from (1) one often meets with in queueing theory, dam theory, risk theory due to the fact that such processes are the natural mathematical models of the problems arising there. The analytical theory of such walks was developed in [1], [2]. Here we apply it to the study of batch arrival queueing systems and to one class of non-Markovian risk processes.

We put

$$f_i(s) = \mathbf{E}e^{-s\alpha_1(i)}, \ \mathbf{P}\{\eta_1(1) = k\} = p_k, \ q(z) = \mathbf{E}z^{\eta_1(2)}.$$

2. Results

Let $\xi(t)$ be the number of customers in the batch arrival system of the type $G^\eta I/G/1$ at the time t.

Theorem 1. *For $m \geq 1$, $\lambda > 0$ the following representation holds true*

$$\int_0^\infty e^{-\lambda t}\mathbf{P}\{\xi(t) = m\}dt = \frac{1}{\lambda}\sum_{k=0}^{m-1} p_m^{(k+1)*}\left(f_1^k(\lambda) - f_1^{k+1}(\lambda)\right)$$

$$+ \frac{1}{f_+(\lambda,\lambda)}\sum_{k=0}^{m}\int_{-\infty}^{+0}\int_0^\infty e^{-\lambda t}\Theta(t-y, m-k, \lambda)dH(t,k)dP_-(y,\lambda),$$

where $f_+(\lambda,\lambda)$, $\Theta(y,k,\lambda)$, $H(y,k), P_-(y,\lambda)$ are some known functions.

Let now we have a risk process of the form

$$\xi(t) = \mu_1(t) - S^{(2)}_{\mu_2(t)} + \xi(0)$$

and $\tau = \inf\{t > 0 : \xi(t) < 0\}$. Then $\Phi(n) = \mathbf{P}\{\tau < \infty\mu\xi(0) = n\}$ is ruin probability, provided the initial capital of the company is equal to n.

Theorem 2. *Let there exist $s_0 < 0$ such that $f_2(-s_0)q(f_1(s_0)) > 1$. Then for some $\varepsilon > 0$*

$$\Phi(n) = K(\nu)f_1^{-n}(-\nu) + o\big(f_1^{-n}(-\nu-\varepsilon)\big)$$

as $n \to \infty$, where $K(\nu)$ is some known constant and $s_0 < -\nu < 0$ is the solution of the equation

$$f_2(-s)q(f_1(s)) = 1.$$

References

[1] Korolyuk, V.S. and Pirliev B. *Random walk on the half-axis on the superposition of two renewal processes*, Ukr. Matem. Zur. **36**, N.4, 1984, 433-436.

[2] Bratiychuk M. S., Kempa W. *Application of the superposition of renewal processes to the study of batch arrival queues,* Queueing systems. **44**, 2003, 51-67.

Mykola S. Bratiychuk
Silesian University of Technology, Gliwice, Poland

Trends in Mathematics, © 2004 Birkhäuser Verlag Basel/Switzerland

Representation of Fixed Points of a Smoothing Transformation

Amke Caliebe

ABSTRACT: *Given a sequence of real random variables $T = (T_1, T_2, ...)$, distributions μ which satisfy the following fixed point equation for distributions are considered:*

$$W \stackrel{\sim}{=} \sum_{j=1}^{\infty} T_j W_j \ ,$$

where $W, W_1, W_2, ...$ have distribution μ and $T, W_1, W_2, ...$ are independent. Only the case $\sum_{j=1}^{\infty} 1_{|T_j|>0} < \infty$ almost surely is regarded in this article. These solutions are known to arise, e.g., in the limiting behaviour of branching processes. Here, such fixed points are characterized as mixtures of infinitely divisible distributions. Depending on the properties of T and of the fixed points in question, it can be shown that the corresponding infinitely divisible distributions can be normal, α-stable ($\alpha < 1$) or degenerate.

1. Introduction

Let $T = (T_1, T_2, ...)$ be a sequence of real random variables with $N := \sum_{j=1}^{\infty} 1_{|T_j|>0} < \infty$ almost surely. The *smoothing transformation* K is defined as

$$K : \mathfrak{D} \to \mathfrak{D}; \qquad K(\mu) \stackrel{\sim}{=} \sum_{j=1}^{\infty} T_j W_j$$

with $\mathfrak{D}$ the space of distributions and $W_1, W_2, ...$ random variables with distribution μ and $T, W_1, W_2, ...$ independent. The random variables T_j are also called *coefficients*.

The object of this article is to characterize fixed points of K. These satisfy the following equation for distributions μ

$$W \stackrel{\sim}{=} \sum_{j=1}^{\infty} T_j W_j \ , \tag{1}$$

where $W, W_1, W_2, ...$ have distribution μ and $T, W_1, W_2, ...$ are independent. A random variable is called a fixed point if its distribution satisfies (1).

Fixed point equation (1) has extensive applications (for an overview see [26], [27], and [32]).Here, only the field of branching processes will be shortly (and not comprehensively) mentioned. Since the first appearance of fixed points in the Galton-Watson process to characterize the limiting behaviour of the generation size [4, 10, 18, 21, 23], they were also discovered in Bellman-Harris and Crump-Mode-Jagers processes [3, 11, 15, 16]. They are of special interest in the case

*Research supported by the German Science Foundation (DFG) Grant RO 498/4-1.

of the branching random walk [5, 6, 7, 8, 9, 30] and of the weighted branching process ([28], who used the expression *marked trees* instead; [33], [34]). Lately, it was shown that fixed points with finite expectation can always be viewed as limiting distributions of the corresponding weighted branching process [14, 25]. A nice recent survey about distributional fixed point equations is given in [1].

Fixed points of Eq. (1) are sometimes regarded as generalized stable distributions. [29] calls them *laws stable by random weighted mean.* If the coefficients T_j are constant then it is known that all fixed points are mixtures of stable distributions (if the closed multiplicative group generated by the coefficients is uncountable; otherwise the situation is slightly more involved) [2]. In the far more complex situation of random coefficients, this result is unknown. In this article, however, some steps are made in that direction. Recently [12], it was proved that fixed points are mixtures of infinitely divisible distributions. Here, for special cases of fixed points these distributions are calculated in more detail (under certain regularity and integrability assumptions):

- For positive coefficients and positive fixed points the infinitely divisible distributions are α-stable with $0 < \alpha < 1$.
- For fixed points with finite, non-zero expectation and positive coefficients the infinitely divisible distributions are constants.
- For fixed points with finite variance and zero expectation the infinitely divisible distributions are normal.

Note that all these distributions are stable. Whether *all* kinds of fixed points are mixtures of α-stable distributions remains still to be investigated.

2. Infinitely Divisible and Stable Distributions

This section displays some basic important features of infinitely divisible and stable distributions as needed in the following. It mainly aims at establishing a consistent notation (cp., e.g., [22]).The experienced reader may skip this section or consult it at convenience. For the results and proofs of this section compare the fundamental works of [19], [24], and [31]; for the representation of stable distributions see also [22].

A probability measure P on $\mathbb{R}$ with characteristic function φ_{P} or a random variable with distribution P is called *infinitely divisible* iff for every $n \in \mathbb{N}$ there is a characteristic function φ_n with $\varphi_{\mathrm{P}} = (\varphi_n)^n$.

Let $\mathcal{B}_0$ denote the Borel σ algebra on $\mathbb{R} \setminus \{0\}$. The function $\varphi : \mathbb{R} \to \mathbb{C}$ is a characteristic function of an infinitely divisible distribution if and only if there exist constants $\gamma \in \mathbb{R}$, $\sigma^2 \in \mathbb{R}_{\geq 0}$ and a (not necessarily finite) measure ν on $\mathcal{B}_0$ with

$$\int_{\mathbb{R}\setminus\{0\}} \frac{u^2}{1+u^2}\nu(\mathrm{d}u) < \infty$$

such that

$$\log \varphi(t) = \mathrm{i}\gamma t - \frac{1}{2}\sigma^2 t^2 + \int_{\mathbb{R}\setminus\{0\}} \left(\mathrm{e}^{\mathrm{i}ut} - 1 - \frac{\mathrm{i}ut}{1+u^2}\right)\nu(\mathrm{d}u) \,. \tag{2}$$

This representation is unique. It is called *Lévy representation* and ν is referred to as the *Lévy measure.* Denote in the subsequent text the measure generating function of ν by

$$Q : \mathbb{R} \setminus \{0\} \to \mathbb{R}, u \mapsto \begin{cases} \nu((-\infty, u]) & : \quad u < 0 \\ -\nu([u, \infty)) & : \quad u > 0 \end{cases}.$$

The *stable distributions* are an important subclass of the class of infinitely divisible distributions. A distribution or random variable with distribution function F is called stable iff for any $a_1, a_2, b_1, b_2 \in \mathbb{R}$, $a_1 > 0$, $a_2 > 0$, there exist $a, b \in \mathbb{R}$, $a > 0$, such that the equality $F(a_1 x + b_1) * F(a_2 x + b_2) = F(ax + b)$ is valid ($*$ denotes the convolution).

Let X be a stable random variable. Then its characteristic function φ_X has the representation

$$\log \varphi_X(t) = \mathrm{i}\bar{\gamma}t - c|t|^\alpha \left(1 - \mathrm{i}\beta \operatorname{sgn}(t)\omega(t, \alpha)\right) \qquad (t \in \mathbb{R}) \tag{3}$$

with parameters $\alpha \in (0, 2]$, $\beta \in [-1, 1]$, $\bar{\gamma} \in \mathbb{R}$, and $c \in \mathbb{R}_{\geq 0}$, and

$$\omega(t, \alpha) = \begin{cases} \tan\left(\frac{1}{2}\alpha\pi\right) & : \quad \alpha \neq 1 \\ -\frac{2}{\pi} \log|t| & : \quad \alpha = 1 \end{cases}.$$

The relation to the parameters of the Lévy representation of infinitely divisible distributions (2) is the following:
$\sigma^2 = 0$ for $\alpha \in (0, 2)$,

$$\bar{\gamma} = \begin{cases} \gamma - \alpha c_1 \int_{-\infty}^0 \frac{u}{(1+u^2)|u|^{1+\alpha}} du - \alpha c_2 \int_0^\infty \frac{u}{(1+u^2)u^{1+\alpha}} du & : \alpha \in (0, 1) \\ \gamma + (c_1 - c_2) \int_0^\infty \left(\frac{\sin u}{u^2} - \frac{1}{u(1+u^2)}\right) du & : \quad \alpha = 1 \\ \gamma + \alpha c_1 \int_{-\infty}^0 \frac{u}{|u|^{1+\alpha}} \left(1 - \frac{u}{1+u^2}\right) du + \alpha c_2 \int_0^\infty \frac{u}{u^{1+\alpha}} \left(1 - \frac{u}{1+u^2}\right) du & : \alpha \in (1, 2) \\ \gamma & : \quad \alpha = 2 \end{cases} \tag{4}$$

and

$$Q(u) = \begin{cases} |u|^{-\alpha}(c_1 1_{u<0} - c_2 1_{u>0}) & : \quad \alpha \in (0, 2) \\ 0 & : \quad \alpha = 2 \end{cases}. \tag{5}$$

c_1 and c_2 are real non-negative numbers with $c_1 + c_2 > 0$, $\beta = \frac{c_2 - c_1}{c_1 + c_2}$, and

$$c = \begin{cases} -(c_1 + c_2) \cos(\pi\alpha/2) \int_0^\infty (e^{-y} - 1)\, y^{-\alpha-1} dy & : \quad \alpha \in (0, 1) \\ (c_1 + c_2)\pi/2 & : \quad \alpha = 1 \\ -(c_1 + c_2) \cos(\pi\alpha/2) \int_0^\infty (e^{-y} - 1 + y)\, y^{-\alpha-1} dy & : \quad \alpha \in (1, 2) \\ \sigma^2/2 & : \quad \alpha = 2 \end{cases}. \tag{6}$$

By the canonical representation (3) it can be seen that stable distributions admit a Lebesgue density.

3. General Representation Result

In this section the general representation result from [12] is given which states that fixed points are mixtures of infinitely divisible distributions. First the corresponding *weighted branching process* is introduced:

Let

$$V := \bigcup_{n \in \mathbb{Z}^+} \mathbb{N}^n \qquad \mathbb{N}^0 := \{\emptyset\}$$

and let $T(v) : \Omega \to \mathbb{R}^{\mathbb{N}}$, $v \in V$, denote independent random variables with $T(v) \stackrel{\sim}{=} T$. For $v = (v_1, v_2, ..., v_n) \in V$ let (for $k \in \mathbb{N}$, $k < n$) $v_{|k} := (v_1, ..., v_k)$

be the restriction to the first k components and $v_{|0} := \emptyset$. On the tree define the random variables or *weights* $L(v) : \Omega \to \mathbb{R}$, $v \in V$, for $v = (v_1, v_2, ..., v_n)$ by

$$L(v) := \prod_{j=1}^{n} T_{v_j}(v_{|j-1}) \qquad L(\emptyset) := 1 .$$

Let $L := (L(v))_{v \in V}$. Using recursion fixed point equation (1) can be expressed for $n \in \mathbb{N}$ as

$$W \stackrel{\sim}{=} \sum_{|v|=n} L(v)W(v) , \tag{7}$$

where $W(v)$, $|v| = n$, are copies of W and L, $W(v)$, $|v| = n$, are independent.

Let W be a solution of fixed point equation (1) and define for $l \in M := \{l \in \mathbb{R}^V : \#\{v \in V : |v| = n, l(v) \neq 0\} < \infty \text{ for all } n \in \mathbb{N}\}$ the random variable

$$Y_n(l) := \sum_{|v|=n} l(v)W(v) , \tag{8}$$

with $W(v)$, $|v| = n$, copies of W and L, $W(v)$, $|v| = n$, independent.

Theorem 3.1. *(Representation; Theorem 1 of* [12]*)*
As $n \to \infty$ assume $\sup_{|v|=n} |L(v)| \to 0$ to hold almost surely. Let W be a solution of fixed point equation (1). For almost every $l \in \mathbb{R}^V$ with respect to P^L there exists an infinitely divisible random variable $Y(l)$ with

$$Y_n(l) \xrightarrow{w} Y(l) .$$

Denote by $\rho(l)$ the characteristic function of $Y(l)$ and let $\gamma(l)$ and $\sigma^2(l)$ be the corresponding constants and $Q(l)$ the corresponding measure generating function of the Lévy representation (2) of $\rho(l)$ (see Section 2). Then ($t \in \mathbb{R}$)

$$\begin{aligned}\varphi_W(t) &= \mathrm{E}\rho(L)(t) \\ &= \mathrm{E}\exp\left(\mathrm{i}\gamma(L)t - \frac{1}{2}\sigma^2(L)t^2 + \int \left(\mathrm{e}^{\mathrm{i}ut} - 1 - \frac{\mathrm{i}ut}{1+u^2}\right) \mathrm{d}Q(L)(u)\right) . \end{aligned} \tag{9}$$

The subsequent lemma shows how to compute Q, σ^2 and γ from the distribution function F_W of a certain fixed point W. Note that only the tail behaviour of F_W is needed.

Lemma 3.2. *(Lemma 6 of* [12]*) As $n \to \infty$ let $\sup_{|v|=n} |L(v)| \to 0$ hold almost surely and assume that W is a fixed point of (1). Let $Q(l)$, $\sigma^2(l)$ and $\gamma(l)$ for almost every $l \in \mathbb{R}^V$ with respect to P^L be as defined in Theorem 3.1. Then for all continuity points u of $Q(l)$*

$$Q(l)(u) = \begin{cases} \lim\limits_{n\to\infty} \sum\limits_{|v|=n} \mathrm{P}(l(v)W \le u) & : \quad u < 0 \\ \lim\limits_{n\to\infty} \sum\limits_{|v|=n} \Big(\mathrm{P}(l(v)W \le u) - 1\Big) & : \quad u > 0 \end{cases} , \tag{10}$$

$$\lim_{\epsilon \to 0} \liminf_{n\to\infty} \sum_{|v|=n} \left(l^2(v) \int_{|u|<\epsilon/|l(v)|} u^2 \,\mathrm{d}F_W(u) - l^2(v) \left(\int_{|u|<\epsilon/|l(v)|} u \,\mathrm{d}F_W(u) \right)^2 \right)$$

$$= \lim_{\epsilon \to 0} \limsup_{n\to\infty} \sum_{|v|=n} \left(l^2(v) \int_{|u|<\epsilon/|l(v)|} u^2 \, \mathrm{d}F_W(u) - l^2(v) \left(\int_{|u|<\epsilon/|l(v)|} u \, \mathrm{d}F_W(u) \right)^2 \right)$$
$$= \sigma^2(l) \qquad (11)$$

and
$$\gamma(\tau)(l) = \lim_{n\to\infty} \sum_{|v|=n} l(v) \int_{|u|<\tau/|l(v)|} u \, \mathrm{d}F_W(u) \,, \qquad (12)$$

where τ and $-\tau$ are continuity points of Q. $\gamma(l)$ can be obtained by

$$\gamma(l) = \gamma(\tau)(l) - \int_{0<|u|<\tau} \frac{u^3}{1+u^2} \mathrm{d}Q(l)(u) + \int_{|u|\geq\tau} \frac{u}{1+u^2} \mathrm{d}Q(l)(u) \,. \qquad (13)$$

4. Positive Solutions

In this section non-negative solutions of fixed point equation (1) are considered. It is always assumed that $T_j \geq 0$ for all $j \in \mathbb{N}$. Furthermore, it is supposed that

$$\mathrm{P}(N = 0 \text{ or } 1) < 1 \qquad \text{and} \qquad \mathrm{P}(\forall j \in \mathbb{N} : T_j = 1 \text{ or } T_j = 0) < 1$$

hold. Otherwise, the situation is easy (see Lemma 1.1 of [27].Often, a stronger assumption is needed:

Assumption 1.
There exists a $\delta > 0$ such that $E\big(\sum_{j=1}^{\infty} T_j\big)^{1+\delta} < \infty$ and $\mathrm{E}N^{1+\delta} < \infty$.

In the investigation of fixed point equation (1) the following function is of major importance (stated here for not necessarily non-negative coefficients)

$$m : \mathbb{R}_{\geq 0} \to \overline{\mathbb{R}}_{\geq 0}; \qquad \gamma \mapsto \mathrm{E} \sum_{j=1}^{\infty} |T_j|^{\gamma} 1_{|T_j|>0} \,.$$

It is easy to show that m is strict convex on $m^{-1}(\mathbb{R}_{>0})$. In particular, there is at most one $\alpha \in \mathbb{R}_{>0}$ such that $m(\alpha) = 1$ and $m'(\alpha) \leq 0$ (e.g., [12, Lemma 1]) Throughout the text it is always assumed that $m(0) > 1$. If $m(0) > 1$ does not hold the situation is trivial ([13, Lemma 2]). Denote

$$\mathcal{P} = \{\mu \in \mathcal{D} \setminus \{\delta_0\} : \mu([0,\infty)) = 1 \text{ and } \mu \text{ is a fixed point of (1)}\}.$$

If W is a random variable then the somewhat sloppy notation $W \in \mathcal{P}$ means that the distribution of W belongs to $\mathcal{P}$.

The fixed point equation (1) is called *non-lattice* if there is no $s > 0$ such that

$$\mathrm{P}(\forall j \in \mathbb{N} : \log T_j \in s\mathbb{Z} \text{ for } T_j > 0) = 1 \,.$$

4.1. Existence

Results regarding the existence and uniqueness of non-negative solutions of fixed point equation (1), under certain integrability condition on the T_j, were given in Theorem 1 in [17] for constant N, and Theorem 1.1 and Corollary 1.5 in [27]:

Theorem 4.1. *(Theorem 1.1 and Corollary 1.5 of [27])*

(i) *Let $\inf_{\beta\in[0,1]} m(\beta) \leq 1$. Then $\mathcal{P} \neq \emptyset$. If $\mathrm{E}\sum_{j=1}^{\infty} T_j \log^+ T_j < \infty$ and $\mathrm{E}N < \infty$ the converse is also true.*

(ii) *Let Assumption 1 hold and assume $\mathcal{P} \neq \emptyset$. According to (i) let α be the unique point of $(0,1]$ such that $m(\alpha)=1$, $m'(\alpha) \leq 0$. If fixed point equation (1) is non-lattice or $\alpha = 1$ then (1) has only one non-negative solution up to multiplicative constants.*

If $W \in \mathcal{P}$ and Assumption 1 holds we denote by α always the unique point in $(0,1]$ such that $m(\alpha)=1$, $m'(\alpha) \leq 0$.

The following is known regarding the tail behaviour ([17, Theorem 2], for constant N; [27, Corollary 1.6]):

Theorem 4.2. *(Corollary 1.6 of [27]) Let Assumption 1 hold and assume that fixed point equation (1) is non-lattice. Let the random variable $W \in \mathcal{P}$ have distribution function F. Assume $\alpha < 1$.*

(i) *If $m'(\alpha) < 0$ then there is a constant $C_1 > 0$ such that*

$$1 - F(u) \sim C_1 u^{-\alpha} \quad \textit{for } u \to \infty\,.$$

(ii) *If $m'(\alpha) = 0$ then there is a constant $C_2 > 0$ such that*

$$1 - F(u) \sim C_2 u^{-\alpha} \log u \quad \textit{for } u \to \infty\,.$$

4.2. The Representation

Let

$$\mathcal{F}_n := \sigma\{T(v) : v \in V, |v| < n\}\,. \tag{14}$$

Define for $\delta > 0$ the operator

$$K_\delta : \mathcal{D} \to \mathcal{D}; \qquad K_\delta(\mu) \cong \sum_{j=1}^{\infty} T_j^\delta X_j\,, \tag{15}$$

where $X_1, X_2, \ldots$ have distribution μ and $T, X_1, X_2, \ldots$ are independent.

That $Y(l)$ of Theorem 3.1 is α-stable in the context of positive fixed points and coefficients is shown in the next theorem. This is the main new result of this article. Thus, it is derived that each fixed point is a mixture of α-stable distributions. Associated results are given in the proof of Theorem 3.1 in [17], in Section II, Proposition 1 in [20], in Section 5 in [27], and in Theorem 2.5 of [28]. The proof of this theorem and of its corollary, which shows the existence of Lebesgue densities, is given in the appendix.

Theorem 4.3. *Let Assumption 1 hold and assume that fixed point equation (1) is non-lattice. Let $\mathcal{P} \neq \emptyset$ and W be the (up to multiplicative constants) unique non-trivial solution of (1). Assume $\alpha < 1$. If C_1 and C_2 are chosen according to Theorem 4.2(i) and (ii), then (Q, σ^2, γ as in the representation of Theorem 3.1):*

(i) *Case $m'(\alpha) < 0$: $\sum_{|v|=n} L^\alpha(v)$ is a non-negative martingale with respect to $(\mathcal{F}_n)_{n\in\mathbb{N}}$ and converges almost surely and in L_1 to a non-negative, L-measurable random variable Z_1 with $EZ_1 = 1$. Z_1 is the (up to multiplicative constants) unique non-trivial, non-negative fixed point of K_α. Furthermore, almost surely*

$$Q(L)(u) = \begin{cases} 0 & : \ u < 0 \\ -u^{-\alpha} C_1 Z_1 & : \ u > 0 \end{cases}\,,$$

$$\sigma^2(L) = 0\,,$$

$$\gamma(L) = \frac{\pi}{2\sin\left(\frac{(1-\alpha)\pi}{2}\right)} C_1 \alpha Z_1\,.$$

(ii) *Case $m'(\alpha) = 0$: $\sum_{|v|=n} L^\alpha(v)\log L(v)$ is a martingale with respect to $(\mathcal{F}_n)_{n\in\mathbb{N}}$ with expectation 0 and converges almost surely to a non-positive, L-measurable random variable Z_2 with $\mathrm{P}(Z_2 < 0) > 0$. Furthermore, almost surely*

$$Q(L)(u) = \begin{cases} 0 & : \quad u < 0 \\ u^{-\alpha} C_2 Z_2 & : \quad u > 0 \end{cases}$$

$$\sigma^2(L) = 0\,,$$

$$\gamma(L) = -\frac{\pi}{2\sin\left(\frac{(1-\alpha)\pi}{2}\right)} C_2 \alpha Z_2\,.$$

(iii) *For almost every $l \in \mathbb{R}^V$ with respect to P^L the random variable $Y(l)$ of Theorem 3.1 is α-stable with $\mathrm{P}^L\,(Y(l)$ is non-degenerate$) > 0$, i.e. φ_W is the integral over characteristic functions of α-stable random variables. In particular:*
Case $m'(\alpha) < 0$: $\varphi_W(t) = \mathrm{E}\exp\left(-N_1 Z_1 |t|^\alpha \left(1 - i\mathrm{sgn}(t)\tan\left(\frac{\alpha\pi}{2}\right)\right)\right)$ with $N_1 = -C_1\cos(\pi\alpha/2)\int_0^\infty \left(e^{-y} - 1\right) y^{-\alpha-1}\,dy$.
Case $m'(\alpha) = 0$: $\varphi_W(t) = \mathrm{E}\exp\left(-N_2 Z_2 |t|^\alpha \left(1 - i\mathrm{sgn}(t)\tan\left(\frac{\alpha\pi}{2}\right)\right)\right)$ with $N_2 = C_2\cos(\pi\alpha/2)\int_0^\infty \left(e^{-y} - 1\right) y^{-\alpha-1}\,dy$.

Let $g : [0,1] \to [0,1];\ t \mapsto \sum_{j=0}^\infty t^j \mathrm{P}(N = j)$ be the generating function of N. Note that, due to Lemma 3.1 of Liu (1998), the existence of a non-trivial non-negative solution W of (1) ensures the existence of a unique fixed point q of g in $[0,1)$ and $\mathrm{P}(W = 0) = q$.

Corollary 4.4. *Under the assumptions of the preceding theorem, W has a Lebesgue density f on $\mathbb{R}\setminus\{0\}$. In particular: If $f_{\alpha,\beta,\bar{\gamma},c}$ denotes the Lebesgue density of a stable distribution with corresponding parameters (see (3)) then, almost surely with respect to the Lebesgue measure,*

(i) *Case $m'(\alpha) < 0$: $f : \mathbb{R}\setminus\{0\} \to \mathbb{R}_{\geq 0};\ t \mapsto \mathrm{E} f_{\alpha,1,M_1 Z_1,N_1 Z_1}(t) 1_{Z_1 \neq 0}$.*
(ii) *Case $m'(\alpha) = 0$: $f : \mathbb{R}\setminus\{0\} \to \mathbb{R}_{\geq 0};\ t \mapsto \mathrm{E} f_{\alpha,1,M_2 Z_2,N_2 Z_2}(t) 1_{Z_2 \neq 0}$.*

Remark 4.1. *Theorem 2.3 of Liu (2001) guarantees the existence of a Lebesgue density if $\mathrm{E}T_\tau^{-\epsilon} 1_{\tau<\infty} < \infty$ for $\tau := \min\{j \in \mathbb{N} : T_j > 0\}$ for some $\epsilon > 0$, also for the lattice case and $\alpha = 1$.*

5. Solutions with Finite Expectation

In this section we always assume that the coefficients T_j are non-negative.

We denote by $\mathcal{E}$ the set of solutions of fixed-point equation (1) with finite non-zero expectation.

5.1. Existence

The next theorem is a direct combination of Corollary 2.1 and Remark 2.1 of [14]. The proof of this corollary applies the result of [30]. The theorem gives necessary and sufficient conditions for the existence of solutions with finite non-zero expectation. It can also be generalized to include the case $\mathrm{P}(N < \infty) < 1$ (see Corollary 2.1 of [14].

Theorem 5.1. *(Cor. 2.1 and Rem. 2.1 of [14]) Suppose that* $\mathrm{E}\sum_{j=1,T_j\neq 0}^{\infty} T_j \log T_j$ *exists and is finite. Assume* $\mathrm{E}T_j < \infty$ *for all* $j \in \mathbb{N}$*. Then,*

$$\mathcal{E} \neq \emptyset \iff m(1) = 1,\ \mathrm{E} \sum_{j=1,T_j\neq 0}^{\infty} T_j \log T_j < 0 \text{ and } \mathrm{E}\left(\sum_{j=1}^{\infty} T_j\right) \log^+ \left(\sum_{j=1}^{\infty} T_j\right) < \infty.$$

5.2. The Representation

For fixed points $W \in \mathcal{E}$ the random variables $Y(l)$ of Theorem 3.1 are degenerate, i.e. constant.

Theorem 5.2. *Assume that* $\mathrm{E}T_j < \infty$ *for all* $j \in \mathbb{N}$ *and that* $\mathcal{E} \neq \emptyset$ *with* $W \in \mathcal{E}$*.*

(i) $m(1) = 1$ *and* $\sum_{|v|=n} L(v)$ *is a non-negative martingale with respect to* $(\mathcal{F}_n)_{n\in\mathbb{N}}$ *(see Eq. (14)) and converges almost surely and in* L_1 *to a non-negative, L-measurable random variable* Z *with* $EZ = 1$*. Furthermore, in the representation of Theorem 3.1, the following statements are true almost surely:*

$$\begin{aligned} Q(L) &\equiv 0\,, \\ \sigma^2(L) &= 0\,, \\ \gamma(L) &= ZEW\,. \end{aligned}$$

(ii) *For almost every* $l \in \mathbb{R}^V$ *with respect to* P^L *the random variable* $Y(l)$ *of Theorem 3.1 is a constant.*

Since $\gamma(L) = ZEW$ for $W \in \mathcal{E}$, Theorem 5.2 also shows the uniqueness of solutions (up to multiplicative constants).

Proof: Assume $ET_j < \infty$ for all $j \in \mathbb{N}$ and that $\mathcal{E} \neq \emptyset$ with $W \in \mathcal{E}$. That $m(1) = 1$ and $\sum_{|v|=n} L(v)$ is a non-negative martingale with respect to $(\mathcal{F}_n)_{n\in\mathbb{N}}$ and converges almost surely to a non-negative, L-measurable random variable Z with $0 < EZ < \infty$, was proved in Theorem 2.2 and Proposition 2.1 of [14]. Because W has finite expectation, certain statements about the tail behaviour are known. These can be used to calculate Q, σ^2 and γ directly from Eqs. (10) - (13). A very similar procedure was performed in detail in part (vi) in the proof of Theorem 2.3 in [14]. □

6. Solutions with Finite Variance

In this section it is assumed that $\mathrm{P}(\sum_{j=1}^{\infty} T_j = 1) < 1$. This is not necessary (Theorem 18 of [13]). However, it clarifies the situation and avoids the distinction between several cases. In this section the assumption $T_j \geq 0$ of Sections 4 and 5 is not necessary.

Denote by $\mathcal{F}^1$, resp. $\mathcal{F}^2$, the set of non-trivial solutions of fixed point equation (1) with finite variance and non-zero, resp. zero, expectation. These cases have to be treated separately. For set $\mathcal{F}^1$ the subsequent Assumption 2 and for set $\mathcal{F}^2$ Assumption 3 is necessary.

Assumption 2. $\mathrm{E}\sum_{j=1}^{\infty} T_j^2 < \infty$ *and* $\left(\sum_{j=1}^{n} T_j\right)_{n\in\mathbb{N}}$ *converges in* L_2.

Assumption 3. $\mathrm{E}\sum_{j=1}^{\infty} T_j^2 < \infty$ *and* $\mathrm{E}\left|\sum_{\substack{j=1\\T_j\neq 0}}^{\infty} T_j^2 \log T_j^2\right| < \infty$.

6.1. Existence

Sufficient and necessary conditions for the existence of solutions with finite variance are given in the next theorem. This is a result of [13]. It does not require that $P(N < \infty) = 1$.

Theorem 6.1. *(Theorem 3 of* [13]*)*

(i) *Suppose that Assumption 2 holds. Then*

$$\mathcal{F}^1 \neq \emptyset \iff \mathrm{E}\sum_{j=1}^{\infty} T_j^2 < 1 \text{ and } \mathrm{E}\sum_{j=1}^{\infty} T_j = 1 .$$

(ii) *Suppose that Assumption 3 holds. Then*

$$\mathcal{F}^2 \neq \emptyset \iff$$

$$\mathrm{E}\sum_{j=1}^{\infty} T_j^2 = 1, \mathrm{E}\sum_{\substack{j=1\\ T_j\neq 0}}^{\infty} T_j^2 \log T_j^2 < 0 \text{ and } \mathrm{E}\left(\left(\sum_{j=1}^{\infty} T_j^2\right)\log^+\left(\sum_{j=1}^{\infty} T_j^2\right)\right) < \infty.$$

(iii) *Suppose that Assumption 2 or 3 holds. If $\mathcal{F}^1 \cup \mathcal{F}^2 \neq \emptyset$, the solution of Eq. (1) with finite variance is unique up to multiplicative constants.*

6.2. The Representation

The subsequent theorem shows the representation in the case $\mathcal{F}^2$. (For $W \in \mathcal{F}^1$ the results for finite expectation apply since $\mathrm{E}W \neq 0$.) Fixed points with finite variance and expectation 0 are mixtures of normal distributions with expectation 0 and a random variance which satisfies a related fixed point equation. Recall the definition of K_δ from Eq. (15).

Theorem 6.2. *Suppose that Assumption 3 holds. Let $\mathcal{F}^2 \neq \emptyset$.*

(i) *$\sum_{|v|=n} L^2(v)$ is a non-negative martingale with respect to $(\mathcal{F}_n)_{n\in\mathbb{N}}$ and converges a.s. and in L_1 to a non-negative, L-measurable random variable Z with $EZ = 1$. Z is the (up to multiplicative constants) unique non-trivial fixed point of K_2 with finite expectation. Furthermore, in the representation of Theorem 3.1, the following statements are true almost surely:*

$$\begin{aligned} Q(L) &\equiv 0 , \\ \sigma^2(L) &= Z \,\mathrm{Var} W , \\ \gamma(L) &= 0 . \end{aligned}$$

(ii) *For almost every $l \in \mathbb{R}^V$ with respect to P^L the random variable $Y(l)$ of Theorem 3.1 is normal distributed, i.e. φ_W is the integral over characteristic functions of normal random variables. In particular:*

$$\varphi_W(t) = \mathrm{E}\exp\left(-\frac{1}{2}t^2 Z \,\mathrm{Var} W\right) \qquad (t \in \mathbb{R}) . \tag{16}$$

Proof: The proof is similar to that of Theorem 5.2. Instead of part (vi) in the proof of Theorem 2.3 in [14], part (vi) of Lemma 16 of [13] is employed. □

Acknowledgements

The author would like to thank Uwe Rösler and Gerold Alsmeyer for stimulating conversations. She is also grateful to two anonymous referees for a careful reading of the manuscript.

Appendix A. Appendix: Proof of Theorem 4.3

Before performing the proof of Theorem 4.3, two preliminary results are stated. Assume $T_j \geq 0$ for all $j \in \mathbb{N}$. For the proof of the first lemma compare Lemma 3.2 in [14]. It was also proved in Lemma 7.2 of [27] and implicitly in Lemma 2.2 of [8] applying a result of [6].

Lemma A.1. *Suppose that there exists an $\alpha \in (0,\infty)$ such that $m(\alpha) = 1$ and a $\delta > 0$ such that $m(\alpha+\delta) < \infty$. Then, as $n \to \infty$,*

$$\sup_{|v|=n} L(v) \longrightarrow 0 \quad \textit{almost surely.}$$

In particular, the result holds if $\mathcal{P} \neq \emptyset$ and Assumption 1 is satisfied.

Corollary A.2. *Suppose that there exists an $\alpha \in (0,\infty)$ such that $m(\alpha) = 1$. Then, for every $\beta > \alpha$, such that there exists a $\delta > 0$ such that $m(\alpha/(\beta-\alpha)+\delta) < \infty$, as $n \to \infty$,*

$$\sum_{|v|=n} L^\beta(v) \longrightarrow 0 \quad \textit{almost surely.}$$

Proof: Choose α and β according to the statement of the corollary. We consider the function

$$\overline{m} : \mathbb{R}_{\geq 0} \to \overline{\mathbb{R}}_{>0}; \qquad \gamma \mapsto \mathrm{E}\sum_{j=1}^{\infty} T_j^{(\beta-\alpha)\gamma} 1_{T_j>0} \,.$$

Then we know that $\overline{m}\left(\frac{\alpha}{\beta-\alpha}\right) = 1$. Therefore, by the preceding lemma as $n \to \infty$,

$$\sup_{|v|=n} L^{(\beta-\alpha)}(v) \longrightarrow 0 \quad \text{almost surely.}$$

Furthermore, $\sum_{|v|=n} L^\alpha(v)$ is a non-negative martingale and converges almost surely. Thus,

$$\sum_{|v|=n} L^\beta(v) \leq \sup_{|v|=n} L^{\beta-\alpha}(v) \sum_{|v|=n} L^\alpha(v) \longrightarrow 0 \quad \text{almost surely as } n \to \infty \,.\square$$

Proof of Theorem 4.3:
Let Assumption 1 hold. Let $W \in \mathcal{P}$, $\alpha < 1$ and fixed point equation (1) be non-lattice. Choose C_1, C_2 according to Theorem 4.2(i) and (ii).
(a) Martingales and Calculation of Q
Case $m'(a) < 0$: $\sum_{|v|=n} L^\alpha(v)$ being a non-negative martingale follows directly thereby guaranteeing the almost sure convergence. Consider the function

$$\overline{m} : \mathbb{R}_{\geq 0} \to \overline{\mathbb{R}}_{>0}; \qquad \gamma \mapsto \mathrm{E}\sum_{j=1}^{\infty} T_j^{\alpha\gamma} 1_{T_j>0}$$

corresponding to the operator K_α. Note that $\overline{m}(\gamma) = m(\alpha\gamma)$, $\overline{m}(1) = 1$ and $\overline{m}'(1) = \alpha m'(\alpha) < 0$. Therefore, $\lim_{n\to\infty} \sum_{|v|=n} L^\alpha(v)$ is a non-trivial fixed point of K_α with mean 1 [30]. Denote by F the distribution function of W and let $u > 0$. By Lemma 3.2 it follows that P^L-almost surely for $l \in \mathbb{R}^V$

$$Q(l)(u) = \lim_{n\to\infty} \sum_{|v|=n} \left(F\left(\frac{u}{l(v)}\right) - 1\right) .$$

Let $\delta > 0$, $\delta < C_1$. Choose by Theorem 4.2(i) a $x_0 \in \mathbb{R}_{>0}$ such that for every $x > x_0$

$$x^{-\alpha}(C_1 - \delta) \le (1 - F(x)) \le x^{-\alpha}(C_1 + \delta) .$$

Due to Lemma A.1 it is possible to choose $n_0 := n_0(l) \in \mathbb{N}$ such that $u/l(v) > x_0$ for each $v \in V$, $|v| > n_0$ (P^L-almost surely). Then for every $n > n_0$

$$-(C_1+\delta)u^{-\alpha} \sum_{|v|=n} l^\alpha(v) \le \sum_{|v|=n} \left(F\left(\frac{u}{l(v)}\right) - 1\right) \le -(C_1-\delta)u^{-\alpha} \sum_{|v|=n} l^\alpha(v).$$

Thus

$$-(C_1+\delta)u^{-\alpha} \lim_{n\to\infty} \sum_{|v|=n} l^\alpha(v) \le Q(l)(u) \le -(C_1-\delta)u^{-\alpha} \lim_{n\to\infty} \sum_{|v|=n} l^\alpha(v).$$

Letting $\delta \to 0$ we obtain P^L-almost surely that

$$Q(l)(u) = -C_1 u^{-\alpha} \lim_{n\to\infty} \sum_{|v|=n} l^\alpha(v)$$

therefore deriving the desired equality.
Case $m'(\alpha) = 0$: It can be proved directly that $\sum_{|v|=n} L^\alpha(v)\log L(v)$ is a martingale. Note that $\mathrm{E}\sum_{j=1}^\infty T_j^\alpha = 1$ and $\mathrm{E}\sum_{j=1}^\infty T_j^\alpha \log T_j = 0$ since $m(\alpha) = 1$ and $m'(\alpha)$=0. The remaining statements are gained analogously to the case $m'(\alpha) < 0$.
□

(b) Calculation of σ^2
Case $m'(\alpha) < 0$: As in part (a) let $\delta > 0$, $\delta < C_1$ and choose $x_0 \in \mathbb{R}$ such that for every $x > x_0$

$$x^{-\alpha}(C_1 - \delta) \le (1 - F(x)) \le x^{-\alpha}(C_1 + \delta) .$$

Then let $\epsilon > 0$ and choose $n_0 := n_0(l) \in \mathbb{N}$ such that $\epsilon/l(v) > x_0$ for each $v \in V$, $|v| > n_0$ (P^L-almost surely). By partial integration we obtain for each $v \in V$, $|v| > n_0$,

$$\begin{aligned}
&\int_0^{\epsilon/l(v)} u^2 \,\mathrm{d}F(u) \\
&= \epsilon^2 l^{-2}(v) F\left(\frac{\epsilon}{l(v)}\right) - 2\int_0^{x_0} uF(u)\,\mathrm{d}u + 2\int_{x_0}^{\epsilon/l(v)} u(1-F(u))\,\mathrm{d}u - 2\int_{x_0}^{\epsilon/l(v)} u\,\mathrm{d}u \\
&\le l^{-2+\alpha}(v)\epsilon^{2-\alpha}\left(\frac{2}{2-\alpha}(C_1+\delta) - (C_1-\delta)\right) + x_0^2 - 2\int_0^{x_0} uF(u)\,\mathrm{d}u \\
&- \frac{2}{2-\alpha}(C_1+\delta)x_0^{2-\alpha} .
\end{aligned}$$

Applying Eq. (11) gives

$$\begin{aligned}
\sigma^2(l) &\le \lim_{\epsilon\to 0} \limsup_{n\to\infty} \sum_{|v|=n} l^2(v) \int_0^{\epsilon/l(v)} u^2\,\mathrm{d}F(u) \\
&\le \lim_{\epsilon\to 0} \limsup_{n\to\infty} \Bigg(\epsilon^{2-\alpha}\left(\frac{2}{2-\alpha}(C_1+\delta) - (C_1-\delta)\right) \sum_{|v|=n} l^\alpha(v)
\end{aligned}$$

$$+\left(x_0^2 - 2\int_0^{x_0} uF(u)\,\mathrm{d}u - \frac{2}{2-\alpha}(C_1+\delta)x_0^{2-\alpha}\right)\sum_{|v|=n} l^2(v)\Bigg).$$

Corollary A.2 guarantees that the last sum goes to 0 for $n \to \infty$ (P^L-almost surely). Since $\sum_{|v|=n} l^\alpha(v)$ converges for $n \to \infty$, this results in $\sigma^2(l) = 0$ P^L-almost surely.
In the case $m'(\alpha) = 0$ the arguments are similar. Note that now $\overline{m}'(1) = 0$ and therefore $\lim_{n\to\infty}\sum_{|v|=n} L^\alpha(v) = 0$ almost surely by [30]. □

(c) Calculation of γ
Case $m'(\alpha) < 0$: Let $\tau \in \mathbb{R}_{>0}$. With the same reasoning as in part (b) we obtain for each $\delta > 0$ almost surely

$$\gamma(\tau)(L) \le \tau^{1-\alpha}\left(\frac{1}{1-\alpha}(C_1+\delta) - (C_1-\delta)\right)\lim_{n\to\infty}\sum_{|v|=n} L^\alpha(v),$$

$$\gamma(\tau)(L) \ge \tau^{1-\alpha}\left(\frac{1}{1-\alpha}(C_1-\delta) - (C_1+\delta)\right)\lim_{n\to\infty}\sum_{|v|=n} L^\alpha(v).$$

For $\delta \to 0$ we obtain ($Z_1 := \lim_{n\to\infty}\sum_{|v|=n} L^\alpha(v)$)

$$\gamma(\tau)(L) = \frac{\alpha}{1-\alpha}C_1 Z_1 \tau^{1-\alpha}.$$

By using the representation of $Q(L)$ in part (a) and Eq. (13), we derive almost surely

$$\gamma(L) = \gamma(\tau)(L) - C_1\alpha Z_1\left(\int_0^\tau \frac{u^{2-\alpha}}{1+u^2}\,\mathrm{d}u - \int_\tau^\infty \frac{u^{-\alpha}}{1+u^2}\,\mathrm{d}u\right).$$

$\tau \to 0$ results in $\gamma(L) = C_1\alpha Z_1 \int_0^\infty \frac{u^{-\alpha}}{1+u^2}\,\mathrm{d}u = C_1\alpha Z_1 \frac{\pi}{2\sin\left(\frac{(1-\alpha)\pi}{2}\right)}$.

The proof of case $m'(\alpha) = 0$ is similar. Observe that if $\lim_{n\to\infty}\sum_{|v|=n} L^\alpha(v)$ $\log L(v) = 0$ almost surely, the solution W would be trivial contrary to the assumption. □

(d) Proof of (iii)
Parts (a), (b) and (c) together with with Eqs. (3) - (6) show that $Y(l)$ has P^L-almost surely a stable distribution and $\alpha(L) = \alpha$, $\beta(L) = 1$, $c_1(L) = 0$ and $\bar{\gamma}(L) = 0$. Furthermore, $c(L) = -c_2(L)\cos(\pi\alpha/2)\int_0^\infty (e^{-y}-1)\,y^{-\alpha-1}dy$ with $c_2(L) = C_1 Z_1$ if $m'(\alpha) < 0$ and $c_2(L) = C_2 Z_2$ if $m'(\alpha) = 0$.□

Proof of Corollary 4.4:
Let Assumption 1 hold and assume that fixed point equation (1) is non-lattice. Let $\mathcal{P} \ne \emptyset$ and W be the (up to multiplicative constants) unique non-trivial solution of (1). Assume $\alpha < 1$.
Case (i): From Theorem 4.3 (i) it follows that

$$\varphi_W(t) = \mathrm{P}(Z_1 = 0) + \mathrm{E}\exp\left(\mathrm{i}M_1 Z_1 t - N_1 Z_1 |t|^\alpha\left(1 - \mathrm{i}\,\mathrm{sgn}(t)\tan\left(\frac{\alpha\pi}{2}\right)\right)\right)1_{Z_1>0}$$

$$= \mathrm{P}(Z_1 = 0) + \int \mathrm{E}\left(f_{\alpha,1,M_1Z_1,N_1Z_1}(u)1_{Z_1>0}\right)\exp(\mathrm{i}ut)\mathrm{d}u.$$

The right hand side is the characteristic function of a distribution with point mass $P(Z_1 = 0)$ at 0 and Lebesgue density f on $\mathbb{R}\setminus\{0\}$ as stated in the lemma. Applying the uniqueness of the characteristic function completes the proof.
Case (ii): The proof is along the same lines as the proof of case (i).□

References

[1] ALDOUS, D. J. AND BANDYOPADHYAY, A. (2004). A survey of max-type recursive distributional equations. Preprint, arXiv:math.PR/0401388.

[2] ALSMEYER, G. AND ROESLER, U. (2003). A stochastic fixed-point equation related to weighted branching with deterministic weights. *Bericht 10/03-S* Angewandte Mathematik, FB 10, University of Münster, http://wwwmath.uni-muenster.de/inst/Statistik/alsmeyer/Publikationen.html.

[3] ATHREYA, K. B. AND NEY, P. E. (1972). *Branching Processes.* Springer, Berlin.

[4] ATHREYA, K. B. (1971). On the absolute continuity of the limit random variable in the supercritical Galton-Watson branching process. *Proc. Amer. Math. Soc.* **30,** 563–565.

[5] BIGGINS, J. D. (1977a). Martingale convergence in the branching random walk. *J. Appl. Prob.* **14,** 25–37.

[6] BIGGINS, J. D. (1977b). Chernoff's theorem in the branching random walk. *J. Appl. Prob.* **14,** 630–636.

[7] BIGGINS, J. D. AND GREY, D. R. (1979). Continuity of limit random variables in the branching random walk. *J. Appl. Prob.* **16,** 740–749.

[8] BIGGINS, J. D. AND KYPRIANOU, A. E. (1997). Seneta-Heyde norming in the branching random walk. *Ann. Probab.* **25,** 337–360.

[9] BIGGINS, J. D. AND KYPRIANOU, A. E. (2004). Fixed points of the smoothing transform; the boundary case, Preprint, http://www.shef.ac.uk/~st1jdb/tsttbc.html.

[10] BINGHAM, N. H. AND DONEY, R. A. (1974). Asymptotic properties of supercritical branching processes I: The Galton-Watson process. *Adv. Appl. Prob.* **6,** 711–731.

[11] BINGHAM, N. H. AND DONEY, R. A. (1975). Asymptotic properties of supercritical branching processes II: Crump-Mode and Jirina processes. *Adv. Appl. Prob.* **7,** 66–82.

[12] CALIEBE, A. (2003). Symmetric fixed points of a smoothing transformation. *Adv. Appl. Prob.* **35,** 377–394.

[13] CALIEBE, A. AND RÖSLER, U. (2003). Fixed points with finite variance of a smoothing transformation. *Stochastic Process. Appl.* **107,** 105–129.

[14] CALIEBE, A. AND RÖSLER, U. (2004). Fixed points of a smoothing transformation with finite expectation: closing a gap. Preprint, http://www-computerlabor.math.uni-kiel.de/stochastik/caliebe

[15] CRUMP, K. AND MODE, C. J. (1968). A general age-dependent branching process (I). *J. Math. Anal. Appl.* **24,** 497–508.

[16] CRUMP, K. AND MODE, C. J. (1969). A general age-dependent branching process (II). *J. Math. Anal. Appl.* **25,** 8–17.

[17] DURRETT, R. AND LIGGETT, T. (1983). Fixed points of the smoothing transformation. *Z. Wahrscheinlichkeitsth.* **64,** 275–301.

[18] GEIGER, J. (2000). A new proof of Yaglom's exponential limit law. In *Algorithms, Trees, Combinatorics and Probability,* eds. D. Gardy and A. Mokkadem. Trends in Mathematics, Birkhäuser, Basel, pp. 245–249.

[19] GNEDENKO, B. V. AND KOLMOGOROV, A. N. (1968). *Limit Distributions for Sums of Independent Random Variables.* Addison-Wesley, Reading.

[20] GUIVARC'H, Y. (1990). Sur une extension de la notion de loi semi-stable. *Ann. Inst. Henri Poincaré - Probabilités et Statistiques* **26,** 261–285.

[21] HARRIS, T. E. (1948). Branching processes. *Ann. Math. Stat.* **19**, 474–494.

[22] HALL, P. (1981). A comedy of errors: the canonical form for a stable characteristic function. *Bull. London Math. Soc.* **13**, 23–27.

[23] HEYDE, C. C. (1970). Extension of a result of Seneta for the super-critical Galton-Watson process. *Ann. Math. Stat.* **41**, 739–742.

[24] IBRAGIMOV, I. A. AND LINNIK, YU. V. (1971). *Independent and Stationary Sequences of Random Variables.* Wolters-Noordhoff, Groningen.

[25] IKSANOV, A. M. (2004). Elementary fixed points of the BRW smoothing transforms with infinite number of summands. Preprint.

[26] LIU, Q. (1997). Sur une équation fonctionnelle et ses applications: une extension du théorème de Kesten-Stigum concernant des processus de branchement. *Adv. Appl. Prob.* **29**, 353–373.

[27] LIU, Q. (1998). Fixed points of a generalized smoothing transformation and applications to the branching random walk. *Adv. Appl. Prob.* **30**, 85–112.

[28] LIU, Q. (2000). On generalized multiplicative cascades. *Stochastic Process. Appl.* **86**, 263–286.

[29] LIU, Q. (2001). Asymptotic properties and absolute continuity of laws stable by random weighted mean. *Stochastic Process. Appl.* **95**, 83–107.

[30] LYONS, R. (1997). A simple path to Biggins' martingale convergence for branching random walk. In *Classical and Modern Branching Processes*, eds. K.B. Athreya and P. Jagers. IMA Volumes in Math. and its Appl., vol. 84, Springer, Berlin, pp. 217–221.

[31] PETROV, V. V. (1975). *Sums of Independent Random Variables.* Springer, Berlin.

[32] ROESLER, U. (1992). A fixed point theorem for distributions. *Stochastic Process. Appl.* **42**, 195–214.

[33] ROESLER, U. (1993). The weighted branching process. In *Dynamics of complex and irregular systems (Bielefeld, 1991),* Bielefeld Encounters in Mathematics and Physics VIII, World Science Publishing, River Edge, NJ, pp. 154–165.

[34] ROESLER, U., TOPCHII, V. AND VATUTIN, V. A. (2002). Convergence rate for stable Weighted Branching Processes. In *Mathematics and Computer Science II,* eds. B. Chauvin, P. Flajolet, A. Mokkadem. Trends in Mathematics, Birkhäuser, Basel, pp. 441–453.

Amke Caliebe
Institut für Medizinische Informatik und Statistik, Universitätsklinikum Schleswig-Holstein, Campus Kiel, Brunswiker Str. 10, D-24105 Kiel, Germany (corresponding address) and
Mathematisches Seminar, Christian-Albrechts-Universität zu Kiel, Ludewig-Meyn-Str. 4, D-24098 Kiel, Germany
caliebe@math.uni-kiel.de

Trends in Mathematics, © 2004 Birkhäuser Verlag Basel/Switzerland

Stochastic Fixed Points for the Maximum

Peter Jagers and Uwe Rösler

ABSTRACT: *We consider stochastic fixed point equations*

$$X \stackrel{\mathcal{D}}{=} \sup_i T_i X_i$$

in $X \geq 0$ for known $T = (T_1, T_2, \ldots)$. The rvs $T, X_i, i \in \mathbb{N}$ are independent and X_i distributed as X. We present a systematic approach in order to find solutions using the monotonicity of the corresponding operator. These equations come up in the natural setting of weighted trees with finite or countable many branches. Examples are in branching processes and the analysis of algorithms (for parallel computing).

The above supremum equation is equivalent to

$$F(t) = E \prod_i F(t/T_i)$$

for distribution functions. In case of a characteristic exponent α solutions are known via the Laplace or Fourier transform of solutions to the stochastic fixed point equation

$$W \stackrel{\mathcal{D}}{=} \sum_i T_i W_i.$$

We shall show, that there are more fixed point for the supremum equation than to the sum equation.

The prominent example is the water cascade problem. We obtain a whole class of such solutions exploiting stochastic monotonicity.

1. Introduction

In the mathematical abstract setting we are interested in random variables X with values in $[0, \infty]$ satisfying a fixed point equation

$$X \stackrel{\mathcal{D}}{=} \bigvee_i T_i X_i. \tag{1}$$

The rvs $T, X_i, i \in \mathbb{N}$ are independent and the positive X_i have the same distribution as X. $\vee$ denotes the supremum. The distribution of $T = (T_1, T_2, \ldots)$ is known. Notice, that we require no dependence assumption on the factors $0 \leq T_i, i \in \mathbb{N}$.

The above equation is a fixed point equation for probability measures on the extended positive reals $[0, \infty]$. Define the map K by

$$K(\mu) := \mathcal{L}(\vee_i T_i X_i) \tag{2}$$

as above and X_i having distribution μ on $[0,\infty]$. Here $\mathfrak{L}$ denotes the distribution of a rv. In terms of distribution functions F the operator K satisfies

$$K(F)(t) := E\prod_i F\left(\frac{t}{T_i}\right). \tag{3}$$

Notice, that dealing with the supremum in (1) is equivalent to dealing with the infimum using the formulation

$$\frac{1}{X} \stackrel{\mathfrak{D}}{=} \bigwedge \frac{1}{T_i}\frac{1}{X_i}. \tag{4}$$

Taking logarithms the equation (1) is equivalent to

$$\ln X \stackrel{\mathfrak{D}}{=} \vee_i (\ln T_i + \ln X_i) \tag{5}$$

replacing the multiplicative group $((0,\infty),\cdot)$ by the additive $(\mathbb{R},+)$.

Besides the mathematical importance of the above fixed points in itself they appear naturally in the setting of probability theory on weighted trees. Consider a weighted branching process with positive factors, i.e., iid positive rvs $T(v) = (T_1(v), T_2(v), \ldots)$, $v \in V = \mathbb{N}^*$ with values in $R^{\mathbb{N}}$ and every vertex v of the tree V with countable many branches has a random weight (=length) $L(v)$ recursively given by $L(vi) = T_i(v)L(v)$, $L(\emptyset) = 1$. The largest weight $L_n = \sup_{|v|=n} L(v)$ in the $n-$th generation satisfies in distribution the equation

$$K(\mathfrak{L}(L_n)) \stackrel{\mathfrak{D}}{=} \mathfrak{L}(L_{n+1}).$$

The easiest example are branching processes [2] with factors T_i either 0 or 1. $L(v) = 1$ corresponds to the particle v is alive and $L_n = 1$ corresponds to at least one living particle in $n-$th generation. The distribution of L_n converges to the measure $q\delta_0 + (1-q)\delta_1$ where δ_x denotes the point measure on x and q is the extinction probability of the branching process.

Another example are branching random walks, [3], the particles split like a branching process and have a displacement like a random walk on $\mathbb{R}$. This is a weighted branching process on the positive reals, using the additive group instead of the multiplicative. The largest (=most right) particle has the position L_n on $\mathbb{R}$.

Other examples of (1) came up in the analysis of algorithms. Suppose parallel computing of a problem following a certain order of the partial tasks according to the tree structure. The result is available, if the last task, let us say in $n-$th generation, is done. For large n we obtain in the limit a solution of a fixed point equation involving the supremum. From that we may obtain estimates for the time spend by the algorithm.

Many solutions of (3) are known in the setting of weighted branching processes, especially for positive coefficients. The stochastic fixed point equation

$$X \stackrel{\mathfrak{D}}{=} \sum_i T_i X_i, \tag{6}$$

in X is well studied. Taking Laplace transforms $\varphi(t) = Ee^{-tX}$ or Fourier transforms for symmetric X we find

$$\varphi(t) = E\prod_i \varphi(tT_i). \tag{7}$$

The function $x \mapsto \varphi(1/x)$ is a fixed point of (3).

A crucial role for the existence of solutions of (6) plays the characteristic exponent α defined by $E\sum_i T_i^\alpha = 1$. We will not stress this connection, just give some literature. Durrett and Liggett [5], Liu [7], Lyons, Pemantle, Peres [8] on the existence of positive solutions, Caliebe [4] Rösler [13] for symmetric solutions. In a broader context, especially for stochastic algorithms, these equations were considered using the contraction method with various metrics on the space of measures, [12], [9], [11].

However the equation 1 has more solutions than those obtainable by equation (7) respectively (6). In section 4 and 5 we present in our main Theorem 5.1 a new class of solutions including cases without a characteristic exponent α.

For the time being let us present a concrete example in branching processes, water cascading. Suppose every edge (respectively knot) of a binary tree gets a value 0 or 1 by independent, Ber(p) distributed rvs. Now we pour water into the root of the tree.

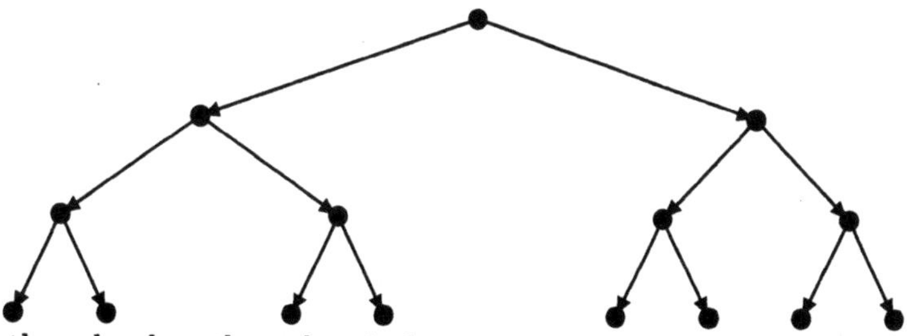

If the edge has the value 0 the water passes immediately through and if the edge has the value 1 then the water has one unit time delay. (Otherwise the water has infinite speed.) At what time can we expect the first water in the n-th generation? And passes the water immediately down through all generations?

In our additive formulation every knot gets as weight the sum of all the 0 and 1 along the path to the root. The infimum L_n of the weights in the n-th generation corresponds to the time of the first arrival of water in the n-th generation. We obtain the recursive equation for the distribution

$$L_{n+1} \stackrel{\mathfrak{D}}{=} (L_n^1 + T_1) \wedge (L_n^2 + T_2)$$

where L_n^1, L_n^2, T_1, T_2 are independent, L_n^1, L_n^2 have the same distribution as L_n and T_1, T_2 are independent Bernoulli distributed to the parameter p as above. If $2p$ is strictly smaller than 1 then with strict positive probability water will run immediately in all generations. This implies L_n converges pointwise to some rv L and $P(L = 0) > 0$. Otherwise the probability is 0 and L_n converges to infinity.

Concerning the probability $P(L = 0)$, the argument is easy by identifying the instantaneous appearance of water with the survival probability of a branching process. A 0 for the edge (v, vi) corresponds exactly to a living particle at vi, given a living particle at v. The reproduction distribution for the branching process is $p_0 = 2p$, $p_2 = q^2$, $p_1 = 2pq$ where $q = 1 - p$. This identifies the probability of instantaneous appearance of water as the survival probability of the branching process. However, the probability that the first water appears in all generations at least within time k is not so easy to calculate. We provide the result in the last section via a fixed point equation

$$L \stackrel{\mathfrak{D}}{=} (L^1 + T_1) \wedge (L^2 + T_2)$$

L^1, L^2, T_1, T_2 independent, $L^1 \sim L^2 \sim L$, $T_1 \sim T_2 \sim \text{Ber}(q)$.

The first section states the problem mathematically precise in the setting of weighted branching processes. Section 3 excludes some not so interesting cases. Section 4 provides in Theorem 4.1 the main monotonicity argument in general and Section 4 gives a more detailed study, Theorem 5.1, on fixed points under the additional restriction of all factors bounded by 1. The last section provides more detailed results especially to the water cascade problem.

There are related recent papers prepared after finishing this work by Neininger and Rüschendorf [6] and another one by Aldous and Bandyopadhyay [1].

2. Mathematical setting

Let $(\Omega, \mathcal{A}, P)$ be a probability space sufficiently large for our purposes. We suppress $\omega \in \Omega$ whenever possible. Let $(V, (G, \cdot), T, L)$ be a weighted branching process [12] on the multiplicative group $G = (0, \infty)$. For simplicity we take the adjoint grave $\triangle$ ($\triangle \cdot g = g \cdot \triangle = \triangle$ for all $g \in G \cup \{\triangle\}$) as the point 0. The set $V := \bigcup_{n \geq 0} \mathbb{N}^n$ denotes the vertices of a tree with countable many branches and a root $\emptyset$. $T(v) = (T_1(v), T_2(v), \ldots)$, $v \in V$ are independent and identical distributed random variables with values in $([0, \infty))^{\mathbb{N}}$.

Define recursively $L(v) : \Omega \to G \cup \{\triangle\} = [0, \infty)$, $v \in V$, by $L(\emptyset) \equiv 1$ and

$$L(vi) = T_i(v) L(v).$$

Notice $L(i) = T_i(\emptyset)$. An equivalent definition is $L(v) = \prod_{j=1}^n T_{v_j}(v_{|j-1})$ for a vertex $v = (v_1, \ldots, v_n) \in \mathbb{N}^n$ of length $|v| = n \in \mathbb{N}$. The symbol $v_{|j} := (v_1, v_2, \ldots, v_j)$ denotes the restriction of v to the first j coordinates.

Let V_n denote the set of vertices of length n. Throughout the paper we assume the random measure

$$\nu_n := \sum_{v \in V_n,\ L(v) \in G} \delta_{L(v)}$$

to be a Radon measure on $G = (0, \infty)$, taking finite values on all compact sets (relative to the induced topology). We use $\nu := \nu_1$ and $\nu(1) := \nu(\{1\})$, $\|\nu\| := \nu(G)$, which might be ∞. Notice ν is a random measure and also $\nu(1), \|\nu\|$ is random.

Define

$$\begin{aligned} L_n &:= \sup\{L(v) \mid v \in V_n\} \\ L_n^i &:= \sup\{\frac{L(v)}{T_i(\emptyset)} \mid v \in V_{n+1},\ v_1 = i\ \} \end{aligned}$$

where $\frac{L(v)}{T_i(\emptyset)}$ is the intuitive writing for the well defined expression $\prod_{j=2}^{n+1} T_{v_j}(v_{|j-1})$. We shall use the intuitive writing whenever possible.

The L-rvs satisfy the following recursive equation

$$L_{n+1} = \bigvee_{i \in \mathbb{N}} L(i) L_n^i \tag{8}$$

as rvs. Notice $L(i) = T(\emptyset), L_n^i, i \in \mathbb{N}$ are independent and all L_n^i have the same distribution as the L_n. The symbol $\bigvee$ denotes the supremum.

Define the map K acting on probabilities measures on $[0,\infty]$ to probability measures on $[0,\infty]$ via

$$K(\mu) \stackrel{\mathcal{D}}{=} \mathcal{L}(\bigvee_{i\in\mathbb{N}} T_i X_i), \tag{9}$$

$\mathcal{L}$ denoting the distribution of a rv. The rv $T = (T_1, T_2, \ldots)$ (on a not specified probability space) has the same distribution as the $T(v)$ above. The rvs $X_i, i \in \mathbb{N}$, (on the same not specified probability space) have all the distribution μ. The rvs $T, X_i, i \in \mathbb{N}$, are independent. The symbol $\stackrel{\mathcal{D}}{=}$ denotes equality in distribution. We use the convention $0 \cdot \infty = 0$ and $x \cdot \infty = \infty$ otherwise.

Notice that (8) rewrites in the weaker form $K(\delta_1)$ is the distribution $\mathcal{L}(L_1)$ of L_1 and in general $K(\mathcal{L}(L_n)) = \mathcal{L}(L_{n+1})$.

On the space of probability measures introduce the stochastic order via

$$\mu \preceq \nu \Leftrightarrow \int f d\mu \leq f d\nu$$

for all positive increasing functions f. This order is equivalent to $F_\mu \geq F_\nu$ pointwise for the distribution functions. Actually the set of probability measures on $[0,\infty]$ with the stochastic order is a lattice and K acts on this space via a lattice operation. For simplicity we shall work with distribution functions instead of the lattice of measures.

Let $\mathcal{C}$ be the set of all increasing, right continuous functions $F : (0,\infty) \to [0,1]$. For notational reasons we use $F(\infty) = 1$. There is a bijective map of probability measures on $[0,\infty]$ and elements of $\mathcal{C}$. Therefore we call an element of $\mathcal{C}$ a distribution function. The map K acts, recall the bijection, on $\mathcal{C}$ to $\mathcal{C}$

$$K(F)(t) = E\prod_{i\in\mathbb{N}} F(\frac{t}{T_i}), \tag{10}$$

$t \in (0,\infty)$. The n-th iterate of K writes

$$K^n(F)(t) = E\prod_{v\in V_n} F\left(\frac{t}{L(v)}\right) = Ee^{\int \ln F(\frac{t}{y})\nu_n(dy)}. \tag{11}$$

If F_n denotes the distribution function of L_n then $K(F_n) = F_{n+1}$.

We have chosen here right continuous distribution functions. These are naturally more suitable for the supremum. Considering the infimum, the left continuous distribution functions are more suitable, since the corresponding map $\overline{K}$ (apply the map $x \in G$ to its inverse $\frac{1}{x}$)

$$\overline{K}(\overline{F})(t) = E\prod_{i\in\mathbb{N}} \overline{F}(tT_i)$$

with $\overline{F}(t) := 1 - F(t-) := 1 - \lim_{s\nearrow t} F(s)$.

In our case of a Radon measure ν we could actually take left or right versions of F, and the fixed point equation rewrites

$$K(F)(t-) = Ee^{\int \ln F(\frac{t-}{y})\nu(dy)}.$$

We sometimes use the operator K also for increasing left continuous functions, if appropriate. The discussion for left or right continuous functions would disappear using the lattice setting of the measures.)

Remark: We consider here the supremum of all $L(v)$, $v \in V_n$. The argument for the infimum is the same using the map $[0,\infty] \ni x \to \frac{1}{x} \in [0,\infty]$. Instead of T_i consider $\frac{1}{T_i}$ for $T_i \geq 0$. Now the grave is ∞ and we would use the convention $0 \cdot \infty = \infty$. The fixed point equation (1) rewrites to

$$\frac{1}{X} \stackrel{\mathcal{D}}{=} \bigwedge \frac{1}{T_i} \frac{1}{X_i}.$$

A translation of $F \in \mathcal{C}$ with respect to $a > 0$ is the function $F_a \in \mathcal{C}$ defined by $F_a(x) = F(ax)$.

Lemma 2.1. *The set $\mathcal{C}$ is convex and closed under translations, finite sums, finite or countable products and decreasing limits.*

Let $T_1, T_2, \ldots$ be positive rvs. The map $K : \mathcal{C} \to \mathcal{C}$ from (10) is well defined, is monotone and σ-continuous from above.

If $F_n \in \mathcal{C}$ is a pointwise increasing sequence and $P(\prod_i F_1\left(\frac{t}{T_i}\right) > 0) = 1$ for all positive t then KF_n increases to $K \lim_n F_n$ at all continuity points of $K \lim_n F_n$.

Proof: The right continuous increasing function are the upper semi continuous increasing functions. The set of upper semi continuous increasing functions is convex and closed with respect to translations, to finite sums and to decreasing limits. Using the positivity of the functions the set is closed to finite products. Since the functions are bounded by 1 the set is closed to countable products.

The map K from $\mathcal{C}$ to some function space is well defined. The map is order preserving, using the pointwise order on function space. The function $K : \mathcal{C} \to \mathcal{C}$ is well defined, since $\lim_{x \searrow t} K(F)(x) = K(F)(t)$ by the Monotone Convergence Theorem.

The monotone operator K is called σ-continuous from above if

$$\mathcal{C} \ni F_n \searrow_n F \Rightarrow K(F_n) \searrow_n K(F) \in \mathcal{C}.$$

The Monotone Convergence Theorem implies the σ-continuity of K from above.

The σ-continuity from below follows also by the Monotone Convergence Theorem. The condition is needed to apply the theorem. q.e.d.

Using left continuous versions for an increasing sequence F_n we obtain $\lim_n F_n$ is a left continuous increasing function and $\lim_n KF_n = K \lim_n F_n$. In terms of stochastic convergence of measures there would be no distinction between left and right continuous versions necessary.

Corollary 2.2. *If $\|\nu\|$ is finite almost everywhere then for some $F \in \mathcal{C}$*

$$K(F)(t-) = E \prod_{i \in \mathbb{N}} F\left(\frac{t-}{T_i}\right). \tag{12}$$

for all $t > 0$ is equivalent to F being a fixed point.

Proof: An a.s. finite $\|\nu\|$ implies the condition in order to apply the Monotone Convergence Theorem. The rest is now obvious from the previous lemma. q.e.d.

3. The set of fixed points

In this section we study the set $\mathcal{F}$ of fixed points of K in some generality. Obviously the constant function 1 is always a fixed point. We call this the trivial fixed point and suppress it if possible. We first treat completely the case $\|\nu\| \leq 1$ a.e. the case

$\nu(1) = \|\nu\|$ a.e. , and the case $P(\nu(t,\infty) = \infty) > 0$ for some $t > 0$. Afterwards all these cases are excluded and we collect general results on $\mathfrak{F}$.

Case $\|\nu\| \leq 1$. The case $\|\nu\| \leq 1$ corresponds, after renaming, to the equation

$$X \stackrel{\mathfrak{D}}{=} TX. \tag{13}$$

Proposition 3.1. *Assume* $\|\nu\| \leq 1$.

If $\nu(1) = 1$ *then any probability measure on* $[0,\infty]$ *is a fixed point of* K.

Otherwise, if $P(\|\nu\| = 0) > 0$ *then* K *has only the trivial fixed point and if* $P(\|\nu\| = 0) = 0$ *then the fixed points of* K *are* δ_0 *and* δ_∞.

Proof: The first statement is easy.

For notational simplicity we use T instead of $T_1 \geq 0$ and use $T_2 = 0 = T_3 = \dots$. If $P(T=0)$ is strict positive then the equation (13) implies

$$P(X=0) = P(T=0) + P(T \neq 0)P(X=0),$$

which in turn implies $X \equiv 0$.

Now let T be not identical 1 and take values in $(0,\infty)$. The probability measures δ_0 and δ_∞ are solutions of (13). Suppose the equation (13) has another solution F. The function G defined by

$$G(t) := E(X\mathbb{I}_{X\leq t}) = \int_{[0,t]} ydF(y)$$

has the following properties:

- G is right continuous and increasing.
- $\lim_{t\to 0} \frac{G(t)}{t} = 0$
- $\lim_{t\to\infty} \frac{G(t)}{t} = 0$
- $\frac{G(t)}{t} = E\frac{G(\frac{t}{T})}{\frac{t}{T}}$

The first statement is easy. The second follows by $G(t) \leq t\int_{(0,t]} dF(y) = t(F(t) - F(0+))$. The third statement follows by

$$\begin{aligned}\lim_{t\to\infty} \frac{G(t)}{t} &= \lim_t \left[\frac{1}{t}\int_0^{t_0} ydF(y) + \frac{1}{t}\int_{t_0}^t ydF(y)\right] \\ &\leq \lim_t (F(t) - F(t_0)) = \lim_{t\to\infty} F(t) - F(t_0) \to_{t_0\to\infty} 0.\end{aligned}$$

The fourth statement follows by

$$G(t) = E(X\mathbb{I}_{X\leq t}) = E(TX\mathbb{I}_{TX\leq t}) = E(TE(X\mathbb{I}_{TX\leq t} \mid T)) = E(TG(\frac{t}{T})).$$

Let t_1, t_2 be the infimum and supremum of the set of values, where the right continuous function $t \mapsto \frac{G(t)}{t}$ takes its supremum. (Notice the points of discontinuity are of the first kind and the function increases at every point of discontinuity. Therefore $0 < t_1 \leq t_2 < \infty$ exist and are well defined.) We obtain

$$\frac{G(t_i)}{t_i} = E\frac{G(\frac{t_i}{T})}{\frac{t_i}{T}},$$

$i = 1,2$. This implies $\frac{t_i}{T}$ is a.e. a supremum of the function $t \mapsto \frac{G(t)}{t}$. But this implies $T \leq 1$ for $i = 1$ and $T \geq 1$ for $i = 2$. But $T \equiv 1$ was excluded. q.e.d.

Case $\nu(1) = \|\nu\|$ and not $\|\nu\| \leq 1$ a.e..

The case $\nu(1) = \|\nu\|$ corresponds to values 0 or 1 of the factors $T_i, i \in \mathbb{N}$. The corresponding weighted branching process is a branching process. The fixed point equation $KF = F$ rewrites

$$KF(t) = EF^{\nu(1)}(t). \tag{14}$$

Possible values of $F(t)$ are the fixed points of the generating function

$$[0,1] \ni s \mapsto f(s) = \sum_{k=0}^{\infty} p_k s^k \tag{15}$$

with $p_k := P(\nu(1) = k)$. (Notice $\nu(1) < \infty$.) There are at most two solutions (f the identity is excluded). One solution is always the value 1. There is another solution in $[0,1)$ iff $E\nu(1) > 1$.

Proposition 3.2. *Assume $P(\|\nu\| \leq 1) < 1$ and $\nu(1) = \|\nu\| < \infty$. The fixed points of K are all functions in $\mathcal{C}$ taking as values only the fixed points of the generating function f (15).*

Case $P(\nu(t,\infty) = \infty) > 0$ for some $0 < t$.

Proposition 3.3. *Assume $P(\nu(t,\infty) = \infty) > 0$ for some $0 < t$. Let q be the smallest fixed point of $[0,1] \ni s \mapsto Es^{\nu(0,\infty)}$. Then the fixed points of K are all functions in $\mathcal{C}$ taking as values only the fixed points of $[0,1] \ni s \mapsto Es^{\nu(0,\infty)}$.*

Proof: Using $Y := \nu(0,\infty)$ and excluding $X \equiv 0$ the value $P(X=0) < 1$ is a fixed point of $s \mapsto Es^Y$ since

$$\begin{aligned} E(\mathbb{I}_{X=0}) &= E(\mathbb{I}_{Y<\infty}\mathbb{I}_{\vee T_i X_i = 0} + E(\mathbb{I}_{Y=\infty}\mathbb{I}_{\vee T_i X_i=0}) \\ &= E(\mathbb{I}_{Y<\infty}\mathbb{I}_{\vee T_i X_i=0}) = E(P^Y(X=0)). \end{aligned}$$

Let $c \in (0,\infty]$ be the essential supremum of X. Then

$$\begin{aligned} E\mathbb{I}_{X<c} &= E(\mathbb{I}_{Y<\infty}\mathbb{I}_{\vee T_i X_i<c} + E(\mathbb{I}_{Y=\infty}\mathbb{I}_{\vee T_i X_i<c})) \\ &= E(\mathbb{I}_{Y<\infty}\mathbb{I}_{\vee T_i X_i<c}) = E(P^Y(X<c)). \end{aligned}$$

is again a fixed point of $[0,1) \ni s \mapsto Es^Y$. There is at most one fixed point. This leads to the possible solutions $\mathcal{L}(X) = q\delta_0 + (1-q)\delta_c$, $c \in [0,\infty]$. All these are solutions. q.e.d.

We collect now a few general results on the set $\mathfrak{F}$ of fixed points excluding the previous cases.

Proposition 3.4. *Any multiplicative translate $(0,\infty) \ni t \mapsto F(at)$, $a > 0$ of a fixed point $F \in \mathfrak{F}$ is a fixed point.*

The right side limit $F(0+) = \lim_{t\to 0} F(t)$ at 0 of a fixed point F solves the equation

$$c = Ec^{\|\nu\|} \tag{16}$$

in c. (Use here the convention $0^0 = 1$.)

There is at most one solution of equation (16) for $c \in [0,1)$. There is a solution iff $E\|\nu\| > 1$.

The function identical to a constant $c \in [0,1]$ is a fixed point iff c solves (16).

Proof: The first statement is obvious.
For any solution $F \in \mathfrak{F}$ we have

$$F(0+) = \lim_{t\to 0} F(t) = \lim_t KF(t) = K \lim_t F(t) = E\prod_i F(0+) = EF^{\|\nu\|}(0+).$$

Define the generating function g of $\|\nu\|$, $g(s) := Es^{\|\nu\|}$ for $s \in [0,1]$. The fixed points of g are exactly the solutions of equation (16). If $P(\|\nu\| < \infty) = 1$ then the statement on fixed points of g is well known in the literature of branching processes [2]. The same arguments apply in the general case, including $P(\|\nu\| < \infty) < 1$. We skip the details. q.e.d.

Proposition 3.5. *Let $\|\nu\| < \infty$ a.e. If the measure $E\nu$ has no discrete mass then any fixed point is continuous. If the measure $E\nu$ has no discrete mass besides in 1 then any fixed point is continuous besides a possible jump to 1. If there is a jump to 1 at t then $P(\sup_i T_i \leq 1) = 1$, $E\nu(1) > 1$ and $F(t-)$ is the unique fixed point of the generating function f (15) in $[0,1)$.*

Proof: Let D be the set of discontinuities of the fixed point F. D is at most countable, since the fixed point F is increasing and can have at most countable many points of discontinuity of the first kind. For the following estimate note that for fixed t the set of all $\omega \in \Omega$ satisfying $\frac{t}{T_i(\omega)} \in D$ if $T_i(\omega) \neq 1$ is a null set relative to the measure $E\nu$,

$$\sum_i P(t \in DT_i,\ T_i \neq 1) = E\nu(t(D^{-1}\backslash\{1\})) = 0.$$

Therefore and by Corollary 2.2 we obtain

$$\begin{aligned} F(t) &= E\prod_i F\left(\frac{t}{T_i}\right) = EF^{\nu(1)}(t) \prod_{\{i|T_i\neq 1\}} F\left(\frac{t}{T_i}\right) \\ &= EF^{\nu(1)}(t) \prod_{\{i|T_i\neq 1\}} F\left(\frac{t-}{T_i}\right) \end{aligned}$$

and recall (12)

$$F(t-) = EF^{\nu(1)}(t-) \prod_{\{i|T_i\neq 1\}} F\left(\frac{t-}{T_i}\right)$$

In case $\nu(1) = 0$ we are done. Otherwise both values $F(t)$ and $F(t-)$ satisfy the equation

$$s = Es^{\nu(1)}b$$

with $s \in [0,1]$ and $b = \prod_{\{i|T_i\neq 1\}} F\left(\frac{t-}{T_i}\right)$.

Consider the function $[0,1] \ni s \mapsto g(s) = Es^{\nu(1)}b$. The function g is convex. In case $P(0 < \nu(1) < \infty,\ b > 0) = 0$ the function g is independent of s. Therefore g has exactly one fixed point and $F(t) = F(t-)$. In case $P(0 < \nu(1) < \infty,\ b > 0) > 0$ the function g is strict convex. Therefore g has at most two fixed points. By $g(0) \geq 0$ and $g(1) = Eb < 1$ there is at most one fixed point in $[0,1)$. If $F(t) < 1$ we are done. It remains the case $F(t-) < F(t) = 1$. This implies $g(1) = Eb = 1$ and consequently $b \equiv 1$ and $\sup_i T_i \leq 1$ a.e. $F(t-)$ is the unique fixed point of g in $[0,1)$. q.e.d.

4. Fixed points by monotonicity

Like in potential theory the monotonicity of K will provide some fixed points. We determine the fixed points in case of $\sup_i T_i \leq 1$.

We always exclude the cases $\|\nu\| \leq 1$ and $\nu(1) = \|\nu\|$.

Theorem 4.1. *Suppose some $F \in \mathcal{C}$ satisfies $F \geq KF$. Then the sequence K^nF is pointwise decreasing in n to some fixed point of K.*

Suppose some $F \in \mathcal{C}$ satisfies $F \leq KF$ and $P(\prod_i F(\frac{t}{T_i}) > 0) = 1$ for all $t > 0$ or $\|\nu\| < \infty$ a.e.. Then the sequence K^nF, $n \in \mathbb{N}$, is pointwise increasing in n to the left continuous version of some fixed point of K.

Proof: Since $F \geq KF$ we obtain by induction $K^nF \geq K^{n+1}F$. The sequence K^nF decreases pointwise to some function $G \in \mathcal{C}$. By the σ-continuity from above we obtain

$$K(G) = K(\lim_n K^nF) = \lim_n K(K^nF) = \lim_n K^{n+1}F = G.$$

The argument for $F \leq KF$ is analog. The imposed condition provides the interchangeability of the limit. Start with the left continuous versions of F. Then K^nF is left continuous and the limit $G = \lim_n K^nF$ is left continuous. The right continuous version of G is a fixed point of K in $\mathcal{C}$. q.e.d.

Using the stochastic order $\preceq_s$ on probability measures the above lemma reads: If $K(\mu) \preceq_s \mu$ ($\mu \preceq_s K(\mu)$) for a probability measure on $[0,\infty]$ then the sequence $K^n\mu$, $n \in \mathbb{N}$, decreases (increases) in stochastic order to a fixed point of K.

The set $\mathcal{F}$ is partially ordered with the stochastic order. Define the infimum $\wedge_{\mathcal{F}}$ and supremum $\vee_{\mathcal{F}}$ in $\mathcal{F}$ by

$$F \wedge_{\mathcal{F}} G := \lim_n K^n(F \wedge G)$$

$$F \vee_{\mathcal{F}} G := \lim_n K^n(F \vee G).$$

Lemma 4.1. *Assume $P(\|\nu\| > 1) > 0$ and $\|\nu\| < \infty$ a.e.. Then $(\mathcal{F}, \leq, \vee_{\mathcal{F}}, \wedge_{\mathcal{F}})$ is a lattice. This lattice has a minimal element given by the function $\equiv 0$ and a maximal element given by the function $\equiv 1$. The lattice is complete, i.e., every (bounded) set has a minimal upper bound and a maximal lower bound in $\mathcal{F}$.*

We skip the details.

5. Factors bounded by 1

In this section we determine the fixed points in case of $\sup_i T_i \leq 1$. We always exclude the cases $\|\nu\| \leq 1$ and $\nu(1) = \|\nu\|$.

Let $(Z_n)_n$ be a branching process with offspring distribution $p_k := P(\nu(1) = k)$, $k \in \mathbb{Z}^+$. Let $q_n := P(Z_n = 0)$ be the extinction probability of Z_n and $q = \lim_n q_n$ be the extinction probability of the branching process $(Z_n)_n$. Notice $f(q_n) = q_{n+1}$ where f is the generating function $f(s) = \sum_{k=0}^{\infty} p_k s^k$ of $(p_k)_k$.

Let $F_0 := 1_{[1,\infty)}$ and $F_n := K(F_{n-1})$. We obtain $F_n = K^n(F_0)$ by iteration.

Theorem 5.1. *Assume $\|\nu\| < \infty$ a.e. and $\sup_i T_i \leq 1$.*

(i) The sequence F_n of right continuous distribution functions increases pointwise to some function F. The right continuous version of F is a fixed point of K.

(ii) $q_n = F_n(1-)$ and $q = F(1-)$.

(iii) $F = F_0$ *is equivalent to* $P(\nu(1) = 0) = 0$.
(iv) The support of F *is the positive real line unless* $\sup_i T_i < 1$ *a.e.*
(v) If $\sup\{T_i \mid T_i < 1\} \leq c < 1$ *a.e. then* $q_n = F_n(t)$ *and* $q = F(t)$ *for any* $t \in [c, 1)$.
(vi) $F \equiv 1$ *is equivalent to* $q = 1$.

Proof: i) For $t \geq 1$

$$K(F_0)(t) = E\prod_i F_0\left(\frac{t}{T_i}\right) = 1 = F_0(t)$$

and for $t < 1$ is $KF_0(t) \geq 0 = F_0(t)$. By iteration we obtain $F_{n+1} = K(F_n) \geq K(F_{n-1}) = F_n$. The sequence F_n increases pointwise to some function F. Lemma 4.1 provides the fixed point property for the right continuous version.

ii) We show this by induction. The induction starts with $F_0(1-) = 0 = q_0$ and the induction step n to $n + 1$ is

$$F_{n+1}(1-) = KF_{n+1}(1-) = E\prod_i F_n\left(\frac{1-}{T_i}\right) = EF_n^{\nu(1)}(1-) = f(q_n) = q_{n+1}.$$

Since F_n increases to F pointwise we obtain $F(1-) = q$.

iii) $F = F_0$ is equivalent to $q = 0$ and this is equivalent to $\nu(1) \geq 1$ a.e..

iv) Let t_0 be the left endpoint of the support of F. Assume $0 < t_0 < 1$. Then

$$0 = F(t_0-) = KF(t_0-) = E\prod_i F\left(\frac{t_0-}{T_i}\right) = EF^{\nu(1)}(t_0-)b = E\mathbb{I}_{\nu(1)=0}b,$$

where $b = \prod_{T_i<1} F(\frac{t_0-}{T_i})$ is appropriate. We obtain $P(\sup_i T_i < 1) = 0$ and consequently $\nu(1) = \|\nu\|$. But this case was excluded. Consequently t_0 is 0 or 1.

v) For $c \leq t < 1$ we obtain

$$F(t) = KF(t) = E\prod_i F^{\nu(1)}(t) = f(F(t)).$$

$F(t)$ is a fixed point of f and therefore either q or 1. By $F(1-) = q$ we conclude $F(t) = q$. If $q = 1$ then consider the infimum t_1 of all times t satisfying $F(t) = 1$. By the above $t_1 < 1$. If $0 < t_1 < 1$ then the function $t \mapsto F(t_1 t)$ is a fixed point and larger or equal than F_0. By the monotonicity of K we conclude $F(t_1\cdot) \geq F(\cdot)$. Therefore $F(t_1 t_1) \geq F(t_1) = 1$ in contradiction to the definition of t_1. The only way out is $t_1 = 0$ and $F \equiv 1$.

vi) $F \equiv 1$ implies $q = 1$. Now the reverse. Assume first $E\nu(1) < 1$. Then there is an $\epsilon > 0$ such that $E\nu((1-\epsilon, 1]) < 1$. Define the factors

$$\tilde{T}_i := \begin{cases} T_i & \text{if} \quad T_i \leq 1 - \epsilon \\ 1 & \quad\quad \text{else} \end{cases}$$

Then $L_n = \sup_{v\in V_n} L(v) \leq \tilde{L}_n$. Since F_n is the distribution function of L_n we obtain $F_n \geq \tilde{F}_n$ and in the limit $F \geq \tilde{F}$. By the last partial claim $\tilde{L}_n$ and L_n converge to 0 and $\tilde{F} \equiv 1 \equiv F$.

We come now to the case $E(\nu(1)) = 1$. (The given proof would work also for the case $E\nu(1) < 1$.) The rv τ gives every infinite path $v \in \mathbb{N}^{\mathbb{N}}$ the smallest natural numbers, such that the weight of $v_{|n}$ is strictly smaller than 1,

$$\tau(v) := \inf\{n \mid L(v_{|n}) < 1\}.$$

(τ is stopping time in an appropriate sense for trees.) Let X be a solution of (1). Define independent copies $X(v), v \in V$. Then by induction for every $n \in \mathbb{N}$

$$X \stackrel{\mathcal{D}}{=} \vee_{v \in A_n} L(v) X(v)$$

where $A_n = \{v_{|\tau \wedge n} \mid v \in V_n\}$. And again by induction

$$X \stackrel{\mathcal{D}}{=} \vee_{v \in A} L(v) X(v)$$

where $A = \limsup_n A_n$. Notice A is still countable and the measure $\hat{\nu} = \sum_{v \in A} \delta_{L(v)}$ is a.e. a Radon measure. (For the last statement use a finite extinction time for a critical branching process.) Interpreting the $L(v), v \in A$, as factors in the last equality we conclude by our results, case $E\hat{\nu}(1) = 0$, that X is identically 0. q.e.d.

Analogously to the construction via an increasing sequence of distribution functions we can use a decreasing sequence (and right continuous versions). Let q be the extinction probability as above and define

$$G_0 = \mathbb{I}_{[1,\infty)} + q 1_{(0,1)}.$$

Lemma 5.1. *Assume $\|\nu\| < \infty$ and $\sup_i T_i \leq 1$.*
$K^n G_0$ decreases pointwise to a fixed point of K. This fixed point is F as above.

Proof: Let $t < 1$.

$$\begin{aligned} K(G_0)(t) &= E \prod_i G_0\left(\frac{t}{T_i}\right) \\ &= E G_0^{\nu(1)}(t) \prod_{\{i \mid T_i < 1\}} G_0\left(\frac{t}{T_i}\right) \\ &\leq E q^{\nu(1)} 1 = q. \end{aligned}$$

This proves $G_0 \geq K G_0$. The sequence $K^n G_0$ is decreasing and converges pointwise to some function called G. By $K^n G_0 \geq K^n F_0$ we obtain $G \geq F$.

Assume $G > F$ which can happen only for $0 < q < 1$. Let t_2 be the smallest real satisfying $G = F$ on $[t_2, 1)$. We consider first the case $\sup\{T_i \mid T_i < 1\} \leq c < 1$ a.e.. for a constant c. Then for any $ct_2 \leq t < t_2$ the values $F(t)$ and $G(t)$ are fixed points of the map $s \mapsto E s^{\nu(1)} b$ with the rv $b = \prod_{T_i < 1} F\left(\frac{t}{T_i}\right) = \prod_{T_i < 1} G\left(\frac{t}{T_i}\right)$. Like for branching processes consider the function $g(s) := E c^{\nu(1)} b$. This is a convex function in $0 \leq s < 1$. It is not difficult to establish $g(0) > 0$ and $g(1) < 1$ in our setting. Therefore $s = g(s)$ has exact one solution in $[0, 1)$. This implies $F(t) = G(t)$ in contradiction to the choice of t_2 and the assumption.

Now we consider the general case without the restriction. For $\epsilon > 0$ consider the factors

$$\tilde{T}_i := \begin{cases} T_i & \text{if} \quad T_i \leq 1 - \epsilon \\ 1 & \text{else} \end{cases}$$

$$\underline{T}_i := \begin{cases} T_i \wedge (1 - \epsilon) & \text{if} \quad T_i < 1 \\ 1 & \text{else} \end{cases}$$

Then $\underline{F} \geq F \geq \tilde{F}$ and $\underline{G} \geq G \geq \tilde{G}$. By $\underline{F} = \underline{G}$ and $\tilde{F} = \tilde{G}$ it suffices to show $\underline{F}$ and $\overline{F}$ converge to F as ϵ decreases to 0. For this it suffices to show $\underline{F}_n$ and $\tilde{F}_n$ converge to F_n as ϵ decreases to 0 for any given n. However this is easy considering the corresponding rvs $L_n = \sup_{v \in V_n} L(v)$ and analogously $\underline{L}_n, \tilde{L}_n$. q.e.d.

Corollary 5.2. *Assume* $\|\nu\| < \infty$ *and* $\sup_i T_i \leq 1$.

Besides the trivial fixed point 1 *there is in case* $q = 1$ *none and in case* $q < 1$ *up to translation only one fixed point* F *satisfying* $F(t) = 1$ *for some* $t \in \mathbb{R}^+$.

6. Water cascades example

We return now to the toy example from the beginning. The state space is the **additive** group of integers. Let $V = \cup_{n=0}^{\infty}\{1,2\}^n$ denote the knots of a binary tree with root $\emptyset$. Let $T_i(v) : \Omega \to \{0,1\},\ v \in V, i \in \{1,2\}$ be independent identical distributed random variables with Bernoulli distribution to the parameter p. Define the weight $L(v)$ of a knot v recursively by

$$L(vi) = L(v) + T_i(v)$$

and $L(\emptyset) = 0$. The weight $L(v)$ represents the number of all 1's on the path from the knot v to the root. (The edge (v, vi) carries a 1 iff $T_i(v) = 1$.) More precisely

$$L(v) = \sum T_{w_{|w|}}(w_{||w|-1})$$

where the sum is over all ancestors $w \neq v$ of v. Define $L_n := \inf_{|v|=n} L(v)$. Then L_n satisfies the backward equation [10]

$$L_{n+1} \stackrel{\mathcal{D}}{=} (L_n^1 + T_1) \wedge (L_n^2 + T_2).$$

Here L_n^1, L_n^2, T_1, T_2 are independent, L_n^1, L_n^2 have the same distributions as L_n and T_1, T_2 are Bernoulli(p) random variables.

In the multiplicative setting we consider the multiplicative group $\{2^n \mid n \in \mathbb{Z}\}$. Taking the map $x \mapsto e^{-x}$ and the notation $\tilde{L}_n = e^{-L_n}$ and $\tilde{T}_i(v) = e^{-T_i(v)}$ we obtain the backward equation

$$\tilde{L}_{n+1} \stackrel{\mathcal{D}}{=} (\tilde{T}_1\tilde{L}_n^1) \vee (\tilde{T}_2\tilde{L}_n^2), \tag{17}$$

the standard form used in our setting.

Proposition 6.1. *The distribution of* $\tilde{L}_n$ *is given by* $F_n = K^n F_0,\ n \in \mathbb{N}$, *where* $F_0 = \mathbb{I}_{[1,\infty)}$. *The limiting distribution* F *is non degenerate iff* $2p < 1$ *and in that case given by*

$$F(t) = \begin{cases} 1 & \text{if} \quad t \geq 1 \\ a_n & \text{if} \quad t \in [e^{-n}, e^{-n+1}) \end{cases}$$

The values a_n *and* $a_0 = 1$ *satisfy the recursive equation*

$$a_n = (1-p)a_n^2 + pa_{n-1}^2$$

and are uniquely determined by this equation.

Proof: A short calculation shows $F_1 = KF_0$ corresponds to the distribution of L_1. By induction we obtain F_n is the distribution function of L_n. The limiting distribution is not trivial, Lemma 5.1, if and only if the extinction probability is strictly less than 1. This is equivalent to $E(1-T_1)+E(1-T_2) > 1$ which provides the criterium $2p < 1$. Via induction on n show F has the above form and the a_n are determined (uniquely) by the equation. We skip the details. q.e.d.

We considered here rvs (=water delays) on the edges. The argument with delays on the knots is, after reformulation, a special case.

References

[1] D. Aldous and A. Bandyopadhyay *A Survey of Max-Type Recursive Distributional Equations.* http://front.math.ucdavis.edu/math.PR/0401388 (2004).

[2] K. Athreya and P. Ney, *Branching Processes.* Springer, Berlin 1972.

[3] J.D. Biggins *Martingale convergence in the branching random walk.* Journal of Applied Probability **14**, 25-37, (1977).

[4] A. Caliebe *Symmetric fixed points of a smoothing transformation.* Advances in Applied Probability **35** 377-394 (2003).

[5] R. Durrett and M. Liggett *Fixed points of the smoothing transformation.* Zeitschrift für Wahrscheinlichkeitstheorie und verwandte Gebiete **64**, 275-301, (1983).

[6] R. Neininger and L. Rüschendorf *Analysis of algorithms by the contraction method: additive and max-recursive sequences.* Preprint 2003.

[7] Qu. Liu *Fixed points of a generalized smoothing transformation and applications to the branching random walk.* Advances Applied Probability **30**, 85-112 (1998).

[8] R. Lyons, R. Pemantle and Y. Peres *Conceptual proofs of of LlogL criteria for mean behavior of branching processes.* Annals of Probability **25**, 1125-1138 (1995).

[9] S.T. Rachev and L. Rüschendorf *A new ideal metric with applications to multivariate stable limit theorems.* Probability Theory and related fields, **94**, 163-187 (1992).

[10] U. Rösler *A fixed point theorem for distributions.* Stochastic Processes Appl. **42**, 195-214 (1992).

[11] U. Rösler, V. Topchii and V. Vatutin *Convergence conditions for the weighted branching process.* Discrete Mathematics, **12** nr.1 (2000).

[12] U. Rösler and L. Rüschendorf *The contraction method for recursive algorithms.* Algorithmica **29**, 3-33 (2001).

[13] U. Rösler *A fixed point equation for distributions.* Berichtsreihe des Mathematischen Seminars Kiel, Christian-Albrechts-Universität zu Kiel, www.numerik.uni-kiel/reports/ Bericht 98-7, (1998).

Peter Jagers
Mathematical Statistics
Chalmers University of Technology
S-412 96 Göteborg
Sweden
jagers@math.chalmers.se

Uwe Rösler
Mathematisches Seminar
Christian-Albrecht-Universität zu Kiel
Ludewig-Meyn-Strasse 4
24098 Kiel
Germany
roesler@math.uni-kiel.de

Trends in Mathematics, © 2004 Birkhäuser Verlag Basel/Switzerland

The Number of Descents in Samples of Geometric Random Variables

Arnold Knopfmacher* and Helmut Prodinger†

ABSTRACT: *For words of length n, generated by independent geometric random variables, we consider the mean and variance, and thereafter the distribution of the number of descents in the words. The cases of strict and weak descents are both considered. We also study the position and height of the first descent as well as the size of the greatest descent over all words of length n.*

1. Introduction

Let X denote a geometrically distributed random variable, i. e. $\mathbb{P}\{X = k\} = pq^{k-1}$ for $k \in \mathbb{N}$ and $q = 1 - p$. The combinatorics of n geometrically distributed independent random variables $X_1, \dots, X_n$ has attracted recent interest, especially because of applications in computer science. We mention just two areas, the skip list [1, 13, 16, 8] and probabilistic counting [3, 6, 7, 9].

One of the first combinatorial questions investigated for words $a_1 \dots a_n$, with the letters a_i independently generated according to the geometric distribution, was the number of left-to-right maxima in [14] . In [10] the study of left-to-right maxima was continued, but now the parameters studied were the mean value and mean position of the r-th maximum.

In this article we study descents in a string of n geometrically distributed independent random letters. For example in $w = 221114431$ we have 3 strict and 7 weak descents (equality is included). In the sequel we denote by $D_n(w)$ the number of descents (strict or weak as indicated) in the word w, where w is of length n.

In section 2 we study the mean and variance of $D_n(w)$. Thereafter, in section 3 we study the distribution of the number of descents, which turns out to be Gaussian. Subsequently, in sections 4 and 5 we study the average position and height of the first descent. Finally, in section 6 we determine the expected value of the size of the greatest descent in a word of length n. By reversing the strings we see that exactly the same results hold for *ascents* as for descents in such words. Our treatment to some extent parallels the study of runs in geometric random variables by Grabner and the present authors in [4].

We make use of the following abbreviations: $p = 1 - q$, $Q = 1/q$, $L = \log Q$. Words have length n.

*This material is based upon work supported by the National Research Foundation under grant number 2053740

†This material is based upon work supported by the National Research Foundation under grant number 2053748

2. The number of descents

As always in the setting of geometrically distributed random variables, we have two choices: to use the '>' or the '≥' sign. Let us do the instance of strict descents first.

2.1. Strict descents

In order to determine the mean and variance of the number of strict descents we will make use of the following decomposition of the set of all words. Here $\{\geq \mathbf{k}\}$ denotes the set $\{k, k+1, \ldots\}$; for a given set A we denote

$$A^+ = \bigcup_{k=1}^{\infty} A^k, \quad A^* = \varepsilon \cup A^+,$$

where ε stands for the empty word. We decompose the set of all words according to runs of 1's, separated by words consisting of larger digits only:

$$\{\geq \mathbf{1}\}^* = \mathbf{1}^*\big(\{\geq \mathbf{2}\}^+\mathbf{1}^+\big)^*\{\geq \mathbf{2}\}^+(\varepsilon + \mathbf{1}^+) + \mathbf{1}^*. \tag{1}$$

We consider a probability generating function $F(z,u)$, where z labels the number of random variables (or length of the word, if you wish), and u counts the number of descents. We should always have $F(z,1) = \frac{1}{1-z}$, which we might use as a check, and a replacement of z by qz, if we increase all letters by 1.

$$F(z,u) = \frac{1}{1-pz}\frac{1}{1-[F(qz,u)-1]\frac{pzu}{1-pz}}[F(qz,u)-1]\Big(\frac{pzu}{1-pz}+1\Big)+\frac{1}{1-pz}. \tag{2}$$

Now we differentiate it w.r.t. u, plug in $u=1$, set $G(z) = \frac{\partial}{\partial u}F(z,1)$, and get

$$G(z) = G(qz)\frac{(1-qz)^2}{(1-z)^2} + \frac{pqz^2}{(1-z)^2}.$$

We set $H(z) = (1-z)^2G(z)$ and get $H(z) = H(qz) + pqz^2$. Comparing coefficients, we see that

$$[z^2]H(z) = \frac{pq}{1-q^2} = \frac{q}{1+q},$$

and that the other coefficients are zero. Consequently,

$$H(z) = \frac{q}{1+q}z^2 \quad \text{and} \quad G(z) = \frac{q}{1+q}\frac{z^2}{(1-z)^2}.$$

Theorem 2.1. *The expected value of the number of descents is given for $n \geq 1$ by*

$$[z^n]G(z) = (n-1)\,\frac{q}{1+q}. \tag{3}$$

Now that we see such a simple result, we will be tempted to look for a simple proof as well. Indeed, since the expectation is additive, we will have $n-1$ times the expectation between two adjacent random variables. This quantity is

$$\sum_{i>j\geq 1} pq^{i-1}pq^{j-1} = \frac{q}{1+q}.$$

If we want to compute the variance, such a simple argument seems to be out of reach. Henceforth, we differentiate (2) twice, and use the notation $V(z) = \frac{\partial^2}{\partial u^2}F(z,1)$.

We see

$$V(z) = V(qz)\frac{(1-qz)^2}{(1-z)^2} + \frac{2q^2(1-q)z^3\big(q(1+q)+(1-q^2)z-q^2z^2\big)}{(1+q)^2(1-z)^3(1-qz)},$$

or, with $W(z) = (1-z)^2V(z)$,

$$W(z) = W(qz) - \frac{2(1-q)q^2z(1+(1+q)z+z^2)}{(1+q)^2} + \frac{2q^2}{(1+q)^2}\Big[\frac{1}{1-z} - \frac{1}{1-qz}\Big].$$

From this we see that (with $w_n := [z^n]W(z)$) for $n \geq 4$,

$$w_n = q^n w_n + \frac{2q^2}{(1+q)^2}\big[1-q^n\big], \quad \text{or} \quad w_n = \frac{2q^2}{(1+q)^2}.$$

Furthermore we see that $w_3 = \dfrac{2q^3(1-q)}{(1+q)(1-q^3)}$ and $w_0 = w_1 = w_2 = 0$. Hence

$$V(z) = \frac{2q^3(1-q)}{(1+q)(1-q^3)}\frac{z^3}{(1-z)^2} + \frac{2q^2}{(1+q)^2}\frac{z^4}{(1-z)^3}$$

and

$$[z^n]V(z) = \frac{q^2}{(1+q)^2}(n-2)(n-3) + \frac{2q^3(1-q)}{(1+q)(1-q^3)}(n-2).$$

Adding the expectation and subtracting the square of the expectation, we obtain the variance.

Theorem 2.2. *The variance of the number of strict descents is given for $n \geq 1$ by*

$$\frac{q(1-q)(1+q^3)}{(1+q)^3(1-q^3)}n - \frac{q(1-3q+q^2)}{(1+q)^2(1+q+q^2)}. \tag{4}$$

2.2. Weak descents

We will use the same decomposition (1), but with a slightly different translation into generating functions.

Without explicitly mentioning it, we will use the same functions as before, adjusted to the new meaning.

$$F(z,u) = \Big(1+\frac{pz}{1-pzu}\Big)\frac{1}{1-[F(qz,u)-1]\frac{pzu}{1-pzu}}[F(qz,u)-1]\Big(\frac{pzu}{1-pzu}+1\Big)$$
$$+1+\frac{pz}{1-pzu}. \tag{5}$$

Therefore

$$G(z) = G(qz)\frac{(1-qz)^2}{(1-z)^2} + \frac{pz^2}{(1-z)^2}$$

and $H(z) = H(qz) + pz^2$. Consequently,

$$H(z) = \frac{1}{1+q}z^2 \quad \text{and} \quad G(z) = \frac{1}{1+q}\frac{z^2}{(1-z)^2}.$$

Theorem 2.3. *The expected value of the number of weak descents is given for $n \geq 1$ by*

$$[z^n]G(z) = (n-1)\,\frac{1}{1+q}. \tag{6}$$

A simple explanation is also available as in the other instance, and therefore omitted.— Now we turn to the variance. We see

$$V(z) = V(qz)\frac{(1-qz)^2}{(1-z)^2} + \frac{2(1-q)z^3(1+q+(1-q^2)qz-q^3z^2)}{(1+q)^2(1-z)^3(1-qz)},$$

or, with $W(z) = (1-z)^2 V(z)$,

$$W(z) = W(qz) - \frac{2(1-q)z(1+(1+q)z+q^2z^2)}{(1+q)^2} + \frac{2}{(1+q)^2}\Big[\frac{1}{1-z} - \frac{1}{1-qz}\Big].$$

From this we see that (with $w_n := [z^n]W(z)$) for $n \geq 4$,

$$w_n = q^n w_n + \frac{2}{(1+q)^2}\big[1-q^n\big], \quad \text{or} \quad w_n = \frac{2}{(1+q)^2}.$$

Furthermore we see that $w_3 = \dfrac{2(1-q)}{(1+q)(1-q^3)}$ and $w_0 = w_1 = w_2 = 0$. Hence

$$V(z) = \frac{2(1-q)}{(1+q)(1-q^3)}\frac{z^3}{(1-z)^2} + \frac{2}{(1+q)^2}\frac{z^4}{(1-z)^3}$$

and

$$[z^n]V(z) = \frac{1}{(1+q)^2}(n-2)(n-3) + \frac{2(1-q)}{(1+q)(1-q^3)}(n-2).$$

Adding the expectation and subtracting the square of the expectation, we obtain the variance.

Theorem 2.4. *The variance of the number of weak descents is given for $n \geq 1$ by*

$$\frac{q(1-q)(1+q^3)}{(1+q)^3(1-q^3)}n - \frac{q(1-3q+q^2)}{(1+q)^2(1+q+q^2)}. \tag{7}$$

We note that this quantity is the same as in the instance of the strict descents.

3. Distribution of the number of descents

In this section we prove a central limit theorem for the distribution of the number of strict descents. In order to do this, we have to extract further information from the functional equation (2). We observe that the terms on the right-hand side are all simple rational functions, except for the terms containing $F(qz,u)$. From the definition of $F(z,u)$ it is clear that $F(z,u)$ can be written as

$$F(z,u) - 1 = \sum_{n\geq 1} z^n f_n(u)$$

for polynomials $f_n(u)$ with $\deg f_n = n$, whose coefficients are positive and ≤ 1. Therefore we have

$$|f_n(u)| \leq \frac{|u|^{n+1}-1}{|u|-1}, \quad \text{for } |u| > 1. \tag{8}$$

Using $q < 1$ and (8) we obtain that $F(qz,u)$ is holomorphic in $|z| < \frac{1}{\sqrt{q}}$, $|u| < \frac{1}{\sqrt{q}}$. Since for $u = z = 1$ we have

$$1 - (F(q,1)-1)\frac{p}{1-p} = 0, \quad \text{and} \quad \frac{\partial}{\partial z}\Big(1 - (F(qz,u)-1)\frac{pzu}{1-pz}\Big)\Big|_{z=1,u=1} \neq 0$$

there exists a function $f(u)$ holomorphic in a neighbourhood of $u = 1$ such that $z = f(u)^{-1}$ solves

$$1 - \big(F(qz,u) - 1\big)\frac{pzu}{1-pz} = 0$$

and satisfies $f(1) = 1$. Furthermore, $|f(e^{it})| < 1$ for $0 < |t| < \varepsilon$ for some $\varepsilon > 0$ by an application of Rouché's theorem. Thus we can write

$$F(z,u) - 1 = \frac{g(z,u)}{1 - f(u)z} + R(z,u), \tag{9}$$

where $g(z,u)$ and $R(z,u)$ are holomorphic in $|z| < 1+\delta$, $|u-1| < \delta$ for some $\delta > 0$. Now we are in the general framework of Hwang's quasi-power theorem (cf. [5]) and can deduce the following theorem.

Theorem 3.1. *The number of strict descents in words of length n produced by independent geometric random variables obeys a central limit law, more precisely*

$$\mathbb{P}\left(D_n(w) \le \frac{q}{1+q}n + t\sqrt{\frac{q(1-q)(1+q^3)}{(1+q)^3(1-q^3)}}\sqrt{n}\right) = \Phi(t) + \mathcal{O}(n^{-\frac{1}{2}}). \tag{10}$$

Similarly we can show

Theorem 3.2. *The number of weak descents in words of length n produced by independent geometric random variables obeys a central limit law, more precisely*

$$\mathbb{P}\left(D_n(w) \le \frac{1}{1+q}n + t\sqrt{\frac{q(1-q)(1+q^3)}{(1+q)^3(1-q^3)}}\sqrt{n}\right) = \Phi(t) + \mathcal{O}(n^{-\frac{1}{2}}). \tag{11}$$

Remarks In [12] Louchard and Prodinger find an explicit bivariate generating function for samples of geometric random variables with k strictly ascending runs. Since the number of weak descents is one less than the number of strictly ascending runs, we may deduce from their result that in the case of weak descents

$$F(z,u) = \frac{(1-u)P}{1-uP}, \quad \text{where} \quad P = \prod_{n\ge 0}(1 + pz(1-u)q^n). \tag{12}$$

To verify this directly one must show that (12) is the solution of the functional equation (5) that satisfies $F(0,u) = 1$. This follows easily by substituting (12) in (5) using $F(qz,u) = \frac{(1-u)P}{1+pz(1-u)-uP}$.

Similarly, in the case of strict descents one easily verifies that the solution of the functional equation (2) that satisfies $F(0,u) = 1$ is

$$F(z,u) = \frac{1-u}{P-u}, \quad \text{where now} \quad P = \prod_{n\ge 0}(1 - pz(1-u)q^n). \tag{13}$$

In both (12) and (13) we may let $q \to 1$ and obtain as expected the bivariate exponential generating function for the Eulerian numbers

$$A(z,u) = \frac{1-u}{e^{(u-1)z} - u}.$$

4. The position of the first descent

4.1. Strict descents

We use the following variation of the number of descents decomposition for the set of all words:

$$\{\geq \mathbf{1}\}^* = \mathbf{1}^*\{\geq \mathbf{2}\}^+\big(\mathbf{1}^+\{\geq \mathbf{2}\}^+\big)^*(\varepsilon + \mathbf{1}^+) + \mathbf{1}^*. \tag{14}$$

We consider a probability generating function $F(z,u)$, where z labels the number of random variables, and u marks the initial position of the first descent. In working with this decomposition it is convenient to adopt the convention that a nondecreasing word has a descent at its end.

$$F(z,u) = \frac{1}{1-pzu}[F(qz,u)-1]\frac{1}{1-[F(qz,1)-1]\dfrac{pz}{1-pz}}\frac{1}{1-pz} + \frac{1}{1-pzu}, \tag{15}$$

which simplifies to

$$F(z,u) = \frac{1}{1-pzu}[F(qz,u)-1]\frac{1-qz}{1-z} + \frac{1}{1-pzu}. \tag{16}$$

Now we differentiate it w.r.t. u, plug in $u=1$, set $G(z) = \frac{\partial}{\partial u}F(z,1)$, and get

$$G(z) = \frac{1}{1-z}G(qz)\frac{1-qz}{1-pz} + \frac{pz}{1-pz}.$$

We set $H(z) = (1-z)G(z)$ and get $(1-pz)H(z) = H(qz) + pz$. Comparing coefficients, we see that $h_n := [z^n]H(z)$ satisfies $h_1 = 1$ and for $n \geq 2$, $h_n - ph_{n-1} = q^n h_n$. By iterating this recurrence, $h_n = p^n/(q;q)_n$, with the notation $(a;q)_n = (1-a)(1-aq)\dots(1-aq^{n-1})$. Consequently,

$$H(z) = \sum_{n\geq 1}\frac{p^n z^n}{(q;q)_n} \quad \text{and} \quad G(z) = \frac{1}{1-z}\sum_{n\geq 1}\frac{p^n z^n}{(q;q)_n} = \frac{1}{1-z}\prod_{i\geq 0}\frac{1}{1-pzq^i},$$

by one of Euler's partition identities.

Our convention gives a contribution of n for the position of the first descent in each nondecreasing word, whereas 0 might be more natural. If we prefer the latter convention then we should subtract n times the probability $p^n/(q;q)_n$ of a nondecreasing word from $[z^n]G(z)$ to compute the average.

Theorem 4.1. *The average position of the first descent is given for $n \geq 1$ by*

$$\sum_{j=1}^{n}\frac{p^j}{(q;q)_j} - \frac{p^n}{(q;q)_n}n = \sum_{j=1}^{n}\frac{1}{[j]_q!} - \frac{n}{[n]_q!}. \tag{17}$$

(We used the notation $[n]_q! = (q;q)_n/p^n$.) Note that as $q \to 1$ we find that the average position of the first descent in a permutation equals $\sum_{j=1}^{n}\frac{1}{j!} - \frac{n}{n!} \to e-1$, as $n \to \infty$, in agreement with results of Knuth [11]. As $q \to 0$, we obtain a word of all $\mathbf{1}$'s and hence 0 for the average position of the first descent as expected.

Previously, using a different approach, Prodinger [15] studied the length of the r-th ascending run in a geometrically distributed word. The length of the first weakly (strictly) ascending run corresponds to the position of the first strict (weak) descent, respectively.

4.2. Weak descents

We use the same decomposition (14) for the set of all words.

We consider a probability generating function $F(z,u)$, where z labels the number of random variables, and u marks the initial position of the first weak descent. Again it is convenient to adopt a convention that a strictly increasing word has a weak descent at its end. We split up our decomposition into the cases of zero or one **1**, or two or more **1**'s. In the latter case a weak descent has already occurred with initial position equal to 1.

$$\begin{aligned}F(z,u) &= (1+pzu)\frac{[F(qz,u)-1]}{1-[F(qz,1)-1]\frac{pz}{1-pz}}\frac{1}{1-pz}\\ &\quad+\frac{p^2z^2u}{1-pz}\frac{[F(qz,1)-1]}{1-[F(qz,1)-1]\frac{pz}{1-pz}}\frac{1}{1-pz}+\Big(1+\frac{pzu}{1-pz}\Big),\end{aligned} \tag{18}$$

which simplifies to

$$F(z,u) = (1+pzu)[F(qz,u)-1]\frac{1-qz}{1-z}+\frac{p^2z^2u}{1-pz}\frac{qz}{1-z}+1+\frac{pzu}{1-pz}. \tag{19}$$

Now we differentiate it w.r.t. u, plug in $u=1$, set $G(z)=\frac{\partial}{\partial u}F(z,1)$, and get

$$G(z)=\frac{1}{1-z}\Big(G(qz)(1-qz)(1+pz)+pz\Big).$$

We set $H(z)=(1-z)G(z)$ and get $H(z)=H(qz)(1+pz)+pz$. By comparing coefficients we find that

$$H(z)=\sum_{n\ge1}\frac{p^nq^{\binom{n}{2}}z^n}{(q;q)_n}\quad\text{and}\quad G(z)=\frac{1}{1-z}\sum_{n\ge1}\frac{p^nq^{\binom{n}{2}}z^n}{(q;q)_n}=\frac{1}{1-z}\prod_{i\ge0}(1+pzq^i),$$

by another of Euler's partition identities.

Since our convention gives a contribution of n rather than 0 for the position of the first descent in each increasing word, we subtract n times the probability $p^nq^{\binom{n}{2}}/(q;q)_n$ of a strictly increasing word from $[z^n]G(z)$ to compute the average.

Theorem 4.2. *The average position of the first descent is given for $n\ge1$ by*

$$\sum_{j=1}^{n}\frac{p^jq^{\binom{j}{2}}}{(q;q)_j}-\frac{p^nq^{\binom{n}{2}}}{(q;q)_n}n=\sum_{j=1}^{n}\frac{q^{\binom{j}{2}}}{[j]_q!}-\frac{nq^{\binom{n}{2}}}{[n]_q!}. \tag{20}$$

Note that as $q\to1$ we find again that the average position of the first descent in a permutation equals $\sum_{j=1}^{n}\frac{1}{j!}-\frac{n}{n!}\to e-1$, as $n\to\infty$. As $q\to0$, we obtain a word of all **1**'s and hence 1 for the average position of the first weak descent as expected.

5. The height of the first descent

5.1. Strict descents

We stay with decomposition (14) and consider a probability generating function $F(z,u)$, where now u marks the initial height of the first descent. Once again we adopt the convention that a nondecreasing word has a descent at its end.

$$F(z,u)=\frac{u}{1-pz}[F(qz,u)-1]\frac{1}{1-[F(qz,1)-1]\frac{pz}{1-pz}}\frac{1}{1-pz}+\Big(1+\frac{pzu}{1-pz}\Big), \tag{21}$$

which simplifies to

$$F(z,u)=\frac{u}{1-pz}[F(qz,u)-1]\frac{1-qz}{1-z}+1+\frac{pzu}{1-pz}. \tag{22}$$

Now we differentiate it w.r.t. u, plug in $u=1$, set $G(z)=\frac{\partial}{\partial u}F(z,1)$, and get

$$G(z)=\frac{1}{1-z}G(qz)\frac{1-qz}{1-pz}+\frac{z}{1-z}.$$

We set $H(z)=(1-z)G(z)$ and get $(1-pz)H(z)=H(qz)+z(1-pz)$. This leads to

$$H(z)=z+\frac{q}{p}\sum_{n\ge1}\frac{p^nz^n}{(q;q)_n}\quad\text{and}\quad G(z)=\frac{z}{1-z}+\frac{q}{p}\frac{1}{1-z}\sum_{n\ge1}\frac{p^nz^n}{(q;q)_n}.$$

Theorem 5.1. *The average initial height of the first descent is given for $n\ge1$ by*

$$\frac{q}{p}\sum_{j=1}^{n}\frac{p^j}{(q;q)_j}+1=\frac{q}{p}\sum_{j=1}^{n}\frac{1}{[j]_q!}+1. \tag{23}$$

As $q\to1$ the sum tends to infinity indicating that the average height of the first descent in a permutation grows as a function of the size of the permutation. As $q\to0$, we obtain a word of all **1**'s and hence 1 for the average height of the first strict descent as expected from our convention that such a word has a strict decent at the end.

5.2. Weak descents

Again we use decomposition (14) and consider a probability generating function $F(z,u)$, where now u marks the initial height of the first weak descent. Once again we adopt the convention that a strictly increasing word has a descent at its end.

We split up our decomposition into the cases of zero or one **1**, or two or more **1**'s. In the latter case a weak descent has already occurred with initial height equal to 1:

$$\begin{aligned}F(z,u)&=u(1+pz)\frac{[F(qz,u)-1]}{1-[F(qz,1)-1]\frac{pz}{1-pz}}\frac{1}{1-pz}\\&\quad+\frac{p^2z^2u}{1-pz}\frac{[F(qz,1)-1]}{1-[F(qz,1)-1]\frac{pz}{1-pz}}\frac{1}{1-pz}+1+\frac{pzu}{1-pz}.\end{aligned} \tag{24}$$

Again we simplify, differentiate w.r.t. u, plug in $u=1$, set $G(z)=\frac{\partial}{\partial u}F(z,1)$, and get

$$G(z)=\frac{1}{1-z}\Big(G(qz)(1-qz)(1+pz)+z\Big).$$

We set $H(z)=(1-z)G(z)$ and get $H(z)=H(qz)(1+pz)+z$. This leads to

$$H(z)=\frac{1}{p}\sum_{n\ge1}\frac{p^nq^{\binom{n}{2}}z^n}{(q;q)_n}\quad\text{and}\quad G(z)=\frac{1}{p}\frac{1}{1-z}\sum_{n\ge1}\frac{p^nq^{\binom{n}{2}}z^n}{(q;q)_n}.$$

Theorem 5.2. *The average initial height of the first weak descent is given for $n \geq 1$ by*

$$\frac{1}{p}\sum_{j=1}^{n}\frac{p^j q^{\binom{j}{2}}}{(q;q)_j} = \frac{1}{p}\sum_{j=1}^{n}\frac{q^{\binom{j}{2}}}{[j]_q!}. \tag{25}$$

As $q \to 1$ we again find that the average height of the first descent in a permutation grows with n. As $q \to 0$, we obtain a word of all **1**'s and hence obtain 1 for the average height of the first weak descent as expected.

6. The maximum size of a descent

We study the following parameter X:

$$X := \max\{x_i - x_{i+1} \mid 1 \leq i < n \ \&\ x_i \geq x_{i+1}\}; \tag{26}$$

it is consistent to assign the value 0 if no pair of consecutive letters satisfies the condition. It is more convenient to think from right to left, and we may still call it X; the statistics are the same:

$$X := \max\{x_{i+1} - x_i \mid 1 \leq i < n \ \&\ x_i \leq x_{i+1}\}.$$

Fix a parameter h and consider all words with $X \leq h$. That is, that for all $1 \leq i < n$, we must have that $x_{i+1} \in \{1, 2, \ldots, x_i + h\}$.

The adding-a-new-slice technique works here: Let $f_n(u)$ be such that $[u^i]f_n(u)$ is the probability that a word of length n has $X \leq h$ and last letter equal to i. Then $f_1(u) = \sum_{i\geq 1} pq^{i-1}u^i = \frac{pu}{1-qu}$, and we have the replacement rule

$$u^i \longrightarrow pu + pqu^2 + \cdots + pq^{i+h-1}u^{i+h} = pu\frac{1-(qu)^{i+h}}{1-qu}.$$

Consequently,

$$f_{n+1}(u) = \frac{pu}{1-qu}f_n(1) - \frac{pq^h u^{h+1}}{1-qu}f_n(qu). \tag{27}$$

Now define $F(u) = F(z,u) = \sum\limits_{n\geq 1} z^n f_n(u)$. Then, upon summing,

$$\sum_{n\geq 1} z^{n+1} f_{n+1}(u) = \frac{pu}{1-qu}\sum_{n\geq 1} z^{n+1} f_n(1) - \frac{pq^h u^{h+1}}{1-qu}\sum_{n\geq 1} z^{n+1} f_n(qu)$$

or

$$F(z,u) - zf_1(u) = \frac{puz}{1-qu}F(z,1) - \frac{pq^h u^{h+1} z}{1-qu}F(z,qu).$$

Now we iterate the functional equation

$$F(u) = \frac{puz}{1-qu} + \frac{puz}{1-qu}F(1) - \frac{pq^h u^{h+1} z}{1-qu}F(qu). \tag{28}$$

Setting $u = 1$ and collecting terms:

$$F(1) = \sum_{k\geq 1}\frac{(-1)^{k-1}p^k q^{\binom{k}{2}(h+1)} z^k}{(q)_k} + F(1)\sum_{k\geq 1}\frac{(-1)^{k-1}p^k q^{\binom{k}{2}(h+1)} z^k}{(q)_k}.$$

It is better now to include the empty word and thus work with

$$1+F(1)=\Big(1-\sum_{k\ge1}\frac{(-1)^{k-1}p^kq^{\binom{k}{2}(h+1)}z^k}{(q)_k}\Big)^{-1}. \tag{29}$$

We must study the dominant zeros ρ_h of

$$1-\sum_{k\ge1}\frac{(-1)^{k-1}p^kq^{\binom{k}{2}(h+1)}z^k}{(q)_k}=1-\sigma_h(z)=0. \tag{30}$$

Bootstrapping is the appropriate technique in this instance:

$$1-z+\frac{q^{h+1}z^2}{1+q}\sim 0.$$

Set $\rho_h=1+\varepsilon_h$, then $\varepsilon_h\sim\frac{q^{h+1}}{1+q}$. We have $1+F(1)\sim\dfrac{A_h}{1-z/\rho_h}$ with

$$A_h=\frac{1}{\rho_h\sigma'(\rho_h)}\sim 1\Big/\sum_{k\ge1}\frac{(-1)^{k-1}kp^kq^{\binom{k}{2}(h+1)}}{(q)_k}. \tag{31}$$

So the probability that $X\le h$ is approximated by

$$A_h\rho^{-n}\sim A_h\Big(1-\frac{q^{h+1}}{1+q}\Big)^n\sim A_h\exp\Big(-\frac{q^{h+1}n}{1+q}\Big).$$

As one can check, $A_h\sim 1$, and thus we may work with $\exp(-\frac{q^{h+1}n}{1+q})$.

We now apply the Mellin transform (cf. [2]) to the function $f(t)=\sum_{h\ge0}(1-\exp(-t\frac{q^{h+1}}{1+q}))$. This yields the transformed function

$$f^*(s)=\Gamma(s)\frac{(1+q)^s}{1-q^s},\quad\text{for }-1<\Re s<0.$$

Application of the Mellin inversion formula, shifting the line of integration to the right and collecting (negative) residues yields

$$\begin{aligned}f(t)&=\frac{1}{2\pi i}\int_{-\frac12-i\infty}^{-\frac12+i\infty}\Gamma(s)\frac{(1+q)^s}{1-q^s}t^{-s}\,ds\\&=-\sum_{k\in\mathbb{Z}}\operatorname{Res}f^*(s)t^{-s}\big|_{s=\frac{2k\pi i}{L}}-\frac{1}{2\pi i}\int_{\frac12-i\infty}^{\frac12+i\infty}\Gamma(s)\frac{(1+q)^s}{1-q^s}t^{-s}\,ds.\end{aligned} \tag{32}$$

The residues are easily computed to be

$$-\operatorname{Res}\Gamma(s)\frac{(1+q)^s}{1-q^s}t^{-s}\Big|_{s=0}=\log_Q t+\frac{\gamma}{L}-\frac{\log(Q+1)}{L}+\frac12, \tag{33}$$

$$-\operatorname{Res}\Gamma(s)\frac{(1+q)^s}{1-q^s}t^{-s}\Big|_{s=\frac{2k\pi i}{L}}=-\frac1L\Gamma\Big(\frac{2k\pi i}{L}\Big)\Big(\frac{t}{Q+1}\Big)^{-\frac{2k\pi i}{L}}. \tag{34}$$

This yields

Theorem 6.1. *The mean value of the size of the greatest descent $M_n(w)$ in a string of n geometric random variables satisfies*

$$\mathbb{E}M_n(w) \sim \log_Q n + \frac{\gamma}{L} - \frac{\log(Q+1)}{L} + \frac{1}{2} - \delta\Big(\log_Q\Big(\frac{n}{Q+1}\Big)\Big), \tag{35}$$

where $\delta(x)$ is a periodic function of mean zero, period one and small magnitude, with Fourier series

$$\delta(x) = \frac{1}{L}\sum_{k\neq 0}\Gamma\Big(-\frac{2k\pi i}{L}\Big)e^{2k\pi i x}.$$

References

[1] L. Devroye. A limit theory for random skip lists. *Advances in Applied Probability*, 2:597–609, 1992.

[2] P. Flajolet, X. Gourdon, and P. Dumas. Mellin transforms and asymptotics: Harmonic sums. *Theoretical Computer Science*, 144:3–58, 1995.

[3] P. Flajolet and G. N. Martin. Probabilistic counting algorithms for data base applications. *Journal of Computer and System Sciences*, 31:182–209, 1985.

[4] P. J. Grabner, A. Knopfmacher and H. Prodinger. Combinatorics of geometrically distributed random variables: Run statistics. *Theoret. Comp. Sci.*, 297: 261–270, 2003.

[5] H.-K. Hwang. On convergence rates in the central limit theorems for combinatorial structures. *European Journal of Combinatorics*, 19:329–343, 1998.

[6] P. Kirschenhofer and H. Prodinger. On the analysis of probabilistic counting. In E. Hlawka and R. F. Tichy, editors, *Number-theoretic Analysis*, volume 1452 of *Lecture Notes in Mathematics*, pages 117–120, 1990.

[7] P. Kirschenhofer and H. Prodinger. A result in order statistics related to probabilistic counting. *Computing*, 51:15–27, 1993.

[8] P. Kirschenhofer and H. Prodinger. The path length of random skip lists. *Acta Informatica*, 31:775–792, 1994.

[9] P. Kirschenhofer, H. Prodinger, and W. Szpankowski. Analysis of a splitting process arising in probabilistic counting and other related algorithms. *Random Structures and Algorithms*, 9:379–401, 1996.

[10] A. Knopfmacher and H. Prodinger. Combinatorics of geometrically distributed random variables: Value and position of the rth left-to-right maximum. *Discrete Math.*, 226:255-267, 2001.

[11] D. E. Knuth. *The art of computer programming*, volume 3: Sorting and Searching. Addison-Wesley, 1973. Second edition, 1998.

[12] G. Louchard and H. Prodinger. Ascending runs of sequences of geometrically distributed random variables: a probabilistic analysis. *Theoretical Computer Science*, 304:59–86, 2003.

[13] T. Papadakis, I. Munro, and P. Poblete. Average search and update costs in skip lists. *BIT*, 32:316–332, 1992.

[14] H. Prodinger. Combinatorics of geometrically distributed random variables: Left-to-right maxima. *Discrete Mathematics*, 153:253–270, 1996.

[15] H. Prodinger. Combinatorics of geometrically distributed random variables: Length of ascending runs. Proceedings of LATIN'2000, *Lecture Notes in Comp. Sci.* 1776, Gonnet, Panario and Viola (Eds.), 473–482, 2000.

[16] W. Pugh. Skip lists: a probabilistic alternative to balanced trees. *Communications of the ACM*, 33:668–676, 1990.

Arnold Knopfmacher
The John Knopfmacher Centre for Applicable Analysis and Number Theory
University of the Witwatersrand, P. O. Wits
2050 Johannesburg, South Africa
arnoldkn@cam.wits.ac.za

Helmut Prodinger
The John Knopfmacher Centre for Applicable Analysis and Number Theory
Department of Mathematics
University of the Witwatersrand, P. O. Wits
2050 Johannesburg, South Africa
helmut@maths.wits.ac.za

Trends in Mathematics, © 2004 Birkhäuser Verlag Basel/Switzerland

Large Deviations for Cascades and Cascades of Large Deviations

Alain Rouault

ABSTRACT: *In a Mandelbrot's multiplicative cascade on* $[0,1]$, *let* r *be the number of cells, and* Z_r^n *the mass of the measure at the height* n *in the* r*-ary tree. Most of the known results deal with the limit* $n \to \infty$ *with* r *fixed. In some previous papers, we began the study of* $r \to \infty$ *when* n *is fixed, showing a law of large numbers and describing a new large deviations phenomenon when the cascade generator is bounded. Actually, there are self-similar rate functions at geometrical scales, reflecting the multiplicative structure of the process. Here we describe the behavior of a typical cascade, knowing that* Z_r^n *is given different from its mean* 1. *The main result is a Sanov type theorem, at different rates, leading to a Gibbs conditioning principle.*

1. Introduction

Cascades are a very popular mathematical model for rainfall, internet packet traffic, market prices, etc. The literature is now very broad since the foundation (Kolmogorov [12], Mandelbrot [17], until more recent improvements. Extensive bibliographies are in Barral [1] or Liu [13], and Ossiander and Waymire [18] for statistical aspects.

Let $W \geq 0$ a random variable of mean 1. The *cascade generators* are given by a family $\{W_{\mathbf{i}}\}$ of independent copies of W, indexed by the set of all finite sequences $\mathbf{i} = i_1 \cdots i_n, n \geq 1$ of positive integers. We are interested in the random variable

$$Z_r^n = r^{-n} \sum_{1 \leq i_1, \ldots, i_n \leq r} W_{i_1} W_{i_1 i_2} \cdots W_{i_1 \cdots i_n}$$

which is the mass of the cascade at height n. Let $\mathfrak{F}_n$ be the σ-field generated by all the $W_{\mathbf{i}}$ for those $\mathbf{i}$ of height less or equal to n. For r fixed, $\{Z_r^n\}_n$ is a $\{\mathfrak{F}_n\}_n$ nonnegative martingale. Let Z_r^∞ be its a.s limit when $n \to \infty$. It is a.s. strictly positive if $\log r > EW \log W$ (Kahane-Peyrière [11]).

Another asymptotics consists in letting $r \to \infty$ with n fixed (finite or infinite). For $n = 1$ it is the classical sum of i.i.d. random variables. For $n \geq 2$, the summands are dependent: the terms corresponding to $\mathbf{i}$ and $\mathbf{j}$ have in their products a number of common random variables equal to the height of the last common ancestor $\mathbf{i} \wedge \mathbf{j}$ of $\mathbf{i}$ and $\mathbf{j}$. Notice that it is also possible to write $Z_r^n = r^{-1} \sum_{k=1}^r W_k \ (Z_r^{n-1} \circ \Theta_k)$ a.s., where Θ_k is the tree shifted at the node of label k, and Z_r^n appears as a sum of r i.i.d. random variables, with a distribution depending on r. In [15], Liu and Rouault and in [14], Liu et al. extended results known for $n = 1$, in particular the law of large numbers. For $n \leq \infty$ fixed, they proved that $\lim_{r \to \infty} Z_r^n = 1$ a.s. (remember $EW = 1$). It means roughly that only the first level contributes to the limit. It is also true for other types of results (CLT, LIL).

In the study of large deviations, a new phenomenon occurs under the assumption

$$\overline{w} := \text{ess sup } W < \infty . \tag{1}$$

Let Λ be the cumulant generating function:

$$\Lambda(\theta) = \log \mathfrak{E} e^{\theta W} \quad , \ \theta \in \mathfrak{R},$$

and Λ^* be its Legendre dual:

$$\Lambda^*(y) = \sup_{\theta \in \mathfrak{R}} \theta y - \Lambda(\theta) \quad , \ y \in \mathfrak{R}.$$

It is well known that Λ^* is finite for $y \in (\underline{w}, \overline{w})$ where $\underline{w} :=$ ess inf W, and infinite for $y \notin [\underline{w}, \overline{w}]$.

In [15] it is proved that the family of distributions of Z_r^∞ satisfies the following tail estimates, at different rates. Roughly, every level contributes to large deviations, at an appropriate scale and appropriate rate, with a kind of self-similarity, justifying the title of the present paper.

Theorem 1.1. *([15]) If assumption (1) holds, then for any $k \geq 0$ and any $x \in (\overline{w}^k, \overline{w}^{k+1}]$,*

$$\lim_{r\to\infty} r^{-(k+1)} \log \mathfrak{P}(Z_r^\infty \geq x) = -\Lambda^*(x\overline{w}^{-k}) .$$

Actually, at intermediate levels (n fixed finite) we have a similar result (see *i)* below). The result for the first range $(1, \overline{w})$ (corresponding to $k = 0$) may be extended to the case $\overline{w} = \infty$ under the assumption

$$\forall \tau > 0 \quad \mathfrak{E} \exp \tau W < \infty , \tag{2}$$

(see *ii)* below). The proof of *i)* is similar to the proof of [15] Theorem 1.4. We could also use [16]. Part *ii)* is a consequence of [14].

Proposition 1.2. i) *If assumption (1) holds, then for any $0 \leq k < n$ and $x \in (\overline{w}^k, \overline{w}^{k+1}]$,*

$$\lim_{r\to\infty} r^{-(k+1)} \log \mathfrak{P}(Z_r^n \geq x) = -\Lambda^*(x\overline{w}^{-k}).$$

ii) *If assumption (2) holds, then for any $x > 1$,*

$$\lim_{r\to\infty} r^{-1} \log \mathfrak{P}(Z_r^n \geq x) = -\Lambda^*(x) .$$

It should be clear that the same results hold when one considers left tails, i.e. $\mathfrak{P}(Z_r^n \leq x)$ for $x < 1$. For the analog of *i)* we put no assumption, and for *ii)* we assume $\underline{w} > 0$.

A quite natural question is then the description of a typical cascade of branching number r (large), of mass approximately $a \neq 1$. We use large deviations techniques for which we recall some definitions.

Let $a_r, r \in \mathfrak{N}$ an increasing sequence of positive real numbers with $\lim_{r\to\infty} a_r = \infty$. We say that a sequence of probability measures $(\mathcal{P}_r)$ on a regular Hausdorff space $(\mathcal{X}, \mathcal{B}(\mathcal{X}))$ satisfies a Large Deviation Principle (LDP) with rate function I at scale a_r if I is a lower semi-continuous function $I : \mathcal{X} \to [0, \infty]$ such that

a) For any closed subset F of $\mathcal{X}$

$$\limsup_{r\to\infty} \frac{1}{a_r} \log \mathcal{P}_r(F) \leq - \inf_{x\in F} I(x)$$

b) For any open subset G of $\mathcal{X}$

$$\limsup_{r\to\infty} \frac{1}{a_r} \log \mathcal{P}_r(G) \leq - \inf_{x\in G} I(x)\,.$$

The rate function is good if for every c the level set $\{x : I(x) \leq c\}$ is compact. The sequence $(\mathcal{P}_r)$ is said to be exponentially tight at scale a_r if for every $L > 0$ there is a compact set K_L such that $\limsup_r \frac{1}{a_r} \log \mathcal{P}_r(K_L^c) < -L$.

As a first answer to the question of typical cascade, our main result (Theorem 4.2) is a Gibbs conditioning principle. If the cascade generator W satisfies (1) , then for $a \in (\bar{w}^{k-1}, \bar{w}^k)$, and n fixed, the distribution of $(W_{i_1}, \cdots, W_{i_1\cdots i_n})$ conditioned on $Z_r^n \in [a-\eta, a+\eta]$ converges weakly to $\delta_{\bar{w}}^{\otimes(k-1)} \otimes Q^\theta \otimes Q^{\otimes(n-k)}$ as $r \to \infty$ followed by $\eta \to 0$. Here δ_c denotes the Dirac mass at c, Q^θ is obtained by tilting Q to get mean $a\bar{w}^{-(k-1)}$ instead of mean 1. It is a consequence of a Sanov type theorem (Theorem 2.1) giving the LDP for the empirical distribution:

$$\mathcal{L}_r^n = \frac{1}{r^n} \sum_{1\leq i_1,\dots,i_n\leq r} \delta_{W_{i_1},\dots,W_{i_1\cdots i_n}}\,. \tag{3}$$

This empirical distribution is a random measure on $[0,\infty)^n$. In Section 3, we use the contraction principle to study again the masses Z_r^n, and in Section 4, we deduce the typical behavior along a fixed branch. It would be interesting to have a more general Gibbs principle, taking into account thè limiting distribution along a fixed number (> 1) of branches, like in Corollary 7.3.5 of [4].

The structure of our problem[1] leads us to conditioning at stage n upon the σ-field $\mathcal{F}_{n-1}$, and use induction. In particular, we have to argue about conditional LDP, or in other words, we have to manage mixture of distributions which satisfy LDP. Various authors have studied this topic, Dinwoodie-Zabell [6] for exchangeable vectors (and more recently Trashorras [20]), Chaganthy [3] for statistical purposes, Grunwald for statistical mechanics ([10] section 2), Finnoff for evolutionary games [9], and Biggins for random graphs in [2]. Here, we have tried an (almost) self-contained treatment of this question adapted to our model.

Let us give some notations used in the sequel. First, Σ is a Polish space in Section 2, and is $\mathfrak{R}^+$ in the end of Section 3 (Corollary 3.2) and in Section 4. We denote by $bm(\Sigma)$ the set of bounded Borel functions from Σ to $\mathfrak{R}$. Let $\mathcal{M}_1(\Sigma)$ be the set of probability measures on Σ equipped with the weak convergence denoted by $\Rightarrow$. If $f \in bm(\Sigma)$ and $\nu \in \mathcal{M}_1(\Sigma)$ we denote as usual,

$$< f, \nu > = \int_\Sigma f d\nu\,.$$

The relative entropy or Kullback information between two probability measures α and β in $\mathcal{M}_1(\Sigma)$ is defined by:

$$H(\alpha\|\beta) = \int_\Sigma \left(\log \frac{d\alpha}{d\beta}\right) d\alpha$$

whenever α is absolutely continuous with respect to β. Otherwise $H(\alpha\|\beta) = +\infty$ (see [4] Appendix D3). We will use the following variational formula ([7] Lemma 1.4.3 p.36):

$$H(\alpha\|\beta) = \sup\{< \varphi, \alpha > - \log < e^\varphi, \beta > :\ \varphi \in \mathcal{C}_b(\Sigma)\} \tag{4}$$

[1]This model involves a dependence between strings of variables different from that of U-empirical measures (see [8]).

For $\nu \in \mathcal{M}_1(\Sigma^n)$ and $1 \le j \le n$, let ν_j be the projection of ν on the first j coordinates, and let $\pi\nu = \nu_{n-1}$.

The random variable W is Σ valued, and its distribution is denoted by Q. In Section 3 and 4, we assume that $\mathfrak{E}W = 1$. For $j \ge 1$ let $Q^j = Q^{\otimes j}$ (and $\nu \otimes Q^0 = \nu$ for every measure ν).

For $f : \Sigma^n \to \mathfrak{R}$ let $\pi f = \mathfrak{E} f(\cdot, W)$ or in other words $\pi f(\mathbf{x}) = \int_\Sigma f(\mathbf{x}, y) Q(dy)$.

2. A Sanov type theorem

In the whole section we fix $n < \infty$. We consider $\mathfrak{L}_r^n$ as (random) element of $\mathcal{M}_1(\Sigma^n)$ equipped with the weak topology. Let S be the support of Q.

Theorem 2.1. a) *The family $\{\mathfrak{P}(\mathfrak{L}_r^n \in .)\}_r$ satisfies the LDP at scale r with good rate function*

$$\begin{aligned} \mathfrak{I}_1^n(\nu) &= H(\nu_1 \parallel Q) \text{ if } \nu = \nu_1 \otimes Q^{n-1} \\ &= +\infty \quad \text{otherwise}. \end{aligned} \tag{5}$$

b) *If S is compact, then for every $2 \le k \le n$, the family $\{\mathfrak{P}(\mathfrak{L}_r^n \in .)\}_r$ satisfies the LDP in $\mathcal{M}_1(S^n)$ at scale r^k with rate function*

$$\begin{aligned} \mathfrak{I}_k^n(\nu) &= H(\nu_k \parallel \nu_{k-1} \otimes Q) \text{ if } \nu = \nu_k \otimes Q^{n-k} \\ &= +\infty \quad \text{otherwise}. \end{aligned} \tag{6}$$

For $n = 1$, a) is the classical Sanov theorem ([4] p.261). For $n > 1$, our strategy uses conditioning, LDP for conditional distributions and induction. As usual, we need exponential tightness, provided by the following lemma. Its proof involves some important formulas used in the sequel.

Lemma 2.2. a) *If $n \ge 1$, $\{\mathfrak{P}(\mathfrak{L}_r^n \in .)\}_r$ is exponentially tight at scale r.*
b) *For $n \ge 2$, and for every sequence $\{\xi_r\}_r \in \mathcal{M}_1(\Sigma^{n-1})$ such that $\xi_r \Rightarrow \xi$, $\{\mathfrak{P}(\mathfrak{L}_r^n \in \cdot \,|\, \mathfrak{L}_r^{n-1} = \xi_r)\}_r$ is exponentially tight at scale r.*

To prove this lemma, we use a slightly modified version of the criteria of exponential tightness of Deuschel-Stroock Lemma 3.2.7 p.67 of [5].

Lemma 2.3. *If $(\gamma_r)_r$ is a family of random probability measures on a Polish space $\mathcal{Y}$, then the family $\{\mathfrak{P}(\gamma_r \in \cdot)\}_r$ is exponentially tight at scale a_r as soon as there is a tight family of probability measures $(\rho_r)_r$ such that for every $\varphi \in bm(\mathcal{Y})$*

$$\mathfrak{E} \exp < \varphi, a_r \gamma_r > \le < \exp \varphi, \rho_r >^{a_r} . \tag{7}$$

Proof of Lemma 2.3: We give a construction similar to that of Deuschel-Stroock's (see also Dembo-Zeitouni's Lemma 6.2.6 [4]). For every $L \ge 1$, we have to find a compact set $C_L \in \mathcal{M}_1(\mathcal{Y})$ such that for every $r \ge 1$

$$\mathfrak{P}(\gamma_r \notin C_L) \le 2 \exp -L a_r . \tag{8}$$

For $\ell \ge 1$ let $\epsilon_\ell = e^{-\ell^2} 2^{-\ell}$. Since $\{\rho_r\}_r$ is tight, one can find a compact K_ℓ such that $\rho_r(K_\ell) \ge 1 - \epsilon_\ell$ for every r. We claim that $C_L = \cap_{\ell \ge L} \{\nu : \nu(K_\ell) \ge 1 - \ell^{-1}\}$ suits. Thanks to the Portmanteau theorem, the set C_L is closed; since it is tight, it is compact by Prohorov's theorem. We have for $f_\ell = \ell^2 + \ell \log 2$ on K_ℓ^c and 0 on K_ℓ

$$\mathfrak{P}(\gamma_r(K_\ell) < 1 - \ell^{-1}) \le \mathfrak{E}(\exp < f_\ell, a_r \gamma_r >) (2e^\ell)^{-a_r} .$$

Now, from (7) $\mathfrak{E}(\exp < f_\ell, a_r\gamma_r >) \le < \exp f_\ell, \rho_r >^{a_r}$ and from the definition of K_ℓ,

$$< \exp f_\ell, \rho_r >= \rho_r(K_\ell) + e^{\ell^2} 2^\ell (1 - \rho_r(K_\ell)) \le 2 .$$

With the three last displays and the union of events bound we get $\mathfrak{P}(\gamma_r \notin C_L) \le \sum_{\ell \ge L} e^{-\ell r^n}$, which yields (8). □

Proof of Lemma 2.2: For $n = 1$, the statement a) is classical ([4] Lemma 6.2.6). We assume from now $n > 1$ and S is unbounded (otherwise it is trivial). Let us begin with a display often used in the sequel. If $h \in bm(\Sigma^n)$, we have

$$\mathfrak{E}\big(\exp < r^n h, \mathcal{L}_r^n > \mid \mathcal{F}_{n-1}\big) = \exp < r^n \tilde{h}, \mathcal{L}_r^{n-1} > , \tag{9}$$

where $\tilde{h}(x_1, x_2, ..., x_{n-1}) = \log \mathfrak{E} \exp h(x_1, x_2, ..., x_{n-1}, W)$, and in particular

$$r^{-n} \log \mathfrak{E}\left(\exp < r^n h, \mathcal{L}_r^n > \mid \mathcal{L}_r^{n-1} = \mu\right) = < \log \mathfrak{E} \exp h(\cdot, W), \mu > . \tag{10}$$

a) Let us prove by induction on n that for every $\varphi \in bm(\Sigma^n)$

$$\mathfrak{E} \exp < \varphi, r\mathcal{L}_r^n > \le \{\mathfrak{E} \exp \varphi(W_1, .., W_n)\}^r , \tag{11}$$

which is (7) with $\rho_r = Q^{\otimes n}$, $a_r = r$. For $n = 1$, we have actually an equality. Assume that (11) holds at level $n - 1$. From (9)

$$\mathfrak{E} \exp < \varphi, r\mathcal{L}_r^n >= \mathfrak{E} \exp < \tilde{\varphi}, r\mathcal{L}_r^{n-1} >$$

with $\tilde{\varphi}(x) = r^{n-1} \log \mathfrak{E} \exp\left(r^{1-n}\, \varphi(x, W)\right)$. We apply (11) to $\tilde{\varphi} \in bm(\Sigma^{n-1})$:

$$\mathfrak{E} \exp < \tilde{\varphi}, r\mathcal{L}_r^{n-1} > \le \left[\mathfrak{E} \exp \tilde{\varphi}(W_1, \cdots, W_{n-1})\right]^r \tag{12}$$

$$= \left(\mathfrak{E}\left[\left(\mathfrak{E}\{\exp r^{1-n}\varphi(W_1, \cdots, W_n) | W_1, \cdots, W_{n-1}\}\right)^{r^{n-1}}\right]\right)^r$$

$$\le \left(\mathfrak{E}\left[\mathfrak{E}\{\exp \varphi(W_1, \cdots, W_n) | W_1, \cdots, W_{n-1}\}\right]\right)^r \tag{13}$$

where the last display comes from Jensen's inequality with the convex function $x \mapsto x^{r^{n-1}}$. This yields (11) at level n.

b) Let $\varphi \in bm(\Sigma^n)$. Applying (10) with $h = r^{1-n}\varphi$ we get

$$\begin{aligned} r^{-1} \log \mathfrak{E}\big(\exp < \varphi, r\mathcal{L}_r^n > |\mathcal{L}_r^{n-1} = \xi_r\big) &= r^{n-1} < \log \mathfrak{E} \exp r^{1-n}\varphi(., W), \xi_r > \\ &\le r^{n-1} \log < \mathfrak{E} \exp r^{1-n}\varphi(., W), \xi_r > \\ &\le \log < \mathfrak{E} \exp \varphi(., W), \xi_r > \end{aligned}$$

where we use twice Jensen's inequality, (concavity of logarithm and convexity of $x \mapsto x^{r^{n-1}}$). If $\xi_r \Rightarrow \xi$, the family $(\xi_r \otimes Q)_r$ is tight, so we have (7) with $\rho_r = \xi_r \otimes Q$, $a_r = r$. □

The large deviations for conditional distribution are ruled by the following lemma. Let $\chi(a|b) = 0$ if $a = b$ and $\chi(a|b) = \infty$ if $a \ne b$. For $\xi \in \mathcal{M}_1(\Sigma^{n-1})$ and $f \in bm(\Sigma^n)$ let

$$\lambda_\xi(f) := < \log \mathfrak{E} \exp f(\cdot, W) \ , \ \xi > .$$

Lemma 2.4. *Let us assume* $\xi_r \Rightarrow \xi$.

a) *For every* $k < n$, *the family* $\{\mathfrak{P}\left(\mathcal{L}_r^n \in \cdot \mid \mathcal{L}_r^{n-1} = \xi_r\right)\}_r$ *satisfies the LDP at scale* r^k *with good rate function* $\chi(. \mid \xi \otimes Q)$.

b) *The family* $\{\mathfrak{P}(\mathcal{L}_r^n \in \cdot \mid \mathcal{L}_r^{n-1} = \xi_r)\}_r$ *satisfies the LDP at scale* r^n *with good rate function*

$$\mathfrak{J}_\xi(\nu) := \sup\{< f, \nu > -\lambda_\xi(f) \; ; \; f \in bm(\Sigma^n)\} \,. \tag{14}$$

Moreover

$$\begin{aligned} \mathfrak{J}_\xi(\nu) &= H(\nu \parallel \xi \otimes Q) \;\; \textit{if}\; \nu_{n-1} = \xi \\ &= +\infty \;\; \textit{otherwise.} \end{aligned} \tag{15}$$

Proof of Lemma 2.4: From Lemma 2.2 our family is exponentially tight at scale r hence at scale r^k, for every $k \geq 1$. Owing to Corollary 4.5.27 of Baldi's theorem ([4] p.160), it is then sufficient to study the limits of the cumulant generating functionals. Let us first notice two well known limits used in the sequel. For $g \in bm(\Sigma)$

$$\lim_{a\to 0} a^{-1} \log \mathfrak{E} \exp ag(W) = \mathfrak{E} g(W) \tag{16}$$

$$\lim_{a\to\infty} a^{-1} \log \mathfrak{E} \exp ag(W) = \operatorname{essup} g(W)\,. \tag{17}$$

a) From (9), we have for $k < n$

$$\begin{aligned} r^{-k} \log \mathfrak{E}\big(\exp < r^k f, \mathcal{L}_r^n > \mid \mathcal{L}_r^{n-1} = \xi\big) = \\ = < r^{n-k} \log \mathfrak{E} \exp \left[r^{k-n} f(\cdot, W)\right] \,,\, \xi > \end{aligned}$$

and its limit, as $r \to \infty$ is $< f, \xi \otimes Q > =: < \pi f, \xi >$ (see (16). Moreover it is possible to replace in the above display ξ by a sequence $\xi_r \Rightarrow \mu$ without change in the conclusion. We get the LDP with rate function

$$\nu \mapsto \sup\{< f, \nu > - < f, \mu \otimes Q > \; ; \; f \in bm(\Sigma^n)\} = \chi(. \mid \xi \otimes Q)\,.$$

b) From (10)

$$\begin{aligned} \lim_r r^{-n} \log \mathfrak{E}\left(\exp < r^n f, \mathcal{L}_r^n > \mid \mathcal{L}_r^{n-1} = \xi_r\right) = \\ = < \log \mathfrak{E} \exp f(\cdot, W) \,,\, \xi >= \lambda_\xi(f)\,, \end{aligned} \tag{18}$$

which gives the LDP with rate function $\mathfrak{J}_\xi$ given by (14).

Let us prove that $\mathfrak{J}_\xi$ has the non variational expression (15). If f involves only the first $n-1$ coordinates, i.e. $f(\mathbf{x}, y) = \psi(\mathbf{x})$ then

$$< f, \nu > -\lambda_\xi(f) = < \psi, \nu_{n-1} > - < \psi, , \xi >\,,$$

and the supremum taken on ψ is infinite if $\nu_{n-1} \neq \xi$. So

$$\mathfrak{J}_\xi(\nu) = +\infty \;\; \text{if}\; \nu_{n-1} \neq \xi\,. \tag{19}$$

Now if $\nu_{n-1} = \xi$ we may write $\nu = \xi \otimes \bar{\nu}$ where $\bar{\nu}$ is a regular probability kernel so that

$$\mathfrak{J}_\xi(\nu) = \sup_f \int_{\Sigma^{n-1}} \left(\int_\Sigma f(\mathbf{x}, y) \bar{\nu}(dy|\mathbf{x}) - \log \mathfrak{E} \exp f(\mathbf{x}, W) \right) \xi(d\mathbf{x})\,.$$

Applying (4) with $\alpha = \bar{\nu}(\cdot|\mathbf{x})$ and $\beta = Q$, we get

$$\mathfrak{J}_\xi(\nu) \leq \int_{\Sigma^{n-1}} H\left(\bar{\nu}(.|\mathbf{x}) \parallel Q\right) \xi(d\mathbf{x}) = H\left(\xi \otimes \bar{\nu} \parallel \xi \otimes Q\right)\,. \tag{20}$$

Applying again (4) but with $\alpha = \xi \otimes \bar{\nu} = \nu$ and $\beta = \xi \otimes Q$ we obtain

$$H\left(\xi \otimes \bar{\nu} \parallel \xi \otimes Q\right) = \sup_g \left\{ < g, \nu > - \log < e^g, \xi \otimes Q > \right\}.$$

Jensen's inequality yields $\log < e^g, \xi \otimes Q > \geq \lambda_\xi(g)$ which leads to

$$H\left(\xi \otimes \bar{\nu} \parallel \xi \otimes Q\right) \leq \mathfrak{J}_\xi(\nu).$$

With (20) and (19) this yields (15). □

Remark 2.1. *This implies uniform conditional LDPs, i.e. that for every closed set F (resp. open G), every $\epsilon > 0$ and every μ one can find an open neighborhood V_μ of μ such that, from the one hand, for $k < n$*

$$\limsup_r \sup_{\xi \in V_\mu} r^{-k} \log \mathfrak{P}\left(\mathcal{L}_r^n \in F \mid \mathcal{L}_r^{n-1} = \xi\right) \leq -\inf_{\nu \in F} \chi(\nu|\mu \otimes Q) + \epsilon \tag{21}$$

$$\liminf_r \inf_{\xi \in V_\mu} r^{-k} \log \mathfrak{P}\left(\mathcal{L}_r^n \in G \mid \mathcal{L}_r^{n-1} = \xi\right) \geq -\inf_{\nu \in G} \chi(\nu|\mu \otimes Q) - \epsilon \tag{22}$$

and from the other hand

$$\limsup_r \sup_{\xi \in V_\mu} r^{-n} \log \mathfrak{P}\left(\mathcal{L}_r^n \in F \mid \mathcal{L}_r^{n-1} = \xi\right) \leq -\inf_{\nu \in F} \mathfrak{J}_\mu(\nu) + \epsilon \tag{23}$$

$$\liminf_r \inf_{\xi \in V_\mu} r^{-n} \log \mathfrak{P}\left(\mathcal{L}_r^n \in G \mid \mathcal{L}_r^{n-1} = \xi\right) \geq -\inf_{\nu \in G} \mathfrak{J}_\mu(\nu) - \epsilon \tag{24}$$

Proof of Theorem 2.1: As previously said, the statement a) is known.

1) If S is compact, we prove by induction that for $n \geq 1$

$$\{\mathfrak{P}(\mathcal{L}_r^n \in .)\}_r \text{ satisfies the LDP at scale } r^n \text{ with rate function } \mathfrak{I}_n^n. \tag{25}$$

Assume that the statement (25) is true for $n-1$. Let us first prove the lower bound of the (unconditioned) LDP at level n. Let G be an open neighborhood of ν such that $\mathfrak{I}_n^n(\nu) < \infty$ and let $\epsilon > 0$. We have

$$\mathfrak{P}\left(\mathcal{L}_r^n \in G\right) \geq \inf_{\xi \in \pi G \cap V_{\pi\nu}} \mathfrak{P}\left(\mathcal{L}_r^n \in G \mid \mathcal{L}_r^{n-1} = \xi\right) \mathfrak{P}\left(\mathcal{L}_r^{n-1} \in \pi G \cap V_{\pi\nu}\right) \tag{26}$$

and from (24)

$$\begin{aligned} \liminf_r r^{-n} \log \inf_{\xi \in \pi G \cap V_{\pi\nu}} \mathfrak{P}\left(\mathcal{L}_r^n \in G \mid \mathcal{L}_r^{n-1} = \xi\right) \geq \\ \geq -\inf_{\rho \in G} \mathfrak{J}_{\pi\nu}(\rho) - \epsilon \geq -\mathfrak{J}_{\pi\nu}(\nu) - \epsilon. \end{aligned} \tag{27}$$

where $\pi G = \{\pi\lambda, \lambda \in G\}$ and $V_{\pi\nu}$ is a neighborhood of $\pi\nu$. It remains to find a lower bound for $r^{-n} \log \mathfrak{P}(\mathcal{L}_r^{n-1} \in \pi G \cap V_{\pi\nu})$. Before using the LDP at order $n-1$ we need a very simple lemma which is surely known but for which we don't know any reference.

Lemma 2.5. *If S is a compact metric space and P a probability measure on S with full support. Then each non empty open subset $\mathcal{O}$ of $\mathcal{M}_1(S)$ contains a probability measure absolutely continuous with respect to P.*

Proof of Lemma 2.5: Every probability measure on S may be approximated in the Lévy metric by a weighted finite sum of atoms. By linearity, it is then sufficient to work with an open neighbourhood of a Dirac mass δ_{x_0} for $x_0 \in S$. Now, if $B_j := \{x \in S : d(x, x_0) < j^{-1}\}$ we have $P(B_j) > 0$ for every j, by definition of the support of P. The probability measure $\xi_j = P(\cdot|B_j)$ is absolutely continuous with respect to P and $\xi_j \Rightarrow \delta_{x_0}$ as $j \to \infty$ (apply the Portmanteau theorem [4] p.356). □

End of the proof of Theorem 2.1 Thanks to the above lemma, we can pick some qQ^{n-1} in $G' := \pi G \cap V_{\pi\nu}$, then for every $M > 0$ we truncate q into $\tilde{q}_M = (M^{-1} \vee q) \wedge M$ and set $q_M =< \tilde{q}_M, Q^{n-1} >^{-1} \tilde{q}_M$. The dominated convergence theorem implies $q_M Q^{n-1} \Rightarrow qQ^{n-1}$ as $M \to \infty$. Choosing M large enough to get $\bar\nu := q_M Q^{n-1} \in G'$, we deduce

$$\mathfrak{I}_{n-1}^{n-1}(\bar\nu) = H(\bar\nu_{n-1} \parallel \bar\nu_{n-2} \otimes Q) \leq M \log M^2 < \infty\,.$$

The LDP at order $n-1$ (statement (25) for $n-1$) yields

$$\liminf_r r^{1-n} \log \mathfrak{P}\left(\mathfrak{L}_r^{n-1} \in \pi G \cap V_{\pi\nu}\right) \geq -\inf\left\{\mathfrak{I}_{n-1}^{n-1}(\xi)\ ;\ \xi \in \pi G \cap V_{\pi\nu}\right\} \geq$$

$\geq -M \log M^2$ for every $M > 0$, so that $\liminf_r r^{-n} \log \mathfrak{P}\left(\mathfrak{L}_r^{n-1} \in \pi G \cap V_{\pi\nu}\right)$ is 0. Gathering (26) and (27) we get the good lower bound.

For the upper bound, we remark that for every Borel subset A of $\mathcal{M}_1(S^n)$,

$$\begin{aligned}\mathfrak{P}\left(\mathfrak{L}_r^n \in A\right) &= \int_{\pi A} \mathfrak{P}\left(\mathfrak{L}_r^n \in A \mid \mathfrak{L}_r^{n-1} = \mu\right) P\left(\mathfrak{L}_r^{n-1} \in d\mu\right) \\ &\leq \sup_{\mu \in \pi A} \mathfrak{P}\left(\mathfrak{L}_r^n \in A \mid \mathfrak{L}_r^{n-1} = \mu\right)\,, \qquad (28)\end{aligned}$$

where $\pi A = \{\pi\xi : \xi \in A\}$.

Since S is compact it is enough to consider only F compact. Following the method of [4] p.150-151, we consider for every $\nu \in F$ a convenient neighborhood A_ν of ν and get an upper bound for $\sup_{\mu\in\pi A} \mathfrak{P}\left(\mathfrak{L}_r^n \in A \mid \mathfrak{L}_r^{n-1} = \mu\right)$, and then make a finite covering of F. The details are left to the reader.

Let us remark that the rate function is

$$\mathcal{J}_{\pi\nu}(\nu) = \int_{\Sigma^{n-1}} H(\nu(.|\mathbf{x}) \parallel Q)\nu_{n-1}(d\mathbf{x}) = H(\nu \parallel \nu_{n-1} \otimes Q) = \mathfrak{I}_n^n(\nu)\,. \qquad (29)$$

We conclude that (25) holds at every level.

2) Let us now consider intermediate levels. Fix $k \geq 1$ and if $k > 1$ assume that S is compact. We show by induction that for $n \geq k$

$$\{\mathfrak{P}(\mathfrak{L}_r^n \in .)\}_r \text{ satisfies the LDP at scale } r^k \text{ with rate function } \mathfrak{I}_k^n\,. \qquad (30)$$

The statement is true for k, from (25) above. Assume that it is true for $n-1$. From (9), we have $r^{-k} \log \mathfrak{E}\big(\exp < r^k f, \mathfrak{L}_r^n > \mid \mathfrak{L}_r^{n-1} = \mu\big) = < r^{n-k} \log \mathfrak{E} \exp\left[r^{k-n} f(\cdot, W)\right], \mu >$ and its limit, as $r \to \infty$ is $< f, \mu \otimes Q > =: < \pi f, \mu >$ (see (16). Moreover it is possible to replace in the above display μ by a sequence $\xi_r \Rightarrow \mu$ without changing the conclusion. With the same argument as above, we get a conditional LDP with rate function $\chi(.|\mu \otimes Q)$. Then we use the same line of argument as in 1). For the proof of the upper bound, there are two cases. For $k \geq 2$ we take into account that S is compact as above. For $k = 1$ we take into account Lemma 2.2 a). Thanks to the statement (30) for $n-1$, we obtain a (unconditional) LDP of rate function

$$\widetilde{\mathfrak{I}}_k^n(\nu) := \inf_{\mu \in \mathcal{M}_{n-1}} \chi(\nu|\mu \otimes Q) + \mathfrak{I}_k^{n-1}(\mu) \qquad (31)$$

Now $\widetilde{\mathfrak{I}}_k^n(\nu) < \infty$ forces $\nu = \mu \otimes Q$ with $\mathfrak{I}_k^{n-1}(\mu) < \infty$. By (6) $\mu = \mu_k \otimes Q^{n-1-k}$, hence $\nu = \mu_k \otimes Q^{n-k}$ and in particular $\nu_k = \mu_k$. Moreover, in this case, (31) yields $\widetilde{\mathfrak{I}}_k^n(\nu) = \mathfrak{I}_k^{n-1}(\mu) = H(\nu_k \parallel \nu_{k-1} \otimes Q) = \mathfrak{I}_k^n(\nu)$, which says that the statement (30) holds for n.

We have ended the proof of Theorem 2.1. □

Remark 2.2. 1) *The rate function $\mathfrak{I}_k^n$ is very similar to the rate function of the empirical measure of k-tuples in a sample of size r of i.i.d. copies of W (as it appears in* [4] *Theorem 6.5.12).*

2) *Formula (31) is of the same type as Lemma 2.3 in* [9], *formula (2) in* [2], *or formula (2.4) in* [3].

3. Contraction and LDP for masses

We first study linear functionals of the empirical measure, and then take the particular case of masses. We assume in this section that Σ is compact.

Proposition 3.1. *Fix n, $k \in \{1, \cdots, n\}$ and $f \in \mathcal{C}(\Sigma^n, \mathfrak{R})$. The family of distributions of $\{< f, \mathcal{L}_r^n >\}_r$ satisfies the LDP at scale r^k and good rate function*

$$I_k(c) = \inf\{\mathcal{J}_n^k(\nu) \ : \nu \in \mathcal{M}_1(\Sigma^n) \ , \ < f, \nu > = c\} . \tag{32}$$

Moreover

$$I_1(c) = \sup_{\theta \in \mathfrak{R}} \{\theta c - \log \int_\Sigma \exp(\theta \ \pi_{n-1} f(y)) \, Q(dy)\} , \tag{33}$$

$$I_k(c) = \sup_{\theta \in \mathfrak{R}} \{\theta c - \log \sup_{\mathbf{x} \in \Sigma^{k-1}} \int_\Sigma \exp(\theta \ \pi_{n-k} f(\mathbf{x}, y)) \, Q(dy)\} \quad \text{if } k \geq 2 . \tag{34}$$

Proof: The first claim is a consequence of the contraction principle, since $\nu \mapsto < f, \nu >$ is continuous. To prove (33), notice that $\mathcal{J}_1^n(\nu) = H(\nu_1 \parallel Q)$ if $\nu = \nu_1 \otimes Q$, and then $< f, \nu > = < \pi_{n-1} f, \nu_1 >$.

Fix $k \geq 2$. Let $\ell^*(c)$ the right hand side of (34) so that we have to prove

$$I_k = \ell^* . \tag{35}$$

Set, for $\mathbf{x} \in \Sigma^{k-1}, \lambda \in \mathfrak{R}$, $\xi \in \mathcal{M}_1(\Sigma^{k-1})$

$$\Lambda_{\mathbf{x}}(\lambda) \ := \ \log \int_\Sigma \exp(\lambda \ \pi_{n-k} f(\mathbf{x}, y)) \, Q(dy) \tag{36}$$

$$\Lambda_\xi(\lambda) \ := \ < \Lambda_.(\lambda), \xi > \text{ and } \Lambda_\xi^*(c) := \sup_\lambda (\lambda c - \Lambda_\xi(\lambda)) , \tag{37}$$

From the obvious equality $\sup\{\Lambda_\xi(\lambda) \ ; \ \xi \in \mathcal{M}_1(\Sigma^{k-1})\} = \sup\{\Lambda_{\mathbf{x}}(\lambda) \ ; \ \mathbf{x} \in \Sigma^{k-1}\}$, and the minimax theorem (the ξ set is compact - see [4] p. 151) we deduce $\ell^*(c) = \sup_\lambda (\lambda c - \sup_\xi \Lambda_\xi(\lambda)) = \sup_\lambda \inf_\xi (\lambda c - \Lambda_\xi(\lambda)) = \inf_\xi \Lambda_\xi^*(c)$. Let $\nu \in \mathcal{M}_1(\Sigma^n)$ such that $< f, \nu > = c$ and $\mathcal{J}_k^n(\nu) = c$. Then $\nu = \nu_k \otimes Q^{n-k}$ and $< g, \nu_k > = c$ where $g := \pi_{n-k} f$. Applying once more (4), we have for $\mathbf{x} \in \Sigma^{k-1}$,

$$H(\nu_k(.|\mathbf{x}) \parallel Q) \geq \lambda \int_\Sigma g(\mathbf{x}, y) \nu_k(dy|\mathbf{x}) - \Lambda_{\mathbf{x}}(\lambda) . \tag{38}$$

If ν_{k-1} satisfies $\int_{\Sigma^{k-1} \times \Sigma} g(\mathbf{x}, y) \nu_{k-1}(d\mathbf{x}) \nu_k(dy|\mathbf{x}) = c$, an integration of (38) and a maximization in λ yield

$$\int_{\Sigma^{k-1}} H(\nu_k(.|\mathbf{x}) \parallel Q) \, \nu_{k-1}(d\mathbf{x}) \geq \Lambda_{\nu_{k-1}}^*(c) ,$$

and taking infimum on ν_{k-1} we get $I_k(c) \geq \ell^*(c)$. Let us prove the reverse inequality. Fix $\xi \in \mathcal{M}_1(\Sigma^{k-1})$ and assume $\Lambda_\xi^*(c) < \infty$. The function Λ_ξ is convex, infinitely differentiable and strictly increasing, and

$$\Lambda_\xi'(\infty) = \int_{\Sigma^{k-1}} \operatorname{essup} g(\mathbf{x}, W) \xi(d\mathbf{x}) \ , \ \ \Lambda_\xi'(-\infty) = \int_{\Sigma^{k-1}} \operatorname{essinf} g(\mathbf{x}, W) \xi(d\mathbf{x}) .$$

If $c \in \left(\Lambda'_\xi(-\infty), \Lambda'_\xi(\infty)\right)$, let λ the unique solution of $\Lambda'_\xi(\lambda) = c$. This λ reaches the supremum in (37). For $\eta(dy|\mathbf{x}) = \exp\left(\lambda g(\mathbf{x}, y) - \Lambda_\mathbf{x}(\lambda)\right)\ Q(dy)$

$$\int_{\Sigma^{k-1}} H\left(\eta(.|\mathbf{x}) \parallel Q\right) \xi(d\mathbf{x}) = \Lambda^*_\xi(c)\,. \tag{39}$$

For fixed ξ, we get the inequality

$$\Lambda^*_\xi(c) \geq \inf\left\{\int_{\Sigma^{k-1}} H\left(\eta(.|\mathbf{x}) \parallel Q\right) \xi(d\mathbf{x})\ ;\ \int_{\Sigma^{k-1}\times\Sigma} g(\mathbf{x}, y)\xi(d\mathbf{x})\eta(dy|\mathbf{x}) = c\right\}. \tag{40}$$

If $c \notin [\Lambda'_\xi(-\infty), \Lambda'_\xi(\infty)]$, we have $\Lambda^*_\xi(c) = \infty$ and (40) holds. If $c = \Lambda'_\xi(\infty) < \infty$, we have $\Lambda^*_\xi(c) = -\int_{\Sigma^{k-1}} \log \mathfrak{P}(g(\mathbf{x}, W) = c)\ \xi(d\mathbf{x}) < \infty$, so that $\mathfrak{P}(g(\mathbf{x}, W) = c) \neq 0$ for ν a.e. $\mathbf{x}$ and choosing

$$\nu_k(dy|\mathbf{x}) = \mathbb{1}_{f(\mathbf{x},y)=c}\{\mathfrak{P}(f(\mathbf{x}, W) = c)\}^{-1}\ Q(dy)$$

we see that (39) holds in this case (and also if $c = \Lambda'_\xi(-\infty)$). Henceforth (40) holds in all cases. Taking infimum on ξ we get $\ell^*(c) = \inf_\xi \Lambda^*_\xi(c) \geq I_k(c)$ and finally equation (35) holds, which ends the proof of Proposition 3.1. □

In the particular case of masses, we assume (1) and take $\Sigma = [0, \bar{w}]$. We denote $p_n(\mathbf{x}) := x_1 x_2 \cdots x_n$ for $\mathbf{x} \in \Sigma^n$ and take $f = p_n$ in the above proposition. This gives the following result, which recovers Proposition 1.2.

Corollary 3.2. 1) *The family of distributions of Z_r^n satisfies the LDP at scale r with good rate function Λ^*.*

2) *For $2 \leq k \leq n$ the family of distributions of Z_r^n satisfies the LDP at scale r^k with good rate function I_k^n given by*

$$\begin{aligned} I_k^n(x) &= 0 && \text{if } x \in [\underline{w}^{k-1}, \bar{w}^{k-1}] \\ I_k^n(x) &= \Lambda^*(x\bar{w}^{-(k-1)}) && \text{if } x \in [\bar{w}^{k-1}, \bar{w}^k) \\ I_k^n(x) &= \Lambda^*(x\underline{w}^{-(k-1)}) && \text{if } x \in [\underline{w}^k, \underline{w}^{k-1}) \\ I_k^n(x) &= +\infty && \text{if } x \notin [\underline{w}^k, \bar{w}^k]\,. \end{aligned} \tag{41}$$

Proof: Start from $\pi_{n-k} f(\mathbf{x}, y) = x_1 \cdots x_{k-1} y$. With the notation of (36) we have $\Lambda_\mathbf{x}(\lambda) = \Lambda(x_1 x_2 \cdots x_{k-1}\lambda)$. This gives $\sup_{\mathbf{x}\in S^{k-1}} \Lambda_\mathbf{x}(\lambda) = \Lambda(\bar{w}^{k-1}\lambda)$ if $\lambda > 0$ and $\Lambda(\underline{w}^{k-1}\lambda)$ si $\lambda < 0$. □

4. Gibbs conditioning principle

We want to characterize typical behaviors along a given branch, by means of the Gibbs conditioning principle, ([19] or Chapter 7.3 in [4]). Mass plays the role of energy. For $\delta > 0$, let $A_\delta = \{\nu \in \mathcal{M}_1(\Sigma^n) : |< p_n, \nu > - a| \leq \delta\}$ be the "energy constraint". The following lemma identifies the Gibbs states. We assume (1).

Let $F_a = \{\nu \in \mathcal{M}_1(\mathfrak{R}^n)\ : < p_n, \nu >= a\}$. For $\theta \in \mathfrak{R}$ let $Q_\theta(dy) := e^{\theta y - \Lambda(\theta)}$ $Q(dy)$, whose mean is $\Lambda'(\theta)$. For every $x \in (\underline{w}, \bar{w})$ let $\theta(x)$ be the unique solution of $\Lambda'(\theta) = x$.

Lemma 4.1. *1) When $\underline{w} < a < \bar{w}$ and $n \geq 1$, $\nu_n^{(a)} = Q_{\theta(a)} \otimes Q^{n-1}$ is the unique solution of*

$$\inf\{\mathfrak{I}_1^n(\nu); \nu \in F_a\} = \mathfrak{I}_1^n(\nu_n^{(a)})\,. \tag{42}$$

2) When $a \in [\bar{w}^{k-1}, \bar{w}^k)$ *for* $k \geq 2$, $\nu_n^{(a)} = \delta_{\bar{w}}^{k-1} \otimes Q_{\theta(a\bar{w}^{-(k-1)})} \otimes Q^{n-k}$ *is the unique solution of*

$$\inf\{\mathfrak{I}_k^n(\nu); \nu \in F_a\} = \mathfrak{I}_k^n(\nu_n^{(a)}) . \tag{43}$$

3) When $a \in [\underline{w}^k, \underline{w}^{k-1})$, $\nu_n^{(a)} = \delta_{\underline{w}}^{k-1} \otimes Q_{\theta(a\underline{w}^{-(k-1)})} \otimes Q^{n-k}$ *is the unique solution of (43).*

Here is the main result of this section.

Theorem 4.2. *Fix* $n > 0$ *and* $k \leq n$. *Let* Γ *be an open neighbourhood of* $\nu_n^{(a)}$

$$\limsup_{\delta \to 0} \limsup_{r} \ r^{-k} \log \mathfrak{P}(\mathcal{L}_r^n \notin \Gamma \mid \mathcal{L}_r^n \in A_\delta) < 0 .$$

Consequently, for any $i_1, i_2, \cdots, i_n$, *the distribution of* $(W_{i_1}, \cdots, W_{i_1 \cdots i_n})$ *conditioned upon* $Z_r^n \in [a-\delta, a+\delta]$ *converges to* $\nu_n^{(a)}$, *as* $r \to \infty$, *followed by* $\delta \to 0$.

Proof of Lemma 4.1: To simplify, we treat only the second case and set $\tilde{a} = a\bar{w}^{-(k-1)}$, $\tilde{\theta} = \theta(\tilde{a})$, $\tilde{\nu} = \nu_n^{(a)}$. From the previous section, we know that

$$\mathfrak{I}_k^n(\tilde{\nu})) = H(Q_{\tilde{\theta}} \parallel Q) = \tilde{\theta}\tilde{a} - \Lambda(\tilde{\theta}) .$$

So $\tilde{\nu}$ is a solution. Let $\mu \in F_a$ another one. Since $\mathfrak{I}_k^n(\mu) < \infty$ we have $\mu = \mu_k \otimes Q^{n-k}$ and

$$\mathfrak{I}_k^n(\mu) = H\big(\mu_k \parallel \mu_{k-1} \otimes Q\big) = H\big(\mu_k \parallel \mu_{k-1} \otimes Q_{\tilde{\theta}}\big) + \tilde{\theta} \int_{\Sigma^k} x_k \mu_k(d\mathbf{x}) - \Lambda(\tilde{\theta}) .$$

Now since μ_k is absolutely continuous with respect to $\mu_{k-1} \otimes Q$, the support of $\mu_k(dy|\mathbf{x})$ is a subset of S. Since $\mathfrak{I}_k^n(\mu) < \infty$ we know from Theorem 2.1 b) that $\mu = \mu_k \otimes Q^{n-k}$, which gives

$$a = < p_n, \mu > = < p_k, \mu_k > \leq \ \bar{w}^{k-1} \int_{\Sigma^k} x_k \mu_k(d\mathbf{x}) . \tag{44}$$

Since $\tilde{\theta} > 0$, we get $\Lambda^*(\tilde{a}) - H\left(\mu_k \parallel \mu_{k-1} \otimes Q_{\tilde{\theta}}\right) \geq \tilde{\theta} a \bar{w}^{-(k-1)} - \Lambda(\tilde{\theta}) = \Lambda^*(\tilde{a})$ which implies $\mu_k = \mu_{k-1} \otimes Q_{\tilde{\theta}}$. Carrying into (44) gives $\bar{w}^{k-1} = < p_{k-1}, \mu_{k-1} >$ which forces $\mu_{k-1} = \delta_{\bar{w}^{k-1}}$. □

Proof of Theorem 4.2: Clearly,

$$\limsup_{\delta \to 0} \limsup_{r} \ r^{-k} \log \mathfrak{P}(\mathcal{L}_r^n \notin \Gamma \mid \mathcal{L}_r^n \in A_\delta) \leq$$

$$\leq \lim_{\delta \to 0} \limsup_{r} \ r^{-k} \log \mathfrak{P}(\mathcal{L}_r^n \in \Gamma^c \cap \ A_\delta) - \lim_{\delta \to 0} \liminf_{r} \ r^{-k} \log \mathfrak{P}(\mathcal{L}_r^n \in A_\delta) .$$

From the upper bound of the LDP (Theorem 2.1 b)), the first term of the right hand side is less than $-\lim_{\delta \to 0} \inf\left\{H(\mu|Q^{\otimes n}) \ ; \ \mu \in \Gamma^c \cap \ A_\delta\right\}$. Since the sets $\Gamma^c \cap \ A_\delta$ are closed and nested (see [4] Lemma 4.1.6), the bound is equal to $-\inf\left\{H(\mu|Q^{\otimes n}) \ ; \ \mu \in \Gamma^c \cap \ F_a\right\}$, which is strictly smaller than $-I_k(a)$. For the second term, we apply the lower bound of the LDP to A_δ so that $\liminf_r \ r^{-k} \log \mathfrak{P}(\mathcal{L}_r^n \in A_\delta) \geq -\inf\{H(\mu|Q^{\otimes n}) \ ; \ \mu \in intA_\delta\} \geq -H(\nu_n^{(a)} \| Q^{\otimes n}) = -I_k(a)$. □

Acknowledgement. I thank Wendelin Werner for asking me this question.

References

[1] J. Barral. Generalized vector multiplicative cascades. *Adv. in Appl. Probab.*, 33 (4): 874–895, 2001.

[2] J.D. Biggins. Large deviations for mixtures. Available at http://www.shef.ac.uk/~st1jdb/.

[3] N.R. Chaganthy. Large deviations for joint distributions and statistical applications. *Sankhya*, 59:147–166, 1997.

[4] A. Dembo and O. Zeitouni. *Large Deviations Techniques and Applications*. Springer, 2nd edition, 1998.

[5] J.D. Deuschel and D.W. Stroock. *Large deviations*. Academic Press, 1989.

[6] I.H. Dinwoodie and S.L. Zabell. Large deviations for exchangeable random vectors. *Ann. Probab.*, pages 1147–1166, 1992.

[7] P. Dupuis and R. Ellis. *A weak convergence approach to the theory of large deviations*. Wiley, 1997.

[8] P. Eichelsbacher and U. Schmock. Large deviations of U-empirical measures in strong topologies and applications. *Ann. Inst. H. Poincaré Probab. Statist.*, 38(5):779–797, 2002.

[9] W. Finnoff. Integration of large-deviation kernels and applications to large deviations for evolutionary games. *Probab. Theory Relat Fields*, 122:141–162, 2002.

[10] M. Grunwald. Sanov results for Glauber spin-glass dynamics. *Probab. Theory Relat Fields*, 106:187–232, 1996.

[11] J.-P. Kahane and J. Peyrière. Sur certaines martingales de Benoit Mandelbrot. *Advances in Math.*, 22(2):131–145, 1976.

[12] A. N. Kolmogorov. The local structure of turbulence in incompressible viscous fluid for very large Reynolds numbers. *Proc. Roy. Soc. London Ser. A*, 434(1890):9–13, 1991. Translated from the Russian by V. Levin, Turbulence and stochastic processes: Kolmogorov's ideas 50 years on.

[13] Q. Liu. On generalized multiplicative cascades. *Stochastic Process. Appl.*, 86(2):263–286, 2000.

[14] Q. Liu, E. Rio, and A. Rouault. Limit theorems for Multiplicative Processes. *J. of Theoret. Probab.*, 16:971–1014, 2003.

[15] Q. Liu and A. Rouault. Limit theorems for Mandelbrot's multiplicative cascades. *Annals of Appl. Probab.*, 10(1):218–239, 2000.

[16] M. Löwe. Iterated large deviations. *Stat. and Probab. Letters*, 26(3):219–223, February 1996.

[17] B. Mandelbrot. Multiplications aléatoires itérées et distributions invariantes par moyenne pondérée aléatoire: quelques extensions. *C. R. Acad. Sci. Paris Sér. A*, 278:355–358, 1974.

[18] M. Ossiander and E.C. Waymire. Statistical estimation for multiplicative cascades. *Ann. Statist.*, 28(6):1533–1560, 2000.

[19] D.W. Stroock and O. Zeitouni. Microcanonical distributions, Gibbs states, and the equivalence of ensembles. In Durrett R. and H. Kesten, editors, *Festschrift in Honour of F.Spitzer*, pages 399–424. Birkhauser, 1991.

[20] J. Trashorras. Large deviations for a triangular array of exchangeable random variables. *Ann. Inst. H. Poincaré Probab. Statist.*, 38(5):649–680, 2002.

Alain Rouault
LAMA, Université de Versailles-Saint-Quentin
rouault@math.uvsq.fr

Trends in Mathematics, © 2004 Birkhäuser Verlag Basel/Switzerland

Partitioning with Piecewise Constant Eigenvectors

Christiane Takacs

ABSTRACT: *In the present paper we investigate the partitioning properties of piecewise constant eigenvectors of matrices describing the mutual positions of points. We discuss differences arising from choosing different matrices and present an example where one has to be careful with the selection of the appropriate eigenvectors.*

Let $V \neq \emptyset$ be a finite set of vertices, e.g. points in some space, with $|V| = n$. In order to describe the mutual positions of the points we introduce the distance matrix $D = (d_{ij})_{i,j \in V}$, where d_{ij} is the (Euclidean) distance of the points i and j. We assume D to be non-negative and symmetric. Secondly, let $W = (w_{ij})_{i,j \in V}$ be the similarity matrix, where w_{ij} is the similarity of the points i and j, and $w_{ij} = 0$ means that i and j are completely different. We assume W to be non-negative and symmetric. The matrix W can be derived from D by e.g. $w_{ij} = e^{-\sigma d_{ij}}$ with some $\sigma > 0$.
Let $d_i = \sum_{j=1}^n d_{ij}$ and $\Delta = \operatorname{diag}(d_1, \ldots, d_n)$ and assume Δ to be regular with the scaling $\sum_i d_i = 1$, further let $w_i = \sum_{j=1}^n w_{ij}$ and $\Omega = \operatorname{diag}(w_1, \ldots, w_n)$ and assume Ω to be regular with $\sum_i w_i = 1$. From the computational viewpoint the matrix W has to be preferred because in applications it is often a sparse matrix. But since my initial contact to the problem was a question concerning the clustering properties of piecewise constant eigenvectors of the distance matrix, I consider this, too.
The object $(V, V \times V, D)$ (resp. $(V, V \times V, W)$) is interpreted as a weighted graph where the points i and j are connected by an edge with weight d_{ij} (resp. w_{ij}). Because of the symmetry of the matrices D and W the graphs are simple and undirected. In the context of electric networks the weight w_{ij} corresponds to the conductance of the edge (i, j). The graph corresponds to a finite and reversible Markov chain with transition probabilities w_{ij}/w_i. This Markov chain performs a transition from one vertex to another with a probability proportional to the conductance of the connecting edge weight. A weight d_{ij} could be interpreted as some sort of electrical resistance, but we will not use this here, instead we interpret d_{ij} as some (crazy) conductance, too.
Now, given the distances or similarities of points one wants to identify clusters of points, i.e. group together points which are near or similar, and put cuts between groups of points which are apart or dissimilar. In the literature there are a lot of algorithms (see [2], [6], [7] for a survey) finding an optimal partition. The spectral methods have in common that they use piecewise almost constant eigenvectors as indicators for an underlying partition. The goal of the present paper is to compare different partitioning criteria and to get a better understanding of their features.

1. Criteria for Partitioning

For $A, B \subset V$ let $d_{AB} := \sum_{i \in A} \sum_{j \in B} d_{ij}$ resp. $w_{AB} := \sum_{i \in A} \sum_{j \in B} w_{ij}$. The prior quantifies the distance between the sets A and B, whereas the latter represents their similarity. Due to our scaling we have $d_{VV} = 1$ and $w_{VV} = 1$. Let $|A|$ be the number of points in A. In the following we introduce criteria for finding a partition (A, B) of the set V.

The average association criterion maximizes $\text{AAss}(A,B) := \frac{w_{AA}}{|A|} + \frac{w_{BB}}{|B|}$.
The average distance criterion minimizes $\text{ADist}(A,B) := \frac{d_{AA}}{|A|} + \frac{d_{BB}}{|B|}$.
The normalized association criterion maximizes $\text{NAss}(A,B) := \frac{w_{AA}}{w_{AV}} + \frac{w_{BB}}{w_{AV}}$.
The normalized distance criterion minimizes $\text{NDist}(A,B) := \frac{d_{AA}}{d_{AV}} + \frac{d_{BB}}{d_{BV}}$.

Note that each criterion optimizes some average quantity within the partitioning subsets, and thus can easily be generalized to partitions with more than two sets. Different features of a set $A \subset V$ are covered by

$$\text{AAss}\, A := \frac{w_{AA}}{|A|}, \ \text{NAss}\, A := \frac{w_{AA}}{w_{AV}},$$

the mean similarity within A and the proportion of the similarity between A and V covered by the similarities within A,

$$\text{ADist}\, A := \frac{d_{AA}}{|A|}, \ \text{NDist}\, A := \frac{d_{AA}}{d_{AV}},$$

the mean distance within A and the proportion of the distance between A and V covered by the distances within A.

Note that we have $0 \le \text{AAss}\, A \le \text{NAss}\, A \le 1$ and $0 \le \text{ADist}\, A \le \text{NDist}\, A \le 1$. The average quantities are average for $A = V$ but may be smaller or larger for smaller sets A. The normalized quantities are usually small for small sets A and maximal for $A = V$. For a partition $(A_1, \ldots, A_k)$ of V we define

$$\text{NAss}\,(A_1, \ldots, A_k) := \text{NAss}\, A_1 + \ldots + \text{NAss}\, A_k$$

and observe that $0 < \text{NAss}\,(A_1, \ldots, A_k) \le k$. Since this estimate depends on k and since usually $\text{NAss}\, A$ is small for small sets A there is no canonical strategy when comparing optima for partitions with different numbers of elements.

Remark 1.1. *One could also optimize quantities* between *the partitioning subsets. In* [5] *the criterion* NCut *is minimized, where*

$$\text{NCut}\,(A,B) := \frac{w_{AB}}{w_{AV}} + \frac{w_{AB}}{w_{AV}} = 2 - \text{NAss}\,(A,B).$$

Thus the NCut *criterion is equivalent to the* NAss *criterion. For the normalized distance cut criterion* $\frac{d_{AB}}{d_{AV}} + \frac{d_{AB}}{d_{BV}} \to \max$, *we also have*

$$\frac{d_{AB}}{d_{AV}} + \frac{d_{AB}}{d_{BV}} = 2 - \text{NDist}\,(A,B)$$

and thus equivalence to the NDist *criterion.*
These equivalences suggest how to generalize cut criteria to partitions of V *with more than two elements, namely e.g.*

$$\begin{aligned} \text{NCut}\,(A_1, \ldots, A_k) \quad : \quad &= (1 - \text{NAss}\, A_1) + \ldots + (1 - \text{NAss}\, A_k) \\ &= \frac{w_{A_1\overline{A_1}}}{w_{A_1 V}} + \ldots + \frac{w_{A_k\overline{A_k}}}{w_{A_k V}}. \end{aligned}$$

2. Features of vectors

For $A \subset V$ let $\mathbf{1}_A \in \mathbb{R}_n$ be a vector with $(\mathbf{1}_A)_i = 1$ for $i \in A$ and 0 otherwise. Since $|A| = \mathbf{1}_A^t\mathbf{1}_A$, $w_{AA} = \mathbf{1}_A^t W\mathbf{1}_A$, $w_{AV} = \mathbf{1}_A^t W\mathbf{1} = \mathbf{1}_A^t\Omega\mathbf{1} = \mathbf{1}_A^t\Omega\mathbf{1}_A$ we have the representations

$$\operatorname{AAss} A = \frac{\mathbf{1}_A^t W\mathbf{1}_A}{\mathbf{1}_A^t\mathbf{1}_A},\ \operatorname{NAss} A = \frac{\mathbf{1}_A^t W\mathbf{1}_A}{\mathbf{1}_A^t\Omega\mathbf{1}_A}$$

and analogous expressions for ADist A and NDist A. Therefore we define the quantities for column vectors $\mathbf{x} \in\mathbb{R}_n$. So let

$$\operatorname{AAss}\mathbf{x} := \frac{\mathbf{x}^t W\mathbf{x}}{\mathbf{x}^t\mathbf{x}}, \qquad \operatorname{NAss}\mathbf{x} := \frac{\mathbf{x}^t W\mathbf{x}}{\mathbf{x}^t\Omega\mathbf{x}},$$
$$\operatorname{ADist}\mathbf{x} := \frac{\mathbf{x}^t D\mathbf{x}}{\mathbf{x}^t\mathbf{x}}, \qquad \operatorname{NDist}\mathbf{x} := \frac{\mathbf{x}^t D\mathbf{x}}{\mathbf{x}^t\Delta\mathbf{x}}.$$

Note that each definition coincides with the original definition for $\mathbf{x} = \mathbf{1}_A$. From the definition

$$\operatorname{AAss}\mathbf{x} = \frac{\sum_{i,j} w_{ij}x_ix_j}{\sum_i x_i^2}$$

we see that AAss $\mathbf{x}$ is large if for large w_{ij}, i.e. high similarity of the points i and j, the entries x_i and x_j have the same sign. The optimal $\mathbf{x}$ has the same sign everywhere and maximum weight at highly connected vertices. Both features can be used for clustering. In contrast we observe that

$$\operatorname{ADist}\mathbf{x} = \frac{\sum_{i,j} d_{ij}x_ix_j}{\sum_i x_i^2}$$

is small (negative) if for large distances of points the corresponding entries of $\mathbf{x}$ have different signs. For NAss and NDist there are similar argumentations, but the following theorem clarifies their features.

Theorem 2.1. *We have*

$$\begin{aligned}\operatorname{NAss}\mathbf{x} &= 1 - \frac{\sum_{i<j} w_{ij}(x_i - x_j)^2}{\sum_i w_ix_i^2}\\ \operatorname{NDist}\mathbf{x} &= 1 - \frac{\sum_{i<j} d_{ij}(x_i - x_j)^2}{\sum_i d_ix_i^2}.\end{aligned}$$

Proof: The first assertion is proved by

$$\begin{aligned}\frac{\sum_{i<j} w_{ij}(x_i - x_j)^2}{\sum_i w_ix_i^2} &= \frac{1}{2}\frac{\sum_{i,j} w_{ij}(x_i - x_j)^2}{\sum_{i,j} w_{ij}x_i^2}\\ &= \frac{\sum_{i,j} w_{ij}(x_i - x_j)^2}{\sum_{i,j} w_{ij}(x_i^2 + x_j^2)}\\ &= 1 - 2\frac{\sum_{i,j} w_{ij}x_ix_j}{2\sum_i w_ix_i^2}\\ &= 1 - \operatorname{NAss} x.\end{aligned}$$

For the second assertion change w to d.

For a normed vector $\mathbf{x}$ the quantity $1 - \mathrm{NAss}\,\mathbf{x}$ calculates the sum of squares of the differences of $\mathbf{x}$ along the edges of the graph. This expression is small, i.e. NAss large, if $\mathbf{x}$ has large differences only where similarities between vertices are small. $\mathrm{NAss}\,\mathbf{x}$ is maximal for $\mathbf{x} = \mathbf{1}$, but this is not useful for partitioning. For $1 - \mathrm{NDist}\,\mathbf{x}$ on the other hand we argue that it is large, i.e. $\mathrm{NDist}\,x$ is small, if $\mathbf{x}$ has large differences where distances between vertices are large. While the Ass–criteria stress the similarities, the Dist–criteria stress the distances.

3. Partitioning with vectors

This section links the features of vectors to the criteria for partitioning. For $A, B \subset V$ with $A \cap B = \emptyset$ we introduce vectors χ_{AB} which indicate the sets A and B by different signs and are zero otherwise.
With $\chi_{AB} := \frac{1}{\mathbf{1}_A^t \mathbf{1}_A}\mathbf{1}_A - \frac{1}{\mathbf{1}_B^t \mathbf{1}_B}\mathbf{1}_B$ we have

$$\begin{aligned} \mathrm{AAss}\,(A \cup B) + \mathrm{AAss}\,\chi_{AB} &= \mathrm{AAss}\,(A, B) \\ \mathrm{ADist}\,(A \cup B) + \mathrm{ADist}\,\chi_{AB} &= \mathrm{ADist}\,(A, B)\,. \end{aligned}$$

With $\chi_{AB} := \frac{1}{\mathbf{1}_A^t \Omega \mathbf{1}_A}\mathbf{1}_A - \frac{1}{\mathbf{1}_B^t \Omega \mathbf{1}_B}\mathbf{1}_B$ we have

$$\mathrm{NAss}\,(A \cup B) + \mathrm{NAss}\,\chi_{AB} = \mathrm{NAss}\,(A, B)\,.$$

With $\chi_{AB} := \frac{1}{\mathbf{1}_A^t \Delta \mathbf{1}_A}\mathbf{1}_A - \frac{1}{\mathbf{1}_B^t \Delta \mathbf{1}_B}\mathbf{1}_B$ we have

$$\mathrm{NDist}\,(A \cup B) + \mathrm{NDist}\,\chi_{AB} = \mathrm{NDist}\,(A, B)\,.$$

Proof: Since all proofs are essentially the same we only prove the third property. As a consequence of $A \cap B = \emptyset$ we have

$$\begin{aligned} \chi_{AB}^t \Omega \chi_{AB} &= \frac{1}{(\mathbf{1}_A^t \Omega \mathbf{1}_A)^2}\mathbf{1}_A^t \Omega \mathbf{1}_A + \frac{1}{(\mathbf{1}_B^t \Omega \mathbf{1}_B)^2}\mathbf{1}_B^t \Omega \mathbf{1}_B \\ &= \frac{1}{\mathbf{1}_A^t \Omega \mathbf{1}_A} + \frac{1}{\mathbf{1}_B^t \Omega \mathbf{1}_B}. \end{aligned}$$

Now we calculate

$$\begin{aligned} \chi_{AB}^t W \chi_{AB} &= \frac{1}{(\mathbf{1}_A^t \Omega \mathbf{1}_A)^2}\mathbf{1}_A^t W \mathbf{1}_A + \frac{1}{(\mathbf{1}_B^t \Omega \mathbf{1}_B)^2}\mathbf{1}_B^t W \mathbf{1}_B \\ &\quad - \frac{1}{\mathbf{1}_A^t \Omega \mathbf{1}_A}\frac{1}{\mathbf{1}_B^t \Omega \mathbf{1}_B}\left(\mathbf{1}_A^t W \mathbf{1}_B + \mathbf{1}_B^t W \mathbf{1}_A\right). \end{aligned}$$

Since

$$\begin{aligned} \mathbf{1}_A^t W \mathbf{1}_B + \mathbf{1}_B^t W \mathbf{1}_A &= \mathbf{1}_A^t W \mathbf{1}_{A\cup B} - \mathbf{1}_A^t W \mathbf{1}_A + \mathbf{1}_B^t W \mathbf{1}_{A\cup B} - \mathbf{1}_B^t W \mathbf{1}_B \\ &= \mathbf{1}_{A\cup B}^t W \mathbf{1}_{A\cup B} - \mathbf{1}_A^t W \mathbf{1}_A - \mathbf{1}_B^t W \mathbf{1}_B. \end{aligned}$$

we have

$$\begin{aligned} \chi_{AB}^t W \chi_{AB} &= \mathrm{NAss}\,A\left(\frac{1}{\mathbf{1}_A^t \Omega \mathbf{1}_A} + \frac{1}{\mathbf{1}_B^t \Omega \mathbf{1}_B}\right) \\ &\quad + \mathrm{NAss}\,B\left(\frac{1}{\mathbf{1}_A^t \Omega \mathbf{1}_A} + \frac{1}{\mathbf{1}_B^t \Omega \mathbf{1}_B}\right) - \frac{\mathbf{1}_{A\cup B}^t W \mathbf{1}_{A\cup B}}{\mathbf{1}_A^t \Omega \mathbf{1}_A\ \mathbf{1}_B^t \Omega \mathbf{1}_B} \end{aligned}$$

and we arrive at

$$\begin{aligned}\operatorname{NAss}\chi_{AB} &= \frac{\chi_{AB}^t W \chi_{AB}}{\chi_{AB}^t \Omega \chi_{AB}} \\ &= \operatorname{NAss} A + \operatorname{NAss} B - \frac{\mathbf{1}_{A\cup B}^t W \mathbf{1}_{A\cup B}}{\mathbf{1}_A^t \Omega \mathbf{1}_A + \mathbf{1}_B^t \Omega \mathbf{1}_B}.\end{aligned}$$

This together with

$$\frac{\mathbf{1}_{A\cup B}^t W \mathbf{1}_{A\cup B}}{\mathbf{1}_A^t \Omega \mathbf{1}_A + \mathbf{1}_B^t \Omega \mathbf{1}_B} = \frac{\mathbf{1}_{A\cup B}^t W \mathbf{1}_{A\cup B}}{\mathbf{1}_{A\cup B}^t \Omega \mathbf{1}_{A\cup B}} = \operatorname{NAss}(A \cup B)$$

completes the proof.

If (A, B) is a partition of V we have with appropriate χ_{AB}

$$\begin{aligned}\operatorname{AAss}(A,B) &= \operatorname{AAss} V + \operatorname{AAss}\chi_{AB} \\ \operatorname{ADist}(A,B) &= \operatorname{ADist} V + \operatorname{ADist}\chi_{AB} \\ \operatorname{NAss}(A,B) &= 1 + \operatorname{NAss}\chi_{AB} \\ \operatorname{NDist}(A,B) &= 1 + \operatorname{NDist}\chi_{AB}.\end{aligned}$$

Since the first terms on the right side do not depend on the partition, the left sides are optimized by optimizing the second terms on the right side. Thus the sum of two Rayleigh quotients on the left side is actually reduced to the one on the right. If we want to partition V into two sets, we use a vector looking like χ_{AB} for some partition (A, B) such that its AAss or NAss is maximal or its ADist or NDist is minimal.

If $\mathbf{x}$ is an eigenvector of W (resp. D) with eigenvalue λ we have

$$\begin{aligned}\operatorname{AAss}\mathbf{x} &= \frac{\mathbf{x}^t W \mathbf{x}}{\mathbf{x}^t \mathbf{x}} = \frac{\mathbf{x}^t(\lambda \mathbf{x})}{\mathbf{x}^t \mathbf{x}} = \lambda, \\ \operatorname{ADist}\mathbf{x} &= \frac{\mathbf{x}^t D \mathbf{x}}{\mathbf{x}^t \mathbf{x}} = \frac{\mathbf{x}^t(\lambda \mathbf{x})}{\mathbf{x}^t \mathbf{x}} = \lambda.\end{aligned}$$

If $\mathbf{x}$ is an eigenvector of $\Omega^{-1}W$ (resp. $\Delta^{-1}D$) with eigenvalue λ we have

$$\begin{aligned}\operatorname{NAss}\mathbf{x} &= \frac{\mathbf{x}^t W \mathbf{x}}{\mathbf{x}^t \Omega \mathbf{x}} = \frac{\mathbf{x}^t \Omega \Omega^{-1} W \mathbf{x}}{\mathbf{x}^t \Omega \mathbf{x}} = \frac{\mathbf{x}^t \Omega(\lambda \mathbf{x})}{\mathbf{x}^t \Omega \mathbf{x}} = \lambda, \\ \operatorname{NDist}\mathbf{x} &= \frac{\mathbf{x}^t D \mathbf{x}}{\mathbf{x}^t \Delta \mathbf{x}} = \frac{\mathbf{x}^t \Delta(\lambda \mathbf{x})}{\mathbf{x}^t \Delta \mathbf{x}} = \lambda.\end{aligned}$$

So if an eigenvector of W or $\Omega^{-1}W$ with the largest (possible) eigenvalue looks like some χ_{AB}, the partition (A, B) is optimal with respect to the corresponding criterion among all 2-partitions. The same is true for eigenvectors with smallest eigenvalues of D or $\Delta^{-1}D$.

In [5] the above argumentation for NAss is the basis of a partitioning algorithm using eigenvectors. The algorithm is analyzed with respect to its running time and experimentally tested for several images. The simultaneous use of k eigenvectors is proposed in [5], too. The latter algorithm is modified, analyzed and tested in [4].

But the above argumentation can be made even more precise.

4. Partitioning with eigenvectors

On the one hand the eigenbasis of an appropriate matrix is used for optimization of a Rayleigh quotient, but on the other hand, as we will see, the eigenvectors themselves have additional clustering properties. The matrices D and W are symmetric and therefore diagonalizable. For $\Delta^{-1}D$ and $\Omega^{-1}W$ this is a consequence of the following (elementary) lemma and the successive remarks, which also hold for W replaced by D and Ω by Δ.

Lemma 4.1. *The following assertions are equivalent.*

(i) $\mathbf{x}$ *is a right eigenvector of* $\Omega^{-1}W$ *with the eigenvalue* λ.
(ii) $\mathbf{x}$ *is a right eigenvector of* $I - \Omega^{-1}W$ *with the eigenvalue* $1-\lambda$.
(iii) $(\Omega\mathbf{x})^t$ *is a left eigenvector of* $\Omega^{-1}W$ *with the eigenvalue* λ.
(iv) $\Omega^{1/2}\mathbf{x}$ *is a right eigenvector of the symmetric matrix* $\Omega^{-1/2}W\Omega^{-1/2}$ *with the eigenvalue* λ.
(v) $\Omega^{1/2}\mathbf{x}$ *is a right eigenvector of the symmetric matrix* $\Omega^{-1/2}\left(\Omega - W\right)\Omega^{-1/2}$ *with the eigenvalue* $1-\lambda$.

Since the Laplacian $\Omega - W$ is symmetric and positive semidefinite, the matrix $\Omega^{-1/2}\left(\Omega - W\right)\Omega^{-1/2}$ is symmetric and positive semidefinite, too. Therefore there exist n linearly independent (orthogonal) eigenvectors with non-negative eigenvalues. Because of the equivalence of 1. and 5. the matrix $\Omega^{-1}W$ is diagonalizable, too, and all eigenvalues are bounded from above by 1. Note that the vector $\mathbf{1}$ is a right eigenvector of $\Omega^{-1}W$ corresponding to the eigenvalue 1.

Theorem 4.1. *Let* $(\mathbf{z}_1,\lambda_1),\ldots,(\mathbf{z}_n,\lambda_n)$ *be the eigenpairs of* W *(resp.* $\Omega^{-1}W$*), where* $\mathbf{z}_i^t\mathbf{z}_j = 0$ *(resp.* $\mathbf{z}_i^t\Omega\mathbf{z}_j = 0$*) for* $i \neq j$*. For* $\mathbf{x} = \sum_{i=1}^n \alpha_i z_i$ *we have*

$$\mathrm{AAss}\,\mathbf{x} = \frac{\sum_{i=1}^n \alpha_i^2\lambda_i}{\sum_{i=1}^n \alpha_i^2} \quad \textit{(resp. } \mathrm{NAss}\,\mathbf{x} = \frac{\sum_{i=1}^n \alpha_i^2\lambda_i}{\sum_{i=1}^n \alpha_i^2}\textit{)}.$$

We see that, if $\mathbf{x}$ is represented as a linear combination of eigenvectors of W (resp. $\Omega^{-1}W$), AAss x (resp. NAss x) is a convex combination, with the squares of the linear factors as coefficients, of the corresponding eigenvalues. To maximize AAss $\mathbf{x}$ (resp. NAss $\mathbf{x}$), where $\mathbf{x}$ looks like some χ_{AB}, one will choose the $\mathbf{x}$ such that it can be combined from eigenvectors with the largest possible eigenvalues. Since the eigenvector of $\Omega^{-1}W$ with the largest possible eigenvalue is useless for combining a χ_{AB}, i.e. $\mathbf{1}^t\Omega\chi_{AB} = 0$, one will take the eigenvector to the second largest eigenvalue. If this almost looks like some χ_{AB} its coefficient in the convex combination of NAss χ_{AB} will be large and the partition (A, B) has a good chance to be optimal.

When we want to minimize ADist (resp. NDist) for some partition and we have an eigenvector of D (resp. $\Delta^{-1}D$) with a small, i.e. negative, eigenvalue looking like some χ_{AB}, its coefficient in the convex combination of NAss χ_{AB} will be large and the partition (A, B) has a good chance to be optimal.

Theorem 4.2. *Let* $(\mathbf{z}_1,\lambda_1),\ldots,(\mathbf{z}_n,\lambda_n)$ *be the eigenpairs of* W *(resp.* $\Omega^{-1}W$*), where* $\mathbf{z}_i^t\mathbf{z}_j = 0$ *(resp.* $\mathbf{z}_i^t\Omega\mathbf{z}_j = 0$*) for* $i \neq j$*. If* $(A_1,\ldots,A_k)$ *is a partition of* V *such that the indicator vectors* $\mathbf{1}_{A_i}$ *are within the linear span of* $(\mathbf{z}_1,\ldots,\mathbf{z}_m)$*, then* $\mathrm{AAss}\,(A_1,\ldots,A_k)$ *(resp.* $\mathrm{NAss}\,(A_1,\ldots,A_k)$*) is a* k*-fold convex combination of* $\lambda_1,\ldots,\lambda_m$.

Proof: Since each indicator vector $\mathbf{1}_{A_i}$ linearly combines from the eigenvectors, its AAss $\mathbf{1}_{A_i}$ is a (1–fold) convex combination of $\lambda_1, \ldots, \lambda_m$. Thus

$$\text{AAss}\,(A_1, \ldots, A_k) = \sum_{i=1}^{k} \text{AAss}\,\mathbf{1}_{A_i}$$

is a k–fold convex combination.

So for finding more than two clusters simultaneously one will choose few piecewise (almost) constant eigenvectors with large eigenvalues. Then the indicated partition has a good chance to be optimal. A completely analogous argumentation holds for the eigenvectors of D and $\Delta^{-1}D$.

The following theorem, which is intimately related to proposition 2 of [4], gives necessary and sufficient conditions for a matrix to have piecewise constant eigenvectors, and thus tells us when the above algorithms are exact.

Theorem 4.3. *Let $M \in \mathbb{R}^n_n$ be a matrix and $(A_1, \ldots, A_k)$ a partition of its index set $\{1, \ldots, n\}$. Then we have:*
M has the eigenvalues $\lambda_1, \ldots, \lambda_k$ and the corresponding right eigenvectors are linearly independent and piecewise constant on $(A_1, \ldots, A_k)$
iff for all $r, s = 1, \ldots, k$ we have

$$\sum_{\sigma \in A_s} m_{i\sigma} = \sum_{\sigma \in A_s} m_{j\sigma} \text{ for all } i, j \in A_r$$

and

$$(\bar{m}_{rs})_{r,s=1,\ldots,k} \text{ where } \bar{m}_{rs} = \sum_{\sigma \in A_s} m_{i\sigma} \text{ for } i \in A_r$$

can be diagonalized with the same eigenvalues $\lambda_1, \ldots, \lambda_k$.

Proof: In this proof let $V := (\mathbf{v}_1, \ldots, \mathbf{v}_k)$ be the matrix of the eigenvectors of M corresponding to the eigenvalues $(\lambda_1, \ldots, \lambda_k)$ and $\Lambda := \text{diag}\,(\lambda_1, \ldots, \lambda_k)$. Let $R = (\mathbf{1}_{A_s})_{s=1,\ldots,k} \in \mathbb{R}^k_n$ be a matrix whose $s - th$ column is the indicator function of A_s and $W \in \mathbb{R}^n_k$ a matrix whose $s-th$ row is a probability measure concentrated on A_s. Note that $WR = I$.
"$\Rightarrow$" Since each eigenvector $\mathbf{v}$ is piecewise constant on A_s, it has a representation $\mathbf{v} = R\mathbf{u}$ where $\mathbf{u} \in \mathbb{R}_k$. The matrix $U := (\mathbf{u}_1, \ldots, \mathbf{u}_k)$ is a square matrix with $V = RU$ and, since V has k linearly independent columns, U is regular. Putting all together we have

$$\begin{aligned} MV &= V\Lambda \\ MRU &= RU\Lambda \\ WMRU &= U\Lambda \\ WMR &= U\Lambda U^{-1}. \end{aligned}$$

Since the $i - th$ row of MR equals the $i - th$ sums of the entries of M within the partition, i.e. $(MR)_i = \left(\sum_{\sigma \in A_s} m_{i\sigma}\right)_{s=1\ldots,k}$, and the probability measures in W are arbitrary, but the above right side does not depend on it, we have

$$(MR)_i = (MR)_j \text{ for } i, j \in A_r,\ r = 1, \ldots, k.$$

Thus the quantities $\bar{m}_{rs} := \sum_{\sigma \in A_s} m_{i\sigma}$ where $i \in A_r$ are uniquely defined by

$$(\bar{m}_{rs})_{r,s=1,\ldots,k} = W\ M\ R = U\ \Lambda\ U^{-1},$$

i.e. $(\bar{m}_{rs})_{r,s=1,\dots,k}$ can be diagonalized with the eigenvalues $\lambda_1,\dots,\lambda_k$.
"$\Leftarrow$" Let $\mathbf{u}$ be an eigenvector of $(\bar{m}_{rs})_{r,s=1,\dots,k}$ with eigenvalue λ. First we show that $R\mathbf{u}$ is an eigenvector of M with eigenvalue λ. We have

$$(\bar{m}_{rs})_{r,s=1,\dots,k}\,\mathbf{u} = WMR\mathbf{u} = \lambda\mathbf{u}.$$

Since W is arbitrary the rows of the matrix MR are constant in each element of the partition and consequently

$$MR\mathbf{u} = \lambda R\mathbf{u}.$$

Thus the matrix M has k linearly independent eigenvectors $R\mathbf{u}_1,\dots,R\mathbf{u}_k$ which are piecewise constant on $(A_1,\dots,A_k)$.

When we have piecewise constant eigenvectors of D the preceding theorem tells us that points are similar if their distances to the sets of the partition indicated by piecewise constant eigenvectors are equal. So the mean distances from each point to the classes have to be equal to the mean distances between the corresponding classes, i.e.

$$\frac{d_{iA_s}}{|A_s|} = \frac{d_{A_rA_s}}{|A_r|\,|A_s|} \text{ for all } i \in A_r,\ r = 1,\dots,k$$

Note that the mean is taken with respect to the number of points. If the piecewise constant eigenvectors correspond to the minimal eigenvalues, ADist will be minimal.

For $\Delta^{-1}D$ the above theorem tells us that points are similar if their distance-ratios are equal, i.e.

$$\frac{d_{iA_s}}{d_{iV}} = \frac{d_{jA_s}}{d_{jV}} \text{ for all } i,j \in A_r,\ r = 1,\dots,k.$$

Note that due to the normalizing property of $\frac{1}{d_{iV}}$ this requirement is less strict than the above. It explains why points in the margin are more likely to be cut off by the eigenvectors of D. To optimize NDist we use the eigenvectors of $\Delta^{-1}D$ corresponding to the smallest eigenvalues.

Points belong to the same class with respect to piecewise constant eigenvectors of W if their similarities to the other classes indicated by piecewise constant eigenvectors are equal. So the mean similarity from each point to each class has to be equal to the mean similarity between the corresponding classes, i.e.

$$\frac{w_{iA_s}}{|A_s|} = \frac{w_{A_rA_s}}{|A_r|\,|A_s|} \text{ for all } i \in A_r,\ r = 1,\dots,k.$$

This criterion requires a very high similarity within the classes and thus favors small classes. It is also quite likely to cut off marginal points (see [5] for experiments). To optimize AAss we use the eigenvectors of W corresponding to the largest eigenvalues. If for example $w_{A_rA_s} = 0$ for $r \neq s$, we obviously have k classes, but points within one class are similar with respect to the present criterion only if they have the same similarity within the class. This strictness is weakened in the following NAss setting.

For points in one class with respect to piecewise constant eigenvectors of $\Omega^{-1}W$ their similarity-ratios are equal, i.e.

$$\frac{w_{iA_s}}{w_{iV}} = \frac{w_{jA_s}}{w_{jV}} \text{ for all } i,j \in A_r,\ r = 1,\dots,k.$$

Note that the similarity-ratio $\frac{w_{iA_s}}{w_{iV}}$ equals the transition probability that the Markov chain on the $W-$weighted graph moves from the state i to class A_s. Since these transition probabilities have to be equal for all points in each class, the Markov chain admits a factorization with respect to the partition $(A_1, \ldots, A_k)$. A factorization is possible if, for example, the state space of the Markov chain consists of k non-communicating classes. Then the eigenvalues corresponding to the piecewise constant eigenvectors are all equal to one. In most applications we will find only piecewise almost constant eigenvectors with eigenvalues near one indicating that the graph corresponds to an almost reducible Markov chain. This situation arises when the graph consists of several classes of very similar points, where the classes themselves are far apart. Because of perturbation theory these classes are found by the eigenvectors of $\Omega^{-1}W$ very well (see [1] and [3] for failure estimates under certain conditions). To optimize NAss we use (almost) piecewise constant eigenvectors of $\Omega^{-1}W$ corresponding to large eigenvalues. Since by lemma 4.1 each right eigenvector of $\Omega^{-1}W$ corresponds to a left eigenvector with the same sign structure and the same eigenvalue, which can be interpreted as a (sub)invariant signed measure, a large eigenvalue tells us that from the signed measure there is only a small part which is cancelled by one step of the Markov chain. Thus the Markov chain tends to stay within the sets of the partition. But sometimes the change in the subinvariant measure is small compared to the number of transitions between the sets, this occurs when the changes themselves cancel out each other.

5. Oscillations of Eigenvectors

For certain arrangements of points in space, i.e. points along lines, circles or band structures in general, one observes optimizing eigenvectors of $\Omega^{-1}W$ changing gradually along these figures. This is a consequence of theorem 2.1, i.e.

$$1 - \mathrm{NAss}\,\mathbf{x} = \frac{\sum_{i<j} w_{ij}\,(x_i - x_j)^2}{\sum_i w_i x_i^2}$$

and $\mathrm{NAss}\,\mathbf{x} = \lambda$ for an eigenvector $\mathbf{x}$ with eigenvalue λ. Since for band structures the w_{ij} are large for neighboring points and small otherwise, the non-constant optimal vector $\mathbf{x}$ changes gradually along the band structure. The next optimal vector $\mathbf{x}'$ does almost the same but has a phase shift or changes faster and so on. If there are parallel structures or concentric circles we also observe piecewise (almost) constant eigenvectors. The corresponding eigenvalues depend on the similarity between the parallel structures. The latter is a consequence of theorem 4.3, since each line is, with the exception of the points near the boundary, an equivalence class of the corresponding Markov chain.

There is a similar argumentation for the eigenvalues of $\Delta^{-1}D$ with the difference that in this case the large distances together with big changes of $\mathbf{x}$ dominate the sum, where the w's are replaced by d's.

The above arguments can be made rigorous for a simplified model without parallel structures.

Lemma 5.1. *A matrix $M = (m_{ij})_{i,j=1,\ldots,n}$ with $m_{i'j'} = m_{ij}$ for $|i'-j'| \in \{|i-j|, n-|i-j|\}$ has the eigenvectors*

$$\begin{aligned} \mathbf{u}_\nu \text{ with } (\mathbf{u}_\nu)_j &= \cos\left(j\frac{2\pi\nu}{n}\right), \text{ where } \nu = 0,\ldots,\left[\frac{n-1}{2}\right], \\ \mathbf{v}_\nu \text{ with } (\mathbf{v}_\nu)_j &= \sin\left(j\frac{2\pi\nu}{n}\right), \text{ where } \nu = 1,\ldots,\left[\frac{n}{2}\right]. \end{aligned}$$

The corresponding eigenvalues are

$$\lambda_\nu = \sum_{i=0}^{n-1} m_{1i+1} \cos\left(i\frac{2\pi\nu}{n}\right).$$

Proof: The k–th line of

$$M \cdot \mathbf{u}_\nu = \begin{pmatrix} \mu_0 & \mu_1 & \mu_2 & \cdots & \mu_2 & \mu_1 \\ \mu_1 & \mu_0 & \mu_1 & \ddots & \ddots & \mu_2 \\ \mu_2 & \mu_1 & \ddots & \ddots & \ddots & \vdots \\ \vdots & \ddots & \ddots & \ddots & \mu_1 & \mu_2 \\ \mu_2 & \ddots & \ddots & \mu_1 & \mu_0 & \mu_1 \\ \mu_1 & \mu_2 & \cdots & \mu_2 & \mu_1 & \mu_0 \end{pmatrix} \cdot \begin{pmatrix} \cos\left(\frac{2\pi\nu}{n}\right) \\ \cos\left(2\frac{2\pi\nu}{n}\right) \\ \vdots \\ \vdots \\ \cos\left((n-1)\frac{2\pi\nu}{n}\right) \\ \cos\left(n\frac{2\pi\nu}{n}\right) \end{pmatrix}$$

simplifies to

$$\begin{aligned} &\mu_0 \cos\left(k\tfrac{2\pi\nu}{n}\right) + \textstyle\sum_{i\geq 1} \mu_i \left(\cos\left(\tfrac{(k-i)2\pi\nu}{n}\right) + \cos\left(\tfrac{(k+i)2\pi\nu}{n}\right)\right) \\ &= \cos\left(k\tfrac{2\pi\nu}{n}\right)\left(\mu_0 + 2\textstyle\sum_{i\geq 1} \mu_i \cos\left(i\tfrac{2\pi\nu}{n}\right)\right) \\ &= \cos\left(k\tfrac{2\pi\nu}{n}\right)\left(m_{11} + \textstyle\sum_{i=1}^{n-1} m_{1i+1} \cos\left(i\tfrac{2\pi\nu}{n}\right)\right), \end{aligned}$$

which proves the assertion concerning the $\mathbf{u}_\nu$. The assertion concerning the $\mathbf{v}_\nu$ is proved in the same way by use of $\sin t = \cos(t - \pi/2)$.

If the matrix D (resp. W) is of this form, then so is $\Delta^{-1/2} D \Delta^{-1/2}$ (resp. $\Omega^{-1/2} W \Omega^{-1/2}$). Note that for all involved matrices we have $\mu_0, \mu_1, \mu_2, \ldots \geq 0$ and in our setting the diagonal elements of Δ (resp. Ω) are all equal to $1/n$. Since by lemma 4.1 (in the (D, Δ) version) a right eigenvector $\mathbf{x}$ of $\Delta^{-1} D$ arises from the eigenvector $\Delta^{1/2}\mathbf{x}$ of $\Delta^{-1/2} D \Delta^{-1/2}$ we have the same eigenpairs for $\Delta^{-1} D$ and for $\Delta^{-1/2} D \Delta^{-1/2}$. The matrix D has the same eigenvectors, too, but the eigenvalues are different. The eigenpairs of $\Omega^{-1} W$ and $\Omega^{-1/2} W \Omega^{-1/2}$ coincide, W has the same eigenvectors but different eigenvalues.
A distance matrix D of this form arises when we arrange points at the successive vertices of a regular n–gon and label them according to this arrangement. If w_{ij} is derived from d_{ij}, the matrix W has the same structure. Thus by the above theorem in such an arrangement we have oscillating eigenvectors, where the corresponding eigenvalues appear twice.

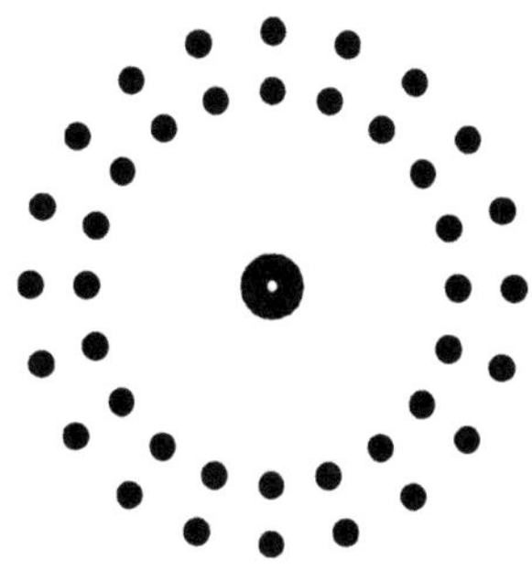

FIGURE 1. Set of points in the plane

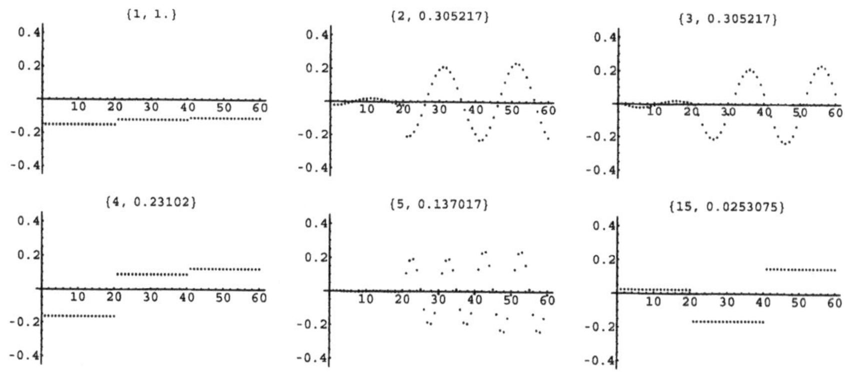

FIGURE 2. Some eigenvectors and eigenvalues of $\Omega^{-1/2}W\Omega^{-1/2}$

Example 5.2. *Oscillating and piecewise constant eigenvectors*
If we want to separate the central points (1-20) of the picture from the outer circles (21-40, 41-60), the first three eigenvectors are of no use, only the fourth eigenvector does the correct separation. together with the first and the $15-th$ *eigenvector we have the complete splitting of the concentric circles.*

6. Conclusions

Piecewise constant eigenvectors of any of the considered matrices are quite well suited for partitioning. Eigenvectors of normalized matrices have better interpretations. Uncritical application of, say sign based, partitioning with the first eigenvectors may produce nonsense in presence of band structures. Whether there are situations, where the distance based partition, despite of its disadvantage of a non-sparse matrix, is preferable, has to be investigated.

References

[1] Deuflhard P., Huisinga W., Fischer A., Schütte Ch. (2000), Identification of almost invariant aggregates in reversible nearly uncoupled Markov chains, Linear Algebra and its Applications 315, 36-59

[2] Kannan R., Vempala S., Vetta A. (2000), On Clusterings - Good, Bad and Spectral, Proceedings of 41st Annual Symposium on Foundations of Computer Science

[3] Ng J., Jordan M., Weiss Y. (2001), On Spectral Clustering: Analysis and an Algorithm, Advances in Neural Information Processing Systems 14: Proceedings

[4] Meila M., Shi J. (2001), A Random Walks View of Spectral Segmentation, Proceedings of the International Workshop on AI and Statistics

[5] Shi J. , Malik J. (2000), Normalized Cuts and Image Segmentation, IEEE Transactions on Pattern Analysis and Machine Intelligence (PAMI)

[6] Seary A, Richards W. (2003), Spectral Methods for Analyzing and Visualizing Networks: An Introduction. In *Dynamic Social Network Modeling and Analysis: Workshop Summary and Papers.*

[7] Weiss Y. (1999), Segmentation using eigenvectors: A unifying view. In *International Conference on Computer Vision*

Christiane Takacs
Department of Stochastics, University of Linz, Austria
christiane.takacs@jku.at

Trends in Mathematics, © 2004 Birkhäuser Verlag Basel/Switzerland

Yaglom Type Limit Theorem for Branching Processes in Random Environment

Vladimir Vatutin and Elena Dyakonova

ABSTRACT: *Let Z_n be the number of particles at moment n in a branching process in random environment with iid probability generating functions. Imposing Spitzer's condition on the associated random walk and applying the quenched approach we find the asymptotics of the survival probability of the process up to moment n as $n \to \infty$ and prove a Yaglom type limit theorem (under appropriate scaling) for Z_n conditioned on $Z_n > 0$.*[1]

1. Introduction and main results

A branching process in random environment (BPRE) is a natural generalization of the simple Galton-Watson process. At each generation the offspring distribution is chosen according to a probability distribution determined by the environmental sequence, $\{\pi_n\}$. In the present paper we use the so-called quenched approach to study the asymptotic behavior of the survival probability up to moment n and the size of the population at moment n conditioned on the survival of the process up to this moment for a wide class of BPRE's that includes, in particular, the classical critical BPRE's. Our results generalize and refine the corresponding statements established in [7] and [8].

Let $\{Z_n,\ n \in \mathsf{N}_0\}$ be a BPRE in which the states of the environment are specified by infinite-dimensional vectors $\{\pi_n,\ n \in \mathsf{N}_0\}$, where

$$\pi_n = \left\{\pi_n^{(0)}, \pi_n^{(1)}, \pi_n^{(2)}, \ldots\right\},\ \pi_n^{(i)} \geq 0,\ \sum_{i=0}^{\infty} \pi_n^{(i)} = 1,\ n \in \mathsf{N}_0 = \{0, 1, 2, \ldots\}, \qquad (1)$$

and the tuples π_n are independent and identically distributed. Given the environment the evolution of the BPRE is described by the relations

$$Z_0 = 1,\ \mathbf{E}\left[s^{Z_{n+1}} \mid f_0, f_1, \ldots, f_n; Z_0, Z_1, \ldots, Z_n\right] = \left(f_n(s)\right)^{Z_n}$$

in which

$$f_n(s) = \sum_{i=0}^{\infty} \pi_n^{(i)} s^i \qquad (2)$$

is the offspring generating function of particles of the n-th generation.

Let $X_k := \ln f_{k-1}^{'}(1),\ \eta_k := f_{k-1}^{''}(1)\left(f_{k-1}^{'}(1)\right)^{-2},\ k \in \mathsf{N} = \{1, 2, \ldots\}$; $X \stackrel{d}{=} X_1,\ S_0 := 0,\ S_n := X_1 + \ldots + X_n,\ n \geq 1$.

In the sequel $\{S_n\}_{n \geq 0}$ is called the *associated* random walk. We suppose that the associated random walk satisfies **Spitzer's condition**:

---

[1]Supported in part by grants RFBR 02-01-00266, RFBR-DFG 02-01-04002 and Russian Scientific School 1758.2003.1

Condition $A1$. There exists $0 < \rho < 1$ such that

$$\frac{1}{n}\sum_{m=1}^{n}\mathbf{P}(S_m > 0) \to \rho, \quad n \to \infty,$$

or, what is the same (see [4]) $\lim_{n\to\infty}\mathbf{P}(S_n > 0) = \rho$.

Condition $A1^*$. The distribution of X is absolutely continuous and condition $A1$ holds.

Our next assumption is related with truncated moments of the distributions π_n :

$$\Theta(a) := \sum_{k=a}^{\infty} k^2\pi^{(k)}/\Big(\sum_{k=0}^{\infty} k\pi^{(k)}\Big)^2, \quad a \in \mathsf{N}_0.$$

To formulate the restrictions we impose on $\Theta(a)$ some additional functions characterizing $\{S_n\}_{n\geq 0}$ are needed.

Let $\gamma_0 := 0, \quad \gamma_{j+1} := \min(n > \gamma_j : S_n < S_{\gamma_j})$ and $\Gamma_0 := 0, \quad \Gamma_{j+1} := \min(n > \Gamma_j : S_n > S_{\Gamma_j})$, $j \geq 0$, be the strict descending and ascending ladder epochs of $\{S_n\}_{n\geq 0}$. Introduce the functions

$$V(x) := \sum_{j=0}^{\infty}\mathbf{P}(S_{\gamma_j} \geq -x), \quad x > 0, \quad V(0) = 1, \quad V(x) = 0, \quad x < 0,$$

and

$$U(x) := \sum_{j=0}^{\infty}\mathbf{P}(S_{\Gamma_j} \leq x), \quad x > 0, \quad U(0) = 1, \quad U(x) = 0, \quad x < 0.$$

If condition A1 holds then S_n is an oscillating random walk. Hence, by Lemma 1 in [6] $U(x)$ and $V(x)$ are harmonic functions, that is,

$$\mathbf{E}U(x - X) = U(x), \quad \mathbf{E}V(x + X) = V(x), \quad x \geq 0. \tag{3}$$

Condition $A2$. There are numbers $\varepsilon > 0$ and $a \in \mathsf{N}_0$ such that

$$\mathbf{E}[(\log^+\Theta(a))^{\frac{1}{\rho}+\varepsilon}] < \infty \quad \text{and} \quad \mathbf{E}[V(X)(\log^+\Theta(a))^{1+\varepsilon}] < \infty, \tag{4}$$

$$\mathbf{E}[(\log^+\Theta(a))^{\frac{1}{1-\rho}+\varepsilon}] < \infty \quad \text{and} \quad \mathbf{E}[U(-X)(\log^+\Theta(a))^{1+\varepsilon}] < \infty. \tag{5}$$

Denote by $\mathbf{L} = \{\mathfrak{L}\}$ the set of all proper probability laws $\mathfrak{L}(\xi)$ of nonnegative random variables ξ, and let $\mathbf{L}^+ \subset \mathbf{L}$ be the set of the probability laws in $\mathbf{L}$ strictly concentrated on $(0, \infty)$: $\mathbf{P}(\xi > 0) = 1$. Let, further, $\mathbf{\Phi} = \{\Phi\}$ and $\mathbf{\Phi}^+ \subset \mathbf{\Phi}$ be the metric spaces of the Laplace transforms $\Phi(\lambda) = \int_0^\infty e^{-\lambda x}\mathfrak{L}(dx)$, $\lambda \in [0, \infty)$, of the laws from $\mathbf{L}$ and $\mathbf{L}^+$, respectively, endowed with the metric

$$d(\Phi_1, \Phi_2) = \sup_{1\leq\lambda\leq 2}|\Phi_1(\lambda) - \Phi_2(\lambda)|.$$

Since Laplace transforms are completely determined by their values in any interval of the positive half-line, convergence $\Phi_n \to \Phi$ as $n \to \infty$ in metric d is equivalent to weak convergence $\mathfrak{L}_n \to \mathfrak{L}$ of the respective probability laws.

Put, finally,

$$f_{k,n}(s) := f_k(f_{k+1}(\ldots(f_{n-1}(s))\ldots)), \ 0 \leq k \leq n-1, \ f_{n,n}(s) := s,$$
$$f_{n,0}(s) := f_{n-1}(f_{n-2}(\ldots(f_0(s))\ldots)), \ n \geq 1.$$

Throughout the paper the symbols $\mathbf{E}$ and $\mathbf{P}$ are used to denote the expectation and probability with respect to the measure over the environment (with some obvious exceptions causing no confusion) while the symbols $\mathbf{E}_{\pi}$, $\mathbf{P}_{\pi}$ always stand for the conditional expectation and probability given the environment $\pi = (\pi_1, \pi_2, ..., \pi_n, ...)$.

Let $\tau(n) := \min\{k \in [0,n] : S_j \geq S_k, \ j = 0, 1, \ldots, n\}$ be the left-most point of the interval $[0,n]$ at which the minimal value of S_j, $j = 0, 1, ..., n$ is attained.

Theorem 1.1. *Let conditions $A1^*$ and $A2$ be valid. Then the distributions of the random variables*

$$\zeta_{0,n} := e^{-S_{\tau(n)}} \mathbf{P}_{\pi}(Z_n > 0) = e^{-S_{\tau(n)}} (1 - f_{0,n}(0)), \ n \in \mathsf{N}_0, \tag{6}$$

weakly converge as $n \to \infty$ to the distribution of a random variable $\zeta \in [0,1]$ being positive with probability 1.

Denote

$$\mathbf{m}_n^* := \mathbf{E}_{\pi}[Z_n \mid Z_n > 0] = e^{S_n}/(1 - f_{0,n}(0)), \quad \widehat{Y}_n := Z_n/\mathbf{m}_n^*$$

and set

$$\Phi_n(\lambda, \pi) = \Phi_n(\lambda) := \mathbf{E}_{\pi}\left[e^{-\lambda \widehat{Y}_n} \mid Z_n > 0\right].$$

Theorem 1.2. *Let conditions $A1^*$ and $A2$ be valid. Then*

$$\Phi_n \Rightarrow \Phi, \quad n \to \infty, \tag{7}$$

where $\Phi(\lambda)$, $\lambda \in [0, \infty)$, is the Laplace transform of a probability law belonging to $\mathbf{L}^+$ with probability 1 and the symbol $\Rightarrow$ stands for the weak convergence of probability measures on the Borel sets of the metric space $\mathbf{\Phi}$.

Corollary 1.3. *If, under the conditions of Theorem 1.2, the offspring generating functions of the individuals of each generation are fractional-linear:*

$$f_n(s) = 1 - \frac{p_n}{1 - q_n} + \frac{p_n s}{1 - q_n s}, \quad 0 < p_n + q_n < 1, \ p_n, q_n \geq 0, \quad n \in \mathsf{N}_0,$$

then

$$\Phi(\lambda) = \frac{1}{1 + \lambda}. \tag{8}$$

The last result is an analogue of the corresponding Yaglom-type limit theorem for the ordinary critical Galton-Watson processes. It seems, however, that (8) is not valid in the general situation (see [2] for a relevant discussion of this problem).

Remark 1. Our results generalize those in [7] and [8] in which much stronger conditions on the characteristics of the branching processes are imposed. Moreover, here we prove that the limit law corresponding to Φ has no atom at zero with probability 1. This problem remained unsolved in the mentioned papers.

Remark 2. For the first time branching processes in random environment meeting conditions $A1$ and $A2$ were investigated in [1]. The mentioned paper deals with the annealed approach and studies the same problems which we analyze here under the quenched approach.

2. Change of measures

Let $\{S_k\}_{k \geq 0}$ be the associated random walk. Denote $M_n := \max_{0 \leq k \leq n} S_k$, $L_n := \min_{0 \leq k \leq n} S_k = S_{\tau(n)}$, $\Gamma = \min\{n \geq 1 : S_n \geq 0\}$, $\gamma = \min\{n \geq 1 : S_n < 0\}$ and set

$$m_n(x) := \mathbf{P}(L_n \geq -x), \qquad \widetilde{m}_n(x) := \mathbf{P}(M_n \leq x), \qquad x \geq 0.$$

In what follows the symbols c, c_i, $i = 1, 2, \ldots$ denote positive constants which may be different in different formulas.

Lemma 2.1. *(compare, for instance, with [8] or [3], Section 8.9) Under condition A1 there exist functions $l_1(n)$ and $l_2(n)$ slowly varying at infinity and $l_1(n)l_2(n) \sim \pi/\sin\pi\rho$, $n \to \infty$, and absolute constants $c_1 > 0$, $c_2 > 0$ such that for all $n \geq 1$ and all $x \in [0,\infty)$*

$$m_n(x) \leq c_1 V(x)/\left(n^{1-\rho} l_1(n)\right), \qquad \widetilde{m}_n(x) \leq c_2 U(x)/\left(n^{\rho} l_2(n)\right).$$

In addition, for any fixed $x \in (0,\infty)$ as $n \to \infty$

$$m_n(x) \sim V(x)/\left(n^{1-\rho} l_1(n)\right), \qquad \widetilde{m}_n(x) \sim U(x)/\left(n^{\rho} l_2(n)\right). \tag{9}$$

If condition A1 holds then*

$$\mathbf{P}(\gamma > n) \sim 1/\left(n^{1-\rho} l_1(n)\right), \qquad \mathbf{P}(\Gamma > n) \sim 1/\left(n^{\rho} l_2(n)\right). \tag{10}$$

Denote by Π the set of all infinite-dimensional vectors (1) and by Σ the natural σ-algebra generated by the subsets of Π. In view of (2) the offspring generating function $f_n(s)$ of particles of the n-th generation may be treated as a random vector π_n on the measurable space (Π,Σ). The joint distribution of the tuple $f_0(s), f_1(s), ..., f_{n-1}(s)$ is specified by the product-measure $\mathbf{P}^n = \mathbf{P}\times\mathbf{P}\times\cdots\times\mathbf{P}$ on the measurable space

$$(\Pi^n,\Sigma^n) = (\Pi,\Sigma)\times(\Pi,\Sigma)\times\cdots\times(\Pi,\Sigma).$$

The infinite sequence $f_0(s), f_1(s), ..., f_{n-1}(s), ...$ is a random element specified on the measurable space $(\Pi^\infty,\Sigma^\infty)$ (where Σ^∞ is a completion of the infinite direct product $\Sigma\times\Sigma\cdots$ to a σ-algebra) and distributed according to the measure $\mathbf{P}^\infty$ which is obtained by extension of the sequence $\{\mathbf{P}^n\}_{n\geq 1}$ to $(\Pi^\infty,\Sigma^\infty)$. Such a measure exists and is unique by Ionescu Tulcea theorem on extending a measure and the existence of a random sequence. If $\mathbf{P}(\pi_n^{(0)} = 1) = 0$ (and we assume this condition throughout the paper), then $X_{n+1} := \log\left(\Sigma_{i=0}^{\infty} i\pi_n^{(i)}\right) = \log f_n'(1), n \in \mathsf{N}_0$, are well-defined random variables on the probability space $(\Pi^\infty,\Sigma^\infty,\mathbf{P}^\infty)$.

Let $\{f_n^-\}_{n\geq 0}$ and $\{f_n^+\}_{n\geq 0}$ be two independent sequences (realizations) of the random environment and let $\{S_n^-\}_{n\geq 0}$ and $\{S_n^+\}_{n\geq 0}$ be the corresponding associated random walks. Later on any characteristics or random variables related with $\{f_n^-\}_{n\geq 0}$ and $\{f_n^+\}_{n\geq 0}$, are supplied with the symbols $-$ or $+$, respectively. For instance, we write $L_n^+ = \min_{0\leq j\leq n} S_j^+$, $\Gamma^- = \min\{n \geq 1 : S_n^- \geq 0\}$ and $\gamma^+ = \min\{n \geq 1 : S_n^+ < 0\}$. Set $A_{k,p} = \{\Gamma^- > k, \gamma^+ > p\}$. For any bounded and $(\Pi^k\times\Pi^p,\Sigma^k\times\Sigma^p)$ – measurable function $\psi : \Pi^k\times\Pi^p \to \mathbf{R}$ let

$$\begin{aligned}&\widehat{\mathbf{E}}\left[\psi\left(f_0^-,...,f_{k-1}^-;f_0^+,...,f_{p-1}^+\right)\right]\\ &\quad := \mathbf{E}\left[\psi\left(f_0^-,...,f_{k-1}^-;f_0^+,...,f_{p-1}^+\right)U\left(-S_k^-\right)V(S_p^+)I\{A_{k,p}\}\right].\end{aligned} \tag{11}$$

For $S_k^- \leq 0$ and $S_p^+ \geq 0$ we have

$$\mathbf{E}\left[U\left(-S_{k+1}^-\right)\mid S_k^-\right] = U\left(-S_k^-\right), \quad \mathbf{E}\left[V(S_{p+1}^+)\mid S_p^+\right] = V(S_p^+).$$

Hence it follows that the definition of $\widehat{\mathbf{E}}[\bullet]$ is consistent for all k and p. Moreover, if $\psi \equiv \psi^-\left(f_0^-,...,f_{k-1}^-\right)$(or $\psi \equiv \psi^+\left(f_0^+,...,f_{p-1}^+\right)$), i.e., if only the first k (the last p) variables of ψ are essential then by (3) and independency of the sequences $\{f_n^-\}_{n\geq 0}$ and $\{f_n^+\}_{n\geq 0}$

$$\begin{aligned}\mathbf{E}[\,\psi^-(f_0^-,\ \ ...,\ \ &f_{k-1}^-)U\left(-S_k^-\right)V(S_p^+)I\{A_{k,p}\}]\\ &= \mathbf{E}\left[\psi^-\left(f_0^-,...,f_{k-1}^-\right)U\left(-S_k^-\right)I\{\Gamma^- > k\}\right]\end{aligned}$$

$$=: \quad \mathbf{E}^{-}\left[\psi^{-}\left(f_0^{-},...,f_{k-1}^{-}\right)\right], \tag{12}$$

where $I\{A\}$ is the indicator of the event A, and

$$\begin{aligned}\mathbf{E}[\,\psi^{+}(f_0^{+}, \quad ..., \quad & f_{p-1}^{+})\,U\left(-S_k^{-}\right)V(S_p^{+})I\left\{A_{k,p}\right\}] \\ = \quad & \mathbf{E}\left[\psi^{+}\left(f_0^{+},...,f_{p-1}^{+}\right)V(S_p^{+})I\left\{\gamma^{+} > p\right\}\right] \\ =: \quad & \mathbf{E}^{+}\left[\psi^{+}\left(f_0^{+},...,f_{p-1}^{+}\right)\right]. \end{aligned} \tag{13}$$

Relations (12), (13) and (11) specify in a natural way measures $\mathbf{P}^{\pm}$ on $(\Pi^{\infty}, \Sigma^{\infty})$ and $\widehat{\mathbf{P}} = \mathbf{P}^{-} \times \mathbf{P}^{+}$ on $(\Pi^{\infty} \times \Pi^{\infty}, \Sigma^{\infty} \times \Sigma^{\infty})$. These measures are defined for any Borel sets $A_0, ..., A_{k-1}, B_0, ..., B_{p-1} \subset \Pi$ by taking

$$\psi^{-}\left(f_0^{-},...,f_{k-1}^{-}\right) := I\left(f_i^{-} \in A_i, i = 0,...,k-1\right),$$

$$\psi^{+}\left(f_0^{+},...,f_{p-1}^{+}\right) := I\left(\, f_j^{+} \in B_j, j = 0,...,p-1\right)$$

and $\psi := \psi^{-} \times \psi^{+}$ in (12), (13) and (11), respectively. In turn, $\mathbf{P}^{-}$ and $\mathbf{P}^{+}$ induce in a natural way the corresponding measures on $\{S_k^{-}\}_{k\geq 0}$ and $\{S_k^{+}\}_{k\geq 0}$ for which we keep the same notation $\mathbf{P}^{-}$ and $\mathbf{P}^{+}$. In addition, we use the symbols $\mathfrak{L}^{-}$ and $\mathfrak{L}^{+}$ to denote the laws generated by the measures $\mathbf{P}^{-}$ and $\mathbf{P}^{+}$. The next lemma shows the importance of $\mathbf{P}^{-}$, $\mathbf{P}^{+}$ and $\widehat{\mathbf{P}}$.

Lemma 2.2. *If condition A1 holds then for any bounded and* $\left(\Pi^{k} \times \Pi^{p}, \Sigma^{k} \times \Sigma^{p}\right)$*–measurable function* $\psi : \Pi^{k} \times \Pi^{p} \to \mathbf{R}$

$$\lim_{\min(n,r)\to\infty} \mathbf{E}\left[\psi\left(f_0^{-},...,f_{k-1}^{-}; f_0^{+},...,f_{p-1}^{+}\right) \mid A_{n,r}\right]$$
$$= \widehat{\mathbf{E}}\left[\psi\left(f_0^{-},...,f_{k-1}^{-}; f_0^{+},...,f_{p-1}^{+}\right)\right]. \tag{14}$$

Proof. For $k \leq n$ and $p \leq r$ we have

$$\mathbf{E}\left[\psi\left(f_0^{-},...,f_{k-1}^{-}; f_0^{+},...,f_{p-1}^{+}\right) \mid A_{n,r}\right]$$
$$= \frac{\mathbf{E}\left[\psi\left(f_0^{-},...,f_{k-1}^{-}; f_0^{+},...,f_{p-1}^{+}\right) I\left\{A_{k,p}\right\} \widetilde{m}_{n-k}^{-}\left(-S_k^{-}\right) m_{r-p}^{+}\left(S_p^{+}\right)\right]}{\mathbf{P}(\Gamma^{-} > n)\mathbf{P}(\gamma^{+} > r)}. \tag{15}$$

It follows from Lemma 2.1 that for any fixed k, m and $x \geq 0$

$$\lim_{n\to\infty} \frac{\widetilde{m}_{n-k}^{-}(x)}{\mathbf{P}(\Gamma^{-} > n)} = U(x), \quad \lim_{r\to\infty} \frac{m_{r-m}^{+}(x)}{\mathbf{P}(\gamma^{+} > r)} = V(x).$$

Moreover, there exists a constant $C > 0$ such that for all $x \geq 0$ and all $n \geq k$ and $r \geq p$

$$\frac{\widetilde{m}_{n-k}^{-}(x)}{\mathbf{P}(\Gamma^{-} > n)} \leq CU(x) \quad \text{and} \quad \frac{m_{r-p}^{+}(x)}{\mathbf{P}(\gamma^{+} > r)} \leq CV(x).$$

Hence using the dominated convergence theorem and passing to the limit in (15) as $\min(n,r) \to \infty$ we get (14).

The lemma is proved.

It will be convenient for us to use Lemma 2.2 in a wider context. To this aim we need the following generalization of Lemma 2.5 in [1].

Let $\mathfrak{F}_n$ be the σ-algebra generated by the random variables $f_0, f_1, \ldots, f_{n-1}$; $Z_0 = 1, Z_1, \ldots, Z_n$ and $\mathfrak{F} = \bigvee_n \mathfrak{F}_n$ be the completion of $\bigvee_n \mathfrak{F}_n$ to a σ-algebra. The σ-algebras $\mathfrak{F}^{-}$ and $\mathfrak{F}^{+}$ are defined in a similar way.

Lemma 2.3. *Let condition A1* be valid and let $T_{l,p}$, $l,p \in \mathsf{N}$ be a tuple of uniformly bounded random variables such that $T_{l,p}$ is measurable with respect to the σ-algebra $\mathfrak{F}_l^- \times \mathfrak{F}_p^+$ for any pair l,p. Then*

$$\lim_{\min(n,r)\to\infty} \mathbf{E}[T_{l,p} \mid A_{n,r}] = \hat{\mathbf{E}}[T_{l,p}]. \tag{16}$$

More generally, if the array $\{T_{n,r},\ n,r \in \mathsf{N}\}$ consists of uniformly bounded random variables adapted to the filtration $\mathfrak{F}^- \times \mathfrak{F}^+$ and $\lim_{\min(n,r)\to\infty} T_{n,r} =: T$ exists $\hat{\mathbf{P}}$–a.s., then

$$\lim_{\min(n,r)\to\infty} \mathbf{E}[T_{n,r} \mid A_{n,r}] = \hat{\mathbf{E}}[T]. \tag{17}$$

Proof. Relation (16) can be proved the same way as Lemma 2.2. To demonstrate (17) observe that for any numbers $\sigma > 1$ and $l \le \min(n,r)$, $l \in \mathsf{N}$

$$\begin{aligned}
|\mathbf{E}[T_{n,r} - T_{l,l} \mid A_{\sigma n,\sigma r}]| &\le \mathbf{E}\left[|T_{n,r} - T_{l,l}| \frac{\widetilde{m}^-_{(\sigma-1)n}(-S_n^-)}{\widetilde{m}^-_{\sigma n}(0)} \frac{m^+_{(\sigma-1)r}(S_r^+)}{m^+_{\sigma r}(0)} I\{A_{n,r}\}\right] \\
&\le c\Big(\frac{\sigma}{\sigma-1}\Big)^{\rho}\Big(\frac{\sigma}{\sigma-1}\Big)^{1-\rho} \mathbf{E}\left[|T_{n,r} - T_{l,l}| U(-S_n^-) V(S_r^+) I\{A_{n,r}\}\right] \\
&= \frac{c\sigma}{\sigma-1} \hat{\mathbf{E}}[|T_{n,r} - T_{l,l}|].
\end{aligned}$$

Hence by the conditions of the lemma and the dominated convergence theorem we get

$$\limsup_{l\to\infty} \limsup_{\min(n,r)\to\infty} \left|\mathbf{E}[T_{n,r} - T_{l,l} \mid A_{\sigma n,\sigma r}]\right| = 0.$$

In particular,

$$\mathbf{E}[T_{n,r} I\{A_{\sigma n,\sigma r}\}] = \big(\hat{\mathbf{E}}[T] + o(1)\big)\mathbf{P}(A_{\sigma n,\sigma r}).$$

Consequently,

$$\begin{aligned}
&|\mathbf{E}[T_{n,r} I\{A_{n,r}\}] - \hat{\mathbf{E}}[T]\mathbf{P}(A_{n,r})| \\
&\quad \le |\mathbf{E}[T_{n,r} I\{A_{\sigma n,\sigma r}\}] - \hat{\mathbf{E}}[T]\mathbf{P}(A_{\sigma n,\sigma r})| + c\big|\mathbf{P}(A_{\sigma n,\sigma r}) - \mathbf{P}(A_{n,r})\big| \\
&\qquad \le \Big(o(1) + c\big((1-\sigma^{-\rho})\sigma^{\rho-1} + (1-\sigma^{1-\rho})\big)\Big)\mathbf{P}(A_{n,r}),
\end{aligned}$$

since by Lemma 2.1

$$\begin{aligned}
\big|\mathbf{P}(A_{\sigma n,\sigma r}) - \mathbf{P}(A_{n,r})\big| &= \big|\mathbf{P}(\Gamma^- > n\sigma)\mathbf{P}(\gamma^+ > \sigma r) - \mathbf{P}(\Gamma^- > n)\mathbf{P}(\gamma^+ > r)\big| \\
&\le \big|\mathbf{P}(\Gamma^- > n\sigma) - \mathbf{P}(\Gamma^- > n)\big|\mathbf{P}(\gamma^+ > \sigma r) \\
&\quad + \big|\mathbf{P}(\gamma^+ > r) - \mathbf{P}(\gamma^+ > \sigma r)\big|\mathbf{P}(\Gamma^- > n) \\
&\le c((1-\sigma^{-\rho})\sigma^{\rho-1} + (1-\sigma^{1-\rho}))\mathbf{P}(A_{n,r}).
\end{aligned}$$

Therefore,

$$\mathbf{E}[T_{n,r} I\{A_{n,r}\}] - \hat{\mathbf{E}}[T]\,\mathbf{P}(A_{n,r}) = o(\mathbf{P}(A_{n,r})),$$

as $\min(n,r) \to \infty$ which is equivalent to (17).

Remark 3. It follows easily from the proof of Lemma 2.3 that under condition *A1**

$$\limsup_{\min(l,p)\to\infty} \limsup_{\min(n,r)\to\infty} \left|\mathbf{E}[T_{n,r} - T_{l,p} \mid A_{n,r}]\right| = 0. \tag{18}$$

Put

$$\chi_j(s) := \frac{1}{1-f_j(s)} - \frac{1}{f_j'(1)(1-s)}, \; 0 \le s \le 1.$$

In the subsequent arguments we essentially use the following known results.

Lemma 2.4. (*see, for instance,* [5]). *Let* $f_j \neq 1, 0 \le j \le n-1$. *Then for any* $0 \le s < 1$ *and* $0 \le m \le n-1$

$$\frac{1}{1-f_{m,n}(s)} = \frac{e^{-S_n+S_m}}{1-s} + \sum_{j=m}^{n-1} \eta_{j,n}(s)\, e^{-S_j+S_m}, \tag{19}$$

where

$$\eta_{j,n}(s) := \chi_j(f_{j+1,n}(s)) \le \eta_{j+1} = \frac{f_j''(1)}{(f_j'(1))^2}. \tag{20}$$

If, in addition, $f_n(s)$ *are fractional-linear for all* $n = 0, 1, \ldots,$ *then*

$$\eta_{j,n}(s) = \eta_{j+1}/2, \qquad j = 0, 1, \ldots, n-1.$$

Lemma 2.5. [1] *If conditions A1 and (4) are valid, then*

$$\sum_{j=0}^{\infty} \eta_{j+1} e^{-S_j} < \infty \qquad \mathbf{P}^+ - a.s.$$

If conditions A1 and (5) are valid, then

$$\sum_{j=0}^{\infty} \eta_j e^{S_j} < \infty \qquad \mathbf{P}^- - \; a.s.$$

3. Proofs of the main results

Proof of Theorem 1.1. Clearly, to demonstrate the theorem it suffices to establish the existence of a random variable ζ such that

$$\hat{\mathbf{P}}(\zeta \in (0,1]) = 1 \quad \text{and} \quad \lim_{n\to\infty} \mathbf{E}[g(\zeta_{0,n})] = \hat{\mathbf{E}}[g(\zeta)]$$

for any function $g(x)$ that is bounded and uniformly continuous in $x \in \mathbf{R}$.

We have

$$\mathbf{E}[g(\zeta_{0,n})] = \sum_{k=0}^{n} \mathbf{E}[g(\zeta_{0,n}) \mid \tau(n) = k]\mathbf{P}(\tau(n) = k). \tag{21}$$

It is known ([9], Ch. IV, § 20) that under condition A1

$$n^{-1}\tau(n) \xrightarrow{w} \tau, \quad n \to \infty \tag{22}$$

(here the symbol $\xrightarrow{w}$ means convergence in distribution), where τ is a random variable having the generalized arc-sine distribution with parameter ρ. The limiting distribution is absolutely continuous on (0,1) and has no atoms at points 0 and 1. Thus, for any $\varepsilon > 0$ there exists $\delta \in (0, 1/2)$ such that $\mathbf{P}(\tau(n) \notin [n\delta, n(1-\delta)]) < \varepsilon$ for all sufficiently large n. Hence, to establish the desired result it suffices to show the existence of a random variable ζ such that

$$\lim_{\min(k,n-k)\to\infty} \mathbf{E}[g(\zeta_{0,n}) \mid \tau(n) = k] = \hat{\mathbf{E}}[g(\zeta)] \tag{23}$$

and then to apply the dominated convergence theorem to (21).

We proceed to prove (21). Rewriting $\zeta_{0,n}$ as

$$\zeta_{0,n} = \left(1 - f_{0,\tau(n)}(f_{\tau(n),n}(0))\right)e^{-S_{\tau(n)}},$$

we get

$$\begin{aligned}\mathbf{E}[g(\zeta_{0,n}) \,|\, \tau(n)=k] &= \mathbf{E}[g((1-f_{0,k}(f_{k,n}(0)))e^{-S_k}) \,|\, \tau(n)=k] \\ &= \mathbf{E}[T_{k,n-k} \,|\, A_{k,n-k}],\end{aligned}$$

where $T_{k,n-k} := g\left((1-f^-_{k,0}(f^+_{0,n-k}(0)))e^{-S^-_k}\right)$ and $\{f^-_n\}_{n\geq 0}$ and $\{f^+_n\}_{n\geq 0}$ are independent sequences of offspring generating functions with $S^\pm_k = \sum_{j=0}^{k-1}\log(f^\pm_j)'(1)$.

Set $\zeta^-_k(s) := (1-f^-_{k,0}(s))e^{-S^-_k}$ and $\eta^-_{k,0}(s) := \chi^-_{k-1}(f^-_{k-1,0}(s))$, $s \in [0,1)$, $k \in \mathsf{N}$. Since $\zeta^-_{k+1}(s) \leq \zeta^-_k(s), k=1,2,...,$ the limit $\lim_{k\to\infty}\zeta^-_k(s) := \zeta^-(s)$ exists $\hat{\mathbf{P}}-$ a.s. By Lemmas 2.4, 2.5 and relations

$$\begin{aligned}\frac{1}{\zeta^-(s)} &= \lim_{k\to\infty}\frac{1}{\zeta^-_k(s)} = \lim_{k\to\infty}\frac{e^{S^-_k}}{1-f^-_{k,0}(s)} = \lim_{k\to\infty}\left(\frac{1}{1-s}+\sum_{j=1}^{k}\eta^-_{j,0}(s)e^{S^-_j}\right) \\ &= \frac{1}{1-s}+\sum_{j=1}^{\infty}\eta^-_{j,0}(s)e^{S^-_j} \leq \frac{1}{1-s}+\sum_{j=1}^{\infty}\eta^-_j e^{S^-_j} < \infty \quad \mathbf{P}^- -\text{ a.s.} \end{aligned}\tag{24}$$

the limiting function $\zeta^-(s)$ is $\hat{\mathbf{P}}$–a.s. uniformly continuous in s from any interval contained in $[0,1)$. For the same reason $\lim_{n-k\to\infty} f^+_{0,n-k}(0) := q^+$ exists $\hat{\mathbf{P}}-$ a.s. and, moreover, $\hat{\mathbf{P}}(q^+<1)=1$, since S^+_i tends to infinity $\mathbf{P}^+-$ a.s. as $i\to\infty$ and, therefore,

$$\begin{aligned}\frac{1}{1-q^+} = \lim_{i\to\infty}\frac{1}{1-f^+_{0,i}(0)} &= \lim_{i\to\infty}\left(e^{-S^+_i}+\sum_{j=0}^{i-1}\eta^+_{j,i}(0)\,e^{-S^+_j}\right) \\ &\leq \sum_{j=0}^{\infty}\eta^+_{j+1}e^{-S^+_j} < \infty \qquad \mathbf{P}^+-\text{a.s.}\end{aligned}\tag{25}$$

Hence it follows that $\zeta := \lim_{\min(k,n-k)\to\infty}\zeta^-_k(f^+_{0,n-k}(0)) = \zeta^-(q^+)$ exists $\hat{\mathbf{P}}-$ a.s. and $\hat{\mathbf{P}}(\zeta\in(0,1])=1$. It is clear that ζ is a random variable possessing all the properties claimed in Theorem 1.1 and, in addition, $\lim_{\min(k,n-k)\to\infty}T_{k,n-k} = g(\zeta)$ $\hat{\mathbf{P}}-$ a.s. Applying now Lemma 2.3 we obtain

$$\lim_{\min(k,n-k)\to\infty}\mathbf{E}[T_{k,n-k}\,|\,A_{k,n-k}] = \hat{\mathbf{E}}[g(\zeta)],$$

proving (23) and, as a result, Theorem 1.1.

Proof of Theorem 1.2. Evidently,

$$\Phi_n(\lambda) = \mathbf{E}_\pi\left[e^{-\lambda\hat{Y}_n}\mid Z_n>0\right] = 1-\frac{1-f_{0,n}(e^{-\lambda/\mathbf{m}^*_n})}{1-f_{0,n}(0)}$$

with $1/\mathbf{m}^*_n = (1-f_{0,n}(0))e^{-S_n} = \zeta_{0,n}e^{S_{\tau(n)}-S_n}$ and, therefore, for any function $G:\mathbf{\Phi}\to\mathbf{R}$ bounded and uniformly continuous on $\mathbf{\Phi}$ with respect to metric d and any integer $k\in[0,n]$

$$\mathbf{E}[G(\Phi_n(\bullet))\,|\,\tau(n)=k] = \mathbf{E}[G(R_{k,n-k}(\bullet))\,|\,A_{k,n-k}],$$

where

$$
\begin{aligned}
R_{k,n-k}(\lambda) &:= 1-\frac{1-f_{k,0}^{-}(f_{0,n-k}^{+}(e^{-\lambda/\mathbf{m}_n^*}))}{1-f_{k,0}^{-}(f_{0,n-k}^{+}(0))} \\
&= 1-\frac{\zeta_k^{-}(f_{0,n-k}^{+}(\exp\{-\lambda\zeta_k^{-}(f_{0,n-k}^{+}(0))e^{-S_{n-k}^{+}}\}))}{\zeta_k^{-}(f_{0,n-k}^{+}(0))}. \qquad (26)
\end{aligned}
$$

Thus, as in Theorem 1.1 it suffices to show that

$$\lim_{\min(k,n-k)\to\infty} R_{k,n-k}(\lambda) =: \Phi(\lambda), \quad \lambda > 0, \qquad (27)$$

exists $\hat{\mathbf{P}}-$ a.s., since (27) implies, evidently, the respective $\hat{\mathbf{P}}-$ a.s. convergence of the Laplace transforms $R_{k,n-k}(\bullet)$ to $\Phi(\bullet)$ as elements of $\mathbf{\Phi}$. Recalling Theorem 1.1 we conclude that it is enough to establish the existence of the limit $\lim_{\min(k,n-k)\to\infty}\zeta_k^{-}(f_{0,n-k}^{+}(\exp\{-\lambda\zeta_k^{-}(f_{0,n-k}^{+}(0))e^{-S_{n-k}^{+}}\}))$ $\hat{\mathbf{P}}-$ a.s. In order to demonstrate this fact consider first the random function

$$\theta_{k,n-k}(\lambda) := \zeta_k^{-}(f_{0,n-k}^{+}(\exp\{-\lambda e^{-S_{n-k}^{+}}\})).$$

Let Z_i^+, $i \in \mathsf{N}_0$ be a branching process generated by the sequence $f_0^+, f_1^+, ..., f_i^+, \ldots$ and let $\Sigma_i = \Sigma\left(Z_0^+, Z_1^+, ..., Z_i^+ \,|f_0^+, f_1^+, .., f_{i-1}^+\right)$ be the minimal σ-algebra generated by $Z_0^+, Z_1^+, ..., Z_i^+$ given $f_0^+, f_1^+, .., f_{i-1}^+$. Since $\mathbf{E}_{\pi^+}\left[Z_{i+1}^+\,|Z_i^+\right] = e^{X_{i+1}^+}Z_i^+$, it follows that the stochastic sequence $\left(Z_i^+e^{-S_i^+}, \Sigma_i\right)$, $i \in \mathsf{N}$ is a nonnegative martingale. We know that

$$\mathbf{E}_{\pi^+}\left[\exp\{-\lambda Z_i^+e^{-S_i^+}\}\right] = f_{0,i}^+\left(\exp\left\{-\lambda e^{-S_i^+}\right\}\right), \quad \lambda \in [0,\infty).$$

Hence, for any $\lambda > 0$

$$\lim_{i\to\infty} f_{0,i}^+\left(\exp\left\{-\lambda e^{-S_i^+}\right\}\right) =: Q^+(\lambda) = Q^+(\lambda, \pi^+)$$

exists $\widehat{\mathbf{P}}-$ a.s. Moreover, $\widehat{\mathbf{P}}(Q^+(0)=1)=1$ in view of

$$\mathbf{P}_{\pi^+}(Z_i^+e^{-S_i^+} > x) \le x^{-1}\mathbf{E}_{\pi^+}[Z_i^+e^{-S_i^+}] = x^{-1} \to 0, \quad x\to\infty,$$

uniformly in $i \in \mathsf{N}$. Observe also that $\widehat{\mathbf{P}}(Q^+(\lambda)<1)=1$ for any $\lambda>0$, since S_j^+ tends to infinity $\mathbf{P}^+-$ a.s. as $j\to\infty$ and, as a result,

$$
\begin{aligned}
\frac{1}{1-Q^+(\lambda)} &= \lim_{i\to\infty}\frac{1}{1-f_{0,i}^+\left(\exp\left\{-\lambda e^{-S_i^+}\right\}\right)} \\
&= \lim_{i\to\infty}\left(\frac{e^{-S_i^+}}{1-\exp\left\{-\lambda e^{-S_i^+}\right\}} + \sum_{j=0}^{i-1}\eta_{j,i}^+\left(\exp\left\{-\lambda e^{-S_i^+}\right\}\right)e^{-S_j^+}\right) \\
&\le \frac{1}{\lambda} + \sum_{j=0}^{\infty}\eta_{j+1}^+e^{-S_j^+} < \infty \qquad \mathbf{P}^+ - \text{a.s.}
\end{aligned}
$$

Thus, for any $\lambda \ge 0$

$$\lim_{\min(k,n-k)\to\infty}\theta_{k,n-k}(\lambda) = \zeta^-(Q^+(\lambda)) \qquad \mathbf{P}^+ - \text{a.s.}$$

The function $Q^+(\lambda) = Q^+(\lambda, \pi)$, being the Laplace transform of a nonnegative nondegenerate random variable, is uniformly continuous in $\lambda \in [0, \infty)$. Hence we deduce by Theorem 1.1 that

$$\begin{aligned}\Phi(\lambda) &:= \lim_{\min(k,n-k)\to\infty} R_{k,n-k}(\lambda) \\ &= 1 - \frac{\lim_{\min(k,n-k)\to\infty} \zeta_k^-(f_{0,n-k}^+(e^{-\lambda/\mathbf{m}_n^*}))}{\lim_{\min(k,n-k)\to\infty} \zeta_k^-(f_{0,n-k}^+(0))} \\ &= 1 - \frac{\zeta^-(Q^+(\lambda\zeta^-(q^+)))}{\zeta} = 1 - \frac{\zeta^-(Q^+(\lambda\zeta))}{\zeta}, \quad \lambda \geq 0, \end{aligned} \tag{28}$$

exists $\widehat{\mathbf{P}}-$ a.s.. Moreover, this convergence is $\widehat{\mathbf{P}}-$ a.s. uniform in $\lambda \in [1, 2]$. Thus, if $\min(k, n-k) \to \infty$ then $R_{k,n-k}(\bullet)$ converge $\widehat{\mathbf{P}}-$ a.s. to $\Phi(\bullet)$ as elements of $\mathbf{\Phi}$. Consequently, for any function $G : \mathbf{\Phi} \to \mathbf{R}$ bounded and uniformly continuous on $\mathbf{\Phi}$ with respect to metric d we have

$$\lim_{\min(k,n-k)\to\infty} \mathbf{E}[G(\Phi_n(\bullet)) \,|\, \tau(n) = k] = \hat{\mathbf{E}}[G(\Phi(\bullet))]. \tag{29}$$

Hence, in view of (22) we get

$$\lim_{n\to\infty} \mathbf{E}[G(\Phi_n(\bullet))] = \lim_{n\to\infty} \sum_{k=0}^{n} \mathbf{E}[G(\Phi_n(\bullet)) \,|\, \tau(n) = k]\mathbf{P}(\tau(n) = k) = \hat{\mathbf{E}}[G(\Phi(\bullet))].$$

Thus, $\Phi_n \Rightarrow \Phi$, $n \to \infty$.

It remains to check that the probability law with Laplace transform $\Phi(\lambda)$ has no atom at zero $\hat{\mathbf{P}}-$a.s. and, therefore, belongs to $\mathbf{L}^+$ with probability 1. Clearly, it suffices to prove that $\lim_{\lambda\to\infty} \Phi(\lambda) = 0 \quad \hat{\mathbf{P}}-$ a.s. To this aim let $W^+ = \lim_{i\to\infty} Z_i^+ e^{-S_i^+}$. It is shown in the proof of Proposition 3.1 in [1] that $\{W^+ > 0\} = \{Z_i^+ > 0 \text{ for all } i\}$ $\mathbf{P}^+-$ a.s. Thus,

$$q^+ = \mathbf{P}_{\pi^+}(\lim_{i\to\infty} Z_i^+ = 0) = \mathbf{P}_{\pi^+}(W^+ = 0) = \lim_{\lambda\to\infty} Q^+(\lambda) \quad \mathbf{P}^+ - \text{a.s.}$$

From this equality, Theorem 1.1, the continuity of $\zeta^-(s)$ in $s \in [0, 1)$ and (28) it is easy to deduce that

$$\lim_{\lambda\to\infty} \Phi(\lambda) = 1 - \frac{\zeta^-(q^+)}{\zeta} = 0 \qquad \hat{\mathbf{P}} - \text{a.s.}.$$

This completes the proof of Theorem 1.2.

Proof of Corollary 1.3. The proof of this corollary repeats almost literally the respective arguments used in [7] to establish (8) under stronger conditions and we give the needed arguments for completeness only.

It follows from (24) and Lemmas 2.4 and 2.5 that

$$\zeta^-(s) = \left(\frac{1}{1-s} + \frac{1}{2}\sum_{j=1}^{\infty} \eta_j^- e^{S_j^-} \right)^{-1} =: \left(\frac{1}{1-s} + \xi^- \right)^{-1} \qquad \widehat{\mathbf{P}} - \text{a.s.}$$

In particular,

$$\zeta = \zeta^-(q^+) = \left(\frac{1}{1-q^+} + \xi^- \right)^{-1}, \quad \zeta^-(Q^+(\lambda\zeta)) = \left(\frac{1}{1 - Q^+(\lambda\zeta)} + \xi^- \right)^{-1}.$$

Since $S_j^+ \to \infty$ $\widehat{\mathbf{P}}$-a.s. as $j \to \infty$, applying Lemma 2.4 to the fractional-linear offspring functions gives

$$\frac{1}{1-q^+} = \lim_{i\to\infty}\left(e^{-S_i^+} + \frac{1}{2}\sum_{j=0}^{i-1}\eta_{j+1}^+ e^{S_j^+}\right) = \frac{1}{2}\sum_{j=0}^{\infty}\eta_{j+1}^+ e^{S_j^+} \qquad \widehat{\mathbf{P}} - \text{a.s.}$$

and

$$\begin{aligned}\frac{1}{1-Q^+(\lambda\zeta)} &= \lim_{\min(k,n-k)\to\infty}\left(\frac{e^{-S_k^+}}{1-\exp\left\{-\lambda\zeta_{0,n}e^{-S_k^+}\right\}} + \frac{1}{2}\sum_{j=0}^{k-1}\eta_{j+1}^+ e^{S_j^+}\right)\\ &= \frac{1}{\lambda\zeta} + \frac{1}{2}\sum_{j=0}^{\infty}\eta_{j+1}^+ e^{S_j^+} = \frac{1}{\lambda\zeta} + \frac{1}{1-q^+} \qquad \widehat{\mathbf{P}} - \text{a.s.}\end{aligned}$$

Hence we obtain

$$\Phi(\lambda) = 1 - \frac{\zeta^-(Q^+(\lambda\zeta))}{\zeta} = 1 - \frac{1}{\zeta}\left(\frac{1}{\lambda\zeta} + \frac{1}{1-q^+} + \xi^-\right)^{-1} = \frac{1}{1+\lambda}$$

as required.

References

[1] Afanasyev V.I., Geiger J., Kersting G., and Vatutin V.A. *Criticality for branching processes in random environment.* submitted in Annals of Probability, 2003.

[2] Athreya K.B. and Karlin S. *Branching processes with random environments, II: limit theorems.* Ann. Math. Statist., 42:1843–1858, 1971.

[3] Bingham N.H., Goldie C.M., and Teugels J.L. *Regular variation.* Cambridge University Press, Cambridge, 1987.

[4] Doney R.A. *Spitzer's condition and the ladder variables in random walks.* Probab. Theory Relat. Fields 101: 577–580, 1995.

[5] Geiger J. and Kersting G. *The survival probability of a critical branching process in random environment.* Theory Probab. Appl., 45:607–615, 2000.

[6] Hirano K. *Determination of the limiting coefficient for exponential functionals of random walks with positive drift.* J. Math. Sci. Univ. Tokyo, 5:299–332, 1998.

[7] Vatutin V.A. and Dyakonova E.E. *Reduced branching processes in random environment.* In: Mathematics and Computer Science II: Algorithms, Trees, Combinatorics and Probabilities (Ed. B.Chauvin, P.Flajolet, D.Gardy, A.Mokkadem), Basel - Boston - Berlin: Birkhäuser, 455–467, 2002.

[8] Vatutin V.A. and Dyakonova E.E. *Galton-Watson branching processes in random environment, I: limit theorems.* Theory Probab. Appl., 48:274–300, 2003. (In Russian.)

[9] Spitzer F. *Principles of random walk.* Princeton NJ, Toronto - New York - London, 1964.

Vladimir Vatutin and Elena Dyakonova
Steklov Mathematical Institute
Gubkin street, 8
117966, Moscow
Russia
e-mail: vatutin@mi.ras.ru, elena@mi.ras.ru

Trends in Mathematics, © 2004 Birkhäuser Verlag Basel/Switzerland

Two-Dimensional Limit Theorem for a Critical Catalytic Branching Random Walk

Valentin Topchii and Vladimir Vatutin

ABSTRACT: *A continuous time branching random walk on the lattice $\mathbb{Z}$ is considered in which individuals may produce children at the origin only. Assuming that the underlying random walk is symmetric and the offspring reproduction law is critical we study the asymptotic behavior of the joint distribution of the number of individuals at the origin and outside the origin at moment t as $t \to \infty$ given that there are individuals at the origin at this moment* [1].

1. Introduction

In [1] and [2] we investigated the following modification of a standard branching random walk on $\mathbb{Z}$. The population is initiated at time $t = 0$ by a single particle. Being outside the origin the particle performs similar a continuous time underlying random walk on $\mathbb{Z}$ with infinitesimal transition matrix

$$A = |a(x,y)|_{x,y\in\mathbb{Z}}, \qquad a(0,0) < 0,$$

until the moment when it hits the origin. At the origin it spends an exponentially distributed time with parameter 1 and then either jumps to a point $y \neq 0$ with probability $-(1-\alpha)a(0,y)a^{-1}(0,0)$ or dies with probability α producing just before the death a random number of children ξ in accordance with offspring generating function

$$f(s) \stackrel{def}{=} \mathbf{E}s^{\xi} = \sum_{k=0}^{\infty} f_k s^k. \tag{1}$$

At the birth moment the newborn particles are located at the origin and from this point they begin their own branching random walks behaving independently and stochastically the same as the parent individual.

We assume that the process is initiated at time $t = 0$ by a single individual located at the origin and impose the following restrictions on the characteristics of the process:

Hypothesis (I): The underlying random walk on $\mathbb{Z}$ is symmetric, irreducible and homogeneous ($a(x,y) = a(0,y-x) \stackrel{def}{=} a(y-x)$) with $a(x) \geq 0$ if $x \neq 0$, $a(0) < 0$,

$$\sum_{x\in\mathbb{Z}} a(x) = 0, \quad \text{and} \quad b^2 \stackrel{def}{=} \sum_{x\in\mathbb{Z}} x^2 a(x) < \infty.$$

Hypothesis (II): The offspring process is critical ($f'(1) = 1$) and $\sigma^2 \stackrel{def}{=} f''(1) \in (0,\infty)$.

---

[1]The first author is supported in part by the grants RFBR 03-01-00045, Russian Scientific School 2139.2003.1; the second author is supported in part by the grants RFBR 02-01-00266, Russian Scientific School 1758.2003.1, and INTAS 03-51-5018

$$b^2 \stackrel{def}{=} \sum_{x\in\mathbb{Z}} x^2 a(x) < \infty \quad \text{and} \quad \sigma^2 \stackrel{def}{=} f''(1) \in (0,\infty).$$

Let $\zeta(t)$ denote the number of particles in the process located at time t at the origin, $\mu(t)$ denote the number of particles in the process at time t outside the origin, and let $\eta(t) = \zeta(t) + \mu(t)$ be the total number of individuals at the process at moment t. The aim of the present paper is to establish a conditional limit theorem for a suitable normalized vector $(\zeta(t),\, \mu(t))$ given that $\zeta(t) > 0$.

The model we investigate here was considered in a wider context in [3]-[6]. In the mentioned papers integral equations were deduced (in a more general situation of random walks on the lattice $\mathbb{Z}^d$, $d \geq 1$) for the generating functions of $\zeta(t,x)$, the number of particles located at moment t at point $x \in \mathbb{Z}^d$, and $\eta(t)$, the total size of the population at time t, and the asymptotic behavior of the moments of these random variables was investigated. The present paper, being a continuation of papers [1] and [2], gives for the case $d = 1$ more detailed information about the properties of the process under study.

Put

$$K \stackrel{def}{=} \frac{2^{3/4}}{\sigma\pi^{1/4}}\sqrt{\frac{b(1-\alpha)}{\alpha}} \quad \text{and} \quad c_q \stackrel{def}{=} \pi K^2 = \frac{2\sqrt{2\pi}b(1-\alpha)}{\sigma^2\alpha}.$$

In the sequel we need the following results established in [1] and [2].

Theorem 1.1. [1] *Let hypotheses* (I) *and* (II) *be valid. Then, as* $t \to \infty$,

$$Q(t) \stackrel{def}{=} \mathbf{P}(\mu(t) > 0) \sim \mathbf{P}(\eta(t) > 0) \sim Kt^{-1/4},$$
$$q(t) \stackrel{def}{=} \mathbf{P}(\zeta(t) > 0) \sim \frac{c_q}{\sqrt{t}\ln t},$$

and for any $s \in [0,1]$

$$\lim_{t\to\infty} \mathbf{E}\left[s^{\mu(t)} \,|\, \eta(t) > 0\right] = \lim_{t\to\infty} \mathbf{E}\left[s^{\eta(t)} \,|\, \eta(t) > 0\right] = 1 - \sqrt{1-s}.$$

Theorem 1.2. [2] *Let hypotheses* (I) *and* (II) *be valid. Then for any* $\lambda \in [0,\infty)$

$$\lim_{t\to\infty} \mathbf{E}\left\{\exp\left\{-\frac{\lambda\zeta(t)}{\mathbf{E}\{\zeta(t)\,|\,\zeta(t)>0\}}\right\}\middle|\, \zeta(t) > 0\right\} = \frac{1}{\lambda+1}.$$

Theorems 1.1 and 1.2 provide no information about the number of individuals outside the origin given that there are individuals at the origin. In the present paper we fill this gap and prove a theorem describing, in particular, the asymptotic behavior of $\mu(t)$ given that $\zeta(t) > 0$. To formulate our main result we need additional notation. Let $\varphi(\lambda_1)$ be a continuous positive solution of the equation

$$\varphi(\lambda_1) = 1 - \alpha\sigma^2\lambda_1^{1/2}\int_0^1 \varphi^2(\lambda_1(1-v))dv^{1/2},\ \lambda_1 \geq 0 \tag{2}$$

(according to Lemma 7 in [9] such a solution exists and is unique). Let, further, $D(\lambda_2)$ be a bounded solution of the equation

$$D(\lambda_2) = 1 - \alpha\sigma^2\lambda_2\int_0^1 \frac{D(\lambda_2\sqrt{y})\varphi(\lambda_2^2 y)}{\sqrt{y(1-y)}}dy,\ \lambda_2 \geq 0, \tag{3}$$

which is unique and positive as is shown later in the paper.

Theorem 1.3. *Let hypotheses* (I) *and* (II) *be valid and, in addition,*

$$h''(x) = \sigma^2 + o(|\ln^{-1} x|) \text{ as } x \to +0.$$

Then for all $\lambda_1, \lambda_2 \geq 0$

$$\lim_{t\to\infty} \mathbf{E}\left\{ \exp\left\{ -\frac{\lambda_1 \zeta(t)}{\mathbf{E}\{\zeta(t)\,|\,\zeta(t)>0\}} \right\} \exp\left\{ -\frac{\lambda_2 \mu(t)}{c_p\sqrt{t}} \right\} \middle| \zeta(t) > 0 \right\} = \frac{D(\lambda_2)}{1+\lambda_1},$$

where $c_p^{-1} \stackrel{def}{=} b\sqrt{2\pi}(1-\alpha)$.

Thus, under our scaling the numbers of individuals at the origin and outside the origin are independent at that moments (rather rare as follows from Theorem 1.1) when $\zeta(t) > 0$.

Before passing to the proof of Theorem 1.3 we say a few words about the asymptotic behavior of the scaling function $\mathbf{E}\{\zeta(t)\,|\,\zeta(t) > 0\}$. To this aim we temporarily forget that our random walk has a point of catalyst and consider an ordinary random walk on $\mathbb{Z}$ satisfying Hypothesis (I). Assume that the random walk starts at the origin at time $t = 0$ and let τ_1 be the time spent by the walking particle at the origin until it leaves the origin and let τ_2 be the time spent by this particle outside the origin until the first return to the origin. Set $G_1(t) \stackrel{def}{=} \mathbf{P}(\tau_1 \leq t) = 1 - e^{-t}$, $G_2(t) \stackrel{def}{=} \mathbf{P}(\tau_2 \leq t)$ and

$$G_3(t) \stackrel{def}{=} \alpha G_1(t) + (1-\alpha) G_1 * G_2(t), \tag{4}$$

where $*$ is the convolution symbol. It is shown in [1] that the function $P(t) \stackrel{def}{=} \mathbf{E}\zeta(t)$ has the representation

$$P(t) = (1 - G_1(\cdot)) * U(t) \tag{5}$$

where

$$U(t) \stackrel{def}{=} \sum_{k=0}^{\infty} G_3^{*k}(t) \tag{6}$$

with $G_3^{*0}(t) = 1$ and $G_3^{*k}(t)$, $k \geq 1$, the k-th convolution of $G_3(t)$ with itself.

Moreover, the following statement is valid:

Lemma 1.4. [1] *$P(t)$ is monotone decreasing and $P(t) \sim c_p t^{-1/2}$ as $t \to \infty$.*

Combining this statement with Theorem 1.1 we see that

$$\mathbf{E}\{\zeta(t)\,|\,\zeta(t) > 0\} = \frac{\mathbf{E}\zeta(t)}{\mathbf{P}\{\zeta(t) > 0\}} = \frac{P(t)}{q(t)} \sim c^* \ln t, \qquad t \to \infty,$$

where $c^* \stackrel{def}{=} c_p c_q^{-1} = \alpha\sigma^2(4\pi)^{-1} b^{-2}(1-\alpha)^{-2}$. As a result we get that given $\zeta(t) > 0$ the number of individuals at the origin is order $\ln t$ (up to a random multiplier) while the number of particles outside the origin is of order $\sqrt{t}$.

2. Branching random walk and Bellman-Harris processes

Similarly to [1] and [2] we prove Theorem 1.3 by introducing an auxiliary Bellman-Harris branching process $(Z_1(t), Z_2(t))$ with two types of particles (see [7, 8]), where by $Z_i(t)$, $i = 1, 2$ we denote the number of individuals of type i in this process at time t.

Let

$$F_i(t;s_1,s_2)=\mathbf{E}\left\{s_1^{Z_1(t)}s_2^{Z_2(t)}|Z_i(0)=1\right\},\qquad i=1,2,$$

be the probability generating function of the number of individuals of both types given that the process is initiated at time zero by a single individual of type i.

The critical Bellman-Harris process with two types of individuals we are interested in is described as follows. A particle of the first type has life length distribution $G_1(t)=\mathbf{P}(\tau_1\le t)=1-e^{-t}$, $t\ge 0$, and dying produces offspring of two types in accordance with probability generating function $f_1(s_1,s_2)=\alpha f(s_1)+(1-\alpha)s_2$, that is it produces with probability αf_k exactly $k\ge 0$ particles of the first type and with probability $1-\alpha$ exactly one particle of the second type (recall the definition of $f(s)$ in (1)). The life length distribution of a particle of the second type is $G_2(t)=\mathbf{P}(\tau_2\le t)$ (that is coincides in distribution with the time spend outside the origin by the parent individual of the catalytic branching random walk under investigation until the first return to the origin provided that the initial individual is located at point 0 at time $t=0$ and it does not produce children during its first stay at 0). Dying a particle of the second type produces offspring in accordance with probability generating function $f_2(s_1,s_2)=s_1$, that is, it produces exactly one particle of the first type and nothing else.

This Bellman-Harris process is critical and indecomposable since the maximal in absolute value eigenvalue (the Perron root) of its mean matrix

$$M\stackrel{def}{=}\left\|\frac{\partial f_i}{\partial s_j}(1,1)\right\|_{i,j=1,2}=\begin{pmatrix}\alpha & 1-\alpha\\ 1 & 0\end{pmatrix}$$

equals 1 and all the elements of M^2 are positive.

It is not difficult to understand that

$$(Z_1(t),Z_2(t))\stackrel{distr}{=}(\zeta(t),\mu(t)). \tag{7}$$

Recall that under our assumptions on the form of $f_i(s_1,s_2)$, $i=1,2$, we have (see [8], Chapter VIII, §1, or [7])

$$\begin{aligned}F_1(t;s_1,s_2)&=s_1(1-G_1(t))\\&+\int_0^t(\alpha f(F_1(t-u;s_1,s_2))+(1-\alpha)F_2(t-u;s_1,s_2))dG_1(u),\\F_2(t;s_1,s_2)&=s_2(1-G_2(t))+\int_0^t F_1(t-u;s_1,s_2)dG_2(u).\end{aligned}$$

Using the second of these equalities in the first and introducing $Q(t;s_1,s_2)\stackrel{def}{=}1-F_1(t;s_1,s_2)$ we see that

$$Q(t;s_1,s_2)=(1-s_1)(1-G_1(t))+(1-s_2)(1-\alpha)(1-G_2(\cdot))*G_1(t)$$
$$+\int_0^t Q(t-u;s_1,s_2)dG_3(u)-\alpha\int_0^t h(Q(t-u;s_1,s_2))dG_1(u) \tag{8}$$

where $h(x)\stackrel{def}{=}f(1-x)-(1-x)$ and $G_3(t)$ is from (4).

Solving renewal equation (8) we obtain by (5) and (6)

$$\begin{aligned}Q(t;s_1,s_2)&=(1-s_1)P(t)+(1-\alpha)(1-s_2)(1-G_2(\cdot))*G_1*U(t)\\&-\alpha\int_0^t h(Q(t-u;s_1,s_2))d\left(G_1*U(u)\right).\end{aligned} \tag{9}$$

Note that

$$(1-\alpha)(1-G_2(\cdot))*G_1(t) = (1-\alpha)G_1(t)-(1-\alpha)G_2*G_1(t) = G_1(t)-G_3(t)$$

and, consequently,

$$\begin{aligned}(1-\alpha)(1 \quad &- \quad G_2(\cdot))*G_1*U(t) = G_1*U(t)-G_3*U(t)\\ &= \quad G_3^0(t)+G_1*U(t)-U(t) = 1-(1-G_1(\cdot))*U(t)\\ &= \quad 1-P(t).\end{aligned}$$

Since $G_1(t) = 1-e^{-t}$ it follows that $(G_1*U(t))' = (1-G_1(\cdot))*U(t) = P(t)$. Using the identities above we rewrite (9) as follows

$$\begin{aligned}Q(t;s_1,s_2) &= (1-s_1)P(t)+(1-P(t))(1-s_2)\\ &\quad -\alpha\int_0^t h(Q(t-u;s_1,s_2))P(u)du. \qquad (10)\end{aligned}$$

Put $Q(t;s,1) \stackrel{def}{=} q(t;s)$. Observe that $q(t) = \mathbf{P}(\zeta(t)>0) = q(t;0)$. Setting $s_1 = s$ and $s_2 = 1$ in (10) leads to the equality

$$q(t;s) = (1-s)P(t)-\alpha\int_0^t h(q(t-u;s))P(u)du.$$

Denote

$$s_1(t) = s_1(t;\lambda_1) \stackrel{def}{=} \exp\left\{-\frac{\lambda_1 q(t)}{P(t)}\right\} = \exp\left\{-\frac{\lambda_1}{\mathbf{E}\{\zeta(t)\,|\zeta(t)>0\}}\right\},$$

$$s_2(t) = s_2(t;\lambda_2) \stackrel{def}{=} \exp\left\{-\frac{\lambda_2}{c_p\sqrt{t}}\right\},$$

$$\begin{aligned}\Delta(t;\lambda_1,\lambda_2) \stackrel{def}{=} q(t;0) \quad &- \quad q(t;s_1(t;\lambda_1))\\ &+ \quad (Q(t;s_1(t;\lambda_1),s_2(t;\lambda_2))-Q(t;0,s_2(t;\lambda_2))).\end{aligned}$$

It is not difficult to check that

$$\begin{aligned}H_t(\lambda_1,\lambda_2) &\stackrel{def}{=} \mathbf{E}\left[s_1^{Z_1(t)}(t)\,s_2^{Z_2(t)}(t)\,\middle|Z_1(t)>0\right]\\ &= \frac{F(t;s_1(t),s_2(t))-F(t;0,s_2(t))}{P(Z_1(t)>0)}\\ &= \frac{Q(t;0,s_2(t))-Q(t;s_1(t),s_2(t))}{q(t)} = \Theta_t(\lambda_1)-\Psi_t(\lambda_1,\lambda_2), \qquad (11)\end{aligned}$$

where

$$\Theta_t(\lambda_1) \stackrel{def}{=} \frac{q(t)-q(t;s_1(t))}{q(t)}, \qquad \Psi_t(\lambda_1,\lambda_2) \stackrel{def}{=} \frac{\Delta(t;\lambda_1,\lambda_2)}{q(t)}.$$

By Theorem 1.2 and (7)

$$\lim_{t\to\infty}\Theta_t(\lambda_1) = \frac{1}{1+\lambda_1}. \qquad (12)$$

Hence, to prove Theorem 1.3 it is necessary to find $\lim_{t\to\infty}\Psi_t(\lambda_1,\lambda_2)$. To solve this problem we need two following basic representations which can be deduced from equation (10) by rather tedious and complicated arguments too lengthy to be given in the present paper:

Lemma 2.1. *Under the conditions of Theorem* 1.3

$$\Psi_t(\lambda_1,\lambda_2) = \alpha\sigma^2\frac{\lambda_2}{1+\lambda_1}\int_0^1 \frac{\varphi(\lambda_2^2 y)}{\sqrt{y(1-y)}}dy - \alpha\sigma^2\lambda_2\int_\varepsilon^{1-\varepsilon}\frac{\Psi_{ty}(\lambda_1,\lambda_2\sqrt{y})\varphi(\lambda_2^2 y)}{\sqrt{y(1-y)}}dy + r_{0,\varepsilon}(t;\lambda_1,\lambda_2),$$

where

$$\overline{\lim_{\varepsilon\to+0}}\;\overline{\lim_{t\to\infty}}\sup_{(\lambda_1,\lambda_2)\in\mathcal{D}}|r_{0,\varepsilon}(t;\lambda_1,\lambda_2)| = 0$$

for any bounded set $\mathcal{D}\subseteq[0,\infty)\times[0,\infty)$, *or, in view of* (11) *and* (12),

$$H_t(\lambda_1,\lambda_2) = \frac{1}{1+\lambda_1} - \alpha\sigma^2\lambda_2\int_\varepsilon^{1-\varepsilon}\frac{H_{ty}(\lambda_1,\lambda_2\sqrt{y})\varphi(\lambda_2^2 y)}{\sqrt{y(1-y)}}dy + r_\varepsilon(t;\lambda_1,\lambda_2), \quad (13)$$

where

$$\overline{\lim_{\varepsilon\to+0}}\;\overline{\lim_{t\to\infty}}\sup_{(\lambda_1,\lambda_2)\in\mathcal{D}}|r_\varepsilon(t;\lambda_1,\lambda_2)| = 0.$$

Our next goal is to show that $H_t(\lambda_1,\lambda_2)\to D(\lambda_2)(1+\lambda_1)^{-1}$ as $t\to\infty$ where $D(\lambda_2)$ solves equation (3). However, first we need to establish that a solution of equation (3) exists and is unique.

The uniqueness problem is partially solved by means of the following two lemmas the first of which can be demonstrated by standard methods and we omit its proof.

Lemma 2.2. *In the domain* $\lambda_2>0$ *there are no bounded solutions of the equation*

$$\psi(\lambda_2) = \alpha\sigma^2\lambda_2\int_0^1\psi(\lambda_2\sqrt{y})\frac{\varphi(\lambda_2^2 y)}{\sqrt{y(1-y)}}dy$$

other than $\psi(\lambda_2)=0$.

Lemma 2.3. *Equation* (3) *has at most one solution on the half-line* $\lambda_2\geq 0$. *Moreover, in the range* $\lambda_2\in[0,2\rho)$, *where* $\rho=\left(2\alpha\pi\sigma^2\right)^{-1}$, *there exists exactly one bounded positive solution of* (3).

Sketch of the proof. The first part of Lemma 2.3 is a corollary of Lemma 2.2. To demonstrate the second part we let $C[0,2\rho)$ be the set of continuous functions on $[0,2\rho)$ and for functions $T(\bullet)\in C[0,2\rho)$ define an operator $\mathfrak{L}: C[0,2\rho)\mapsto C[0,2\rho)$ by the equality

$$\mathfrak{L}T(\lambda_2) \stackrel{def}{=} 1-\alpha\sigma^2\lambda_2\int_0^1\frac{T(\lambda_2\sqrt{y})\varphi(\lambda_2^2 y)}{\sqrt{y(1-y)}}dy.$$

Setting $T^1(\lambda_2)\equiv 1$, defining $T^{n+1}(\lambda_2)=\mathfrak{L}T^n(\lambda_2)$, $n\in\mathbb{N}$, and recalling that $0\leq\varphi(x)\leq 1$ if $x\geq 0$ by (2), it is not difficult to check by induction that for $\lambda_2\in[0,2\rho)$

$$0\leq T^n(\lambda_2)\leq 1, \qquad |T^{n+1}(\lambda_2)-T^n(\lambda_2)|\leq(\lambda_2/2\rho)^n.$$

Hence the second part of the lemma follows.

Lemma 2.4. *Under the conditions of Theorem* 1.3 *for all* $\lambda_1\geq 0$ *and* $\lambda_2\in[0,\rho]$

$$\lim_{t\to\infty}H_t(\lambda_1,\lambda_2) = \frac{D(\lambda_2)}{1+\lambda_1}.$$

Proof. It is easy to see that in the range of λ_1 and λ_2 we are interested in $H(\lambda_1,\lambda_2) \stackrel{def}{=} D(\lambda_2)(1+\lambda_1)^{-1}$ satisfies the relation

$$H(\lambda_1,\lambda_2) = \frac{1}{1+\lambda_1} - \alpha\sigma^2\lambda_2 \int_0^1 \frac{H(\lambda_1,\lambda_2\sqrt{y})\varphi(\lambda_2^2 y)}{\sqrt{y(1-y)}} dy. \tag{14}$$

Moreover, using Lemma 2.3 one can show that there is no $H(\lambda_1,\lambda_2)$ other than $D(\lambda_2)(1+\lambda_1)^{-1}$ which solves equation (14) for $(\lambda_1,\lambda_2) \in [0,\infty)\times[0,\rho]$. By (13) it is not difficult to deduce that for any $\varepsilon \in (0,1/2)$

$$|H_t(\lambda_1,\lambda_2) - H(\lambda_1,\lambda_2)| \le \alpha\sigma^2\lambda_2 \int_\varepsilon^{1-\varepsilon} |H_{ty}(\lambda_1,\lambda_2\sqrt{y}) - H(\lambda_1,\lambda_2\sqrt{y})| \times$$
$$\times \frac{\varphi(\lambda_2^2 y)}{\sqrt{y(1-y)}} dy + r_\varepsilon(t;\lambda_1,\lambda_2),$$

where

$$\overline{\lim_{\varepsilon\to+0}}\, \overline{\lim_{t\to\infty}} \sup_{(\lambda_1,\lambda_2)\in\mathcal{D}} |r_{\varepsilon}(t;\lambda_1,\lambda_2)| = 0$$

for any $\mathcal{D} \subseteq [0,\infty)\times[0,\rho]$. Setting

$$M_t(\lambda_1;\lambda_2) = \sup_{0\le z\le\lambda_2} |H_t(\lambda_1,z) - H(\lambda_1,z)|, \qquad \mathcal{M}_T(\lambda_1;\rho) = \sup_{t\ge T} M_t(\lambda_1;\rho)$$

it is easy to see that for all sufficiently large t and all $\lambda_1 \ge 0$ and $\lambda_2 \in [0,\rho]$

$$\begin{aligned} |H_t(\lambda_1,\lambda_2) - H(\lambda_1,\lambda_2)| &\le g(\varepsilon) + 2^{-1}\sup_{y\ge\varepsilon} M_{ty}(\lambda_1;\rho) \\ &= g(\varepsilon) + 2^{-1}\mathcal{M}_{t\varepsilon}(\lambda_1;\rho), \end{aligned}$$

where $g(\varepsilon)$ is a function in ε such that $\lim_{\varepsilon\downarrow 0} g(\varepsilon) = 0$. Therefore, as $t \ge t_0(\varepsilon)$

$$M_t(\lambda_1;\lambda_2) \le g(\varepsilon) + 2^{-1}\mathcal{M}_{t\varepsilon}(\lambda_1;\rho).$$

Hence

$$\mathcal{M}_T(\lambda_1;\rho) \le g(\varepsilon) + 2^{-1}\mathcal{M}_{T\varepsilon}(\lambda_1;\rho)$$

for all sufficiently large T. Setting $\mathcal{M}(\lambda_1;\rho) = \lim_{T\to\infty}\mathcal{M}_T(\lambda_1;\rho)$ we get

$$\mathcal{M}(\lambda_1;\rho) \le g(\varepsilon) + 2^{-1}\mathcal{M}(\lambda_1;\rho).$$

Since $\varepsilon > 0$ may be selected arbitrary small, it follows that $\mathcal{M}(\lambda_1;\rho) = 0$ proving the lemma.

3. Proof of Theorem 1.3

From Lemma 2.4, relations (12), (11), (7), and the uniqueness theorem for Laplace transforms we conclude that there exists a function $W(\lambda_1,\lambda_2)$ analytic in the domain $\mathcal{R}e\lambda_1 > 0$, $\mathcal{R}e\lambda_2 > 0$ such that

$$\lim_{t\to\infty} \mathbf{E}\left\{\exp\left\{-\frac{\lambda_1\zeta(t)}{\mathbf{E}\{\zeta(t)\,|\,\zeta(t)>0\}}\right\}\exp\left\{-\frac{\lambda_2\mu(t)}{c_p\sqrt{t}}\right\}\middle|\,\zeta(t)>0\right\} = W(\lambda_1,\lambda_2)$$

for all λ_1,λ_2: $\mathcal{R}e\lambda_1 > 0$, $\mathcal{R}e\lambda_2 > 0$. In particular,

$$W(\lambda_1,\lambda_2) = \frac{D(\lambda_2)}{1+\lambda_1} \quad \text{for} \quad \lambda_1,\lambda_2 \in [0,\infty)\times[0,\rho].$$

Letting $\lambda_1 = 0$ we see that there exists a unique analytic extension of $D(\lambda_2)$ to the domain $\Re e\lambda_2 > 0$. We keep the same symbol $D(\lambda_2)$ for this extension. Clearly, $D(\lambda_2)$ is the Laplace transform of a non-degenerate distribution on the positive half- line and, in particular,

$$W(\lambda_1, \lambda_2) = \lim_{t\to\infty} H_t(\lambda_1, \lambda_2) = \frac{D(\lambda_2)}{1+\lambda_1}, \quad \lambda_1, \lambda_2 \in [0,\infty) \times [0,\infty).$$

Thus, the proof of Theorem 1.3 will be finished, if we establish that $D(\lambda_2)$ solves (3) on the positive half-line. Note that we have proved in Lemma 2.2 that if a solution of (3) exists then it is unique. Evidently, $0 \le H_t(\lambda_1, \lambda_2) \le 1$ and $0 \le \varphi(x) \le 1$ for $x \ge 0$. Therefore,

$$0 \le \frac{H_t(\lambda_1, \lambda_2)\varphi(\lambda_2^2 y)}{\sqrt{y(1-y)}} \le \frac{1}{\sqrt{y(1-y)}}, \quad \text{for all} \quad y \in (0,1), \ \lambda_2 \ge 0.$$

Recalling that $D(\lambda_2)$, being a Laplace transform, is monotone decreasing in the domain $\lambda_2 > 0$ and applying the dominated convergence theorem to (13) we obtain

$$D(\lambda_2) = 1 - \alpha\sigma^2\lambda_2 \int_{\varepsilon}^{1-\varepsilon} \frac{D(\lambda_2\sqrt{y})\varphi(\lambda_2^2 y)}{\sqrt{y(1-y)}} dy + r_{0,\varepsilon}(\lambda_1, \lambda_2), \tag{15}$$

where

$$\overline{\lim_{\varepsilon\to+0}} \sup_{(\lambda_1,\lambda_2)\in\mathfrak{D}} |r_{0,\varepsilon}(\lambda_1, \lambda_2)| = 0.$$

In view of $0 \le D(\lambda_2) \le 1$ for all $\lambda_2 > 0$ we may apply the dominated convergence theorem to (15) as $\varepsilon \downarrow 0$. As a result we get that $D(\lambda_2)$ indeed solves (3) on the positive half-line.

Theorem 1.3 is proved.

References

[1] V. A. Topchii, V. A. Vatutin, E. B. Yarovaya, *Catalytic branching random walk and queueing systems with random number of independent servers.* Theory of Probability and Mathematical Statistics, v. 69, 158-172 (2003).

[2] V. A. Topchii, V. A. Vatutin, *Individuals at the origin in the critical catalytic branching random walk.* Discrete Mathematics & Theoretical Computer Science, v. 6, 325-332 (2003).

[3] S. Albeverio, L. V. Bogachev, *Branching random walk in a catalytic medium. I. Basic equations.* Positivity, v. 4, 41-100 (2000).

[4] S. Albeverio, L. V. Bogachev, E. B. Yarovaya, *Asymptotics of branching symmetric random walk on the lattice with a single source.* C.-R.-Acad.-Sci.-Paris-Ser.-I-Math., v. 326, 975-980 (1998).

[5] L. V. Bogachev, E. B. Yarovaya, *A limit theorem for a supercritical branching random walk on $\mathbb{Z}^d$ with a single source.* Russian-Math.-Surveys, v. 53, 1086-1088 (1998).

[6] L. V. Bogachev, E. B. Yarovaya, *Moment analysis of a branching random walk on a lattice with a single source.* Dokl.-Akad.-Nauk, v. 363, 439-442 (1998). (In Russian.)

[7] T. E. Harris, (1963) *The Theory of Branching Processes* Springer-Verlag, Berlin; Prentice-Hall, Inc., Englewood Cliffs, N.J.

[8] B. A. Sewastjanow, (1974) *Verzweigungsprozesse.* Akademie-Verlag, Berlin.

[9] V. A. Vatutin, *Critical Bellman-Harris branching processes starting with a large number of particles.* Mat.-Zametki, v. 40, 527-541 (1986). (In Russian.)

Valentin Topchii
Omsk Branch of Sobolev Institute of Mathematics
Pevtcova Street, 13
644099 Omsk
Russia
e-mail: topchij@iitam.omsk.net.ru

Vladimir Vatutin
Steklov Mathematical Institute
Gubkin Street, 8
119991 Moscow
Russia
e-mail: vatutin@mi.ras.ru

Part VI

Combinatorial Stochastic Processes

Trends in Mathematics, © 2004 Birkhäuser Verlag Basel/Switzerland

A Combinatorial Approach to Jumping Particles II: General Boundary Conditions

Enrica Duchi and Gilles Schaeffer

ABSTRACT: *We consider a model of particles jumping on a row of cells, called in physics the one dimensional totally asymmetric exclusion process with open boundaries (TASEP). From the point of view of combinatorics a remarkable feature of this Markov chain is that Catalan numbers are involved in several entries of its stationary distribution.*

In a companion paper, we gave a combinatorial interpretation and a simple proof of these observations in the simplest case where the particles enter, jump and exit at the same rate. To do this we revealed a second row in which particles travel backward and defined on these two row configurations a Markov chain with uniform stationary distribution which is a covering of the TASEP.

In this paper we show how to deal with general rates. The covering chain is still defined on two row configurations, but its stationary distribution is not uniform anymore. Instead it is described in terms of two natural combinatorial statistics.

1. Jumping particles

We shall consider a model of jumping particles on a row of n cells that was studied by and large since the early 90's in physics, under the name *one dimensional totally asymmetric exclusion process with boundaries*, or TASEP for short. Roughly speaking, the TASEP consists of black particles entering a row of n cells from an infinite reservoir on the left hand side and randomly hopping to the right with the simple exclusion rule that each cell may contain only one particle.

FIGURE 1. An informal illustration of the TASEP.

The TASEP is usually defined as a continuous-time Markov process on a finite set of configurations of particles on a line. We shall use an alternative definition as a finite state Markov chain —with discrete time— which is more convenient for our combinatorial purpose. One could insist on calling our chain the TASEC, with "C" for chain instead of "P" for process, but as we will argue later, there is no need for this distinction. Another cosmetic modification we allow ourselves consists in putting a white particle in each empty cell, to make explicit the left-right symmetry of the system.

1.1. The basic system, or TASEP

A *basic configuration* is a row of n cells, each containing either one black particle or one white particle (see Figure 2). These cells are delimited by $n+1$ walls: the left border (or wall 0), the ith separation wall for $i = 1, \ldots, n-1$, and the right border (or wall n).

FIGURE 2. A basic configuration with $n = 10$ cells.

The TASEP, which we shall sometimes refer to as the *basic system*, is a Markov chain $S_{\alpha\beta\gamma}$ defined for any three parameters α, β and γ in the interval $[0,1]$ on the set of basic configurations. From time t to $t+1$, the chain evolves from the basic configuration $\tau = S_{\alpha\beta\gamma}(t)$ to a basic configuration $\tau' = S_{\alpha\beta\gamma}(t+1)$ as follows:

- A wall i is chosen uniformly at random among the $n+1$ walls, and then may become *active* with probability $\lambda(i)$, with $\lambda(i) = \alpha$ for $i = 1, \ldots, n-1$, $\lambda(0) = \beta$ and $\lambda(n) = \gamma$.
- If the wall did not become active, then nothing happens: $\tau' = \tau$.
- Otherwise the configuration may change near the active wall:
  a. If the active wall is not on the border ($i \in \{1, \ldots, n-1\}$) and has a black particle on its left and a white one on its right, then these two particles swap: $\bullet|\circ \to \circ|\bullet$.
  b. If the active wall is the left border ($i = 0$) and the leftmost cell contains a white particle, then the white particle leaves the system and it is replaced by a black particle: $|\circ \to |\bullet$.
  c. If the active wall is the right border ($i = n$) and the rightmost cell contains a black particle, then the black particle leaves the system and it is replaced by a white particle: $\bullet| \to \circ|$.
  d. Otherwise the configuration is left unchanged.

The four cases a, b, c, d define an application ϑ from the set of configurations with an active wall into the set of configurations, and, in terms of this application, the chain can be described as: let $I(t)$ be a sequence of independent uniform random variables on $\{0, \ldots, n\}$, and set

$$S_{\alpha\beta\gamma}(t+1) = \begin{cases} \vartheta(S_{\alpha\beta\gamma}(t), I(t)) & \text{with probability } \lambda(I(t)), \\ S_{\alpha\beta\gamma}(t) & \text{otherwise,} \end{cases}$$

where $\lambda(i) = \alpha$ for $i \in \{1, \ldots, n-1\}$, $\lambda(0) = \beta$, and $\lambda(n) = \gamma$.

As illustrated by Figure 3, during the evolution of the system, black particles travel from left to right while white particles do the opposite. As already mentioned, one can equivalently think about white particles as empty cells. See Figure 10 for the entire system with $n = 3$.

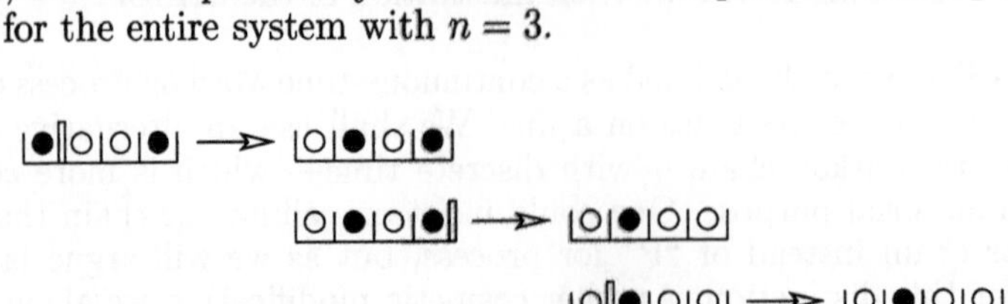

FIGURE 3. A possible evolution, with $n = 4$. The active wall triggering each transition is indicated.

1.2. A remarkable stationary distribution

Among many results on the TASEP, Derrida *et al.* [1, 3] proved the following nice property of the system in the case $\alpha = \beta = \gamma = 1$. First,

$$\text{Prob}(S_{111}(t) \text{ contains 0 black particles}) \underset{t\to\infty}{\longrightarrow} \frac{1}{C_{n+1}}, \tag{1}$$

where $C_{n+1} = \frac{1}{n+2}\binom{2n+2}{n+1}$ is the $(n+1)$th Catalan number. More generally, for all $0 \leq k \leq n$,

$$\text{Prob}(S_{111}(t) \text{ contains } k \text{ black particles}) \underset{t\to\infty}{\longrightarrow} \frac{\frac{1}{n+1}\binom{n+1}{k}\binom{n+1}{n-k}}{C_{n+1}}. \tag{2}$$

The finite state Markov chain S_{111} is clearly ergodic so that the previous limits are in fact the probabilities of the same events in the unique stationary distribution of the chain [6]. More generally, Derrida *et al.* provided expressions for the stationary probabilities of $S_{\alpha\beta\gamma}$. Since their original work a number of papers have appeared, providing alternative proofs and further results on correlations, time evolutions, etc. It should be moreover stressed that the model we presented is a special case among the many existing variants of asymmetric exclusion processes. See for instance the article [4] for recent advances and a bibliography. However, the remarkable apparition of Catalan numbers is not easily understood from the proofs in the physics literature. As far as we know, these proofs rely either on a *matrix ansatz*, or on a *Bethe ansatz*, both being then proved by a recursion on n.

In a previous paper [5], we proposed a combinatorial proof of Formulas (1) and (2) based on a combinatorial interpretation of the stationary distribution of S_{111}. The aim of the present paper is to give a combinatorial derivation of the general stationary distribution of $S_{\alpha\beta\gamma}$.

1.3. The complete system

The main ingredient we introduced in [5] to study the TASEP consisted in a new Markov chain X_{111} on a set Ω_n^0 of complete configurations, that satisfies two main requirements. On the one the stationary distribution of the basic chain S_{111} can be simply expressed in terms of that of the chain X_{111}. On the other hand the stationary behavior of the chain X_{111} is easy to understand. The complete configurations that we introduced for this purpose are made of two rows of n cells containing black and white particles. The first requirement was met by imposing that in the first row, the chain X_{111} simulates the chain S_{111}, *i.e.* X_{111} is a covering chain of S_{111}. The second requirement was met by adequately choosing the complete configurations and the transition rules so that X_{111} has clearly a uniform stationary distribution. In this paper we shall proceed in an analogous way and construct a complete chain $X_{\alpha\beta\gamma}$ on Ω_n^0, that will allow us to describe the stationary distribution of the basic chain $S_{\alpha\beta\gamma}$.

A *complete configuration* of Ω_n^0 is a pair of rows of n cells satisfying the following constraints:

(i) The balance condition: The two rows contain together n black and n white particles.

(ii) The positivity condition: On the left hand side of any vertical wall there are no more white particles than black ones.

An example of complete configuration is given in Figure 4. In view of Formulas (1) and (2), one first reason to introduce these complete configurations is that the cardinal of Ω_n^0 is $\frac{1}{n+2}\binom{2n+2}{n+1}$, and that, for all $0 \leq k \leq n$, the cardinal of the set

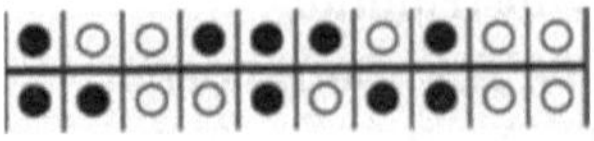

FIGURE 4. A complete configuration with $n = 10$.

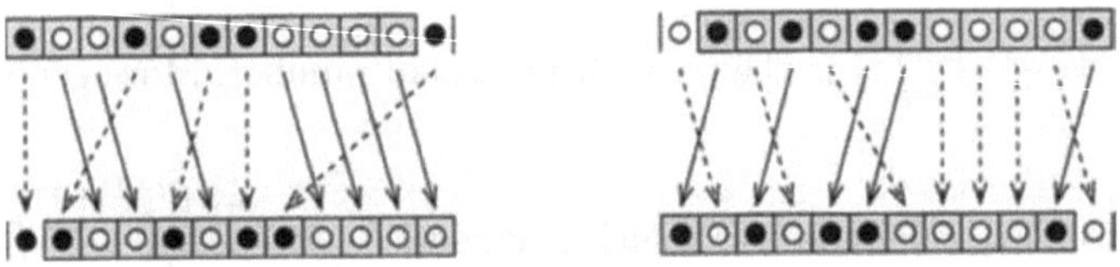

FIGURE 5. A white sweep and a black sweep.

$\Omega^0_{k,m}$ of complete configurations with k black and $m = n - k$ white particles on the top row is $\frac{1}{n+1}\binom{n+1}{k}\binom{n+1}{n-k}$. These formulas can be obtained in many ways, for instance using the cycle lemma (see [5]), or through one-to-one correspondences between complete configurations and bicolored Motzkin paths with n steps, or Dyck paths with $2n+2$ steps [8, Chap. 6]. Yet another classical way to obtain them is using generating functions, as we shall do in Section 2.

The Markov chain $X_{\alpha\beta\gamma}$ on Ω^0_n will be defined in terms of an application T from the set $\Omega^0_n \times \{0, \dots, n\}$ to the set Ω^0_n. This application, which we already used in [5], is derived in Section 3 as the first component of a fundamental bijection $\bar{T}$ and can be conveniently described as follows. Given a complete configuration ω and an active wall i, the actions of T on the first row of ω do not depend on the second row, and mimic the application ϑ describing the evolution of the Markov chain $S_{\alpha\beta\gamma}$ in the cases a, b, c and d of the description of the basic TASEP. In particular in the top row, black particles travel from left to right and white particles from right to left. As opposed to that, in the bottom row, T moves black and white particles backward. In order to describe this, we first introduce the concept of sweep (see Figure 5):

- A *white sweep* between walls i_1 and i_2 consists in all white particles of the bottom row and between walls i_1 and i_2 simultaneously hopping to the right (some black particles thus being displaced to the left in order to fill the gaps).
- A *black sweep* between walls i_1 and i_2 consists in all black particles of the bottom row and between walls i_1 and i_2 simultaneously hopping to the left (some white particles thus being displaced to the right in order to fill the gaps).

Next, around the active wall i, we distinguish the following walls: if $i \neq 0$, let $j_1 < i$ the leftmost wall such that there are only white particles in the top row between walls j_1 and $i-1$; if $i \neq n$, let $j_2 > i$ be the rightmost wall such that there are only black particles in the bottom row between walls $i+1$ and j_2.

With these definitions, we are in position to describe completely the application T. Given a configuration $\omega \in \Omega^0_n$ and an active wall $i \in \{0, \dots, n\}$, the cases a, b c and d of the basic chain describe the first row of $T(\omega, i)$, and they are complemented as follows to give the second row:

a. Depending whether the particle on the bottom right of the ith wall in ω is black or white, a white sweep occurs between j_1 and i, or a black one between $i+1$ and j_2+1 (see Figure 6).

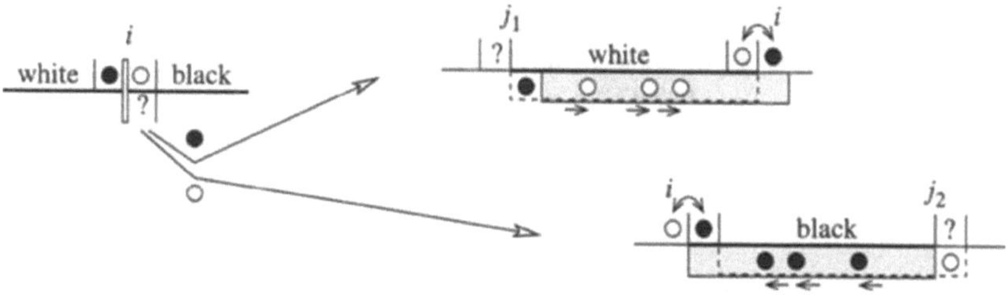

FIGURE 6. Sweeps occurring below the transition $(b|w \to w|b)$.

b. The leftmost column of ω consists of a $\left|\substack{\circ\\ \bullet}\right|$-column. These two particles exchange (in agreement with the rule for the top row), and a black sweep occurs between the left border and wall j_2+1.

c. The rightmost column of ω consists of a $\left|\substack{\bullet\\ \circ}\right|$-column. These two particles exchange (in agreement with the rule for the top row), and a white sweep occurs between wall j_1 and the right border.

d. As in the top row, nothing happens in the bottom row.

The Markov chain $X_{\alpha\beta\gamma}$ is the Markov chain on the set Ω_n^0 of complete configurations that is defined from the application T exactly as the TASEP is described from ϑ: the evolution rule from time t to $t+1$ consists in choosing $i = I(t)$ uniformly at random in $\{0, \dots, n\}$ and setting

$$X_{\alpha\beta\gamma}(t+1) = \begin{cases} T(X_{\alpha\beta\gamma}(t), i) \text{ with probability } \lambda(i), \\ X_{\alpha\beta\gamma}(t) \text{ otherwise,} \end{cases}$$

where $\lambda(i) = \alpha$ for $i \in \{1, \dots, n-1\}$, $\lambda(0) = \beta$, and $\lambda(n) = \gamma$.

By construction, the Markov chains $S_{\alpha\beta\gamma}$ and $X_{\alpha\beta\gamma}$ are related by

$$S_{\alpha\beta\gamma} \equiv \operatorname{top}(X_{\alpha\beta\gamma}),$$

where $\operatorname{top}(\omega)$ denotes the top row of a complete configuration ω, and the $\equiv$ is intended as identity in law. An appealing interpretation from a combinatorial point of view is that we have revealed a circulation of the particles, that use the bottom row to travel backward and implement the infinite reservoirs.

1.4. The stationary distribution of the complete system

In order to express the stationary distribution of the chain $X_{\alpha\beta\gamma}$, we introduce two combinatorial statistics and use them to associate a weight $q(\omega)$ to each complete configuration.

By definition, a complete configuration ω is a concatenation of four types of columns $\left|\substack{\bullet\\ \bullet}\right|$, $\left|\substack{\bullet\\ \circ}\right|$, $\left|\substack{\circ\\ \bullet}\right|$ and $\left|\substack{\circ\\ \circ}\right|$, subject to the balance and positivity conditions. Observe that the concatenation of two complete configurations of Ω_i^0 and Ω_j^0 with $i+j=n$ yields a complete configuration of Ω_n^0. Let us call *prime* a configuration that cannot be decomposed in this way. A complete configuration ω can be uniquely written as a concatenation $\omega = \omega_1 \cdots \omega_m$ of prime configurations. These prime factors can be of three types: $\left|\substack{\bullet\\ \circ}\right|$-columns, $\left|\substack{\circ\\ \bullet}\right|$-columns, and *blocks* of the form $\left|\substack{\bullet\\ \bullet}\right|\omega'\left|\substack{\circ\\ \circ}\right|$ with ω' a complete configuration. The inner part ω' of a block $\omega = \left|\substack{\bullet\\ \bullet}\right|\omega'\left|\substack{\circ\\ \circ}\right|$ is referred to as its *inside*.

Now, given a complete configuration ω, let us assign labels to some of the black and white particles of its bottom row: a white particle is labeled x if it is not in a block, and a black particle is labeled y if it is not in the inside of a block

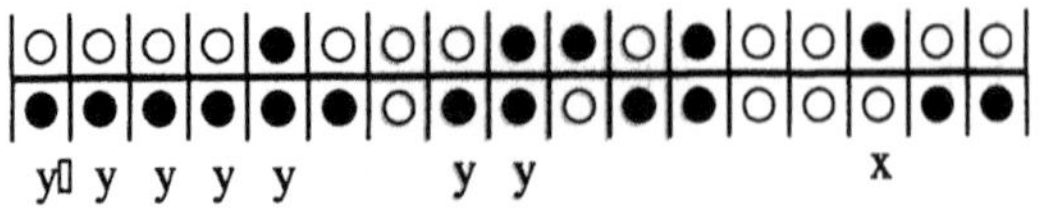

FIGURE 7. A configuration ω with weight $q(\omega) = \alpha^8\beta^{10}\gamma^{16}$. Labels are indicated below particles.

and if on its left hand side, all white particles belong to some block. Let us denote by $n_x(\omega)$ and $n_y(\omega)$ respectively the number of labels of type x and the number of labels of type y in the configuration ω. Then the *weight* of a configuration ω is defined as

$$q(\omega) = \beta^n\gamma^n\left(\frac{\alpha}{\beta}\right)^{n_y(\omega)}\left(\frac{\alpha}{\gamma}\right)^{n_x(\omega)} = \alpha^{n_x(\omega)+n_y(\omega)}\beta^{n-n_y(\omega)}\gamma^{n-n_x(\omega)}.$$

For instance, the weight of the configuration of Figure 7 is $\alpha^8\beta^{10}\gamma^{16}$, and more generally the weight is a monomial with total degree $2n$.

Theorem 1.1. *The Markov chain $X_{\alpha\beta\gamma}$ is ergodic and has the following stationary distribution:*

$$Prob(X_{\alpha\beta\gamma}(t) = \omega) \underset{t\to\infty}{\longrightarrow} \frac{q(\omega)}{Z_n}, \qquad \text{where } Z_n = \sum_{\omega'\in\Omega_n^0} q(\omega'),$$

where $q(\omega)$ is the previously defined weight on complete configurations.

In particular for $\alpha = \beta = \gamma = 1$, $q(\omega) = 1$ for all configurations and we recover uniformity as in [5],

$$Prob(X_{111}(t) = \omega) \underset{t\to\infty}{\longrightarrow} \frac{1}{|\Omega_n^0|} = \frac{1}{C_n}.$$

The Markov chain $X_{\alpha\beta\gamma}$ is clearly aperiodic, and the fact that it is irreducible follows from the irreducibility of S_{111} done in [5]. This granted, it is sufficient for the proof of Theorem 1.1 to show that the distribution induced by the weights q is stationary. We shall use an alternate description of T, which rely on the following result proved in Section 3.

Theorem 1.2. *There exists a bijection $\bar{T}$ from $\Omega_n^0 \times \{0,\dots,n\}$ onto itself such that*

- *the application T is the first component of $\bar{T}$: for all ω and i, $\bar{T}(\omega,i) = (\omega',j) \Rightarrow T(\omega,i) = \omega'$,*
- *the bijection $\bar{T}$ transports weights: for all $(\omega',j) = \bar{T}(\omega,i)$,*

$$\lambda(j)\,q(\omega') = \lambda(i)\,q(\omega),$$

where $\lambda(i) = \alpha$ for $i \in \{1,\dots,n-1\}$, $\lambda(0) = \beta$ and $\lambda(n) = \gamma$.

In order to see that the distribution induced by q is stationary, we assume that

$$\text{Prob}(X_{\alpha\beta\gamma}(t) = \omega) = \frac{q(\omega)}{Z_n}, \qquad \text{for all } \omega \in \Omega_n^0,$$

and try to compute $\text{Prob}(X_{\alpha\beta\gamma}(t+1) = \omega')$.

Recall that $I(t)$ denotes the wall chosen at time t, so that $X_{\alpha\beta\gamma}(t+1) = T(X_{\alpha\beta\gamma}(t), I(t))$, and define $J(t+1)$ by $\bar{T}(X_{\alpha\beta\gamma}(t), I(t)) = (X_{\alpha\beta\gamma}(t+1), J(t+1))$

if $I(t)$ became active, or by $J(t+1) = I(t)$ otherwise. Then, in view of the first point in Theorem 1.2,

$$\mathrm{Prob}\big(X_{\alpha\beta\gamma}(t+1) = \omega'\big) \;=\; \sum_{j=0}^{n} \mathrm{Prob}\big(X_{\alpha\beta\gamma}(t+1) = \omega',\; J(t+1) = j\big).$$

Now, by definition of the Markov chain $X_{\alpha\beta\gamma}$, for all ω' and j,

$$\begin{aligned}\mathrm{Prob}\big(X_{\alpha\beta\gamma}(t+1) = \omega',\, J(t+1) = j\big) & \\ &= \lambda(i)\cdot \mathrm{Prob}\big(X_{\alpha\beta\gamma}(t) = \omega,\; I(t) = i\big) \\ &\quad + (1-\lambda(j))\cdot \mathrm{Prob}\big(X_{\alpha\beta\gamma}(t) = \omega',\; I(t) = j\big),\end{aligned}$$

where $(\omega, i) = \bar{T}^{-1}(\omega', j)$. Since the random variable $I(t)$ is uniform on $\{0, \ldots, n\}$, we get

$$\begin{aligned}\mathrm{Prob}\big(X_{\alpha\beta\gamma}(t+1) = \omega',\, J(t+1) = j\big) & \\ &= \lambda(i)\cdot \frac{q(\omega)}{Z_n}\frac{1}{n+1} \;+\; (1-\lambda(j))\cdot \frac{q(\omega')}{Z_n}\frac{1}{n+1}.\end{aligned}$$

But according to the second point in Theorem 1.2, $\lambda(i)q(\omega) = \lambda(j)q(\omega')$ so that the terms involving λ cancel. Finally

$$\mathrm{Prob}(X_{\alpha\beta\gamma}(t+1) = \omega') \;=\; \sum_{j=0}^{n} \frac{q(\omega')}{Z_n}\frac{1}{n+1} \;=\; \frac{q(\omega')}{Z_n},$$

and this completes the proof that the distribution induced by q is stationary.

1.5. From the complete to the basic system.

The relation $S_{\alpha\beta\gamma} \equiv \mathrm{top}(X_{\alpha\beta\gamma})$ now allows to derive from Theorem 1.1 a combinatorial interpretation for the basic system.

Theorem 1.3. *Let top(ω) denote the top row of a complete configuration ω. Then for any initial configurations $S_{\alpha\beta\gamma}(0)$ and $X_{\alpha\beta\gamma}(0)$ with $top(X_{\alpha\beta\gamma}(0)) = S_{\alpha\beta\gamma}(0)$, and any basic configuration r,*

$$Prob(S_{\alpha\beta\gamma}(t) = r) \;=\; Prob(top(X_{\alpha\beta\gamma}(t)) = r) \;\underset{t\to\infty}{\longrightarrow}\; \frac{1}{Z_n} \sum_{\{\omega\in\Omega_n^0 \mid top(\omega)=r\}} q(\omega).$$

In particular, in the case $\alpha = \beta = \gamma = 1$, we recover

$$Prob(S_{\alpha\beta\gamma}(t) = r) \;\underset{t\to\infty}{\longrightarrow}\; \frac{|\{\omega \in \Omega_n^0 \mid top(\omega) = r\}|}{|\Omega_n^0|},$$

which is yields

$$Prob(S_{\alpha\beta\gamma}(t) \text{ contains } k \text{ black particles}) \;\underset{t\to\infty}{\longrightarrow}\; \frac{|\Omega_{k,m}^0|}{|\Omega_n^0|} \;=\; \frac{\frac{1}{n+1}\binom{n+1}{k}\binom{n+1}{m}}{C_{n+1}}.$$

As discussed in Section 4 this interpretation sheds a new light on some recent results of Derrida *et al.* connecting the TASEP to Brownian excursions [2].

1.6. Continuous-time descriptions of the TASEP

In the physics literature, the TASEP is usually described in the following terms. The time is continuous, and one consider walls where a move can take place: at any time, wall i has probability $\lambda(i)dt$ to trigger a move $\omega \to \vartheta(\omega, i)$ during time interval dt (the rate $\lambda(i)$ is defined as previously).

Following the probabilistic literature [7], one can give an formulation which is equivalent to the previous one, but already closer to ours. In this description, each wall waits for an independent exponential random time with rate 1 before waking up (in other terms, the probability that wall i will still be sleeping in t seconds is e^{-t}). When wall i wakes up, it has probability $\lambda(i)$ to become active. If this is the case, then the transition $\omega \to \vartheta(\omega, i)$ is applied to the current configuration ω. In any case the wall falls again asleep, restarting its clock again.

This continuous-time TASEP is now easily coupled to the Markov chain $S_{\alpha\beta\gamma}$. Let the time steps of $S_{\alpha\beta\gamma}$ correspond to the succession of moments at which a wall wakes up. Then in both versions, the index of next wall to wake up is at any time an uniform random variable on $\{0, \ldots, n\}$, and when a wall wakes up the transition probabilities are identical. This implies that the stationary distribution of the continuous-time TASEP and its Markov chain replica are identical.

1.7. Outline of the rest of the paper.

In Section 2 an approach to compute explicit quantities from the combinatorial interpretation is briefly exposed. Theorem 1.2 is proved in Section 3. Finally some concluding remarks are gathered in Section 4.

2. Enumeration

Let us introduce the weighted generating function of complete configurations with respect to their length:

$$Z(t; u, v) = \sum_{n\geq 0} \sum_{\omega \in \Omega_n^0} u^{n_x(\omega)} v^{n_y(\omega)} t^n, \qquad \text{so that} \quad Z(\beta\gamma t; \tfrac{\alpha}{\gamma}, \tfrac{\alpha}{\beta}) = \textstyle\sum_{n\geq 0} Z_n t^n.$$

The decomposition of a configuration at its first block yields

$$Z(t; u, v) = 1 + tvZ(t; u, v) + tuZ(t; 1, v) + t^2 vZ(t; 1, 1)Z(t; u, v).$$

Solving this equation yields

$$Z(t; u, v) \;=\; \frac{2 - u - v + uv - 2tuv - (u + v - uv)\sqrt{1 - 4t}}{2(1 - u + tu^2)(1 - v + tv^2)}.$$

Extracting coefficients in this expression allow to recover for instance a formula for Z_n. One could also have taken into account the number of particles in the top row in the equation. We do not pursue on this line since Z_n was obtained by other ways and largely studied as a function of α, β, γ in the physical literature.

3. The bijection $\bar{T}$

The aim of this section is to prove Theorem 1.2, thus giving an alternate description of the transformation T and a case by case analysis of its action on the weight q.

We need the following properties, the verification of which is left to the reader.

Property 3.1. *Let ω be a complete configuration belonging to Ω_n, then we have the following structure properties:*

1. *In a local configuration* $|{\bullet \atop ?}|{\circ \atop \bullet}|$ *the black particle in the bottom row never contributes a label* y.
2. *The white particle in the bottom row of a* $|{\circ \atop \circ}|$*-column never has a label* x.

and movement properties:

i. *The deletion/insertion of a* $|{\circ \atop \bullet}|$*-column does not change the labels of other particles.*
ii. *The deletion/insertion of a pair* ${}^{\bullet}|_{\circ}$ *taking the form* $|{\bullet \atop ?}|{\circ \atop \circ}| \leftrightarrow |{\circ \atop ?}|$ *does not change the labels of other particles.*

From now on in this section, (ω, i) denotes an element of the current class, and (ω', j) its image by $\bar{T}$. In the pairs (ω, i) and (ω', j), i and j refer to walls of the configurations ω and ω', and i is called the active wall of ω. Following the notations of Section 1, when $i \neq 0$, we also consider $j_1 < i$ the smallest integer such that in the top row of ω all cells between walls j_1 and $i-1$ contain white particles. Symmetrically, when $i \neq n$, we consider $j_2 > i$ the largest integer such that in the top row of ω all cells between walls $i+1$ and j_2 contain black particles.

To define the bijection $\bar{T}$ and prove Theorem 1.2 we shall partition the set $\Omega_n \times \{0, \ldots, n\}$ into classes $A_{a'}$, $A_{a'_s}$, $A_{a''}$, $A_{a''_s}$, A_b, A_{b_s}, A_c, A_{c_s}, A_d, and describe, for each class A_z, its image $B_z = \bar{T}(A_z)$ under the action of $\bar{T}$ and the corresponding variation of the weight q:

The active wall of ω separates in the top row a black particle P and a white particle Q. Then in the top row the particles P and Q swap. In the bottom row, the sweep that occurs depends on the type of the particle R that is below Q in ω (see Figure 8):

$A_{a'}$ The particle R is black and the wall j_1 is different from 0. Then $j = j_1$ and, in the bottom row, a white sweep occurs between walls j and i. The new configuration ω' belongs to Ω_n. Indeed ω' can also be described as obtained from ω by moving a $|{\circ \atop \bullet}|$-column from the right of the ith wall to the right of the j_1th. But moving a $|{\circ \atop \bullet}|$-column has no effect on the positivity constraint.

Now we want to compare $q(\omega)$ and $q(\omega')$. According to Property 3.1.1 the particle R does not contribute a label y neither in ω nor in ω'. Moreover, according to Property 3.1.i, the displacement of the $|{\circ \atop \bullet}|$-column does not affect labels of other particles. Hence $q(\omega) = q(\omega')$, in agreement with $\lambda(i) = \lambda(j)$.

The image $B_{a'}$ of the class $A_{a'}$ consists of pairs (ω', j), $j > 0$, with a $|{\circ \atop \bullet}|$-column on the right hand side of the jth wall of ω' and such that the sequence of white particles on the right hand side of the jth wall in the top row is followed by a black particle.

$A_{a'_s}$ The particle R is black and the wall $j_1 = 0$. Then $j = 0$ and a white sweep occurs between walls 0 and i. The new configuration ω' still belongs to Ω_n since it is again obtained from ω by moving a $|{\circ \atop \bullet}|$-column from the right of the ith wall to the right of the wall 0.

As opposed to the previous case, Property 3.1.1 applies only to ω: in ω', the displaced $|{\circ \atop \bullet}|$-column is the leftmost one, so that its black particle contributes a supplementary y label. Therefore $q(\omega') = q(\omega)\frac{\alpha}{\beta}$, in agreement with $\lambda(i) = \alpha$, $\lambda(0) = \beta$.

The image $B_{a'_s}$ consists of pairs $(\omega', 0)$ such that the top row starts on the left by a non-empty sequence of white particles followed by a black one.

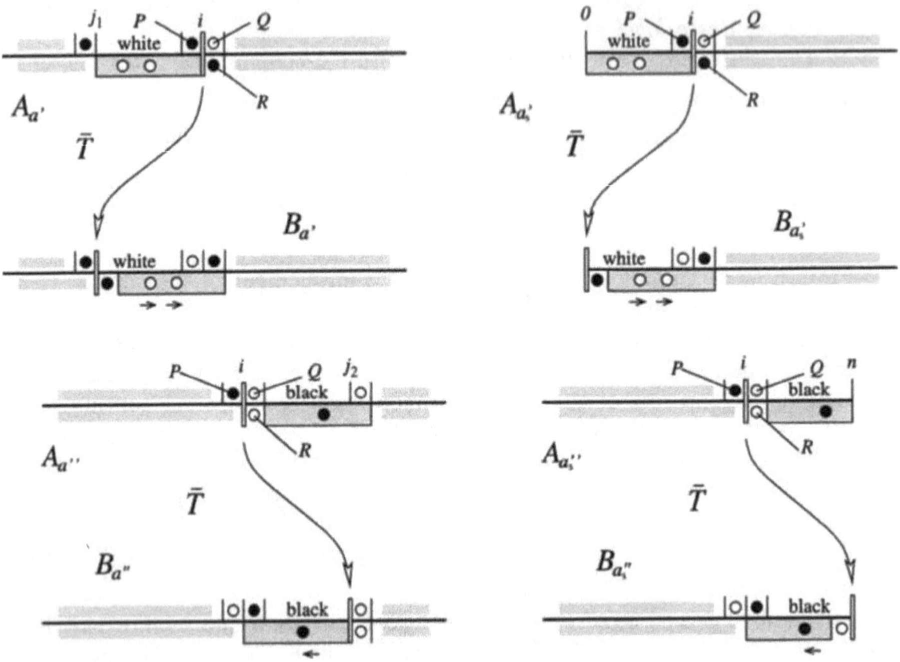

FIGURE 8. Jump moves in the case $\bullet|\circ \to \circ|\bullet$.

$A_{a''}$ The particle R is white and the wall j_2 is different from n. Then $j = j_2$ and, in the bottom row, a black sweep occurs between walls $i+1$ and $j+1$. The new configuration ω' belongs to Ω_n. Indeed ω' can be described as obtained from ω by moving a ${}^\bullet|_\circ$-diagonal from the ith wall to the j_2th wall: this movement has no effect on the positivity constraint.

From Property 3.1.2 we see that the particle R does not contribute a label y neither in ω nor in ω', and from Property 3.1.ii the displacement of the ${}^\bullet|_\circ$-diagonal does not affect labels of other particles. Hence $q(\omega) = q(\omega')$, in agreement with $\lambda(i) = \lambda(j)$.

The image $B_{a''}$ of the class $A_{a''}$ consists of pairs (ω', j) with a $|{}^{\circ}_{\circ}|$-column on the right hand side of the jth wall and such that the sequence of black particles on the left hand side of the jth wall in the top row is followed by a white particle.

$A_{a''_s}$ The particle R is white and the wall j_2 is equal to n. Then $j = n$ and, in the bottom row, a black sweep occurs between walls $i+1$ and n. The new configuration ω' still belongs to Ω_n since it is obtained from ω by removing a ${}^\bullet|_\circ$-diagonal around the ith wall and inserting a $|{}^{\bullet}_{\circ}|$-column to the left of the n-th wall.

As opposed to the previous case, Property 3.1.2 applies only to ω: the inserted $|{}^{\bullet}_{\circ}|$-column is the rightmost one, so that its white particle contributes an x label. Therefore $q(\omega') = q(\omega)\frac{\alpha}{\gamma}$, in agreement with $\lambda(i) = \alpha$ and $\lambda(n) = \gamma$.

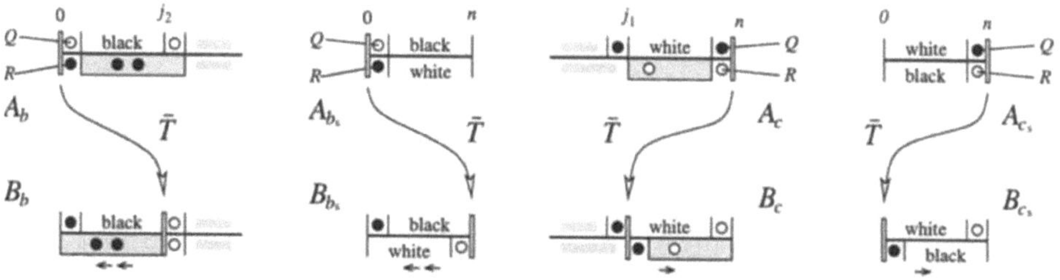

FIGURE 9. Active left border and active right border with respectively a black and a white particle in the top row.

The image $B_{a''_s}$ of the class $A_{a''_s}$ consists of pairs (ω', n) such that there is a sequence of black particles on the left hand side of the nth wall in the top row, followed by a white particle.

The active wall of ω is the left border with a white particle Q on its right in the top row. Again, the cell under Q must contain a black particle R (see Figure 9). First the two particle Q and R exchange to form a $|{}^{\bullet}_{\circ}|$-column. Then we have two cases:

A_b the wall j_2 is not n. Then $j = j_2$ and, in the bottom row, a black sweep occurs between walls 1 and $j+1$. The configuration ω' belongs to Ω_n. Indeed no black particle moves to the right.

Equivalently, the new configuration ω' is obtained by inserting a ${}^{\bullet}|_{\circ}$-diagonal at the wall j_2 and deleting the first $|{}^{\circ}_{\bullet}|$-column. According to Properties 3.1.i and 3.1.ii only the labels of the displaced particles are affected. Since the deleted $|{}^{\circ}_{\bullet}|$-column is the leftmost, it contributes a y label in ω. As opposed to that, Property 3.1.2 forbids the ${}^{\bullet}|_{\circ}$-diagonal to contribute a x label. Therefore $q(\omega') = q(\omega)\frac{\beta}{\alpha}$, in agreement with $\lambda(0) = \beta$ and $\lambda(j) = \alpha$.

The image B_b consist of pairs (ω', j) with a $|{}^{\circ}_{\circ}|$-column on the right of the jth wall of ω' and such that the sequence of black particles on left of the jth wall in the top row ends at the border.

A_{b_s} the wall j_2 is equal to n. Then $j = n$ and, in the bottom row, a black sweep occurs between walls 1 and n. Finally, $q(\omega') = q(\omega)\frac{\beta}{\alpha}\frac{\alpha}{\gamma} = q(\omega)\frac{\beta}{\gamma}$, in agreement with $\lambda(0) = \beta$ and $\lambda(n) = \gamma$.

The image B_{b_s} is reduced to the configuration with all black particles in the top row.

The active wall of ω is the right border with a black particle Q on its left in the top row. The cell under Q must contain a white particle R (see Figure 9): First the particles Q and R exchange to form a $|{}^{\circ}_{\bullet}|$-column. There are then two cases:

A_c the wall j_1 is different from 0. Then $j = j_1$ and in the bottom row, a white sweep occurs between walls j and $n-1$. The configuration ω' belongs to Ω_n since the transformation amounts to moving and flipping a $|{}^{\bullet}_{\circ}|$-column.

Equivalently, ω' is obtained by inserting a $|{}^{\circ}_{\bullet}|$-column at the left of the wall n and deleting the $|{}^{\circ}_{\bullet}|$-column on the right of j_1. According to Property 3.1.i only the labels of displaced particles can be affected. Since the deleted $|{}^{\bullet}_{\circ}|$-column is the rightmost column, its white particle contributes an x label in ω. As opposed to that, Property 3.1.1 forbids the $|{\circ \atop \bullet}|$-column to contribute a label in ω'. Therefore $q(\omega') = q(\omega)\frac{\gamma}{\alpha}$, in agreement with $\lambda(n) = \gamma$ and $\lambda(j) = \alpha$.

The image B_c consist of pairs (ω', j) with a $|{\circ \atop \bullet}|$-column on the right hand side of the jth wall of ω' and such that the sequence of white particles on the right hand side of the jth wall in the top row ends at the right border.

A_{c_s} the wall j_1 is equal to 0. Then $j = 0$ and in the bottom row, a white sweep occurs between walls 0 and $n-1$. Finally, $q(\omega') = q(\omega)\frac{\gamma}{\alpha}\frac{\alpha}{\beta} = q(\omega)\frac{\gamma}{\beta}$, in agreement with $\lambda(n) = \gamma$ and $\lambda(0) = \beta$.

The image B_{b_s} is reduced to the configuration with all white particles in the top row.

A_d: This class contains all the remaining cases. On these pairs the application $\bar{T}$ has no effect, that is, for $(\omega, i) \in A_d$, $\bar{T}(\omega, i) = (\omega, i)$. In particular the weights are left unchanged.

The observation that image classes $\{B_{a'}, B_{a'_s}, B_{a''}, B_{a''_s}, B_b, B_{b_s}, B_c, B_{c_s}, B_d\}$ form a partition of $\Omega_n \times \{0, \ldots, n\}$ completes the proof of Theorem 1.2.

4. Conclusions and links to Brownian excursions

The starting point of the paper [5] was a "combinatorial ansatz": the stationary distribution of the TASEP can be expressed in terms of Catalan numbers hence should have a nice combinatorial interpretation. As we have seen in the present paper, our approach is natural enough to extend to more general TASEP.

We do not claim that our combinatorial interpretation is of any physical relevance. However, as already pointed out in [5], apart from explaining the occurrence of "magical" Catalan numbers in the problem, it sheds a new light on the recent results of Derrida *et al.* [2] connecting the TASEP with Brownian excursion. More precisely, using explicit calculations, Derrida *et al.* show that when $\alpha = \beta = \gamma = 1$, the density of black particles in configurations of the TASEP can be expressed in terms of a pair (e_t, b_t) of independent processes, a Brownian excursion e_t and a Brownian motion b_t. In our interpretation these two quantities appear at the discrete level, associated to each complete configuration ω of Ω_n^0:

- The role of the Brownian excursion is played for ω by the halved differences $e(i) = \frac{1}{2}(B(i) - W(i))$ between the number of black and white particles sitting on the left of the ith wall, for $i = 0, \ldots, n$. By definition of complete configurations, $(e(i))_{i=0,\ldots,n}$ is a discrete excursion, that is, $e(0) = e(n) = 0$, $e(i) \geq 0$ and $|e(i) - e(i-1)| \in \{0, 1\}$, for $i = 0, \ldots, n$.
- The role of the Brownian motion is played for ω by the differences $b(i) = B_{top}(i) - B_{bot}(i)$ between the number of black particles sitting in the top and in the bottom row, on the left of the ith wall, for $i = 0, \ldots, n$. This quantity $(b(i))_{i=0,\ldots,n}$ is a discrete walk, with $|b(i) - b(i-1)| \in \{0, 1\}$ for $i = 0, \ldots, n$.

Since $e(i) + b(i) = 2B_{top}(i) - i$, these quantities allow to describe the cumulated number of black particles in the top row of a complete configuration. Accordingly, the density in a given segment (i, j) is

$$\frac{B_{top}(j) - B_{top}(i)}{j - i} = \frac{1}{2} + \frac{e(j) - e(i)}{2(j - i)} + \frac{b(j) - b(i)}{2(j - i)}.$$

This is a discrete version of the quantity considered by Derrida *et al.* in [2].

Now the two walks $e(i)$ and $b(i)$ are correlated since one is stationary when the other is not, and vice versa: $|e(i) - e(i-1)| + |b(i) - b(i-1)| = 1$. Given ω, let $I_e = \{\alpha_1 < \ldots < \alpha_p\}$ be the set of indices of $|{\bullet \atop \bullet}|$- and $|{\circ \atop \circ}|$-columns, and $I_b = \{\beta_1 < \ldots < \beta_q\}$ the set of indices of $|{\bullet \atop \circ}|$- and $|{\circ \atop \bullet}|$-columns $(p + q = n)$. Then the walk $e'(i) = e(\alpha_i) - e(\alpha_{i-1})$ is the excursion obtained from e by ignoring stationary steps, and the walk $b'(i) = b(\beta_i) - b(\beta_{i-1})$ is obtained from b in the same way. Conversely given a simple excursion e' of length p, a simple walk b' of length q and a subset I_e of $\{1, \ldots, p+q\}$ of cardinal p, two correlated walks e and b, and thus a complete configuration ω can be uniquely reconstructed. The consequence of this discussion is that the uniform distribution on Ω_n^0 (which is stationary for $\alpha = \beta = \gamma = 1$) corresponds to the uniform distribution of triples (I_e, e', b') where given I_e, e' and b' are independent.

A direct computation shows that in the large n limit, with probability exponentially close to 1, a random configuration ω is described by a pair (e', b') of walks of roughly equal lengths $n/2 + O(n^{1/2+\varepsilon})$. In particular up to multiplicative constants the normalized pairs $(\frac{e'(tn/2)}{n^{1/2}}, \frac{b'(tn/2)}{n^{1/2}})$ and $(\frac{e(tn)}{n^{1/2}}, \frac{b(tn)}{n^{1/2}})$ both converge to the same pair (e_t, b_t) of independent processes, with e_t a standard Brownian excursion and b_t a standard Brownian walk.

We thus obtained a combinatorial interpretation of the apparition of the pair (e_t, b_t) in the TASEP at $\alpha = \beta = \gamma = 1$. How do these considerations extend to other TASEP? The case α much smaller than β and γ essentially reduces to $\alpha = \beta = \gamma = 1$ on a system of length $n - 2$ (with border cells acting as reservoirs). The case β and γ smaller than α appears more interesting to consider: at a rough level, the weights force $e(t)$ to spend more time at the value exactly zero and favors negative value of $b(t)$. The derivation of the associated continuum quantities in this case could be of some interest.

Another challenge raised by our approach is to give an explicit construction of a continuum TASEP by taking the limit of the Markov chain $X_{\alpha\beta\gamma}$, viewed as a Markov chain on pairs of walks. An appealing way to give a geometric meaning to the transitions in the continuum limit could be to use a representation in terms of parallelogram polyominoes, where the process $e(t)$ (or e_t in the continuum limit) describes the width of the polymonino and $b(t)$ (or b_t in the continuum limit) describes its vertical displacement.

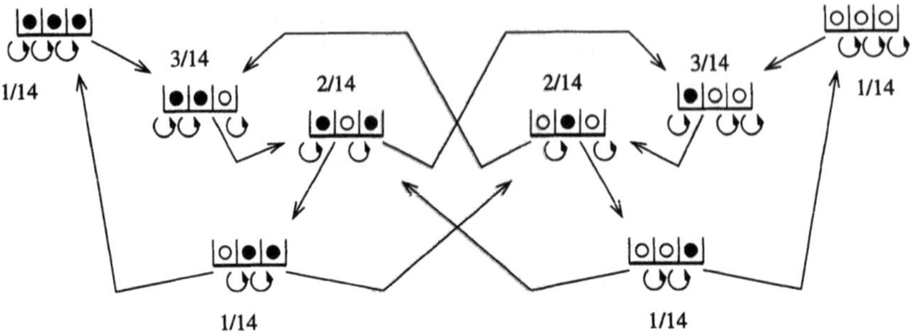

FIGURE 10. The basic configurations for $n = 3$ and transitions between them. The start point of each arrow indicates the wall triggering the transition. The numbers are the stationary probabilities.

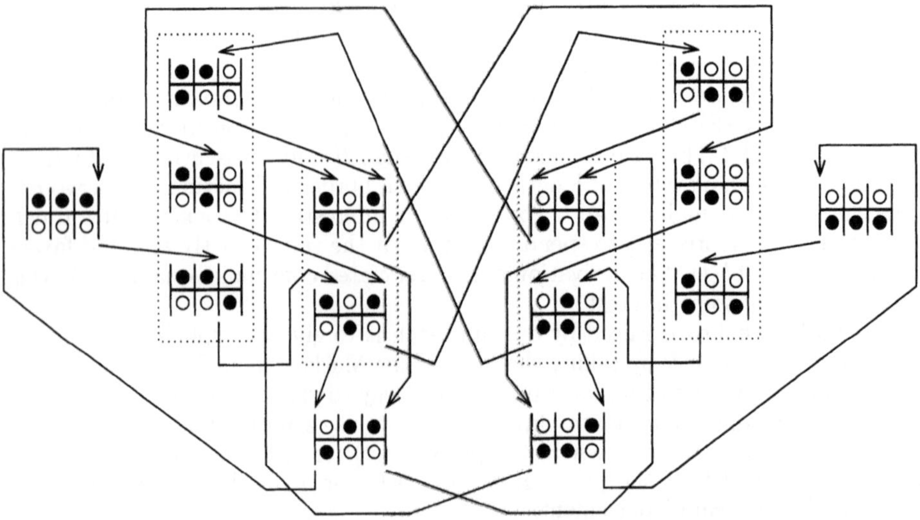

FIGURE 11. The 14 complete configurations for $n = 3$ and transitions between them. The start point of each arrow indicates the wall triggering the transition (loop transitions are not indicated). Stationary probabilities are uniform (equal to 1/14) since each configuration has equal in and out degrees. Ignoring bottom rows reduces this Markov chain to the chain of Figure 10.

References

[1] B. Derrida, E. Domany, and D. Mukamel. An exact solution of a one dimensional asymmetric exclusion model with open boundaries. *J. Stat. Phys.*, 69:667–687, 1992.

[2] B. Derrida, C. Enaud, and J. L. Lebowitz. The asymmetric exclusion process and Brownian excursions. Available electronically as `arXiv:cond-mat/0306078`.

[3] B. Derrida, M.R. Evans, V. Hakim, and V. Pasquier. Exact solution of a one-dimensional asymmetric exclusion model using a matrix formulation. *J. Phys. A: Math.*, 26:1493–, 1993.

[4] B. Derrida, J. L. Lebowitz, and E. R. Speer. Exact large deviation functional of a stationary open driven diffusive system: the asymmetric exclusion process. Available electronically as `arXiv:cond-mat/0205353`.

[5] E. Duchi and G. Schaeffer. A combinatorial approach to jumping particles I: maximal flow regime. *In proceedings of FPSAC'04*, 2004.

[6] O. Häggström. *Finite Markov Chains and Algorithmic Applications.* Cambridge University Press, 2002.

[7] T. M. Liggett. *Interacting Particle Systems.* Springer, New York, 1985.

[8] R. Stanley. *Enumerative Combinatorics, volume II.* Cambridge University Press, 1999.

Enrica Duchi
CAMS, EHESS, 87, bd Raspail, 75006 Paris, France
Enrica.Duchi@ehess.fr

Gilles Schaeffer
LIX, CNRS - École Polytechnique, 91128 Palaiseau, France
Gilles.Schaeffer@lix.polytechnique.fr

Trends in Mathematics, © 2004 Birkhäuser Verlag Basel/Switzerland

Stochastic Deformations of Sample Paths of Random Walks and Exclusion Models

Guy Fayolle and Cyril Furtlehner

ABSTRACT: *This study in centered on models accounting for stochastic deformations of sample paths of random walks, embedded either in $\mathbb{Z}^2$ or in $\mathbb{Z}^3$. These models are immersed in multi-type particle systems with exclusion. Starting from examples, we give necessary and sufficient conditions for the underlying Markov processes to be reversible, in which case their invariant measure has a Gibbs form. Letting the size of the sample path increase, we find the convenient scalings bringing to light phase transition phenomena. Stable and metastable configurations are bound to time-periods of limiting deterministic trajectories which are solution of nonlinear differential systems: in the example of the ABC model, a system of Lotka-Volterra class is obtained, and the periods involve elliptic, hyper-elliptic or more general functions. Lastly, we discuss briefly the contour of a general approach allowing to tackle the transient regime via differential equations of Burgers' type.*

1. Introduction

We are interested in models describing evolution of sample paths of random walks, when they are submitted to random local deformations involving possibly several links. Roughly speaking, given a finite sample path, say of size N, forming a not necessarily closed curve, the problem will be to characterize the evolution of an associated family $\{\mathbb{Y}_i, i = 1, \ldots, N\}$ of Markov processes in the thermodynamic limit as $N \to \infty$. This requires to guess and to find the interesting scalings.

In a previous study [7], we considered random walks on a square lattice, deformations involved pairs of links and occurred at the epochs of Poisson jump processes in continuous time (see section 2 for a more exact definition). The analysis was carried out by means of an explicit mapping, which led to view the system as a coupling of two exclusion processes.

Starting from a number of observations, we intend to hint in this paper that the model in [7] can indeed be cast into a broader class, the ultimate goal being to propose methods of wide applicability concerning the following questions:

- conditions ensuring Gibbs states and explicit forms of the corresponding invariant measures;
- steady-state equations in the thermodynamic limit as $N \to \infty$, and their solutions in the case of Gibbs states, but also in situations involving permanent currents;
- hydrodynamic and transient equations, when N is sufficiently large, yielding thus a complete picture of the evolution.

Generalizations of the model in $\mathbb{Z}^2$ can follow two natural trends. First, in modifying the construction of the random walk. Indeed, in the square lattice, we dealt with a 4-letter alphabet. Considering instead a finite alphabet of l letters is then tantamount to constructing random walks with oriented links, whose affixes are multiples of $\frac{2k\pi}{l}, k = 0, \ldots, l-1$. The case $l = 2$ corresponds to the simple exclusion process in $\mathbb{Z}$, and $l = 3$ yields the so-called *ABC model*.

Another possible extension is to relax the constraint that the walk lives in $\mathbb{Z}^2$ and to define a stochastic deformation process in higher dimension. In the sequel, we shall restrict ourselves to some paradigms in $\mathbb{Z}^2$ and $\mathbb{Z}^3$.

In section 2, we define a class of two-dimensional models, together with related patterns in $\mathbb{Z}^3$, in terms of exclusion particle systems. Section 3 is devoted to stochastic reversibility of the Markov processes of interest and to the Gibbs form of their invariant measure. In section 4, the non-symmetric classical ABC model is solved (fundamental scaling, phase transitions, classification of stable configurations) through the analysis of a Lotka-Volterra differential system. The concluding section 5 gives a brief overview of ongoing research about large scale dynamics, nonequilibrium and transient regimes.

2. Model descriptions via exclusion particle systems

Our main objective in this section is to show how the evolution of the sample paths of random walks can be fruitfully described by means of particle exclusion processes.

Beforehand, to avoid repetition and clumsy notation, let us emphasize that we shall only deal with jump Markov processes in continuous time. So implicitly the word *transition rate* will always refer to some underlying generator. Also, N will always stand for the size of the sample path, or equivalently the number of its links.

2.1. Preliminaries

In $\mathbb{Z}^1$, the simple exclusion model coincides with the well known KPZ system (see e.g. [9]), which represents a fluctuating and eventually growing interface. This system is coded by a sequence of binary variables $\{\tau_j\}, j = 1, \ldots, N$, depending on whether a particle is present or not, with asymmetric jump rates. This system has been extensively studied. In particular, the invariant measure has been obtained in a closed matrix form solution, for fairly arbitrary parameters and boundary conditions [3]. Large scale dynamics has also been analyzed [14], showing Burgers' equations [1].

2.2. 2-dimensional models

1) The triangular lattice and the ABC model. Here the evolution of the random walk is restricted to the triangular lattice. Each link (or step) of the walk is either 1, $e^{2i\pi/3}$ or $e^{4i\pi/3}$, and quite naturally will be said to be of type A, B and C, respectively. This corresponds to the so-called *ABC model*, since there is a coding by a 3-letter alphabet. The set of *transitions* (or reactions) is given by

$$AB \underset{p^+}{\overset{p^-}{\leftrightarrows}} BA, \qquad BC \underset{q^+}{\overset{q^-}{\leftrightarrows}} CB, \qquad CA \underset{r^+}{\overset{r^-}{\leftrightarrows}} AC, \tag{1}$$

where the introduced rates are fixed, but not necessarily equal. Also we impose *periodic boundary conditions* on the sample paths. This model was first introduced in [4] in the context of particles with exclusion, and a Gibbs form corresponding to reversibility has been found in [5] in some cases.

2) The square lattice and coupled exclusion model. This model was introduced in [7] to analyze stochastic deformations of a walk in the square lattice, and it will be referred to from now on as the $\{\tau_a\tau_b\}$ model. Assuming links counterclockwise oriented, we the following transitions can take place.

$$AB \overset{\lambda^{ab}}{\underset{\lambda^{ba}}{\rightleftarrows}} BA, \qquad BC \overset{\lambda^{bc}}{\underset{\lambda^{cb}}{\rightleftarrows}} CB, \qquad CD \overset{\lambda^{cd}}{\underset{\lambda^{dc}}{\rightleftarrows}} DC, \qquad DA \overset{\lambda^{da}}{\underset{\lambda^{ad}}{\rightleftarrows}} AD,$$

$$AC \overset{\gamma^{ac}}{\underset{\delta^{bd}}{\rightleftarrows}} BD, \qquad BD \overset{\gamma^{bd}}{\underset{\delta^{ca}}{\rightleftarrows}} CA, \qquad CA \overset{\gamma^{ca}}{\underset{\delta^{db}}{\rightleftarrows}} DB, \qquad DB \overset{\gamma^{db}}{\underset{\delta^{ac}}{\rightleftarrows}} AC.$$

We studied a rotation invariant version of this model when

$$\begin{cases} \lambda^+ \overset{\text{def}}{=} \lambda^{ab} = \lambda^{bc} = \lambda^{cd} = \lambda^{da}, \\ \lambda^- \overset{\text{def}}{=} \lambda^{ba} = \lambda^{cb} = \lambda^{dc} = \lambda^{ad}, \\ \gamma^+ \overset{\text{def}}{=} \gamma^{ac} = \gamma^{bd} = \gamma^{ca} = \gamma^{db}. \\ \gamma^- \overset{\text{def}}{=} \delta^{ac} = \delta^{bd} = \delta^{ca} = \delta^{db}. \end{cases}$$

Define the mapping $(A, B, C, D) \to (\tau^a, \tau^b) \in \{0, 1\}^2$, such that

$$\begin{cases} A \to (0, 0), \\ B \to (1, 0), \\ C \to (1, 1), \\ D \to (0, 1). \end{cases}$$

Then the dynamics can be formulated in terms of coupled exclusion processes. The evolution of the sample path is represented by a Markov process with state space the $2N$ binary random variables $\{\tau_j^a\}$ and $\{\tau_j^b\}$, $j = 1, \ldots, N$, taking value 1 if a particle is present and 0 otherwise. The jump rates to the right $(+)$ or to the left $(-)$ are given by

$$\begin{cases} \lambda_a^{\pm}(i) = \bar{\tau}_i^b \bar{\tau}_{i+1}^b \lambda^{\mp} + \tau_i^b \tau_{i+1}^b \lambda^{\pm} + \bar{\tau}_i^b \tau_{i+1}^b \gamma^{\mp} + \tau_i^b \bar{\tau}_{i+1}^b \gamma^{\pm}, \\ \lambda_b^{\pm}(i) = \bar{\tau}_i^a \bar{\tau}_{i+1}^a \lambda^{\pm} + \tau_i^a \tau_{i+1}^a \lambda^{\mp} + \bar{\tau}_i^a \tau_{i+1}^a \gamma^{\pm} + \tau_i^a \bar{\tau}_{i+1}^a \gamma^{\mp}. \end{cases} \tag{2}$$

Notably, one sees the jump rates of a given sequence are locally conditionally defined by the complementary sequence.

3) An extended stochastic clock model. We propose an extension of the preceding model for an arbitrary two-dimensional regular graph. To this end, consider a random walk composed of oriented links, the affixes of which take values $\omega_k = \exp(2ik\pi/n), k = 1, \ldots, n$, the n-th roots of unity. There are two different situations.

(a) $n = 2p + 1$ is odd. Then the walk cannot have a *fold* of two successive links, so that local displacements of edges can only be performed by exchanging two successive links. Let $X = \{X^1, \ldots, X^n\}$, denote the particle types viewed as letters of an alphabet, and let $\{\lambda^{kl}\}$ be the transition rates. The set of reactions is defined by

$$X^k X^l \overset{\lambda^{kl}}{\underset{\lambda^{lk}}{\rightleftarrows}} X^l X^k, \quad k \in [1,\, 2p+1],\, k \neq l.$$

These rules provide an extension of the ABC model, which we shall discuss in detail in section 3.

(b) When $n = 2p$ is even, the grammar is altered when two successive links fold, so that this elementary transition amounts merely to a rotation of the fold of angle $\pm\frac{2\pi}{n}$ (instead of a π rotation which would occur when exchanging the two links). The situation is thus slightly more complicated and the set of reactions is now given by

$$\begin{cases} X^k X^l \underset{\lambda^{lk}}{\overset{\lambda^{kl}}{\rightleftarrows}} X^l X^k, & k = 1, \ldots, n, \quad l \neq k + p, \\ X^k X^{k+p} \underset{\delta^{k+1}}{\overset{\gamma^k}{\rightleftarrows}} X^{k+1} X^{k+p+1}, & k = 1, \ldots, n, \end{cases} \tag{3}$$

where $k + p$ is taken modulo n. It is worth noting that γ^k (resp. δ^k) concern folds rotating in the counterclockwise direction (resp. clockwise) and that the number of letters of each type is no longer conserved. In other words, odd models give rise to pure diffusions with eventual drifts, when even models are truly reaction-diffusion models.

2.3. 3-dimensional generalizations of the coupled exclusion model on a diamond lattice

2.3.1. Elementary deformations Actually, the diamond lattice formulation of stochastic deformations in $\mathbb{Z}^3$ provides several straightforward generalization of the 2-dimensional $\{\tau_a\tau_b\}$ model. Indeed, between two nodes there are $8 = 2^3$ possible links (the jumps of the sample path). Let (τ_a, τ_b, τ_c) be the vector of binary components corresponding to a displacement in each direction, where a, b and c denote here the three particle families (letters).

As in the 2-dimensional model, elementary deformations consist in exchanging between neighbouring sites the value of one of the binary components. By construction this model gives a kind of geometric decoupling between the three types of particles. In fact all possible existing stochastic coupling result solely from the conditional transition rates $\lambda_a^{\pm}(\tau_b, \tau_c)$, $\lambda_b^{\pm}(\tau_c, \tau_a)$, and $\lambda_c^{\pm}(\tau_a, \tau_b)$, according to the various possible models. In order to obtain some non-trivial dynamical effect, we couple the three systems of particles, keeping in mind the possibility of getting back the 2-dimensional $\{\tau_a\tau_b\}$ model under certain conditions. This is the subject of the next two paragraphs.

2.3.2. A linear coupled exclusion model Here we propose a coupling which is linear with respect to the complementary fields, via the following intensities:

$$\begin{cases} \lambda_a^{\pm}(i) = \lambda \pm 2(\tau_i^b - \tau_i^c)\mu_a, \\ \lambda_b^{\pm}(i) = \lambda \pm 2(\tau_i^c - \tau_i^a)\mu_b, \\ \lambda_c^{\pm}(i) = \lambda \pm 2(\tau_i^a - \tau_i^b)\mu_c. \end{cases}$$

Suppose for a while $\mu_c = 0$ and $\mu_a = \mu_b = \mu$. Then the sequence $\{tau_c\}$ remains disordered, which means the marginal law of τ_i^c is $\frac{1}{2}$. As for the subsystem $\{\tau_a, \tau_b\}$, up to random contributions $(\tau_i^c - \frac{1}{2})\mu$, the transition rates $\lambda_a^{\pm}$ and $\lambda_b^{\pm}$ correspond to a particular definition of the $\{\tau_a\tau_b\}$ model. Hence, in the limit, we obtain a kind of *disordered* $\{\tau_a\tau_b\}$ model.

2.3.3. A NON-LINEAR COUPLED EXCLUSION MODEL It is also expected to recover the $\tau_a\tau_b$ model when one of the components is completely frozen in an ordered phase, taking for example $\tau_i^c = 0$ for $i = 1, \dots, N$. This situation is fulfilled by choosing the following non-linear couplings

$$\begin{cases} \lambda_a^{\pm}(i) = \lambda \pm (\tau_i^b - \bar{\tau}_i^b)(\tau_i^c - \bar{\tau}_i^c)\mu_a, \\ \lambda_b^{\pm}(i) = \lambda \pm (\tau_i^c - \bar{\tau}_i^c)(\tau_i^a - \bar{\tau}_i^a)\mu_b, \\ \lambda_c^{\pm}(i) = \lambda \pm (\tau_i^a - \bar{\tau}_i^a)(\tau_i^b - \bar{\tau}_i^b)\mu_c. \end{cases}$$

2.4. Boundary conditions

Although they live in $\mathbb{Z}^2$ or $\mathbb{Z}^3$, let us emphasize that all the objects considered throughout this study are curves, hence one-dimensional dynamical systems.

Hence, for any given sample path of size N, there are two links referred to as site 1 and N, at which *boundary conditions* have to be specified. We shall consider only *periodic* boundary conditions: this means essentially the system is *invariant* under circular permutation of the sites. Consequently, certain geometric quantities locally conserved will remain also globally conserved. For example, the distance between the two extremities (not necessarily distinct) of a curve remains constant, so that closed curves will stay closed for ever.

3. Reversibility, Gibbs measures and equilibrium states

In the sequel, our goal will be to find exact scalings permitting to derive phase transition conditions, for deformation processes of the class defined in section 2, as $N \to \infty$. In fact, for the sake of shortness, we shall restrict ourselves to case studies where the sample paths have a *Gibbs invariant measure*, but this is by no means a necessary constraint, as commented in section 5.

Consider a system having a state space of the form $\mathcal{W} = K^S$, where $K$and S are finite sets. The following notion, adapted from [12], will be sufficient for our purpose.

Definition 3.1. *For any finite set $R \in S$, let $\{V_R\}$ be a collection of real numbers. A probability measure ζ on $\mathcal{W}$ is said to be a* Gibbs state *or a* Gibbs measure *relative to the potential $\{V_R\}$ if , for all $w = (w_1, \dots, w_{|S|}) \in \mathcal{W}$,*

$$\zeta(w) = \frac{1}{Z} \exp\left[\sum_R V_R \prod_{i \in R} w_i\right],$$

where Z is a normalizing constant.

Conditions will now be established, which are either of a geometrical nature or bear directly on the transition rates, for the processes of interest to be *reversible*: detailed balance equations hold and they indeed suffice to equilibrate all possible cycles in the state space (see e.g. [10]). In that case a potential does exist and the invariant measure can be explicitly expressed as a Gibbs measure.

3.1. Odd alphabet

For 2-dimensional walks, when the size n of the alphabet is odd, as already noticed in section 2.2, there is a dual point of view saying that a n-species particle system moves in a one-dimensional lattice: there is exactly one particle per site and the transition rates λ^{kl} correspond to exchanges of a particle k with a particle l between adjacent sites. In a very different context, this model was proposed in [5], from which we extract some results pertaining to our topic.

Up to a slight abuse in the notation, we let $X_i^k \in \{0,1\}$ denote the binary random variable representing the occupation of site i by a letter of type k. The state of the system is represented by the array $\mathbb{X} \stackrel{\text{def}}{=} \{X_i^k, i=1,\ldots,N; k=1,\ldots,n\}$ of size $N \times n$. Then the invariant measure of the associated Markov process is given by

$$P(\mathbb{X}) = \frac{1}{Z}\exp[-\mathcal{H}(\mathbb{X})], \tag{4}$$

where

$$\mathcal{H}(\mathbb{X}) = \sum_{i<j}\sum_{k,l} \alpha^{kl} X_i^k X_j^l, \tag{5}$$

Z is the normalizing constant, and

$$\alpha^{kl} - \alpha^{lk} = \log\frac{\lambda^{kl}}{\lambda^{lk}},$$

provided that the following condition holds

$$\sum_{k\neq l} \alpha^{kl} N_k = 0. \tag{6}$$

Indeed, a typical balance equation reads

$$\frac{P[\ldots, X_i^k = 1, X_{i+1}^l = 1, \ldots]}{P[\ldots, X_i^l = 1, X_{i+1}^k = 1, \ldots]} = \frac{\lambda^{lk}}{\lambda^{kl}} = \exp(\alpha^{lk} - \alpha^{kl}), \tag{7}$$

and relation (6) proceeds directly from enforcing the above measure to be invariant by translation.

3.2. Even alphabet

When the cardinal of the alphabet is even, say $n \stackrel{\text{def}}{=} 2p$, the situation is rendered a bit more involved due possible rotations of consecutive folded links. There is no longer conservation law for the number of letters in each class, and one should instead introduce the quantities

$$\Delta^k \stackrel{\text{def}}{=} N^{k+p} - N^k, \quad k = 1, \ldots, p-1,$$

which represent the differences between populations of links with opposite directions. Moreover, as a rule, some non-trivial cycles in the state-space are not balanced (see figure 1), unless transition rates satisfy additional conditions. This gives rise to the next theorem.

Theorem 3.2. *Assume $n = 2p$ and periodic boundary conditions. Then the system is* reversible *if and only if the following conditions are imposed on the rates and on the particles numbers:*

$$(i) \quad \prod_{k=l}^{l+p-1} \frac{\gamma^k}{\delta^{k+1}} = 1, \forall l = 1, \ldots, n$$

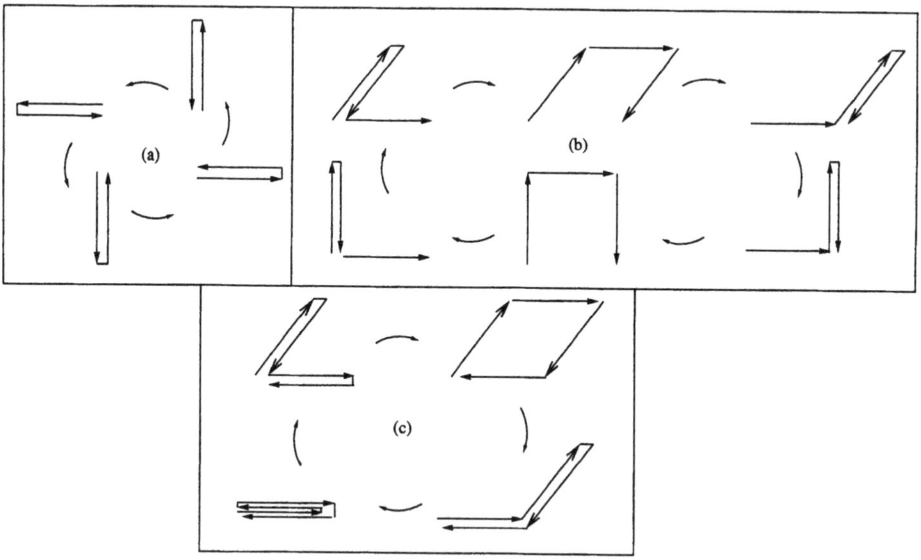

FIGURE 1. Elementary cycles: fold (a), 3-link motion (b), square loop (c).

$$(ii) \quad \frac{\lambda^{kl}}{\lambda^{lk}} \frac{\lambda^{k+p,\,l}}{\lambda^{l,\,k+p}} = 1, \quad \forall k, l = 1 \dots n,\ k \neq l,$$

$$(iii) \quad \sum_{l \neq k+p} \Delta^l \log \frac{\lambda^{kl}}{\lambda^{lk}} = 0, \quad k = 1, \dots, n.$$

The result relies on the next lemma.

Lemma 3.3. *In the case of periodic boundary conditions, if the invariant measure has a Gibbs form given by (4) and (5), then the following relationships must hold:*

$$(iv) \quad \begin{cases} \alpha^{kl} - \alpha^{lk} = \log \dfrac{\lambda^{kl}}{\lambda^{lk}}, & k = 1, \dots, n,\ l \neq k + p; \\ \alpha^{k+1,\,k+p+1} - \alpha^{k,\,k+p} = \log \dfrac{\gamma^k}{\delta^{k+1}}, & k = 1, \dots, n; \end{cases}$$

(v) *there exists a constant* $\alpha \in \mathbb{R}$ *such that*

$$\alpha^{kl} + \alpha^{k+p,l} = \alpha^{lk} + \alpha^{l,k+p} = \alpha, \ \forall k, l = 1, \dots, n;$$

$$(vi) \quad \sum_{l=1}^{p-1} (\alpha^{kl} - \alpha^{lk}) \Delta^l = 0, \quad k = 1, \dots, n.$$

Proof: We only present the main steps. Condition (iv) in the lemma comes from a balance equation of type (7). The case $l = k + p$ corresponds to adjacent folded links, and, after setting

$$U(k,i) \stackrel{\text{def}}{=} \sum_l \sum_{j > i+1} \left(\alpha^{k+1,\,l} + \alpha^{k+p+1,\,l} - \alpha^{kl} - \alpha^{k+p,\,l}\right) X_j^l,$$

$$V(k,i) \stackrel{\text{def}}{=} \sum_l \sum_{j<i} (\alpha^{l,\,k+1} + \alpha^{l,\,k+p+1} - \alpha^{lk} - \alpha^{l,\,k+p}) X_j^l,$$

equation (7) has to be replaced by

$$\frac{P[\ldots, X_i^k = 1, X_{i+1}^{k+p} = 1, \ldots]}{P[\ldots, X_i^{k+1} = 1, X_{i+1}^{k+p+1} = 1, \ldots]} = \frac{\gamma^k}{\delta^{k+1}}$$

$$= \exp\left[\alpha^{k+p+1,\,k+1} - \alpha^{k,\,k+p} - U(k,i) - V(k,i)\right],$$

which leads to impose condition (v).

To take into account the invariance by translation, let σ denote a circular permutation σ among the indices, such that $\sigma(i) = 1 + i \mod (N)$. Consider

$$\mathcal{H}_\sigma(\mathbb{X}) = \sum_{1 \geq k,l \geq N} \sum_{i<j} \alpha^{kl} X_{\sigma(i)}^k X_{\sigma(j)}^l,$$

the resulting energy obtained after applying permutation σ. Then

$$\mathcal{H}_\sigma(\mathbb{X}) = \mathcal{H}(\mathbb{X}) + \sum_{k=1}^{N} X_1^k \sum_{l=1}^{N} (\alpha^{kl} - \alpha^{lk}) N^l,$$

where N^l is the number of links l. Since $N^{l+p} - N^l$, $l = 1, \ldots, p$ is conserved (but N^l is not), the rule (v) leads to the translational invariance in the form of (vi).

As to reversibility, one can check that (i) and (ii) are necessary to equilibrate the cycles depicted in figure 1. Moreover, the cycle condition imposed by a circular permutation is exactly given by (iii). Indeed, this cycle is performed by transporting one particle through the system from site 1 to site N: during this operation, a tagged particle X^k will encounter the N^l particles corresponding to all other species $l \neq k$ and the resulting transition weight is then given by $\prod_{l \neq k, k+p} \left[\frac{\lambda^{kl}}{\lambda^{lk}}\right]^{N^l}$, which in turn, by using (ii), amounts to condition (iii) after taking the logarithm. These three conditions are in fact sufficient to determine the parameters $\{\alpha^{kl}\}$ in order to solve $(iv), (v), (vi)$, thus ensuring reversibility.

4. Steady state of the ABC system in the thermodynamic limit

As an illustration, we will present a detailed analysis of the thermodynamic equilibrium situation in the case of the ABC model. When there are three particle species, the form $\mathcal{H}$ of (5) comes to

$$\mathcal{H}(\{\mathbb{X}\}) = \sum_{i<j} \alpha^{ab} A_i B_j + \alpha^{bc} B_i C_j + \alpha^{ca} C_i A_j, \tag{8}$$

where the constants $\alpha^{ab}, \alpha^{bc}, \alpha^{ca}$ take the values

$$\alpha^{ab} = \log \frac{p^+}{p^-}, \quad \alpha^{bc} = \log \frac{q^+}{q^-}, \quad \alpha^{ca} = \log \frac{r^+}{r^-},$$

and the constraints (6) now become

$$\frac{N_A}{N_B} = \frac{\alpha^{bc}}{\alpha^{ca}}, \quad \frac{N_B}{N_C} = \frac{\alpha^{ca}}{\alpha^{ab}}, \quad \frac{N_C}{N_A} = \frac{\alpha^{ab}}{\alpha^{bc}}. \tag{9}$$

4.1. Scaling and Lotka-Volterra equations

In the example of the ABC model [2], we have at hand an explicit analytic expression for the invariant measure. In fact, our claim is that equations at steady state in the thermodynamic limit can be derived by using a method proposed in [7] for the square lattice model, which a priori *does not require any explicit knowledge* of the invariant measure, which is most of the time untractable.

We shall only sketch the main lines of argument in the case of the ABC model. The number of sites N is fixed and we are interested in the steady state behaviour as $t \to \infty$.

For the sake of shortness, let $[A, B, C] \stackrel{\text{def}}{=} \{(A_i, B_i, C_i), i = 1, \ldots, N\}$ denote the $3N$-dimensional boolean vector representing the state of the Markov process. Combining conditionings and couplings, the approach relies on the construction of a stochastic *iterative rocking-scheme*, to generate the global invariant measure of the system.

The algorithm constructs a convergent sequence of random kernels $Q^{(n)}, n \geq 0$.

S1 Set $n = 0$ and take the system in some fixed state $[A, B, C]^{(0)} = [A, B, C]$.

S2 At step n, assume we are given a 3-tuple $[A, B, C]^{(n)}$, corresponding to an admissible configuration (with respect to the ABC model). To construct $[A, B, C]^{(n+1)}$, one allows each particle family to evolve as an *independent* one-dimensional exclusion process until reaching equilibrium, with conditional transition rates *compatible* with the rates of the original process introduced in section 2.2. For instance, with regard to type A particles *(omitting sometimes the indexing by n for the sake of brevity)*, we have

$$\begin{cases} \lambda_a^+(i) = p^+ B_i + r^- C_i + \Gamma\, \overline{B}_i \overline{C}_i, \\ \lambda_a^-(i+1) = p^- B_i + r^+ C_i + \Gamma\, \overline{B}_i \overline{C}_i, \end{cases}$$

these expressions being similar to the form of (2) in the $\{\tau_a \tau_b\}$ model. Γ is a non-zero quantity taking into account the exclusion at site i in the following sense: when $\overline{B}_i \overline{C}_i = 1$, then necessarily $A_i = 1$. In the free process, letting q_i^a denote the random variable representing the conditional probability of finding a particle A_i at site i, we have

$$\frac{u_{i+1}^{(n)}}{u_i^{(n)}} = \frac{\lambda_a^+(i)}{\lambda_a^-(i+1)}, \quad \text{with } u_i^{(n)} \stackrel{\text{def}}{=} \frac{q_i^a}{1 - q_i^a}.$$

Using the boolean nature of B_i and C_i, we can write

$$\log \frac{u_{i+1}^{(n)}}{u_i^{(n)}} = \alpha^{ca} C_i \; - \; \alpha^{ab} B_i, \tag{10}$$

with analogous quantities $v_i^{(n)}$ and $w_i^{(n)}$ for B and C species respectively.

S3 Clearly, the equilibrium states thus obtained are not all admissible, since at a site i, several particles of different types may coexist. Therefore, we only retain configurations satisfying the constraint $A_i^{(n+1)} + B_i^{(n+1)} + C_i^{(n+1)} =$

1, for all i, the distribution of which, after some algebra, takes the form

$$Q([(A,B,C)]^{(n+1)} \mid [(A,B,C)]^{(n)}) = \frac{1}{Z}\prod_{i=1}^{N}[u_i^{(n)}A_i^{(n+1)} + v_i^{(n)}B_i^{(n+1)} + w_i^{(n)}C_i^{(n+1)}],$$

where Z is a normalizing constant. Then **do** $n \leftarrow n+1$ and **go to S2**.

Let $r_i^{(n)} \stackrel{\text{def}}{=} u_i^{(n)} + v_i^{(n)} + w_i^{(n)}$. Then, by means of transfer and coupling theorems, one can prove the existence of the random vectors

$$(r_i^a, r_i^b, r_i^c) = \lim_{n\to\infty}\left(\frac{u_i^{(n)}}{r_i^{(n)}}, \frac{v_i^{(n)}}{r_i^{(n)}}, \frac{w_i^{(n)}}{r_i^{(n)}}\right), \quad i \geq 0.$$

Fundamental scaling We introduce the so-called *fundamental scaling* defined by

$$\alpha^{bc} = \frac{\alpha}{N} + o\Big(\frac{1}{N}\Big), \quad \alpha^{ca} = \frac{\beta}{N} + o\Big(\frac{1}{N}\Big), \quad \alpha^{ab} = \frac{\gamma}{N} + o\Big(\frac{1}{N}\Big),$$

where α, β, γ are three positive real constant.

Limiting equations Set from now on $x \stackrel{\text{def}}{=} i/N$, for $1 \leq i \leq N$. Then one can show the weak limits

$$\rho^a(x) = \lim_{N\to\infty} r_{xN}^a, \qquad \rho^b(x) = \lim_{N\to\infty} r_{xN}^b, \qquad \rho^c(x) = \lim_{N\to\infty} r_{xN}^c$$

exist and satisfy the system of deterministic differential equations

$$\begin{cases} \dfrac{\partial \rho_a}{\partial x} = \rho_a(\beta\rho_c - \gamma\rho_b), \\ \dfrac{\partial \rho_b}{\partial x} = \rho_b(\gamma\rho_a - \alpha\rho_c), \\ \dfrac{\partial \rho_c}{\partial x} = \rho_c(\alpha\rho_b - \beta\rho_a), \end{cases} \tag{11}$$

with the crucial constraints due to the periodic boundary conditions

$$\rho_u(x+1) = \rho_u(x), \quad \forall u = a, b, c. \tag{12}$$

Proof: Starting from (10), one applies the law of large numbers and ergodic theorems to justify the approximation of finite sums by Riemann integrals, as $N \to \infty$.

It is amusing to see that (11) belongs to the class of generalized Lotka-Volterra systems. The original Lotka-Volterra model was the simplest model of predator-prey interactions, proposed independently by Lotka (1925) and Volterra (1926), see for instance [13]. Nonetheless, in our case the world is less cruel and all particle types are treated on an equal footing...!

Now the form of (11) lends itself to a reasonably explicit solution in terms of special functions. The first step is to remark the existence of two level surfaces

$$\rho_a + \rho_b + \rho_c = 1, \tag{13}$$

$$\rho_a^\alpha \rho_b^\beta \rho_c^\gamma = \kappa, \tag{14}$$

where (13) follows at once from (9) and κ is a constant of motion to be determined. Using (13) to eliminate ρ_c, we rewrite (14) as

$$\rho_a^\alpha \rho_b^\beta (1 - \rho_a - \rho_b)^\gamma = \kappa. \tag{15}$$

The change of functions $u = \rho_a + \rho_b$ and $v = \beta\rho_a - \alpha\rho_b$, yields the first order nonlinear differential equation

$$\frac{du}{dx} = (1-u)v(u), \tag{16}$$

where $v(u)$ satisfies the equation

$$(\alpha u + v)^{\alpha}(\beta u - v)^{\beta}(1-u)^{\gamma} = \kappa(\alpha+\beta)^{\alpha+\beta}. \tag{17}$$

Formally, $u(x)$ can be expressed as

$$x = \int_{u(0)}^{u(x)} \frac{du}{(1-u)v(u)}. \tag{18}$$

It appears that the ratios in (9) are rational for all finite N, but, since we have let $N, N_A, N_B, \to \infty$, they might become arbitrary real numbers in the interval $[0,1]$. When they are rational, (16) is a polynomial equation, and then $u(x)$ is in some sense just a bit more general that an hyperelliptic function, since, after some algebra, we are left with integrals of the form

$$\int \sqrt{\left[s^2 - a^2(1-s)^{-\frac{p}{q}}\right]}\,ds,$$

where p and q stand for positive integers. The particular case $\alpha = \beta = \gamma$ is rather simple, since then

$$\left(\frac{du}{dx}\right)^2 = (1-u)[u^2(1-u) - \kappa],$$

showing u to be a standard Jacobi elliptic function.

For any given κ, the system (11) admits of a unique solution. In particular, there is *always* a degenerate solution crumbled to the fixed point

$$\widetilde{\rho}_a = \frac{\alpha}{\alpha+\beta+\gamma}, \quad \widetilde{\rho}_b = \frac{\beta}{\alpha+\beta+\gamma}, \quad \widetilde{\rho}_c = \frac{\gamma}{\alpha+\beta+\gamma}, \tag{19}$$

and corresponding to the constant

$$\widetilde{\kappa} \stackrel{\text{def}}{=} \frac{\alpha^{\alpha}\beta^{\beta}\gamma^{\gamma}}{(\alpha+\beta+\gamma)^{\alpha+\beta+\gamma}}.$$

The purpose of the next paragraph is to discriminate between solutions of (11) and (12), in order to relate them to admissible limit-points of Φ_N, as $N \to \infty$.

4.2. Stability, fundamental period and phase transition

To catch a rough qualitative insight into the solution of (11), a standard approach relies on a linearization of the right-hand side around the fixed point (19). This yields a linear differential system, whose matrix

$$\frac{1}{\alpha+\beta+\gamma}\begin{bmatrix} 0 & -\alpha\gamma & \alpha\beta \\ \beta\gamma & 0 & -\alpha\beta \\ -\beta\gamma & \alpha\gamma & 0 \end{bmatrix},$$

has three eigenvalues 0 and $\lambda_{\pm} = \pm i\sqrt{\alpha\beta\gamma(\alpha+\beta+\gamma)}$, and the trajectories are located on an ellipsoid. But, since the above eigenvalues are purely imaginary, it is well known that no conclusion can be drawn as for the original system, which might be of a quite different nature. However, it is pleasant to see that the modulus of the nonzero eigenvalues plays in fact a crucial role as shown in the next theorem.

Theorem 4.1. *Let* $s \stackrel{\text{def}}{=} \alpha+\beta+\gamma$ *and* $\eta \stackrel{\text{def}}{=} \frac{s}{3}$. *The limit* $\Phi \stackrel{\text{def}}{=} \lim_{N\to\infty} \Phi_N$ *of the ABC model is deterministic and there is a second order phase transition phenomenon. There exits a critical value*

$$\eta_c \stackrel{\text{def}}{=} \frac{2\pi}{3\sqrt{\widetilde{\rho}_a\widetilde{\rho}_b\widetilde{\rho}_c}},$$

such that if $\eta > \eta_c$ *then there are closed non-degenerate trajectories of (11) satisfying (12), with period* $T(\kappa_p) = \frac{1}{p}$, $p \in \{1,\ldots,[\frac{\eta}{\eta_c}]\}$. *The only admissible stable* Φ *corresponds*

- *either to the trajectory associated with* κ_1 *if* $\eta > \eta_c$;
- *or to the degenerate one consisting of the single point (19) if* $\eta \le \eta_c$.

The proof involves a forest of technicalities and we only sketch the main lines of argument. The first step is to switch to polar coordinates

$$\begin{cases} u_a \stackrel{\text{def}}{=} \widetilde{\rho}_a - \dfrac{\alpha}{s} = r\cos\theta, \\ u_b \stackrel{\text{def}}{=} \widetilde{\rho}_b - \dfrac{\beta}{s} = r\sin\theta. \end{cases}$$

Rewrite (14) as

$$H(r,\theta) = \log\kappa, \tag{20}$$

with

$$H(r,\theta) \stackrel{\text{def}}{=} \alpha\log\big[r\cos\theta + \frac{\alpha}{s}\big] + \beta\log\big[r\sin\theta + \frac{\beta}{s}\big] + \gamma\log\big[\frac{\gamma}{s} - r(\cos\theta+\sin\theta)\big],$$

and let $r(\theta,\kappa)$ be the single root in r of (20). Then θ satisfies the differential equation

$$\frac{d\theta}{dx} = G(\theta,\kappa), \tag{21}$$

where

$$\begin{aligned} G(\theta,\kappa) &\stackrel{\text{def}}{=} \frac{1}{s}\big[\beta(\alpha+\gamma)\cos\theta + \alpha\beta\sin^2\theta + \alpha(\beta+\gamma)\sin^2\theta\big] \\ &+ r(\theta,\kappa)\cos\theta\sin\theta\big[(\beta+2\alpha+2\gamma)\cos\theta + (\alpha+2\beta+2\gamma)\sin\theta\big]. \end{aligned}$$

Letting $T(\kappa)$ be the period of the orbit, we have

$$T(\kappa) = \int_0^{2\pi} \frac{d\theta}{G(\theta,\kappa)}.$$

The second important step relies on the monotonic behaviour of $T(\kappa)$ with respect to the parameter κ, yielding the inequality

$$T(\kappa) \ge T(\widetilde{\kappa}).$$

Observing that

$$r(\theta,\widetilde{\kappa}) = 0, \quad \forall\theta \in [0,2\pi],$$

we can write, by (21), $T(\widetilde{\kappa})$ as a contour integral on the unit circle, namely

$$T(\widetilde{\kappa}) = -4is\oint_\Gamma \frac{1}{z^2[\gamma(\beta-\alpha)-2i\alpha\beta] + 2z[2\alpha\beta+\gamma(\alpha+\beta)] + \gamma(\beta-\alpha)+2i\alpha\beta},$$

or, after a simple calculus,

$$T(\widetilde{\kappa}) = 2\pi\sqrt{\frac{s}{\alpha\beta\gamma}},$$

which leads precisely to the critical value η_c announced in the theorem.

5. Perspectives

This paper is the continuation of [7], but is certainly an intermediate step. For the sake of shortness, we did restrict ourselves to the thermodynamics of the ABC model. Actually, our goal is to analyze the dynamics of random curves evolving in $\mathbb{Z}^m$ (no spatial constraints) or in $\mathbb{Z}_+^m$, when they warp under the action of some stochastic deformation grammar.

In [6], this project will be carried out in the framework of large scale dynamics for exclusion processes, and it will mainly address the points listed hereafter.

More on thermodynamic equilibrium: The trick to derive limiting differential systems amounts essentially to writing conditional flow equations on suitable sample paths, even in the presence of particle currents. These equations involve functionals of Markov and they enjoy special features encountered in many systems. It might also be interesting to note that most of the Lotka-Volterra equations can be explained in the light of the famous urns of Ehrenfest.

Phase transition: There exists a global interpretation by means of a free energy functional with two components: the entropy of the system, and the algebraic area enclosed by the curve. It turns out that the contention between these two quantities yield, after taking limits $\lim_t \to \infty \lim_N \to \infty$ (in that order), either stretched deterministic curves or Brownian objects when the scaling is of central limit-type.

Transient regime: Our claim is that time-dependent behaviour can be treated along the same ideas, up to technical subtleties, by means of a numerical scheme based on the conservation of particle currents. This should yield a system of Burgers equations, extending those obtained in [7] for the symmetric $\{\tau_a\tau_b\}$ model, which had the form

$$\begin{cases} \dfrac{\partial \rho^a(x,t)}{\partial t} &= D\dfrac{\partial^2 \rho^a(x,t)}{\partial x^2} - 2D\eta\dfrac{\partial}{\partial x}\left[\rho^a(1-\rho^a)(1-2\rho^b)\right](x,t), \\ \dfrac{\partial \rho^b(x,t)}{\partial t} &= D\dfrac{\partial^2 \rho^b(x,t)}{\partial x^2} + 2D\eta\dfrac{\partial}{\partial x}\left[\rho^b(1-\rho^b)(1-2\rho^a)\right](x,t). \end{cases}$$

References

[1] J. BURGERS, *A mathematical model illustrating the theory of turbulences*, Adv. Appl. Mech., 1 (1948), pp. 171–199.

[2] M. CLINCY, B. DERRIDA, AND M. EVANS, *Phase transition in the ABC model*, Phys. Rev. E, 67 (2003), pp. 6115–6133.

[3] B. DERRIDA, M. EVANS, V. HAKIM, AND V. PASQUIER, *Exact solution for 1d asymmetric exclusion model using a matrix formulation*, J. Phys. A: Math. Gen., 26 (1993), pp. 1493–1517.

[4] M. EVANS, D. P. FOSTER, C. GODRÈCHE, AND D. MUKAMEL, *Spontaneous symmetry breaking in a one dimensional driven diffusive system*, Phys. Rev. Lett., 74 (1995), pp. 208–211.

[5] M. EVANS, Y. KAFRI, M. KODUVELY, AND D. MUKAMEL, *Phase Separation and Coarsening in one-Dimensional Driven Diffusive Systems*, Phys. Rev. E., 58 (1998), p. 2764.

[6] G. FAYOLLE AND C. FURTLEHNER, *Stochastic deformations of random walks and exclusion models. Part II: Gibbs states and dynamics in $\mathbb{Z}^2$ and $\mathbb{Z}^3$*. In preparation.

[7] G. FAYOLLE AND C. FURTLEHNER, *Dynamical Windings of Random Walks and Exclusion Models. Part I: Thermodynamic limit in $\mathbb{Z}^2$*, Journal of Statistical Physics, 114 (2004), pp. 229–260.

[8] O. KALLENBERG, *Foundations of Modern Probability*, Springer, second edition ed., 2001.

[9] M. KARDAR, G. PARISI, AND Y. ZHANG, *Dynamic scaling of growing interfaces*, Phys. Rev. Lett., 56 (1986), pp. 889–892.

[10] F. P. KELLY, *Reversibility and stochastic networks*, John Wiley & Sons Ltd., 1979. Wiley Series in Probability and Mathematical Statistics.

[11] R. LAHIRI, M. BARMA, AND S. RAMASWAMY, *Strong phase separation in a model of sedimenting lattices*, Phys. Rev. E, 61 (2000), pp. 1648–1658.

[12] T. M. LIGGETT, *Interacting Particle Systems*, vol. 276 of Grundlehren der mathematischen Wissenschaften, Springer-Verlag, 1985.

[13] J. MURRAY, *Mathematical Biology*, vol. 19 of Biomathematics, Springer-Verlag, second ed., 1993.

[14] H. SPOHN, *Large Scale Dynamics of Interacting Particles*, Springer, 1991.

Guy Fayolle
INRIA Rocquencourt – Domaine de Voluceau BP 105
78153 Le Chesnay, France.
Guy.Fayolle@inria.fr

Cyril Furtlehner
INRIA Rocquencourt – Domaine de Voluceau BP 105
78153 Le Chesnay, France.
Cyril.Furtlehner@inria.fr

Trends in Mathematics, © 2004 Birkhäuser Verlag Basel/Switzerland

A Markov Chain Algorithm for Eulerian Orientations of Planar Triangular Graphs

Johannes Fehrenbach and Ludger Rüschendorf

ABSTRACT: *On the set of Eulerian orientations of a planar Eulerian graph a natural Markov chain is defined and is shown to converge to the uniform distribution. For the class of planar triangular graphs this chain is proved to be rapidly mixing. The proof uses the path coupling technique of Bubley and Dyer (1997) and the comparison result of Randall and Tetali (1998). For the class of planar triangular graphs our result improves essentially the mixing rate result from Mihail and Winkler (1996) for general Eulerian graphs. As consequence we obtain a faster polynomial randomized approximation scheme for counting the number of Euler orientations.*

1. Introduction

An undirected, connected graph $G = (V, E)$ is called Euler graph if all vertices have even degree. A Eulerian orientation X of G is an orientation of the edges of G such that for each vertex $v \in V$ the set of edges directed towards v and the set of edges directed out of v have the same cardinality, i.e., with $E^-(v) := \{e = (w, v) \in X \mid w \in V\}$ and $E^+(v) := \{e = (v, w) \in X \mid w \in V\}$ holds: $|E^+(v)| = |E^-(v)|$, for all $v \in V$.

Counting the number of Euler orientations is relevant to some problems in statistical physics. Welsh (1990) observed that the crucial partition function of the *ice-type model* is equal to the number of Eulerian orientations of some underlying Eulerian graphs. It has also been observed that the counting problem for Eulerian orientations corresponds to evaluating the Tutte polynomial, which encodes important information on the graph, at the point $(0, -2)$. It is not difficult to construct a Eulerian orientation in polynomial time. The corresponding exact counting problem of all Eulerian orientations is however $\#P$-complete as was established in Mihail and Winkler (1996). Mihail and Winkler (1996) also proved that counting of Eulerian orientations of G can be reduced to counting the perfect matchings of a related graph G'. Thus by the randomized approximation scheme (RAS) of Jerrum and Sinclair (1989) this yields a polynomial RAS for the orientations. The mixing time for this scheme is of the class $O\left((n')^3 m'(n' \log n' + \log \epsilon^{-1})\right)$, ϵ the approximation error, $n' = |V'|$, $m' = |E'|$, $G' = (V', E')$. Since by the construction of G' in Mihail and Winkler (1996) $n' \geq nm$, $m' \geq m^2$ one gets mixing times of considerable high polynomial order.

In this paper, which is based on the dissertation of Fehrenbach (2003), we introduce a natural direct Markov chain on the set of Eulerian orientations and prove that in the case of planar triangular graphs one gets a considerable lower mixing rate order and thus one obtains a faster randomized scheme. A related construction and mixing rate result for sampling Eulerian orientations has been given for bounded Cartesian lattices with specified boundaries in Luby, Randall, and Sinclair (2001).

Our mixing rate result uses the path coupling method of Bubley and Dyer (1997). For an ergodic Markov chain $\mathcal{M}$ with transition matrix $P = (p_{ij})$ on a finite set Ω a (Markov-)coupling is a stochastic process $(X_t, Y_t)_{t\in\mathbb{N}}$ on $\Omega \times \Omega$ such that for all $x, y, z \in \Omega$ and $t \in \mathbb{N}$

$$\begin{aligned} P(X_{t+1} = x \mid X_t = y, Y_t = z) &= p_{yx} \\ P(Y_{t+1} = x \mid X_t = y, Y_t = z) &= p_{zx} \end{aligned}$$

and $X_t = Y_t$ implies $X_{t+1} = Y_{t+1}$. Then by the coupling lemma

$$\left\|P^{X_t} - P^{Y_t}\right\| \le P(X_t \neq Y_t), \tag{1}$$

where $\| \ \|$ is the variation norm.

Let $(X_t, Y_t)_{t\in\mathbb{N}}$ be a coupling and let $\delta : \Omega \times \Omega \to \mathbb{N}$ be a metric, such that for some $\beta \le 1$ and all t

$$E(\delta(X_{t+1}, Y_{t+1}) \mid (X_t, Y_t) = (x, y)) \le \beta\delta(x, y). \tag{2}$$

Let $\tau(\epsilon)$ be the mixing time of the Markov chain for the approximation error ϵ, then

$$\tau(\epsilon) \le \frac{\log\left(\delta(\Omega)\epsilon^{-1}\right)}{1-\beta} \qquad \text{if } \beta < 1. \tag{3}$$

If $\beta = 1$ and if for some $\alpha > 0$, $P\left(\delta(X_{t+1}, Y_{t+1}) \neq \delta(x, y) \mid (X_t, Y_t) = (x, y)\right) \ge \alpha$, for all t and all $x, y \in \Omega$, then

$$\tau(\epsilon) \le \frac{e\delta(\Omega)^2}{\alpha} \log \epsilon^{-1}, \tag{4}$$

where $\delta(\Omega) = \max\{\delta(x, y) \mid x, y \in \Omega\}$ is the diameter of Ω (see Dyer and Greenhill (1998), Aldous (1983)).

The path coupling method is a technique which simplifies the construction of a coupling on all of $\Omega \times \Omega$ that satisfies condition (2). It was introduced in Bubley and Dyer (1997). The following formulation is from Dyer and Greenhill (1998).

Let $S \subset \Omega \times \Omega$ be a set of transitions such that for all $x, y \in \Omega$ there exists a path $x = z_0, z_1, \ldots, z_r = y$ for x to y with transitions $(z_i, z_{i+1}) \in S$, $\forall i < r$. If $(x, y) \to (X', Y')$ is an $\mathcal{M}$-coupling for all $(x, y) \in S$, then an extension can be defined via the path in S for any state $(X_t, Y_t) = (x, y)$. One obtains thus a sequence $Z_0', \ldots, Z_r'$ and a coupling $(X_{t+1}, Y_{t+1})_{t\in\mathbb{N}}$ on $\Omega \times \Omega$ with $X_{t+1} = Z_0'$ and $Y_{t+1} = Z_r'$. For a function $\varphi : S \to \mathbb{N}_0$ we define a metric $\delta(x, y) := \min \sum_{i=0}^{r-1} \varphi(z_i, z_{i+1})$, the minimum taken over all paths from x to y in S. If for some $\beta \le 1$

$$E(\delta(X', Y') \mid (X, Y) = (x, y)) \le \beta\delta(x, y) \text{ for all } (x, y) \in S, \tag{5}$$

then

$$E(\delta(X_{t+1}, Y_{t+1}) \mid (X_t, Y_t) = (x, y)) \le \beta\delta(x, y) \text{ for all } (x, y) \in \Omega \times \Omega. \tag{6}$$

For the comparison of mixing times τ_1, τ_2 of two ergodic, reversible Markov chains $\mathcal{M}_1, \mathcal{M}_2$ with the same stationary distribution π and transition functions p_1, p_2 an effective method has been developed by Randall and Tetali (1998). Let $T_i = \{(x, y) \in \Omega \times \Omega \mid p_i(x, y) > 0\}$, $i = 1, 2$, be the graphs of $\mathcal{M}_1, \mathcal{M}_2$ and let $\Gamma = \{\gamma_{x,y}; (x, y) \in T_2\}$ be a set of *canonical* paths $\gamma_{x,y}$ in T_1 from x to y for any pair

$(x, y) \in T_2$. For any $(w, z) \in T_1$ let $\Gamma(w, z) = \{(x, y) \in T_2; (w, z) \in \gamma_{x,y}\}$ be the set of all canonical paths in Γ which contain the edge (w, z). Finally, we denote by

$$A(\Gamma) := \max_{(w,z)\in T_1} \frac{1}{\pi(\omega)p_1(w, z)} \sum_{(x,y)\in\Gamma(w,z)} \pi(x)p_2(x, y)|\gamma_{x,y}|, \tag{7}$$

where $\gamma_{x,y}$ is the length of the path $\gamma_{x,y}$, the comparison measure. Then the following holds for all $\epsilon \in \left(0, \frac{1}{2}\right)$:

$$\tau_1(\epsilon) \leq A(\Gamma)\frac{4\tau_2(\epsilon)}{\log((2\epsilon)^{-1})}\left(\log\hat{\pi}^{-1} + \log\epsilon^{-1}\right) \tag{8}$$

where $\hat{\pi} = \min_{x\in\Omega} \pi(x)$.

This comparison result will be applied in section 3 to determine bounds for the mixing time of a natural Markov chain by comparison with a chain which is simpler to analyse.

2. A Markov chain on Eulerian orientations of planar graphs

Let $G = (V, E)$ be a planar, undirected, connected Euler graph and let $EO(G)$ denote the set of Eulerian orientations. Let $F(G)$ denote the open domains generated by the embedding of G in $\mathbb{R}^2$. The inner domains are bounded while exactly one domain—the outer domain—is unbounded. For $\alpha \in F(G)$ and a Eulerian orientation X we denote by X^α the edges in the boundary of α directed according to X. The inversion of an edge $e = (v, w) \in X$ is defined as $\overline{e} := (w, v)$. Similarly for $C \subset X$ we define the inversion $\overline{C} := \{\overline{e}; e \in C\}$. The following construction of a Markov chain $\mathcal{M}_0(G) = (X_t)_{t\in\mathbb{N}}$ on $EO(G)$ is quite natural.

Markov chain $\mathcal{M}_0(G)$ on $EO(G)$: To define the transition of $\mathcal{M}_0(G)$ from $X_t = x \in EO(G)$, we use two steps:

1) Let $\Lambda \in F(G)$ be a randomly sampled domain
2) If $\Lambda = \alpha$ and with $C := x^\alpha$ define

$$x' := \begin{cases} (x - C) \cup \overline{C} & \text{if } (x - C) \cup \overline{C} \in EO(G) \\ x & \text{else.} \end{cases} \tag{9}$$

Then define

$$X_{t+1} = \begin{cases} x' & \text{with probability } \frac{1}{2} \\ x & \text{else.} \end{cases} \tag{10}$$

The corresponding transition matrix is denoted by P_0. The inversion of the edges in x^α for $\alpha \in F(G)$ is called α-transition.

Thus the Markov chain randomly chooses domains of the planar graph and inverts with probability $\frac{1}{2}$ the orientation of the boundary if possible.

To determine an initial state S_G of the Markov chain let $G^* = (V^*, E^*)$ be the dual graph with node set V^* the set of domains $F(G)$ and where two domains α, α' are connected in V^* if α, α' have a common boundary in G. For more details on planar graphs see Diestel (1996). G^* is a bipartite graph. Let V_1^*, V_2^* denote the partition of V^* with corresponding partition F_1, F_2 of $F(G)$. W.l.g. let the outer domain of G be in F_2. To define the initial state S_G of the Markov chain, the edges of G are oriented in such a way that for any $\alpha \in F_1$ the edges in S_g^α are oriented clockwise. This defines in fact a Eulerian orientation since any node $v \in V$ in the boundary of α is final node of two edges in the boundary of α which are oriented clockwise. Thus one edge points into v while the other edge points out of v. For each edge towards v thus there exists an edge out of v and thus $|E^+(v)| = |E^-(v)|$ for all $v \in V$. Since the domains with v in their boundary belong cyclically to F_1 or F_2, the edges of v in S_G are cyclically oriented out of v or towards v (see Figure 2.1). We use this Euler orientation S_G as initial state of the chain. The analysis of the mixing time however will be independent of the initial state. The following

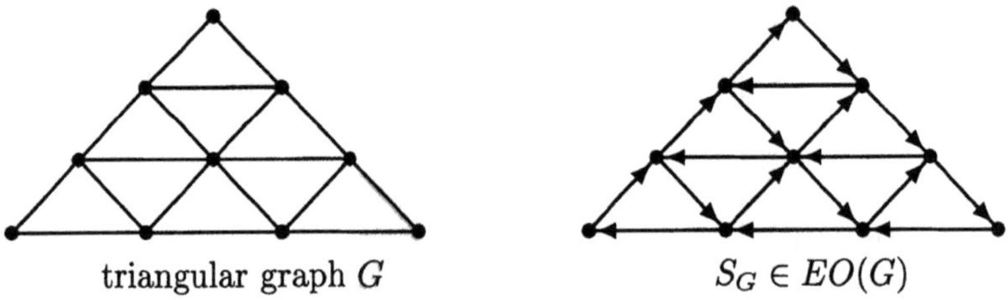

Figure 2.1 The domains with clockwise orientation define F_1. F_2 consists of the domains which are oriented counterclockwise.

property of the initial state S_G will be used in the following.

Lemma 2.1. *If $C \subset S_G$ is a simple clockwise oriented circle in G, then C^* is a minimal cut in G^*, which decomposes V^* in two components Z_1^* and Z_2^*. If the outer domain of G belongs to Z_2^*, then all nodes in Z_1^* which are final nodes of an edge in C^* are in V_1^*.*

Proof: The property that C^* is a minimal cut is stated in Diestel (1996, Proposition 3.10). For $v \in V$ a node in the circle C let k be the number of edges e on v with dual edge $e^* \in Z_1^*$. k is even since by construction of S_G the edges on v cyclically are out of v or towards v. Therefore, for the domains α, β on both edges on v in C with α^*, $\beta^* \in Z_1^*$ holds α^*, $\beta^* \in V_1^*$ (see Figure 2.2). □

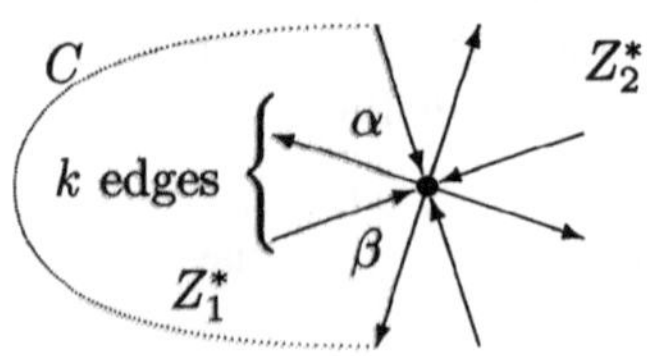

Figure 2.2 orientation S_G. The edges on v are cyclically *in* or *out* v. Thus k is even.

Similarly, if C as in Lemma 2.1 is counterclockwise oriented then the end nodes of edges in Z_1^* are in V_2^*.

Theorem 2.2. *Let $G = (V, E)$ be a planar undirected, connected Euler graph. Then the Markov chain $\mathcal{M}_0(G)$ defined in (10) is ergodic. The stationary distribution π of $\mathcal{M}_0(G)$ is the uniform distribution on $EO(G)$.*

Proof: By the construction of $\mathcal{M}_0(G)$ the transition matrix is symmetric $p_0(x, y) = p_0(y, x)$ for $x, y \in EO(G)$. Thus $\mathcal{M}_0(G)$ is reversible with respect to the uniform distribution on $E(G)$. By (10) it is also aperiodic. The main part of the proof is to establish irreducibility. We will prove by induction on $|F(G)|$ that there is a path of any $x \in EO(G)$ to the initial state S_G. This implies by symmetry irreducibility of $\mathcal{M}_0(G)$.

1) $|F(G)| = 2$.
   Then G has two Euler orientations S_G and $\bar{S}_G$. A transition from $\bar{S}_G$ to S_G has by (9) positive probability.
2) **Induction step:** $|F(G)| = \ell > 2$.
   Let $x \in EO(G)$. Let in the first case $x = \bar{S}_G$. Choosing in the first construction step of $\mathcal{M}_0(G)$ iteratively points from the set F_1 one gets a finite path from x to S_G. The result is independent of the sequence since the inversion of the orientation of edges on a domain in F_1 does not influence edges on other domains in F_1. No edge is in the boundary of two domains in F_1 (see also Figure 2.1).

   If $x \neq \bar{S}_G$ then we consider $\tilde{G} = (\tilde{V}, \tilde{E})$ with $\tilde{E} = \{e \in E : e_{S_G} \neq e_x\}$ and $\tilde{V} = \{v \in V : v$ is final node of an edge in $\tilde{E}\}$; e_{S_G}, e_x denote the orientations of e by S_G resp. x. We first assume that $\tilde{G}$ is connected and define $\tilde{S}_G = \{e_{S_G} = (v, w) \in S_G : \{v, w\} \in \tilde{E}\}$, $\tilde{X} = \{e_x = (v, w) \in x : \{v, w\} \in \tilde{E}\}$ the corresponding orientations of $\tilde{G}$. Then $\tilde{G}$ is a Euler graph and $\tilde{S}_G$, $\tilde{x} \in EO(\tilde{G})$. Since $x \neq \tilde{S}_G$ we have $|F(G)| > |F(\tilde{G})|$ and, therefore, from the assumption of the induction there is a path $(C_i)_{0 \leq i \leq k}$ from $\tilde{X} = C_0$ to $\tilde{S}_G = C_k$ w.r.t. $\mathcal{M}_0(\tilde{G})$. As consequence one gets also a path from x to S_G in $\mathcal{M}_0(G)$.

   Let the transition from C_0 to C_1 be an α-transition for some $\alpha \in F(\tilde{G})$. If $\alpha \in F(G)$ then this transition of $\mathcal{M}_0(\tilde{G})$ is also a transition of $\mathcal{M}_0(G)$. If $\alpha \notin F(G)$ then the edges of $\tilde{x}^\alpha$ are a simple directed circle in $\tilde{x}$, which corresponds to a minimal cut in G^*. We denote the component which does not contain the outer domain of G^* by T^*. The edges dual to those in T^* are in X and in S_G identically oriented. Therefore, all nodes in T^* which are final nodes of edges in $\tilde{x}^\alpha$ are by Lemma 2.1 in the same partition of G^* say in V_1^*. Since in S_G the edges in S_G^β are directed circles for any $\beta \in F(G)$, also the edges in x^γ are directed circles in x for nodes γ^* in $T^* \cap V_2^*$. Thus starting from x for all these γ the γ-transitions can be made and as a result the edges at all $\alpha \in F(G)$ with $\alpha^* \in T^* \cap V_2^*$ form circles which get an inverse orientation. Finally the orientation of all edges $e \in E$ with $e^* \in T^*$ has two times been inverted, while the edges in $\tilde{x}^\alpha$ have one time been inverted. Thus the transition of C_0 to C_1 in $\mathcal{M}_0(\tilde{G})$ has been transferred to a sequence of transitions in $\mathcal{M}_0(G)$. Iterating this procedure for all the other transitions we finally obtain a path from x to S_G.

   If $\tilde{G}$ has several components, then this procedure can be applied to all of these producing finally a path in $\mathcal{M}_0(G)$ and establishes irreducibility. Thus the Markov chain $\mathcal{M}_0(G)$ is ergodic and the uniform distribution of $EO(G)$ is the unique stationary distribution of $\mathcal{M}_0(G)$. □

3. Rapid mixing of $\mathcal{M}_0(G)$ for triangular graphs

The aim of this section is to prove the rapid mixing property of the Markov chain $\mathcal{M}_0(G)$ of section 2 for the case of planar, triangular graphs. A graph G is called a triangular graph if the boundary of any inner domain contains exactly three nodes (cf. the example in Figure 2.1). Let $\tau_0(\epsilon)$ denote the mixing time of $\mathcal{M}_0(G)$ for approximation error $\varepsilon > 0$.

Theorem 3.1. *Let $G = (V, E)$ be a Eulerian, planar triangular graph. Then the mixing time $\tau_0(\epsilon)$ of the Markov chain $\mathcal{M}_0(G)$ is polynomially bounded and*

$$\tau_0(\epsilon) \leq 3072n^3(\log \epsilon^{-1})(3n + \log \epsilon^{-1}). \tag{11}$$

To establish the bound in (11) we introduce a modified Markov chain for planar Eulerian triangle graphs with extended transitions which is easier to analyse (see also the proof of Theorem 3.5)

Lemma 3.2. *Let $G = (V, E)$ be a planar, Eulerian triangular graph and $x \in EO(G)$. If $\alpha \in F(G)$ and x^α is not a directed circle, then α has at most one neighbour domain $\beta \in F(G)$ such that $x^\alpha \oplus x^\beta$ is a directed circle.*

Proof: If x^α is not a directed circle, then there is an edge $e \in x^\alpha$ such that $(x^\alpha - \{e\}) \cup \{\bar{e}\}$ is a directed circle. This edge e is not contained in any simple directed circle in x together with any of the other two edges form x^α. Therefore, only the neighbour domain β, separated from α by the edge e, can possibly satisfy that $x^\alpha \oplus x^\beta$ is a circle, see Figure 3.3 for the corresponding part of G.

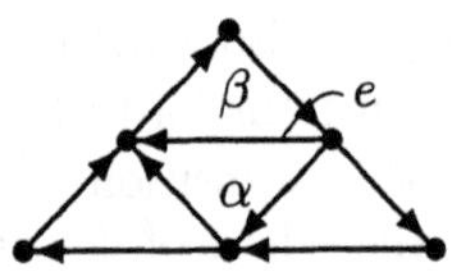

Figure 3.3 □

If for a Eulerian orientation x it is not possible to invert the edges of a domain α then there can be one (but at most one) neighbour domain β such that $x^\alpha \oplus x^\beta$ is a directed circle. The following modification $\mathcal{M}'_0(G)$ of $\mathcal{M}_0(G)$ allows the Markov chain to invert circles of this kind.

Definition 3.3. *Let $G = (V, E)$ be a planar Eulerian triangular graph. The Markov chain $\mathcal{M}'_0(G) = (X_t)_{t\in\mathbb{N}}$ on $EO(G)$ is defined by the following transitions of $\mathcal{M}'_0(G)$ from $X_t = x \in EO(G)$:*

1) *Let $\Lambda \in F(G)$ be a randomly sampled domain*
2) *For $\Lambda = \alpha$ and if $C := x^\alpha$ is a directed circle then define $x' := (x - C) \cup \bar{C}$. If C is not a directed circle and if $\beta \in F(G)$ is a neighbour domain of α such that $C' = x^\alpha \oplus x^\beta$ is a directed circle, then set*

$$X' := \begin{cases} (x - C') \cup \bar{C}' & \text{with probability } \frac{1}{6} \\ x & \text{else} \end{cases} \tag{12}$$

If no neighbour β of this type exists then set $X' := x$.

3) *Define* $X_{t+1} := \begin{cases} X' & \text{with probability } \frac{1}{2} \\ x & \text{else.} \end{cases}$

Let $P'_0 = (p'_0(\cdot,\cdot))$ denote the transition matrix of $\mathcal{M}'(G)$. If $X' \neq x$ in the construction above then we speak of transitions of type 1 and type 2 corresponding to the use of simple circles or combined circles in step 2).

Theorem 3.4. *Let $G = (V, E)$ be a Eulerian, planar, triangular graph. Then the Markov chain $\mathcal{M}'_0(G)$ is ergodic and the stationary distribution π of $\mathcal{M}'_0(G)$ is the uniform distribution on $EO(G)$.*

Proof: Obviously $\mathcal{M}_0'(G)$ is irreducible and aperiodic since $\mathcal{M}_0(G)$ has these properties. To prove reversibility, let $x, y \in EO(G)$, $x \neq y$ with $p_o'(x,y) > 0$. Then $p_0'(x,y) = p_0'(y,x)$ by the proof of Theorem 2.2 for those pairs x, y with $p_0(x,y) > 0$. If $p_0(x,y) = 0$, then for some neighbour domains α, β holds, that x^α is not a directed circle and $y = (x - C) \cup \bar{C}$ with $C = x^\alpha \oplus x^\beta$. We have that $p_0'(x,y) = \frac{1}{|F(G)|-1} \cdot \frac{1}{12}$. Thus we have a situation as in Figure 3.4. One sees directly

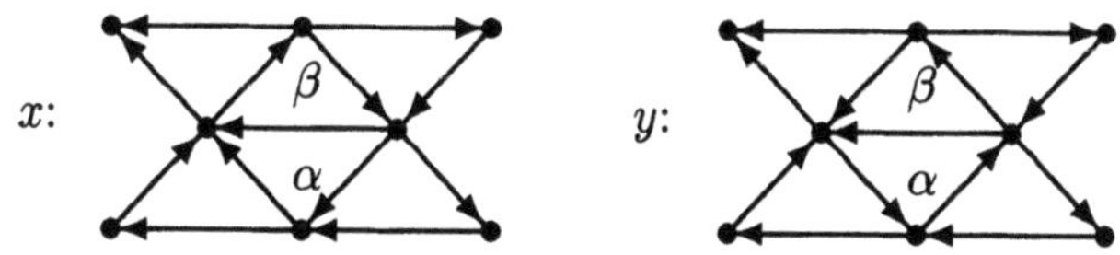

Figure 3.4

that x^β and y^α are directed circles while x^α and y^β are not directed circles. If in one step transition of the Markov chain $\mathcal{M}_0'(G)$ one draws in step 1) β then in step 2) one obtains $C' = y^\beta \oplus y^\alpha$ and thus $Y' = x$ with probability $\frac{1}{6}$. Together we obtain $p_0'(y,x) = \frac{1}{|F(G)|-1} \cdot \frac{1}{12} = p_0'(x,y)$.

The mixing time τ_0' of $\mathcal{M}_0'(G)$ can be efficiently bounded for this Markov chain in the case of triangular graphs.

Theorem 3.5. *Let $G = (V, E)$ be a Eulerian, planar triangular graph. Then the mixing time τ_0' of $\mathcal{M}_0'(G)$ is polynomially bounded in $n = |V|$:*

$$\tau_0'(\epsilon) \leq \lceil 32en^3 \rceil \lceil \log \epsilon^{-1} \rceil \text{ for all } \epsilon \in (0,1). \tag{13}$$

Proof: We apply the path coupling method with

$$S := \{(x,y) \in \Omega \times \Omega : \exists \text{ inner domain } \alpha \text{ such that } y = (x - x^\alpha) \cup \overline{x^\alpha}\},$$

where $\Omega = EO(G)$. The irreducibility of $\mathcal{M}_0(G)$ implies that for any $(x,y) \in \Omega$ there exists a path $(z_i)_{0 \leq i \leq r}$ with $(z_i, z_{i+1}) \in S$ for all $i < r$ and $z_0 = x$, $z_r = y$. Define $\varphi(x,y) = 1$ for all $(x,y) \in S$. Then $\delta(x,y)$ is the length of the shortest path from x to y in S. We have to determine the coupling on S.

For $(x_1, y_1) \in S$ with $y_1 = (x_1 - x_1^\alpha) \cup \overline{x_1^\alpha}$ for some inner domain $\alpha \in F(G)$ let β, γ, λ denote the neighbour domains of α. To construct the coupling (X_2, Y_2) starting from $(X_1, Y_1) = (x_1, y_1)$ choose in step 1 for both pairs the same inner domain $\kappa \in F(G)$. After construction of x_2 the coupling transition step of y_1 is given as follows:

1. case: $\kappa \notin \{\alpha, \beta, \gamma, \lambda\}$.
In this case holds $x_1^\kappa = y_1^\kappa$ as well as $x_1^\kappa \oplus x_1^{\kappa'} = y_1^\kappa \oplus y_1^{\kappa'}$ for all neighbours κ' of x. Therefore, in step 2 for the transition of x_1 w.r.t. $\mathcal{M}_0'(G)$ we have the same situation as for y_1 and thus may choose a transition from y_1 to y_2 parallel to that of x_1 to x_2. y_2 differs from x_2 only in the orientation of the edges in x_1^α. Thus we obtain

$$(x_2, y_2) \in S \text{ and } \delta(x_2, y_2) = \delta(x_1, y_1) = 1. \tag{14}$$

2. case: $\kappa = \alpha$.
Then by step 2) we obtain $X_1' = y_1$ and $Y_1' = x_1$. Defining

$$(X_2, Y_2) = \begin{cases} (x_1, x_1) & \text{with probability } \frac{1}{2} \\ (y_1, y_1) & \text{with probability } \frac{1}{2} \end{cases}$$

we obtain a.s.

$$\delta(X_2, Y_2) = \delta(x_1, y_1) - 1 = 0. \tag{15}$$

There are two extreme situations to consider.

3. case: $\kappa \in \{\beta, \gamma, \lambda\}$.
Without loss of generality let $\kappa = \beta$. The first possible extreme situation is this: x_1^β is no directed circle, but for some neighbour β' of β, $\beta' \neq \alpha$, $C = x_1^\beta \oplus x_1^{\beta'}$ is an oriented circle. Also assume that y_1^β is no directed circle but $C'' = y_1^\beta \oplus y_1^{\beta''}$ for some neighbour β'' of β, $\beta'' \notin \{\alpha, \beta'\}$ is a directed circle. Then define

$$(X_2, Y_2) = \begin{cases} (x_1', y_1') & \text{with probability } \frac{1}{12}, \\ (x_1, y_1) & \text{with probability } \frac{11}{12}, \end{cases}$$

where $x_1' = (X_1 - C') \cup \bar{C}'$, $y_1' = (Y_1 - C'') \cup \bar{C}''$. Then for this worst case we obtain

$$\delta(X_2, Y_2) = \begin{cases} \delta(x_1, y_1) + 4 & \text{with probability } \frac{1}{12}, \\ \delta(x_1, y_1) & \text{with probability } \frac{11}{12}. \end{cases} \tag{16}$$

Since $\delta(x_1, y_1) = 1$ we obtain from (14)–(16):

$$\begin{aligned} &E(\delta(X_2, Y_2) \mid (X_1, Y_1) = (x_1, y_1)) - \delta(x_1, y_1) \\ \leq\ & \frac{1}{|F(G)| - 1} \left((|F(G)| - 5) \cdot 0 + 1 \cdot (-1) + 3 \cdot \frac{1}{12} \cdot 4 \right) \\ =\ & \frac{1}{|F(G)| - 1} (-1 + 1) = 0 \end{aligned} \tag{17}$$

and thus $E(\delta(X_2, Y_2) | (X_1, Y_1) = (x_1, y_1)) \leq \delta(x_1, y_1)$. Note that this situation allows to choose the transition of type 2 in Definition 3.3 with probability at most $\frac{1}{12}$.

The second extreme situation to consider is this: x_1^β is no directed circle but $C = x_1^\beta \oplus x_1^\alpha$ is a directed circle. Simultaneously y_1^β is a directed circle. Then the transitions of type 2 in x_1 and of type 1 in y_1 yield the same result $x_1' = y_1'$; the first transition occurs with probability $\frac{1}{12}$, the second with probability $\frac{1}{2}$. As a result we obtain in this second situation

$$\delta(X_2, Y_2) = \begin{cases} \delta(x_1, y_1) - 1 & \text{with probability } \frac{1}{12} \\ \delta(x_1, y_1) + 1 & \text{with probability } \frac{1}{2} - \frac{1}{12} \\ \delta(x_1, y_1) & \text{with probability } \frac{1}{2} \end{cases}$$

and, therefore,

$$\begin{aligned} &E(\delta(X_2, Y_2) \mid (X_1, Y_1) = (x_1, y_1)) - \delta(x_1, y_1) \\ =\ & \frac{1}{|F(G)| - 1} \Bigg((|F(G)| - 5) \cdot 0 + 1 \cdot (-1) \\ & \qquad + 3 \cdot \left(\frac{1}{12}(-1) + \left(\frac{1}{2} - \frac{1}{12} \right) \cdot (+1) + \frac{1}{2} \cdot 0 \right) \Bigg) \\ =\ & \frac{1}{|F(G)| - 1} (-1 + 1) = 0. \end{aligned}$$

In this second situation the transition of type 2 has to be chosen with probability at least $\frac{1}{12}$. So both of these situations lead to the transition probability $\frac{1}{12}$. This second situation is also the reason that the original Markov chain has to be modified by allowing transitions of type 2. For the Markov chain $\mathcal{M}_0(G)$ the contraction condition is not fulfilled in this situation.

As a result we obtain that the assumption of the path coupling theorem in (5) is fulfilled with $\beta = 1$ and thus we obtain a coupling on $\Omega \times \Omega$.

In the proof of Theorem 2.2 it was shown that any x in Ω is connected with S_G by a path of length $\leq |F(G)| - 1$. Thus we obtain $\delta(\Omega) = \max_{x,y \in \Omega} \delta(x, y) \leq 2(|F(G)| - 1)$.

From case 2 we obtain for $(x_1, y_1) \in S$

$$\alpha := P(\delta(X_2, Y_2) \neq \delta(x_1, y_1) \mid (X_1, Y_1) = (x_1, y_1)) \geq (|F(G)| - 1)^{-1}. \tag{18}$$

This estimate extends by an induction argument to any pair $(x_1, y_1) \in \Omega^2$. Thus (4) implies

$$\tau_0'(\epsilon) \leq \lceil 4e(|F(G)| - 1)^3 \rceil \lceil \log \epsilon^{-1} \rceil. \tag{19}$$

If a denotes the number of edges in the outer domain of G, then using that G is triangular we obtain with $n = |V|$, $m = |E|$, $\ell = |F(G)|$

$$3(\ell - 1) + a = 2m. \tag{20}$$

The Euler polyeder formula for planar graphs gives

$$n - m + \ell = 2. \tag{21}$$

Thus we obtain $n + \ell - \frac{3}{2}(\ell - 1) - \frac{a}{2} = 2$. This implies

$$\ell < 2n \tag{22}$$

and again by Euler's polyeder formula

$$m < 3n. \tag{23}$$

From (19) we conclude

$$\tau_0'(\epsilon) \leq \lceil 4e\ell^3 \rceil \lceil \log \epsilon^{-1} \rceil \leq \lceil 32en^3 \rceil \lceil \log \epsilon^{-1} \rceil.$$

From the polynomial bound for the mixing time τ_0' of $\mathcal{M}_0'(G)$ we next derive the mixing time bound for τ_0 by means of the comparison method of Randall and Tetali (1998).

Proof of Theorem 3.1: In order to compare the mixing time $\tau_0(\epsilon)$ with $\tau_0'(\epsilon)$ from $\mathcal{M}_0'(G)$ we have to construct a set Γ of canonical paths $\gamma_{x,y}$ in $\mathcal{M}_0(G)$ for each transition (x, y) in $\mathcal{M}_0'(G)$, $\Gamma = \{\gamma_{x,y}; (x, y) \text{ a transition in } \mathcal{M}_0'(G)\}$.

1. case: (x, y) is a type 1 transition.
Then for some $\alpha \in F(G)$, $y = (x - C) \cup \bar{C}$ with $C = x^\alpha$. This is also a transition in $\mathcal{M}_0(G)$ and thus we define $\gamma_{x,y} = (x, y)$.

2. case: (x, y) is a type 2 transition.
Then $y = (x - C) \cup \bar{C}$ with $C = x^\alpha \oplus x^\beta$ for some inner neighbour domains $\alpha, \beta \in F(G)$. C is a directed circle while x^α is not a directed circle. Therefore, x^β is also a directed circle. Therefore, the type 2 transition (x, y) can be replaced by two type 1 transitions, first in the domain β and then in the domain α. These two transitions define the path $\gamma_{x,y}$. Together we obtain the set of canonical paths Γ.

We have to bound the comparison measure $A(\Gamma)$. Let $(w, z) \in T_1$ be a transition of $\mathcal{M}_0(G)$, i.e., $z = (w - C) \cup \bar{C}$, $C = w^\alpha$ for some inner domain $\alpha \in F(G)$. If β is a neighbour domain to α, then w^β and z^β can not be both directed circles since one common edge of these transitions is inverted. Therefore, the transition (w, z) is contained for each neighbour β of α in at most one path $\gamma_{x,y} \in \Gamma$ and the transition (x, y) in $\mathcal{M}'_0(G)$ is then of type 2. Thus we obtain $|\Gamma(w, z)| \leq 4$ and, therefore,

$$\begin{aligned} A(\Gamma) &\leq \max_{(w,z)\in T_1} \sum_{(x,y)\in\Gamma(w,z)} \frac{p'_0(x,y)}{p_0(w,z)} |\gamma_{x,y}| \\ &\leq \max_{(w,z)\in T_1} 2|\Gamma(w,z)| \leq 8. \end{aligned}$$

Since the stationary distribution π of $\mathcal{M}_0(G)$ is the uniform distribution on $EO(G)$ we obtain $\widehat{\pi}^{-1} \leq 2^m$ with $m = |E|$. Also by (23) $m < 3n$ and thus $\widehat{\pi}^{-1} \leq 2^{3n}$. This implies by the comparison result in (8):

$$\begin{aligned} \tau_0(\epsilon) &\leq \frac{4\log((\widehat{\pi}\epsilon)^{-1})}{\log((2\epsilon)^{-1})} 8\tau'_0(\epsilon) \leq 32\frac{\log(2^{3n}\epsilon^{-1})}{\log((2\epsilon)^{-1})}[32en^3]\lceil\log\epsilon^{-1}\rceil \\ &\leq 3072n^3\lceil\log\epsilon^{-1}\rceil(3n + \log\epsilon^{-1}) \end{aligned}$$

for all $\epsilon \in (0, \frac{1}{2})$. □

Remarks 3.6. *a) The mixing rates τ'_0, τ_0 are given in Theorems 3.1 and 3.5 only in terms of $n = |V|$ the number of nodes. τ'_0 is of the order n^3 while τ_0 is of the order n^4. The effective mixing time of the chains could be much lower since the estimation by the path coupling method is uniform and thus too much concentrated on the worst case. In comparison our rates are much better than the rates obtained by the general method of Mihail and Winkler (1996) which are of an order $\geq n^{10}$. In consequence the standard associated randomized approximation scheme (RAS) of our specialized Markov chain yields improved estimates for the number of Euler orientations.*

b) It is an open problem whether the chain $\mathcal{M}_0(G)$ in section 2 is rapidly mixing for all (or a large class of) planar graphs. The rapid mixing property of this chain has been established also for the Euler orientations in Cartesian grids in the dissertation of Fehrenbach (2003).

References

[1] Aldous, D. (1983). *Random walks on finite groups and rapidly mixing Markov chains*, Volume 986 of *Lecture Notes in Mathematics*, pp. 243–297. Springer.

[2] Bubley, R. and M. Dyer (1997). Path coupling: a technique for proving rapid mixing in Markov chains. In *Proceedings of the 38th Annual IEEE Symposium on foundations of Computer Science (FOCS)*, pp. 223–231. IEEE Computer Society Press.

[3] Diestel, R. (1996). *Graphentheorie*. Springer.

[4] Dyer, M. and C. Greenhill (1998). A more rapidly mixing Markov chain for graph colorings. *Random Structure and Algorithms 13*, 285–317.

[5] Fehrenbach, J. (2003). *Design und Analyse stochastischer Algorithmen auf kombinatorischen Strukturen.* PhD thesis, University of Freiburg.

[6] Jerrum, M. and A. Sinclair (1989). Approximating the permanent. *SIAM Journal on Computing 18*, 1149–1178.

[7] Luby, M., D. Randall, and A. Sinclair (2001). Markov chain algorithms for planar lattice structures. *SIAM Journal on Computing 31*, 167–192.

[8] Mihail, M. and P. Winkler (1996). On the number of Eulerian orientations of a graph. *Algorithmica 16*, 402–414.

[9] Randall, D. and P. Tetali (1998). *Analyzing Glauber dynamics by comparison of Markov chains*, Volume 1380 of *Lecture Notes in Computer Science*, pp. 292–304. Springer.

[10] Welsh, D. (1990). The computational complexity of some classical problems from statistical physics. In G. Grimmett and J. Hammersley (Eds.), *Disorder in Physical Systems*, pp. 307–321. Oxford University Press.

[11] Welsh, D. (1998). *Complexity: Knots, Colourings, and Counting*, Volume 186 of *LMS Lecture Note Series*. Cambridge University Press.

Johannes Fehrenbach
Department of Mathematics, University of Freiburg, Eckerstr. 1, 79104 Freiburg, Germany

Ludger Rüschendorf
Department of Mathematics, University of Freiburg, Eckerstr. 1, 79104 Freiburg, Germany
ruschen@stochastik.uni-freiburg.de

Trends in Mathematics, © 2004 Birkhäuser Verlag Basel/Switzerland

Regenerative Composition Structures: Characterisation and Asymptotics of Block Counts

Alexander Gnedin
(based on joint work with Jim Pitman and Marc Yor)

A regenerative composition structure is a sequence $(\mathcal{C}_n)$ of random compositions of integers $n = 1, 2, \ldots$ which satisfies two conditions:

- sampling consistency: if n identical balls are distributed into an ordered series of boxes according to $(\mathcal{C}_n)$, then a distributional copy of $\mathcal{C}_{n-1}$ is obtained by discarding one of the balls picked uniformly at random, and then deleting an empty box in case one is created.
- the first-part deletion invariance: for all $n > r \geq 1$, given that the first part of $\mathcal{C}_n$ is r, the remaining composition of $n-r$ is distributed like $\mathcal{C}_{n-r}$.

The first condition is an ordered version of the defining property of Kingman's partition structures, while the second generalises a deletion property of the partition structure associated with the Ewens sampling formula (ESF). Next representation is the principal characterisation result.

Theorem 14. [3] *For a measure* ν *on* $]0,1]$ *with finite first moment and* $\mathrm{d} \geq 0$ *let*

$$\Phi(n:r) = \binom{n}{r} \int_0^1 x^r (1-x)^{n-r} \widetilde{\nu}(\mathrm{d}x)\,, \ 1 \leq r \leq n\,, \qquad \Phi(n) = \sum_{r=1}^{n} \Phi(n:r).$$

Then for

$$q(n:r) = \frac{\Phi(n:r)}{\Phi(n)}$$

the function on compositions

$$p(\lambda_1, \ldots, \lambda_\ell) = \prod_{j=1}^{\ell} q(\Lambda_j : \lambda_j)$$

where $\Lambda_j = \lambda_j + \ldots + \lambda_\ell$ *is the distribution of a regenerative composition structure. The parameterisation of distributions by* $(\widetilde{\nu}, \mathrm{d})$ *is unique subject to the normalisation* $\Phi(1) = 1$.

There is a paintbox-type sampling scheme which constructs a composition structure associated with $(\widetilde{\nu}, \mathrm{d})$. Let (S_t) be a subordinator with drift d and the Lévy measure obtained from $\widetilde{\nu}$ by the transform $x \mapsto -\log(1-x)$. Let $\widetilde{\mathcal{R}} \subset [0,1]$ be the closed range of the process $\widetilde{S}_t = 1 - \exp(-S_t)$, and let $u_1, u_2, \ldots$ be independent sample from the uniform distribution. Call u_i, u_j *blocked* if the interval spanned on these points does not intersect $\widetilde{\mathcal{R}}$. The composition $\mathcal{C}_n$ appears as the record of clusters of blocked points among the first n sample points, from left to right.

There is a direct characterisation of regenerative composition structures in terms of arrangements of special (unordered) partition structures. Call a nonnegative function $d(\lambda, r)$ depending on partition λ and a part $r \in \lambda$ a *deletion kernel* if $\sum_{r\in\lambda} d(\lambda, r) = 1$. For each λ such a kernel defines a subpartition of λ obtained by deleting a random part. Call a partition structure $(\mathcal{P}_n)$ *deletion invariant* if for all $n > r \geq 1$, given that the deleted part of $\mathcal{P}_n$ is r, the remaining partition of $n - r$ is distributed like $\mathcal{P}_{n-r}$. Starting with some partition λ, repeated application of the deletion rule encoded in d yields an ordering of the parts of λ.

Theorem 15. [4] *For a partition structure with distribution p, invariant with respect to some deletion kernel d, the arrangement of parts in the order of deletion yields a regenerative composition structure. The distribution of such a partition structure is described then by the formula*

$$p(\lambda_1, \dots, \lambda_\ell) = \sum_\sigma \prod_{j=1}^{\ell} q(\Lambda_{\sigma(j)} : \lambda_{\sigma(j)})$$

where the sum expands over all distinct compositions $(\lambda_{\sigma(1)}, \dots, \lambda_{\sigma(\ell)})$ derived from partition λ by arranging the parts in some order, and

$$q(n : r) = \sum_{\{\lambda \vdash n:\, r\in\lambda\}} d(\lambda, r)p(\lambda).$$

Particular choice of the measure $\widetilde{\nu}[x, 1] = x^{-\alpha}(1 - x)^\theta$ with $\alpha \in [0, 1]$, $\theta > 0$ yields a regenerative composition structure whose associated partition structure belongs to the Ewens-Pitman two-parameter family (and for other values of parameters no regenerative realisation is possible). For example, the case $\alpha = 0$, $\theta > 0$ corresponds to the GEM paintbox, the case $\alpha > 0$, $\theta = 0$ to the paintbox induced by the zero set of a Bessel process, and the case $\alpha = \theta$ is associated with a Bessel bridge (in the last case there is a further characterisation by reversibility or by the invariance of distribution of $\mathcal{C}_n$ under permutations of parts). Partition structures from the two-parameter family are deletion-invariant with respect to the kernel

$$d(\lambda, r) = \frac{1}{n} \frac{(n - r)\tau + r(1 - \tau)}{1 - \tau + (\ell - 1)\tau},$$

(the formula distinguishes among the equal parts), and the deletion property is characteristic (the ESF being a special case $\tau = 0$).

A novel example of composition structure is provided by the formula [1]

$$p(\lambda) = \frac{n!}{[\theta]_n} \prod_j \frac{1}{\lambda_j h(\Lambda_j)}, \quad \text{where } h(r) = \sum_{j=1}^{r} \frac{1}{j + \theta - 1},$$

which corresponds to a gamma-like subordinator with $\widetilde{\nu}(dx) = x^{-1}(1-x)^{\theta-1}$, $\theta > 0$.

Let K_n be the number of parts of $\mathcal{C}_n$ and let $K_{n,r}$ be the multiplicity of part r. Next is a summary of available results regarding these functionals. Suppose $\mathsf{d} = 0$. When $\widetilde{\nu}$ is finite, under very mild assumptions the distribution of K_n is close to normal, with both expectation and variance of the order $\log n$ [2], and $K_{n,r}$ is bounded for each r as $n \to \infty$. When $\widetilde{\nu}[x, 1] \sim \ell(1/x)x^{-\alpha}$ as $x \downarrow 0$ for slowly varying ℓ and $0 < \alpha \leq 1$, we have $K_n/(\ell(n)n^\alpha)$ strongly converging to the exponential functional of a subordinator (the case $\alpha = 1$ is special), and a similar joint convergence result holds for $K_{n,r}$'s [5]. A very interesting case, intermediate between the above two, is the case of gamma-type subordinators with $\widetilde{\nu}[x,1] \sim -\log x$: in this case K_n is again asymptotically normal, with expectation and variance growing like $\log^2 n$ and $\log^3 n$, respectively, and $(K_{n,1}, K_{n,2}, \ldots)$ is jointly approximately normal [6].

References

[1] A. V. Gnedin. Three sampling formulas, *Comb. Probab. Comp.*, 13: 185–193, 2004.

[2] A. V. Gnedin. The Bernoulli sieve, *Bernoulli*, 10: 79–96, 2004.

[3] A. V. Gnedin and J. Pitman. Regenerative composition structures, *Ann. Probab.*, 2004 (to appear)

[4] A.V. Gnedin and J. Pitman. Regenerative partition structures, (paper in progress)

[5] A. V. Gnedin, J. Pitman and M. Yor. Asymptotic laws for compositions derived from transformed subordinators, available at arXive.

[6] A. V. Gnedin, J. Pitman and M. Yor. Asymptotic laws for regenerative composition structures: gamma subordinators and the like, available at arXive.

Alexander Gnedin
Utrecht University, The Netherlands

Trends in Mathematics, © 2004 Birkhäuser Verlag Basel/Switzerland

Random Walks on Groups With a Tree-Like Cayley Graph

Jean Mairesse and Frédéric Mathéus

ABSTRACT: *We consider a transient nearest neighbor random walk on a group G with finite set of generators Σ. The pair (G, Σ) is assumed to admit a natural notion of normal form words which are modified only locally when multiplied by generators. The basic examples are the free products of a finitely generated free group and a finite family of finite groups, with natural generators. We prove that the harmonic measure is Markovian and can be completely described via a finite set of polynomial equations. It enables to compute the drift, the entropy, the probability of ever hitting an element, and the minimal positive harmonic functions of the walk. The results extend to monoids. In several simple cases of interest, the set of polynomial equations can be explicitly solved, to get closed form formulas for the drift, the entropy,... Various examples are treated: the modular group $\mathbb{Z}/2\mathbb{Z} \star \mathbb{Z}/3\mathbb{Z}$, the Hecke groups $\mathbb{Z}/2\mathbb{Z} \star \mathbb{Z}/k\mathbb{Z}$, the free products of two isomorphic cyclic groups $\mathbb{Z}/k\mathbb{Z} \star \mathbb{Z}/k\mathbb{Z}$, the braid group B_3, and Artin groups with two generators.*

1. Introduction

The properties of the harmonic measure associated with a random walk on a finitely generated free group have been studied by many authors [6, 14, 16, 21]. A remarkable result is that for nearest neighbor random walks, the harmonic measure is Markovian. If the probability defining the random walk does not only charge the generators but is of finite support instead, then the result fails to be true, see [14]. In other words, having a Markovian harmonic measure is a property which depends not only on the group G but also on the chosen set of generators Σ. Here, we prove the existence of a Markovian harmonic measure for a whole class of pairs (G, Σ). We then use the property for computational purposes.

A pair (G, Σ) formed by a group (group law $*$, unit element 1_G) and a finite set of generators is called *0-automatic* if the set of words $L(G, \Sigma) = \{u_1 \cdots u_k \mid \forall i, u_i * u_{i+1} \notin \Sigma \cup 1_G\}$ is a cross-section of G.

Consider a group $G = \mathbb{F}(S) \star G_1 \star \cdots \star G_k$ which is a free product of a finitely generated free group and a finite family of finite groups, also called *plain* group. Consider the *natural* (but not necessarily minimal) set of generators $\Sigma = S \sqcup S^{-1} \sqcup_i G_i \backslash \{1_{G_i}\}$. Then the pair (G, Σ) is 0-automatic. Now consider an arbitrary 0-automatic pair (G, Σ). Then G is isomorphic to a plain group. This is not the end of the story since Σ may be different (in particular, larger) than the natural set of generators of the group seen as a free product. (For an example of this, see the end of §3.) And what is relevant in our context is the pair group-generators rather that the group itself, since the generators form the support of the measure defining the nearest neighbor random walk. Hence it seems appropriate and convenient to coin a specific term as "0-automatic pairs" to highlight the notion.

Apart from the free group with one generator and the group $\mathbb{Z}/2\mathbb{Z}\star\mathbb{Z}/2\mathbb{Z}$, all the plain groups are non-amenable. It follows that any random walk living on the whole group is transient and has a strictly positive drift (see [11] and [26, Chapter 1.B] for details).

Consider a transient nearest neighbor random walk $(X_n)_n$ on a 0-automatic pair (G, Σ). Let $(Y_n)_n$ be the corresponding random walk on normal form words. The harmonic measure is the law of $Y_\infty = \lim_n Y_n$. We prove that this harmonic measure is Markovian. The transition probabilities of this Markov chain are the unique solutions of a set of polynomial equations of degree 2, that we call the Traffic Equations. We can then compute the drift and the entropy of the random walk. Some additionnal work is needed to compute the probability of ever hitting a generator, which leads to a computational description of the minimal positive harmonic functions. Mutatis mutandis, the results extend to monoids.

All the random walks considered here belong to the general setting of random walks on regular languages studied in [15]. In [15], local limit theorems are proved. Also, our random walks can be viewed as random walks on a tree with finitely many cone types in the sense of [19], but with one-step moves at distance 1 and 2. In [19], it is proved, among other things, that the harmonic measure is Markovian for nearest neighbor random walks. See also [23]. The method of proof is different from the one we use, see the discussion in §5.2.

A natural question is whether the Traffic Equations can be solved "explicitly", in order to get for instance a closed form formula for the drift or the entropy. This is feasible in many situations. We treat completely the following cases: the general nearest neighbor random walk on the modular group $\mathbb{Z}/2\mathbb{Z} \star \mathbb{Z}/3\mathbb{Z}$, three one-parameter families of random walks on $\mathbb{Z}/3\mathbb{Z} \star \mathbb{Z}/3\mathbb{Z}$, a one-parameter family of random walks on the three strands braid group B_3, and the simple random walks on $\mathbb{Z}/k\mathbb{Z}\star\mathbb{Z}/k\mathbb{Z}$, on the Hecke groups $\mathbb{Z}/2\mathbb{Z}\star\mathbb{Z}/k\mathbb{Z}$, and on the Artin groups with two generators.

None of these computations appeared in the literature before. For the few examples of non-elementary explicit computations previously available, see [5, 6, 16, 19, 20, 21]. Nevertheless, there exists an alternative potential method for the effective computation of the drift (not the entropy) which is due to Sawyer and Steger [21]. In this approach, the drift is expressed as a functional of the *first-passage generating series* of the random walk. The simplest of our computational results can also be retrieved using this method. We detail and discuss this method in §6.1. The Sawyer-Steger method links the problem of computing the drift with the problem of computing the generating series of transition probabilities. Concerning the latter problem, there exists an important literature, especially for random walks on free groups and free products, see [2, 3, 25] and [26, Sections II.9 and III.17].

The usual motivation for studying generating functions of transition probabilities is to get central or local limit theorems. For much more material on this and on other aspects of random walks on discrete infinite groups only touched upon here (like boundary theory), see [13, 24, 26] and the references there.

The results, presented here without proofs, are exposed in full details in [17, 18].

Let $\mathbb{N}$ be the set of non-negative integers. If μ is a measure on a group $(G,*)$, then μ^{*n} is the n-fold convolution product of μ, that is the image of the product measure $\mu^{\otimes n}$ by the product map $G\times\cdots\times G\to G$, $(g_1,\ldots,g_n)\mapsto g_1*g_2*\cdots*g_n$. The symbol $\sqcup$ is used for the disjoint union of sets.

2. Random walks on groups

Given a set Σ, the free monoid it generates is denoted by Σ^*. As usual, Σ is called the *alphabet*, the elements of Σ and Σ^* are called respectively the *letters* and *words*. The empty word is denoted by 1_{Σ^*}. The *length* (number of letters) of a word u is denoted by $|u|_\Sigma$.

Consider a finitely generated group $(G,*)$ with unit element 1_G. Let Σ be a finite set of generators of G (with $1_G\notin\Sigma$ and $u\in\Sigma \implies u^{-1}\in\Sigma$). Denote by $\pi:\Sigma^*\to G$ the monoid morphism which associates to a word $a_1\cdots a_k$ the group element $a_1*\cdots*a_k$. The *length with respect to* Σ of a group element u is:

$$|u|_\Sigma=\min\{k \mid u=s_1*\cdots*s_k, s_i\in\Sigma\}\,. \tag{1}$$

A word $u\in\Sigma^*$ is a *geodesic* if $|u|_\Sigma=|\pi(u)|_\Sigma$.

The Cayley graph $\mathfrak{X}(G,\Sigma)$ of a group G with respect to a set of generators Σ is the directed graph with G as the set of nodes and with an arc from u to v if $u^{-1}*v\in\Sigma$. It is often convenient to view $\mathfrak{X}(G,\Sigma)$ as a labelled graph with set of labels Σ (with $u\xrightarrow{a}v$ if $u*a=v$). Observe that $|u|_\Sigma$ is the geodesic distance from 1_G to u in the Cayley graph.

Let $G_1\star G_2$ be the free product of two groups G_1 and G_2. Roughly, the elements of $G_1\star G_2$ are the finite alternate sequences of elements of $G_1\backslash\{1_{G_1}\}$ and $G_2\backslash\{1_{G_2}\}$, and the group law is the concatenation with simplication. More rigorously, the definition is as follows. Set $S=G_1\sqcup G_2$. Let $\sim$ be the least congruence on S^* containing the relations: $\forall u,v\in S^*, \forall i\in\{1,2\},\ u1_{G_i}v\sim uv,\quad \forall a,b,c\in G_i$, s.t. $c=a*b,\ uabv\sim ucv$. The quotient monoid $(S^*/\sim)$ is a group called the *free product* $G_1\star G_2$.

Let μ be a probability distribution over Σ. Consider the Markov chain on the state space G with one-step transition probabilities given by: $\forall g\in G, \forall a\in\Sigma,\ P_{g,g*a}=\mu(a)$. This Markov chain is called the *random walk* (associated with) (G,μ). It is a *nearest neighbor* random walk: one-step moves occur between nearest neighbors in the Cayley graph $\mathfrak{X}(G,\Sigma)$. When $\mu(s)=1/|\Sigma|$ for all $s\in\Sigma$, we say that the random walk is *simple*.

Let $(x_n)_n$ be a sequence of i.i.d. r.v.'s distributed according to μ. Set

$$X_0=1,\ X_{n+1}=X_n*x_n=x_0*x_1*\cdots*x_n\,. \tag{2}$$

The sequence $(X_n)_n$ is a *realization* of the walk (G,μ). The law of X_n is μ^{*n}.

Drift, entropy. The first step in understanding the asymptotic behaviour of X_n consists in studying the length $|X_n|_\Sigma$ as $n\to\infty$. Since $|uv|_\Sigma\le|u|_\Sigma+|v|_\Sigma$, Guivarc'h [11] observed that a simple corollary of Kingman's subadditive ergodic theorem is the existence of a constant $\gamma\in\mathbb{R}_+$ such that *a.s.* and in L^p, for all $1\le p<\infty$,

$$\lim_{n\to\infty}|X_n|_\Sigma/n=\gamma\,. \tag{3}$$

We call γ the *drift* of the random walk. Intuitively, γ is the speed of escape to infinity of the walk.

Another quantity of interest is the entropy. The *entropy* of a probability measure μ with finite support S is defined by $H(\mu) = -\sum_{x\in S}\mu(x)\log[\mu(x)]$. The *entropy of the random walk* (G,μ), introduced by Avez [1], is

$$h = \lim_n \frac{H(\mu^{*n})}{n} = \lim_n -\frac{1}{n}\log \mu^{*n}(X_n)\,, \tag{4}$$

a.s. and in L^p, for all $1 \le p < \infty$. The existence of the limits as well as their equality follow again from Kingman's subadditive ergodic theorem.

3. Zero-automaticity

Let G be a group with finite set of generators Σ. A language L of Σ^* is a *cross-section* of G (over the alphabet Σ) if the restriction of π to L defines a bijection, that is if every element of G has a unique representative in L. A word of L is then called a *normal form* word. The map $\varphi : G \to L$ which associates to a group element its unique representative in L is called the *normal form map.*

Definition 3.1. *Let G be a group with finite set of generators Σ. Define the sets*

$$\forall a\in\Sigma,\ \mathit{Next}(a)=\{b\in\Sigma \mid a*b\notin\Sigma\cup\{1_G\}\},\ \mathit{Prev}(a)=\{b\in\Sigma\mid b*a\notin\Sigma\cup\{1_G\}\}. \tag{5}$$

and the language of Σ^*,

$$\begin{aligned} L(G,\Sigma) &= \{u_1\cdots u_k \mid \forall i\in\{2,\dots,k\},\ u_i\in \mathit{Next}(u_{i-1})\} \qquad (6)\\ &= \{u_1\cdots u_k \mid \forall i\in\{1,\dots,k-1\},\ u_i\in \mathit{Prev}(u_{i+1})\}\,. \end{aligned}$$

We say that the pair (G,Σ) is 0-automatic *if $L(G,\Sigma)$ is a cross-section of G.*

Here are some consequences of Definition 3.1. First, let $\varphi : G \to L(G,\Sigma)$ be the normal form map. Then: $\forall g\in G$ s.t. $\varphi(g) = u_1\cdots u_k$, $\forall a\in\Sigma$,

$$\varphi(g*a) = \begin{cases} u_1\cdots u_{k-1} & \text{if } a = u_k^{-1} \\ u_1\cdots u_{k-1}v & \text{if } u_k*a = v\in\Sigma \\ u_1\cdots u_{k-1}u_k a & \text{if } u_k*a\notin\Sigma\cup 1_G \end{cases}, \tag{7}$$

and the analog for $\varphi(a*g)$ also holds.

Second, the language $L(G,\Sigma)$ is regular and recognized by the following automaton: Set of states: $\Sigma\cup 1_G$, initial state: 1_G, final states: $\Sigma\cup 1_G$; Transitions: $a \xrightarrow{b} b$ if $a*b\notin\Sigma\cup 1_G$.

Third, the Cayley graph $\mathfrak{X}(G,\Sigma)$ has a tree-like structure. In particular, $\mathfrak{X}(G,\Sigma)$ has uniform node-connectivity 1, that is, the removal of any node disconnects the graph, see Figure 6.

Fourth, all the group elements have a unique geodesic representative with respect to Σ, and the set of these geodesic representatives is precisely $L(G,\Sigma)$.

Fifth, the set of simple circuits going through the node 1_G in $\mathfrak{X}(G,\Sigma)$ is finite. In fact, this last property is equivalent to the property that $L(G,\Sigma)$ be a cross-section, see [12].

Here are examples of 0-automatic pairs:

- Let G be a finite group. Then $(G, G\backslash\{1_G\})$ is 0-automatic.
- Let $\mathbb{F}(\Sigma)$ be the free group generated (as a group) by Σ. Set $\tilde\Sigma = \Sigma\sqcup\Sigma^{-1}$. Then $(\mathbb{F}(\Sigma),\tilde\Sigma)$ is 0-automatic.
- Let (G_1,Σ_1) and (G_2,Σ_2) be 0-automatic. Then $(G_1\star G_2, \Sigma_1\sqcup\Sigma_2)$ is 0-automatic.

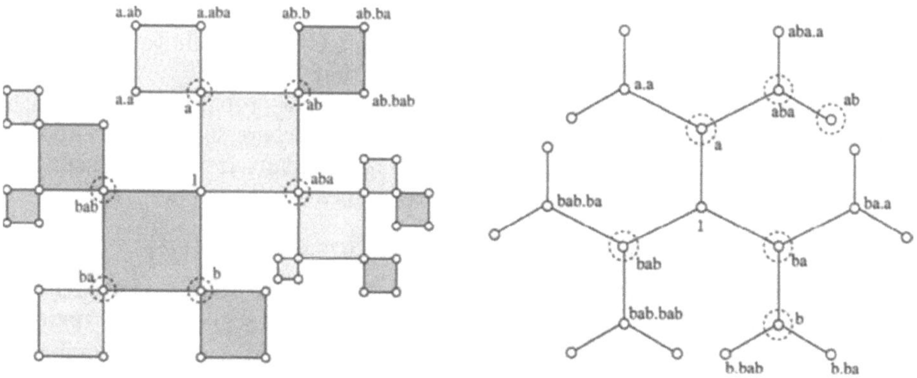

Figure 3.1 The Cayley graph of G with respect to $\{a, a^{-1}, b, b^{-1}\}$ (left), and $\{a, a^{-1}, ba\}$ (right).

Following [12], define a *plain group* as the free product of a finitely generated free group and a finite family of finite groups. Let $G = \mathbb{F}(S) \star G_1 \star \cdots \star G_k$ be a plain group. Then $\Sigma = S \sqcup S^{-1} \sqcup_i \Sigma_i$, $\Sigma_i = G_i \backslash \{1_{G_i}\}$, is a set of generators for G that we call the *natural* generators. It follows from the above that (G, Σ) is 0-automatic. The sets $\text{Next}(\cdot), \text{Prev}(\cdot)$, defined in (5), can be explicited:

$$\forall a \in S \sqcup S^{-1},\ \text{Next}(a) = \text{Prev}(a) = \Sigma \backslash \{a^{-1}\},\ \forall a \in \Sigma_i,\ \text{Next}(a) = \text{Prev}(a) = \Sigma \backslash \Sigma_i.$$

A 0-automatic pair (G, Σ) is precisely a *unique factorization pair* in Stallings [22], with the additional assumption that Σ be finite. Using the results from [22], we get:

Proposition 3.2. *Let (G, Σ) be a 0-automatic pair. Then G is isomorphic to a plain group.*

The proof of Stallings is constructive and provides more precise information. Let φ be any isomorphism from G to the plain group $\widetilde{G}$. Then there exists a set S of natural generators of $\widetilde{G}$ such that $S \subset \varphi(\Sigma)$. Note that the inclusion may be strict. Also, for $u \in \varphi(\Sigma)$, we have $|u|_S = 1, 2$, or 3.

Plain groups are hyperbolic in the sense of Gromov [10] and automatic in the sense of Epstein & al [7]. Besides, (G, Σ) is 0-automatic iff $(\Sigma, L(G, \Sigma))$ is an automatic pair (in the sense of [7]) satisfying the 0-fellow traveller property. This is our justification for the chosen terminology.

Consider the group $G = \langle\, a, b \mid abab = 1 \,\rangle$. (This is the Artin group $A_4 = \langle\, a, b \mid abab = baba \,\rangle$ quotiented by its center.) Set $\Sigma = \{a, b, ab = (ab)^{-1}, ba = (ba)^{-1}, aba = b^{-1}, bab = a^{-1}\}$. Then (G, Σ) is a 0-automatic pair. Here, $\text{Next}(x) = \{a, ab, aba\}$ if $x \in \{a, ba, aba\}$ and $\text{Next}(x) = \{b, ba, bab\}$ if $x \in \{b, ab, bab\}$.

Now, the group G is isomorphic to $\mathbb{F}(a) \star \{1, ba\} \sim \mathbb{Z} \star \mathbb{Z}/2\mathbb{Z}$, see Figure 3. Set $S = \{a, a^{-1} = bab, ba\}$ for the corresponding set of natural generators. We have for instance $|b|_S = 2$ and $|ab|_S = 3$. Concentrating on the right of Figure 3, it is not obvious that (G, Σ) is 0-automatic.

4. Random walks on zero-automatic pairs

From now on, the setting is the following one. Let G be an infinite group with a finite set of generators Σ, with $\Sigma = \Sigma^{-1}$, such that (G, Σ) is 0-automatic. Let $L(G, \Sigma)$ and φ be defined as in (6) and (7). Let μ be a probability on Σ which generates the whole group, that is, $\cup_n \text{supp}(\mu^{*n}) = G$, where supp is the support of the measure. We consider the random walk (G, μ), assumed to be transient.

Define the set of *infinite normal form* words $L^\infty \subset \Sigma^{\mathbb{N}}$ by

$$L^\infty = \{u = u_0 u_1 \cdots u_k \cdots \in \Sigma^{\mathbb{N}} \mid \forall i \in \mathbb{N}, u_{i+1} \in \text{Next}(u_i)\}\,. \tag{8}$$

A word belongs to L^∞ iff all its finite prefixes belong to $L(G, \Sigma)$. Consider the natural action $\Sigma \times L^\infty \to L^\infty, (a, \xi) \mapsto a \cdot \xi$, with $a \cdot \xi = a\xi_0\xi_1 \cdots$ if $a \in \text{Prev}(\xi_0)$, $a \cdot \xi = (a * \xi_0)\xi_1 \cdots$ if $a * \xi_0 \in \Sigma$, and $a \cdot \xi = \xi_1\xi_2 \cdots$ if $a = \xi_0^{-1}$. Equip $\Sigma^{\mathbb{N}}$ with the Borel σ-algebra associated with the product topology. This induces a σ-algebra on L^∞. Given a measure ν^∞ on L^∞ and $a \in \Sigma$, define the measure $a\nu^\infty$ by: $\int f(\xi) d(a\nu^\infty)(\xi) = \int f(a \cdot \xi) d\nu^\infty(\xi)$. A probability measure ν^∞ on L^∞ is *invariant* if

$$\nu^\infty(\cdot) = \sum_{a \in \Sigma} \mu(a)[a\nu^\infty](\cdot)\,. \tag{9}$$

Proposition 4.1. *Let $(X_n)_n$ be a realization of the random walk (G, μ) and set $Y_n = \varphi(X_n)$. There exists a r.v. Y^∞ valued in L^∞ such that a.s.*

$$\lim_{n \to \infty} Y_n = Y^\infty\,,$$

meaning that the length of the common prefix between Y_n and Y^∞ goes to infinity a.s. Let μ^∞ be the distribution of Y^∞. The measure μ^∞ is invariant and is the only invariant probability on L^∞. We call it the harmonic measure *of (G, μ). The drift and the entropy of the random walk are given by:*

$$\gamma = \sum_{x \in \Sigma} \mu(x) \Big[-\mu^\infty(x^{-1}\Sigma^{\mathbb{N}}) + \sum_{y \in Next(x)} \mu^\infty(y\Sigma^{\mathbb{N}}) \Big]\,, \tag{10}$$

$$h = -\sum_{x \in \Sigma} \mu(x) \int \log\Big[\frac{dx^{-1}\mu^\infty}{d\mu^\infty}(\xi)\Big] d\mu^\infty(\xi)\,, \tag{11}$$

where $dx^{-1}\mu^\infty / d\mu^\infty$ is the Radon-Nikodym derivative of $x^{-1}\mu^\infty$ with respect to μ^∞.

In the context of the free group, this is proved for instance in [16, Theorem 1.12, Theorem 4.10, Corollary 4.5]. The proofs adapt easily to the present setting. Several of the key arguments go back to [8], see [16] for precise references.

Intuitively, the harmonic measure μ^∞ gives the direction in which $(X_n)_n$ goes to infinity.

5. Markovian harmonic measure

Define $\mathring{\mathcal{B}} = \{x \in \mathbb{R}^\Sigma \mid \forall i, x_i > 0, \sum_i x_i = 1\}$. Consider $r \in \mathring{\mathcal{B}}$. Define the matrix P of dimension $\Sigma \times \Sigma$ by

$$P_{u,v} = \begin{cases} r(v) \big(\sum_{x \in \text{Next}(u)} r(x)\big)^{-1} & \text{if } v \in \text{Next}(u) \\ 0 & \text{otherwise} \end{cases}\,. \tag{12}$$

It is the transition matrix of a Markov Chain on the state space Σ, which can be proved to be irreducible. For convenience, set for all $a \in \Sigma$, $s(a) = \sum_{x\in \text{Next}(a)} r(x)$. Let $(U_n)_n$ be a realization of the Markov chain with transition matrix P and starting from U_1 such that $P\{U_1 = x\} = r(x)$. Set $U^\infty = \lim_n U_1 \cdots U_n$, and let ν^∞ be the distribution of U^∞. Clearly the support of ν^∞ is included in L^∞. For $u_1 \cdots u_k \in L(G,\Sigma)$, we have

$$\begin{aligned} \nu^\infty(u_1\cdots u_k\Sigma^{\mathbb{N}}) &= r(u_1)P_{u_1,u_2}\cdots P_{u_{k-1},u_k} \\ &= r(u_1)\frac{r(u_2)}{s(u_1)}\cdots\frac{r(u_k)}{s(u_{k-1})} = \frac{r(u_1)}{s(u_1)}\frac{r(u_2)}{s(u_2)}\cdots\frac{r(u_{k-1})}{s(u_{k-1})}r(u_k)\,. \end{aligned} \tag{13}$$

We call ν^∞ the *Markovian multiplicative* probability measure associated with r.

Observe that the measure ν^∞ is in general non-stationary with respect to the translation shift $\tau : \Sigma^{\mathbb{N}} \to \Sigma^{\mathbb{N}}, (x_n)_n \mapsto (x_{n+1})_n$. Indeed, the distribution of the first marginal is r which is in general different from the stationary distribution of P.

5.1. The main theorem

The *Traffic Equations* associated with (G,μ) are defined by: $\forall a \in \Sigma$,

$$x(a) = \mu(a)\sum_{u\in\text{Next}(a)} x(u) + \sum_{u*v=a}\mu(u)x(v) + \sum_{u\in\text{Prev}(a)}\mu(u^{-1})\frac{x(u)}{\sum_{v\in\text{Next}(u)}x(v)}x(a). \tag{14}$$

We are now ready to state the main result.

Theorem 5.1. *Let (G,Σ) be 0-automatic. Let μ be a probability measure on Σ such that $\cup_n \text{supp}(\mu^{*n}) = G$ and such that the random walk (G,μ) is transient. Then the Traffic Equations (14) have a unique solution $x \in \mathring{\mathcal{B}}$. The harmonic measure of the random walk is the Markovian multiplicative measure associated with x.*

The harmonic measure is not stationary in general. However, we have the following result:

Proposition 5.2. *Let H be a finite group and let $(G_i)_{i\in I}$ be a finite family of copies of H. Let π_i be the isomorphism between G_i and H. Let ν be a probability measure on $H\backslash\{1_H\}$. Consider the free product $G = \star_{i\in I}G_i$ and let μ be the probability measure on $\Sigma = \sqcup_i G_i\backslash\{1_{G_i}\}$ defined by: $\forall g \in G_i\backslash\{1_{G_i}\}$, $\mu(g) = \nu\circ\pi_i(g)/|I|$. Then the harmonic measure of (G,μ) is stationary and ergodic.*

Consider now a random walk (G,μ) where $G = G_1 \star G_2$ is the free product of two arbitrary finite groups. Then the harmonic measure μ^∞ is 2-stationary, meaning that: $\forall u \in L(G,\Sigma), \forall k \in \mathbb{N}$, $\mu^\infty(u\Sigma^{\mathbb{N}}) = \mu^\infty(\Sigma^{2k}u\Sigma^{\mathbb{N}})$.

Starting from (10)-(11) and using Theorem 5.1, we obtain a simple formula for the drift:

$$\gamma = \sum_{a\in\Sigma}\mu(a)\big[-r(a^{-1}) + \sum_{b\in\text{Next}(a)} r(b)\big]\,. \tag{15}$$

and for the entropy:

$$h = -\sum_{a\in\Sigma}\mu(a)\Big[\log\big[\frac{1}{q(a^{-1})}\big]r(a^{-1}) + \sum_{b,a*b\in\Sigma}\log\big[\frac{q(a*b)}{q(b)}\big]r(b) + \log[q(a)]r(\text{Next}(a))\Big]$$

where $r(\text{Next}(a)) = \sum_{b\in\text{Next}(a)} r(b)$ and $q(a) = r(a)/r(\text{Next}(a))$.

In particular, if the probabilities $\mu(a)$ are algebraic numbers, then the drift and the entropy are algebraic numbers.

5.2. Harmonic functions

For all $u \in G$, define $q(u) = P\{\exists n \mid X_n = u\}$, the probability of ever reaching u. If $\varphi(u) = u_1 \cdots u_k \in L(G,\Sigma)$, by the strong Markov property, we have $q(u) = q(u_1)q(u_2)\cdots q(u_k)$. Therefore, all we need to compute are the quantities $q(a), a \in \Sigma$. These quantities satisfy the equations: $\forall a \in \Sigma$,

$$q(a) \quad = \quad \mu(a) + \sum_{u*v=a} \mu(u)q(v) + q(a) \sum_{c\in\Sigma\setminus\Sigma_a} \mu(c)q(c^{-1}) \,. \qquad (16)$$

Proposition 5.3. *The Equations (16) characterize $(q(a))_{a\in\Sigma}$. Let r be the unique solution to the Traffic Equations in $\mathring{\mathfrak{B}}$. We have: $\forall a \in \Sigma$, $q(a)=r(a)/[\sum_{b\in Next(a)} r(b)]$. The harmonic measure satisfies: $\forall u_1 \cdots u_k \in L(G,\Sigma)$, $\mu^\infty(u_1\cdots u_k\Sigma^{\mathbb{N}}) = q(u_1)\cdots q(u_{k-1})r(u_k)$.*

Consider now a free product of finite groups $G_1\star\cdots\star G_k$. Set $\Sigma_i = G_i\setminus\{1_{G_i}\}$, and $q(\Sigma_i) = \sum_{a\in\Sigma_i} q(a)$. We have: $\forall a \in \Sigma_i$, $r(a) = q(a)/[1+q(\Sigma_i)]$. At last, for a finitely generated free group $\mathbb{F}(S)$, we have: $\forall a \in S \sqcup S^{-1}$, $r(a) = q(a)/(1+q(a))$.

For a general 0-automatic pair, there is no simple formula giving r as a function of q. We come back to this point at the end of §5.2.

Specializing Proposition 5.3 to the free product of two groups, we obtain an unexpected identity:

$$q(\Sigma_1)q(\Sigma_2) = 1\,.$$

In words, the average number of visited elements in Σ_1 is the inverse of the average number of visited elements in Σ_2. It is the identity used to prove the last part of Proposition 5.2.

The knowledge of $q(.)$ enables to determine explicitly the minimal harmonic functions. A *positive harmonic* function is a function $f : G \to \mathbb{R}_+$ such that $\forall u \in G$, $\sum_{a\in\Sigma} f(u*a)\mu(a) = f(u)$. A positive harmonic function f is *minimal* if $f(1_G) = 1$ and if for any positive harmonic function g such that $f \geq g$, there exists $c \in \mathbb{R}+$ such that $f = cg$.

For $g \in G$, set $\Gamma(g) = \sum_{i=0}^{\infty} \mu^{*i}(g)$ (the Green function), and define the map $K_g : G \to \mathbb{R}+$ by

$$K_g(x) = \frac{\Gamma(x^{-1}*g)}{\Gamma(g)} = \frac{q(x^{-1}*g)}{q(g)}\,.$$

The right-hand equality is obtained by observing that $\Gamma(v) = q(v)\Gamma(1_G)$ for all v.

For $\xi \in L^\infty$, define $K_\xi : G \to \mathbb{R}+$ by $K_\xi = \lim_n K_{\pi(\xi[n])}$, where $\xi[n]$ is the length n prefix of ξ. Set $\xi = \xi_0\xi_1\cdots$. For $x \in G$ with $\varphi(x) = x_0\cdots x_{n-1} \in L(G,\Sigma)$, set $k = |\varphi(x)\wedge\xi|_\Sigma$, the length of the longest joint prefix of $\varphi(x)$ and ξ. We have

$$K_\xi(x) = \begin{cases} [q(x_k^{-1})\cdots q(x_{n-1}^{-1})]/[q(\xi_0)\cdots q(\xi_{k-1})] & \text{if } x_k^{-1} \in \text{Prev}(\xi_k) \\ [q(x_k^{-1}*\xi_k)q(x_{k+1}^{-1})\cdots q(x_{n-1}^{-1})]/[q(\xi_0)\cdots q(\xi_k)] & \text{otherwise} \end{cases}\,.$$

Proposition 5.4. *The minimal positive harmonic functions are the functions K_ξ, $\xi \in L^\infty$.*

The set $\{K_\xi,\ \xi \in L^\infty\}$ forms the Martin boundary of the random walk.

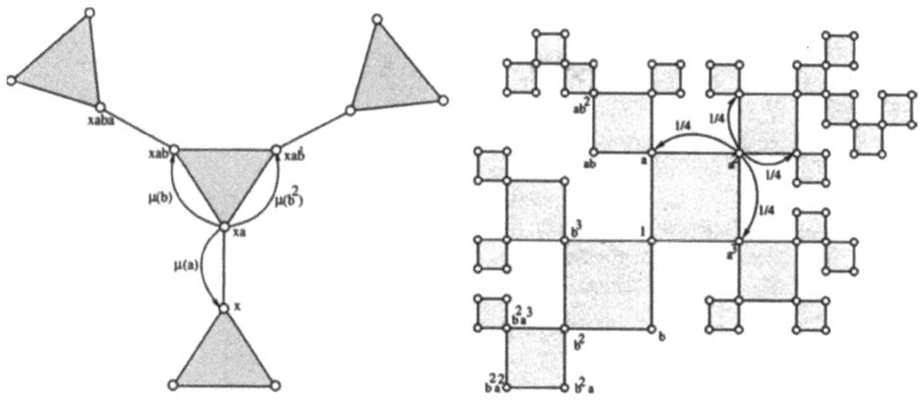

Figure 6.2 A nearest neighbor random walk on $\mathbb{Z}/2\mathbb{Z}\star\mathbb{Z}/3\mathbb{Z}$ (left), and the simple random walk on $\mathbb{Z}/4\mathbb{Z}\star\mathbb{Z}/4\mathbb{Z}$ (right).

Comparison with the literature. The importance of the Equations (16) in q is well-known. In the seminal paper of Dynkin & Malyutov [6], these equations are explicitely solved in the free group case, and the harmonic functions are then derived as above. In [21], the authors prove that μ^∞ is Markovian for the free group as follows: they use the expression for q obtained in [6], they define r as in the second part of Proposition 4.1, and then prove that the measure defined by $\nu^\infty(u_1\cdots u_k\Sigma^{\mathbb{N}}) = q(u_1)\cdots q(u_{k-1})r(u_k)$ is the harmonic measure. For trees of finite cone types [19], the proof that the harmonic measure is Markovian is also centered around the analog of the Equations (16). The series version of (16) is the main ingredient in the Sawyer & Steger approach [21] for computing γ, see §6.1. See also [15, 19]. At last, the series version of (16) can be used for *finite range* random walks on free groups to get qualitative results (central limit theorem, asymptotic type of $P\{X_n = 1_G\}$), see Lalley [14].

Here the proof that μ^∞ is Markovian is different and based on the Traffic Equations (14) instead of the Equations (16). This is a more direct path and the only way to proceed in the general case since we cannot retrieve a solution to the Traffic Equations from a solution to (16).

6. Explicit computations

In Theorem 5.1, the harmonic measure is completely determined via the vector r which is itself the solution of an explicit finite set of polynomial equations. In small or simple examples, it is possible to go further, that is, to solve these equations to get closed form formulas for the harmonic measure, the drift, the entropy, or the harmonic functions. It is the program that we now carry out for several specific and interesting cases. In §6.2-§6.6, we compute explicitly r, and we illustrate by providing closed form formulas for the drift. The computations have been carried out using Maple and Mathematica.

We first discuss alternative existing methods for computing the drift. (They do not work for computing the entropy for instance.)

6.1. Comparison with other methods for computing the drift

Let $G = G_1 \star G_2$ be a free product of two finite groups. Set $\Sigma_i = G_i \backslash \{1_{G_i}\}$ and $\Sigma = \Sigma_1 \sqcup \Sigma_2$. Let μ be a probability measure on Σ such that: $\forall i, \forall x \in \Sigma_i,\ \mu(x) = \mu(\Sigma_i)/\#\Sigma_i$. In words, μ is uniform on each of the two groups. Consider the random walk (G, μ). Here, computing the drift becomes elementary and does not require knowing that the harmonic measure is Markovian.

Set $p = \mu(\Sigma_1), k_1 = \#\Sigma_1$, and $k_2 = \#\Sigma_2$. Denote by $i \in \{1,2\}$, the set of elements of G whose normal form representative ends with a letter in Σ_i. When we are far from the unit element 1_G, the random walk on G induces a Markov chain on $\{1,2\}$ with transition matrix:

$$P = \begin{bmatrix} p(k_1-1)/k_1 & p/k_1 + 1 - p \\ (1-p)/k_2 + p & (1-p)(k_2-1)/k_2 \end{bmatrix}.$$

Let π be the stationary distribution, that is $\pi P = \pi, \pi(1) + \pi(2) = 1$. By the Ergodic Theorem for Markov Chains, we have $\lim_n [\, P\{X_n \in 1\}\, ,\, P\{X_n \in 2\}\,] = \pi$. The value of the drift follows readily:

$$\gamma = \lim_n \frac{1}{n} \sum_{i=0}^{n-1} E[X_{i+1} - X_i] + \frac{X_0}{n} = E_\pi[X_1 - X_0] = \frac{2p(1-p)(k_1k_2 - 1)}{(1-p)k_1 + pk_2 + k_1k_2}.$$

Now assume that $G = G_1 \star \cdots \star G_k$, where the G_i are finite groups, and assume that $\forall i, \forall x \in \Sigma_i = G_i \backslash \{1_{G_i}\},\ \mu(x) = \mu(\Sigma_i)/\#\Sigma_i$. Then each of the finite groups can be collapsed into a single node, and the random walk (G, μ) projects into a nearest neighbor randow walk on a tree with k cone types. Therefore, the formulas for the drift given in [19] apply.

With the exception of (22), none of the formulas obtained in §6.2-§6.6 corresponds to the above two situations.

Now let us discuss the Sawyer-Steger method [21]. It was developped for the free group but adapts to the present situation. Let (G, Σ) be a 0-automatic pair. For $g \in G$, define the r.v. $\tau(g) = \min\{n \mid X_n = g\}$ (with $\tau(g) = \infty$ if g is not reached). Define the *first-passage generating series* $S \in \mathbb{R}[[y, z]]$ by:

$$S(y,z) = \sum_{k \in \mathbb{N}} y^k \sum_{|g|_\Sigma = k} \sum_{n \in \mathbb{N}} P\{\tau(g) = n\} z^n. \tag{17}$$

Let S_y and S_z denote the partial derivatives of S with respect to y and z. Adapting the results in [21, Theorem 2.2 and Section 6] (see also [19, Section 6]), one obtains the following formula for the drift:

$$\gamma = S_y(1,1)/S_z(1,1). \tag{18}$$

Now assume, for simplicity, that $G = G_1 \star \cdots \star G_k$ is a free product of finite groups. Set $\Sigma_i = G_i \backslash \{1_{G_i}\}$ and $\Sigma = \sqcup_i \Sigma_i$. For $u \in G$, define the series $q(u,z) = \sum_{n \in \mathbb{N}} P\{\tau(u) = n\} z^n$. Observe that $q(u,1) = q(u)$, the probability of ever reaching u, defined in §5.2. In particular, if one encapsulates the Equations (16) as $q(a) = \Phi_a(q)$, then $q(a,z) = z\Phi_a(q(z))$. Using this last set of Equations, the corresponding set of Equations for the derivatives $dq(a,z)/dz$, and playing around with the Equations (17) and (18), we get:

$$\gamma = \frac{\sum_{i=1}^k q_i/(1+q_i)^2}{\sum_{i=1}^k q_i'/(1+q_i)^2}, \qquad q_i = \sum_{u \in \Sigma_i} q(u,1), \quad q_i' = \sum_{u \in \Sigma_i} \Big[\frac{dq(u,z)}{dz}\Big]_{|z=1}. \tag{19}$$

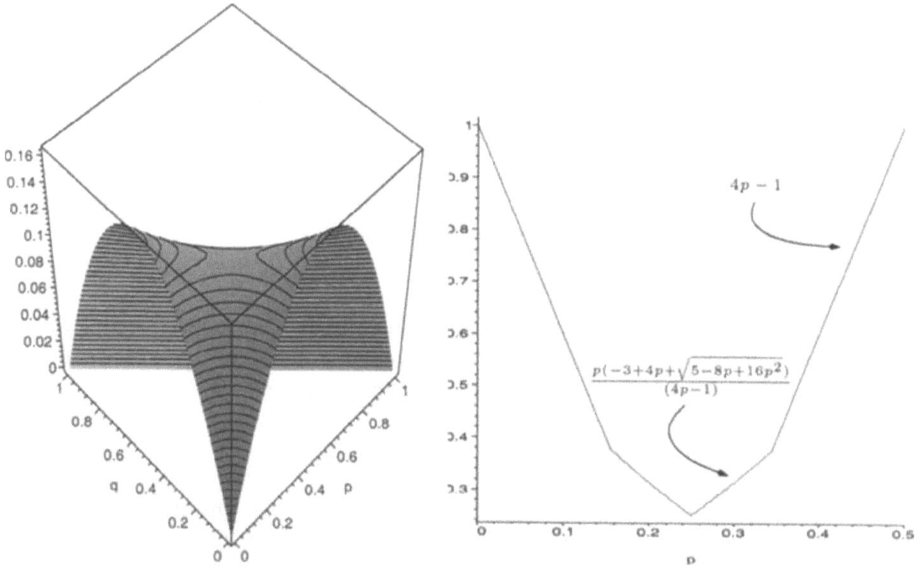

Figure 6.3 The drift of $(\mathbb{Z}/2\mathbb{Z} \star \mathbb{Z}/3\mathbb{Z}, \mu)$ as a function of $p = \mu(b)$ and $q = \mu(b^2)$ (left), and the drift of (B_3, μ) as a function of $p = \mu(a) = \mu(b) = 1/2 - \mu(a^{-1}) = 1/2 - \mu(b^{-1})$ (right).

This formula is more complicated than the one obtained in (15). The reason for this is easy to understand. There is much more information in the series S than what is relevant for computing γ. In particular, all the 'transient' behavior of the walk is encoded into it. Our approach centers around the knowledge that μ^∞ is Markovian. It allows to compute γ via a simple 'equilibrium' argument: $\mu * \mu^\infty = \mu^\infty$. This is a more direct path. Consequently, it gives more chances to solve the equations to get a closed form formula. As an exercise, we tried to retrieve the results for γ in §6.2-§6.6 using (19). We succeeded in two cases: formulas (21) and (23). On the other hand, the results in §6.4-§6.6 seem totally out of reach.

6.2. Random walks on $\mathbb{Z}/2\mathbb{Z} \star \mathbb{Z}/3\mathbb{Z}$

The group $\mathbb{Z}/2\mathbb{Z} \star \mathbb{Z}/3\mathbb{Z}$ is isomorphic to the modular group $\mathrm{PSL}(2, \mathbb{Z})$ (i.e. the group of 2x2 matrices with integer entries and determinant 1, quotiented by $\pm\mathrm{Id}$). Let a and b be the respective generators of $\mathbb{Z}/2\mathbb{Z}$ and $\mathbb{Z}/3\mathbb{Z}$. A possible representation of the group is

$$a = \begin{bmatrix} 0 & -1 \\ 1 & 0 \end{bmatrix}, \qquad b = \begin{bmatrix} 1 & -1 \\ 1 & 0 \end{bmatrix}. \tag{20}$$

Quoting [4, Chapter II.B]: "The modular group is one of the most important groups in mathematics".

Consider a general nearest neighbor random walk $(\mathbb{Z}/2\mathbb{Z} \star \mathbb{Z}/3\mathbb{Z}, \mu)$. Set $\mu(a) = r, \mu(b) = p, \mu(b^2) = q$. The Traffic Equations can be solved explicitly.

The resulting value of the drift is:

$$\gamma = \frac{2r\left(pq - p - q + \sqrt{(p^2+q^2)(3+(r+p)^2+(r+q)^2) + 2pq(2r+1)}\right)}{(r+p)^2+(r+q)^2 - pq + 2} . \quad (21)$$

For instance, the drift is maximized for $r = z_0, p = 1 - z_0, q = 0$ (or $r = z_0, p = 0, q = 1 - z_0$)), where z_0 is the root of $[z^6 + 12z^4 - 4z^3 + 47z^2 - 48z + 12]$ whose numerical value is $0.490275\cdots$. The corresponding numerical value of the drift is $\gamma_{\max} = 0.163379\cdots$. This was not a priori obvious!

6.3. Random walks on $\mathbb{Z}/3\mathbb{Z} \star \mathbb{Z}/3\mathbb{Z}$

Consider the free product $\mathbb{Z}/3\mathbb{Z}\star\mathbb{Z}/3\mathbb{Z}$ with a and b being the respective generators of the two cyclic groups. Consider the probability μ such that $\mu(a) = \mu(a^2) = p, \mu(b) = \mu(b^2) = q = 1/2 - p$. Solving the Traffic Equations yields:

$$r(a) = r(a^2) = \frac{1+2p}{6}, \quad r(b) = r(b^2) = \frac{1-p}{3} .$$

We have $\mu^\infty(u_1u_2\cdots u_k\Sigma^{\mathbb{N}}) = r(u_1)(1/2)^{k-1}$, i.e. μ^∞ is the uniform measure except for the distribution of the initial element of the normal form. The drift is

$$\gamma = 4pq = 2p(1-2p) . \quad (22)$$

Assume now that $\mu(a) = \mu(b) = p, \mu(a^2) = \mu(b^2) = q = 1/2 - p$. We obtain:

$$r(a) = r(b) = \frac{4p - 3 + \sqrt{16p^2 - 8p + 5}}{4(4p-1)},$$
$$r(a^2) = r(b^2) = \frac{4p + 1 - \sqrt{16p^2 - 8p + 5}}{4(4p-1)}.$$

The harmonic measure is the Markovian multiplicative measure associated with r. It is stationary, see Proposition 5.2. The drift is

$$\gamma = -\frac{1}{4} + \frac{1}{4}\sqrt{16p^2 - 8p + 5} . \quad (23)$$

At last, consider the case $\mu(a) = p, \mu(a^2) = q$, and $\mu(b) = \mu(b^2) = (1-p-q)/2$. Solving explicitly the Traffic Equations is feasible but provides formula which are too lengthy to be reproduced here. However, for the drift, several simplifications occur, and we obtain the following formula:

$$\gamma = 2(1-p-q)\sqrt{(p^2+q^2+pq)/(p^2+q^2-2pq+3)} .$$

For the general nearest neighbor random walk on $\mathbb{Z}/3\mathbb{Z}\star\mathbb{Z}/3\mathbb{Z}$, we did not succeed in solving completely the Traffic Equations.

6.4. The simple random walk on $\mathbb{Z}/2\mathbb{Z}\star\mathbb{Z}/k\mathbb{Z}$

The groups $\mathbb{Z}/2\mathbb{Z}\star\mathbb{Z}/k\mathbb{Z}$ are known as the Hecke groups. We consider the random walk $(\mathbb{Z}/2\mathbb{Z}\star\mathbb{Z}/k\mathbb{Z}, \mu)$ with $\mu(a) = \mu(b) = \mu(b^{-1}) = 1/3$.

Consider the applications $G_n : [0,1] \to \mathbb{R}, n \in \mathbb{N}$, defined by

$$G_0(x) = \frac{1}{4} + \frac{x}{2}, \quad G_1(x) = x, \quad \forall n \geq 2, \ G_n(x) = \frac{8(1-x)}{3-2x}G_{n-1}(x) - G_{n-2}(x) .$$

For instance, $G_4(x) = (1552x^4 - 4416x^3 + 4296x^2 - 1600x + 165)/(2x-3)^3$.

Theorem 6.1. *For $k \geq 3$, the equation $G_{k-1}(x) = x$ has a unique solution in $(0, 1/2)$ that we denote by y_k. The harmonic measure of (G, μ) is the Markovian multiplicative measure associated with r: $r(a) = G_0(y_k)$, $\forall i \in \{1, \ldots, k-1\}$, $r(b^i) = G_i(y_k)$. The drift is $\gamma_k = (1 - 2y_k)/3$. It is strictly increasing in k and $\lim_k \gamma_k = 2/9$.*

Here is a table of the first values of γ, given either in closed form or numerically when no closed form could be found. Set $\mathbb{Z}_k = \mathbb{Z}/k\mathbb{Z}$.

	$\mathbb{Z}_2 \star \mathbb{Z}_3$	$\mathbb{Z}_2 \star \mathbb{Z}_4$	$\mathbb{Z}_2 \star \mathbb{Z}_5$	$\mathbb{Z}_2 \star \mathbb{Z}_6$	$\mathbb{Z}_2 \star \mathbb{Z}_7$	$\mathbb{Z}_2 \star \mathbb{Z}_8$
γ	$2/15$	$(\sqrt{7}-1)/9$	$(2\sqrt{61}-4)/57$	0.21341...	0.21792...	0.22010...

6.5. The simple random walk on $\mathbb{Z}/k\mathbb{Z} \star \mathbb{Z}/k\mathbb{Z}$

Consider the free product $G_1 \star G_2 = \mathbb{Z}/k\mathbb{Z} \star \mathbb{Z}/k\mathbb{Z}$. Set $\Sigma_1 = G_1 \backslash \{1_1\} = \{a, \ldots, a^{k-1}\}$, $\Sigma_2 = G_2 \backslash \{1_2\} = \{b, \ldots, b^{k-1}\}$, and $\Sigma = \Sigma_1 \sqcup \Sigma_2$. Consider the simple random walk $(\mathbb{Z}/k\mathbb{Z} \star \mathbb{Z}/k\mathbb{Z}, \mu)$ with $\mu(a) = \mu(b) = \mu(a^{-1}) = \mu(b^{-1}) = 1/4$. See Figure 6.

Consider the applications $F_n : [0, 1] \to \mathbb{R}, n \in \mathbb{N}$, defined by

$$F_0(x) = 1, \quad F_1(x) = x, \quad \forall n \geq 2, \; F_n(x) = 2(2 - x)F_{n-1}(x) - F_{n-2}(x) \,. \quad (24)$$

Theorem 6.2. *For $k \geq 3$, the equation $F_k(x) = 1$ has a unique solution in $(0, 1)$ that we denote by x_k. The harmonic measure of (G, μ) is the Markovian multiplicative measure associated with r: $\forall i \in \{1, \ldots, k-1\}$, $r(a^i) = r(b^i) = F_i(x_k)/2$. The drift is $\gamma_k = (1 - x_k)/2$. It is strictly increasing in k and $\lim_k \gamma_k = 1/3$.*

	$\mathbb{Z}_3 \star \mathbb{Z}_3$	$\mathbb{Z}_4 \star \mathbb{Z}_4$	$\mathbb{Z}_5 \star \mathbb{Z}_5$	$\mathbb{Z}_6 \star \mathbb{Z}_6$	$\mathbb{Z}_7 \star \mathbb{Z}_7$	$\mathbb{Z}_8 \star \mathbb{Z}_8$
γ	$1/4$	$(\sqrt{5}-1)/4$	$(\sqrt{13}-1)/8$	0.33085...	0.33251...	0.33306...

6.6. Random walks on Artin Groups with 2 generators

The *Artin group with 2 generators* A_k $(k \geq 3)$ is the group with finite presentation

$$A_k = \langle \, a, b \mid \mathrm{prod}(a, b; k) = \mathrm{prod}(b, a; k) \, \rangle \,, \quad (25)$$

where $\mathrm{prod}(a, b; k) = ababa\ldots$, with k terms in the product on the right-hand side. Observe that $A_3 = B_3$, the well-known braid group over three stands.

Set $\Sigma = \{a, a^{-1}, b, b^{-1}\}$. The pair (A_k, Σ) is not 0-automatic. In particular there is no natural notion of geodesic normal form over the generators Σ. However, we are able to go back to the 0-automatic framework by a series of intermediate steps, that we now list.

Consider the random walk (A_k, μ) where μ is a probability on Σ. Write the group elements in Garside normal form [9]. This requires the switch to a new set of generators. In particular, the Garside normal form is not geodesic for the generators Σ. Let Z be the center of A_k. Consider the induced random walk on A_k/Z, the group quotiented by its center. Assume that $\mu(a) = \mu(b), \mu(a^{-1}) = \mu(b^{-1})$. Show that the induced random walk behaves like a nearest neighbor random walk on $\mathbb{Z}/k\mathbb{Z} \star \mathbb{Z}/k\mathbb{Z}$. Use the results of §6.3 and §6.5. Go back from A_k/Z to A_k. Go back to the natural generators.

In the end, what is lost is an explicit description of the harmonic measure of (A_k, μ), but what remains is an explicit formula for the drift.

Theorem 6.3. *Consider the random walk (B_3, μ) where μ is a probability measure on Σ such that $\mu(a) = \mu(b) = p, \mu(a^{-1}) = \mu(b^{-1}) = 1/2 - p$. The drift with respect to the natural generators $\Sigma = \{a, b, a^{-1}, b^{-1}\}$ is*

$$\gamma(p) = \max\left[1-4p, g(1/2-p), g(p), -1+4p\right], \; g(p) = \frac{p(-3+4p+\sqrt{5-8p+16p^2})}{-1+4p}.$$

Consider the simple random walk (A_k, μ) with $\mu(a) = \mu(a^{-1}) = \mu(b) = \mu(b^{-1}) = 1/4$. Let γ_k be the drift with respect to Σ. We have

$$\gamma_k = \begin{cases} (1-x_k)\left[\, \sum_{i=1}^{j-1} iF_i(x_k) + (j/2)F_j(x_k) \,\right] & \text{if } k = 2j \\ (1-x_k)\left[\, \sum_{i=1}^{j} iF_i(x_k) \,\right] & \text{if } k = 2j+1 \end{cases}, \tag{26}$$

where the F_i were defined in (24). The drift γ_k is strictly increasing in k, and $\lim_k \gamma_k = 1/2$.

Set $u = 7/24 + 11/[24(71+6\sqrt{177})^{1/3}] + (71+6\sqrt{177})^{1/3}/24 = 0.155979\cdots$. The function $\gamma(p)$, represented on Figure 6.2, has several unexpected characteristics: it is linear on the intervals $[0, u]$ and $[1/2-u, 1/2]$ but not on the interval $[u, 1/2-u]$; and it has 3 points of non-differentiability: $u, 1/4, 1/2-u$. Hence, the behavior of the braid model has some kind of phase transitions whose physical interpretation is intriguing.

	A_3	A_4	A_5	A_6	A_7	A_8
γ	$1/4$	$(\sqrt{5}-1)^2/4$	$(\sqrt{13}-1)^2/16$	0.462598...	0.475221...	0.487636...

7. From groups to monoids

A pair (M, Σ) formed by a monoid and a finite set of generators is *0-automatic* if the set of locally reduced words $L(M, \Sigma)$ is a cross-section *and* the analog of (7) holds. A plain monoid together with the natural generators, forms a 0-automatic pair. Define the Traffic Equations as in (14) with the convention that $\mu(u^{-1}) = 0$ if u has no inverse. Then Theorem 5.1 and Proposition 5.2 hold for monoids. (But not Propositions 5.3 and 5.4.)

To illustrate, consider a free product of the form $M = M_a \star M_b \star M_c \star M_d$, where M_i is equal either to $\mathbb{Z}_2$ or to $\mathbb{B}$. Here $\mathbb{B} = \langle\, a \mid a^2 = a \,\rangle$ is the Boolean monoid and $\mathbb{Z}_2 = \mathbb{Z}/2\mathbb{Z}$. Let i be the generator of M_i. We consider the simple random walk on M, that is the random walk defined by $\mu : \mu(i) = 1/4, \forall i$. For $\mathbb{Z}_2 \star \mathbb{Z}_2 \star \mathbb{Z}_2 \star \mathbb{Z}_2$ and $\mathbb{B} \star \mathbb{B} \star \mathbb{B} \star \mathbb{B}$, it is elementary that γ equals respectively 1/2 and 3/4. The other values of the drift are given below.

	$\mathbb{Z}_2 \star \mathbb{Z}_2 \star \mathbb{Z}_2 \star \mathbb{B}$	$\mathbb{Z}_2 \star \mathbb{Z}_2 \star \mathbb{B} \star \mathbb{B}$	$\mathbb{Z}_2 \star \mathbb{B} \star \mathbb{B} \star \mathbb{B}$
γ	$(12+3\sqrt{2})/28$	$(6+\sqrt{3})/12$	$7/10$

The value 7/10 can be obtained by elementary arguments without having to solve the Traffic Equations. But not the other two values.

Acknowledgement. The authors would like to thank both W. Woess and an anonymous referee for pointing out the method of [21] for computing the drift, see §6.1; as well as several other references.

References

[1] A. Avez. Entropie des groupes de type fini. *C. R. Acad. Sci. Paris Sér. A-B*, 275:1363–1366, 1972.

[2] D. Cartwright. Some examples of random walks on free products of discrete groups. *Ann. Mat. Pura Appl. (4)*, 151:1–15, 1988.

[3] D. Cartwright and P. Soardi. Random walks on free products, quotients and amalgams. *Nagoya Math. J.*, 102:163–180, 1986.

[4] P. de la Harpe. *Topics in geometric group theory*. Chicago Lectures in Mathematics. University of Chicago Press, 2000.

[5] Y. Derriennic. Quelques applications du théorème ergodique sous-additif. *Astérisque*, 74:183–201, 1980.

[6] E. Dynkin and M. Malyutov. Random walk on groups with a finite number of generators. *Sov. Math. Dokl.*, 2:399–402, 1961.

[7] D. Epstein, J. Cannon, D. Holt, S. Levy, M. Paterson, and W. Thurston. *Word processing in groups*. Jones and Bartlett, Boston, 1992.

[8] H. Furstenberg. Noncommuting random products. *Trans. Amer. Math. Soc.*, 108: 377–428, 1963.

[9] F. Garside. The braid groups and other groups. *Quart. J. Math. Oxford*, 20:235–254, 1969.

[10] M. Gromov. Hyperbolic groups. In *Essays in group theory*, volume 8 of *Math. Sci. Res. Inst. Publ.*, pages 75–263. Springer, 1987.

[11] Y. Guivarc'h. Sur la loi des grands nombres et le rayon spectral d'une marche aléatoire. *Astérisque*, 74:47–98, 1980.

[12] R. Haring-Smith. Groups and simple languages. *Trans. Amer. Math. Soc.*, 279(1): 337–356, 1983.

[13] V. Kaimanovich. The Poisson formula for groups with hyperbolic properties. *Ann. of Math. (2)*, 152(3):659–692, 2000.

[14] S. Lalley. Finite range random walk on free groups and homogeneous trees. *Ann. Probab.*, 21(4):2087–2130, 1993.

[15] S. Lalley. Random walks on regular languages and algebraic systems of generating functions. In *Algebraic methods in statistics and probability*, volume 287 of *Contemp. Math.*, pages 201–230. Amer. Math. Soc., 2001.

[16] F. Ledrappier. Some asymptotic properties of random walks on free groups. In J. Taylor, editor, *Topics in probability and Lie groups: boundary theory*, number 28 in CRM Proc. Lect. Notes, pages 117–152. American Mathematical Society, 2001.

[17] J. Mairesse. Random walks on groups and monoids with a Markovian harmonic measure. In preparation, 2004.

[18] J. Mairesse and F. Mathéus. Random walks on free products of cyclic groups and on Artin groups with two generators. In preparation, 2004.

[19] T. Nagnibeda and W. Woess. Random walks on trees with finitely many cone types. *J. Theoret. Probab.*, 15(2):383–422, 2002.

[20] S. Nechaev and R. Voituriez. Random walks on three-strand braids and on related hyperbolic groups. *J. Phys. A*, 36(1):43–66, 2003.

[21] S. Sawyer and T. Steger. The rate of escape for anisotropic random walks in a tree. *Probab. Theory Related Fields*, 76(2):207–230, 1987.

[22] J. Stallings. A remark about the description of free products of groups. *Proc. Cambridge Philos. Soc.*, 62:129–134, 1966.

[23] C. Takacs. Random walk on periodic trees. *Electron. J. Probab.*, 2:no. 1, 1–16, 1997.

[24] A. Vershik. Dynamic theory of growth in groups: Entropy, boundaries, examples. *Russ. Math. Surv.*, 55(4):667–733, 2000. Translation from Usp. Mat. Nauk 55(4):59-128, 2000.

[25] W. Woess. Nearest neighbour random walks on free products of discrete groups. *Boll. Un. Mat. Ital. B (6)*, 5(3):961–982, 1986.

[26] W. Woess. *Random walks on infinite graphs and groups*. Number 138 in Cambridge Tracts in Mathematics. Cambridge University Press, 2000.

Jean Mairesse
LIAFA, CNRS-Université Paris 7, case 7014, 2, place Jussieu, 75251 Paris Cedex 05, France. `Jean.Mairesse@liafa.jussieu.fr`

Frédéric Mathéus
LMAM, Université de Bretagne-Sud, Campus de Tohannic, BP 573, 56017 Vannes, France. `Frederic.Matheus@univ-ubs.fr`

Trends in Mathematics, © 2004 Birkhäuser Verlag Basel/Switzerland

Nested Regenerative Sets and Their Associated Fragmentation Process

Philippe Marchal

ABSTRACT: *We give a fractal construction of nested, stable regenerative sets and study the associated inhomogeneous fragmentation process.*

1. Introduction

The purpose of this paper is to introduce an explicit construction of nested, stable regenerative sets and to study the related fragmentation process. The complete construction of regenerative sets on the whole half-line is explained in the next section. However, the intersections of these sets with the interval $[0, 1)$ admit the following description as random fractal sets. Let us introduce the

SPLITTING PROCEDURE. Given $\alpha \in [0, 1]$, this procedure takes an interval $[a, b)$, cuts it into two intervals $[a, U), [U, b)$, where U is uniformly distributed on $[a, b)$, and returns $[a, U), [U, b)$ with probability α and $[a, U)$ with probability $1-\alpha$, independently of U.

Fix $\alpha \in [0, 1]$ and use the splitting procedure to construct a

RANDOM FRACTAL SET. We construct recursively a sequence $(S_n(\alpha), n \geq 0)$ where for each n, $S_n(\alpha)$ is a collection of intervals of $[0, 1)$. Put $S_0(\alpha) = \{[0, 1)\}$. Then to obtain $S_{n+1}(\alpha)$ from $S_n(\alpha)$, apply to each interval of $S_n(\alpha)$ the splitting procedure with parameter α, independently.

Theorem 1.1. *Set*

$$S_\infty(\alpha) = \overline{\cap_{n\geq 0} S_n(\alpha)}$$

Let T be a stable subordinator with index α and let $R = \{T_t, t \geq 0\}$ be its range. Then

$$R \cap [0, 1) \stackrel{d}{=} S_\infty(\alpha)$$

A corollary is the following representation of the generalized arcsine law:

Corollary 1.2. *Let $I_0 = [0, 1]$. Define by induction a sequence of intervals $I_n = [a_n, b_n]$ as follows.*
(i) Let V_n be a uniform random variables on I_n, independent of the past.
(ii) Set $I_{n+1} = [a_n, V_n]$ (resp.$I_{n+1} = [V_n, b_n]$) with probability $1 - \alpha$ (resp. α), independently of the past.

Then the point $I_\infty = \cap I_n$ is distributed according to the generalized arcsine law with parameter α, that is, the law on $[0, 1]$ with density

$$\frac{s^{\alpha-1}(1-s)^{-\alpha}}{\Gamma(\alpha)\Gamma(1-\alpha)} \tag{1}$$

Our main interest is to investigate the following

COUPLED CONSTRUCTION. We can perform the splitting procedure with parameter $\alpha \in [0,1]$ by taking an independent, uniform random variable V on $[0,1]$ and decide to return one or two sets according as $V > \alpha$ or $V \leq \alpha$. Proceeding this way, we see that we obtain a coupled construction of $(S_n(\alpha), n \geq 0)$, and therefore of $S_\infty(\alpha)$, simultaneously for all $\alpha \in [0,1]$, by starting at $\alpha = 1$.

It is clear that $(S_\infty(\alpha), \alpha \in [0,1])$ is a nested family of regenerative sets, i.e. if $\alpha < \alpha'$, $S_\infty(\alpha) \subset S_\infty(\alpha')$. It can be shown [7] that this family has the subordination property, that is, if $\alpha < \alpha'$, $S_\infty(\alpha)$ can be obtained from $S_\infty(\alpha')$ by subordination. See [1] for details on subordination.

Moreover, for each α, the regenerative set $S_\infty(\alpha)$ cuts the interval $[0,1]$ into disjoint intervals. Consider the coupled construction with α going from 1 to 0. We obtain this way a process where the intervals delimited by $S_\infty(\alpha)$ merge together. The subordination property entails [4] that this process is a Markov process known as the Bolthausen-Sznitman coalescent. This process was introduced in [5] for the study of spin glasses. See also [3, 6].

We shall focus here on the time reversal, that is, we let α go from 0 to 1, in which case we obtain a process where the intervals are broken into smaller ones. It turns out that this process has the law of an inhomogeneous fragmentation and our construction enables us to study its dislocation measure.

The whole construction of regenerative sets is explained in the next section. We give some results on random partitions in Section 3 and study more specifically the fragmentation process in Section 4.

2. The construction of stable regenerative sets

We give here the construction of stable, regenerative sets on the positive half-line. It is easy to check that if we consider the intersection of these sets with the interval $[0,1)$, we recover the fractal construction described in the introduction. In particular, Theorem 1.1 is a direct corollary of Theorem 2.1.

2.1. Regenerative sets

For a detailed account on regenerative sets, see for instance [1]. A random subset $\mathcal{R}$ of $\mathbb{R}_+$ is regenerative if and only if $0 \in \mathcal{R}$ and for every $t > 0$, conditionally on $t \in \mathcal{R}$, the subset $\mathcal{R} \cap [t, \infty)$ has the same law as $\mathcal{R} + t$ and is independent of $\mathcal{R} \cap [0,t]$. A classical result tells us that every regenerative set can be viewed as the range of a subordinator.

Moreover, if $\mathcal{R}$ is self-similar, i.e. if $\mathcal{R}$ has the same law as $c\mathcal{R}$ for every $c > 0$, then $\mathcal{R}$ is the range of a stable subordinator. In that case, to identify the index of the subordinator, one can use for instance the generalized arcsine law. For every time t set $g_t = \sup(\mathcal{R} \cap [0,t])$. Then g_1 (and more generally, by self-similarity, g_t/t) is distributed according to the generalized arcsine law with parameter α, i.e. the law whose density is given by (1).

2.2. The construction

Let $\mathcal{N}$ be a Poisson point process on $\mathbb{R}_+ \times \mathbb{R}_+$ with intensity $d\mu = dt \otimes x^{-2}dx$. Call C the (random) support of the measure $\mathcal{N}$. To each point $M \in C$ associate an independent, uniform random variable $U(M)$ on $[0,1]$. Note that almost surely, the coordinates of the points of C are all different, there are infinitely many points in

Figure 2.1 Percolating points of the Poisson point process.

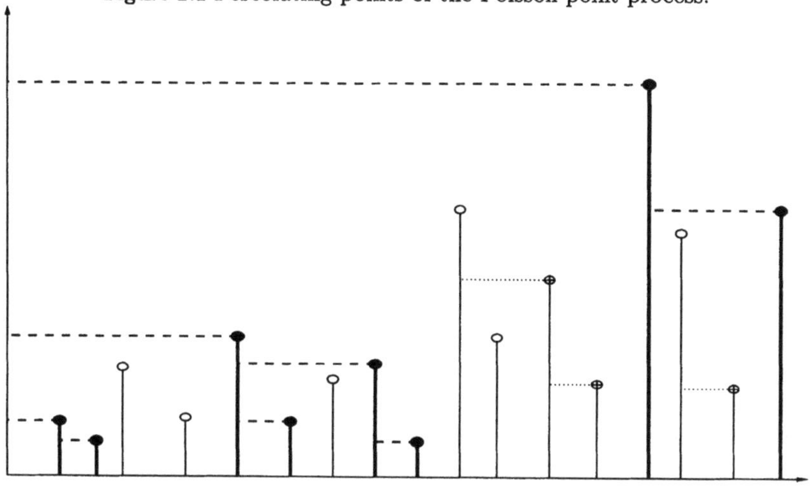

each strip $[s,t]\times\mathbb{R}_+$ ($s<t$) and only finitely many points in each strip $[s,t]\times[y,\infty)$ ($y>0$). We shall implicitly assume all these properties in the sequel (that is, we shall often omit to write "almost surely").

For $s<t\in\mathbb{R}_+$ let $M(s,t)$ be the highest point of C between times s and t. Formally, $M(s,t)$ is the point of coordinates (u,y) such that $M(s,t)\in C$, $s\le u\le t$ and for each point $M'=(u',y')\in C$ with $s\le u'\le t$, one has $y'\le y$.

One can then define $M_1(s,t):=(s_1,y_1)=M(s,t)$ and by induction, $M_{n+1}(s,t):=(s_{n+1},y_{n+1})=M(s_n,t)$. The sequence is finite if and only if there is a point of C with x-coordinate t. We say that $t\in\mathbb{R}_+$ percolates at parameter $\alpha\in[0,1]$ if for each n, $U(M_n(0,t))\le\alpha$. By convention, we say that 0 percolates.

See the figure. The black points are those which percolate, the white ones are those for which $U>\alpha$ and which, consequently, do not percolate, and the crossed ones are those for which $U\le\alpha$ but which do not percolate. The set of percolating points is the closure of the set of x-coordinates of the black points.

Loosely speaking, if $U(M)>\alpha$, M does not percolate. In turn, when $U(M)\le\alpha$, M "looks to the left". If the first vertical line seen by M is either the y-axis or a vertical line of a point which percolates, then M percolates. Otherwise, if this first vertical line seen by M is the line of a point which does not percolate, then M does not percolate. The procedure to determine whether M percolates is well-defined since there is only a finite number of points at a higher altitude than M and on the left of M.

2.3. Regenerativity and stability

Theorem 2.1. *The set $\mathcal{R}_\alpha$ of points percolating at parameter α has the law of the range of a stable subordinator with index α.*

Proof:

(i) First we prove that the set of points percolating at parameter α is regenerative. Indeed, assume that s percolates at parameter α. Remark that for every t, the x-coordinates of the points $M_n(t)$ converge to t almost surely as n increases. Hence for every $t>s$, there exists a minimal $k=k(s,t)$ such that for $n\ge k$, $M_n(0,t)>s$. Then for $n<k$, $M_n(0,t)=M_n(0,s)$ and for $n\ge k$, $M_n(0,t)=M_{n-k}(s,t)$.

Since s percolates, one has $U(M_n(t)) \leq \alpha$ for every $n < k$. Hence conditionally on the event that s percolates, t percolates if and only if for each $n \geq k$, $U(M_{n-k}(s,t)) \leq \alpha$. In other words, the set of times $t > s$ percolating at parameter α is the set of t such that $U(M_n(s,t)) \leq \alpha$ for every n. This is independent of $\mathcal{R}_\alpha \cap [0,s]$ because of the independence properties of the Poisson point process $\mathcal{N}$ and has the same law as $\mathcal{R}_\alpha + s$ because of the translation-invariance of $\mathcal{N}$.

(ii) The self similarity is quite obvious. Indeed, write $\mathcal{N} = \sum_k \delta_{t_k,x_k}$ and set $\widehat{\mathcal{N}} = \sum_k \delta_{ct_k,cx_k}$ for some $c > 0$. Since $\mu(B) = \mu(cB)$ for every Borel set $B \in \mathbb{R}_+ \times \mathbb{R}_+$, $\widehat{\mathcal{N}}$ has the same law as $\mathcal{N}$. Therefore $\mathcal{R}_\alpha$ has the same law as $c\mathcal{R}_\alpha$ for every $c > 0$ and consequently, $\mathcal{R}_\alpha$ is the range of a stable subordinator.

(iii) It remains to prove that the parameter of the subordinator is α. Let $g_1 = \sup\{t \leq 1,\ t \in R_\alpha\}$. Define by induction the sequence $A_n := (u_n, z_n)$ by $A_0 = M_n(0,1)$ and $A_{n+1} = M(0,u_n)$. Then by construction, $g_1 \in [u_{K+1}, u_K]$ if K is defined as the smallest integer such that $U(A_{K+1}) \leq \alpha$. Remark that u_n is the product of n independent, uniform random variables on $[0,1]$ and therefore has density $(-\log t)^{n-1}/(n-1)!$. Hence the distribution of u_K is given by

$$\begin{aligned}\mathbb{P}(u_K \in dt) &= \sum_{n\geq 1} \mathbb{P}(K = n, u_n \in dt)\\ &= \alpha(1-\alpha)\sum_{n\geq 1}(1-\alpha)^{n-1}\frac{(-\log t)^{n-1}}{(n-1)!}dt\\ &= \alpha(1-\alpha)\exp[(\alpha-1)\log t]dt = \alpha(1-\alpha)t^{\alpha-1}dt\end{aligned}$$

and consequently $P(u_K \leq \varepsilon) \sim c\varepsilon^\alpha$ as ε goes to 0. Similar calculations show that $P(u_{K+1} \leq \varepsilon) \sim c'\varepsilon^\alpha$ as ε goes to 0. Hence $P(g_1 \leq \varepsilon) \asymp \varepsilon^\alpha$. Comparing this with the generalized arcsine law, we conclude that the index of the subordinator is α. □

3. Partitions

3.1. Partitions of integers

For more details on random partitions, as well as numerous references, we refer to Pitman's *Cours de St-Flour* [9]. We consider exchangeable random partitions of the set $\mathbb{N}$, i.e. random partitions whose law is invariant under any permutation of a finite number of integers.

An interesting construction of random partitions is the so-called "Chinese restaurant" process [9]. The model is parameterized by two reals α, θ and an initial condition $(a_1, \ldots, a_{k_0})$ where k_0 and $a_1, \ldots, a_{k_0}$ are positive integers. We must have $0 \leq \alpha \leq 1$ and $\theta \geq -k_0\alpha$.

Imagine a restaurant with infinitely many tables, labeled by the integers. There are initially a_1 customers at the first table, a_2 customers at the second table and so on. Then new customers arrive and sit at some table according to the following rule.

Suppose that at a given moment, n customers are seated and occupy k tables, the number of customers at each table being $n_1 \ldots n_k$ respectively with $n_1 + \ldots + n_k = n$. Then the $(n+1)$-th customer sits at table number i, $1 \leq i \leq k$, with probability $(n_i - \alpha)/(n+\theta)$, and at table number $k+1$ with probability $(k\alpha + \theta)/(n+\theta)$.

Associate with this process a partition of $\mathbb{N}$ by saying that i and j are in the same block of the partition if and only if the i-th and j-th customers sit at the same table. Then one can check that this random partition is exchangeable, provided that the first $a_1 + \ldots + a_{k_0}$ customers sit in an exchangeable manner. We shall call an (α, θ)-partition a partition derived from a Chinese restaurant with parameters (α, θ) and initial condition (1), that is, one has initially one customer at the first table and no other customers.

3.2. The partition associated with the construction

We want to show that the Chinese restaurant can be embedded in the construction of Section 2. From now on we restrict ourselves to the the points of C lying in the strip $[0,1] \times \mathbb{R}_+$. Reorder these points by decreasing ordinate, denoting

$$P_1 = (t_1, y_1), \ldots P_n = (t_n, y_n) \ldots$$

with $y_1 > y_2 > \ldots$. Add by convention $P_0 = (1, \infty)$.

Say that P_m is on the left of P_n if $t_m < t_n$.

SEATING RULE. We interpret P_n as the $(n+1)$-th customer.

- The first customer sits at the first table.
- The $(n+1)$-th customer P_n sits at a new table if t_n percolates.
- Otherwise, P_n sits at the same table as the left most customer on the right of P_n.

This seating rule induces a partition $\mathcal{P}_\alpha$ and we have

Theorem 3.1. *The seating rule described above yields the Chinese restaurant with parameters $(\alpha, 0)$ and initial condition $k_0 = 1$, $a_1 = 1$.*

3.3. Proof of Theorem 3.1

For every $n \geq 2$, denote by σ_n the permutation of $\{1, 2 \ldots n\}$ such that

$$t_{\sigma_n(1)} < t_{\sigma_n(2)} \cdots < t_{\sigma_n(n)}$$

Set $s_i^n = t_{\sigma_n(i)}$ for $1 \leq i \leq n$. First we have:

Lemma 3.2. *For every n and every permutation σ of $\{1, \ldots n\}$, conditionally on $\sigma = \sigma_n$, the family of reals $(s_1^n, \ldots s_n^n)$ has the law $\mathcal{L}$ of the increasing reordering of n independent, uniform random variables on $[0,1]$.*

Proof By elementary results on Poisson point processes, the t_i's are independent, uniform random variables on $[0,1]$. Hence $(s_1^n, \ldots s_n^n)$ has the desired law $\mathcal{L}$.

Let σ be a permutation of $\{1, \ldots n\}$ and $\mathcal{L}_\sigma$ be the law of the increasing reordering conditionally on $\sigma_n = \sigma$. Remark that for every permutation τ of $\{1, \ldots n\}$, $(t_1, \ldots t_n)$ has the same law as $(t_{\tau(1)}, \ldots t_{\tau(n)})$. This entails that $\mathcal{L}_\sigma = \mathcal{L}_{\sigma \circ \tau}$ for every τ. Thus $\mathcal{L}_\sigma$ does not depend on σ and equals $\mathcal{L}$. □

Let us introduce some more notation. The restaurant at parameter α induced by our construction is formally a map $f^\alpha : \mathbb{N} \to \mathbb{N}$ depending on the Poisson point process $\mathcal{N}$ and on the variables $(U(P_n), n \geq 1)$, where $f^\alpha(n) = k$ means that the n-th customer sits at the k-th table. Denote by f_n^α the restriction of f^α to $\{1, \ldots n+1\}$. Set $U = (U(P_n), n \geq 1)$ and $U_n = (U(P_m), 1 \leq m \leq n)$.

Remark that for every n, whether or not P_n percolates is determined by (σ_n, U_n), since it only depends on the points of C on the left of P_n and at a higher altitude than P_n. Hence for every n, f_n^α is determined by (σ_n, U_n). We shall write $f_n^\alpha = H_n(\sigma_n, U_n)$. As U_n and σ_n are independent, the lemma entails that conditionally on f_n^α, the variables s_m^n, $1 \le m \le n-1$, are the increasing reordering of $n-1$ independent, uniform random variables on $[0,1]$. In particular, conditionally on f_n^α,

$$\mathbb{P}(t_n \in (s_i^n, s_{i+1}^n)) = 1/n \tag{2}$$

for every $i \le n-1$.

Observe that according to the seating rule, the set of customers sitting at a given table at a given moment is a set of consecutive points if we reorder the customers by their x-coordinates. Moreover, it is easily seen by induction that if $s_m^n, s_{m+1}^n, \dots s_{m'}^n$ is the set of x-coordinates of the customers sitting at the i-th table at time n, then $s_{m'}^n$ percolates, s_i^n does not percolate for $m \le i < m'$ and s_{m-1}^n percolates.

As a consequence, the event that the $(n+1)$-th customer sits at the i-th table is the union of these two events:

- $\{s_m^n \le t_n \le s_{m'}^n\}$
- $\{s_{m-1}^n \le t_n \le s_m^n,\ U(P_n) > \alpha\}$

According to (2), the probability of the first event is $(m'-m)/n$ and the probability of the second event is $(1-\alpha)/n$. Hence conditionally on f_n^α, the probability that the $(n+1)$-th customer sits at the i-th table is

$$\frac{m'-m+1-\alpha}{n} = \frac{n_i-\alpha}{n}$$

where $n_i = m'-m+1$ is the number of customers sitting at table i at time n. This is the probability defining the Chinese restaurant. This being true for every table, we have proved Theorem 3.1. □

4. Fragmentation

In this section, we shall have to consider cadlag functions and we denote as usually

$$f(\alpha-) = \lim_{\beta\to\alpha,\beta<\alpha} f(\beta)$$

4.1. Inhomogeneous fragmentations

We refer to [2] for more details on fragmentation. Let $S^\downarrow$ be the set of decreasing sequences reals $x_1 \ge x_2 \ge \dots$, $\sum x_i \le 1$. We call a partition of mass a random element of $S^\downarrow$. If a partition of integers is exchangeable, it is a classical consequence of De Finetti's theorem that each block B of the partition almost surely has an asymptotic frequency, which is the limit of $|B \cap \{1,2,\dots n\}|/n$. This way one can derive a partition of mass by reordering the asymptotic frequencies of the blocks in decreasing order. In particular, given a Chinese restaurant with parameters (α,θ) and initial condition $(a_1,\dots,a_{k_0})$, we denote the law of the associated partition of mass by

$$M(\alpha,\theta; a_1,\dots,a_{k_0}).$$

Remark that translating Theorem 3 into a result for partitions of mass (rather that partitions of the integers), we recover the Poisson-Dirichlet representation of a partition derived from a stable subordinator, as established in [8].

A general fragmentation process $(X_t, 0 \le t \le 1)$ is a cadlag Markov process such that:

(a) For every t, X_t is an exchangeable partition of integers. For an integer i, we denote by $B_t(i)$ the block containing i in the partition X_t.

(b) If $s < t$, X_t is a finer partition than X_s.

(c) If $B_t(i) \neq B_t(j)$, then conditionally on $(B_t(i), B_t(j))$, $(B_s(i), s \geq t)$ and $(B_s(j), s \geq t)$ are independent.

The set $Dis(i)$ of dislocation times for i is the set of times t such that $B_t(i) \neq B_{t-}(i)$. If $t \in Dis(i)$, consider the set of all blocks of X_t contained in $B_{t-}(i)$ and let $a_1(t,i) > a_2(t,i) \ldots$ be the sequence of their asymptotic frequencies. Let $b_t(i)$ be the asymptotic frequency of $B_{t-}(i)$, set $b_n(t,i) = a_n(t,i)/b(t)$ and

$$b(t,i) = (b_1(t,i), b_2(t,i), \ldots)$$

We say that a fragmentation process $(X_t, 0 \leq t \leq 1)$ is an inhomogeneous fragmentation with dislocation measure $(\mu_t, 0 \leq t \leq 1)$ if for each t, μ_t is a sigma-finite measure on $S^{\downarrow}$ and if for each integer i

$$((t, b(t,i)),\ t \in Dis(i))$$

is a Poisson point process on $[0,1] \times S^{\downarrow}$ with intensity measure ν given by $\nu(dtds) = dt\mu_t(ds)$. We can now state:

Theorem 4.1. *The process $(\mathcal{P}_\alpha, 0 \leq 1 \leq \alpha)$, is an inhomogeneous fragmentation. Its dislocation measure is given by*

$$\mu_\alpha = \frac{1}{1-\alpha} \sum_{n=1}^{\infty} c_n M(\alpha, -\alpha; n, 1)$$

where c_n is given by

$$\sum_{n=1}^{\infty} c_n t^n = \frac{t}{1 + \log(1-t)}$$

Remark As shown in Section 3, the partition obtained at time α is an $(\alpha, 0)$-partition. Moreover it is known [9] that if $\alpha' > \alpha$, one can obtain an $(\alpha', 0)$-partition from an $(\alpha, 0)$-partition by splitting each block of the $(\alpha, 0)$-partition according an independent $(\alpha', -\alpha)$ partition. This suggests that, loosely speaking, one could obtain $\mathcal{P}_{\alpha+}$ from $\mathcal{P}_{\alpha-}$ by an $(\alpha, -\alpha)$-partition. The expression of the dislocation measure in Theorem 4.1 provides a rigorous version of this idea, whereas an $(\alpha, -\alpha)$-partition does not make sense strictly speaking.

4.2. Proof of Theorem 4.1

It is easy to verify that (a), (b) and (c) hold for the process $(\mathcal{P}_\alpha, 0 \leq 1 \leq \alpha)$, using the independence properties of the construction of Section 2. We want to show the Markov property and the Poisson point process representation for the dislocations of a given block. We shall study the evolution of the block containing 1 in the partition $\mathcal{P}_\alpha$, denoted by $B(\alpha)$

Recall that the integers in $B(\alpha)$ correspond to points of the Poisson point process $\mathcal{N}$ and to the additional point $P_0 = (1, \infty)$ (see Section 2.2). We shall identify $B(\alpha)$ with these points. Rank these points by decreasing y-coordinate, denoting them by

$$Q_0(\alpha) = (x_0(\alpha), z_0(\alpha)),\ Q_1(\alpha) = (x_1(\alpha), z_1(\alpha)), \ldots$$

with $\infty = z_0(\alpha) > z_1(\alpha) > \ldots$. Set $H_1(\alpha) = 1$ and by induction $H_{k+1}(\alpha) = \min\{n, x_n(\alpha) < x_{H_k(\alpha)}(\alpha)\}$.

Also, there is natural order on $B(\alpha-)$ by ordering the points in decreasing x-coordinate. Let $\prec_\alpha$ be this order.

By construction, for every k, $U(Q_{H_k(\alpha)}) > \alpha$, and it is easily seen that $U(Q_{H_k(\alpha)})$ is uniformly distributed on $(\alpha, 1)$, independently of the fragmentation up to time α. In turn, for all $n \notin \{H_k(\alpha), k \geq 1\}$, $U(Q_n(\alpha))$ is uniformly distributed on $(0,1)$, independently of the fragmentation up to time α.

Moreover, it is easily seen that for every n, the restriction of $\prec_\alpha$ to $Q_1(\alpha)$, $\ldots Q_n(\alpha)$ is a random total order, uniformly distributed over all possible total orders and independent of the fragmentation up to time α. We shall see that the fragmentation process is determined by $\prec_\alpha$ and the variables $U(Q_n(\alpha))$, and since they are independent of the fragmentation up to time α, this entails the Markov property.

So we have, for every k, $U(Q_{H_k(\alpha)}) > \alpha$, and $U(Q_{H_k(\alpha-)}) \geq \alpha$. There is a dislocation at time α if for some $k \geq 1$, $U(Q_{H_k(\alpha-)}) = \alpha$. In that case, we say that we have a dislocation of index $I(\alpha) = k$ and almost surely, for every $k' \neq k$, $U(Q_{H_{k'}(\alpha-)}) > \alpha$. It is easy to check the following:

Proposition 4.2. *For every $k \geq 1$, let T_k be the set of dislocation times of index k. Then the T_k's are iid Poisson point processes on $[0,1]$ with intensity measure $dx/(1-x)$.*

Proof: This is just due to the fact that, conditionally on $U(P) > \alpha$, $U(P)$ is uniformly distributed on $(\alpha, 1)$. □

If α is a dislocation time, let $\mathcal{P}'(\alpha)$ be the partition of $B(\alpha-)$ into blocks of the partition $\mathcal{P}(\alpha)$.

Proposition 4.3. *Assume that there is a dislocation at time α. Then conditionally on $I(\alpha) = k$ and $H_k(\alpha-) = k'$, the partition $\mathcal{P}'(\alpha)$ has the same law as the partition derived from a Chinese restaurant with parameters $(\alpha, -\alpha)$ and initial condition $k_0 = 2$, $a_1 = k'$, $a_2 = 1$.*

Proof The proof uses the same arguments as Theorem 3.1 Suppose that $I(\alpha) = k$ and $H_k(\alpha-) = k'$. Let $\overline{P} = (x, z)$ be the left-most point percolating at α and such that $z > z_{H_k(\alpha-)}$. If there is no such point, put by convention $\overline{P} = (0, \infty)$.

We view $B(\alpha-)$ as the set of customers of the Chinese restaurant. There are initially two tables and $k'+1$ customers. At the first table, P_0 and $Q_1(\alpha-), \ldots Q_{k'-1}(\alpha-)$ are seated. At the second table, $Q_{k'}(\alpha-)$ is seated.

We also have $k+2$ reals, namely $x, x_1, \ldots x_{k'}, 1$, yielding $k'+1$ intervals, which we denote by $J_1(\alpha), \ldots J_{k'+1}(\alpha)$ from left to right.

Let n be the minimal integer such that $P_n = (t_n, y_n)$ satisfies $t_n > x$ and $y_n < z_{H_k(\alpha-)}$. It is important to notice that we are considering P_n, as defined in Section 3.2, and not Q_n. In particular P_n may not belong to $B(\alpha-)$. Set $Q = P_n$.

By the same arguments as in the proof of Theorem 3, Q lies in one of the $k'+1$ intervals $J_1(\alpha), \ldots J_{k'+1}(\alpha)$ with equal probability. Then one has to distinguish between 5 possibilities.

- $x \in J_1(\alpha)$, and $U(Q) < \alpha$. In that case, $Q \notin B(\alpha-)$.
- $x \in J_1(\alpha)$ and $U(Q) > \alpha$. In that case, Q sits at the second table.
- $x \in J_2(\alpha)$ and $U(Q) < \alpha$. In that case, Q sits at a new table.
- $x \in J_2(\alpha)$, and $U(Q) > \alpha$. Then Q sits at the first table.
- $x \notin (J_1(\alpha) \cup J_2(\alpha))$. Then Q sits at the first table.

Therefore the probabilities to sit at the first table, at the second table or at a new table are proportional to $k' - \alpha$, $1 - \alpha$, α respectively. The same scheme apply for the next customers, which proves the proposition. □

Proposition 4.4. *Assume that there is a dislocation at time* α. *Then*

$$\sum_{k'} \mathbb{P}(H_k(\alpha-) = k' | I(\alpha) = k) t^{k'} = t(-\log(1-t))^{k-1}$$

Proof: If $\prec$ is a total order on $\{1, 2, \ldots n\}$, set $m_1 = 1$ and by induction, $m_{i+1} = \min\{j > m_i, j \prec m_i\}$. Now let $D(k, k')$ be set of total orders $\prec$ on $\{1, 2, \ldots k'\}$ such that $m_k = k'$. As an elementary application of the theory of combinatorial species,

$$\sum_{k' \geq 1} |D(k, k')| \frac{t^{k'}}{k'!} = t(-\log(1-t))^{k-1}$$

Let us apply this to the order $\prec_\alpha$. The event that $I(\alpha) = k$, $H_k(\alpha-) = k'$ is exactly the event that

(A) $U(Q_{H_k(\alpha-)}(\alpha-)) = \alpha$ and for all $i \neq k$, $U(Q_{H_i(\alpha-)}(\alpha-)) > \alpha$.

(B) The order $\prec_\alpha$, restricted to $Q_1(\alpha-), \ldots Q_{k'}(\alpha-)$, lies in $\in D(k, k')$.

On the other hand, the event that α is a dislocation time and that $I(\alpha) = k$ is exactly the event that (A) holds. Since the variables $U(P_n)$ are independent of the Poisson point process $\mathcal{N}$, they are independent of $\prec_\alpha$ and therefore (A) and (B) occur independently. As a consequence, the probability that (A) and (B) hold given that (A) holds is the probability of (B). Finally, remark that the restriction of $\prec_\alpha$ to $Q_1(\alpha-), \ldots Q_{k'}(\alpha-)$ is uniformly distributed over the $k'!$ possible total orders. This proves the proposition. □

Propositions 4.2, 4.3 and 4.4, together with obvious independence properties of $\mathcal{N}$, entail that the process of dislocations of the block containing 1 follow the law induced by the dislocation measure ν.

Moreover, it is easily seen that the law of dislocations for the block containing i is the same for all integers i. This concludes the proof of Theorem 4.1.

References

[1] Bertoin, Jean. *Subordinators: examples and applications.* Lectures on probability theory and statistics (Saint-Flour, 1997), 1–91, Lecture Notes in Math., 1717, Springer, Berlin, 1999.

[2] Bertoin, Jean Homogeneous fragmentation processes. *Probab. Theory Related Fields* **121** (2001), no. 3, 301–318.

[3] Bertoin, Jean; Le Gall, Jean-François. The Bolthausen-Sznitman coalescent and the genealogy of continuous-state branching processes. *Probab. Theory Related Fields* **117** (2000).

[4] Bertoin, Jean; Pitman, Jim. Two coalescents derived from the ranges of stable subordinators. *Electron. J. Probab.* **5** (2000), no. 7.

[5] Bolthausen, E.; Sznitman, A.-S. On Ruelle's probability cascades and an abstract cavity method. *Comm. Math. Phys.* **197** (1998), no. 2, 247–276.

[6] Bovier, Anton and Kurkova, Irina. Derrida's generalized random energy models 4: Continuous state branching and coalescents *Preprint.*

[7] Marchal, P. Regenerative sets, random partitions and the Bolthausen-Sznitman coalescent. *Preprint.*

[8] Perman, Mihael; Pitman, Jim; Yor, Marc. Size-biased sampling of Poisson point processes and excursions. *Probab. Theory Related Fields* **92** (1992), no. 1, 21–39.

[9] Pitman, Jim. *Combinatorial stochastic processes* (Saint-Flour,2002). To appear, Lecture Notes in Math., Springer.

Philippe Marchal
DMA, Ecole Normale Supérieure, 75005 Paris, France.
Philippe.Marchal@ens.fr

Part VII

Applications

Real Numbers with Bounded Digit Averages

Eda Cesaratto and Brigitte Vallée

ABSTRACT: *This paper considers numeration schemes, defined in terms of dynamical systems and studies the set of reals which obey some constraints on their digits. In this general setting, (almost) all sets have zero Lebesgue measure, even though the nature of the constraints and the numeration schemes are very different. Sets of zero measure appear in many areas of science, and Hausdorff dimension has shown to be an appropriate tool for studying their nature. Classically, the studied constraints involve each digit in an independent way. Here, more global conditions are studied, which only provide constraints on each digit prefix. The main example of interest deals with reals whose all the digit prefix averages in their continued fraction expansion are bounded by M. More generally, a weight function is defined on the digits, and the weighted average of each prefix has to be bounded by M. This setting can be translated in terms of random walks where each step performed depends on the present digit, and walks under study are constrained to be always under a line of slope M. We first provide a characterization of the Hausdorff dimension s_M, in terms of the dominant eigenvalue of the weighted operator transfer relative to the dynamical system, in a quite general setting. We then come back to our main example; With the previous characterization at hand and use of the Mellin transform, we exhibit the behaviour of $|s_M - 1|$ when the bound M becomes large. Even if this study seems closely related to previous works in Multifractal Analysis, it is in a sense complementary, because it uses weights on digits which grow faster and deals with different methods. This paper only presents a detailed abstract of this study; it describes the main tools and the sketches of proofs whereas a full paper [6] contains all the proofs of the statements which are briefly described here.*

1. Description of the framework

1.1. Numeration schemes

A numeration process associates to each real number x of the unit interval I a sequence of digits $(m_1(x), m_2(x), \dots, m_n(x), \dots)$ where each m_i belongs to some alphabet $\mathcal{M} \subset \mathbf{N}^\star$, finite or denumerable. We consider here such processes which are defined in terms of dynamical systems. For a readable treatment of dynamic systems of the interval, see [24].

Definition 1. [Good Class] *A dynamical system of the Good Class is defined by four elements*

(i) An alphabet $\mathcal{M}$ included in $\mathbf{N}^\star$, whose elements are called digits.

(ii) A topological partition of $I := [0,1]$ with disjoint open intervals I_m, $m \in \mathcal{M}$, i.e. $[0,1] = \cup_{m\in\mathcal{M}} \overline{I}_m$; the length of the interval I_m is denoted by p_m.

(iii) A mapping ρ which is constant and equal to m on each I_m.

(iv) A mapping T –often called the shift– whose restriction to each $\overline{I}_m$ is a $\mathcal{C}^2$ bijection from $\overline{I}_m$ to I. Let h_m be the inverse branch of T restricted to $\overline{I}_m$. The mappings h_m satisfy the following:

(*a*) [Contraction.] *For each* $m \in \mathcal{M}$, *there exist* η_m, δ_m *with* $0 < \eta_m < \delta_m < 1$ *for which* $\eta_m \le |h'_m(x)| \le \delta_m$ *for* $x \in I$. *The quantity* $\delta := \sup_{m\in\mathcal{M}} \delta_m$ *satisfies* $\delta < 1$.
(*b*) [Bounded Distortion Property]. *There exists a constant* $r > 0$, *called the distortion constant such that* $|h''_m(x)| \le r|h'_m(x)|$ *for all* $m \in \mathcal{M}$ *and for all* $x \in I$.
(*c*) [Convergence on the left of $s = 1$]. *There exists* $\sigma < 1$ *for which the series* $\sum_{m\in\mathcal{M}} p_m^\sigma$ *is convergent. The infimum* σ_0 *of such* σ *is the abscissa of convergence.*

With a system of the Good Class, a representation scheme for real numbers of I is built as follows: We relate to x its trajectory $\mathcal{T}(x) = (x, T(x), T^2(x), ..., T^n(x), ...)$ which can be encoded by the (infinite) sequence of the digits produced by applying the map ρ on each element $T^i(x)$ of the trajectory,

$$(m_1(x), m_2(x), ..., m_n(x), ...), \qquad \text{with } m_i(x) := \rho(T^{i-1}(x)). \tag{1}$$

This framework provides numeration processes where the distribution n–th digit m_n may depend on the whole previous history.
Each branch (or inverse branch) of the n-th iterate of the shift T is called a branch of depth n. It is then associated in a unique way to an n-tuple $\mathbf{m} = (m_1, ..., m_n)$ of length n, and is of the form $h_\mathbf{m} := h_{m_1} \circ h_{m_2} \circ ... \circ h_{m_n}$. The interval $I_\mathbf{m} := h_\mathbf{m}(]0, 1[)$ gathers all the reals x for which the sequence of the first n digits equals $\mathbf{m}$: it is called the fundamental interval relative to $\mathbf{m}$. Its depth equals the length $|\mathbf{m}|$ of prefix $\mathbf{m}$ and its Lebesgue measure denoted by $p_\mathbf{m}$ satisfies $p_\mathbf{m} \le \delta^n$. The set of inverse branches of depth n is exactly $\mathcal{H}^n$, and the set of all the inverse branches of any depth is $\mathcal{H}^\star$. Distortion and contraction properties entail the existence of a constant $L > 0$ such that

$$\frac{1}{L} \le \left|\frac{h'(x)}{h'(y)}\right| \le L, \qquad \text{for any } h \in \mathcal{H}^\star. \tag{2}$$

1.2. Main examples

Here, as we explain next, we focus on dynamical systems relative to an infinite alphabet $\mathcal{M}$. The most classical examples are memoryless sources (of Riemann type) and the Continued Fraction expansion.

Continued fraction expansion. The shift T, also known as the Gauss map, is

$$T(x) = \frac{1}{x} - \lfloor \frac{1}{x} \rfloor \text{ for } x \neq 0, \qquad T(0) = 0. \tag{3}$$

It is relative to the topological partition $I_m = (1/(m+1), 1/m)$. The inverse branch of depth 1 associated to the digit m is the LFT (linear fractional transformation) $h_m(z) = 1/(m + z)$. This map induces the numeration scheme related to continued fraction expansion.

Memoryless dynamical systems. A dynamical system of the Good Class is memoryless when the branches h_m are affine. It is completely defined (up to isomorphism) by the length $p_m = \delta_m$ of each interval I_m of depth one, which equals the probability p_m of emitting m at each step of the process. The fundamental interval can be chosen as $I_m :=]q_m, q_{m+1}[$ with $q_1 = 0$ and $q_m := \sum_{k<m} p_k$. A special type of memoryless source is studied here as a main example: the Riemann type $\mathcal{R}(\alpha)$ (for $\alpha > 1$) where the associated probabilities are

$$p_m^{(\alpha)} := \frac{1}{\zeta(\alpha)} \frac{1}{m^\alpha}. \tag{4}$$

The affine approximation of the Gauss map is the memoryless system relative to the partition $I_m = (1/(m+1), 1/m)$. The length p_m equals $1/(m(m+1))$, and it is of the same type as the system $\mathcal{R}(2)$.

1.3. Elementary constraints on numeration processes

In this setting, it is now classical to study numbers x for which the sequence (1) satisfies some particular constraints. The instance of Cantor sets where the constraint is the same for each digit m_i and only allows a subset $\mathcal{A}$ of possible values is well known. In this case, the set of such constrained numbers $E_{\mathcal{A}}$ has zero measure, and it is thus of great interest to study its Hausdorff dimension. The first study on the subject is relative to numeration in base b and due to Eggleston [8]. The problem is now completely solved when the alphabet is finite. The case when the alphabet is infinite is quite important since it contains a particular case of great interest: the reals whose continued fraction expansion only contains digits m_i less than M. These reals are badly approximable by rationals, and intervene in many contexts of number theory (see [31]). The case of an infinite alphabet (even if the process is memoryless) is a little bit more difficult to deal with, and the set $\mathcal{A}$ of constraints has to be made precise [27, 28]. In a quite general setting (dynamical systems of the Good Class, "open" constraints), the question is solved. The main tool is the constrained transfer operator $\mathbf{H}_{s,\mathcal{A}}$ defined by

$$\mathbf{H}_{s,\mathcal{A}}[f] := \sum_{m\in\mathcal{A}} |h'_m|^s \cdot f\circ h_m.$$

For a dynamical system of the Good Class, for real values of parameter s, and on a convenient functional space, the operator $\mathbf{H}_{s,\mathcal{A}}$ has a unique dominant eigenvalue denoted by $\lambda_{\mathcal{A}}(s)$. When the set $\mathcal{A}$ is "open", there exists a (unique) real $s = \tau_{\mathcal{A}}$ for which $\lambda_{\mathcal{A}}(s) = 1$ and the Hausdorff dimension of $E_{\mathcal{A}}$ equals $\tau_{\mathcal{A}}$.

The particular case of "constrained" continued fractions was extensively studied; the beginners were Jarnik [20], Besicovitch [4] and Good [11]. Then, Cusick, Hirst, Bumby, brought important contributions, and finally Hensley [13, 14] completely solved the problem. In [32], this result was extended to the case of "periodic" constraints.

Another general question of interest is the asymptotic behaviour of $\dim(E_{\mathcal{A}})$ when the constraint becomes weaker (i.e., $\mathcal{A} \to \mathcal{M}$). Then, the Hausdorff dimension tends to 1, and the speed of convergence towards 1 is also an important question. In the case of continued fractions, Hensley [14] studies the case when $\mathcal{A}_M := \{1, 2 \ldots, M\}$ and exhibits the asymptotic behaviour of $\tau_M := \dim(E_{\mathcal{A}_M})$

$$|\tau_M - 1| = \frac{6}{\pi^2}\frac{1}{M} + O\left(\frac{\log M}{M^2}\right), \qquad \text{when } M \to \infty. \tag{5}$$

1.4. Costs and weighted prefix averages

We consider here constraints which are more general than previous ones. They are defined by conditions which only bound all the weighted prefix averages and appear in a natural way in the Multifractal Analysis framework [9, 10].

A digit-cost c relative to a dynamical system (I, T) is a strictly positive function $c : \mathcal{M} \mapsto I\!R^+$ which extends to a function $c : \mathcal{M}^\star \mapsto I\!R^+$ via the additive property

$$\text{for } \mathbf{m} = (m_1, m_2, \ldots, m_n), \qquad c(\mathbf{m}) := \sum_{i=1}^{n} c(m_i). \tag{6}$$

On each trajectory $\mathcal{T}(x)$ encoded by the sequence $(m_1(x), m_2(x), \ldots, m_n(x) \ldots)$ defined in (1), the weighted prefix average of length n is defined as

$$M_n(x) := \frac{1}{n}\sum_{i=1}^{n} c(m_i(x)), \tag{7}$$

and we study here the set F_M of reals for which all the $M_n(x)$ are bounded by M.

1.5. Triples of large growth

The strength of these constraints depends on the relation between the cost c and the occurrence probability of the digits. Consider the (initial) probability distribution $p : k \mapsto p_k$ of digit m_k, together with the limit distribution $\overline{p}$ of the n–th digit m_n which always exists in the Good Class Setting. We are mostly interested in the case of a dynamical system of the Good Class, with an infinite alphabet $\mathcal{M}$, where the sequences $p_m, c(m)$ satisfy

(c1) $m \mapsto p_m$ is decreasing,

(c2) $m \mapsto c(m)$ is increasing to $+\infty$.

The mixed sequences $\pi_n := \min\{p_m;\ c(m) \le n\}, \overline{\pi}_n := \min\{\overline{p}_m;\ c(m) \le n\}$ summarize the balance between the increase of cost c and the decrease of distributions $p, \overline{p}$, and the conditions

(c3) $\qquad \limsup \pi_n^{1/n} = 1, \qquad \text{or} \qquad \limsup \overline{\pi}_n^{1/n} = 1$

(which are equivalent for systems of the Good Class) informally express that the increase of c is faster than the decrease of p. Condition (c3) is equivalent to requiring that the convergence radii of the series $U(z)$, $V(z)$ equal 1, with

$$U(z) := \sum_{n \ge 1} \pi_n z^n, \qquad V(z) := \sum_{m \ge 1} p_m z^{\lfloor c(m) \rfloor}.$$

Definition 2. *Consider a triple (I, T, c) made with a system (I, T) of the Good Class. If it does not satisfy Condition (c3), it is said to be of moderate growth ($\mathcal{GMG}$–setting). If it satisfies Conditions (c1)(c2)(c3), it is said to be of large growth ($\mathcal{GLG}$–setting). When, furthermore, the abscissa of convergence of the Dirichlet series*

$$P(s) := \sum_{m \in \mathcal{M}} c(m)\, p_m^s, \qquad \overline{P}(s) := \sum_{m \in \mathcal{M}} c(m)\, \overline{p}_m^s \tag{8}$$

relative to a triple (I, T, c) of class $\mathcal{GLG}$ equals 1, the triple (I, T, c) is called a boundary triple.

Here, we focus on the $\mathcal{GLG}$–setting. In the boundary case, the (stationary) average $\mu(c)$ of cost c (which equals $\overline{P}(1)$) is infinite,

$$\mu(c) := \sum_{k=1}^{n} c(k) \cdot \overline{p}_k = +\infty. \tag{9}$$

Examples of triples (I, T, c) of $\mathcal{GLG}$–type. The boundary triple $\mathcal{BR}(\alpha)$ related to the Riemann memoryless sources $\mathcal{R}(\alpha)$ defined by

$$p_m^{(\alpha)} := \frac{1}{\zeta(\alpha)} \frac{1}{m^\alpha}, \qquad c(m) = m^{\alpha - 1}, \tag{10}$$

is studied here as a main example. Note that $\mathcal{BR}(2)$ provides an approximate memoryless version of the boundary triple formed with the Gauss map (I, T) together with the cost $c(m) = m$ which will be one of the most interesting example of our study.

1.6. Subsets F_M and statement of the main results

We wish to study the sets F_M defined as follows:

Definition 3. [Set F_M] *Consider a triple (I, T, c) of $\mathcal{GLG}$ type, and denote by $\gamma(c) > 0$ the minimal value of c. For any $M > \gamma(c)$, the set F_M is the set of ordinary reals x of I for which all the weighted averages $M_n(x)$ defined in (7) satisfy $M_n(x) \le M$.*

The set F_M can be also described in terms of random walks: To each real number x, one associates the walk formed with points $(P_i(x))_{i\geq 0}$. One begins with $P_0(x) := (0,0)$, and, at time i, one performs a step $P_i(x) - P_{i-1}(x) := (1, c(m_i(x))$. The set F_M is the set of reals x for which the walk $(P_i(x))_{i\geq 0}$ is always under the line of slope M.
For any $M \in]\gamma(c), \mu(c)[$, the Lebesgue measure of the set F_M equals 0, and we wish to study the Hausdorff dimension s_M of F_M. We obtain three main results.

The first Theorem provides a (mathematical) characterization of the Hausdorff dimension s_M of F_M as a solution of a differential system.

Theorem 1.1. *Consider the set F_M relative to a triple (I,T,c) of $\mathcal{GLG}$-type. Denote by $\mathbf{H}_{s,w}$ the weighted operator relative to the triple (I,T,c) defined by*

$$\mathbf{H}_{s,w}[f] := \sum_{m\in\mathcal{M}} \exp[wc(m)] \cdot |h'_m|^s \cdot f \circ h_m,$$

and by $\Lambda(s,w)$ the logarithm of its dominant eigenvalue when $\mathbf{H}_{s,w}$ acts on $C^1(I)$. Then, for any $\gamma(c) < M < \mu(c)$, there exists a unique pair $(s_M, w_M) \in [0,1] \times]-\infty, 0[$ for which the two relations hold:

$$(\mathcal{S}): \qquad \Lambda(s,w) = Mw, \qquad \frac{\partial}{\partial w}\Lambda(s,w) = M.$$

If, furthermore the second derivative $\Lambda''_{w^2}(s_M, w_M)$ is non zero, then the Hausdorff dimension of F_M equals s_M. The two functions $M \mapsto s_M, M \mapsto w_M$ are analytic at any point $M \in]\gamma(c), \mu(c)[$ for which $\Lambda''_{w^2}(s_M, w_M) \neq 0$.

We now focus on boundary triples: when M tends to $\mu(c) = +\infty$, the dimension s_M tends to 1 and we describe the exact asymptotic behaviour of $|s_M - 1|$. The following two results prove that, in each case (Riemann memoryless sources or Continued Fraction framework), the speed of convergence of s_M towards 1 is exponential in M, and exhibit the precise convergence rate.

Theorem 1.2. *The Hausdorff dimension s_M of the set F_M relative to the boundary Riemann triple $\mathcal{BR}(\alpha)$ satisfies, for $M \to \infty$*

$$|s_M - 1| = \frac{1}{(\alpha-1)\zeta(\alpha)h(\alpha)} \exp[-M(\alpha-1)\zeta(\alpha)] + O(\exp[-M\theta])$$

$$\textit{with} \qquad h(\alpha) = \alpha\frac{\zeta'(\alpha)}{\zeta(\alpha)} - \log\zeta(\alpha), \quad \textit{and any} \quad \theta < 2(\alpha-1)\zeta(\alpha).$$

Theorem 1.3. *The Hausdorff dimension s_M of the set F_M relative to the triple formed with the Euclidean system and the cost $c(m) = m$ satisfies, for $M \to \infty$,*

$$|s_M - 1| = \frac{6}{\pi^2} \cdot 2^{-M}[1 + O(\theta^{-M})] \qquad \textit{for any } \theta < 2.$$

1.7. Relation with Multifractal Analysis

This work is partially related to Multifractal Analysis introduced by Mandelbrot [26]. See also [9, 10]. For a dynamical system (I,T), for any n and any x, the interval $I^{(n)}(x)$ is the fundamental interval of depth n which contains x. Each fundamental interval has two measures, the Lebesgue measure and another measure ν which is defined by the cost c. More precisely, for costs c which give rise to a series $\sum_m \exp[-c(m)] = 1$, the measure ν of a fundamental interval $I_{\mathbf{m}}$ related to the prefix $\mathbf{m} := (m_1, m_2, \ldots m_n)$ is defined by $|\log\nu(I_{\mathbf{m}})| = \sum_{i=1}^n c(m_i)$. In this way, with respect to this measure ν, the numeration process is memoryless and always produces the digit m with probability $\exp[-c(m)]$. In order to compare the two measures, the Lebesgue measure, and the measure ν, Multifractal Analysis introduces the set G_β of real x for which

$$B_n(x) := \left|\frac{\sum_{i=1}^{n} c(m_i(x))}{\log |I^{(n)}(x)|}\right| = \left|\frac{\log \nu(I^{(n)}(x))}{\log |I^{(n)}(x)|}\right| \qquad \text{satisfies } \lim_{n\to\infty} B_n(x) = \beta, \quad (11)$$

and studies the Hausdorff Dimension t_β of the set G_β.
For Dynamical systems of the Good Class, the sequence $-(1/n)\log |I^{(n)}(x)|$ tends almost everywhere to the entropy h, so that the asymptotic behaviour of the two sequences $M_n(x)$ and $hB_n(x)$ defined in (7) and (11) is the same almost everywhere. However, this is only true "almost everywhere", and finally, the relation between the two Hausdorff dimensions s_M and t_β is not so clear, and it is of great interest to compare our result on F_M to the following result on G_β, recently obtained by Hanus, Mauldin and Urbański [12] which we translate in our setting.

Theorem 1.4. [12] *Consider the set G_β relative to a triple (I,T,c) of $\mathcal{GMG}$-type. Suppose furthermore that the cost c satisfies $\sum_m \exp[-c(m)] = 1$. Denote by $\mathbf{H}_{s,w}$ the weighted operator relative to the triple (I,T,c) and by $\Lambda(s,w)$ the logarithm of its dominant eigenvalue. Then, for any β near the value $\beta_0 = \mu(c)/h$, there exists a unique pair $(t,w) = (t_\beta, w_\beta) \in [0,1]\times]-\infty,+\infty[$ for which the two relations hold:*

$$(\mathcal{G}): \qquad \Lambda(t-\beta w, w) = 0, \qquad \frac{\partial}{\partial w}\Lambda(t-\beta w, w) = -\beta\frac{\partial}{\partial s}\Lambda(t-\beta w, w).$$

The Hausdorff dimension of G_β equals t_β. The two functions $\beta \mapsto t_\beta, \beta \mapsto w_\beta$ are analytic when β is near β_0.

Note that, even if the two results (our Theorem 1.1, and the previous Theorem) are of the same spirit and involve the same kind of systems $(\mathcal{F})$ and $(\mathcal{G})$, the result on G_β is obtained in the $\mathcal{GMG}$ setting, while ours is obtained in the $\mathcal{GLG}$ setting. This explains why the methods used cannot be similar: they both deal with the weighted transfer operator $\mathbf{H}_{s,w}$. However, the authors in [12] used analyticity of $(s,w) \to \mathbf{H}_{s,w}$ at $(1,0)$, together with ergodic theorems. These properties are no longer true in the $\mathcal{GLG}$ setting, and we have to introduce other tools, similar to those used in Large Deviations results.

1.8. Main steps of the proof

The two sequences of subsets

$$\mathcal{A}_n(M) := \{\mathbf{m} \in \mathcal{M}^n; c(\mathbf{m}) \le Mn\}, \qquad \mathcal{B}_n(M) := \bigcap_{r=1}^{n} \mathcal{A}_r(M). \qquad (12)$$

are useful for defining the set F_M, in two different ways,

$$F_M = \bigcap_{n\ge 1} \bigcup_{\mathbf{m}\in\mathcal{A}_n(M)} I_{\mathbf{m}}, \quad \text{or} \quad F_M = \bigcap_{n\ge 1} \bigcup_{\mathbf{m}\in\mathcal{B}_n(M)} I_{\mathbf{m}},$$

as means of fundamental intervals $I_{\mathbf{m}}$. There is a close link between these two sequences, due to the next property (proven in the full paper [6]). It is of the same spirit as the so-called "Cyclic Lemma", which is useful in the Random Walk Setting.
Cyclic Lemma. *For any $\mathbf{m} \in \mathcal{A}_n(M)$, there exists a circular permutation τ for which $\tau(\mathbf{m})$ belongs to $\mathcal{B}_n(M)$.*

Furthermore, both sequences will be useful: the sequence $\mathcal{B}_n(M)$, in studies on coverings (Section 2), and the sequence $\mathcal{A}_n(M)$, in studies on transfer operators (Section 3). Section 4 provides the proof of Theorem 1.1, while Section 5 is devoted to proving Theorems 1.2 and 1.3.

2. Hausdorff dimension of sets constrained by their prefixes

We first recall some classical facts about coverings and Hausdorff dimension. The definition of Hausdorff dimension of a given set *a priori* involves all its possible coverings. Here, we introduce a class of sets (the sets which are well-constrained by their prefixes) which contains all the sets F_M relative to triples of large growth. We prove in Proposition 1 that, for such sets, the Hausdorff dimension can be determined via particular coverings, formed with fundamental intervals of fixed depth. For sets F_M, this characterization involves sets $\mathcal{B}_n(M)$ of (12).

2.1. Coverings and Hausdorff dimension

Let E be a subset of I. A covering $\mathcal{J} := (J_\ell)_{\ell\in\mathcal{L}}$ of E is a set of open intervals J_ℓ for which $E \subset \bigcup_{\ell\in\mathcal{L}} J_\ell$. It is said to be finite if $\operatorname{card}\mathcal{L}$ is finite. The diameter of a covering is the real ρ that is the supremum of the lengths $|J_\ell|$. A covering is fundamental [with respect to some dynamical system (I,T)] if its elements J_ℓ are fundamental intervals. For each covering $\mathcal{J}$ of E, the quantity $\Gamma_\sigma(\mathcal{J}) := \sum_{J\in\mathcal{J}} |J|^\sigma$ plays a fundamental rôle in the following.

A subset E of I has zero measure in dimension σ (i.e., $\mu_\sigma(E) = 0$) if for any $\epsilon > 0$, there exists a covering $\mathcal{K}$ of E for which $\Gamma_\sigma(\mathcal{K}) < \epsilon$. A subset E of I has an infinite measure in dimension σ (i.e., $\mu_\sigma(E) = \infty$) if for any $A > 0$, there exists $\rho > 0$, such that, for any covering $\mathcal{K}$ of E of diameter at most ρ, one has $\Gamma_\sigma(\mathcal{K}) > A$.

The Hausdorff dimension of E, denoted by $\dim(E)$ is the unique number d for which $\mu_\sigma(E) = 0$ for any $\sigma > d$ and $\mu_\sigma(E) = +\infty$ for any $\sigma < d$,

$$\dim(E) = \inf\{\sigma;\ \mu_\sigma(E) = 0\} = \sup\{\sigma;\ \mu_\sigma(E) = +\infty\}.$$

2.2. Sets which are well–constrained by their prefixes

We are interested in studying sets of the same type as F_M, and we will consider in this section a class of more general sets which are defined by constraints on their prefixes of any length.

Definition 4.[$\mathcal{WCP}$ sets] *Let (I,T) a dynamical system of the Good Class, and $\mathcal{M}$ its associated alphabet. A subset E is defined by its prefixes if there exists a sequence $\mathcal{M}_\star := (\mathcal{M}_n)_{n\geq 1}$ of non-empty subsets $\mathcal{M}_n \subset \mathcal{M}^n$ (the constraints) for which*

$$E := \bigcap_{n\geq 1} \bigcup_{\mathbf{m}\in\mathcal{M}_n} I_{\mathbf{m}}.$$

The sequence $\mathcal{M}_\star$ is the canonical sequence of E. Moreover, if the sequence $\mathcal{M}_\star$ of constraints satisfies the following four conditions,

(i) For any $n \geq 1$, the set $\mathcal{M}_n$ is finite,

(ii) If $(m_1, ..., m_n) \in \mathcal{M}_n$ then $(m_1, ..., m_{n-1}) \in \mathcal{M}_{n-1}$,

(iii) $\mathcal{M}_{n_1} \times \mathcal{M}_{n_2} \subset \mathcal{M}_{n_1+n_2}$ for all n_1, n_2,

(iv) $\pi_n := \min\{p_{m_n};\ \exists\, (m_1, m_2, \ldots, m_{n-1})$ s. t. $(m_1, m_2, \ldots, m_n) \in \mathcal{M}_n\}$ satisfies $\underline{\lim}\, \pi_n^{1/n} = 1$,

the sequence $\mathcal{M}_\star$ is said to be well–conditioned. In this case, the set E is said to be well–constrained by its prefixes. For each n, the set $\mathcal{J}_n := \{I_{\mathbf{m}}; \mathbf{m} \in \mathcal{M}_n\}$ is *a covering of E, which is finite and fundamental: The sequence $(\mathfrak{J}_n)$ is called the canonical system of coverings of E.*

The following proposition (proven in [6]) shows that the Hausdorff dimension of a set E which is well–constrained by its prefixes can be uniquely characterized via its canonical system of coverings $(\mathfrak{J}_n)$.

Proposition 2.1. *[Characterization of the Hausdorff dimension of a $\mathcal{WCP}$ set via its canonical system of coverings.] Let E be a subset of I, which is well–constrained by its prefixes, and $(\mathfrak{J}_n)_{n\geq 1}$ its canonical system of coverings. Then*

$$\dim(E) = \inf\{\sigma;\quad \sup\{\Gamma_\sigma(\mathfrak{J}_n); n \geq 1\} < \infty\}.$$

Proof. It mainly uses Condition *(iv)* of Definition 4 which is equivalent to a condition introduced by J. Peyrière [29]

$$\forall \alpha > 0, \quad \overline{\lim}\{\tau(J)\cdot|J|^\alpha; J \in \mathfrak{J}_\star,\ |J| \to 0\} \leq 1.$$

Here, for any $J \in \mathfrak{J}_\star$ the quantity $\tau(J)$ relates the length of J to the lengths of fundamental intervals $K \subset J$, when depths $d(K), d(J)$ satisfy $d(K) = d(J)+1$,

$$\tau(J) := \sup\left\{\frac{|J|}{|K|};\ K \subset J, K \in \mathfrak{J}_\star, d(K) = d(J)+1\right\}. \blacksquare$$

We denote by $A_n(M,\sigma)$ and $B_n(M,\sigma)$, the associated Γ_σ quantities (defined in 2.1), related to constraints $\mathcal{A}_n(M), \mathcal{B}_n(M)$ defined in (12),

$$A_n(M,\sigma) := \sum_{\mathbf{m}\in\mathcal{A}_n(M)} p_{\mathbf{m}}^\sigma, \qquad B_n(M,\sigma) := \sum_{\mathbf{m}\in\mathcal{B}_n(M)} p_{\mathbf{m}}^\sigma. \tag{13}$$

The following result summarizes the results of this Section : Its proof uses Proposition 1 when applied to the $\mathcal{WCP}$ set F_M, together with the relation between $A_n(M,s)$ and $B_n(M,s)$ due to the Cyclic Lemma, and Property (2).

Corollary 2.2. *Consider a triple (I,T,c) of $\mathcal{GLG}$–type. For $M > \gamma(c)$, the sequence $\mathcal{B}_n(M)$ defined in (12) is well–conditioned and the set F_M is well–constrained by its prefixes. The Hausdorff dimension of the set F_M satisfies*

$$\dim(F_M) = \inf\{\sigma;\quad \sup_n B_n(M,\sigma) < \infty\} = \inf\{\sigma;\quad \sup_n A_n(M,\sigma) < \infty\}.$$

3. The main tool: the weighted transfer operator

In this Section we introduce our main tool: the weighted transfer operator. In the following section, this operator will provide useful informations on the asymptotic behaviour of sequences $A_n(M,s), B_n(M,s)$. Here, we summarize its main well-known properties, and, more precisely, its dominant spectral properties.

3.1. Transfer operators

Consider a dynamical system (I,T) of the Good Class with a cost c. The weighted transfer operator $\mathbf{H}_{\mathbf{m},s}$ relative to a prefix $\mathbf{m} \in \mathcal{M}^\star$ is defined as

$$\mathbf{H}_{\mathbf{m},s,w}[f](x) := \exp[wc(\mathbf{m})]\cdot|h_{\mathbf{m}}(x)|^s \cdot f\circ h_{\mathbf{m}}(x).$$

Due to the additivity of the cost (6) and the multiplicativity of the derivative, the operators $\mathbf{H}_{\mathbf{m},s,w}$ satisfy a fundamental composition property

$$\mathbf{H}_{\mathbf{m},s,w}\circ\mathbf{H}_{\mathbf{n},s,w} = \mathbf{H}_{\mathbf{n}\cdot\mathbf{m},s,w}.$$

The weighted transfer operator $\mathbf{H}_{s,w}$ is defined as the sum of all $\mathbf{H}_{m,s,w}$ relative to a symbol $m \in \mathcal{M}$, and the composition property entails that the n-th iterate $\mathbf{H}^n_{s,w}$ of $\mathbf{H}_{s,w}$ satisfy

$$\mathbf{H}^n_{s,w} := \sum_{\mathbf{m}\in\mathcal{M}^n} \mathbf{H}_{\mathbf{m},s,w} \qquad \text{for any } n \geq 1.$$

For $w = 0$, the operator $\mathbf{H}_{\mathbf{m},s,w}$ coincides with the classical transfer operator

$$\mathbf{H}_{\mathbf{m},s}[f](x) := |h'_{\mathbf{m}}(x)|^s \cdot f \circ h_{\mathbf{m}}(x), \tag{14}$$

which is closely related to the Lebesgue measure $p_{\mathbf{m}}$ of the fundamental interval $I_{\mathbf{m}}$: The length $p_{\mathbf{m}}$ satisfies $p^s_{\mathbf{m}} = |h_{\mathbf{m}}(0) - h_{\mathbf{m}}(1)|^s = |h'_{\mathbf{m}}(\theta_{\mathbf{m}})|^s$ for some $\theta_{\mathbf{m}} \in]0,1[$, and, Distortion Property (2) entails that, for any $\mathbf{m} \in \mathcal{M}^\star$,

$$L^{-s} \cdot p^s_{\mathbf{m}} \leq \mathbf{H}_{\mathbf{m},s}[1](0) \leq L^s \cdot p^s_{\mathbf{m}}.$$

Then, the sequence

$$D_n(M,s) := \sum_{\mathbf{m}\in A_n(M)} \mathbf{H}_{\mathbf{m},s}[1](0). \tag{15}$$

satisfies $L^{-s} \cdot A_n(M,s) \leq D_n(M,s) \leq L^s \cdot A_n(M,s)$, and provides another characterization of the Hausdorff dimension of F_M,

$$\dim(F_M) = \inf\{s;\ \sup_{n\in\mathbf{N}} D_n(M,s) < \infty\}. \tag{16}$$

3.2. Functional analysis

For a complete treatment of transfer operators, see [1]. Here, the functional space $C^1(I)$ is endowed with the norm $||.||_1$ defined by $||f||_1 := \sup\{|f(t)|; t \in I\} + \sup\{|f'(t)|; t \in I\}$. When $(\sigma := \Re s, \rho := \Re w)$ belongs to the set

$$\mathcal{S} := \{(\sigma,\rho);\ \sum_{m\in\mathcal{M}} \exp[\rho c(m)] \cdot \delta^\sigma_m < \infty\},$$

the norm $||\mathbf{H}_{s,w}||_1$ is bounded and $\mathbf{H}_{s,w}$ acts on $C^1(I)$. For a triple (I,T,c) of $\mathcal{GLG}$, one has

$$\mathcal{S} \supseteq \mathcal{S}_1 \qquad \text{with} \quad \mathcal{S}_1 := \{(\sigma,\rho);\ \sigma \geq 0, \rho < 0\} \cup \{(\sigma,0); \sigma > \sigma_0\}.$$

In addition, at any point (s,w) for which $(\Re s, \Re w)$ is in $\mathrm{Int}(\mathcal{S}_1)$, the operator $\mathbf{H}_{s,w} : C^1(I) \to C^1(I)$ is analytic with respect to both variables. Remark that the analyticity with respect to w is no longer true on the frontier of $\mathcal{S}_1$, when $s > \sigma_0$ and $w = 0$. This is why the $\mathcal{GLG}$-setting is not so easy to handle.

3.3. Spectral properties

For (s,w) in $\mathrm{Int}(\mathcal{S}_1)$, the weighted operator $\mathbf{H}_{s,w}$ satisfies a Perron-Frobenius property: there is a unique dominant eigenvalue $\lambda(s,w)$ which is positive and simple and the corresponding eigenfunction $f_{s,w}$ is positive on I. The corresponding eigenmeasure for the dual operator $\mathbf{H}^\star_{s,w}$ is denoted by $\nu_{s,w}$. The pair $(f_{s,w}, \nu_{s,w})$ is unique via the normalization condition $\int_I f_{s,w} d\nu_{s,w} = 1$. Note that, in the case when the branches are affine, the eigenfunction $f_{s,w}$ equals 1, and the eigenvalue $\lambda(s,w)$ is explicit,

$$\lambda(s,w) = \sum_{m\in\mathcal{M}} \exp[wc(m)] \cdot p^s_m. \tag{17}$$

Furthermore, $\mathbf{H}_{s,w}$ is quasi-compact, and there is a spectral gap which separates $\lambda(s,w)$ from the remainder of the spectrum. Since $\mathbf{H}_{s,w}$ depends analytically on (s,w), perturbation theory applies [22], and entails, for (s,w) sufficiently near $\mathrm{Int}(\mathcal{S}_1)$, the existence of dominant spectral objects which depend analytically on s and w. Finally, the operator decomposes as $\mathbf{H}_{s,w} = \lambda(s,w)\mathbf{P}_{s,w} + \mathbf{N}_{s,w}$. Here, $\mathbf{P}_{s,w}$ defined by $\mathbf{P}_{s,w}[f](x) := f_{s,w}(x) \cdot \nu_{s,w}[f]$ is the projection on the eigenspace, the spectral radius $\rho_{s,w}$ of $\mathbf{N}_{s,w}$ is strictly less than $|\lambda(s,w)|$, and $\mathbf{P}_{s,w} \circ \mathbf{N}_{s,w} =$

$\mathbf{N}_{s,w} \circ \mathbf{P}_{s,w} = 0$. For any α strictly larger than the ratio $\rho_{s,w}/|\lambda(s,w)|$, and for any $f \in \mathcal{C}^1(I), f > 0$, a quasi-powers property holds, and we have

$$\mathbf{H}^n_{s,w}[f](x) = \lambda(s,w)^n \cdot \mathbf{P}_{s,w}[f](x) \cdot [1 + O(\alpha^n)] \tag{18}$$

for $x \in I$ and (s,w) near $\mathrm{Int}(\mathcal{S}_1)$.
For $w = 0$, the weighted operator coincides with the transfer operator, and we omit the index w. For $(s,w) = (1,0)$, the operator $\mathbf{H}_{s,w}$ coincides with the density transformer. Then, the dominant spectral objects of $\mathbf{H}_{1,0}$ satisfy the following:

$$\lambda(1,0) = 1, \quad f_{1,0} = f_1 = \text{stationary density}, \quad \nu_{1,0} = \nu_1 = \text{ Lebesgue measure.}$$

3.4. Influence of Condition $\mathcal{GLG}$ on the derivative of the dominant eigenvalue

For real pairs $(s,w) \in \mathcal{S}_1$, denote by $\Lambda(s,w)$ the logarithm of the dominant eigenvalue, i.e., $\Lambda(s,w) := \log \lambda(s,w)$. Then, since the alphabet $\mathcal{M}$ is infinite, and the cost c is not constant (that is always the case for a cost c of large growth), the map $(s,w) \to \Lambda(s,w)$ is strictly decreasing and strictly convex (see the full paper [6]). The quantity $\Lambda'_s(1,0)$ equals the opposite of the entropy h of the system.
It is important to describe the behaviour of the derivative with respect to w, when $w \to 0^-$. For $s > \sigma_0$ and $w > 0$, the (fundamental) equality

$$\lambda'_w(s,w) = \sum_{m \in \mathcal{M}} c(m) \exp[wc(m)] \int_I |h'_m(t)|^s \cdot f_{s,w} \circ h_m(t)\, d\nu_{s,w}(t). \tag{19}$$

proves that, when $w \to 0^-$,

$$\lim_{w \to 0^-} \lambda'_w(s,w) = \mu(c,s) := \sum_{m \in \mathcal{M}} c(m) \int_I |h'_m(t)|^s \cdot f_s \circ h_m(t)\, d\nu_s(t).$$

This limit is finite if and only if $\overline{P}(s)$ defined in (8) is finite. In all the cases, the derivative $\lambda'_w(1,w)$ admits a limit (finite or infinite) when $w \to 0^-$. This limit is the average value $\mu(c)$ of cost c with respect to the stationary density f_1,

$$\mu(c,1) := \sum_{m \in \mathcal{M}} c(m) \int_{h_m(I)} f_1(t)\, dt = \sum_{m \in \mathcal{M}} c(m)\, \overline{p}_m = \mu(c).$$

4. Hausdorff dimension and dominant eigenvalues

This Section is devoted to proving Theorem 1.1 which relates the Hausdorff dimension to the root of a differential system that involves the dominant eigenvalue of the weighted transfer operator. The proof deals with tools that are often used for proving Large Deviation results, since it strongly uses the Quasi-Powers Theorem together with a well-known technique called "shifting of the mean" [5]. We use the sequence $D_n(s)$ defined in (15) and we begin with Relation (16)

$$\dim(F_M) = \inf\{s;\ \sup_{n \in \mathbf{N}} D_n(M,s) < \infty\}.$$

We wish to relate the sequence $D_n(M,s)$ to the dominant eigenvalue $\lambda(s,w)$ of the weighted operator $\mathbf{H}_{s,w}$. We first describe some useful properties of the function $\Lambda(s,w) := \log \lambda(s,w)$ which are closely related to its strict convexity.

Lemma 4.1. *Consider a triple of $\mathcal{GLG}$–type. For $M < \mu(c)$, denote by $\mathcal{V}_M$ the intersection of $[0,1]$ with the largest neighborhood of $s = 1$ for which $\mu(c,s) > M\lambda(s)$. For any pair (s,M) which satisfies $\gamma(c) < M < \mu(c)$ and $s \in \mathcal{V}_M$, the following is true: (a) there exists a unique $w = \eta(s,M) < 0$ for which $\frac{\partial}{\partial w}\Lambda(s,w)|_{w=\eta} = M$.*

(b) *The function* $]-\infty,0[\to \mathbf{R}$ *that associates to* w *the quantity* $\lambda(s,w)e^{-Mw}$ *attains its minimum at* $w=\eta(s,M)$. *The minimal value is denoted by* $\alpha_M(s)$,

$$\alpha_M(s) := \min\{\exp[-Mw]\cdot\lambda(s,w); w\in]-\infty,0[\}.$$

Proof. Since the function $\Lambda_{[s]}: w\mapsto \log\lambda(s,w)$ is strictly convex, its first derivative $\Lambda'_{[s]}$ is strictly increasing. It then admits a limit when w tends to 0^- or when w tends to $-\infty$. We prove in [6] that $\lim_{w\to-\infty}\Lambda'_{[s]}(w)=\gamma(c)$. On the other hand, we recall that $\lim_{w\to 0^-}\Lambda'_{[s]}(w)=\mu(c,s)/\lambda(s)$ defines a continuous function of s which equals $\mu(c)$ for $s=1$. Then, for any $M<\mu(c)$, there exists a neighborhood of $s=1$ on which $\lim_{w\to 0^-}\Lambda'_{[s]}(w) > M$. ■

Remark. When $\mu(c)=+\infty$, the series $P(s)$ defined in (8) is divergent for any $s\in[0,1]$, and $\mu(c,s)=+\infty$ for any $s\in[0,1]$. Then, one can choose $\mathcal{V}_M := [0,1]$. In the general case, denote by a_M the largest $s\in[0,1]$ for which $\mu(c,s)\le M\lambda(s)$. One can choose $\mathcal{V}_M =]a_M,1]$. Then $\eta(s,M)$ tends to 0 when $s\to a_M$, and $\lim_{s\to a_M}\alpha_M(s)=\lambda(a_M,0)>1$.

The following result relates the sequence $D_n(M,s)$ and $\alpha_M(s)$.

Lemma 4.2. *Consider any pair* (s,M) *which satisfies the following three conditions :*$\gamma(c)<M<\mu(c)$, $s\in\mathcal{V}_M$ *and* $\Lambda''_{w^2}(s,\eta(s,M))\neq 0$. *Then, the sequence* $[D_n(M,s)]^{\frac{1}{n}}$ *admits a limit when* $n\to\infty$ *and this limit equals* $\alpha_M(s)$.

Proof. We relate the sequence $D_n(M,s)$ to the n-th iterate of the weighted transfer operator $\exp[-wM]\cdot\mathbf{H}_{s,w}$. First, note that, for (s,w) in $\mathrm{Int}(\mathcal{S}_1)$,

$$D_n(M,s) := \sum_{\mathbf{m}\in\mathcal{A}_n(M)}\mathbf{H}_{\mathbf{m},s}[1](0)\le e^{-wMn}\cdot\mathbf{H}^n_{s,w}[1](0).$$

and Property (18) ensures the existence of a function $a(s,w)$ such that

$$D_n(M,s)\le a(s,w)\left(\frac{\lambda(s,w)}{e^{Mw}}\right)^n. \tag{20}$$

This is true in particular when w equals the value $\eta(s,M)$ of Lemma 5.1, and

$$\limsup_{n\to\infty}\,[D_n(M,s)]^{1/n}\le\alpha_M(s). \tag{21}$$

We now prove the converse inequality, with classical technics, that are well-known as the shifting of the mean [5]. Remark that $D_n(M,s)$ satisfies

$$D_n(M,s)=\sum_{r\le Mn}p_n^{(s)}(r)\qquad\text{with}\qquad p_n^{(s)}(r):=\sum_{\substack{\mathbf{m}\in\mathcal{M}^n,\\ c(\mathbf{m})=r}}\mathbf{H}_{\mathbf{m},s}[1](0). \tag{22}$$

For $\eta:=\eta(s,M)$ of Lemma 5.1, consider the random variables $Z_n^{(s,\eta)}$ defined by

$$\mathbb{P}[Z_n^{(s,\eta)}=r]:=\frac{1}{\mathbf{H}^n_{s,\eta}[1](0)}e^{\eta r}p_n^{(s)}(r).$$

Their generating functions $\mathbb{E}\left[\exp(w\,Z_n^{(s,\eta)})\right]$ can be expressed in terms of the n-th iterate of the weighted transfer operator $\mathbf{H}_{s,\eta}$ and $\mathbf{H}_{s,w+\eta}$,

$$\mathbb{E}\left[\exp(wZ_n^{(s,\eta)})\right]=\frac{\mathbf{H}^n_{s,w+\eta}[1](0)}{\mathbf{H}^n_{s,\eta}[1](0)}. \tag{23}$$

Since η is strictly negative, there exists a (complex) neighborhood of $w=0$ for which $\Re(w+\eta)<\eta_0<0$. Then, the spectral decomposition holds and entails a quasi-power expression for the moment generating function,

$$\mathbb{E}\left[\exp(wZ_n^{(s,\eta)})\right] = \left(\frac{\lambda(s,w+\eta)}{\lambda(s,\eta)}\right)^n a(s,w,\eta)\cdot[1+O(\alpha^n)] \tag{24}$$

where $a(s,w,\eta)$ is bounded and α is related to the spectral gap of operators $\mathbf{H}_{s,\eta}$ and $\mathbf{H}_{s,w+\eta}$. If η and $w+\eta$ belong to compact sets included in $\Re(w)<0$, we can choose $|\alpha|\le\alpha_0$ uniformly in w. Moreover, the function $U_{[s]}$ defined by $U_{[s]}(w) := \Lambda_{[s]}(w+\eta)-\Lambda_{[s]}(\eta)$ is analytic around $w=0$ because $\Lambda_{[s]}(w)$ is analytic around $\eta<0$. At $w=0$, the derivative $U'_{[s]}(0)$ equals $\Lambda'_{[s]}(\eta)=M$. Furthermore, the second derivative $U''_{[s]}(0)$ is non zero by hypothesis.

We then apply the Quasi-Powers Theorem [17, 18, 19] which proves that the variables $Z_n^{(s,\eta)}$ follow an asymptotic Gaussian law: The probability

$$\Pi_n := \mathbb{P}\left[Z_n^{(s,\eta)}\in[Mn-\sqrt{n},Mn]\right] = \frac{1}{\mathbf{H}^n_{s,\eta}[1](0)} \sum_{Mn-\sqrt{n}\le r\le Mn} p_n^{(s)}(r)e^{\eta r}$$

can be approximated by the corresponding probability of the Gaussian distribution

$$\Pi_n = b(s)+O(\frac{1}{\sqrt{n}}), \qquad \text{with} \qquad b(s) := \frac{1}{\sqrt{2\pi}}\int_{\frac{-1}{\sqrt{U''_s(0)}}}^{0} e^{-\frac{t^2}{2}}\,dt.$$

For large enough n, one has $\Pi_n > b(s)/2$, so that, with (22) and (18),

$$e^{\eta(Mn-\sqrt{n})}D_n(M,s) \ge \Pi_n\cdot\mathbf{H}^n_{s,\eta}[1](0) \ge \frac{b(s)}{2}\,\mathbf{H}^n_{s,\eta}[1](0) \ge d(s,\eta)\,\lambda(s,\eta)^n,$$

for some bounded function $d(s,\eta)$. Therefore,

$$D_n(M,s) \ge \frac{b(s)}{2}\,e^{\eta\sqrt{n}}\,d(s,\eta)\cdot\left(e^{-\eta M}\lambda(s,\eta)\right)^n, \tag{25}$$

and finally

$$\liminf_{n\to\infty}\,[D_n(M,s)]^{1/n} \ge \alpha_M(s).$$

With (21), this ends the proof of the Lemma. ∎

The next Lemma (proven in [6]) describes the main properties of function $\alpha_M(s)$ and ends the proof of Theorem 1.1.

Lemma 4.3. *Let $M\in]\gamma(c),\mu(c)[$ and $s\in\mathcal{V}_M$. The function $s\mapsto\alpha_M(s)$ is well-defined. On the neighborhood $\mathcal{V}_M$, α_M is a continuous function, strictly decreasing, and there exists a unique value $s=s_M\in\mathcal{V}_M$ such that $\alpha_M(s)=1$.*

5. Asymptotic behaviour of s_M when $M\to+\infty$

In this Section, we consider boundary triples (I,T,c). In this case, the Dirichlet series $P(s)$ defined in (8) has an abscissa of convergence $s=1$, is divergent at $s=1$ and the average $\mu(c) := \lim_{w\to 0^-}\Lambda'_w(1,w)$ is infinite, so that the second derivative $\Lambda''_{w^2}(1,w)$ tends to $+\infty$ when $w\to 0^-$. It is then strictly positive when (s,w) is a point of the interior of $\mathcal{S}_1$ sufficiently near $(1,0)$, and Theorem 1.1 can be applied without any restriction.

We wish to describe the asymptotic behaviour of $\dim(F_M)$ when M goes to $+\infty$. Since, in this case, the point (s_M,w_M) tends to $(1,0)$, the second derivative $\Lambda''_{w^2}(s_M,w_M)$ is non zero, and Theorem 1.1 can be applied. We focus on two particular cases: the Boundary Riemann system (Theorem 1.2 in Section 5.2) and the Boundary triple relative to Euclid dynamical system with cost $c(m)=m$ (Theorem 1.3, in Section 5.3).

We first deal with memoryless schemes, where the dominant eigenvalue admits an explicit expression. The Mellin transform is a very useful tool in this context (Lemma 5.2). Then, Lemma 5.4 shows that the memoryless source which "approximates" the continued fraction process provides relevant informations about the dominant eigenvalue of the continued fraction scheme.

5.1. General facts about s_M

Since M tends to $+\infty$, we let $z := \exp(-M)$, together with $s(z) := s_M, w(z) := -w_M$, and we consider the case when $z \to 0^+$. We denote by $\widetilde{\lambda}(s,w) := \lambda(s,-w)$. Then, the pair (s_M, w_M) is a solution of system $(\mathcal{S})$ if and only if $(s(z), w(z))$ is a solution of system

$$(\overline{\mathcal{S}}): \qquad \widetilde{\lambda}(s,w) = z^w, \qquad \frac{\partial}{\partial w}\widetilde{\lambda}(s,w) = \log z \cdot z^w.$$

When z varies in $]\exp[-\gamma(c)], 0[$, these systems define a parameterized curved denoted by $\mathcal{C}$ which is the set of points $(s(z), w(z))$. The maps $z \mapsto w(z), z \mapsto s(z)$ are analytic for $z \in]\exp[-\gamma(c)], 0[$. When $z = 0$, this is no longer true, and we wish to describe the curve for $z \to 0$. The point $(s(z), w(z))$ tends to $(1,0)$, and the following lemma (proven in [6]) describes the behaviour of $s(z) - 1$.

Lemma 5.1. *For $z \to 0^+$, the behaviours of $s'(z)$ and $w(z)$ are related by*

$$s'(z) \asymp -\frac{1}{h}\frac{w(z)}{z}, \qquad |s(z) - 1| \asymp \frac{1}{h}\int_0^z \frac{w(t)}{t}dt,$$

where $h = -\Lambda_s'(1,0)$ is the entropy of the dynamical system (I,T).

We focus now on two particular (boundary) triples: the boundary Riemann triples $\mathcal{BR}(\alpha)$, and the Euclidean dynamical system with $c(m) = m$.

5.2. Study of the boundary Riemann system $\mathcal{BR}(\alpha)$

We now prove Theorem 1.2. Here, we use the explicit expression (17) of the dominant eigenvalue $\lambda(s,w)$ together with two main lemmas. Lemma 5.2 describes the behaviour of the derivative $\lambda_s'(s,w)$ when $(s,w) \to (1,0)$.

Lemma 5.2. *For the boundary Riemann system $\mathcal{B}R(\alpha)$, one has*

$$\zeta(\alpha)^s \frac{\partial}{\partial w}\widetilde{\lambda}(s,w) = \frac{1}{\alpha(s-1)}(w^{\alpha(s-1)/(\alpha-1)} - 1) + \gamma\left(\frac{w^{\alpha(s-1)/(\alpha-1)}}{\alpha-1} - 1\right) + \\ + O(|s-1| + w^{1/2}), \qquad (s,w) \to (1,0). \tag{26}$$

Proof. It uses Mellin transforms. We assume s near 1 and $0 < w < 1$. The function $\frac{\partial}{\partial w}\widetilde{\lambda}(s,w)$ is considered first as a function of w and denoted by $K_s(w)$,

$$K_s(w) := \frac{-1}{\zeta(\alpha)^s}\sum_{m \in \mathcal{M}} m^{\alpha-1}\exp[-m^{\alpha-1}w]\frac{1}{m^{\alpha s}}.$$

$K_s(w)$ is a harmonic sum (also a version of the generalized polylogarithm) and the Mellin transform $K_s^\star$ of function K_s is then explicit,

$$K_s^\star(u) = \frac{-1}{\zeta(\alpha)^s}\cdot\Gamma(u)\cdot\zeta(\alpha s + (u-1)(\alpha-1)).$$

The transform $K_s^\star$ has poles at the nearby points $u = 0$ (due to function Γ) and $u = \alpha(1-s)/(\alpha-1)$ (due to function ζ). and its existence strip is $\Re(u) > \alpha(1-s)/(\alpha-1)$. The Mellin inversion theorem yields, for any $D > \alpha(1-s)/(\alpha-1)$,

$$K_s(w) = \frac{-1}{\zeta(\alpha)^s}\frac{1}{2i\pi}\int_{D-i\infty}^{D+i\infty}\Gamma(u)\zeta(\alpha s + (u-1)(\alpha-1))w^{-u}\,du,$$

and shifting the integral to the left leads to

$$\zeta(\alpha)^s \cdot K(s,w) = \frac{-1}{\alpha-1}\Gamma\left(\frac{\alpha(1-s)}{\alpha-1}\right) w^{\alpha(s-1)/(\alpha-1)} - \zeta(\alpha s - \alpha + 1)$$
$$-\frac{1}{2i\pi}\int_{-1/2-i\infty}^{-1/2+i\infty} \Gamma(u)\zeta(\alpha s + (u-1)(\alpha-1))w^{-u}\, du.$$

By well-known properties of the zeta function (its growth is controlled uniformly in vertical strips), the remainder integral is $O(w^{1/2})$, *uniformly* with respect to s in the stated range (s near 1 and $0 < w < 1$). Consequently, one has

$$\zeta(\alpha)^s \cdot K_s(w) = \frac{-1}{\alpha-1}\Gamma\left(\frac{\alpha(1-s)}{\alpha-1}\right) w^{\alpha(s-1)/(\alpha-1)} - \zeta(\alpha s - \alpha + 1) + O(w^{1/2}).$$

Assume now that $s \to 1$. Both quantities

$$\zeta(\alpha s - \alpha + 1) - \frac{1}{\alpha(s-1)} - \gamma, \qquad \Gamma\left(\frac{\alpha(1-s)}{\alpha-1}\right) + \frac{\alpha-1}{\alpha(s-1)} + \gamma$$

are $O(s-1)$. Thus, when both w and $s-1$ tend to 0, Relation (26) holds. ∎

Lemma 5.3. *For the boundary Riemann system $\mathcal{BR}(\alpha)$, the quantity $(s(z)-1)\log z$ tends to 0 when z tends to 0. Moreover $w \asymp z^{(\alpha-1)\zeta(\alpha)}$ for $z \to 0$.*

Proof. There are three main steps.
Step 1. We first remark that, for any boundary Riemann system $\mathcal{BR}(\alpha)$, the quantity $(s(z)-1)\log z$ is *bounded* when z tends to 0. This is due to the fact that the set F_M contains the set E_K of reals whose digits m are less than K with $K = M^{1/(\alpha-1)}$, together with an easy estimate of the Hausdorff dimension of E_K (see [6]).
Step 2. We now prove that z, w, and $(s-1)$ are polynomially related when $z \to 0$: With Lemmas 5.1 and 5.2, the second equation of $(\overline{\mathcal{S}})$ system, and this last result, we first deduce that $|s-1||\log z|$ tends to 0. Using again relation (26) and the second equation of $(\overline{\mathcal{S}})$ system together with this new fact, we obtain that the difference $1 - w^{\alpha(s-1)/(\alpha-1)}$ tends to 0. This entails that $|s-1||\log w|$ tends to 0, and this proves that $|\log z^{(\alpha-1)\zeta(\alpha)}|$ and $|\log w(z)|$ are equivalent, Then, with Lemma 5.1, this entails that $|s-1| = O(z^{(\alpha-1)\zeta(\alpha)/\rho}) = O(w^{(\alpha-1)\zeta(\alpha)/\rho^2})$ (for any $\rho > 1$) and both $w\log^2 z$ and $|s-1|\log^2 w$ are $O(z^{(\alpha-1)\zeta(\alpha)-\epsilon})$ for any $\epsilon > 0$.
Step 3. Finally, with Lemma 5.2, $w = z^{(\alpha-1)\zeta(\alpha)} + O(z^{2(\alpha-1)\zeta(\alpha)-\epsilon})$ when $z \to 0$. ∎

This ends the Proof of Theorem 1.2. Note that the rôle played by the normalization constant $(\alpha-1)\zeta(\alpha)$ is due to the equality

$$\frac{1}{(\alpha-1)\zeta(\alpha)} = \lim \frac{1}{\log w}\frac{\partial}{\partial w}\lambda(s,-w), \qquad (w \to 0^-, (s-1)\log w \to 0). \qquad (27)$$

5.3. Study of the Euclid Dynamical System with $c(m) = m$

Here, we will prove our Theorem 1.3. We first obtain qualitative results, with a comparison between the Euclid dynamical system and the boundary system $\mathcal{BR}(2)$, which plays the rôle of an approximate model, and is denoted by $\overline{x}$.
Consider a compact set $\mathcal{S}_2 = \{(\sigma,\nu) \in \mathcal{S}_1 : \sigma_2 \le \sigma \le 1, \nu_1 \le \nu \le 0\}$, with $\sigma_2 > \sigma_0$, and $\nu_1 < 0$. The explicit expression (19) proves (see [6]) the existence of strictly positive constants a, b such that, for any $(s,w) \in \mathcal{S}_2$

$$a \cdot \overline{\lambda}'_w(s,w) \le \lambda'_w(s,w) \le b \cdot \overline{\lambda}'_w(s,w). \qquad (28)$$

The first two steps of Lemma 5.3 extend to the Euclidean boundary triple. First, the result due to Hensley [14] and recalled in (5) proves that $|s-1|\log z$ is bounded along $\mathcal{C}$. Then, a step quite similar to Step 2 of Lemma 5.3 shows that $(s-1), w$ and z are polynomially related when $z \to 0$. Then, when $(s(z), w(z))$ tends to $(1,0)$ on the curve $\mathcal{C}$, the quantity $(s-1)\log w$ tends to 0.
In order to extend Step 3 of Lemma 5.3 to our present study, we need an exact equivalent of $(1/\log w)\lambda'_w(s,-w)$, when both w and $(s-1)\log w$ tend to 0, which could be similar to (27). The following Lemma is the main step for proving Theorem 1.3 and shows that $\log 2$ is now the good normalization constant.

Lemma 5.4. *Denote by $f_{s,w}$ the dominant eigenfunction of $\mathbf{H}_{s,w}$ and by $\nu_{s,w}$ the dominant eigenmeasure of the dual $\mathbf{H}^*_{s,w}$. Denote by $F(s,w)$ the quantity*

$$F(s,w) = \sum_{m=1}^{\infty} -m\exp[-wm]\int_I \frac{1}{(m+t)^{2s}}\cdot f_1\left(\frac{1}{m+t}\right)dt$$

which deals with the stationary density f_1. For any constant B, $0 < B < \log 2$, the following holds when $(s(z), w(z))$ tends to $(1,0)$ along the curve $\mathcal{C}$:

(i) $\quad F(s,w) = \dfrac{-1}{\log 2}\log w + O(z^B).$

(ii) $\quad ||f_{s,w} - f_1||_1 = O(z^B), \qquad ||\nu_{s,w} - \nu_1||_1 = O(z^B).$

(iii) $\quad \widetilde{\lambda}'_w(s,w) - F(s,w) = O(z^B),$

Sketch of the proof. For (i), we recall that the stationary density f_1 of the Euclidean System satisfies $(\log 2)\cdot f_1(t) = 1/(1+t)$, so that

$$\int_I \frac{1}{(m+t)^{2s}}\cdot f_1\left(\frac{1}{m+t}\right)dt = \frac{J_m(s)}{\log 2} \qquad \text{with} \quad J_m(s) := \int_I \frac{1}{(m+t)^{2s}}\frac{m+t}{m+1+t}dt$$

Remark that $\quad J_m(s) = \dfrac{1}{m^{2s}}\left(1 + O\left(\dfrac{1}{m}\right)\right) \quad$ with a O uniform for s near 1.

The Mellin transform of $F_s(w) = F(s,w)$ is then explicit and given by

$$(\log 2)\cdot F_s^\star(u) = \Gamma(u)\cdot B(s,u) \qquad \text{with} \quad B(s,u) := \sum_{m\geq 1}\frac{1}{m^{u-1}}J_m(s).$$

Note that $B(s,u) = \zeta(2s+u-1) + C(s,u)$ with $|C(s,u)| \leq K|\zeta(2s+u)|$. Then, the poles of $F_s^\star$ near 0 are $u = 0$ and $u = 2-2s$ with $\mathrm{Res}[B(s,u), u = 2-2s] = 1$. Finally,

$$(-\log 2)\cdot F(s,w) = B(s,0) + \Gamma(2-2s)\cdot w^{2(s-1)} + O(|s-1|) + O(w^{1/2})$$

Now, when s is near 1, one has

$$B(s,0) = \zeta(2s-1) + O(1) = \frac{1}{2(s-1)} + O(1), \qquad \Gamma(2-2s) = -\frac{1}{2(s-1)} + O(1).$$

We now use the same arguments as in Lemma 5.3: When $s(z), w(z))$ tends to $(1,0)$ on the curve $\mathcal{C}$, the quantity $(s-1)\log w$ tends to 0, and the relation proves (i). We omit the proof of (ii) (see [6]). For (iii), the difference of interest is a series whose general term is the product of $-m\exp[-wm]$ by

$$\left(\int_I \frac{1}{(m+t)^{2s}}\cdot f_{s,w}\left(\frac{1}{m+t}\right)d\nu_{s,w}(t) - \int_I \frac{1}{(m+t)^{2s}}\cdot f_1\left(\frac{1}{m+t}\right)d\nu_1 t\right).$$

Then, with distortion property,

$$|\widetilde{\lambda}'_w(s,w) - F(s,w)| = O\left(||f_{s,w} - f_1||_1 + ||\nu_{s,w} - \nu_1||_1\right)\cdot\left(\sum_{m=1}^{\infty} m\exp[-wm]\frac{1}{m^{2s}}\right).$$

The second factor is equal, up to a multiplicative factor, to the derivative $\bar{\lambda}'_w(s,w)$. Relation (28) proves that this factor is $O(\log z)$ when $(s(z), w(z))$ tend to $(1,0)$ on the curve $\mathcal{C}$. Using (ii) ends the proof of (iii).■

End of the proof of Theorem 1.3. We know now that $w \asymp z^{\log 2}$, more precisely that $w = z^{\log 2}[1 + O(z^{\log 2 - \epsilon})]$. We conclude with Lemma 5.1.

6. Conclusions and Open Problems

The main two results of this paper (Theorem 1.1 and Theorem 1.3) are of different nature. Theorem 1.1 is a general result which shows that the Hausdorff dimension of a wide class of sets can be characterized in terms of a solution of a differential system which involves the dominant eigenvalues of the weighted transfer operator. It deals with triples (I, T, c) of large growth, and, it is, in a sense, complementary of the multifractal result of [12] which is only obtained in the case of a triple of moderate growth. Is it possible to obtain our result in the $\mathcal{GMG}$–setting? and the result of [12] in the $\mathcal{GLG}$–setting?
On the other hand, Theorems 1.2 and 1.3 deal with particular cases and provide precise asymptotic estimates for the dimension of set F_M, particularly in the continued fraction context. Theorem 1.3 proves that the characterization given in Theorem 1.1 is useful for effective computations, even for dynamical systems with memory. The main idea is to relate systems with memory to memoryless schemes which approximate them. It is clear that these "approximation" techniques are applicable to more general instances of Dynamical systems.
Finally, Theorem 1.1 together with the results of Hanus, Mauldin and Urbański [12] poses an important question: Is it possible to describe a general framework where systematic proven computation of dimensions can be provided ? In the case when the constraints deal with each digit in an independent way, the algorithm proposed by Daudé, Flajolet, Vallée [7], used in [32] and justified by Lhote [25] provides (in polynomial time) proven numerical values for the Hausdorff dimension. The present case is certainly more difficult, since, now, a system (of two equations) has to be solved, whereas there was previously a unique equation to solve.

We precisely described the sets of *reals* whose continued fraction expansion has all its prefix digit averages less than M. For performing the precise analysis of the Euclidean subtractive algorithm (see [33, 34]), one needs precise information on the set of *rational* numbers whose digits in the continued fraction expansion have an average less than M. This discrete problem is more difficult to solve than the present continuous one. In the case of "fast Euclidean Algorithms", relative to costs of moderate growth, the weighted transfer operator $\mathbf{H}_{s,w}$ is analytic at the reference point $(s, w) = (1, 0)$. Then, Tauberian Theorems or Perron's Formula [33, 34, 2] allow a transfer "from continuous to discrete". Here, it does not seem possible to use directly these tools, due to the non–analyticity of $\mathbf{H}_{s,w}$ at $(1, 0)$.

References

[1] Baladi, V., *Positive Transfer operators and decay of correlations,* Advanced Series in non linear dynamics, World Scientific, 2000.

[2] Baladi, V., and Vallée, B., Distributional analyses of Euclidean algorithms, to appear in *Proceedings of ANALCO'04,* also Les Cahiers du GREYC, Université de Caen, 2004.

[3] BALADI, V., AND VALLÉE, B., Euclidean algorithms are Gaussian, available from the ArXiv, submitted, also Les Cahiers du GREYC 2004.

[4] BESICOVITCH, A.S., Sets of fractional dimensions: On rational approximation to real numbers. *J. London Math. Society*, vol 9, pp. 126-131, 1934.

[5] BILLINGSLEY, P., *Probability and Measure* John Wiley & Sons, 1979

[6] CESARATTO, E. and VALLÉE, B. Hausdorff dimension of real numbers with bounded digit averages, Long version of this paper, les Cahiers du GREYC 2004, Submitted.

[7] DAUDÉ, H., FLAJOLET, P., VALLÉE B., An average-case analysis of the Gaussian Algorithm for lattice reduction *Combinatorics, Probability and Computing* No 6 pp 397–433, 1997.

[8] EGGLESTON, H., The fractional dimension of a set defined by decimal properties. *Quaterly Journal of Mathematics*, Oxford Series, 20, pp. 31-36, 1949.

[9] FALCONER, K., *Fractal Geometry -Mathematical Foundations and Applications*, John Wiley & Sons, New York, 1990.

[10] FALCONER, K., *Techniques in Fractal Geometry*, John Wiley & Sons, New York, 1997.

[11] GOOD, I. J. The fractional dimensional theory of continued fractions, *Math. Proc. Cambridge Philos. Soc. 37 (1941) pp 199-228.*

[12] HANUS, P., MAULDIN, D., URBAŃSKI, M., Thermodynamic formalism and multifractal analysis of conformal infinite iterated function system. *Acta Math. Hungarica*, Vol. 96, pp. 27-98, 2002.

[13] HENSLEY, D., The Hausdorff dimensions of some continued fraction Cantor sets, *Journal of Number theory* 33, (1989) pp 182-198

[14] HENSLEY, D., Continued Fraction Cantor sets, Hausdorff dimension, and functional analysis, *Journal of Number Theory* 40 (1992) pp 336-358.

[15] HENSLEY, D., A polynomial time algorithm for the Hausdorff dimension of a continued fraction Cantor set, *Journal of Number Theory* 58 pp 9–45, 1996.

[16] HENSLEY, D., The statistics of the continued fraction digit sum, *Pacific Journal of Mathematics*, Vol. 192, No2, 2000.

[17] HWANG, H.-K., *Théorèmes limite pour les structures combinatoires et les fonctions arithmétiques,* PhD thesis, Ecole Polytechnique, Dec. 1994.

[18] HWANG, H.-K., Large deviations for combinatorial distributions: I. Central limit theorems, *The Annals of Applied Probability* 6 (1996) 297–319.

[19] HWANG, H.-K., On convergence rates in the central limit theorems for combinatorial structures, *European Journal of Combinatorics* 19 (1998) 329–343.

[20] JARNIK, V., Zur metrischen Theorie der diophantischen Approximationen. *Prace Mat. Fiz.* 36, pp. 91-106, 1928.

[21] JARNIK, V., Über die simultanen diophantischen Approximationen. *Math. Zeit.*, 33, pp. 505-543, 1931.

[22] KATO, T., *Perturbation Theory for Linear Operators,* Springer-Verlag, 1980.

[23] KHINCHIN, A. I., *Continued Fractions,* Dover Publications, Mineola, New York, 1997.

[24] LASOTA, A., MACKEY, M., *Chaos, Fractals and Noise; Stochastic Aspects of Dynamics*, Applied Mathematical Science 97, Springer, 1994.

[25] LHOTE, L., Computation of a Class of Continued Fraction Constants, *Proceedings of ANALCO'04*, to appear

[26] MANDELBROT, B., Intermittent turbulence in self-similar cascades: divergence of high moments and dimension of the carrier, *J. of Fluid Mech.* 62, pp. 331-358, 1974.

[27] MAULDIN, D., URBAŃSKI, M. Dimensions and measures in infinite iterated function systems. *Proc. London Math. Soc.* (3) 73, pp 105–154, 1996.

[28] Mauldin, D., Urbański, M. Conformal iterated function systems with applications to the geometry of continued fractions, *Trans. Amer. Math. Soc.* 351 (1999) pp 4995–5025.

[29] Peyrière, J., Calculs de dimensions de Hausdorff, *Duke Math. J.* Vol. 44., No 3, pp. 591-600, 1977.

[30] Pesin, Y. B., Dimension Theory in dynamical systems: contemporary views and applications, Chicago Lectures in Mathh., The university of Chicago Press (1997).

[31] Shallit, J., Real numbers with bounded partial quotients. A survey. *L'Enseignement Mathématique*, t. 38, pp 151-187, 1992.

[32] Vallée, B., Fractions Continues à contraintes périodiques. *Journal of Number Theory* , 72, pp.183-235, 1998.

[33] Vallée, B., Dynamical Analysis of a Class of Euclidean Algorithms, *Theoretical Computer Science* vol 297/1-3 (2003) 447–486.

[34] Vallée B., Digits and Continuants in Euclidean Algorithms: Ergodic vs Tauberian Theorems. *Journal de Théorie des Nombres de Bordeaux* pp 531–570, 2000.

Eda Cesaratto
Facultad de Ingeniería, Universidad de Buenos Aires, Paseo Colón 850, Buenos Aires, Argentina. ecesara@fi.uba.ar

Brigitte Vallée
CNRS UMR 6072, GREYC, Université de Caen, F-14032 Caen, France
brigitte.vallee@info.unicaen.fr

Trends in Mathematics, © 2004 Birkhäuser Verlag Basel/Switzerland

Large Deviation Analysis of Space-Time Trellis Codes

Adriana Climescu-Haulica

ABSTRACT: *A method for computing the probability of symbol error for space-time trellis codes is developed by means of results from large deviation theory for quadratic Gaussian forms. The issue is that in order to build a space-time trellis code with small probability of error it is sufficient to find a metric for the Viterbi decoding algorithm for which there exists a Cramer's rate function who minimizes a criterion on the transfer function of the trellis system. In particular, this approach gives a theoretical explanation for a simulation result of* [2].

The space-times coding is appropriate to multiple input multiple output antenna transmission. Assume a communication channel with n transmit and m receive antennas. Let $\mathcal{A}$ be the alphabet of information symbols and let $L \in \mathbb{N}^*$ denotes the information stream duration. A space-time encoder is a bijective application $\mathcal{E} : \mathcal{A}^L \to \mathcal{C}$ where $\mathcal{C}$ is a subset of the set of $\mathbb{N}$ valued $n \times L$ matrices and denotes the codewords set. The encoder associated with a trellis code is a finite-memory system: its output depends on the input sequence and on the state of the device. Its behavior is completely described by a signal flow graph admitting a rational transfer function. Figure 1. shows an example of 8 states space-time trellis code and its associated signal flow graph.

At time l the code symbols c_i^l, $1 \leq i \leq n$ associated with an information symbol s^l are transmitted simultaneously from the corresponding antenna.The signal at each receive antenna is a noisy superposition of the channel distorted versions of the n transmitted signals. Assume the same time invariant channel model as one considered in [3], where the construction of space-time trellis codes was first time derived:

$$r_j^l = \sum_{i=1}^{n} h_{ij} c_i^l + z_j^l$$

where r_j^l, is the signal received at instant l by the j-th antenna, $j = 1, \ldots m$, $H = (h_{ij})$ is a $n \times m$ random matrix describing the fading channel whose elements are complex Gaussian random variables, circularly distributed with zero mean and variance one, $\mathbf{c} = (c_i^l)$ denotes the transmitted codeword and each z_j^l is a zero mean, complex Gaussian random variable, circularly distributed.

Let T denotes the transfer function associated with the space time trellis. It is known that

$$\left.\frac{\partial T(x,y)}{\partial x}\right\|_{x=1} = \sum_{d=d_f}^{\infty} A_d y^d \doteq F(y),$$

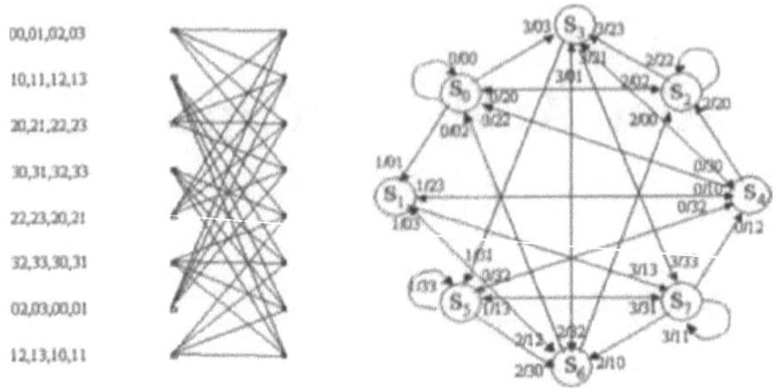

Figure 0.1 Example of space-time trellis code and its state diagram

where d_f denotes the free Hamming distance of the code and A_d is the number of codewords with Hamming weight d. Let E_d denotes the event that the error arise from a path with Hamming weight d. Assuming the trellis is regular and such that there is no path which re-emerges with the all-zero path, having diverged at some point in the past, the probability of an error event satisfies

$$P(E) = P\left(\cup_d E_d\right) = \sum_{d=d_f}^{L} P\left(E_d\right) = \sum_{d=d_f}^{L} A_d P\left(\left(\tilde{\mathbf{c}} \neq \mathbf{c}\right) \cap \left(d_{\mathcal{H}}(\tilde{\mathbf{c}}, \mathbf{c}) = d\right)\right)$$

where $\tilde{\mathbf{c}} = (\tilde{c}_i^l)$ denotes the codeword detected from the received signal by a soft decision Viterbi algorithm, based on the Euclidean metric and $d_{\mathcal{H}}(\cdot,\cdot)$ denotes the Hamming distance. The Euclidean distance between $\tilde{\mathbf{c}}$ and $\mathbf{c}$ is defined as

$$d_{\mathcal{E}}^2\left(\mathbf{c}, \tilde{\mathbf{c}}\right) = \sum_{l=1}^{L} \parallel H^* \left(c_{\cdot}^l - \tilde{c}_{\cdot}^l\right) \parallel^2$$

where $*$ denotes Hermitian transposition. Expressing the Hermitian matrix $K_l = \left(c_{\cdot}^l - \tilde{c}_{\cdot}^l\right)\left(c_{\cdot}^l - \tilde{c}_{\cdot}^l\right)^*$ by means of a unitary matrix U_l as $K_l = U_l \Lambda_l U_l^*$, the Euclidean metric above become

$$d_{\mathcal{E}}^2\left(\mathbf{c}, \tilde{\mathbf{c}}\right) = \sum_{l=1}^{L}\sum_{j=1}^{m} h_{\cdot j}^* K_l h_{\cdot j} = \sum_{l=1}^{L}\sum_{j=1}^{m} \beta(l)_{\cdot j}^* \Lambda_l \beta(l)_{\cdot j}$$

where $\beta(l)_{\cdot j} = U_l^* h_{\cdot j}$ is a complex Gaussian vector with zero mean and covariance matrix identity.It follows that

$$d_{\mathcal{E}}^2\left(\mathbf{c}, \tilde{\mathbf{c}}\right) = \sum_{l=1}^{L}\sum_{j=1}^{m}\sum_{i=1}^{n} \lambda_i^l \mid \beta(l)_{ij} \mid^2$$

with λ_i^l, $i = 1, \cdot n$ the eigenvalues of the matrix K_l and each $\mid \beta(l)_{ij} \mid^2$ a $\chi^2(2)$ random variable. When $d_{\mathcal{H}}(\tilde{\mathbf{c}}, \mathbf{c}) = d$, the Euclidean distance can be expressed as a function of d,

$$d_{\mathcal{E}}^{2}\left(\mathbf{c},\tilde{\mathbf{c}}\right) \mid d=\sum_{l=1}^{L(d)}\sum_{i=1}^{n}\lambda_{i}^{l}(d)B_{i}^{l}(d)$$

with $B_i^l(d)$ a $\chi^2(2m)$ random variable. When d goes to infinity, $(L(d))$ is a sequence of integers increasing to infinity.

Following ideas from [1], a large deviation principle is proved for the quadratic Gaussian form expressed in the above relation,under some constraints on the family of eigenvalues $\left(\lambda_i^l(d)\right)$. This result is also an extension of the Gärtner-Ellis theorem. It follows that there exist a rate function $J:[0,+\infty)\rightarrow I\!R$ such that

$$P\left(\sum_{l=1}^{L(d)}\sum_{i=1}^{n}\lambda_{i}^{l}(d)B_{i}^{l}(d)>\epsilon\right)\approx e^{-dJ(\epsilon)}$$

As

$$P\left((\tilde{\mathbf{c}}\neq\mathbf{c})\cap(d_{\mathcal{H}}(\tilde{\mathbf{c}},\mathbf{c})=d)\right)=P\left(\sum_{l=1}^{L(d)}\sum_{i=1}^{n}\lambda_{i}^{l}(d)B_{i}^{l}(d)>0\right)$$

the conclusion is that

$$P(E)\approx F\left(e^{-J(\epsilon)}\right)\quad\text{for small } \epsilon>0.$$

This result explains that the better performance obtained in [2] for the (8,4,4) Reed Muller code, where the codewords are split evenly among two transmit antenna, than for a usual 1/2 space-time code is due to the behavior of theirs trellis transfer function with respect to the Euclidean norm.

References

[1] B.Bercu, F. Gamboa, and A. Rouault, Large deviations for quadratic forms of stationary Gaussian processes, *Stoch.Proc. and Appl.*, 71 (1997) 75-90

[2] E. Biglieri, G. Taricco, and A.Tulino, Performance of Space-Time Codes for a Large Number of Antennas, *IEEE Trans. Inform. Theory*, vol. 48, July 2002, 1794-1803

[3] V. Tarokh, N.Seshadri, and A.R.Calderbank, Space-Time codes for high data rate wireless communication, *IEEE Trans. Inform. Theory*, vol. 44, March 1998,744-765

Adriana Climescu-Haulica
Laboratoire de Modelisation et Calcul, Institut d'Informatique et Mathématiques Appliquées de Grenoble
51, rue des Mathématiques,38041-Grenoble cedex 9 France
email: adriana.climescu@imag.ca

Trends in Mathematics, © 2004 Birkhäuser Verlag Basel/Switzerland

A Zero-One Law for First-Order Logic on Random Images

David Coupier, Agnès Desolneux, and Bernard Ycart

ABSTRACT: *For a $n \times n$ random image with independent pixels, black with probability $p(n)$ and white with probability $1-p(n)$, the probability of satisfying any given first-order sentence tends to 0 or 1, provided both $p(n)n^{\frac{2}{k}}$ and $(1-p(n))n^{\frac{2}{k}}$ tend to 0 or $+\infty$, for any integer k. The result is proved by computing the threshold function for basic local sentences, and applying Gaifman's theorem.*

1. Introduction

The motivation for this work came for the Gestalt theory of vision (see [3] and references therein), a basic idea of which is that the human eye focuses first on remarkable or unusual features of an image, i.e. features that would have a low probability of occurring if the image were random. Hence the natural question: which properties of a random image have a low or high probability? Here we shall deal with the simplest model for random images:

Definition 1.1. *Let n be a positive integer. Consider the set $X_n = \{1,\ldots,n\}^2$, called the* pixel set. *An* image of size $n \times n$ *is a mapping from X_n to $\{0,1\}$ (white or black). Their set is denoted by E_n. It is endowed with the product of n^2 independent copies of the Bernoulli distribution with parameter p, that will be denoted by $\mu_{n,p}$:*

$$\forall \eta \in E_n\,, \quad \mu_{n,p}(\eta) = \prod_{i,j=1}^{n} p^{\eta(i,j)}(1-p)^{1-\eta(i,j)}\,.$$

A random image of size $n \times n$ and level p, *denoted by $\mathfrak{I}_{n,p}$, is a random element of E_n with distribution $\mu_{n,p}$.*

In other words, a random image of size $n \times n$ and level p is a square image in which all pixels are independent, each being black with probability p or white with probability $1-p$.

We shall use the elementary definitions and concepts of first-order logic on finite models, such as described for instance in Ebbinghaus and Flum [4]. Gaifman's theorem ([8] and [4] p. 31) shows that first-order sentences are essentially local. They can be logically reduced to the appearance of fixed sub-images (precise definitions will be given in section 2). Assume p is fixed. Then as n tends to infinity, any given sub-image of fixed size should appear somewhere in the random image $\mathfrak{I}_{n,p}$, with probability tending to 1: this is the two dimensional version of the well known "typing monkey" paradox. It justifies intuitively that the zero-one law should hold for fixed values of p. Our main result is more general.

Theorem 1.2. *Let $p(n)$ be a function from $\mathbb{N}$ into $[0,1]$ such that:*

$$\forall k = 1, 2, \ldots, \lim_{n\to\infty} n^{\frac{2}{k}} p(n) = 0 \text{ or } +\infty \text{ and } \lim_{n\to\infty} n^{\frac{2}{k}}(1 - p(n)) = 0 \text{ or } +\infty .$$

Let A be a first-order sentence. Then:

$$\lim_{n\to\infty} Prob[\mathfrak{I}_{n,p} \models A] = 0 \text{ or } 1 .$$

Zero-one laws have a long history (cf. Compton [2] for a review and chapter 3 of [4]). The first of them was proved independently by Glebskii et al. [9] and Fagin [6]. It applied to the first-order logic on a finite universe without constraints, and uniform probability. As an example, interpret the elements of E_n as directed graphs with vertex set $\{1, \ldots, n\}$, by putting an edge between i and j if pixel (i, j) is black. Then $\mathfrak{I}_{n,p}$ becomes a random directed graph (or digraph) with edge probability p (see for instance [11, 12], or [1] for a general reference). As a particular case of the Glebskii et al. – Fagin theorem, the zero-one law holds for first-order propositions on random digraphs. However, first-order logic on images is more expressive than on digraphs, since the geometry of images is not preserved in the graph interpretation. For instance, "there exists a horizontal segment of 5 neighboring black pixels" is a first order sentence on images, not on digraphs, where neither "horizontal" nor "neighbor" can be expressed.

The theory of random (undirected) graphs was inaugurated by Erdös and Rényi [5] (see [1, 16] for general references). The zero-one law holds for random graphs with edge probability p, as a consequence of Oberschelp's theorem [13] on parametric classes (see [4] p. 74 or [16] p. 318). At first, zero-one laws were essentially combinatorial, as they applied to the uniform probability on the set of all structures, corresponding to edge probability $p = \frac{1}{2}$ in the case of graphs. It was soon noticed that they also hold for any fixed value of p. But it is well known that random graphs become more interesting by letting $p = p(n)$ tend to 0 as n tends to infinity. A crucial notion for random graphs is the appearance of given subgraphs ([16] p. 309). The threshold function for the appearance of a given subgraph in a random graph is $p(n) = n^{-\frac{v}{e}}$, where v and e are integers. For $p(n) = n^{-\frac{v}{e}}$, the probability of appearance for certain subgraphs does not tend to 0 or 1. Using the extension technique, ([7, 6] and [4] p. 73), Shelah and Spencer [15] made a complete study of those functions $p(n)$ for which the zero-one law holds for random graphs, and proved in particular that it does for $p(n) = n^{-\alpha}$, for any irrational α. Theorem 1.2 is the analogue for random images of Shelah and Spencer's result. To understand why, first notice that the random image model is invariant through exchanging black and white, together with p and $1-p$. Thus we will consider only functions $p(n)$ tending to 0. We shall define precisely the notion of threshold function in section 3, and prove that all threshold functions for patterns are of type $p(n) = n^{-\frac{2}{k}}$: the zero-one law does not hold for these values. For instance, if $p(n)$ is small (resp.: large) compared to n^{-2}, the probability of having at least one black pixel tends to 0 (resp.: 1). But for $p(n) = n^{-2}$, it tends to $1-e^{-1}$. Theorem 1.2 essentially says that the zero-one law holds for any function $p(n)$ which is not a threshold function.

It is worth pointing out here that theorem 1.2 can be extended to other random structures, along two different directions. Firstly, we chose to restrict the study to binary images, using a single unary relation in the language (cf. section 2). With slight modifications of the proofs, and the values of threshold functions, one could introduce a finite set of "color" unary relations, allowing for the coding of multilevel gray or color images. The other possible generalisation concerns the type of graphs. An image is essentially a colored square lattice. The crucial property of that graph for our proof is that there exists a fixed number of vertices at fixed distance of any vertex (balls have bounded cardinality). Our study extends to any family of graphs with bounded balls. For instance, theorem 1.2 also holds for a randomly colored d-dimensional square lattice with n^d points, up to replacing $n^{\frac{2}{k}}$ by $n^{\frac{d}{k}}$ in its statement.

Section 2 is devoted to first-order logic on images. There we shall discuss basic local sentences (definition 2.2 and [4] p. 31), and reduce them to combinations of "pattern sentences" (definition 2.3), showing that a zero-one law holds for all first-order sentences if it holds for basic local or pattern sentences (proposition 2.4). This will trivially imply that theorem 1.2 holds for fixed values of p. The section will end with two examples of (second-order) sentences whose probability under $\mu_{n,\frac{1}{2}}$ tends to $\frac{1}{2}$.

In section 3, we shall define the notion of threshold function (definition 3.2) and prove that all threshold functions for basic local sentences are of type $n^{-\frac{2}{k}}$ (proposition 3.4). Theorem 1.2 easily follows from propositions 2.4 and 3.4.

2. First-order logic for images

We shall follow the notations and definitions in chapter 0 of [4] for the syntax and semantics of first-order logic. The *vocabulary* is the set of *relations* (or predicates). They apply to the *universe* (or domain). In our case the universe will be the pixel set X_n. Image properties will not only be statements on colors of pixels but also about their geometrical arrangement. Our vocabulary will consist of 1 unary and 4 binary relations. The unary relation C is interpreted as the color: Cx means that x is a black pixel and $\neg Cx$ that it is white. Before defining the binary relations, the geometry of X_n needs to be precised.

The pixel set X_n is embedded in $\mathbb{Z}^2$, and naturally endowed with a *graph structure.* In image analysis (see for instance chapter 6 of Serra [14]), the cases most often considered are:

- the 4-*connectivity.* For $i, j > 0$, the neighbors of (i, j) are:

$$(i+1,j),(i-1,j),(i,j+1),(i,j-1)\;.$$

- the 8-*connectivity.* The 4 diagonal neighbors are also included:

$$(i+1,j+1),(i-1,j+1),(i+1,j-1),(i-1,j-1)\;.$$

At this point a few words about the borders are needed. In order to avoid particular cases (pixels having less than 4 or 8 neighbors), we shall impose a periodic boundary, deciding for instance that $(1, j)$ is neighbor with $(n, j), (n, j-1)$, and $(n, j+1)$, so that the graph becomes a regular 2-dimensional torus. Although it may seem somewhat unnatural for images, without that assumption the zero-one law would fail. Consider indeed the (first-order) sentence "there exist 4 black pixels each having only one horizontal neighbor". Without periodic boundary conditions, it applies to the 4 corners, and the probability for a random image $\mathfrak{I}_{n,p}$ to satisfy it is p^4. From now on, the identification $n+1 \equiv 1$ holds for all operations on pixels.

Once the graph structure is fixed, the relative positions of pixels can be described by binary predicates. In the case of 4-connectivity 2 binary predicates suffice, U (up) and R (right): Uxy means that $y = x + (0,1)$ and Rxy that $y = x + (1,0)$. In the case of 8-connectivity, two more predicates must be added,

D_1 and D_2: D_1xy means that $y = x + (1,1)$ and D_2xy that $y = x + (1,-1)$. For convenience reasons, we shall stick to 8-connectivity. Thus the vocabulary of images is the set $\{C, U, R, D_1, D_2\}$. Once the universe and the vocabulary are fixed, the *structures* are particular models of the relations, applied to variables in the domain. To any structure, a graph is naturally associated ([4] p. 26), connecting those pairs of elements $\{x, y\}$ which are such that Sxy or Syx are satisfied, where S is any of the binary relations. Of course only those structures for which the associated graph is the square lattice with diagonals and periodic boundaries will be called images. As usual, the *graph distance* d is defined as the minimal length of a path between two pixels. We shall denote by $B(x,r)$ the ball of center x and radius r:

$$B(x,r) = \{\, y \in X_n \,;\ d(x,y) \leq r \,\}$$

In the case of 8-connectivity, $B(x,r)$ is a square containing $(2r+1)^2$ pixels.

Formulas such as Cx, Uxy, $Rxy\ldots$ are called *atoms*. The *first-order logic* ([4] p. 5) is the set of all formulas obtained by recursively combining first-order formulas, starting with atoms.

Definition 2.1. *The set $\mathfrak{L}_1$ of first-order formulas is defined by:*

(i) All atoms belong to $\mathfrak{L}_1$.
(ii) If A and B are first-order formulas, then $(\neg A)$, $(\forall x Ax)$ and $(A \wedge B)$ also belong to $\mathfrak{L}_1$.

Here are two examples of first-order formulas:

(i) $\forall x, y,\ (Rxy \wedge Uyz) \rightarrow D_1xz$,
(ii) $(\exists y\ (Rxy \wedge Uyz)) \leftrightarrow D_1xz$

Notice that any image satisfies them both: adding the two diagonal relations D_1 and D_2 does not make the language any more expressive. The only reason why the 8-connectivity was preferred here is that the corresponding balls are squares.

We are interested in formulas for which it can be decided if they are true or false for any given image, i.e. for which all variables are quantified. They are called *closed formulas*, or *sentences*. Such a sentence A defines a subset A_n of E_n: that of all images η that *satisfy* A ($\eta \models A$). Its probability for $\mu_{n,p}$ will still be denoted by $\mu_{n,p}(A)$.

$$\mu_{n,p}(A) = Prob[\mathfrak{I}_{n,p} \models A] = \sum_{\eta \models A} \mu_{n,p}(\eta)\ .$$

Gaifman's theorem ([4] p. 31), states that every first-order sentence is equivalent to a boolean combination of *basic local sentences*.

Definition 2.2. *A basic local sentence has the form:*

$$\exists x_1 \ldots \exists x_m \left(\bigwedge_{1 \leq i < j \leq m} d(x_i, x_j) > 2r \right) \wedge \left(\bigwedge_{1 \leq i \leq m} \psi_i(x_i) \right), \tag{1}$$

where:

- *m and r are fixed nonnegative integers,*
- *for all $i = 1, \ldots, m$, $\psi_i(x) \in \mathfrak{L}_1$ is a formula for which only variable x is free (not bound by a quantifier), and the other variables all belong to the ball $B(x,r)$.*

For any x and a fixed radius r, consider now a *complete description* $D(x)$ of the ball $B(x, r)$, i.e. a first-order sentence for which all statements concerning pixels at distance at most r of x are either asserted or negated. There exists a single image I_D of size $(2r+1)\times(2r+1)$, centered at x, satisfying it. Thus $D(x)$ can be interpreted as: "the pattern of pixels at distance at most r of x is I_D".

Definition 2.3. *A pattern sentence has the form:*

$$\exists x_1 \ldots \exists x_m \left(\bigwedge_{1\le i<j\le m} d(x_i, x_j) > 2r \right) \wedge \left(\bigwedge_{1\le i\le m} D_i(x_i) \right), \tag{2}$$

where:

- *m and r are fixed nonnegative integers,*
- *for all $i = 1, \ldots, m$, $D_i(x)$ is a complete description of the ball $B(x, r)$.*

Examples of (interpreted) pattern sentences are:

(i) "there exist 3 black pixels",
(ii) "there exists a 3×3 white square",
(iii) "there exist 3 non overlapping 5×5 white squares with a black pixel on the center".

Figure 2.1 gives another illustration. Obviously, pattern sentences are particular

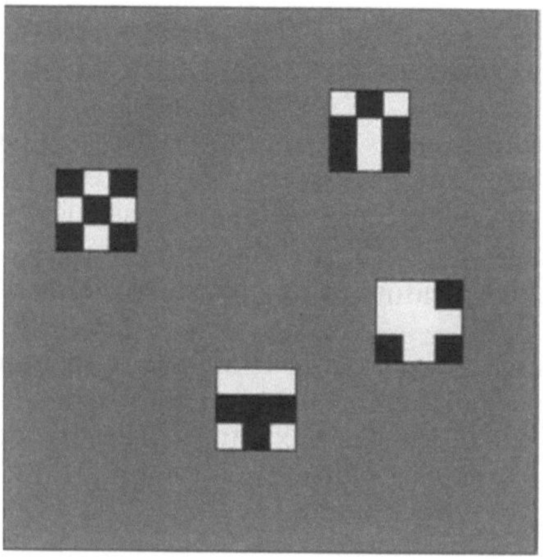

Figure 2.1 Illustration of a pattern sentence, for $m = 4$ and $r = 1$.

cases of basic local sentences. Proposition 2.4 below reduces the proof of zero-one laws for random images to pattern sentences.

Proposition 2.4. *Consider the following three assertions.*

(i) *The probability of any* pattern *sentence tends to* 0 *or* 1.
(ii) *The probability of any* basic local *sentence tends to* 0 *or* 1.
(iii) *The probability of any* first order *sentence tends to* 0 *or* 1.

Then (i) implies (ii) and (ii) implies (iii).

Proof. Observe first that if the probabilities of sentences A and B tend to 0 or 1, then so do the probabilities of $\neg A$ and $A \wedge B$. This follows from elementary properties of probabilities. As a consequence, if the probability of A tends to 0 or 1 for any A in a given family, this remains true for any finite boolean combination of sentences in that family. Thus Gaifman's theorem yields that *(ii)* implies *(iii)*. We shall prove now that every basic local sentence is either unsatisfiable or a finite boolean combination of pattern sentences. Indeed, consider a formula $\psi(x)$ for which only variable x is free, and the other variables all belong to the ball $B(x,r)$. Either it is not satisfiable, or there exists a finite set of $(2r+1)\times(2r+1)$ images (at most $2^{(2r+1)^2}$) which satisfy it. To each of those images corresponds a complete description $D(x)$ which implies $\psi(x)$. So $\psi(x)$ is equivalent to the disjunction of these $D(x)$'s:

$$\psi(x) \leftrightarrow \bigvee_{D(x)\rightarrow\psi(x)} D(x)\,. \tag{3}$$

In formula (1), one can replace each $\psi_i(x_i)$ by a disjunction of complete descriptions. Rearranging terms, one sees that the basic local sentence (1) is itself a finite disjunction of pattern sentences. □

The zero-one law for fixed values of p is an easy consequence of proposition 2.4. Indeed, for fixed p, the probability of any pattern sentence tends to 1. To see why, consider the following sentence:

$$\exists x \left(\bigwedge_{1\leq i\leq m} D_i(x + ((i-1)(2r+1), 0)) \right)\,, \tag{4}$$

interpreted as: "sub-images $I_{D_1}, \ldots, I_{D_m}$ appear in m consecutive, horizontally adjacent balls of radius r". It clearly implies (2). But (4) is equivalent to the appearance of a given sub image on a rectangle of size $(2r+1)\times(m(2r+1))$. This occurs in a random image $\mathcal{I}_{n,p}$ with probability tending to 1 as n tends to infinity. Thus (2) has a probability tending to one of being satisfied by $\mathcal{I}_{n,p}$.

This section ends with two counter-examples of (second-order) sentences the probability of which does not tend to 0 or 1. The first one is "the number of black pixels is even". This is one of the basic examples of second order sentences, that do not belong to first order logic (see [4] example 1.3.4 p. 21 and p. 37). Its probability is

$$\frac{1}{2}\sum_{k=0}^{n^2} \binom{n^2}{k} p^k (1-p)^{n^2-k}(1+(-1)^k) = \frac{1}{2}(1+(1-2p)^{n^2})\,,$$

which tends to $\frac{1}{2}$ for any p such that $0<p<1$.

The second example is more relevant to images. Define a 6-*connected path* as a path where the directions $(-1,1)$ and $(1,-1)$ are forbidden, or more precisely a m-tuple of pixels $(x_1,\ldots,x_m)$, such that for $i = 1,\ldots,m-1$, $x_{i+1} \in x_i \pm \{(1,0),(0,1),(1,1)\}$, and the borders of the image are not crossed (see an illustration on figure 2.2). Consider now the two sentences:

(i) BLR: "there exists a 6-connected path of black pixels from left to right",
(ii) WTB: "there exists a 6-connected path of white pixels from top to bottom".

Some geometrical considerations show that an image satisfies BLR if and only if it does not satisfy WTB (this would not hold for 4- or 8-connected paths: see [14] p. 183). Take now $p = \frac{1}{2}$. Symmetry implies that

$$\mu_{n,\frac{1}{2}}(BLR) = \mu_{n,\frac{1}{2}}(WTB) ,$$

hence both probabilities must be equal to $\frac{1}{2}$.

The sentences BLR and WTB are examples of those properties studied by percolation theory (see Grimmett [10] for a general reference). Actually the random image model that we consider here is a finite approximation of site percolation ([10] p. 24). Using percolation techniques, one can prove that $\mu_{n,p}(BLR)$ tends to 0 if $p < \frac{1}{2}$, to 1 if $p > \frac{1}{2}$.

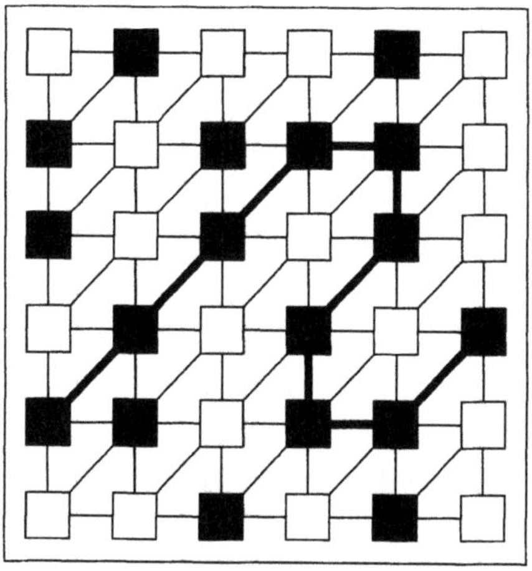

Figure 2.2 A 6-connected path of black pixels from left to right.

3. Threshold functions for basic local sentences

The notions studied in this section have exact counterparts in the theory of random graphs as presented by Spencer [16]. We begin with the asymptotic probability of single pattern sentences, which correspond to the appearance of subgraphs ([16] p. 309).

Proposition 3.1. *Let r and k be two integers such that $0 < k < (2r+1)^2$. Let I be a fixed $(2r+1)\times(2r+1)$ image, with k black pixels and $h = (2r+1)^2 - k$ white pixels. Let $D(x)$ be the complete description of the ball $B(x,r)$ satisfied only by a copy of I, centered at x. Let $\tilde{D}$ be the sentence $(\exists x\ D(x))$. Let $p = p(n)$ be a function from $\mathbb{N}$ to $[0,1]$.*

$$\text{If } \lim_{n\to\infty} n^2 p(n)^k = 0 \text{ then } \lim_{n\to\infty} \mu_{n,p(n)}(\tilde{D}) = 0 . \tag{5}$$

$$\text{If } \lim_{n\to\infty} n^2 p(n)^k (1-p(n))^h = +\infty \text{ then } \lim_{n\to\infty} \mu_{n,p(n)}(\tilde{D}) = 1 . \tag{6}$$

$$\text{If } \lim_{n\to\infty} n^2 (1-p(n))^h = 0 \text{ then } \lim_{n\to\infty} \mu_{n,p(n)}(\tilde{D}) = 0 . \tag{7}$$

Proof. We already noticed the symmetry of the problem: swapping black and white together with p and $1-p$ should leave statements unchanged. In particular the proofs of (5) and (7) are symmetric, and only the former will be given.

For a given x, the probability of occurrence of I in the ball $B(x,r)$ is :

$$\mu_{n,p(n)}(D(x)) = p(n)^k(1-p(n))^h .$$

The pattern sentence $\tilde{D}$ is the disjunction of all $D(x)$'s:

$$\tilde{D} \leftrightarrow \bigvee_{x \in X_n} D(x) .$$

Hence:

$$\mu_{n,p(n)}(\tilde{D}) \leq n^2 p(n)^k (1-p(n))^h ,$$

from which (5) follows.

Consider now the following set of pixels:

$$T_n = \{ (r+1+\alpha(2r+1), r+1+\beta(2r+1)) ,\ \alpha,\beta = 0,\dots, \lfloor \tfrac{n}{2r+1} \rfloor - 1 \} , \qquad (8)$$

where $\lfloor \cdot \rfloor$ denotes the integer part. Call $\tau(n)$ the cardinality of $T(n)$:

$$\tau(n) = \left\lfloor \frac{n}{2r+1} \right\rfloor^2 ,$$

which is of order n^2. Notice that the disjunction of $D(x)'s$ for $x \in T_n$ implies $\tilde{D}$.

$$\bigvee_{x \in T_n} D(x) \rightarrow \tilde{D} .$$

The distance between any two distinct pixels $x, y \in T_n$ is larger than $2r$, and the balls $B(x,r)$ and $B(y,r)$ do not overlap. Therefore the events "$\mathcal{I}_{n,p} \models D(x)$" for $x \in T_n$ are mutually independent. Thus:

$$\begin{aligned} \mu_{n,p(n)}(\tilde{D}) &\geq \mu_{n,p(n)}\left(\bigvee_{x \in T_n} D(x) \right) \\ &= 1 - \left(1 - p(n)^k(1-p(n))^h\right)^{\tau(n)} \\ &\geq 1 - \exp(-\tau(n)p(n)^k(1-p(n))^h) , \end{aligned}$$

hence (6). □

Due to the symmetry of the model, we shall consider from now on that $p(n) < \frac{1}{2}$. Proposition 3.1 shows that the appearance of a given sub-image only depends on its number of black pixels: if $p(n)$ is small compared to $n^{-\frac{2}{k}}$, then no sub image of fixed size, with k black pixels, should appear in $\mathcal{I}(n, p(n))$. If $p(n)$ is large compared to $n^{-\frac{2}{k}}$, all sub-images with k black pixels should appear. Proposition 3.1 does not cover the particular cases $k = 0$ (appearance of a white square) and $k = (2r+1)^2$ (black square). They are easy to deal with. Denote by W (resp.: B) the pattern sentence $(\exists x\ D(x))$, where $D(x)$ denotes the complete description of $B(x,r)$ being all white (resp.: all black). Then $\mu_{n,p(n)}(W)$ always tends to 1 (remember that $p(n) < \frac{1}{2}$). Statements (5) and (6) apply to B, with $k = (2r+1)^2$.

The notion of *threshold function* is a formalisation of the behaviors that have just been described.

Definition 3.2. *Let A be a sentence. A threshold function for A is a function $r(n)$ such that:*

$$\lim_{n\to\infty} \frac{p(n)}{r(n)} = 0 \textit{ implies } \lim_{n\to\infty} \mu_{n,p(n)}(A) = 0 ,$$

and :

$$\lim_{n\to\infty} \frac{p(n)}{r(n)} = +\infty \textit{ implies } \lim_{n\to\infty} \mu_{n,p(n)}(A) = 1 .$$

Notice that a threshold function is not unique. For instance if $r(n)$ is a threshold function for A, then so is $cr(n)$ for any positive constant c. It is customary to ignore this and talk about "the" threshold function of A. For instance, the threshold function for "there exists a black pixel" is n^{-2}.

Proposition 3.1 essentially says that the threshold function for the appearance of a given sub-image I is $n^{-\frac{2}{k}}$, where k is the number of black pixels in I. Proposition 3.4 below will show that the threshold function for a basic local sentence L is $n^{-\frac{2}{k(L)}}$, where $k(L)$ is an integer that we call the *index* of L. Its definition refers to the decomposition (3) of a local property into a finite disjunction of complete descriptions, already used in the proof of proposition 2.4.

Definition 3.3. *Let L be the basic local sentence defined by:*

$$\exists x_1 \ldots \exists x_m \left(\bigwedge_{1\le i<j\le m} d(x_i,x_j) > 2r \right) \wedge \left(\bigwedge_{1\le i\le m} \psi_i(x_i) \right) .$$

If L is not satisfiable, then we shall set $k(L) = +\infty$. If L is satisfiable, for each $i = 1,\ldots,m$, consider the finite set $\{D_{i,1},\ldots,D_{i,d_i}\}$ of those complete descriptions on the ball $B(x_i,r)$ which imply $\psi_i(x_i)$.

$$\psi_i(x_i) \leftrightarrow \bigvee_{1\le j\le d_i} D_{i,j}(x_i) .$$

Each complete description $D_{i,j}(x_i)$ corresponds to an image on $B(x_i,r)$. Denote by $k_{i,j}$ its number of black pixels.

The index *of L, denoted by $k(L)$ is defined by:*

$$k(L) = \max_{i=1}^{m} \min_{j=1}^{d_i} k_{i,j} . \tag{9}$$

The intuition behind definition 3.3 is the following. Assume $p(n)$ is small compared to $n^{-\frac{2}{k(L)}}$. Then there exists i such that none of the $D_{i,j}(x_i)$ can be satisfied, therefore there is no x_i such that $\psi_i(x_i)$ is satisfied, and L is not satisfied. On the contrary, if $p(n)$ is large compared to $n^{-\frac{2}{k(L)}}$, then for all $i = 1,\ldots,m$, $\psi_i(x_i)$ should be satisfied for at least one pixel x_i, and the probability of satisfying L should be large. In other words, $n^{-\frac{2}{k(L)}}$ is the threshold function of L.

Proposition 3.4. *Let L be a basic local property, and $k(L)$ be its index. If L is satisfiable and $k(L) > 0$, then its threshold function is $n^{-\frac{2}{k(L)}}$. If $k(L) = 0$, its probability tends to 1 (for $p(n) < \frac{1}{2}$).*

Proof. Assume L is satisfiable (otherwise its probability is null) and $k(L) > 0$. Let $r(n) = n^{-\frac{2}{k(L)}}$. For $p(n) < \frac{1}{2}$, we need to prove that $\mu_{n,p(n)}$ tends to 0 if $p(n)/r(n)$

tends to 0, and that it tends to 1 if $p(n)/r(n)$ tends to $+\infty$. The former will be proved first.

Consider again the decomposition of L into complete descriptions:

$$L \leftrightarrow \exists x_1 \dots \exists x_m \left(\bigwedge_{1\leq i<j\leq m} d(x_i, x_j) > 2r \right) \wedge \left(\bigwedge_{1\leq i\leq m} \bigvee_{1\leq j\leq d_i} D_{i,j}(x_i) \right) .$$

If $p(n)/r(n)$ tends to 0, there exists i such that:

$$\forall j = 1, \dots, d_i \,, \lim_{n\to\infty} n^2 p(n)^{k_{i,j}} = 0 \,.$$

By proposition 3.1, the probability of $(\exists x \; D_{i,j}(x))$ tends to zero for all $j = 1, \dots, d_i$. Therefore the probability of $(\exists x \, \psi_i(x))$ tends to 0, which implies that $\mu_{n,p(n)}(L)$ tends to 0.

Conversely, for each $i = 1, \dots, m$, choose one of the $D_{i,j}(x)$'s, such that the number of black pixels in the corresponding image is minimal (among all $k_{i,j}$'s). Denote that particular description by $D_i(x)$. Consider now the following pattern sentence, which implies L:

$$\exists x_1 \dots \exists x_m \left(\bigwedge_{1\leq i<j\leq m} d(x_i, x_j) > 2r \right) \wedge \left(\bigwedge_{1\leq i\leq m} D_i(x_i) \right) . \tag{10}$$

As in the proof of proposition 3.1, we shall use the lattice T_n, defined by (8). Remember that its cardinality $\tau(n)$ is of order n^2. The pattern sentence (10) is implied by:

$$\exists x_1 \dots \exists x_m \left(\bigwedge_{1\leq i\leq m} x_i \in T_n \right) \wedge \left(\bigwedge_{1\leq i<j\leq m} x_i \neq x_j \right) \wedge \left(\bigwedge_{1\leq i\leq m} D_i(x_i) \right) . \tag{11}$$

Assume first that $k(L) = 0$. Then necessarily, for each i, the image corresponding to $D_i(x)$ has only white pixels. With $p(n) < \frac{1}{2}$, the probability of observing a $(2r+1) \times (2r+1)$ white image is larger than $\pi = 2^{-(2r+1)^2}$. Since sub-images centered at the points of T_n are independent, the probability of (11) is larger than:

$$1 - \sum_{l=0}^{m-1} \binom{\tau(n)}{l} \pi^l (1-\pi)^{\tau(n)-l} \,,$$

which tends to 1 as n tends to infinity.

Assume now that $k(L) > 0$. The images corresponding to the minimal descriptions D_i need not be all different: renumber *different* descriptions D_i as $D'_1, \dots, D'_{m'}$. Denote by $k(i)$ the number of black pixels of D'_i (hence $k(L) = \max\{k(i)\}$). Let $\pi_i(n)$ be the probability of $D'_i(x)$, for a given x:

$$\pi_i(n) = p(n)^{k(i)} (1 - p(n))^{(2r+1)^2 - k(i)} \,.$$

From the random image $\mathcal{I}_{n,p}$ define the random variable N_i as the number of those pixels $x_i \in T_n$ such that $\mathcal{I}_{n,p}$ is described by $D'_i(x_i)$ on the ball $B(x_i, r)$. Since the different balls do not overlap, N_i has a binomial distribution, with parameters $\tau(n)$ and $\pi_i(n)$. Assuming $p(n)/r(n)$ tends to $+\infty$, it is easy to check that the product $\tau(n)\pi_i(n)$ also tends to infinity. Indeed, $\tau(n)$ is of order n^2, and $r(n)^{k(i)} = n^{-2\frac{k(i)}{k(L)}}$ is at least of order n^{-2}. Therefore the probability that N_i is larger than m tends to

1 for each i, which implies that the probability for all the N_i's to be larger than m also tends to 1. But if all the N_i's are larger than m, then one can be sure that all the $D_i(x_i)$ are satisfied for different centers $x_1, \ldots, x_m$ of the lattice $\mathcal{T}_n$. Therefore $\mathcal{I}_{n,p}$ satisfies (11), hence (10) and L. □

Having characterized the threshold functions of all basic local properties, the proof of theorem 1.2 is now clear. If $p(n)n^{\frac{2}{k}}$ tends to 0 or $+\infty$ for any positive integer k, then by proposition 3.4 the probability of any basic local sentence tends to 0 or 1. This remains true for any boolean combination of basic local sentences (cf. proposition 2.4). By Gaifman's theorem, these boolean combinations cover all first-order sentences. Hence the zero-one law holds for first-order logic.

References

[1] B. Bollobás. *Random Graphs.* Academic Press, London, 1985.

[2] K.J. Compton. 0-1 laws in logic and combinatorics. In I. Rival, editor, *Algorithms and order*, pages 353–383. Kluwer, Dordrecht, 1989.

[3] A. Desolneux, M. Moisan, and J.M. Morel. Meaningful alignments. *Int. J. Computer Vision*, 40(1):7–23, 2000.

[4] H.D. Ebbinghaus and J. Flum. *Finite model theory.* Springer-Verlag, Berlin, 1995.

[5] P. Erdös and A. Rényi. On the evolution of random graphs. *Mat. Kuttató. Int. Közl.*, 5:17–60, 1960.

[6] R. Fagin. Probabilities on finite models. *J. of Symbolic Logic*, 41:50–58, 1976.

[7] H. Gaifman. Concerning measures in first-order calculi. *Israel J. of Mathematics*, 2:1–18, 1964.

[8] H. Gaifman. On local and non local properties. In J. Stern, editor, *Logic Colloquium '81*, pages 105–135. North Holland, Amsterdam, 1982.

[9] Y.V. Glebskii, D.I. Kogan, M.I. Liogonkii, and V.A. Talanov. Range and degree of realizability of formulas in the restricted predicate calculus. *Cybernetics*, 5:142–154, 1969.

[10] G. Grimmett. *Percolation.* Springer-Verlag, New York, 1989.

[11] R.M. Karp. The transitive closure of a random digraph. *Rand. Struct. Algo.*, 1(1):73–94, 1990.

[12] T. Łuczac. The phase transition in the evolution of random digraphs. *J. Graph Theory*, 14(2):217–223, 1990.

[13] W. Oberschelp. Asymptotic 0-1 laws in combinatorics. In D. Jungnickel, editor, *Combinatorial theory*, volume 969 of *L.N. in Mathematics*, pages 276–292. Springer-Verlag, Berlin, 1982.

[14] J. Serra. *Image analysis and mathematical morphology*, volume 1. Academic Press, New York, 1982.

[15] S. Shelah and J. Spencer. Zero-one laws for sparse random graphs. *J. Amer. Math. Soc.*, 1:97–115, 1988.

[16] J. Spencer. Nine lectures on Random Graphs. In P. Bernard, editor, *Ecole d'été de probabilité de Saint-Flour XXI*, volume 1541 of *L.N. in Mathematics*, pages 293–343. Springer-Verlag, New York, 1991.

David Coupier, Agnès Desolneux, and Bernard Ycart
MAP5 CNRS UMR 8145, Université Paris 5, France
{coupier,desolneux,ycart}@math-info.univ-paris5.fr

Trends in Mathematics, © 2004 Birkhäuser Verlag Basel/Switzerland

Coarse and Sharp Transitions for Random Generalized Satisfyability Problems

Nadia Creignou and Hervé Daudé

ABSTRACT: *We study threshold phenomena for random generalized satisfiability problems. These fundamental problems were defined by Schaefer, who gave a complete complexity classification. We give here a complete classification of the nature (coarse or sharp) of the threshold for all generalized satisfiability problems. This new classification is based on easily decidable local properties, and thus provides an exact probabilistic counterpart of Schaefer's complexity result. [T. Schaefer. The complexity of satisfiability problems, in* Proceedings 10th STOC, San Diego (CA, USA), *pages 216–226. Association for Computing Machinery, 1978.]*

1. Introduction

The satisfiability problem (SAT) is the problem of determining whether there exists an assignment of values to the variables of a given Boolean formula which causes it to evaluate to true. While this problem appears in numerous engineering, scientific and operations research applications, it is also unfortunately the prototypical NP-complete problem. Average case analysis and experiments have provided evidence of the existence of a phase transition for the probability of a random formula being satisfiable (see for instance [12]). More precisely, a sharp threshold phenomenon has been observed in the probability of a random k-CNF formula being satisfiable with respect to the ratio of the number of clauses to the number of variables. Experiments have revealed that the critical value of this ratio, c_k, at which the phase transition occurs, coincides with the value at which the average cost of natural solvers for the problem peaks. Therefore it has become clear that one could gather some interesting information in studying the relationship between hardness and random formulae (see [8]).

In order to make general statements about the probabilistic behavior of natural problems, one has to provide a framework which both captures a class of problems which exhibits much of the natural diversity of computational problems and allows a uniform probabilistic analysis. Constraint satisfaction problems are good candidates for that. These problems are interesting in their own right since they occur commonly in practice, in optimization and in settings arising from Artificial Intelligence. In [4] we initiated the study of random generalized satisfiability problems, $\mathsf{SAT}(\mathfrak{F})$, first defined by Schaefer in 1978 [15] (and which correspond to constraint satisfaction problems over the Boolean domain). In such a problem the set of types of constraints that are allowed in the input, $\mathfrak{F}$, is fixed. The major interest of this class of problems is that it has been completely classified in complexity, and thus it has been proved to be an excellent platform to search for a formal basis for empirical observations (see [7] for a survey).

In [4] we gave a first location of the transition for such random problems, in considering a natural probabilistic model: the so-called $G_n(p)$ model arising in random graph theory. Such a model has been independently considered by Molloy [13] for constraint satisfaction over finite domains. In [5] we clarified Friedgut-Bourgain's sharpness criterion [10] for random CSPs and applied it to get a classification of the nature of the threshold for the restricted framework of *symmetric generalized satisfiability problems* (in which, roughly speaking, the truth values 0 and 1 have the same weight). We give now the complete classification of the nature of the threshold for **all generalized satisfiability problems**, together with the scale at which the transition occurs. As in Schaefer's dichotomy theorem, this classification is based on easily decidable local properties on the set of the allowed constraints $\mathcal{F}$. Our main contribution here is twofold. On the one hand we identify all the scales at which a coarse transition can occur. We show that the scale of such a transition is governed by minimal elements. On the other hand we show that supersaturated hypergraphs are a powerful combinatorial tool in to order prove sharpness.

2. Random Boolean Constraint satisfaction problems

2.1. Boolean Constraint satisfaction problems

Consider n Boolean variables $x_1, \ldots, x_n$, a Boolean function f of arity k, $f : \{0,1\}^k \longrightarrow \{0,1\}$. A *constraint generated from* f is given by f and a subset of k indices in $\{1, \ldots, n\}$, $(f, i_1 < \ldots < i_k)$, it is referred to as an application of the function f to $x_1, \ldots, x_n$, $f(x_{i_1}, \ldots, x_{i_k})$. A truth assignment $\Phi : \{x_1, \ldots, x_n\} \longrightarrow \{0,1\}$ satisfies such a constraint if $f(\Phi(x_{i_1}), \cdots, \Phi(x_{i_k})) = 1$.

The set $C_n(f)$ denotes the set of all constraints that can be generated from f over n variables, $\#C_n(f) = \binom{n}{k}$.

Let $\Phi : \{x_1, \ldots, x_n\} \longrightarrow \{0,1\}$ be a truth assignment and $C \in C_n(f)$ be a constraint, $C = (f, i_1 < i_2 < \cdots < i_k)$. We denote by $\Phi(C) = f(\Phi(x_{i_1}), \ldots, \Phi(x_{i_k}))$ and $S_n(C) = \{\Phi \text{ such that } \Phi(C) = 1\}$.

Throughout the paper $\mathcal{F}$ will denote a finite multi-set of constraint functions of fixed arity k over the domain $\{0,1\}$, $\{f_1 \cdots, f_h\}$. The set $C_n(\mathcal{F})$ is the set of all constraints that can be generated over n variables from any constraint function f_i in $\mathcal{F}$, $C_n(\mathcal{F}) = \bigcup_{i=1}^{h} C_n(f_i)$. There are $N_{\mathcal{F}}$ such constraints, $N_{\mathcal{F}} := \#C_n(\mathcal{F}) = \binom{n}{k} \cdot \#\mathcal{F}$.

The set of collections of constraints, also called formulas, from $C_n(\mathcal{F})$ is denoted by $\mathsf{CSP}_n(\mathcal{F})$ (or simply $\mathsf{CSP}(\mathcal{F})$ when no confusion can arise), $\mathsf{CSP}_n(\mathcal{F}) \subseteq \mathcal{P}(C_n(\mathcal{F}))$. We will denote by $\mathsf{SAT}_n(\mathcal{F})$ (or simply $\mathsf{SAT}(\mathcal{F})$) the property for such a collection of constraints of being satisfiable and by $\mathsf{UNSAT}(\mathcal{F})$ the property of being unsatisfiable.

In order to obtain a nontrivial satisfiability property $\mathsf{SAT}(\mathcal{F})$ we restrict our attention to **interesting sets** $\mathcal{F}$, that are sets $\mathcal{F}$ generating no empty constraints, that is for every f in $\mathcal{F}$, $f^{-1}(1) \neq \emptyset$, and in which there exist g_0 and g_1 such that $g_0(\vec{0}) = 0$ and $g_1(\vec{1}) = 0$. These interesting sets correspond to non-0-valid and non-1-valid sets in Schaefer's terminology.

Example 1.

3-SAT: : *for* $0 \leq i \leq 7$, *let* f_i *be the ternary Boolean function such that* $f_i^{-1}(0)$ *corresponds to the dyadic expansion of* i *(for instance* $f_0^{-1}(0) = \{(0,0,0)\}$ *and thus in disjunctive form* $f_0(x,y,z) = (x \vee y \vee z)$*, and* $f_5^{-1}(0) = \{(1,0,1)\}$ *and thus* $f_5(x,y,z) = (\bar{x} \vee y \vee \bar{z})$*). Then,* 3-SAT $=$ SAT$(\{f_0, \cdots, f_7\})$.

k-XOR-SAT: : *let* $h_0(x_1, \cdots, x_k) = x_1 \oplus \cdots \oplus x_k$ *and* $h_1(x_1, \cdots, x_k) = x_1 \oplus \cdots \oplus x_k \oplus 1$*. Then,* k-XOR-SAT $=$ SAT$(\{h_0, h_1\})$.

For any fixed $\mathfrak{F}$, we are interested in studying the probability that a collection of constraints in $C_n(\mathfrak{F})$ is unsatisfiable. Such a collection (or formula) s can be seen as an element of $\{0,1\}^{N_\mathfrak{F}}$ and we will consider the model, analogous to the so-called $G_n(p)$ model of random graph theory, in which each constraint appears in s independently with probability p. For $0 \leq p \leq 1$, let μ_p be the probability measure defined on $\{0,1\}^{N_\mathfrak{F}}$ by $\mu_p(\{s\}) = (1-p)^{N-w(s)}p^{w(s)}$, where $w(s)$ denotes the size of s ($w(s) = \#\{i \ / \ s_i = 1\}$). We will study the asymptotic behavior of $\mu_p(\mathsf{UNSAT}(\mathfrak{F}))$ when $N_\mathfrak{F}$ (or equivalently n) tends to infinity. As a function of p, $\mu_p(\mathsf{UNSAT}(\mathfrak{F}))$ defines an increasing one-to-one correspondence from $[0,1]$ onto $[0,1]$ (see [3]). We will make precise the abrupt change of $\mu_p(\mathsf{UNSAT}(\mathfrak{F}))$ from near 0 to near 1 in considering, for any $c \in [0,1]$, the critical probability $p_c(n,\mathfrak{F})$ defined by $\mu_{p_c}(\mathsf{UNSAT}(\mathfrak{F})) = c$. Indeed, in our model $N_\mathfrak{F} \cdot p$ is the average number of constraints in a random formula. Discussing on the type of constraint functions allowed in $\mathfrak{F}$, we will give the scale of $N_\mathfrak{F} \cdot p_c$ as a function of n. For any set $\mathfrak{F}$, after having made precise the scale at which a transition from satisfiability to unsatisfiability occurs, we will classify the nature, coarse or sharp, of this transition. Let us recall that one says that $\mathsf{UNSAT}(\mathfrak{F})$ has a *sharp transition* if for every $\varepsilon \in]0,1/2]$ the ratio $\frac{p_{1-\varepsilon}(n)-p_\varepsilon(n)}{p_{1/2}(n)}$ tends to 0 as n tends to infinity. If for some $\varepsilon > 0$, $\underline{\lim} \frac{p_{1-\varepsilon}(n)-p_\varepsilon(n)}{p_{1/2}(n)} > 0$, then one says that $\mathsf{UNSAT}(\mathfrak{F})$ has a *coarse transition*.

2.2. Minimal elements

Gathering some useful results on minimal elements will play an essential role in the analysis of the nature of the threshold.

Definition 2.1. [Minimal unsatisfiable collection] *A collection of constraints* $m \in$ CSP$(\mathfrak{F})$ *is said to be* minimal *for* UNSAT$(\mathfrak{F})$ *if* $m \in$ UNSAT$(\mathfrak{F})$ *and for all* m' *strictly contained in* m, $m' \in$ SAT$(\mathfrak{F})$.

For every collection of constraints s, $Var(s)$ denotes the number of distinct variables occurring in s. Let $\mathcal{M}_\mathfrak{F}$ denote the set of minimal formulas for UNSAT$(\mathfrak{F})$. Lower bound of the scale of the phase transition is obtained via minimal elements by the following formula. Let

$$M_r = \#\{m \text{ minimal for } \mathsf{UNSAT}(\mathfrak{F}), w(m) = r\}.$$

Since for every function f in $\mathfrak{F}$, $f^{-1}(1)$ is not empty we have:

$$\mu_p(\mathsf{UNSAT}(\mathfrak{F})) \leq \sum_{r \geq 2} M_r \cdot p^r. \tag{1}$$

Observe that every m in $\mathcal{M}_\mathfrak{F}$ verifies

$$Var(m) \leq (k-1)w(m) + 1.$$

This inequality provides a first upper bound for M_r. Indeed, in observing that $N_{\mathfrak{F}} = \#\mathfrak{F} \cdot \frac{(n)_k}{k!}$, standard bounds for binomial coefficients give:

$$M_r \leq \binom{n}{(k-1)\cdot r+1}\binom{\binom{(k-1)\cdot r+1}{k}\cdot \#\mathfrak{F}}{r} \leq n \cdot \left(\frac{N_{\mathfrak{F}}\cdot k \cdot \exp(k)}{n}\right)^r. \quad (2)$$

Naturally the vacuity of the two following sets will play a central role for getting a lower bound of the scale of the phase transition.

$$\mathcal{M}_{\mathfrak{F}}^{(1)} = \{m \in \mathcal{M}_{\mathfrak{F}} \;/\; Var(m) = (k-1)w(m)+1\},$$

$$\mathcal{M}_{\mathfrak{F}}^{(2)} = \{m \in \mathcal{M}_{\mathfrak{F}} \;/\; Var(m) = (k-1)w(m)\}.$$

Moreover, it turns out that the vacuity of these two sets is determined by two simple local properties of $\mathfrak{F}$ that we define next.

Definition 2.2. *A Boolean function $f : \{0,1\}^k \longrightarrow \{0,1\}$* strongly depends on one component *if there exist $\varepsilon = 0$ or 1, $1 \leq i \leq k$ such that for all $(a_1, \ldots, a_k)$ in $\{0,1\}^k$, $f(a_1,\ldots,a_k) = 1$ implies $a_i = \varepsilon$.*

Definition 2.3. *A Boolean constraint function $f : \{0,1\}^k \longrightarrow \{0,1\}$* strongly depends on a 2-XOR-relation *if there exist two indices $1 \leq i \neq j \leq k$ such that, for all $(a_1,\ldots,a_k)$ in $\{0,1\}^k$, $f(a_1,\ldots,a_k) = 1$ implies $a_i \oplus a_j = 1$.*

Notation 1. *In the sequel $(P1)$ and $(P2)$ will denote the following properties:*

(P1): $\mathfrak{F}$ does not contain any function strongly depending on one component.

(P2): $\mathfrak{F}$ does not contain any function strongly depending on a 2-XOR-relation.

If $\mathfrak{F}$ contains a function u depending on one component, without loss of generality suppose that $u(a_1,\ldots,a_k) = 1$ implies $a_k = 0$. Then, let us consider a formula s of size $(k+1)$ formed by k constraints whose associated constraint function is u and that have no common variables, together with the constraint applying the non-0-valid function g_0 to the set of the last indices of each of the previous constraints. For example $s = (u, 1 < \cdots < k), \cdots, (u, k^2-k+1 < \cdots < k^2), (g_0, k < 2k \cdots < k^2)$. Such a "comb"-formula s is UNSAT and every UNSAT subformula s' of s must contain the g_0-constraint, thus verifying $\#Var(s') = (k-1)w(s')+1$. This shows that if if $(P1)$ does not hold then $\mathcal{M}_{\mathfrak{F}}^{(1)} \neq \emptyset$. The converse is also true and was proved in [4].

If $\mathfrak{F}$ contains a function v depending on a 2-XOR relation, without loss of generality suppose that $v(a_1,\ldots,a_k) = 1$ implies $a_1 \oplus a_k = 1$. Then, let us consider the formula m formed by the three following constraints (whose associated constraint function is v) $C_1 = (v, 1 < \cdots < k), C_2 = (v, 1 < k+1 < k+3 \cdots < 3k-5 < 3k-3)$ and $C_3 = (v, k < k+2 \cdots < 3k-4 < 3k-3)$. Such a "triangle"-formula m is UNSAT and verifies $\#Var(m) = 3(k-1)$. If in addition $\mathfrak{F}$ does not contain any function depending on one component, then m is also minimal UNSAT. Therefore, if $(P1)$ holds but $(P2)$ does not, then $\mathcal{M}_{\mathfrak{F}}^{(1)} = \emptyset$ and $\mathcal{M}_{\mathfrak{F}}^{(2)} \neq \emptyset$. The converse is also true and was proved in [4].

Finally we have proved the following result:

Proposition 2.1. *Let $\mathfrak{F}$ be a multi-set of constraint functions.*

(i) $\mathcal{M}_{\mathfrak{F}}^{(1)} \neq \emptyset$ if and only if $(P1)$ does not hold.

(ii) $\mathcal{M}_{\mathfrak{F}}^{(1)} = \emptyset$ and $\mathcal{M}_{\mathfrak{F}}^{(2)} \neq \emptyset$ if and only if $(P1)$ holds but $(P2)$ does not.

As we will see next this proposition is a first step to establish a clear link between the types of allowed constraints, $\mathfrak{F}$, and the nature of the transition from $\mathsf{SAT}(\mathfrak{F})$ to $\mathsf{UNSAT}(\mathfrak{F})$.

3. Coarse thresholds

Theorem 3.1. *Let $\mathfrak{F}$ be a multi-set of constraint functions. If $\mathfrak{F}$ contains a function depending on one component, i.e., if $(P1)$ does not hold, let $l_{\mathfrak{F}} = \min\{w(m) \;/\; m \in \mathcal{M}_{\mathfrak{F}}^{(1)}\}$, then for every $0 \leq \varepsilon \leq 1$, and n large enough:*

$$\frac{\varepsilon^{1/2}}{l_{\mathfrak{F}} \cdot k \exp(k)} \cdot n^{(l_{\mathfrak{F}}-1)/l_{\mathfrak{F}}} \leq N_{\mathfrak{F}} \cdot p_{\varepsilon}(n) \leq \varphi^{-1}(2\varepsilon) \cdot n^{(l_{\mathfrak{F}}-1)/l_{\mathfrak{F}}},$$

where $\varphi(x) = 1 - \exp(-(\frac{x \cdot k!}{\#\mathfrak{F}})^{l_{\mathfrak{F}}} \cdot \frac{1}{((k-1)l_{\mathfrak{F}}+1)!})$.

Therefore, $\mathsf{SAT}(\mathfrak{F})$ has a coarse transition occurring at the scale $n^{(l_{\mathfrak{F}}-1)/l_{\mathfrak{F}}}$.

Proof: According to the discussion preceding Proposition 2.1, $l_{\mathfrak{F}}$ is well defined and that $2 \leq l_{\mathfrak{F}} \leq (k+1)$. In the following, since no confusion can arise $l_{\mathfrak{F}}$ will be simply denoted by l. The lower bound for the scale of the transition is obtained from (1). Indeed,

$$\mu_p(\mathsf{UNSAT}(\mathfrak{F})) \leq \sum_{r \leq l-1} M_r \cdot p^r + \sum_{r \geq l} M_r \cdot p^r. \tag{3}$$

If $r \leq l-1$, then

$$M_r \leq \binom{n}{(k-1) \cdot r} \binom{\binom{(k-1)\cdot r}{k} \cdot \#\mathfrak{F}}{r} \leq \Big(\frac{N_{\mathfrak{F}} \cdot k \cdot \exp(k)}{n}\Big)^r.$$

If $r \geq l$, then

$$M_r \leq \binom{n}{(k-1) \cdot r + 1} \binom{\binom{(k-1)\cdot r+1}{k} \cdot \#\mathfrak{F}}{r} \leq n \cdot \Big(\frac{N_{\mathfrak{F}} \cdot k \cdot \exp(k)}{n}\Big)^r.$$

Hence, if $\dfrac{N_{\mathfrak{F}} \cdot p \cdot k \cdot \exp(k)}{n} = \dfrac{\varepsilon^{\frac{1}{2}}}{l \cdot n^{\frac{1}{l}}}$, (3) gives

$$\mu_p(\mathsf{UNSAT}(\mathfrak{F})) \leq \varepsilon \cdot (\frac{l-2}{l^2} + \frac{1}{l^l - l^{l-1}}) \leq \varepsilon \text{ (for every } l \geq 2).$$

Therefore, $N_{\mathfrak{F}} \cdot p_{\varepsilon}(n) \geq \dfrac{\varepsilon^{\frac{1}{2}}}{l \cdot k \exp(k)} \cdot n^{1-\frac{1}{l}}$.

To get the second assertion, let us first choose a minimal UNSAT formula m_0 of weight l, such that $\#Var(m_0) = (k-1)l + 1$, and let us consider the set $\mathcal{M}_0$ formed by the $\binom{n}{(k-1)l+1}$ minimal UNSAT formulas isomorphic to m_0.

For each $m \in \mathcal{M}_0$, let X_m indicate if m is a subformula of s or not. As m is of size l, $\mu_p(X_m = 1) = p^l = o(1)$ when $p = \Theta(n^{\frac{l-1}{l}-k})$. Now observe that:

$$\mu_p(\mathsf{UNSAT}(\mathfrak{F})) \geq 1 - \mu_p(\bigcap_{m \in \mathcal{M}_0} X_m = 0). \tag{4}$$

The Janson Inequality (see for instance [2]) gives:

$$\mu_p(\bigcap_{m\in\mathcal{M}_0} X_m = 0) \le \exp(-\nu + \frac{\sum_{j=1}^{l-1}\Delta_j}{2.\cdot(1-p^l)}),$$

where $\nu = \sum_{m\in\mathcal{M}_0} \mu_p(X_m = 1) = p^l \cdot \binom{n}{(k-1)l+1}$, and for $1 \le j \le l-1$

$$\Delta_j = \sum_{m,\ m'\ s.t.\ w(m\ \cap\ m')=j} \mu_p(X_m = X_{m'} = 1) = \sum_{m,\ m'\ s.t.\ w(m\ \cap\ m')=j} p^{2l-j}$$

When $N_{\mathcal{F}} \cdot p = x.n^{\frac{l-1}{l}}$, first observe that $\nu \sim (\frac{x\cdot k!}{\#\mathcal{F}})^l \cdot \frac{1}{((k-1)l+1)!}$. Second, two elements of $\mathcal{M}_0$ having j common constraints share at least $k+(j-1)(k-1)$ variables, thus if $p = O(n^{\frac{l-1}{l}-k})$, then $\Delta_j = O\Big(\binom{n}{2(k-1)l+2-k-(j-1)(k-1)} \cdot (n^{\frac{l-1}{l}-k})^{2l-j}\Big) = O(n^{-1+\frac{j}{l}}) = O(n^{-1/l})$.

Now, if $x = \varphi^{-1}(2\varepsilon)$, then for n large enough, $1 - \mu_p(\bigcap_{m\in\mathcal{M}_0} X_m = 0) \ge \varphi(x) - \varepsilon = \varepsilon$. From (4), we conclude that $\mu_p(\mathsf{UNSAT}(\mathcal{F})) \ge \varepsilon$, when $N_{\mathcal{F}} \cdot p = \varphi^{-1}(2\varepsilon)\cdot n^{\frac{l-1}{l}}$.

Theorem 3.2. *Let $\mathcal{F}$ be a multi-set of constraint functions. If $\mathcal{F}$ does not contain any function depending on one component but contains a function depending on a* 2-XOR-*relation, i.e., if (P1) holds but (P2) does not, let $l_{\mathcal{F}} = \min\{w(m)\ /\ m \in \mathcal{M}_{\mathcal{F}}^{(2)}\}$, then for every $0 \le \varepsilon \le 1$, n large enough:*

$$\frac{\varepsilon^{\frac{1}{2}}}{2k\exp(k)} \cdot n \le N_{\mathcal{F}} \cdot p_\varepsilon(n) \le \psi^{-1}(2.\varepsilon)\cdot n,$$

where $\psi(x) = 1 - \exp(-(\frac{x\cdot k!}{\#\mathcal{F}})^{l_{\mathcal{F}}} \cdot \frac{1}{((k-1)l_{\mathcal{F}})!})$.

Therefore, $\mathsf{SAT}(\mathcal{F})$ *has a coarse transition occurring at the scale n.*

Proof(sketch): Note that $l_{\mathcal{F}}$ is well defined from Proposition 2.1 and that $2 \le l_{\mathcal{F}} \le 3$. As above the lower bound of the scale of the phase transition is obtained from the second point of Proposition 2.1. Here we get, that if $\dfrac{N_{\mathcal{F}}\cdot p\cdot k\cdot\exp(k)}{n} = \dfrac{\varepsilon^{\frac{1}{2}}}{2}$, then $\mu_p(\mathsf{UNSAT}(\mathcal{F})) \le \varepsilon$.

For the upper bound, when $N_{\mathcal{F}}\cdot p = x.n$, the parameters in Janson inequality verify $\nu \sim (\frac{x\cdot k!}{\#\mathcal{F}})^l \cdot \frac{1}{((k-1)l)!}$ and $\Delta_j = O(n^{-1})$, and thus the conclusion follows. □

4. Sharp thresholds

According to the previous section, in order to get a complete classification for the nature of the threshold for $\mathsf{SAT}(\mathcal{F})$ it remains to deal with the following conjecture, which we stated in [4].

Conjecture 4.1. [4] *Let $\mathcal{F}$ be a multi-set of constraint functions. If $\mathcal{F}$ does not contain any constraint function strongly depending on one component or on a* 2XOR-*relation, then* $\mathsf{SAT}(\mathcal{F})$ *exhibits a sharp threshold.*

In [5] we gave a sharpness criterion specified for random CSPs.

Theorem 4.1. [5] *Let $\mathcal{F}$ be a multi-set of constraint functions. If the three following conditions are verified, then the monotone property* $\mathsf{SAT}(\mathcal{F})$ *has a sharp threshold.*

(*D*0) *For each* $c \in (0,1)$, $p_c(n) = O(n^{1-k})$.

(*D*1) *For every* m *minimal for* $\mathsf{UNSAT}(\mathfrak{F})$, $\#Var(m) \leq (k-1)\cdot w(m) - 1$.

(*D*2) *For each* $c \in (0,1)$, *for each* t, *for all* $\delta = (\delta_1, \ldots, \delta_t) \in \{0,1\}^t$, *and all* $\gamma > 0$

$$\mu_{p_c(n)}(s \notin Q_\delta, \#\mathcal{A}_\delta(s) \geq \gamma \cdot n^{k-1}) = o(1),$$

Q_δ denoting the property for $s \in \mathsf{CSP}(\mathfrak{F})$ of having no satisfying assignment with $x_1 = \delta_1, \ldots, x_t = \delta_t$,

$\mathcal{A}_\delta(s)$ denoting, for $s \notin Q_\delta$, the set of constraints C having at least one variable in $\{x_1, \ldots, x_t\}$ and such that $s \cup \{C\} \in Q_\delta$.

If both (*P*1) and (*P*2) hold, then the two first conditions are verified (see [4]). Verifying the last condition is the most challenging task. In [5] we verified it in the restricted framework of *symmetric constraints* (for which, roughly speaking, the truth values 0 and 1 have the same weight). The general case presented here will require the use of a powerful combinatorial tool: supersaturated hypergraphs. We use an ordered version of a theorem of Erdös and Simonovits about supersaturated hypergraphs [9], which has already been used for obtaining sharp threshold results for Ramsey properties of random graphs by Friedgut and Krivelevich [11].

Let us consider h-uniform hypergraphs. Let $K_h^{(h)}(m, \ldots, m)$ denote the following generalization of the complete bipartite graph. Fix h disjoint sets of vertices $V_1, \ldots V_h$, where each V_i has m elements and take all those h-tuples $(x_1, \ldots, x_h)$ such that for every i, $x_i \in V_i$. We are considering graphs on the set of vertices $\{x_1, \ldots, x_n\}$. We say that two disjoint subsets of vertices A and B verify $A < B$ if for all x_i in A and all x_j in B we have $i < j$. Now we define an *ordered copy of* $K_h^{(h)}(m, \ldots, m)$ as an h-uniform hypergraph defined as above with $V_1 < \ldots < V_h$. Thus, the ordered version of the theorem from Erdös and Simonovits about supersaturated uniform hypergraphs [9, Corollary 2, page 184] can be stated as follows.

Theorem 4.2. [9] *Given $c > 0$, there exists a $c' > 0$ such that if an h-uniform hypergraph G^n over n vertices $\{x_1, \ldots, x_n\}$ has at least cn^h hyperedges, then it contains at least $c'n^{hm}$ ordered copies of $L = K_h^{(h)}(m, \ldots, m)$.*

This result enables us to prove the following:

Lemma 4.3. *If* (*P*1) *holds, so does* (*D*2).

Proof:

For more readability we will perform the proof in the special case $k = 3$, it will be clear that it is extendable to any $k \geq 3$. Recall that

$$\begin{aligned}\mathcal{A}_\delta(s) \;=\; & \{C = (g, i_1 < i_2 < i_3) \;/\; i_1 \leq t, g \in \mathfrak{F} \text{ such that} \\ & s \cup C \text{ has no satisfying assignment with } x_1 = \delta_1, \ldots, x_t = \delta_t\},\end{aligned}$$

and that we have to prove that if (*P*1) holds, then for all $\gamma > 0$:

$$\mu_{p_c(n)}(s \notin Q_\delta, \#\mathcal{A}_\delta(s) \geq \gamma' \cdot n^2) = o(1).$$

It is sufficient to prove that for every $\varepsilon = 0$ or 1, for every $f \in \mathfrak{F}$ and all $\gamma' > 0$:

$$\mu_{p_c(n)}(s \notin Q_\delta, \#\mathcal{A}_\delta^{\varepsilon,f}(s) \geq \gamma' \cdot n^2) = o(1),$$

where

$$\mathcal{A}_\delta^{\varepsilon,f}(s) = \{C = (f, i_1 \leq t < i_2 < i_3) \text{ such that } s \cup C \in Q_\delta \text{ and } \delta_{i_1} = \varepsilon\}.$$

The strategy will be as follows. First, for $s \notin Q_\delta$, let us consider the following set:

$$\mathcal{B}_\delta(s) = \{(C_1 = (g, i_1 < i_2 < i_3), C_2 = (g', i_4 < i_5 < i_6)) \\ \text{such that } g, g' \in \mathcal{F}, s \cup \{C_1, C_2\} \in Q_\delta\}.$$

We know that the probability that $\mathcal{B}_\delta(s)$ is dense in the set of conjunctions of two clauses is negligible (see [5, Lemma 5.2]):

$$\text{For all } \nu > 0, \mu_{p_c(n)}(s \notin Q_\delta, \#\mathcal{B}_\delta(s) \geq \nu \cdot n^6) = o(1).$$

Therefore, in order to prove our lemma we will prove that for every $\gamma' > 0$, there exists some $\nu > 0$ such that:

$$\mu_{p_c(n)}(s \notin Q_\delta, \#\mathcal{A}_\delta^{\varepsilon,f}(s) \geq \gamma' \cdot n^2) \leq \mu_{p_c(n)}(s \notin Q_\delta, \#\mathcal{B}_\delta(s) \geq \nu \cdot n^6), \quad (5)$$

thus proving the lemma.

Hence, the trick is to provide a relationship between the cardinality of $\mathcal{A}_\delta^{\varepsilon,f}(s)$ and the one of $\mathcal{B}_\delta(s)$. First recall that $\mathcal{F}$ is interesting, that is for every $\varepsilon = 0, 1$, there exists $g_\varepsilon \in \mathcal{F}$ such that $g_\varepsilon(1-\varepsilon, 1-\varepsilon, 1-\varepsilon) = 0$. Moreover by assumption f does not depend on one component, therefore there exists $\alpha, \beta \in \{0,1\}$ such that $f(\varepsilon, \alpha, \beta) = 1$. These two values α and β, together with the corresponding functions g_α and g_β will be of use later on.

With $\mathcal{A}_\delta^{\varepsilon,f}(s)$ we associate a graph $G_\delta(s)$: the set of vertices is $\{x_{t+1}, \ldots, x_n\}$, and for each constraint $C = (f, i_1 \leq t < i_2 < i_3) \in \mathcal{A}_\delta^{\varepsilon,f}(s)$ we create the edge $\{x_{i_2}, x_{i_3}\}$. Consider in $G_\delta(s)$ an ordered copy of the complete bipartite graph $K_{3,3}$, whose bipartition $A = \{x_{j_1}, x_{j_2}, x_{j_3}\} \cup B = \{x_{j_4}, x_{j_5}, x_{j_6}\}$ verifies $A < B$. Then, we claim that $\{(g_\alpha, j_1 < j_2 < j_3), (g_\beta, j_4 < j_5 < j_6)\} \in \mathcal{B}_\delta(s)$. Indeed, in order to get a contradiction suppose that $s' = s \cup (g_\alpha, j_1 < j_2 < j_3) \cup (g_\beta, j_4 < j_5 < j_6) \notin Q_\delta$. Then, s' has a satisfying assignment I with $x_1 = \delta_1, \ldots, x_t = \delta_t$. By the choice of g_α, I assigns at least one of the literals out of $\{x_{j_1}, x_{j_2}, x_{j_3}\}$ to α, w.l.o.g let us suppose that $I(x_{j_1}) = \alpha$. In the same way we can suppose that $I(x_{j_4}) = \beta$. Thus, I satisfies the constraint $f(\varepsilon, x_{j_1}, x_{j_4})$ since by assumption $f(\varepsilon, \alpha, \beta) = 1$. This contradicts the fact that $\{x_{j_1}, x_{j_4}\}$ is an edge from G_δ. Now, if $G_\delta(s)$ is dense, then according to Theorem 4.2 (with $h = 2$ and $m = 3$) there exists $\nu > 0$ such that $G_\delta(s)$ contains at least $\nu \cdot n^6$ ordered copies of the complete bipartite graph $K_{3,3}$. Therefore, the one-to-one correspondence we have established between the ordered copies of $K_{3,3}$ in $G_\delta(s)$ and $\mathcal{B}_\delta(s)$ proves that if $\#\mathcal{A}_\delta^{\varepsilon,f}(s) \geq \gamma' \cdot n^2$, then $\#\mathcal{B}_\delta(s) \geq \nu \cdot n^6$ for some $\nu > 0$. Therefore, we have proved (5), the desired inequality.

The proof can be extended to any $k \geq 3$. In the general case $\mathcal{B}_\delta(s)$ is formed with conjunctions of $(k-1)$ k-clauses. The graph $G_\delta(s)$ is then a $(k-1)$-uniform hypergraph that contains at least $\gamma \cdot n^{k-1}$ hyperedges and from which every ordered copy of $K_{k-1}^{(k-1)}(k, \ldots, k)$ provides an element of $\mathcal{B}_\delta(s)$.

Finally we have proved the following theorem, which completes our classification.

Theorem 4.4. *Let $\mathcal{F}$ be a multi-set of constraint functions. If $\mathcal{F}$ does not contains any constraint function strongly depending on one component or on a 2XOR-relation, i.e., if both $(P1)$ and $(P2)$ hold, then* SAT($\mathcal{F}$) *exhibits a sharp threshold at scale n.*

5. Conclusion

We have proved that generalized satisfiability problems provide a robust and challenging framework to take a unified look at the relationship between complexity and probabilistic behavior. Besides capturing the usual satisfiability problems like k-SAT, k-XOR-SAT, 1-in-k-SAT and Not-All-Equal-k-SAT, this model –where $\mathcal{F}$ is a multi-set– also captures the problems $(2+p)$-SAT (see [14]), XOR2SAT, 1-in-$(2+p)$-SAT and Not-All-Equal-$(2+p)$-SAT (see [16]), which interpolate smoothly between P and NP by mixing together a polynomial and an NP-complete problems. On one hand our classification re-proves that $(2+p)$-SAT exhibits a sharp threshold (see [1]) and shows that XOR2SAT, in which 3-clauses and 3-XOR-clauses are mixed, also does. On the other hand, for $p < 1$, it proves that 1-in-$(2+p)$-SAT and Not-All-Equal-$(2+p)$-SAT have both a coarse threshold.

Of course, the techniques developed here can be used to establish sharpness results for CSPs over finite domains (not only Boolean), see for example [6]. The classification of the nature of the transition for a large class of random CSPs over finite domains remains an interesting and challenging task.

References

[1] D. Achlioptas, L. Kirousis, E. Kranakis and D. Krizanc. Rigorous results for random (2+p)-SAT. *Theoretical Computer Science*, 265(1-2):109–129, 2001

[2] N. Alon and J. Spencer. The Probabilistic method. Wiley, New-York, 1992.

[3] B. Bollobás. *Random graphs.* Academic Press, 1985.

[4] N. Creignou and H. Daudé. Generalized satisfiability problems: Minimal elements and Phase transitions. *Theoretical Computer Science* 302(1-3): 417–430, 2003.

[5] N. Creignou and H. Daudé. Combinatorial sharpness criterion and phase transition classification for random CSPs. In *Proceedings of the 6th International Symposium on Theory and Applications of Satisfiability Testing*, SAT'2003 Santa Margherita, pages 81–87, 2003. (To appear in a full version in Information and Computation).

[6] N. Creignou, H. Daudé and J. Franco. A sharp threshold for the Renamable Horn and the q-Horn properties. Submitted for publication, 2004.

[7] N. Creignou, S. Khanna and M. Sudan. *Complexity classifications of Boolean constraint satisfaction problems.* SIAM Monographs on discrete mathematics and applications, 2001.

[8] O. Dubois, R. Monasson, B. Selman and R. Zecchina: Editorial. *Theoretical Computer Science*, 265(1-2), 2001.

[9] P. Erdös and M. Simonovits. Supersaturated graphs and hypergraphs. *Combinatorica* 3(2): 181–192, 1982

[10] E. Friedgut and an appendix by J. Bourgain. Sharp thresholds of graph properties, and the k-sat problem. *Journal of the A.M.S.*, 12(4):1017–1054, 1999.

[11] E. Friedgut and M. Krivelevich. Sharp thresholds for Ramsey properties, of random graphs. *Random structures and algorithms*, 17(1):1–19, 2000.

[12] S. Kirkpatrick and B. Selman. Critical behavior in the satisfiability of random Boolean expressions. *Science*, 264:1297–1301,1994.

[13] M. Molloy. Models for random constraint satisfaction problems. *SIAM Journal on Computing*, 32(4):935–949, 2003.

[14] R. Monasson, R. Zecchina, S. Kirkpatrick, B. Selman and L. Troyansky. 2+p-SAT: Relation of typical-case complexity to the nature of the phase transition. *Random structures and algorithms*, 15(3-4):414–435, 1999.

[15] T.J. Schaefer. The complexity of satisfiability problems. In *Proceedings 10th STOC, San Diego (CA, USA)*, pages 216–226. Association for Computing Machinery, 1978.

[16] T. Walsh. The interface between P and NP: COL, XOR, NAE, 1-in-k, and Horn SAT. In *Proceedings of the Eighteenth national conference on Artificial intelligence*, AAAI'2002 Edmonton, pages 695–700, 2002.

Nadia Creignou
LIF, UMR CNRS 6166, Université de la Méditerranée, 163, avenue de Luminy, 13 288 Marseille, France
creignou@lif.univ-mrs.fr

Hervé Daudé
LATP, UMR CNRS 6632, Université de Provence, 39, rue Joliot-Curie, 13 453 Marseille, France
daude@gyptis.univ-mrs.fr

Trends in Mathematics, © 2004 Birkhäuser Verlag Basel/Switzerland

Stochastic Chemical Kinetics with Energy Parameters

Guy Fayolle, Vadim Malyshev, and Serguei Pirogov

ABSTRACT: *We introduce new models of energy redistribution in stochastic chemical kinetics with several molecule types and energy parameters. The main results concern the situations when there are product form measures. Using a probabilistic interpretation of the related Boltzmann equation, we find some invariant measures explicitly and we prove convergence to them.*

1. Introduction

Metabolic pathways in molecular biology are chains or networks of chemical reactions providing redistribution of energy, in particular synthesis of ATP molecules, universal energy stocks in cells. Here we elaborate simple models of energy redistribution. According to a classical approximation, the energy of a molecule can be subdivided in two parts: internal (chemical) energy and kinetic energy. The model is the following.

Assume that there are V molecule types $v \in \{1, \ldots, V\}$, $n_v(t)$ molecules for each type v at time t. Types v can be interpreted as chemical substances with different formulas, different isomers of the same formula, or even as different energy levels (spectrum) of the same molecule.

The total number of molecules $M = \sum_v n_v(t)$ will be conserved. A molecule may be characterized by a pair $(v, T), v = 1, \ldots, V$, where $T \in R_+$ is the kinetic energy of the molecule. Then each molecule of type v at time t has energy

$$E(t) = I(v) + T(t),$$

where $I(v)$ is the internal (or chemical) energy of any molecule of type v, $T(t)$ being the kinetic energy of a concrete molecule at time t. Thus, for any v, t, $I(v)$ are fixed numbers and $T(t)$ are random.

We use the approach usually referred to as *stochastic chemical kinetics*. It appeared in physical papers, see [5], but was also explored also by mathematicians for many models with small V, (see e.g. the reviews [7, 4]). However these models did not consider any energy parameter. Independently of this, Kac [3] considered a beautiful model with mean field collisions. Deeper results in this model appear even recently, see [1]. However, in Kac's model molecules were characterized only by kinetic energies, that is $V = 1$. Our model can be considered as a mixture of these two: there are molecule types and energy parameter.

The plan of the paper is as follows. In section 2, we introduce our probabilistic microscopic model and provide the corresponding Boltzmann type equation. Proof of the finite microtime scaling limit convergence to this equation uses standard technical tools and will be published elsewhere. In section 3 we get deeper results for the one type case with uniform scattering: find invariant measures and prove convergence of the Boltzmann equation for large macrotime. In section 4 we provide many examples, with similar results for multi-type models.

2. Finite time scaling limit

Unless otherwise stated, we consider a system of binary reactions of the form $A+B \to C+D$. We assume energy conservation and random momentary collisions, that is when a pair of different molecules $(v,T),(v',T')$ collide at time t then a new pair $(v_1,U),(v_1',U')$ appears at time $t+0$, so that

$$I(v)+T+I(v')+T' = I(v_1)+U+I(v_1')+U'.$$

Obviously, the reaction is possible only if

$$I(v)+T+I(v')+T' \geq I(v_1)+I(v_1'). \tag{1}$$

We define the following continuous time Markov chain. The state is an array of V vectors $((v,T_i),i=1,\ldots,n_v), v \in V$. Thus, their total length $M=\sum_{v=1}^{V} n_v$ is conserved, but not necessarily n_v. The order of components in each vector $((v,T_i),i=1,\ldots,n_v)$ does not play any role, so that we will consider only functions symmetric in the vector coordinates.
On the time interval $(t,t+dt)$, each pair of molecules $(v,T),(v',T')$ has a collision with probability $\frac{1}{M}\alpha_{vv'}(T,T')dt$. The functions $\alpha_{vv'}(x,y)$ are assumed to be bounded and smooth on R_+^2. As a result of this collision, some pair $(v_1,U),(v_1',U')$ appears, provided that condition (1) holds for at least one pair (v_1,v_1'). Otherwise nothing occurs. The distribution of the new pair is defined by the rules listed hereafter. For any v_1,v_1',v,v',T,T', the conditional densities

$$P((v_1,U),v_1'|(v,T),(v',T')) \geq 0$$

are supposed to satisfy the following properties.

(i) If

$$I(v)+T+I(v')+T' < I(v_1)+I(v_1'),$$

then

$$P((v_1,U),v_1'|(v,T),(v',T')) = 0.$$

(ii) For any v,v',v_1,v_1',T,T', the density function

$$f(U) = P((v_1,U),v_1'|(v,T),(v',T'))$$

is defined on the interval $I=[0,I_v+T+I_{v'}+T'-I(v_1)-I(v_1')]$ and

$$\sum_{v_1,v_1'} \int_I P((v_1,U),v_1'|(v,T),(v',T'))dU = 1.$$

Thus the distribution of the triple (v_1,U,v_1') is entirely defined by

$$P((v_1,U),v_1'|(v,T),(v',T'))$$

and $U' = I(v)+T+I(v')+T'-(I(v_1)+U+I(v_1'))$.
Hence, for V finite sets $\{T_{v,1},\ldots,T_{v,n_v}\}, v=1,\ldots,V$, we have defined a Markov process on R_+^M, which will be denoted by $\mathfrak{L}_M$. It is worth remarking that, when the total energy U is fixed, $\mathfrak{L}_M$ has a compact state space. Then, under some nondegeneracy conditions on α and P, this Markov chain for fixed M approaches, as $t \to \infty$, its unique stationary distribution $\pi^{(M)}(U)$. Our goal will be to study,

under some conditions, the scaling limit $M \to \infty$ for fixed t, and also the large time limit $t \to \infty$.

Let $n_v^{(M)}(A,t)$ denote the number of type v molecules at time t having kinetic energy T in the set $A \subset R_+$. In the limit $M \to \infty$ we have to impose initial conditions at time zero

$$\lim_{M\to\infty} \frac{n_v^{(M)}(A,0)}{M} = \int_A \rho_v(x,0)dx,$$

for some nonnegative functions $\rho_v(x,0), \sum_v \int_{R_+} \rho_v(x,0)dx = 1$, called *concentrations.* Our goal is to prove that, as $M \to \infty$, the sequence of Markov processes $\mathfrak{L}_M$ converges to some deterministic evolution $\mathfrak{L}$ of the concentrations. We state now our first result.

Theorem 2.1. *For any A and t, there exist deterministic limits (in probability)*

$$\lim_{M\to\infty} \frac{n_v^{(M)}(A,t)}{M} = \int_A \rho_v(x,t)dx,$$

where the $\rho_v(x,t)$'s are some non-negative functions satisfying the following Boltzmann type equations

$$\begin{aligned}\frac{d\rho_{v_1}(x,t)}{dt} = \sum_{v,v',v_1'} \int_{R+} &\big[\alpha_{vv'}(y,z)P\big((v_1,x),v_1'|(v,y),(v',z)\big)\rho_v(y,t)\rho_{v'}(z,t) \\ &- \alpha_{vv'}(x,z)P\big((v,y),v'|(v_1,x),(v_1',z)\big)\rho_{v_1}(x,t)\rho_{v_1'}(z,t)\big]dydz,\end{aligned} \tag{2}$$

with the initial condition $\rho_v(x,0)$.

Other reaction types Quite similarly one can consider other types of reactions. For example consider the reaction $A \to B + C$. In this case on the time interval $(t, t+dt)$ each molecule (v,T) with probability $\alpha_v(T)dt$ is transformed to two molecules (note that the scaling is different here). The distribution of the products $(v_1,U),(v_1',U')$ is defined by similar kernels $P((v_1,U),v_1'|(v,T))$ under the condition

$$I_{v_1} + I_{v_1'} \leq I_v + T.$$

3. One type case

3.1. Probabilistic interpretation

We consider in this section the particular situation with only one molecule type v. It will be also assumed that the rates $\alpha(T,T') = \alpha_{vv}(T,T') = \alpha$ and the conditional probabilities $P(U|T,T')$ are uniform on the interval $[0, T+T']$. It turns out that the limiting stationary distribution can be found explicitly. Indeed, equation (2) can be rewritten as

$$\frac{\partial\rho(x,t)}{\partial t} = \alpha \int_x^\infty \frac{ds}{s} \int_0^s \rho(u,t)\rho(s-u,t)du - \alpha\rho(x,t). \tag{3}$$

[Similar equations appeared in [2] in a different context]. Now one can guess a fixed point: it is $\rho(x) = \beta e^{-\beta x}$, but it also can be obtained from a very clear probabilistic picture.

Let us consider *finite particle dynamics*, that is the chain $\mathfrak{L}_M$, the states of which are finite subsets of R_+ with M elements.

Take first the case $M = 2$. Define the chains $\mathcal{L}_2(U)$ as the restriction of $\mathcal{L}_2$ on states with total energy U. Then the chains $\mathcal{L}_2(U)$ are irreducible and nilpotent: that is, already after the first jump we get the stationary distribution $\pi_2(U)$, with T uniformly distributed on $[0, U]$ and $T' = U - T$. Hence, for any initial condition, $\mathcal{L}_2$ is a mixture of $\mathcal{L}_2(U)$. We see that, for any density $f(U)$, the measure

$$\int_{R+} \pi_2(U) f(U) dU$$

is an invariant measure for $\mathcal{L}_2$. Indeed one of these invariant measures is of greatest interest to us. Let the random vector (ξ_1, ξ_2) on R_+^2 be defined by the measure $\mu_{2,\beta}$, such that the two random variables ξ_1, ξ_2 on R_+ be i.i.d with density $\rho(x) = \beta \exp(-\beta x)$. Consider a new random vector (η_1, η_2), where η_1 is picked at random on the interval $[0, \xi_1 + \xi_2]$ and $\eta_2 = \xi_1 + \xi_2 - \eta_1$. This defines a transformation of measures $\mu'_{2,\beta} = W\mu_{2,\beta}$. In fact we have the following

Lemma 3.1. *The measure $\mu_{2,\beta}$ is invariant with respect to W, that is*

$$\mu'_{2,\beta} = \mu_{2,\beta}. \tag{4}$$

Proof: Immediate, since the density of $\xi = \xi_1 + \xi_2$ is $\beta^2 x \exp(-\beta x)$. Then picking a random point on the interval $[0, x]$ yields $\mu_{2,\beta}$, whence equality (4) follows.

In addition, (4) gives

$$\rho(u) = \int_u^\infty \frac{dx}{x} \int_0^x \rho(y)\rho(x - y)dy,$$

which is exactly the stationary form of equation (3).

For $M \geq 3$, the Markov chain $\mathcal{L}_M$ has also irreducible components $\mathcal{L}_M(U)$, consisting of all states $(T_1, \ldots, T_M)$ with $T_1 + \cdots + T_M = U$. For fixed M and U the invariant measure of the chain $\mathcal{L}_M(U)$ is the uniform measure on the simplex $T_1 + \cdots + T_M = U$. An invariant measure on $\mathcal{L}_M$ can be found as follows.

Take M independent particles, having each density $\beta e^{-\beta x}$ on R_+ and let $\mu_{M,\beta}$ denote their joint distribution.

Lemma 3.2. *The measure $\mu_{M,\beta}$ is invariant for $\mathcal{L}_M$.*

Proof: It follows from the previous lemma, because the generator of $\mathcal{L}_M$ is the sum of generators corresponding to all pairs $(i, j), i, j = 1, \ldots, M, i \neq j$.

Remark 3.1. *One can show that $\mathcal{L}_M(U)$ is reversible, by using the classical Kolmogorov's reversibility criteria for Markov with transition rates $\lambda_{\alpha\beta}$, namely*

$$\lambda_{\alpha_1\alpha_2}\lambda_{\alpha_2\alpha_3} \cdots \lambda_{\alpha_k\alpha_1} = \lambda_{\alpha_1\alpha_k}\lambda_{\alpha_k\alpha_{k-1}} \cdots \lambda_{\alpha_2\alpha_1}.$$

See related questions in [8].

3.2. Convergence for Boltzmann equation

According to the above section, we have

$$\lim_{M\to\infty} \lim_{t\to\infty} \frac{n_v^{(M)}(A,t)}{M} = \int_A \beta e^{-\beta x} dx.$$

We will consider now the quantity $\lim_{t\to\infty} \lim_{M\to\infty}$.

Theorem 3.3. *For Boltzmann equation (2), for any initial condition $\rho(x, 0)$, we have*

$$\lim_{t\to\infty} \rho(x,t) = \beta e^{-\beta x}, \quad x \geq 0 \tag{5}$$

Proof: The sketch is the following. First, we prove in the next subsection, under more general assumptions, that any initial distribution converges to some fixed point. Secondly, we will show that there is only one-dimensional manifold of fixed points, namely $\beta e^{-\beta x}, 0 < \beta < \infty$. This will conclude the proof, since β itself is uniquely determined by the initial mean energy

$$T(0) = \lim_{M\to\infty} \frac{1}{M}\sum_{i=1}^{M} T_i(0) = \frac{1}{\beta}.$$

3.3. Local equilibrium condition

We come back here to an arbitrary number of types. We will say that a positive function $f(v,x)$ on $V \times R_+$ with $\sum_v \int f(v,x)dx = C < \infty$, satisfies a *local equilibrium condition* (LE) if, for any γ, γ_1,

$$\sum_{\gamma',\gamma_1'} (w(\gamma,\gamma_1|\gamma',\gamma_1')f(\gamma')f(\gamma_1') - w(\gamma',\gamma_1'|\gamma,\gamma_1)f(\gamma)f(\gamma_1)) = 0, \tag{6}$$

where we use the notation

$$\gamma = (v,x), \quad \sum_\gamma = \sum_v \int dx,$$

and

$$w(\gamma,\gamma_1|\gamma',\gamma_1') =$$
$$\alpha_{v'v_1'}(x',x_1')P\big((v,x),v_1|(v',x'),(v_1',x_1')\big)\delta(x_1-(x'+x_1'+I_{v'}+I_{v_1'}-x-I_v-I_{v_1})).$$

One can assume $C = 1$. Then, in the one type case, this is tantamount to saying that $\mathfrak{L}_2$ has the invariant product form distribution $f(x)f(y)$.

The *fixed point condition* (FP)

$$\sum_{\gamma_1,\gamma',\gamma_1'} (w(\gamma,\gamma_1|\gamma',\gamma_1')f(\gamma')f(\gamma_1') - w(\gamma',\gamma_1'|\gamma,\gamma_1)f(\gamma)f(\gamma_1)) = 0, \tag{7}$$

valid for any γ, follows immediately from (6).

We shall say that $f(\gamma)$ satisfies a *detailed balance condition* (DB) whenever

$$w(\gamma,\gamma_1|\gamma',\gamma_1')f(\gamma')f(\gamma_1') - w(\gamma',\gamma_1'|\gamma,\gamma_1)f(\gamma)f(\gamma_1) = 0, \tag{8}$$

for any $\gamma, \gamma', \gamma_1, \gamma_1'$. In the above one type example, DB condition holds if one chooses

$$f_0 = \beta e^{-\beta x},$$

for any positive β. Note that DB$\rightarrow$LE$\rightarrow$FP.

Let us define the *relative entropy* of f with respect to f_0, assuming both f and f_0 are positive. Farther on, f_0 will be fixed and therefore omitted in the notation, so that

$$H(f) \equiv H(f,f_0) = \sum_\gamma f(\gamma)\log\frac{f_0(\gamma)}{f(\gamma)}. \tag{9}$$

Theorem 3.4. *Assume that there exists some $f_0(\gamma) > 0$ satisfying the local equilibrium condition. Then for any initial $f(\gamma,0)$ with $H(f(.,0))$ finite, the function $f(\gamma) = f(\gamma,t)$, that is the solution of equation (2), does satisfy*

$$\frac{dH(f)}{dt} \geq 0.$$

Moreover, as $t \to \infty$, $f(\gamma, t)$ tends, to some fixed point f_∞ which depends in general on the initial data $f(\gamma, 0)$. LE condition holds for any stationary solution f, that is for any fixed point of (2).

Proof: The integrability of $\frac{df(\gamma)}{dt}$ follows from (2), so that the following conservation law holds

$$\sum_\gamma \frac{df(\gamma)}{dt} = 0. \tag{10}$$

Differentiating (9) and using (10), we get

$$\frac{dH(f)}{dt} = \sum_\gamma \frac{df(\gamma)}{dt} \log\left[\frac{f_0(\gamma)}{f(\gamma)}\right].$$

We rewrite condition (6) as

$$\sum_{\gamma', \gamma_1'} w(\gamma, \gamma_1 | \gamma', \gamma_1') \frac{f_0(\gamma') f_0(\gamma_1')}{f_0(\gamma) f_0(\gamma_1)} = \sum_{\gamma', \gamma_1'} w(\gamma', \gamma_1' | \gamma, \gamma_1),$$

and set for the sake of shortness $\int \equiv \sum_{\gamma, \gamma_1, \gamma', \gamma_1'}$. Then, for any function $f(\gamma)$, we have

$$\int w(\gamma, \gamma_1 | \gamma', \gamma_1') \frac{f_0(\gamma') f_0(\gamma_1')}{f_0(\gamma) f_0(\gamma_1)} f(\gamma) f(\gamma_1) = \int w(\gamma', \gamma_1' | \gamma, \gamma_1) f(\gamma) f(\gamma_1),$$

or, after a change of variables,

$$\int w(\gamma', \gamma_1' | \gamma, \gamma_1) f(\gamma) f(\gamma_1) = \int w(\gamma', \gamma_1' | \gamma, \gamma_1) \frac{f_0(\gamma) f_0(\gamma_1)}{f_0(\gamma') f_0(\gamma_1')} f(\gamma') f(\gamma_1').$$

Let $\varphi(\gamma) = \log\left[\frac{f_0(\gamma)}{f(\gamma)}\right]$. Then

$$\begin{aligned}
\frac{dH(f)}{dt} &= \sum_\gamma \frac{df(\gamma)}{dt} \varphi(\gamma) \\
&= \int \varphi(\gamma) \left[w(\gamma, \gamma_1 | \gamma', \gamma_1') f(\gamma') f(\gamma_1') - w(\gamma', \gamma_1' | \gamma, \gamma_1) f(\gamma) f(\gamma_1) \right] \\
&= \int [\varphi(\gamma) - \varphi(\gamma')] w(\gamma, \gamma_1 | \gamma', \gamma_1') f(\gamma') f(\gamma_1') \\
&= \frac{1}{2} \int [\varphi(\gamma) + \varphi(\gamma_1) - \varphi(\gamma') - \varphi(\gamma_1')] w(\gamma, \gamma_1 | \gamma', \gamma_1') f(\gamma') f(\gamma_1').
\end{aligned}$$

Set for a while

$$\begin{cases}
\xi = \dfrac{f_0(\gamma) f_0(\gamma_1) f(\gamma') f(\gamma_1')}{f(\gamma) f(\gamma_1) f_0(\gamma') f_0(\gamma_1')}, \\[2ex]
\alpha = \dfrac{f_0(\gamma') f_0(\gamma_1')}{f_0(\gamma) f_0(\gamma_1)},
\end{cases}$$

so that

$$\log \xi = \varphi(\gamma) + \varphi(\gamma_1) - \varphi(\gamma') - \varphi(\gamma_1').$$

Then

$$\frac{dH(f)}{dt} = \frac{1}{2} \int \alpha \xi \log \xi \, w(\gamma, \gamma_1 | \gamma', \gamma_1') f(\gamma) f(\gamma_1).$$

On the other hand, from the LE condition,

$$\int \alpha\xi w(\gamma,\gamma_1|\gamma',\gamma_1')f(\gamma)f(\gamma_1) = \int \alpha w(\gamma,\gamma_1|\gamma',\gamma_1')f(\gamma)f(\gamma_1),$$

which yields

$$\frac{dH(f)}{dt} = \frac{1}{2}\int (\xi\log\xi - \xi + 1)\alpha w(\gamma,\gamma_1|\gamma',\gamma_1')f(\gamma)f(\gamma_1) \geq 0,$$

since $\xi\log\xi - \xi + 1 > 0$ if $\xi > 0$, due to the convexity of $\xi\log\xi$.

Assume now that for some $f_0 > 0$ the local equilibrium condition holds. Then it holds also for any other stationary solution f, i.e. satisfying $\frac{df}{dt} = 0$. In fact, note that $\frac{dH(f)}{dt} > 0$ if $f(\gamma)f(\gamma_1) > 0, w(\gamma,\gamma_1|\gamma',\gamma_1') > 0$ and $\xi \neq 1$. Also, if f is a stationary solution of equation (2) then $\frac{df(\gamma)}{dt} = 0$ and hence $\frac{dH(f)}{dt} = 0$. It follows that for any $\gamma,\gamma_1,\gamma',\gamma_1'$ such that $f(\gamma)f(\gamma_1) > 0$ and $w(\gamma,\gamma_1|\gamma',\gamma_1') > 0$ we have $\xi = 1$, that is

$$\frac{f(\gamma')f(\gamma_1')}{f(\gamma)f(\gamma_1)} = \frac{f_0(\gamma')f_0(\gamma_1')}{f_0(\gamma)f_0(\gamma_1)}$$

On the other hand, if $\frac{df(\gamma)}{dt} = 0, f(\gamma) = 0$ and $w(\gamma,\gamma_1|\gamma',\gamma_1') = 0$, then we get $f(\gamma')f(\gamma_1') = 0$ as a consequence of equation (2). Thus, for any γ,γ_1, equation (6) holds. Any solution $f(t)$ of the equation (2) as $t \to \infty$ tends to some stationary solution f_∞, which depends in general on the initial data $f(0)$. In fact, from the proof of theorem 2.1, it follows that f is a stationary solution, i.e. $\frac{df}{dt} = 0$, if and only if $\frac{dH(f)}{dt} = 0$ (provided that (2) holds). This means that $H(f)$ is a Lyapunov function. Consequently, the expected result follows from the general theory of Lyapunov functions and the proof of the theorem is terminated.

3.4. Fixed points and conservation laws

Now we will prove that, for any two fixed points f_0, f, the function $\log\frac{f}{f_0}$ is an additive conservation law. Consider the equation

$$\frac{f(\gamma')f(\gamma_1')}{f(\gamma)f(\gamma_1)} = \frac{f_0(\gamma')f_0(\gamma_1')}{f_0(\gamma)f_0(\gamma_1)}. \tag{11}$$

For $f_0 = 1$, we have

$$\frac{f(\gamma')f(\gamma_1')}{f(\gamma)f(\gamma_1)} = 1, \tag{12}$$

which shows that $\log f$ is an additive conservation law. Vice versa, if there is a set J of additive conservation laws such that

$$\eta_j(\gamma) + \eta_j(\gamma_1) = \eta_j(\gamma') + \eta_j(\gamma_1'), j \in J,$$

then, for any constants c, c_j,

$$f(\gamma) = c\prod_{j\in J}\exp(c_j\eta_j(\gamma))$$

is a solution of (12). Note that additive conservation laws form a linear space. Thus we have proved that any solution of (12) has this form. In the general case (that is if $f_0 \neq 1$), we have

$$\frac{f}{f_0} = c\prod_{j\in J}\exp(c_j\eta_j(\gamma)).$$

It is worth noticing that a nonzero additive conservation law for the chain $\mathfrak{L}_M$ is in fact unique, if the chains $\mathfrak{L}_M(U)$ are irreducible, for all U.

4. Invariant measures for multi-type models

Here we will analyze some cases with $V > 1$, when there exists an invariant measure having a product form.

4.1. Binary reactions without type change

Let for any $v = 1, \ldots, V$ a density $\rho_v(x) > 0$ on R_+ be given. Assume only reactions $v, w \to v, w$ are possible, so that the n_v's are conserved. Then one can introduce finite particle Markov chains $\mathfrak{L}_{n_1,\ldots,n_V}$. Suppose in addition that, for any couple of types (v, w),

$$\alpha_{vw}(T, T') = \alpha_{vw}(T + T'),$$

which means that the rates depend only on the sum of energies.
We need the following definition. Fix a pair (v, w) of types and let ξ_v, ξ_w be independent random variables with joint density $\rho_v(x)\rho_w(y)$. Denote

$$P_{\rho_v\rho_w} = P_{\rho_v\rho_w}(x, y|T) = P(\xi_v = x, \xi_w = y|\xi_v + \xi_w = T)$$

the corresponding conditional distributions, which will be called *canonical kernels* corresponding to the density array $(\rho_1, \ldots, \rho_V)$.
Let $\xi_{v,i}, i = 1, \ldots, n_v$, stand for the energy of the i-th particle of type v.

Theorem 4.1. *Fix an array $\rho_1, \ldots, \rho_V$ and let a system of $\frac{V(V+1)}{2}$ reactions with canonical kernels $P_{\rho_v\rho_w}$ be given. Then, for any $n_1, \ldots, n_V$, the invariant measures of $\mathfrak{L}_{n_1,\ldots,n_V}$ are such that the random variables $\xi_{v,i}$ have independent distributions equal to ρ_v. In the thermodynamic limit, for any initial concentrations of types $(c_1, \ldots, c_V)$ (here the concentrations of types do not change at all), the invariant energy distribution is unique and given by the independent densities ρ_v. Moreover, for any initial energy distribution, there is convergence to this invariant measure. Also, for any array $(\rho_1, \ldots, \rho_V)$ with arbitrary rates $\alpha_{vw}(U)$, there is only one system of kernels for which this array defines an invariant (product form) distribution, these kernels being canonical kernels.*

Proof: Any transition $v, v' \to v, v'$ conserves U and the related measures. Hence, as for the convergence, the argument is similar to that in the previous section. The other statements follow directly from the definitions.

When $\rho_v(x) = \beta e^{-\beta x}$, the kernels are uniform on $[0, T]$, as in the one type case study. Let P^β denote such a kernel. An interesting situation depicted in the next remark arises when

$$\rho_v(x) = \beta_v \exp(-\beta_v x),$$

with different β_v's.

Remark 4.1. *All other cases can be reduced to the simplest one by the following transformation. Given any density $\rho > 0$ and any $\beta > 0$, introduce the one-to-one transformation $U = U(\rho, \beta) : R_+ \to R_+$ such that, for any $x \in R_+$,*

$$\int_0^x \rho(y)dy = \int_0^{Ux} \beta e^{-\beta y} dy.$$

Then

$$P_{vw} = \big(U^{-1}(\rho_v, \beta), U^{-1}(\rho_w, \beta)\big) P^\beta \big(U(\rho_v, \beta), U(\rho_w, \beta)\big).$$

4.2. Unary reactions

Now we want to tackle examples in which the n_v's are not conserved. Then, in general, only

$$\mathfrak{L}^M = \cup_{n_1+\cdots+n_V=M}\mathfrak{L}_{n_1,\ldots,n_V}$$

is a Markov chain. In this subsection, we assume that unary reactions

$$v \to w$$

can take place with rates a_{vw}. Such reactions could be interpreted as isomer to isomer transformations. In case $I_v \geq I_w$ the reaction $v \to w$ always occurs, and kinetic energy T of v-particle becomes kinetic energy $I_v - I_w + T$ of w-particle. The reaction $w \to v$ however occurs only if $T - I_v + I_w \geq 0$, and in this case T kinetic energy T of w-particle becomes the kinetic energy $T - I_v + I_w$ of the v-particle. Consider first the case without binary reactions. Define the following *one-particle* Markov chain: its states are all pairs (v, T), that is $M = 1$. Moreover, assume that there are only two types. Let $I_1 < I_2$. Consider a pair of densities ρ_1, ρ_2, denote ξ_1, ξ_2 corresponding random variables. We call this pair *admissible* if the conditional density of $\xi_1 - (I_2 - I_1)$, on the event $\{\xi_1 > I_2 - I_1\}$, is equal to ρ_2. One example is $\rho_1 = \rho_2 = \beta\exp(-\beta x)$, another being

$$\rho_1(x) = \begin{cases} 0, & \text{for } x < I_2 - I_1; \\ \rho_2(x - I_2 + I_1), & \text{otherwise.} \end{cases}$$

Any invariant measure on $\{1,2\} \times R_+$ can be written as $\pi_1(1,\rho_1) + \pi_2(2,\rho_2)$ with positive coefficients π_i such that $\pi_1 + \pi_2 = 1$. We have for π_1, π_2 the following equations

$$\pi_1 Y_1 a_{12} = \pi_2 a_{21}, \quad Y_1 = \int_{I_2-I_1}^{\infty} \rho_1(x)dx.$$

This case exhibits the highest degree of reducibility, each class containing one or two elements: there is plenty of invariant measures – but this is clearly a very unnatural situation. For an arbitrary M with only two types, we have the product of M chains $\mathfrak{L}^1$, which again leads to a rather unnatural situation.

When there are $V > 2$ types, each class also has a finite number of elements. It is then possible to order the internal energies, assuming for example

$$I_1 \leq I_2 \leq \ldots \leq I_V,$$

and also $a_{vw} > 0, \forall v, w$. If the full energy satisfies $I_m < U < I_{m+1}, m = 1, \ldots, V$ (putting $I_{m+1} = \infty$) then there are no possible jumps to the types $m+1, \ldots, V$, so that the process evolves as a Markov chain $\mathfrak{L}_{1,m}$ with state space $1, \ldots, m$ and rates $a_{vw}, v, w = 1, \ldots, m$. Hence $\mathfrak{L}_{1,m}$ are restrictions of $\mathfrak{L}_{1,V}$. For $m = 1$, it becomes a trivial one-point Markov chain. Let $\pi_{m,v}, v = 1, \ldots, m$ denote the stationary probability of the state v in $\mathfrak{L}_{1,m}$. We have $\pi_{1,1} = 1$.

Note that, if at time 0 the state is $(1, U)$ and U has some density $f(U)$ in $[I_m, I_{m+1}]$, then the stationary distribution is defined by $\pi_{m,v}$ and by the conditional density f of the full energy. Thus everything is defined by the rates a_{vw} and by $f(U)$, that is ρ_1. Moreover, these quantities can be chosen arbitrarily. Setting for the sake of shortness

$$\pi_v = \pi_{V,v},$$

we propose hereafter some examples.

Shifts In this first example we take $\rho_1(x) = 0$ if $x < I_V - I_1$. Then each ρ_v is just a shift of ρ_1.

Reversibility Analogously, a system $(\rho_1, \ldots, \rho_V)$ of densities will be said *admissible* if the following condition holds: for any v the pair (ρ_v, ρ_{v+1}) is admissible. Then it follows that each pair of densities $(\rho_i, \rho_j), i < j$, is admissible.

Theorem 4.2. *If $I_1 < \ldots < I_V$, all $\rho_v(x)$ are strictly positive and the system $(\rho_1, \ldots, \rho_V)$ is admissible then $\mathfrak{L}^M$ is reversible.*

Proof: Let $f_v(U) = \rho_v(U - I_v)$ for $U \geq I_v$ and $f_v(U) = 0$ for $U < I_v$. We suppose the invariant distribution for the chain $\mathfrak{L}^M$ has a product form, each factor being given by $\pi_v f_v(U)$. This means that for each m and for $I_m \leq U < I_{m+1}$

$$\sum_{i=1}^{m} \pi_i f_i(U) a_{ij} = \pi_j f_j(U) \sum_{i=1}^{m} a_{ji},$$

for $j = 1, \ldots, m$. Then admissibility means

$$f_i(U) = \begin{cases} A_i f_1(U), & \text{for } U \geq I_i, \\ f_i(U) = 0, & \text{otherwise.} \end{cases}$$

Hence

$$\sum_{i=1}^{m} \pi_i A_i a_{ij} = \pi_j A_j \sum_{i=1}^{m} a_{ji}, \quad 1 \leq j \leq m.$$

Putting $p_i = \pi_i A_i$, it follows that

$$\sum_{i=1}^{m} p_i a_{ij} = p_j \sum_{i=1}^{m} a_{ji}, \; 1 \leq j \leq m, \quad \forall m = 1, \ldots, V.$$

The comparison of these equations for m and $m+1$ yields

$$p_{m+1} c_{m+1.j} = p_j c_{j,m+1}, \; \forall j = 1, \ldots, m.$$

By induction this yields

$$p_i c_{ij} = p_j c_{ji}, \; \forall i, j,$$

which implies the announced reversibility of $\mathfrak{L}^M$.

Exponent In this third example, we also assume the system $(\rho_1, \ldots, \rho_V)$ of densities is admissible, and moreover that, for some $\rho(T)$ and all v,

$$\rho_v = \rho.$$

Theorem 4.3. *Suppose $V \geq 3$, and that the quantities $I_2 - I_1$ and $I_3 - I_2$ are incommensurable. Then*

$$\rho(T) = \beta \exp(-\beta T),$$

for some $\beta > 0$.

Proof: Admissibility implies that

$$\begin{aligned} \rho_2(T) &= A_2 \rho_1(T + I_2 - I_1), \\ \rho_3(T) &= A_3 \rho_1(T + I_3 - I_1). \end{aligned}$$

If $\rho_1 = \rho_2 = \rho_3 = \rho$ then

$$\rho(T) = A_2 \rho(T + x_2) = A_3 \rho(T + x_3),$$

where $x_i = I_i - I_1, \; i = 2, 3$ are incommensurable. But these last two equations are compatible only if $A_2^{x_3} = A_3^{x_2}$ and $\rho(T) = \beta \exp(-\beta T)$, with

$$\beta = \frac{\log A_2 - \log A_3}{x_2 - x_3}.$$

Energy dependence In the fourth example, the rates $a_{vw} = a_{vw}(T)$ depend on the energy of the input particle v. To construct a model which will be needed later, consider a reversible Markov chain $\mathcal{V}_1$ on $\{1, \ldots, V\}$ with stationary probabilities p_v and rates b_{vw}. Thus

$$p_v b_{vw} = p_w b_{wv}$$

For reactions $v \to w$, define the reaction rates as

$$a_{vw}(U) = \begin{cases} 0, & \text{if } U < I_w, \\ (U - I_w)^{\alpha_w} b_{vw}, & \text{otherwise.} \end{cases}$$

Note that these rates are close to zero if the kinetic energy $T_w = U - I_w$ of the w-particle is close to zero. Letting $f_v(U)$ be the density of the full energy of the v-particle, the reversibility condition writes

$$\pi_v f_v(U) a_{vw}(U) = \pi_w f_w(U) a_{wv}(U), \tag{13}$$

for $U > \max(I_v, I_w)$. We take as density f the shifted Γ-distribution

$$f_v(U) = \begin{cases} \dfrac{\beta^{\nu_v}}{\Gamma(\nu_v)} (U - I_v)^{\nu_v - 1} \exp[-\beta(U - I_v)], \text{ if } U > I_v, \\ 0, \text{otherwise.} \end{cases} \tag{14}$$

Here $\nu_v = \alpha_v + 1$. Then equation (13) becomes

$$\frac{\pi_v \beta^{\nu_v}}{\Gamma(\nu_v)} e^{\beta I_v} b_{vw} = \frac{\pi_w \beta^{\nu_w}}{\Gamma(\nu_w)} e^{\beta I_w} b_{wv} ,$$

showing that the stationary probabilities π_v of type v are equal to (up to a common factor)

$$p_v e^{-\beta I_v} \Gamma(\nu_v) \beta^{-\nu_v}, \tag{15}$$

and the resulting Markov chain is reversible.

4.3. Binary reactions without energy dependence

Let us suppose that a_{vw} do not depend on energies, so that types evolve independently of the energies. Thus at any time t, we will have probabilities $p_t(n_1, \ldots, n_V)$. We will look for cases when there exists an invariant measure on each $\mathcal{L}^M$, defined by probabilities $p(n_1, \ldots, n_V)$, and independent conditional distribution of energies

$$\prod_{v,i} \rho_{v,i},$$

[given $n_1, \ldots, n_V$], defined by densities $\rho_{v.i}(x) = \rho_v(x)$.

Assume all I_v's are equal, but any reaction $v, w \to v', w'$ can occur and let now a reaction $v, w \to v', w'$ be given. We again call

$$P_{\rho_{v'} \rho_{w'}}(x, y|T)$$

the canonical kernel corresponding to the reaction $v, w \to v', w'$ and we denote by $\rho_{vw}(T)$ the density of $\xi_v + \xi_w$.

Theorem 4.4. *Suppose an array $(\rho_1, \ldots, \rho_V)$ of densities is given, satisfying for any binary reaction $v, w \to v', w'$ the conditions*

$$\rho_{vw}(T) = \int_{x+y=T} \rho_v(x) \rho_w(y) dx dy = \int_{x+y=T} \rho_{v'}(x) \rho_{w'}(y) dx dy = \rho_{v'w'}(T).$$

Assume also canonical kernels and that, as $t \to \infty$, the limit of $p_t(n_1, \ldots, n_V)$ exists. Then there is an invariant measure having these densities.

4.4. General binary reaction case

Here the I_v's can be different, but we assume only binary reactions $v, w \to v', w'$ are possible.

4.4.1. COMPLETE FACTORIZATION Denote $\hat{\imath}$ a pair of types (v, w). Thus reaction $v, w \to v', w'$ will be written as $\hat{\imath} \to \hat{\jmath}$, where $\hat{\imath} = (v, w), \hat{\jmath} = (v', w')$. We shall use the analog of the third example with binary reactions. Consider a Markov chain $\mathcal{V}_1 \times \mathcal{V}_1$ on $\{1, \ldots, V\} \times \{1, \ldots, V\}$ with rates $b_{\hat{\imath}\hat{\jmath}}$, such that its stationary distribution be a product form $p_{(v,w)} = p_v p_w$ and the chain be reversible. We define, for *vector particles* $\hat{\imath} = (v, w)$, the energies $I_{\hat{\imath}} = I_v + I_w$ and

$$f_{\hat{\imath}}(U) = (f_v * f_w)(U),$$

where f_v, f_w are given in (14). Thus $f_{\hat{\imath}}(U)$ has also a shifted Γ-distribution with parameters $I_{\hat{\imath}} = I_v + I_w$, $\nu_{\hat{\imath}} = \nu_v + \nu_w$, β. The reversibility condition, with some unspecified stationary probabilities $\pi_{\hat{\imath}}$, writes

$$\pi_{\hat{\imath}} f_{\hat{\imath}}(U) a_{\hat{\imath}\hat{\jmath}}(U) = \pi_{\hat{\jmath}} f_{\hat{\jmath}}(U) a_{\hat{\jmath}\hat{\imath}}(U), \tag{16}$$

where $U > \max(I_{\hat{\imath}}, I_{\hat{\jmath}})$. Letting

$$a_{\hat{\imath}\hat{\jmath}}(U) = \begin{cases} 0, & \text{if } U < I_{\hat{\jmath}}; \\ (U - I_{\hat{\jmath}})^{\alpha_{\hat{\jmath}}} b_{\hat{\imath}\hat{\jmath}} & \text{otherwise.} \end{cases}$$

Here

$$\alpha_{\hat{\jmath}} = \nu_{\hat{\jmath}} - 1 = \nu_{v'} + \nu_{w'} - 1,$$

and the reversibility condition becomes

$$\pi_{\hat{\imath}} \frac{\beta^{\nu_{\hat{\imath}}}}{\Gamma(\nu_{\hat{\imath}})} e^{\beta I_{\hat{\imath}}} b_{\hat{\imath}\hat{\jmath}} = \pi_{\hat{\jmath}} \frac{\beta^{\nu_{\hat{\jmath}}}}{\Gamma(\nu_{\hat{\jmath}})} e^{\beta I_{\hat{\jmath}}} b_{\hat{\jmath}\hat{\imath}}.$$

We are looking for solutions $\pi_{(v,w)} = \pi_v \pi_w$, since we are primarily interested in factorizable invariant distributions. To this end,we assume in addition that, for any binary reaction $v, w \to v', w'$, the condition

$$\nu_v + \nu_w = \nu_{v'} + \nu_{w'}$$

is fulfilled. Then

$$\pi_v \pi_w \beta^{\nu_v + \nu_w} e^{\beta I_v} e^{\beta I_w} b_{\hat{\imath}\hat{\jmath}} = \pi_{v'} \pi_{w'} \beta^{\nu_{v'} + \nu_{w'}} e^{\beta I_{v'}} e^{\beta I_{w'}} b_{\hat{\jmath}\hat{\imath}},$$

and up to a common factor, the solution of this system has the form

$$\pi_v = p_v e^{-\beta I_v} \beta^{-\nu_v} \tag{17}$$

4.4.2. UNARY REACTIONS INCLUDED Let $V = \cup_\alpha V_\alpha$ be a disjoint union of sets V_α of isomers. Thus, we assume that unary reactions $v \to w$ are allowed only if v and w belong to the same V_α. The energy dependence of unary reactions will be defined in the same way as in section 4.2, but additionally we take ν_v to be constant on each V_α, in other words $\nu_v = \nu_w$ for any two isomers $v, w \in V_\alpha$.

We consider the same binary reactions as in section 4.4.1, with the assumption that they are concordant with unary reactions in the following sense: $p_{(v,w)} = p_v p_w$ are such that, for any α, the probabilities p_v have the form given in section 4.2 up to a constant factor.

Theorem 4.5. *If the previous conditions are fulfilled, then formula (17) gives the factorized reversible invariant distribution, both for binary and unary reactions.*

Proof: It suffices to compare the formulae (17) and (15), remarking that the factor $\Gamma(\nu_v)$ in (15) can be omitted, since ν_v is constant on V_α.

References

[1] E. CARLEN, M. CARVALHO AND M. LOSS, *Determination of the spectral gap for Kac's master equation and related stochastic evolutions* (2001), Preprint.

[2] M. ERNST, In *Nonequilibrium Phenomena I. The Boltzmann Equation*, North Holland, 1983.

[3] M. KAC, *Probability and Related Topics in Physical Sciences*, Interscience Publishers, 1958.

[4] A. KALINKIN *Markov branching processes with interaction*, Russian Math. Reviews, vol. 57, No. 2 (2002), pp. 23–84.

[5] M. A. LEONTOVICH, *Main equations of kinetical theory of gases from the random processes point of view*, J. of Experim. and Theor. Physics, vol. 5, No. 3–4 (1935), pp. 211–231.

[6] V. MALYSHEV, S. PIROGOV AND A. RYBKO, *Random walks and chemical networks*, to appear in Moscow Math. J.

[7] D. MCQUARRIE, *Stochastic approach to chemical kinetics*, J. Appl. Prob., 4 (1967), pp. 413–478.

[8] P. WHITTLE, *Systems in Stochastic Equilibrium*, John Wiley. 1986.

Guy Fayolle
INRIA Rocquencourt – Domaine de Voluceau BP 105
78153 Le Chesnay, France. Guy.Fayolle@inria.fr

Vadim Malyshev
INRIA Rocquencourt – Domaine de Voluceau BP 105
78153 Le Chesnay, France. Vadim.Malyshev@inria.fr

Serguei Pirogov
IPPI – Russian Academy of Sciences
19 Bolshoi Karetny – 101447 Moscow, Russia.
Work partially supported by RFBR grant 02-01-01276.

Trends in Mathematics, © 2004 Birkhäuser Verlag Basel/Switzerland

Large Deviations of Hellinger Distance on Partitions

László Györfi

ABSTRACT: *We discuss Chernoff-type large deviation properties of the Hellinger distance on partitions. If H_n denotes the Hellinger distance of the empirical distribution and the distribution restricted to a partition then for small $\epsilon > 0$, $\mathbf{P}\{H_n > \epsilon\} \approx e^{-n(\epsilon^2+o(1))}$, where n is the sample size.*

1. Introduction

We consider the problem of testing an unknown probability density function. The test statistics are derived from dissimilarity measures of probability measures, like φ-divergences introduced by Csiszár (1967). The most important φ-divergences in mathematical statistics and information theory are the total variation distance, the information divergence and the χ^2-divergence. In this paper we consider the Hellinger distance.

Let $\varphi : (0, \infty) \to R$ be a convex function, extended on $[0, \infty)$ by the continuity. If μ and ν are probability measures on $\mathbb{R}^d$ with densities f and g with respect to a dominating measure λ, then the *φ-divergence* of μ and ν is defined by

$$D_\varphi(\mu, \nu) = \int_{\mathbb{R}^d} \varphi\left(\frac{f(x)}{g(x)}\right) g(x)\lambda(dx). \tag{1}$$

We can consider some goodness of fit tests for the hypotheses that the unknown distribution is μ. Suppose that μ is non-atomic. Assume a sample of independent random vectors $X_1, \ldots, X_n$, distributed according to a probability measure μ, and let μ_n denote the empirical measure. Based on a finite partition $\mathcal{P}_n = \{A_{n,1}, \ldots, A_{n,m_n}\}$ of $\mathbb{R}^d$ introduce a test statistics comparing μ and μ_n:

$$D_{\varphi,\mathbf{P}_n}(\mu, \mu_n) = \sum_j \varphi\left(\frac{\mu(A_{n,j})}{\mu_n(A_{n,j})}\right) \mu_n(A_{n,j}). \tag{2}$$

Usually, the hypotheses is accepted if $D_{\varphi,\mathbf{P}_n}(\mu, \mu_n)$ is small. In order to characterize such a test either the limit distribution of $D_{\varphi,\mathbf{P}_n}(\mu, \mu_n)$ should be known, or its large deviation property.

For

$$\varphi(t) = |1 - t|$$

the φ-divergence is called the L_1 error:

$$\|f - g\| := \int |f(x) - g(x)|\lambda(dx).$$

Györfi and van der Meulen (1991) introduced the corresponding test statistic

$$L_n = \sum_{j=1}^{m_n} |\mu(A_{n,j}) - \mu_n(A_{n,j})|.$$

Beirlant et al (2001) proved that if

$$\lim_{n\to\infty} \max_j \mu(A_{n,j}) = 0 \tag{3}$$

and

$$\lim_{n\to\infty} \frac{m_n \log n}{n} = 0. \tag{4}$$

then for all $0 < \epsilon < 2$

$$\lim_{n\to\infty} \frac{1}{n} \log \mathbf{P}\{L_n > \epsilon\} = -g_L(\epsilon), \tag{5}$$

where

$$g_L(\epsilon) = \inf_{0<p<1-\epsilon/2} D(p\|p + \epsilon/2), \tag{6}$$

and

$$D(\alpha\|\beta) = \alpha \log \frac{\alpha}{\beta} + (1-\alpha) \log \frac{1-\alpha}{1-\beta}. \tag{7}$$

It means that

$$\mathbf{P}\{L_n > \epsilon\} = e^{-n(g_L(\epsilon)+o(1))}.$$

For

$$\varphi(t) = -\log t$$

the φ-divergence is denoted by $I(\mu,\nu)$ and is called the *reversed order information divergence* (also called I-divergence, Kullback-Leibler number, relative entropy). The corresponding statistics is called modified likelihood ratio statistics:

$$I_n = \sum_{j=1}^{m_n} \mu_n(A_{n,j}) \log \frac{\mu_n(A_{n,j})}{\mu(A_{n,j})}.$$

Kallenberg (1985), and Quine and Robinson (1985) showed that under (3) and (4), for all $\epsilon > 0$

$$\mathbf{P}\{I_n > \epsilon\} = e^{-n(\epsilon+o(1))}. \tag{8}$$

2. Hellinger distance

For

$$\varphi(t) = (\sqrt{t} - 1)^2$$

the φ-divergence is called the *squared Hellinger distance*, thus

$$\begin{aligned} H^2(\mu,\nu) &= H^2(f,g) \\ &= \int_{\mathbb{R}^d} \left(\sqrt{\frac{f(x)}{g(x)}} - 1\right)^2 g(x)\lambda(dx) \\ &= \int_{\mathbb{R}^d} \left(\sqrt{f(x)} - \sqrt{g(x)}\right)^2 \lambda(dx) \end{aligned}$$

and the corresponding *squared Hellinger statistic* is defined as

$$H_n^2 = \sum_{j=1}^{m_n} \left(\sqrt{\mu(A_{n,j})} - \sqrt{\mu_n(A_{n,j})} \right)^2 .$$

Theorem 2.1. *Assume (3) and (4), and that for all $0 < \epsilon < \sqrt{2}$ the rate function*

$$\lim_{n\to\infty} \frac{1}{n} \log \mathbf{P}\{H_n > \epsilon\} = -g_H(\epsilon) \tag{9}$$

exists. Then

$$\epsilon^2 \le g_H(\epsilon) \tag{10}$$

and

$$g_H(\epsilon) \le g^*(\epsilon), \tag{11}$$

where

$$g^*(\epsilon) = \inf_{0<p<q<1, H((p,1-p),(q,1-q))\ge\epsilon} D(p\|q). \tag{12}$$

Remark 1. According to Theorem 1

$$\epsilon^2 \le g_H(\epsilon) \le g^*(\epsilon).$$

We may have an upper bound on $g^*(\epsilon)$

$$g^*(\epsilon) = \inf_{0<p<q<1, H((p,1-p),(q,1-q))\ge\epsilon} D(p\|q) \le \inf_{0<q<1, H^2((0,1),(q,1-q))\ge\epsilon^2} D(0\|q).$$

Because of

$$H^2((0,1),(q,1-q)) = 2 - 2\sqrt{1-q}$$

and

$$D(0\|q) = -\log(1-q),$$

the minimizer $q = q_\epsilon = 1 - (1-\epsilon^2/2)^2$. Thus

$$g^*(\epsilon) \le -\log(1-q_\epsilon) = -2\log(1-\epsilon^2/2).$$

For $0 < x < 1/2$ we have that $-\log(1-x) \le x + x^2$ therefore for $0 < \epsilon < 1$

$$g^*(\epsilon) \le -2\log(1-\epsilon^2/2) \le \epsilon^2 + \epsilon^4/2.$$

This bound and (10) imply that for small ϵ

$$g_H(\epsilon) \approx \epsilon^2.$$

It means that, for small ϵ,

$$\mathbf{P}\{H_n > \epsilon\} \approx e^{-n(\epsilon^2 + o(1))}.$$

Remark 2. The Hellinger distance is closely related to the L_1 error (cf. Devroye, Györfi (1985)):

$$\begin{aligned} H^2(f,g) &= \int_{\mathbb{R}^d} \left(\sqrt{f(x)} - \sqrt{g(x)}\right)^2 \lambda(dx) \\ &\le \int_{\mathbb{R}^d} \left|\sqrt{f(x)} - \sqrt{g(x)}\right| \left(\sqrt{f(x)} + \sqrt{g(x)}\right) \lambda(dx) \\ &= \int_{\mathbb{R}^d} |f(x) - g(x)| \lambda(dx) \\ &= \|f - g\| \end{aligned}$$

and

$$\|f-g\|^2 \le \int_{\mathbb{R}^d} \left(\sqrt{f(x)}-\sqrt{g(x)}\right)^2 \lambda(dx) \int_{\mathbb{R}^d} \left(\sqrt{f(x)}+\sqrt{g(x)}\right)^2 \lambda(dx)$$
$$= H^2(f,g)(4-H^2(f,g)),$$

therefore

$$H_n^2 \le L_n \le H_n\sqrt{4-H_n^2}.$$

These bounds imply that

$$g_L(\epsilon^2) \le g_H(\epsilon) \le g_L(\epsilon\sqrt{4-\epsilon^2}). \tag{13}$$

For small ϵ,

$$g_L(\epsilon) \approx \epsilon^2/2,$$

therefore the bounds (13) imply that for small ϵ

$$\epsilon^4/2 \le g_H(\epsilon) \le 2\epsilon^2.$$

By numerical calculations, one can verify that for $0 < \epsilon \le 1.26$ the lower bound in (13) is weaker than (10). Next we show that, in general, the upper bound in (13) is weaker than (11), i.e.

$$g^*(\epsilon) \le g_L(\epsilon\sqrt{4-\epsilon^2}).$$

In order to see this put

$$F(\epsilon) = \epsilon\sqrt{4-\epsilon^2}.$$

Then for $\epsilon \le \sqrt{2}$ the function $F(\epsilon)$ is monotonically increasing, therefore

$$\begin{aligned} g^*(\epsilon) &= \inf_{0<p<q<1, H((p,1-p),(q,1-q))\ge\epsilon} D(p\|q) \\ &= \inf_{0<p<q<1, F(H((p,1-p),(q,1-q)))\ge F(\epsilon)} D(p\|q) \\ &\le \inf_{0<p<q<1, L_1((p,1-p),(q,1-q))\ge F(\epsilon)} D(p\|q) \\ &= g_L(F(\epsilon)) \\ &= g_L(\epsilon\sqrt{4-\epsilon^2}). \end{aligned}$$

REMARK 3. The above result obviously can be generalized to a large deviation result for the Hellinger inaccuracy rate $H(f, f_n)$ of the histogram density estimator f_n based on a sample of independent random vectors $X_1, \ldots, X_n$, distributed according to a probability measure μ with density f with respect to the Lebesgue measure λ.
Introducing a partition $\mathcal{P}_n = \{A_{n,j};\ j \ge 1\}$ of $\mathbb{R}^d$ such that $\sup_{j\ge1}\lambda(A_{n,j}) < \infty$, then the *histogram density estimator* is defined by

$$f_n(x) = \frac{\mu_n(A_n(x))}{\lambda(A_n(x))},$$

where $A_n(x) = A_{n,i}$ if $x \in A_{n,i}$. Mimicking the proof of Corollary 1 in Beirlant et al (2001) we can get the following:
Assume that for each sphere S centered at the origin

$$\lim_{n\to\infty} \sup_{A_{n,j}\cap S\neq\emptyset} diam(A_{n,j}) = 0,$$

and there exists a sequence of spheres S_n centered at the origin such that $S_n \uparrow \mathbb{R}^d$ and

$$\lim_{n\to\infty} \text{card}\{A_{n,j} \cap S_n \neq \emptyset\} \frac{\log n}{n} = 0.$$

Then for all $0 < \epsilon < \sqrt{2}$

$$\lim_{n\to\infty} \frac{1}{n} \log \mathbf{P}\{H(f, f_n) > \epsilon\} = -g_H(\epsilon), \tag{14}$$

where $g_H(\epsilon)$ is defined by (9).
Note that in this statement there is no condition on f, and so (14) holds for all f, and the rate function $g_H(\epsilon)$ does not depend on f.

3. Proof

We derive the lower bound using the reversed order information divergence. The squared Hellinger distance can be written as

$$\begin{aligned} H^2(\mu,\nu) &= \int_{\mathbb{R}^d} \left(\sqrt{f(x)} - \sqrt{g(x)}\right)^2 \lambda(dx) \\ &= \int_{\mathbb{R}^d} 2\left(g(x) - \sqrt{f(x)g(x)}\right) \lambda(dx) \\ &= D_{\varphi^*}(\mu,\nu), \end{aligned}$$

where

$$\varphi^*(t) = 2(1-\sqrt{t}).$$

The reversed order information divergence was defined by $\varphi(t) = -\log t$, therefore because of the inequality

$$2(1-\sqrt{t}) \leq -\log t,$$

we have that

$$H^2(\mu,\nu) \leq I(\mu,\nu),$$

which implies that

$$H_n^2 \leq I_n$$

and together with (8) we have (10).

Let $\bar{\mu}_n$ and μ_n^* be the restrictions of μ and μ_n to the partition $\mathcal{P}_n$. In the proof of (11) we shall use the following lemma.

Lemma 3.1. (Sanov (1957), see p. 16 in Dembo, Zeitouni (1992)). *Let Σ be a finite set (alphabet), $\mathcal{L}_n$ be a set of types (possible empirical distributions) on Σ, and let Γ be a set of distributions on Σ. Then*

$$\left|\frac{1}{n} \log \mathbf{P}\{\mu_n^* \in \Gamma\} + \inf_{\tau\in\Gamma\cap\mathcal{L}_n} I(\tau,\bar{\mu}_n)\right| \leq \frac{|\Sigma| \log(n+1)}{n}, \tag{15}$$

where $|\Sigma|$ denotes the cardinality of Σ.

We apply (15) for

$$\Sigma = \{A_{n,1}, \ldots, A_{n,m_n}\}$$

such that

$$\Gamma = \{\tau : H(\bar{\mu}_n, \tau) \geq \epsilon\}.$$

Then, according to (15),

$$\left|\frac{1}{n} \log \mathbf{P}\{H_n \geq \epsilon\} + \inf_{\tau\in\Gamma\cap\mathcal{L}_n} I(\tau,\bar{\mu}_n)\right| \leq \frac{m_n \log(n+1)}{n}$$

and therefore, under (4),

$$-g_H(\epsilon) = \lim_{n\to\infty} \frac{1}{n} \log \mathbf{P}\{H_n > \epsilon\} = -\lim_{n\to\infty} \inf_{\tau\in\Gamma\cap\mathcal{L}_n} I(\tau, \bar{\mu}_n).$$

The distributions in $\mathcal{L}_n$ are possible empirical distributions, having components of the form $\frac{r}{n}$, where r is integer. Because of (3) we have that

$$m_n \to \infty,$$

therefore the continuity of $H(\tau, \bar{\mu}_n)$ and $I(\tau, \bar{\mu}_n)$ implies that

$$g_H(\epsilon) = \lim_{n\to\infty} \inf_{\tau\in\Gamma\cap\mathcal{L}_n} I(\tau, \bar{\mu}_n) = \lim_{n\to\infty} \inf_{H(\tau,\bar{\mu}_n)\geq\epsilon} I(\tau, \bar{\mu}_n). \tag{16}$$

Here

$$I(\tau, \bar{\mu}_n) = \sum_{j=1}^{m_n} \tau(A_{n,j}) \log \frac{\tau(A_{n,j})}{\mu(A_{n,j})}.$$

Put

$$L = \{j : \mu(A_{n,j}) > \tau(A_{n,j})\}$$

and

$$A_n = \cup_{j\in L} A_{n,j}.$$

In order to get an upper bound on $g_H(\epsilon)$, consider a subclass of the distributions such that $\frac{\tau(A_{n,j})}{\mu(A_{n,j})}$ is constant both on L and L^c. Then for such τ,

$$\begin{aligned}
&\sum_{j=1}^{m_n} \left(\sqrt{\mu(A_{n,j})} - \sqrt{\tau(A_{n,j})}\right)^2 \\
= \;& \sum_{j\in L} \left(\sqrt{\mu(A_{n,j})} - \sqrt{\tau(A_{n,j})}\right)^2 + \sum_{j\in L^c} \left(\sqrt{\mu(A_{n,j})} - \sqrt{\tau(A_{n,j})}\right)^2 \\
= \;& \left(\sqrt{\mu(A_n)} - \sqrt{\tau(A_n)}\right)^2 + \left(\sqrt{1-\mu(A_n)} - \sqrt{1-\tau(A_n)}\right)^2 \\
= \;& H^2\left((\tau(A_n), 1-\tau(A_n)), (\mu(A_n), 1-\mu(A_n))\right)
\end{aligned}$$

and

$$\begin{aligned}
&\sum_{j=1}^{m_n} \tau(A_{n,j}) \log \frac{\tau(A_{n,j})}{\mu(A_{n,j})} \\
= \;& \sum_{j\in L} \tau(A_{n,j}) \log \frac{\tau(A_{n,j})}{\mu(A_{n,j})} + \sum_{j\in L^c} \tau(A_{n,j}) \log \frac{\tau(A_{n,j})}{\mu(A_{n,j})} \\
= \;& \tau(A_n) \log \frac{\tau(A_n)}{\mu(A_n)} + (1-\tau(A_n)) \log \frac{1-\tau(A_n)}{1-\mu(A_n)} \\
= \;& D(\tau(A_n)\|\mu(A_n)).
\end{aligned}$$

Thus

$$\begin{aligned}
&\lim_{n\to\infty} \inf_{H(\tau,\bar{\mu}_n)\geq\epsilon} I(\tau, \bar{\mu}_n) \\
\leq \;& \inf_{H((\tau(A_n),1-\tau(A_n)),(\mu(A_n),1-\mu(A_n)))\geq\epsilon} D(\tau(A_n)\|\mu(A_n)) \\
= \;& g^*(\epsilon),
\end{aligned}$$

and we proved (11).

References

[1] Beirlant, J., Devroye, L., Györfi, L. and Vajda, I. (2001): Large deviations of divergence measures on partitions. *J. Statist. Planning Infer.* **93**, 1–16.

[2] Csiszár, I. (1967). Information-type measures of divergence of probability distributions and indirect observations. *Studia Sci. Math. Hungar.*, 2, pp. 299-318.

[3] Dembo, A. and Zeitouni, O. (1992). *Large Deviations Techniques and Applications.* Jones and Bartlett Publishers.

[4] Devroye, L. and Györfi, L. (1985). *Nonparametric Density Estimation: the L_1 View.* Wiley, New York.

[5] Györfi, L. and van der Meulen, E. C. (1991). A consistent goodness–of–fit test based on the total variation distance. In: *Nonparametric Functional Estimation and Related Topics* (G. Roussas, ed.), Kluwer, Boston, 631-646.

[6] Kallenberg, W. C. M.(1985). On moderate and large deviations in multinomial distributions. *Annals of Statistics*, 13, 1554-1580.

[7] Quine, M.P. and Robinson, J. (1985). Efficiencies of chi-square and likelihood ratio goodness-of-fit tests. *Ann. Statist.*, 13, 727-742.

[8] Sanov, I. N. (1957). On the probability of large deviations of random variables. *Mat. Sb.*, 42, pp. 11-44 (English translation in *Sel. Transl. Math. Statist. Prob.*, 1, (1961), pp.213-244).

László Györfi
Department of Computer Science and Information Theory, Technical University of Budapest, 1521 Stoczek u. 2, Budapest, Hungary.
e-mail: gyorfi@szit.bme.hu

Trends in Mathematics, © 2004 Birkhäuser Verlag Basel/Switzerland

Estimation of the Offspring Mean for a General Class of Size-Dependent Branching Processes. Application to Quantitative Polymerase Chain Reaction

Nadia Lalam and Christine Jacob

ABSTRACT: *We first address the problem of estimating the offspring mean for a general class of size-dependent branching processes. Then we apply our results to the particular setting of Quantitative Polymerase Chain Reaction.*

1. Estimation in a general size-dependent setting

We consider a single-type supercritical or near-critical size-dependent branching process $\{N_n\}_n$ such that the offspring mean $m(N_n)$ converges to a limit $m \geq 1$ with a rate of convergence of order N_n^{α}, as the population size N_n grows to ∞, and the offspring variance $\sigma^2(N_n)$ converges to σ^2 at the rate N_n^{β}, where $-1 \leq \beta < 1$. We assume that the offspring mean $m(N) = m + \mu N^{-\alpha} + o(N^{-\alpha})$ depends on an unknown asymptotically identifiable parameter θ_0 that belongs either to the limit model ($\theta_0 = m$) or to the transient model ($\theta_0 = \mu$). When $\theta_0 = m$, $m(N) - m$ is a nuisance part asymptotically negligible relatively to the identifiable part m; when $\theta_0 = \mu$, we assume that m and α are known and $m(N) - (m + \mu N^{-\alpha})$ is the nuisance part asymptotically negligible relatively to the identifiable part $m + \mu N^{-\alpha}$. For ease of presentation, we will denote $m(N)$ by $m_{\theta_0,\nu_0}(N)$ where θ_0 is the true identifiable parameter and ν_0 represents the nuisance term.
We estimate θ_0 on the non-extinction set of the process from the observations $\{N_h, \ldots, N_n\}$, by using the conditional least-squares method weighted by $\{N_{n-1}^{-\gamma}\}_n$. Let $\widehat{\theta}_{n,\gamma,\nu} = \arg\min_\theta \sum_{k=h+1}^n (N_k - m_{\theta,\nu}(N_{k-1})N_{k-1})^2 N_{k-1}^{-\gamma}$, where ν has a given value. We study the asymptotic properties of $\{\widehat{\theta}_{n,\gamma,\nu}\}_n$ according to γ, with either h or $n-h$ remaining constant as $n \to \infty$. The main assumptions to get the consistency concern the asymptotic behavior of the process and namely the existence of a deterministic sequence $\{a_n\}_n$ such that $\lim_{n\to\infty} N_n a_n^{-1} \overset{a.s.}{=} W$ for some random variable W. To get the convergence rate of the estimator, we assume mainly a condition of the Lindeberg type. Let $\alpha_* = 0$ when $\theta_0 = m$ and $\alpha_* = \alpha$ when $\theta_0 = \mu$. Our results are given in the following theorem:

Theorem 1.1. *a) Strong consistency:* $\lim_{n\to\infty} \widehat{\theta}_{n,\gamma,\nu} \overset{a.s.}{=} \theta_0$;
b) Rate of convergence: there exist stochastic sequences $\{P_{n,\gamma,\nu}\}_n$ *and* $\{Q_{n,\gamma,\nu}\}_n$ *such that* $\Phi_{n,\gamma}^{-1}(\widehat{\theta}_{n,\gamma,\nu} - \theta_0) = P_{n,\gamma,\nu}/Q_{n,\gamma,\nu}$, *where* $\lim_{n\to\infty} P_{n,\gamma,\nu} \overset{D}{=} P$ *with* $E[e^{(itP)}]i = E[\exp(-\frac{t^2}{2}\sigma^2 W^{1+\beta+2(1-\alpha_*)-2\gamma})]$, *and* $\lim_{n\to\infty} Q_{n,\gamma,\nu} \overset{a.s.}{=} W^{2(1-\alpha_*)-\gamma}$. *The rate of convergence is*

$$\Phi_{n,\gamma}^{-1} = \left[\sum_{k=h+1}^{n} a_{k-1}^{2(1-\alpha_*)-\gamma}\right] \Big/ \left[\sqrt{\sum_{k=h+1}^{n} a_{k-1}^{1+\beta+2(1-\alpha_*)-2\gamma}}\right]$$

and the best rate is attained for $\gamma = 1 + \beta$.

We prove the consistency by using the minimum contrast method and a strong law of large numbers for martingales [2]. We get the asymptotic distribution of the estimator by applying a Taylor approximation to the conditional least-squares contrast and by using Rahimov's central limit theorem for random sums. For more details, see [4]. Notice that to obtain the consistency, it is sufficient to assume that the offspring variance satisfies $\sigma^2(N) \leq \sigma^2 N^\beta$ and the normalizing sequence $\{a_n\}_n$ may be stochastic (e.g. $a_n = N_n$, for all n).

2. Application to Quantitative Polymerase Chain Reaction (QPCR)

We apply our theoretical results to the particular setting of the QPCR consisting in the *in vitro* amplification of a DNA molecules population and which aims at quantifying the initial amount N_0 of the population. The PCR is largely used in molecular biology since it enables to detect low abundance of DNA and it has many applications (for more details, see [1]).
In the PCR setting, the current quantification method is based on inadequate mathematical models and needs many amplification trajectories to estimate the efficiency in order to quantify N_0, where the efficiency is the probability that a DNA molecule will be duplicated.
The branching process theory is naturally used to model the PCR and recently a size-dependent branching process has been developed by Jagers and Klebaner [3]. We propose a new modelling of a PCR amplification trajectory relying on size-dependent branching processes. Our modelling is based on the concept of saturation and generalizes the efficiency model proposed in [3].
We estimate the reaction efficiency from successive observations of a single trajectory and we study the properties of the estimators at finite distances using simulations and real-time PCR data.

References

[1] Ferré F., 1998, Gene quantification. Ed. Ferré F., Birkhäuser, New-York.

[2] Hall P. & Heyde C. C., 1980, Martingale limit theory and its application, Probability and Mathematical Statistics, Academic Press, New York.

[3] Jagers P. & Klebaner F. C., 2003, Random variation and concentration effects in PCR, *J. Theoret. Biol.*, 224, 299-304.

[4] Lalam N. & Jacob C., 2004, Estimation of the offspring mean in a supercritical or near-critical size-dependent branching process. *Adv. Appl. Prob.* (accepted).

[5] Lalam N., Jacob C. & Jagers P., 2004, Modelling the PCR amplification process by a size-dependent branching process and estimation of the efficiency. *Adv. Appl. Prob.* (accepted).

Nadia Lalam
EURANDOM, P.O. Box 513, 5600 MB Eindhoven, The Netherlands
lalam@eurandom.tue.nl

Christine Jacob
INRA, Laboratoire de Biométrie, 78352 Jouy-en-Josas Cedex, France
cj@banian.jouy.inra.fr

Trends in Mathematics, © 2004 Birkhäuser Verlag Basel/Switzerland

Decidability of Simple Brick Codes

Małgorzata Moczurad and Włodzimierz Moczurad

ABSTRACT: *Bricks are polyominoes with labelled cells. The problem whether a given set of bricks is a code, is undecidable in general. It is open for two-element sets. Here we consider sets consisting of square bricks only. We show that in this setting, the codicity of small sets (two bricks) is decidable, but 15 bricks are enough to make the problem undecidable. Thus the frontier between decidability and undecidability lies somewhere between these two numbers. Additionally we show that the undecidability frontier could be improved to 13 if the Post Correspondence Problem with three pairs should prove undecidable.*

1. Introduction

Let A be a finite alphabet of labels. A *brick* is a partial mapping $k : \mathbf{Z}^2 \to A$, where $\operatorname{dom} k$ is finite and connected. It can be viewed as a polyomino with its cells labelled with the symbols of A. The set of all bricks over A is denoted by $A^{\bowtie}$. Given a set of bricks $X \subseteq A^{\bowtie}$, the set of all bricks tilable with (translated copies of) the elements of X is denoted by $X^{\bowtie}$. Note that we do not allow rotations of bricks. $X \subseteq A^{\bowtie}$ is a *brick code*, if every element of $X^{\bowtie}$ admits exactly one tiling with the elements of X. The *effective alphabet* of $X \subseteq A^{\bowtie}$ is the set of all symbols that appear on bricks in X, i.e., $\bigcup_{k \in X} k(\operatorname{dom} k)$.

The problem whether a given set of bricks (even polyominoes) is a code, is undecidable in general [1, 5]. The problem is open for two-element sets. Here we consider sets consisting of square bricks only. We show that in this setting, the codicity of small sets (two bricks) is decidable, but 15 bricks are enough to make the problem undecidable. Note that apart from the single-label case, the decidability of sets of squares does not depend on the size of the effective alphabet, since a larger alphabet can be "simulated" with two symbols at the expense of the size of the squares.

2. Decidability frontier

Consider sets consisting of just two bricks, each of them being a square (in the geometrical sense). We show that there exists a simple algorithm to verify whether a set of this kind is a brick code.

Proposition 2.1. *Let* $X = \{k, l\} \subseteq A^{\bowtie}$, *where* $\operatorname{dom} k$ *and* $\operatorname{dom} l$ *are squares,* $k \neq l$. *Then* X *is not a brick code iff* k *and* l *have a common rectangular tiler, i.e., there exists a rectangle* $t \in A^{\bowtie}$ *such that* $k, l \in \{t\}^{\bowtie}$.

Proposition 2.2. *If* $k, l \in A^{\bowtie}$ *have a common rectangular tiler, then they have a common square tiler.*

Corollary 2.3. *Let* $X = \{k, l\} \subseteq A^{\bowtie}$, *where* $\operatorname{dom} k$ *and* $\operatorname{dom} l$ *are squares. It is decidable whether* X *is a brick code.*

A Thue system can be reduced to a brick code problem with square bricks. There exists a non-erasing Thue system with an undecidable word problem over a two-letter alphabet with just three relations. This is the smallest example known to us, due to Matiyasevich [4]. We encode this system, including two arbitrary input words v and w, in a set $X_{S,u,v}$ containing 15 square bricks.

Proposition 2.4. *Let (Σ, S) be a Thue system and let $u, v \in \Sigma^*$. Let $X_{S,u,v}$ be the set constructed as described above. The following equivalence holds: $u \equiv_S v$ iff $X_{S,u,v}$ is not a brick code.*

Corollary 2.5. *The codicity problem for sets containing 15 or more square bricks is undecidable.*

The question whether a solution exists for a given Post system S is undecidable in general. The problem is open for $3 \leq |S| \leq 6$, cf. [2]. Using a reduction similar to the Thue-to-brick one and starting with a PCP with three pairs, we obtain a set of 13 square bricks.

Corollary 2.6. *If PCP with three pairs is undecidable, then the codicity problem for sets containing 13 square bricks is undecidable.*

3. Conclusions

(i) If the effective alphabet is trivial, the problem is trivially decidable. (ii) If the effective alphabet contains at least two symbols, we have decidability for two-element sets and undecidability for sets with at least 15 elements. (iii) The undecidability frontier could be improved to 13 if PCP with three pairs should prove undecidable. We are of course interested in finding the exact frontier between decidability and undecidability (cf. [3]).

References

[1] P. Aigrain, D. Beauquier: Polyomino tilings, cellular automata and codicity. *Theoret. Comp. Sci.* 147 (1995) 165–180.

[2] V. Halava, T. Harju, M. Hirvensalo: Binary (generalized) Post Correspondence Problem. *Theoret. Comp. Sci.* 276 (2002) 183–204.

[3] M. Margenstern: Frontiers between decidability and undecidability: a survey. *Theoret. Comp. Sci.* 231 (2000) 217–251.

[4] Yu. Matiyasevich: Word problem for Thue systems with a few relations. LNCS 909 (1995) 39–53.

[5] W. Moczurad: Algebraic and algorithmic properties of brick codes. Doctoral thesis, Jagiellonian University (1999).

[6] W. Moczurad: Brick codes: families, properties, relations. *Intern. J. Comp. Math.* 74 (2000) 133–150.

Małgorzata Moczurad and Włodzimierz Moczurad
Inst. of Computer Science, Jagiellonian Univ., Nawojki 11, 30-072 Kraków, Poland
{mmoczurad,wkm}@ii.uj.edu.pl

Trends in Mathematics, © 2004 Birkhäuser Verlag Basel/Switzerland

A Constrained Version of Sauer's Lemma

Joel Ratsaby

ABSTRACT: *We generalize Sauer's Lemma to finite VC-dimension classes $\mathcal{H}$ of binary-valued functions on $[n] = \{1, \ldots, n\}$ which have a margin of at least N on every element in a sample $S \subseteq [n]$ of cardinality l, where the margin $\mu_h(x)$ of $h \in \mathcal{H}$ on a point $x \in [n]$ is defined as the largest non-negative integer a such that h is constant on the interval $I_a(x) = [x - a, x + a]$.*

1. Introduction

Estimation of the complexity of classes of binary-valued functions has been behind much of recent developments in the of theory learning. In a seminal paper Vapnik & Chervonenkis [9] applied the law of large numbers uniformly over an infinite class $\mathcal{F}$ of binary functions, i.e., indicator functions of sets A in a general domain X, and showed that the complexity of the problem of learning pattern recognition from samples of n randomly drawn examples can be characterized in terms of a combinatorial complexity of $\mathcal{F}$.

This complexity, known as the *growth function* of $\mathcal{F}$ and denoted by $\varphi_{\mathcal{F}}(n)$, counts the maximal number of dichotomies, i.e., binary vectors corresponding to the restriction of functions $f \in \mathcal{F}$ on a finite subset $S \subset X$ of cardinality n, where the maximum runs over all such S. The Vapnik-Chervonenkis dimension of $\mathcal{F}$, denoted as $VC(\mathcal{F})$, plays a crucial role in controlling the rate in which $\varphi_{\mathcal{F}}(n)$ increases with n. Such binary vectors may be viewed as binary-valued functions on a finite domain $[n] \equiv \{1, \ldots, n\}$ and hence form a finite class $\mathcal{H}$ of the same VC-dimension as $\mathcal{F}$.

In this paper we consider finite VC-dimension classes $\mathcal{H}(S)$ of binary-valued functions on $[n]$ which satisfy the constraint of having a large margin on any one set $S \subset X$ where the margin $\mu_h(x)$ of $h \in \mathcal{H}(S)$ on a point $x \in [n]$ is defined as the largest non-negative integer a such that h is constant on the interval $I_a(x) = [x - a, x + a]$. We denote by $\mu_S(h) = \min_{x \in S} \mu_h(x)$. The main result of the paper is an estimate on the cardinality of $\mathcal{H}(S)$.

Part of the motivation behind our work arises from results obtained in recent years which show that learning classes of real-valued functions which are constrained to have a large margin on a training sample yields more-accurate hypotheses. There has thus been significant interest in learning-algorithms which maximize the sample margin as for instance in support vector machines (see Vapnik [8], Cristianini & Shawe-Taylor [3]). We note in passing that samples on which the target function (the one to be learnt) has a large margin are therefore of considerable information worth to a learner and estimates of the complexity of sets of such samples have been recently obtained in Ratsaby [6] with an explicit dependence on the margin parameter and sample size.

The current paper extends Sauer's result on the cardinality of finite VC-dimension classes $\mathcal{H}$ (see Lemma 2.1 below) to classes $\mathcal{H}(S)$ of functions with the above constraint. While we anticipate interesting learning-theoretic consequences emerging from our constrained version of Sauer's result such investigation however is beyond the scope of this paper.

We start by introducing some needed notation.

2. Some notations, definitions and existing results

Let $\mathbb{I}(E)$ denote the indicator function which equals 1 if the expression E is true and 0 otherwise. Let F be a class of functions $f : [n] \to \{0, 1\}$. For a set $A = \{a_1, \ldots, a_k\} \subseteq [n]$ denote by $f_{|A} = [f(a_1), \ldots, f(a_k)]$. F is said to *shatter* A if

$$\left|\{f_{|A} : f \in F\}\right| = 2^k.$$

The Vapnik-Chervonenkis dimension of F, denoted as $VC(F)$, is defined as the cardinality of the largest set shattered by F.

Sauer [4] obtained the following result:

Lemma 2.1. [4] *If the VC-dimension of F is d then*

$$|F| \le \sum_{k=0}^{d} \binom{n}{k}.$$

We note that the bound is tight as for all $d, n \ge 1$ there exist classes $F \subseteq 2^{[n]}$ of VC-dimension d which achieve the equality.

Consider the following definition of functional margin[1] which naturally suits binary-valued functions.

Definition 1. *The* margin $\mu_f(x)$ *of $f \in F$ on an element $x \in [n]$ is the largest non-negative integer a such that f has a constant value of either* 0 *or* 1 *on the interval set* $I_a(x) = \{\max\{1, x - a\}, \ldots, \min\{x + a, n\}\}$.

The sample-margin $\mu_S(f)$ of f on a subset $S \subseteq [n]$ is defined as

$$\mu_S(f) \equiv \min_{x \in S} \mu_f(x).$$

More generally, this definition applies also to classes on other domains X if there is a linear ordering on X.

3. Technical results

We start with the main result of the paper:

Theorem 3.1. *Let $\mathcal{H}$ be a class of binary-valued functions h on $[n]$ having $VC(\mathcal{H}) \le d$. Let $S \subseteq [n]$ have cardinality $l \ge 1$ and consider the subclass $\mathcal{H}(S) \subset \mathcal{H}$ which consists of all functions $h \in \mathcal{H}$ with a margin $\mu_h(x) > N$ only on elements $x \in S$. Then*

$$|\mathcal{H}(S)| \le \beta_d^{(N)}(n - l - N - 1).$$

where $\beta_d^{(N)}(n)$ is defined in Lemma 3.3 below.

[1] For other definitions of margin see for instance [3].

Proof. The condition $\mu_h(x) > N$ implies only two types of functions h are allowed, those which take either a constant-0 value or a constant-1 value over all elements in the interval $I_N(x)$. The condition $\mu_h(x) \leq N$ implies that any function is possible except those taking a constant-0 or a constant-1 value over $I_N(x)$. Hence clearly the first condition is significantly more restrictive.

Since we seek an upper bound on $|\mathcal{H}(S)|$ then we consider among all sets of $[n]$ of cardinality l a set S with the least restrictive constraint, namely, with as few elements x as possible for which $\mu_h(x) > N$. This is achieved by a maximally-packed set $S^* \subset [n]$ of l elements, for instance $S^* = \{1, \ldots, l\}$. It yields a minimal-size region $R = \{1, \ldots, l+N+1\}$ on which every candidate h must take either a constant-0 or constant-1 value, i.e., have a margin larger than N for every $x \in S^*$. This leaves a maximal-size region $[n] \setminus R$ on which the less stringent constraint of having a margin no larger than N must hold[1] for each function h. By Lemma 3.4 (see below), there are no more than $\beta_d^{(N)}(n-l-N-1)$ functions in $\mathcal{H}$ that satisfy the latter. Hence for any $S \subseteq [n]$ of cardinality $|S| = l$, $|\mathcal{H}(S)| \leq \beta_d^{(N)}(n-l-N-1)$. □

We proceed by starting with a few auxiliary lemmas.

Lemma 3.2. *For $N \geq 0$, $n \geq 0$, $0 \leq m \leq n$, let $w_{m,N}(n)$ be the number of standard (one-dimensional) ordered partitions of a nonnegative integer n into m parts each no larger than N. Then*

$$w_{m,N}(n) = \begin{cases} \mathbb{I}(n=0) & \text{if } m = 0 \\ \displaystyle\sum_{i=0,N+1,2(N+1),\ldots}^{n} (-1)^{i/(N+1)} \binom{m}{i/(N+1)} \binom{n-i+m-1}{n-i} & \text{if } m \geq 1. \end{cases}$$

Remark 3.1. *While our interest is in $[n] = \{1, \ldots, n\}$, we allow $w_{m,N}(n)$ to be defined on $n = 0$ for use by Lemma 3.3.*

Proof. The generating function (g.f.) for $w_{m,N}(n)$ is

$$W(x) = \sum_{n \geq 0} w_{m,N}(n) x^n = \left(\frac{1-x^{N+1}}{1-x}\right)^m.$$

When $m = 0$ the only non-zero coefficient is of x^0 and it equals 1 so $w_{0,N}(n) = \mathbb{I}(n=0)$. Let $T(x) = (1-x^{N+1})^m$ and $S(x) = \left(\frac{1}{1-x}\right)^m$. Then

$$T(x) = \sum_{i=0}^{m} (-1)^i \binom{m}{i} x^{i(N+1)}$$

which generates the sequence $t_N(n) = \binom{m}{n/(N+1)}(-1)^{n/(N+1)}\mathbb{I}(n \bmod (N+1) = 0)$. Similarly, for $m \geq 1$, it is easy to show $S(x)$ generates $s(n) = \binom{n+m-1}{n}$. The product $W(x) = T(x)S(x)$ generates their convolution $t_N(n) \star s(n)$, namely,

$$w_{m,N}(n) = \sum_{i=0,N+1,2(N+1),\ldots,}^{n} (-1)^{i/(N+1)} \binom{m}{i/(N+1)} \binom{n-i+m-1}{n-i}. \square$$

[1]Note, it is possible to have a constant value b (where $b \in \{0,1\}$) on $\{1, \ldots, l+N+1\}$ and a value of $1-b$ on the $(l+N+2)^{th}$ element, which is necessary for h in order to have $d_h(x) > N$ for $x \in \{1, \ldots, l\}$ and $d_h(x) \leq N$ for $x \in \{l+1, \ldots, n\}$.

Remark 3.2. *By an alternate proof one obtains a slightly simpler form of*

$$w_{m,N}(n) = \sum_{k=0}^{m} (-1)^k \binom{m}{k} \binom{n+m-1-k(N+1)}{m-1},$$

over $m \geq 1$.

We have two additional lemmas.

Lemma 3.3. *Let the integer* $1 \leq N \leq n$ *and consider the class* F *consisting of all binary-valued functions* f *on* $[n]$ *which take the value* 1 *on no more than* $r \leq n$ *elements of* $[n]$ *and whose margin on any element* $x \in [n]$ *satisfies* $\mu_f(x) \leq N$. *Then*

$$|F| \leq \sum_{k=0}^{r} \sum_{m=1}^{n} c(k, n-k; m, N) \equiv \beta_r^{(N)}(n)$$

where

$$c(k, n-k; m, N) = \left(\sum_{i=(k-m-N+1)_+}^{k-m+1} w_{m-1,2N}(i) \sum_{j=(n-k-m-N+1)_+}^{n-k-m+1} w_{m-1,2N}(j) \right),$$

$(a)_+ = \max\{a, 0\}$ *and* $w_{m,N}(n)$ *is defined in Lemma 3.2.*

Proof: Consider the integer pair $[k, n-k]$, where $n \geq 1$ and $0 \leq k \leq n$. A two-dimensional ordered m-partition of $[k, n-k]$ is an ordered partition into m two-dimensional parts, $[a_j, b_j]$ where $0 \leq a_j, b_j \leq n$ but not both are zero and where $\sum_{j=1}^{m} [a_j, b_j] = [k, n-k]$. For instance, $[2,1] = [0,1] + [2,0] = [1,1] + [1,0] = [2,0] + [0,1]$ are three partitions of $[2,1]$ into two parts (for more examples see Andrews [1]).

Suppose we add the constraint that only a_1 or b_m may be zero while all remaining $a_j, b_k \geq 1$, $2 \leq j \leq m$, $1 \leq k \leq m-1$. Denote any partition that satisfies this as *valid*. For instance, let $k = 2$, $m = 3$ then the m-partitions of $[k, n-k]$ are: $\{[0,1][1,1][1,n-4]\}$,$\{[0,1][1,2][1,n-5]\}$, ...,$\{[0,1][1,n-3][1,0]\}$, $\{[0,2][1,1][1,n-5]\}$, $\{[0,2][1,2][1,n-6]\}$, ..., $\{[0,2][1,n-4][1,0]\}$, ..., $\{[0,n-3][1,1][1,0]\}$. For $[k, n-k]$, let $\mathcal{P}_{n,k}$ be the collection of all valid partitions of $[k, n-k]$.

Let F_k denote all binary functions on $[n]$ which take the value 1 over exactly k elements of $[n]$. Define the mapping $\Pi : F_k \to \mathcal{P}_{n,k}$ where for any $f \in F_k$ the partition $\Pi(f)$ is defined by the following procedure: Start from the first element of $[n]$, i.e., 1. If f takes the value 1 on it then let a_1 be the length of the constant 1-segment, i.e., the set of all elements starting from 1 on which f takes the constant value 1. Otherwise if f takes the value 0 let $a_1 = 0$. Then let b_1 be the length of the subsequent 0-segment on which f takes the value 0. Let $[a_1, b_1]$ be the first part of $\Pi(f)$. Next, repeat the following: if there is at least one more element of $[n]$ which has not been included in the preceding segment, then let a_j be the length of the next 1-segment and b_j the length of the subsequent 0-segment. Let $[a_j, b_j]$, $j = 1, \ldots, m$, be the resulting sequence of parts where m is the total number of parts. Only the last part may have a zero valued b_m since the function may take the value 1 on the last element n of $[n]$ while all other parts, $[a_j, b_j]$, $2 \leq j \leq m-1$, must have $a_j, b_j \geq 1$. The result is a valid partition of $[k, n-k]$ into m parts.

Clearly, every $f \in F_k$ has a unique partition. Therefore Π is a bijection. Moreover, we may divide $\mathcal{P}_{n,k}$ into mutually exclusive subsets V_m consisting of all valid partitions of $[k, n-k]$ having exactly m parts, where $1 \le m \le n$. Thus

$$|F_k| = \sum_{m=1}^{n} |V_m|.$$

Consider the following constraint on components of parts:

$$a_i, b_j \le \begin{cases} N, & i = 1, j = m \\ 2N+1, & 2 \le i \le m, 1 \le j \le m-1. \end{cases} \tag{1}$$

Denote by $V_{m,N} \subset \mathcal{P}_{n,k}$ the collection of valid partitions of $[k, n-k]$ into m parts each of which satisfies this constraint.

Let $F_{k,N} = F \cap F_k$ consist of all functions satisfying the margin constraint in the statement of the lemma and having exactly k ones. Note that f having a margin no larger than N on any element of $[n]$ implies there does not exist a first or last segment of length larger than N on which f takes a constant value. It also means that all other 0-segments or 1-segments must not have a length larger than $2N+1$. Hence the parts of $\Pi(f)$ satisfy (1). Hence, for any $f \in F_{k,N}$, its unique valid partition $\Pi(f)$ must be in $V_{m,N}$. We therefore have

$$|F_{k,N}| = \sum_{m=1}^{n} |V_{m,N}|. \tag{2}$$

By definition of F it follows that

$$|F| = \sum_{k=0}^{r} |F_{k,N}|. \tag{3}$$

Let us denote by

$$c(k, n-k; m, N) \equiv |V_{m,N}| \tag{4}$$

the number of valid partitions of $[k, n-k]$ into exactly m parts whose components satisfy (1). In order to determine $|F|$ it therefore suffices to determine $c(k, n-k; m, N)$.

We next construct the generating function

$$G(t_1, t_2) = \sum_{\alpha_1 \ge 0} \sum_{\alpha_2 \ge 0} c(\alpha_1, \alpha_2; m, N) t_1^{\alpha_1} t_2^{\alpha_2}. \tag{5}$$

For $m \ge 1$,

$$\begin{aligned} G(t_1, t_2) = {} & (t_1^0 + t_1^1 + \cdots + t_1^N)(t_2^1 + t_2^2 + \cdots + t_2^{2N+1})^{\mathbb{I}(m \ge 2)} \\ & \cdot \left((t_1^1 + \cdots + t_1^{2N+1})(t_2^1 + \cdots + t_2^{2N+1})\right)^{(m-2)_+} \\ & \cdot (t_1^1 + \cdots + t_1^{2N+1})^{\mathbb{I}(m \ge 2)} (t_2^0 + t_2^1 + \cdots + t_2^N) \end{aligned} \tag{6}$$

where the values of the exponents of all terms in the first and second factors represent the possible values for a_1 and b_1, respectively. The values of the exponents in the middle $m-2$ factors are for the values of a_j, b_j, $2 \le j \le m-1$ and those in the factor before last and last are for a_m and b_m, respectively. Equating this to (5) implies the coefficient of $t_1^{\alpha_1} t_2^{\alpha_2}$ equals $c(\alpha_1, \alpha_2; m, N)$ which we seek.

The right side of (6) equals

$$t_1^{m-1} t_2^{m-1} \frac{1 - t_1^{N+1}}{1 - t_1} \left(\frac{1 - t_1^{2N+1}}{1 - t_1}\right)^{m-1} \frac{1 - t_2^{N+1}}{1 - t_2} \left(\frac{1 - t_2^{2N+1}}{1 - t_2}\right)^{m-1}. \tag{7}$$

Let

$$U(x) = \frac{1 - x^{N+1}}{1 - x}$$

which is the g.f. for

$$u(n) = \begin{cases} 1, & 0 \le n \le N \\ 0, & n > N \end{cases}$$

and let $W(x) = \left(\frac{1-x^{2N+1}}{1-x}\right)^{m-1}$ generate $w_{m-1,2N}(n)$ which is defined in Lemma 3.2. The product $U(x)W(x)$ is the g.f. for their convolution

$$y(n) = u(n) \star w_{m-1,2N}(n) = \sum_{i=(n-N)_+}^{n} w_{m-1,2N}(i).$$

So (7) becomes

$$\sum_{\alpha_1,\alpha_2 \ge 0} \sum_{i=(\alpha_1-N)_+}^{\alpha_1} \sum_{j=(\alpha_2-N)_+}^{\alpha_2} w_{m-1,2N}(i) w_{m-1,2N}(j) t_1^{\alpha_1+m-1} t_2^{\alpha_2+m-1}. \tag{8}$$

Equating the coefficients of $t_1^{\alpha_1'} t_2^{\alpha_2'}$ in (5) and (8) yields

$$c(\alpha_1', \alpha_2'; m, N) = \sum_{i=(\alpha_1'-m-N+1)_+}^{\alpha_1'-m+1} \sum_{j=(\alpha_2'-m-N+1)_+}^{\alpha_2'-m+1} w_{m-1,2N}(i) w_{m-1,2N}(j),$$

$m - 1 \le \alpha_1', \alpha_2' \le m(2N+1) - 2(N+1)$. Substituting k for α_1', $n-k$ for α_2', combining (2), (3) and (4) yields the result. $\square$

The next lemma extends the result of Lemma 3.3 to classes $\mathcal{H}$ of finite VC-dimension.

Lemma 3.4. *Let $n \ge 1$ and $0 \le d \le n$. Let $\mathcal{H}$ be a class of binary-valued functions h on $[n]$ satisfying $\mu_h(x) \le N$ on any $x \in [n]$ and let $VC(\mathcal{H}) \le d$. Then*

$$|\mathcal{H}| \le \beta_d^{(N)}(n)$$

where $\beta_d^{(N)}$ is defined in Lemma 3.3.

Proof. The proof builds on that of Lemma 7 in Haussler & Long [5] which considered generalizations of the VC-dimension and is done by double induction on n and d.

Start with the case $d = 0$, the bound reduces to $|\mathcal{H}| \le 1$ since $\beta_0^{(N)}(n) \le 1$ when $n \ge 1$. The bound is correct since if $|\mathcal{H}| > 1$ then it implies there are two distinct functions h, g. Let $k \in [n]$ be the element on which they differ. Then the singleton $\{k\}$ is shattered by $\mathcal{H}$ hence the VC-dimension of $\mathcal{H}$ is at least 1 which contradicts the assumption that $d = 0$ hence $|\mathcal{H}| \le 1$ and the lemma holds.

Next, suppose $d = n$. Consider the class F in Lemma 3.3 with $r = n$. Such F consists of all binary-valued functions f on $[n]$ which satisfy the margin constraint $\mu_f(x) \le N$ on every $x \in [n]$. By Lemma 3.3, $|F| \le \beta_n^{(N)}(n)$. Clearly by definition, $\mathcal{H} \subseteq F$. Hence $|\mathcal{H}| \le \beta_n^{(N)}(n)$ as claimed.

Next, suppose $0 < d < n$. Define $\pi : \mathcal{H} \to \{0,1\}^{n-1}$ by $\pi(h) = [h(1), \ldots, h(n-1)]$. Define $\alpha : \pi(\mathcal{H}) \to \{0,1\}$ by $\alpha(u_1, \ldots, u_{n-1}) = \min\{v : \exists h \in \mathcal{H}, h(i) = u_i, h(n) = v, 1 \le i \le n-1\}$. Define $A = \{h \in \mathcal{H} : h(n) = \alpha(h(1), \ldots, h(n-1))\}$ and denote by $A^c = \mathcal{H} \setminus A$. Considering all $h \in \mathcal{H}$, if the minimal value $h(n)$ is 1 then A^c is empty. Otherwise, it is not empty and its members take the value 1 on n.

Make the inductive assumption that the claimed bound holds for all classes $\mathcal{H}$ on any subset of $[n]$ having cardinality $n-1$ and satisfying the margin constraint. Then we claim the following:

Claim 1.

$$|A| \leq \beta_d^{(N)}(n-1).$$

This is proved next: the mapping π is one-to-one on A and the set $\pi(A)$ has VC-dimension no larger than d since any subset of $[n]$ shattered by $\pi(A)$ is also shattered by A which is in $\mathcal{H}$ and $VC(\mathcal{H}) \leq d$. Hence by the induction hypothesis

$$|\pi(A)| \leq \beta_d^{(N)}(n-1)$$

and since π is one-to-one then $|A| = |\pi(A)|$. □

Next, under the same induction hypothesis, we have:

Claim 2.

$$|A^c| \leq \beta_{d-1}^{(N)}(n-1).$$

We prove this next: First we show that $VC(A^c) \leq d-1$. Let $E \subset [n]$ be shattered by A^c and let $|E| = l$. Note that $n \notin E$ since as noted earlier $h(n) = 1$ for all $h \in A^c$. For any $b \in \{0,1\}^{l+1}$ let $h \in A^c$ be such that $h_{|E} = [b_1, \ldots, b_l]$. If $b_{l+1} = 1$ then $h(n) = b_{l+1}$ since all functions in A^c take the value 1 on n. If $b_{l+1} = 0$ then since A^c is non-empty, there exists a $g \in A$ which satisfies $g(i) = h(i)$, $1 \leq i \leq n-1$ and $g(n) = \alpha(h(1), \ldots, h(n-1))$, the latter being $g(n) = 0$. It follows that $E \cap \{n\}$ is shattered by $\mathcal{H}$. But by assumption $VC(\mathcal{H}) \leq d$ and $n \notin E$ hence $|E| \leq d-1$. Since E was chosen arbitrarily then $VC(A^c) \leq d-1$. The same argument as in the proof of Claim 1 applied to A^c using $d-1$ to bound its VC-dimension, obtains the statement of Claim 2. □

From Claims 1 and 2 and recalling the definition of $c(k, n-k; m, N)$ from Lemma 3.3, it follows that

$$\begin{aligned}
|\mathcal{H}| &\leq \beta_d^{(N)}(n-1) + \beta_{d-1}^{(N)}(n-1) \\
&= \sum_{k=0}^{d} \sum_{m=1}^{n-1} c(k, n-k-1; m, N) + \sum_{k=0}^{d-1} \sum_{m=1}^{n-1} c(k, n-k-1; m, N) \\
&= \mathbb{I}(n \leq N) + \sum_{k=1}^{d} \sum_{m=1}^{n-1} c(k, n-k-1; m, N) + \sum_{k=1}^{d} \sum_{m=1}^{n-1} c(k-1, n-k; m, N) \\
&= \mathbb{I}(n \leq N) + \sum_{k=1}^{d} \sum_{m=1}^{n-1} \left(c(k, n-k-1; m, N) + c(k-1, n-k; m, N)\right) \qquad (9)
\end{aligned}$$

where the indicator $\mathbb{I}(n \leq N)$ enters here since in case $k = 0$ the only valid function is the constant-0 on $[n]$ with $n \leq N$. We now have:

Claim 3.

$$\sum_{m=1}^{n} c(k, n-k; m, N) = \sum_{m=1}^{n-1} \left(c(k, n-k-1; m, N) + c(k-1, n-k; m, N)\right). \quad (10)$$

Note that this is a recurrence formula for the number (left hand side of (10)) of valid partitions of $[k, n-k]$ (excluding the case $k=0$) into parts that satisfy (1).

We prove the claim next: given any such partition π_n there is exactly one of four possible ways that it can be constructed by adding a part to a valid two dimensional partition π_{n-1} of $[n-1]$ while still satisfying (1). The first two amount to starting from a partition π_{n-1} of $[k-1, n-k]$ and: (i) adding the part $[1,0]$ algebraically to any existing part in π_{n-1}, e.g., $[x,y]+[1,0]=[x+1,y]$, to obtain a π_n with $[x+1,y]$ as one of the parts (provided that (1) is still satisfied) which yields a total number of parts no larger than $n-1$ or (ii) adding $[1,0]$ to π_{n-1} as a new last part to obtain a π_n (provided it is still valid) with no more than n parts. The remaining two ways amount to starting from a partition π_{n-1} of $[k, n-k-1]$ and acting as before except now adding the part $[0,1]$ instead of $[1,0]$ either algebraically or as a new first part. There are

$$\sum_{m=1}^{n-1} c(k, n-k-1; m, N)$$

valid partitions of $[k, n-k-1]$ and there are

$$\sum_{m=1}^{n-1} c(k-1, n-k; m, N)$$

valid partitions of $[k-1, n-k]$, all satisfying (1). Doing the aforementioned construction to each one of these partitions yields all valid partitions of $[k, n-k]$ that satisfy (1). □

Continuing, the right hand side of (9) becomes

$$\mathbb{I}(n \leq N) + \sum_{k=1}^{d} \sum_{m=1}^{n} c(k, n-k; m, N) = \sum_{k=0}^{d} \sum_{m=1}^{n} c(k, n-k; m, N)$$

which is precisely $\beta_d^{(N)}(n)$. This completes the induction. □

Considering a class $\mathcal{H}$ of VC-dimension d with the margin constraint as defined in Lemma 3.4 then $\beta_d^{(N)}(n)$ is crucial for understanding the effect of the margin parameter N on the cardinality of $\mathcal{H}$. A recent result (cf. [7]) shows that $\beta_d^{(N)}(n)$ exhibits an interesting sharp threshold behavior at a point $N = N^*$, i.e., the number of functions $h \in \mathcal{H}$ that have a margin $\mu_h(x) \leq N$ on every $x \in X$ decreases sharply as N becomes less than N^* and increases sharply as N becomes larger than N^*.

4. Conclusions

The main result of the paper is a bound on the cardinality of a class of finite VC-dimension consisting of binary functions on $[n]$ which have a margin greater than N on a set S of cardinality l. This result generalizes the well known Sauer's Lemma and is analogous to existing bounds on the covering number of classes of *real-valued* functions that have a large-margin on a sample S.

References

[1] Andrews G.E., (1998), *The Theory of Partitions*, Cambridge University Press.

[2] Anthony M., Bartlett P. L., (1999), *Neural Network Learning:Theoretical Foundations*, Cambridge University Press, UK.

[3] Cristianini N. and Shawe-Taylor J., (2000), *An Introduction to Support Vector Machines and other Kernel-based learning methods*, Cambridge University Press, UK.

[4] Sauer N., (1972), On the density of families of sets, *J. Combinatorial Theory (A)*, Vol. 13, pp. 145-147.

[5] Haussler D., Long P.M., (1995), A generalization of Sauer's Lemma, *Journal of Combinatorial Theory (A)*, Vol. 71(2), pp. 219-240.

[6] Ratsaby J., (2004), On the Complexity of Good Samples for Learning, *Proc. Tenth International Computing and Combinatorics Conference(COCOON 2004)*, Jeju Island, Korea, August 2004, Lecture Notes in Computer Science Vol. 3106, Springer-Verlag.

[7] Ratsaby J., (2004), A Sharp Threshold Result for Finite-VC Classes of Large-Margin Functions, Department of Computer Science Technical Report RN/04/06, University College London.

[8] Vapnik V.N., (1998), *Statistical Learning Theory*, Wiley..

[9] Vapnik V. N and Chervonenkis A. Ya. (1971), On the uniform convergence of relative frequencies of events to their probabilities. *Theoret. Probl. Appl.*, Vol. 16, p.264-280.

Joel Ratsaby
Department of Computer Science
University College London
London WC1E 6BT, U.K.

http://www.cs.ucl.ac.uk/staff/J.Ratsaby
J.Ratsaby@cs.ucl.ac.uk

Index

Author Index

MIX
Papier aus verantwortungsvollen Quellen
Paper from responsible sources
FSC® C105338

If you have any concerns about our products,
you can contact us on
ProductSafety@springernature.com

In case Publisher is established outside the EU,
the EU authorized representative is:
Springer Nature Customer Service Center GmbH
Europaplatz 3, 69115 Heidelberg, Germany

Printed by Libri Plureos GmbH
in Hamburg, Germany